Advanced Linear and Matrix Algebra

Nathaniel Johnston

Advanced Linear and Matrix Algebra

 Springer

Nathaniel Johnston
Department of Mathematics and
Computer Science
Mount Allison University
Sackville, NB, Canada

ISBN 978-3-030-52817-1   ISBN 978-3-030-52815-7   (eBook)
https://doi.org/10.1007/978-3-030-52815-7

Mathematics Subject Classification: 15Axx, 97H60, 00-01

This Springer imprint is published by the registered company Springer Nature Switzerland AG
The registered company address is: Gewerbestrasse 11, 6330 Cham, Switzerland

For Devon
…who was very eager at age 2 to contribute to this book:
ndfshfjds kfdshdsf kdfsh kdsfhfdsk hdfsk

Preface

The Purpose of this Book

Linear algebra, more so than any other mathematical subject, can be approached in numerous ways. Many textbooks present the subject in a very concrete and numerical manner, spending much of their time solving systems of linear equations and having students perform laborious row reductions on matrices. Many other books instead focus very heavily on linear transformations and other basis-independent properties, almost to the point that their connection to matrices is considered an inconvenient after-thought that students should avoid using at all costs.

This book is written from the perspective that both linear transformations and matrices are useful objects in their own right, but it is the connection between the two that really unlocks the magic of linear algebra. Sometimes when we want to know something about a linear transformation, the easiest way to get an answer is to grab onto a basis and look at the corresponding matrix. Conversely, there are many interesting families of matrices and matrix operations that seemingly have nothing to do with linear transformations, yet can nonetheless illuminate how some basis-independent objects and properties behave.

This book introduces many difficult-to-grasp objects such as vector spaces, dual spaces, and tensor products. Because it is expected that this book will accompany one of the first courses where students are exposed to such abstract concepts, we typically sandwich this abstractness between concrete examples. That is, we first introduce or emphasize a standard, prototypical example of the object to be introduced (e.g., $\mathbb{R}^n$), then we discuss its abstract generalization (e.g., vector spaces), and finally we explore other specific examples of that generalization (e.g., the vector space of polynomials and the vector space of matrices).

This book also delves somewhat deeper into matrix decompositions than most others do. We of course cover the singular value decomposition as well as several of its applications, but we also spend quite a bit of time looking at the Jordan decomposition, Schur triangularization, and spectral decomposition, and we compare and contrast them with each other to highlight when each one is appropriate to use. Computationally-motivated decompositions like the QR and Cholesky decompositions are also covered in some of this book's many "Extra Topic" sections.

Continuation of *Introduction to Linear and Matrix Algebra*

This book is the second part of a two-book series, following the book *Introduction to Linear and Matrix Algebra* [Joh20]. The reader is expected to be familiar with the basics of linear algebra covered in that book (as well as other introductory linear algebra books): vectors in $\mathbb{R}^n$, the dot product, matrices and matrix multiplication, Gaussian elimination, the inverse, range, null space, rank, and determinant of a matrix, as well as eigenvalues and eigenvectors. These preliminary topics are briefly reviewed in Appendix A.1.

Because these books aim to not overlap with each other and repeat content, we do not discuss some topics that are instead explored in that book. In particular, diagonalization of a matrix via its eigenvalues and eigenvectors is discussed in the introductory book and not here. However, many extensions and variations of diagonalization, such as the spectral decomposition (Section 2.1.2) and Jordan decomposition (Section 2.4) are explored here.

Features of this Book

This book makes use of numerous features to make it as easy to read and understand as possible. Here we highlight some of these features and discuss how to best make use of them.

Notes in the Margin

This text makes heavy use of notes in the margin, which are used to introduce some additional terminology or provide reminders that would be distracting in the main text. They are most commonly used to try to address potential points of confusion for the reader, so it is best not to skip them.

For example, if we want to clarify why a particular piece of notation is the way it is, we do so in the margin so as to not derail the main discussion. Similarly, if we use some basic fact that students are expected to be aware of (but have perhaps forgotten) from an introductory linear algebra course, the margin will contain a brief reminder of why it's true.

Exercises

Several exercises can be found at the end of every section in this book, and whenever possible there are three types of them:

- There are **computational exercises** that ask the reader to implement some algorithm or make use of the tools presented in that section to solve a numerical problem.
- There are **true/false exercises** that test the reader's critical thinking skills and reading comprehension by asking them whether some statements are true or false.
- There are **proof exercises** that ask the reader to prove a general statement. These typically are either routine proofs that follow straight from the definition (and thus were omitted from the main text itself), or proofs that can be tackled via some technique that we saw in that section.

Roughly half of the exercises are marked with an asterisk ($*$), which means that they have a solution provided in Appendix C. Exercises marked with *two* asterisks ($**$) are referenced in the main text and are thus particularly important (and also have solutions in Appendix C).

To the Instructor and Independent Reader

This book is intended to accompany a second course in linear algebra, either at the advanced undergraduate or graduate level. The only prerequisites that are expected of the reader are an introductory course in linear algebra (which is summarized in Appendix A.1) and some familiarity with mathematical proofs. It will help the reader to have been exposed to complex numbers, though we do little more than multiply and add them (their basics are reviewed in Appendix A.3).

The material covered in Chapters 1 and 2 is mostly standard material in upper-year undergraduate linear algebra courses. In particular, Chapter 1 focuses on abstract structures like vector spaces and inner products, to show students that the tools developed in their previous linear algebra course can be applied to a much wider variety of objects than just lists of numbers like vectors in $\mathbb{R}^n$. Chapter 2 then explores how we can use these new tools at our disposal to gain a much deeper understanding of matrices.

Chapter 3 covers somewhat more advanced material—multilinearity and the tensor product—which is aimed particularly at advanced undergraduate students (though we note that no knowledge of abstract algebra is assumed). It could serve perhaps as content for part of a third course, or as an independent study in linear algebra. Alternatively, that chapter is also quite aligned with the author's research interests as a quantum information theorist, and it could be used as supplemental reading for students who are trying to learn the basics of the field.

Sectioning

The sectioning of the book is designed to make it as simple to teach from as possible. The author spends approximately the following amount of time on each chunk of this book:

- **Subsection:** 1–1.5 hour lecture
- **Section:** 2 weeks (3–4 subsections per section)
- **Chapter:** 5–6 weeks (3–4 sections per chapter)
- **Book:** 12-week course (2 chapters, plus some extra sections)

Of course, this is just a rough guideline, as some sections are longer than others. Furthermore, some instructors may choose to include material from Chapter 3, or from some of the numerous in-depth "Extra Topic" sections. Alternatively, the additional topics covered in those sections can serve as independent study topics for students.

Extra Topic Sections

Half of this book's sections are called "Extra Topic" sections. The purpose of the book being arranged in this way is that it provides a clear main path through the book (Sections 1.1–1.4, 2.1–2.4, and 3.1–3.3) that can be supplemented by the Extra Topic sections at the reader's/instructor's discretion. It is expected that many courses will not even make it to Chapter 3, and instead will opt to explore some of the earlier Extra Topic sections instead.

We want to emphasize that the Extra Topic sections are not labeled as such because they are less important than the main sections, but only because they are not prerequisites for any of the main sections. For example, norms and isometries (Section 1.D) are used constantly throughout advanced mathematics, but they are presented in an Extra Topic section since the other sections of this book do not depend on them (and also because they lean quite a bit into "analysis territory", whereas most of the rest of the book stays firmly in "algebra territory").

Similarly, the author expects that many instructors will include the section on the direct sum and orthogonal complements (Section 1.B) as part of their course's core material, but this can be done at their discretion. The subsections on dual spaces and multilinear forms from Section 1.3.2 can be omitted reasonably safely to make up some time if needed, as can the subsection on Gershgorin discs (Section 2.2.2), without drastically affecting the book's flow.

For a graph that depicts the various dependencies of the sections of this book on each other, see Figure ★.

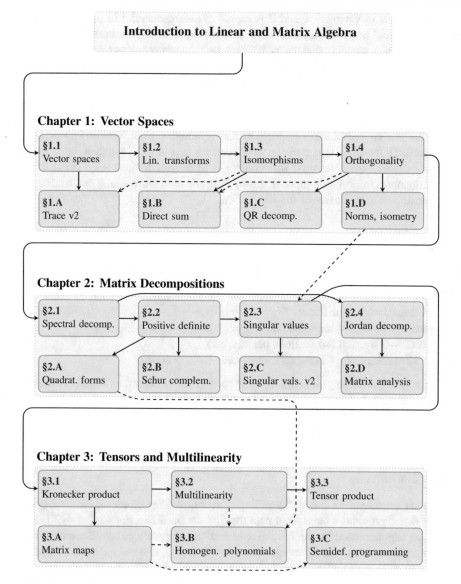

Figure ★: A graph depicting the dependencies of the sections of this book on each other. Solid arrows indicate that the section is required before proceeding to the section that it points to, while dashed arrows indicate recommended (but not required) prior reading. The main path through the book consists of Sections 1–4 of each chapter. The extra sections A–D are optional and can be explored at the reader's discretion, as none of the main sections depend on them.

Acknowledgments

Thanks are extended to Geoffrey Cruttwell, Mark Hamilton, David Kribs, Chi-Kwong Li, Benjamin Lovitz, Neil McKay, Vern Paulsen, Rajesh Pereira, Sarah Plosker, Jamie Sikora, and John Watrous for various discussions that have either directly or indirectly improved the quality of this book.

Thank you to Everett Patterson, as well as countless other students in my linear algebra classes at Mount Allison University, for drawing my attention to typos and parts of the book that could be improved.

Parts of the layout of this book were inspired by the *Legrand Orange Book* template by Velimir Gayevskiy and Mathias Legrand at LaTeXTemplates.com.

Finally, thank you to my wife Kathryn for tolerating me during the years of my mental absence glued to this book, and thank you to my parents for making me care about both learning and teaching.

Sackville, NB, Canada   Nathaniel Johnston

Table of Contents

1. Vector Spaces

> It is my experience that proofs involving matrices can be shortened by 50% if one throws the matrices out.
>
> Emil Artin

$\mathbb{R}^n$ denotes the set of vectors with n (real) entries and $\mathcal{M}_{m,n}$ denotes the set of $m \times n$ matrices.

Our first exposure to linear algebra is typically a very concrete thing—it consists of some basic facts about $\mathbb{R}^n$ (a set made up of lists of real numbers, called vectors) and $\mathcal{M}_{m,n}$ (a set made up of $m \times n$ arrays of real numbers, called matrices). We can do numerous useful things in these sets, such as solve systems of linear equations, multiply matrices together, and compute the rank, determinant, and eigenvalues of a matrix.

When we look carefully at our procedures for doing these calculations, as well as our proofs for why they work, we notice that most of them do not actually require much more than the ability to add vectors together and multiply them by scalars. However, there are many other mathematical settings where addition and scalar multiplication work, and almost all of our linear algebraic techniques work in these more general settings as well.

With this in mind, our goal right now is considerably different from what it was in introductory linear algebra—we want to see exactly how far we can push our techniques. Instead of defining objects and operations in terms of explicit formulas and then investigating what properties they satisfy (as we have done up until now), we now focus on the properties that those familiar objects have and ask what other types of objects have those properties.

For example, in a typical introduction to linear algebra, the dot product of two vectors $\mathbf{v}$ and $\mathbf{w}$ in $\mathbb{R}^n$ is defined by

$$\mathbf{v} \cdot \mathbf{w} = v_1 w_1 + v_2 w_2 + \cdots + v_n w_n,$$

and then the "nice" properties that the dot product satisfies are investigated. For example, students typically learn the facts that

$$\mathbf{v} \cdot \mathbf{w} = \mathbf{w} \cdot \mathbf{v} \quad \text{and} \quad \mathbf{v} \cdot (\mathbf{w} + \mathbf{x}) = (\mathbf{v} \cdot \mathbf{w}) + (\mathbf{v} \cdot \mathbf{x}) \quad \text{for all} \quad \mathbf{v}, \mathbf{w}, \mathbf{x} \in \mathbb{R}^n$$

almost immediately after being introduced to the dot product. In this chapter, we flip this approach around and instead define an "inner product" to be any function satisfying those same properties, and then show that everything we learned about the dot product actually applies to every single inner product (even though many inner products look, on the surface, quite different from the dot product).

© Springer Nature Switzerland AG 2021
N. Johnston, *Advanced Linear and Matrix Algebra*,
https://doi.org/10.1007/978-3-030-52815-7_1

1.1   Vector Spaces and Subspaces

In order to use our linear algebraic tools with objects other than vectors in $\mathbb{R}^n$, we need a proper definition that tells us what types of objects we can consider in a linear algebra setting. The following definition makes this precise and serves as the foundation for this entire chapter. Although the definition looks like an absolute beast, the intuition behind it is quite straightforward—the objects that we work with should behave "like" vectors in $\mathbb{R}^n$. That is, they should have the same properties (like commutativity: $\mathbf{v} + \mathbf{w} = \mathbf{w} + \mathbf{v}$) that vectors in $\mathbb{R}^n$ have with respect to vector addition and scalar multiplication.

More specifically, the following definition lists 10 properties that must be satisfied in order for us to call something a "vector space" (like $\mathbb{R}^3$) or a "vector" (like $(1, 3, -2) \in \mathbb{R}^3$). These 10 properties can be thought of as the answers to the question "what properties of vectors in $\mathbb{R}^n$ can we list *without explicitly referring to their entries?*"

Definition 1.1.1 **Vector Space** $\mathbb{R}$ is the set of real numbers and $\mathbb{C}$ is the set of complex numbers (see Appendix A.3).   Notice that the first five properties concern addition, while the last five concern scalar multiplication.

Let $\mathbb{F}$ be a set of scalars (usually either $\mathbb{R}$ or $\mathbb{C}$) and let $\mathcal{V}$ be a set with two operations called **addition** and **scalar multiplication**. We write the addition of $\mathbf{v}, \mathbf{w} \in \mathcal{V}$ as $\mathbf{v} + \mathbf{w}$, and the scalar multiplication of $c \in \mathbb{F}$ and $\mathbf{v}$ as $c\mathbf{v}$.

If the following ten conditions hold for all $\mathbf{v}, \mathbf{w}, \mathbf{x} \in \mathcal{V}$ and all $c, d \in \mathbb{F}$, then $\mathcal{V}$ is called a **vector space** and its elements are called **vectors**:

    a) $\mathbf{v} + \mathbf{w} \in \mathcal{V}$ (closure under addition)
    b) $\mathbf{v} + \mathbf{w} = \mathbf{w} + \mathbf{v}$ (commutativity)
    c) $(\mathbf{v} + \mathbf{w}) + \mathbf{x} = \mathbf{v} + (\mathbf{w} + \mathbf{x})$ (associativity)
    d) There exists a "zero vector" $\mathbf{0} \in \mathcal{V}$ such that $\mathbf{v} + \mathbf{0} = \mathbf{v}$.
    e) There exists a vector $-\mathbf{v}$ such that $\mathbf{v} + (-\mathbf{v}) = \mathbf{0}$.

    f) $c\mathbf{v} \in \mathcal{V}$ (closure under scalar multiplication)
    g) $c(\mathbf{v} + \mathbf{w}) = c\mathbf{v} + c\mathbf{w}$ (distributivity)
    h) $(c + d)\mathbf{v} = c\mathbf{v} + d\mathbf{v}$ (distributivity)
    i) $c(d\mathbf{v}) = (cd)\mathbf{v}$
    j) $1\mathbf{v} = \mathbf{v}$

Remark 1.1.1 **Fields and** **Sets of Scalars**   It's also useful to know that every field has a "0" and a "1": numbers such that $0a = 0$ and $1a = a$ for all $a \in \mathbb{F}$.

Not much would be lost throughout this book if we were to replace the set of scalars $\mathbb{F}$ from Definition 1.1.1 with "either $\mathbb{R}$ or $\mathbb{C}$". In fact, in many cases it is even enough to just explicitly choose $\mathbb{F} = \mathbb{C}$, since oftentimes if a property holds over $\mathbb{C}$ then it automatically holds over $\mathbb{R}$ simply because $\mathbb{R} \subseteq \mathbb{C}$.

However, $\mathbb{F}$ more generally can be any "field", which is a set of objects in which we can add, subtract, multiply, and divide according to the usual laws of arithmetic (e.g., $ab = ba$ and $a(b + c) = ab + ac$ for all $a, b, c \in \mathbb{F}$)—see Appendix A.4. Just like we can keep $\mathbb{R}^n$ in mind as the standard example of a vector space, we can keep $\mathbb{R}$ and $\mathbb{C}$ in mind as the standard examples of a field.

We now look at several examples of sets that are and are not vector spaces, to try to get used to this admittedly long and cumbersome definition. As our first example, we show that $\mathbb{R}^n$ is indeed a vector space (which should not be surprising—the definition of a vector space was designed specifically so as to mimic $\mathbb{R}^n$).

Example 1.1.1

Euclidean Space is a Vector Space

Show that $\mathbb{R}^n$ is a vector space.

Solution:

We have to check the ten properties described by Definition 1.1.1. If $\mathbf{v}, \mathbf{w}, \mathbf{x} \in \mathbb{R}^n$ and $c, d \in \mathbb{R}$ then:

a) $\mathbf{v} + \mathbf{w} \in \mathbb{R}^n$ (there is nothing to prove here—it follows directly from the definition of vector addition in $\mathbb{R}^n$).

b) We just repeatedly use commutativity of real number addition:

> v_j, w_j, and x_j denote the j-th entries of $\mathbf{v}$, $\mathbf{w}$, and $\mathbf{x}$, respectively.

$$\mathbf{v} + \mathbf{w} = (v_1 + w_1, \ldots, v_n + w_n) = (w_1 + v_1, \ldots, w_n + v_n) = \mathbf{w} + \mathbf{v}.$$

c) This property follows in a manner similar to property (b) by making use of associativity of real number addition:

$$(\mathbf{v} + \mathbf{w}) + \mathbf{x} = (v_1 + w_1, \ldots, v_n + w_n) + (x_1, \ldots, x_n)$$
$$= (v_1 + w_1 + x_1, \ldots, v_n + w_n + x_n)$$
$$= (v_1, \ldots, v_n) + (w_1 + x_1, \ldots, w_n + x_n) = \mathbf{v} + (\mathbf{w} + \mathbf{x}).$$

d) The zero vector $\mathbf{0} = (0, 0, \ldots, 0) \in \mathbb{R}^n$ clearly satisfies $\mathbf{v} + \mathbf{0} = \mathbf{v}$.

e) We simply choose $-\mathbf{v} = (-v_1, \ldots, -v_n)$, which indeed satisfies $\mathbf{v} + (-\mathbf{v}) = \mathbf{0}$.

f) $c\mathbf{v} \in \mathbb{R}^n$ (again, there is nothing to prove here—it follows straight from the definition of scalar multiplication in $\mathbb{R}^n$).

g) We just expand each of $c(\mathbf{v} + \mathbf{w})$ and $c\mathbf{v} + c\mathbf{w}$ in terms of their entries:

$$c(\mathbf{v} + \mathbf{w}) = c(v_1 + w_1, \ldots, v_n + w_n)$$
$$= (cv_1 + cw_1, \ldots, cv_n + cw_n) = c\mathbf{v} + c\mathbf{w}.$$

h) Similarly to property (g), we just notice that for each $1 \le j \le n$, the j-th entry of $(c+d)\mathbf{v}$ is $(c+d)v_j = cv_j + dv_j$, which is also the j-th entry of $c\mathbf{v} + d\mathbf{v}$.

i) Just like properties (g) and (h), we just notice that for each $1 \le j \le n$, the j-th entry of $c(d\mathbf{v})$ is $c(dv_j) = (cd)v_j$, which is also the j-th entry of $(cd)\mathbf{v}$.

j) The fact that $1\mathbf{v} = \mathbf{v}$ is clear.

We should keep $\mathbb{R}^n$ in our mind as the prototypical example of a vector space. We will soon prove theorems about vector spaces in general and see plenty of exotic vector spaces, but it is absolutely fine to get our intuition about how vector spaces work from $\mathbb{R}^n$ itself. In fact, we should do this every time we define a new concept in this chapter—keep in mind what the standard example of the new abstractly-defined object is, and use that standard example to build up our intuition.

> For all fields $\mathbb{F}$, the set $\mathbb{F}^n$ (of ordered n-tuples of elements of $\mathbb{F}$) is also a vector space.

It is also the case that $\mathbb{C}^n$ (the set of vectors/tuples with n complex entries) is also a vector space, and the proof of this fact is almost identical to the argument that we provided for $\mathbb{R}^n$ in Example 1.1.1. In fact, the set of all *infinite* sequences of scalars (rather than finite tuples of scalars like we are used to) is also a vector space. That is, the **sequence space**

$\mathbb{N}$ is the set of natural numbers. That is, $\mathbb{N} = \{1, 2, 3, \ldots\}$.

$$\mathbb{F}^{\mathbb{N}} \stackrel{\text{def}}{=} \big\{ (x_1, x_2, x_3, \ldots) : x_j \in \mathbb{F} \text{ for all } j \in \mathbb{N} \big\}$$

is a vector space with the standard addition and scalar multiplication operations (we leave the proof of this fact to Exercise 1.1.7).

To get a bit more comfortable with vector spaces, we now look at several examples of vector spaces that, on the surface, look significantly different from these spaces of tuples and sequences.

Example 1.1.2

The Set of Matrices is a Vector Space

Show that $\mathcal{M}_{m,n}$, the set of $m \times n$ matrices, is a vector space.

Solution:

We have to check the ten properties described by Definition 1.1.1. If $A, B, C \in \mathcal{M}_{m,n}$ and $c, d \in \mathbb{F}$ then:

a) $A + B$ is also an $m \times n$ matrix (i.e., $A + B \in \mathcal{M}_{m,n}$).

The "(i,j)-entry" of a matrix is the scalar in its i-th row and j-th column. We denote the (i,j)-entry of A and B by $a_{i,j}$ and $b_{i,j}$, respectively, or sometimes by $[A]_{i,j}$ and $[B]_{i,j}$.

b) For each $1 \le i \le m$ and $1 \le j \le n$, the (i,j)-entry of $A + B$ is $a_{i,j} + b_{i,j} = b_{i,j} + a_{i,j}$, which is also the (i,j)-entry of $B + A$. It follows that $A + B = B + A$.

c) The fact that $(A + B) + C = A + (B + C)$ follows similarly from looking at the (i,j)-entry of each of these matrices and using associativity of addition in $\mathbb{F}$.

d) The "zero vector" in this space is the zero matrix O (i.e., the $m \times n$ matrix with every entry equal to 0), since $A + O = A$.

e) We define $-A$ to be the matrix whose (i,j)-entry is $-a_{i,j}$, so that $A + (-A) = O$.

f) cA is also an $m \times n$ matrix (i.e., $cA \in \mathcal{M}_{m,n}$).

g) The (i,j)-entry of $c(A + B)$ is $c(a_{i,j} + b_{i,j}) = ca_{i,j} + cb_{i,j}$, which is the (i,j)-entry of $cA + cB$, so $(A + B) = cA + cB$.

h) Similarly to property (g), the (i,j)-entry of $(c + d)A$ is $(c + d)a_{i,j} = ca_{i,j} + da_{i,j}$, which is also the (i,j)-entry of $cA + dA$. It follows that $(c + d)A = cA + dA$.

i) Similarly to property (g), the (i,j)-entry of $c(dA)$ is $c(da_{i,j}) = (cd)a_{i,j}$, which is also the (i,j)-entry of $(cd)A$, so $c(dA) = (cd)A$.

j) The fact that $1A = A$ is clear.

Since $\mathbb{R} \subset \mathbb{C}$, real entries are also complex, so the entries of B might be real.

If we wish to emphasize which field $\mathbb{F}$ the entries of the matrices in $\mathcal{M}_{m,n}$ come from, we denote it by $\mathcal{M}_{m,n}(\mathbb{F})$. For example, if we say that $A \in \mathcal{M}_{2,3}(\mathbb{R})$ and $B \in \mathcal{M}_{4,5}(\mathbb{C})$ then we are saying that A is a 2×3 matrix with real entries and B is a 4×5 matrix with complex entries. We use the briefer notation $\mathcal{M}_{m,n}$ if the choice of field is unimportant or clear from context, and we use the even briefer notation $\mathcal{M}_n$ when the matrices are square (i.e., $m = n$).

Typically, proving that a set is a vector space is not hard—all of the properties either follow directly from the relevant definitions or via a one-line proof. However, it is still important to actually verify that all ten properties hold, especially when we are first learning about vector spaces, as we will shortly see some examples of sets that look somewhat like vector spaces but are not.

Example 1.1.3

The Set of Functions is a Vector Space

Show that the set of real-valued functions $\mathcal{F} \overset{\text{def}}{=} \{f : \mathbb{R} \to \mathbb{R}\}$ is a vector space.

Solution:

Once again, we have to check the ten properties described by Definition 1.1.1. To do this, we will repeatedly use the fact that two functions are the same if and only if their outputs are always the same (i.e., $f = g$ if and only if $f(x) = g(x)$ for all $x \in \mathbb{R}$). With this observation in mind, we note that if $f, g, h \in \mathcal{F}$ and $c, d \in \mathbb{R}$ then:

a) $f + g$ is the function defined by $(f + g)(x) = f(x) + g(x)$ for all $x \in \mathbb{R}$. In particular, $f + g$ is also a function, so $f + g \in \mathcal{F}$.

b) For all $x \in \mathbb{R}$, we have

$$(f + g)(x) = f(x) + g(x) = g(x) + f(x) = (g + f)(x),$$

so $f + g = g + f$.

c) For all $x \in \mathbb{R}$, we have

$$((f + g) + h)(x) = (f(x) + g(x)) + h(x)$$
$$= f(x) + (g(x) + h(x)) = (f + (g + h))(x),$$

so $(f + g) + h = f + (g + h)$.

d) The "zero vector" in this space is the function 0 with the property that $0(x) = 0$ for all $x \in \mathbb{R}$.

e) Given a function f, the function $-f$ is simply defined by $(-f)(x) = -f(x)$ for all $x \in \mathbb{R}$. Then

$$(f + (-f))(x) = f(x) + (-f)(x) = f(x) - f(x) = 0$$

for all $x \in \mathbb{R}$, so $f + (-f) = 0$.

f) cf is the function defined by $(cf)(x) = cf(x)$ for all $x \in \mathbb{R}$. In particular, cf is also a function, so $cf \in \mathcal{F}$.

g) For all $x \in \mathbb{R}$, we have

$$(c(f + g))(x) = c(f(x) + g(x)) = cf(x) + cg(x) = (cf + cg)(x),$$

so $c(f + g) = cf + cg$.

h) For all $x \in \mathbb{R}$, we have

$$((c + d)f)(x) = (c + d)f(x) = cf(x) + df(x) = (cf + df)(x),$$

so $(c + d)f = cf + df$.

i) For all $x \in \mathbb{R}$, we have

$$(c(df))(x) = c(df(x)) = (cd)f(x) = ((cd)f)(x),$$

so $c(df) = (cd)f$.

j) For all $x \in \mathbb{R}$, we have $(1f)(x) = 1f(x) = f(x)$, so $1f = f$.

> All of these properties of functions are trivial. The hardest part of this example is not the math, but rather getting our heads around the unfortunate notation that results from using parentheses to group function addition and also to denote inputs of functions.

In the previous three examples, we did not explicitly specify what the addition and scalar multiplication operations were upfront, since there was an obvious choice on each of these sets. However, it may not always be clear what these operations should actually be, in which case they have to be explicitly defined before we can start checking whether or not they turn the set into a vector space.

In other words, vector spaces are a package deal with their addition and scalar multiplication operations—a set might be a vector space when the operations are defined in one way but not when they are defined another way. Furthermore, the operations that we call addition and scalar multiplication might look nothing like what we usually call "addition" or "multiplication". All that matters is that those operations satisfy the ten properties from Definition 1.1.1.

Example 1.1.4

A Vector Space With Weird Operations

Let $\mathcal{V} = \{x \in \mathbb{R} : x > 0\}$ be the set of positive real numbers. Show that $\mathcal{V}$ is a vector space when we define addition $\oplus$ on it via usual multiplication of real numbers (i.e., $\mathbf{x} \oplus \mathbf{y} = xy$) and scalar multiplication $\odot$ on it via exponentiation (i.e., $c \odot \mathbf{x} = x^c$).

Solution:

Once again, we have to check the ten properties described by Definition 1.1.1. Well, if $\mathbf{x}, \mathbf{y}, \mathbf{z} \in \mathcal{V}$ and $c, d \in \mathbb{R}$ then:

a) $\mathbf{x} \oplus \mathbf{y} = xy$, which is still a positive number since x and y are both positive, so $\mathbf{x} \oplus \mathbf{y} \in \mathcal{V}$.

b) $\mathbf{x} \oplus \mathbf{y} = xy = yx = \mathbf{y} \oplus \mathbf{x}$.

c) $(\mathbf{x} \oplus \mathbf{y}) \oplus \mathbf{z} = (xy)z = x(yz) = \mathbf{x} \oplus (\mathbf{y} \oplus \mathbf{z})$.

d) The "zero vector" in this space is the number 1 (so we write $\mathbf{0} = 1$), since $\mathbf{0} \oplus \mathbf{x} = 1x = x = \mathbf{x}$.

e) For each vector $\mathbf{x} \in \mathcal{V}$, we define $-\mathbf{x} = 1/x$. This works because $\mathbf{x} \oplus (-\mathbf{x}) = x(1/x) = 1 = \mathbf{0}$.

f) $c \odot \mathbf{x} = x^c$, which is still positive since x is positive, so $c \odot \mathbf{x} \in \mathcal{V}$.

g) $c \odot (\mathbf{x} + \mathbf{y}) = (xy)^c = x^c y^c = (c \odot \mathbf{x}) \oplus (c \odot \mathbf{y})$.

h) $(c + d) \odot \mathbf{x} = x^{c+d} = x^c x^d = (c \odot \mathbf{x}) \oplus (d \odot \mathbf{x})$.

i) $c \odot (d \odot \mathbf{x}) = (x^d)^c = x^{cd} = (cd) \odot \mathbf{x}$.

j) $1 \odot \mathbf{x} = x^1 = x = \mathbf{x}$.

In this example, we use bold variables like $\mathbf{x}$ when we think of objects as vectors in $\mathcal{V}$, and we use non-bold variables like x when we think of them just as positive real numbers.

The previous examples illustrate that vectors, vector spaces, addition, and scalar multiplication can all look quite different from the corresponding concepts in $\mathbb{R}^n$. As one last technicality, we note that whether or not a set is a vector space depends on the field $\mathbb{F}$ that is being considered. The field can often, but not always, be inferred from context.

Before presenting an example that demonstrates why the choice of field is not always obvious, we establish some notation and remind the reader of some terminology. The **transpose** of a matrix $A \in \mathcal{M}_{m,n}$ is the matrix $A^T \in \mathcal{M}_{n,m}$ whose (i,j)-entry is $a_{j,i}$. That is, A^T is obtained from A by reflecting its entries across its main diagonal. For example, if

$$A = \begin{bmatrix} 1 & 2 & 3 \\ 4 & 5 & 6 \end{bmatrix} \quad \text{then} \quad A^T = \begin{bmatrix} 1 & 4 \\ 2 & 5 \\ 3 & 6 \end{bmatrix}.$$

Similarly, the **conjugate transpose** of $A \in \mathcal{M}_{m,n}(\mathbb{C})$ is the matrix

$A^* \in \mathcal{M}_{n,m}(\mathbb{C})$ whose (i,j)-entry is $\overline{a_{j,i}}$, where the horizontal line denotes complex conjugation (i.e., $\overline{a+ib} = a - ib$). In other words, $A^* \stackrel{\text{def}}{=} \overline{A}^T$. For example, if

Complex conjugation is reviewed in Appendix A.3.2.

$$A = \begin{bmatrix} 1 & 3-i & 2i \\ 2+3i & -i & 0 \end{bmatrix} \quad \text{then} \quad A^* = \begin{bmatrix} 1 & 2-3i \\ 3+i & i \\ -2i & 0 \end{bmatrix}.$$

Finally, we say that a matrix $A \in \mathcal{M}_n$ is **symmetric** if $A^T = A$ and that a matrix $B \in \mathcal{M}_n(\mathbb{C})$ is **Hermitian** if $B^* = B$. For example, if

$$A = \begin{bmatrix} 1 & 2 \\ 2 & 3 \end{bmatrix}, \quad B = \begin{bmatrix} 2 & 1+3i \\ 1-3i & -4 \end{bmatrix}, \quad \text{and} \quad C = \begin{bmatrix} 1 & 2 \\ 3 & 4 \end{bmatrix},$$

then A is symmetric (and Hermitian), B is Hermitian (but not symmetric), and C is neither.

Example 1.1.5

Is the Set of Hermitian Matrices a Vector Space?

Let $\mathcal{M}_n^H$ be the set of $n \times n$ Hermitian matrices. Show that if the field is $\mathbb{F} = \mathbb{R}$ then $\mathcal{M}_n^H$ is a vector space, but if $\mathbb{F} = \mathbb{C}$ then it is not.

Solution:

Before presenting the solution, we clarify that the field $\mathbb{F}$ does not specify what the entries of the Hermitian matrices are: in both cases (i.e., when $\mathbb{F} = \mathbb{R}$ and when $\mathbb{F} = \mathbb{C}$), the entries of the matrices themselves can be complex. The field $\mathbb{F}$ associated with a vector space just determines what types of scalars are used in scalar multiplication.

To illustrate this point, we start by showing that if $\mathbb{F} = \mathbb{C}$ then $\mathcal{M}_n^H$ is *not* a vector space. For example, property (f) of vector spaces fails because

$$A = \begin{bmatrix} 0 & 1 \\ 1 & 0 \end{bmatrix} \in \mathcal{M}_2^H, \quad \text{but} \quad iA = \begin{bmatrix} 0 & i \\ i & 0 \end{bmatrix} \notin \mathcal{M}_2^H.$$

That is, $\mathcal{M}_2^H$ (and by similar reasoning, $\mathcal{M}_n^H$) is not closed under multiplication by complex scalars.

On the other hand, to see that $\mathcal{M}_n^H$ *is* a vector space when $\mathbb{F} = \mathbb{R}$, we check the ten properties described by Definition 1.1.1. If $A, B, C \in \mathcal{M}_n^H$ and $c, d \in \mathbb{R}$ then:

a) $(A+B)^* = A^* + B^* = A + B$, so $A + B \in \mathcal{M}_n^H$.

d) The zero matrix O is Hermitian, so we choose it as the "zero vector" in $\mathcal{M}_n^H$.

e) If $A \in \mathcal{M}_n^H$ then $-A \in \mathcal{M}_n^H$ too.

f) Since c is real, we have $(cA)^* = cA^* = cA$, so $cA \in \mathcal{M}_n^H$ (whereas if c were complex then we would just have $(cA)^* = \overline{c}A$, which does not necessarily equal cA, as we saw above).

All of the other properties of vector spaces follow immediately using the same arguments that we used to show that $\mathcal{M}_{m,n}$ is a vector space in Example 1.1.2.

In cases where we wish to clarify which field we are using for scalar multiplication in a vector space $\mathcal{V}$, we say that $\mathcal{V}$ is a vector space "over" that field. For example, we say that $\mathcal{M}_n^H$ (the set of $n \times n$ Hermitian matrices) is a vector space over $\mathbb{R}$, but not a vector space over $\mathbb{C}$. Alternatively, we refer to the field in question as the **ground field** of $\mathcal{V}$ (so the ground field of $\mathcal{M}_n^H$, for example, is $\mathbb{R}$).

Despite the various vector spaces that we have seen looking so different on the surface, not much changes when we do linear algebra in this more general setting. To get a feeling for how we can prove things about vector spaces in general, we now prove our very first theorem. We can think of this theorem as answering the question of why we did not include some other properties like "$0\mathbf{v} = \mathbf{0}$" in the list of defining properties of vector spaces, even though we did include "$1\mathbf{v} = \mathbf{v}$". The reason is simply that listing these extra properties would be redundant, as they follow from the ten properties that we did list.

Theorem 1.1.1

Zero Vector and Additive Inverses via Scalar Multiplication

Suppose $\mathcal{V}$ is a vector space and $\mathbf{v} \in \mathcal{V}$. Then

a) $0\mathbf{v} = \mathbf{0}$, and
b) $(-1)\mathbf{v} = -\mathbf{v}$.

Proof. To show that $0\mathbf{v} = \mathbf{0}$, we carefully use the various defining properties of vector spaces from Definition 1.1.1:

$$
\begin{aligned}
0\mathbf{v} &= 0\mathbf{v} + \mathbf{0} && \text{(property (d))} \\
&= 0\mathbf{v} + \big(0\mathbf{v} + (-(0\mathbf{v}))\big) && \text{(property (e))} \\
&= (0\mathbf{v} + 0\mathbf{v}) + \big(-(0\mathbf{v})\big) && \text{(property (c))} \\
&= (0 + 0)\mathbf{v} + \big(-(0\mathbf{v})\big) && \text{(property (h))} \\
&= 0\mathbf{v} + \big(-(0\mathbf{v})\big) && (0 + 0 = 0 \text{ in every field}) \\
&= \mathbf{0}. && \text{(property (e))}
\end{aligned}
$$

Proving things about vector spaces will quickly become less tedious than in this theorem—as we develop more tools, we will find ourselves referencing the defining properties (a)–(j) less and less.

Now that we have $0\mathbf{v} = \mathbf{0}$ to work with, proving that $(-1)\mathbf{v} = -\mathbf{v}$ is a bit more straightforward:

$$
\begin{aligned}
\mathbf{0} &= 0\mathbf{v} && \text{(we just proved this)} \\
&= (1 - 1)\mathbf{v} && (1 - 1 = 0 \text{ in every field}) \\
&= 1\mathbf{v} + (-1)\mathbf{v} && \text{(property (h))} \\
&= \mathbf{v} + (-1)\mathbf{v}. && \text{(property (j))}
\end{aligned}
$$

It follows that $(-1)\mathbf{v} = -\mathbf{v}$, which completes the proof. ∎

From now on, we write vector subtraction in the usual way $\mathbf{v} - \mathbf{w}$ that we are used to. The above theorem ensures that there is no ambiguity when we write subtraction in this way, since it does not matter if $\mathbf{v} - \mathbf{w}$ is taken to mean $\mathbf{v} + (-\mathbf{w})$ or $\mathbf{v} + (-1)\mathbf{w}$.

1.1.1 Subspaces

It is often useful to work with vector spaces that are contained within other vector spaces. This situation comes up often enough that it gets its own name:

Definition 1.1.2

Subspace

If $\mathcal{V}$ is a vector space and $\mathcal{S} \subseteq \mathcal{V}$, then $\mathcal{S}$ is a **subspace** of $\mathcal{V}$ if $\mathcal{S}$ is itself a vector space with the same addition, scalar multiplication, and ground field as $\mathcal{V}$.

It turns out that checking whether or not something is a subspace is much simpler than checking whether or not it is a vector space. We already saw this somewhat in Example 1.1.5, where we only explicitly showed that four vector space properties hold for $\mathcal{M}_n^H$ instead of all ten. The reason we could do this is that $\mathcal{M}_n^H$ is a subset of $\mathcal{M}_n(\mathbb{C})$, so it inherited many of the properties of vector spaces "for free" from the fact that we already knew that $\mathcal{M}_n(\mathbb{C})$ is a vector space.

It turns out that even checking four properties for subspaces is overkill—we only have to check two:

Theorem 1.1.2

Determining if a Set is a Subspace

Let $\mathcal{V}$ be a vector space over a field $\mathbb{F}$ and let $\mathcal{S} \subseteq \mathcal{V}$ be non-empty. Then $\mathcal{S}$ is a subspace of $\mathcal{V}$ if and only if the following two conditions hold:

a) If $\mathbf{v}, \mathbf{w} \in \mathcal{S}$ then $\mathbf{v} + \mathbf{w} \in \mathcal{S}$, and   (closure under addition)

b) if $\mathbf{v} \in \mathcal{S}$ and $c \in \mathbb{F}$ then $c\mathbf{v} \in \mathcal{S}$.   (closure under scalar mult.)

Proof. For the "only if" direction, properties (a) and (b) in this theorem are properties (a) and (f) in Definition 1.1.1 of a vector space, so of course they must hold if $\mathcal{S}$ is a subspace (since subspaces are vector spaces).

For the "if" direction, we have to show that all ten properties (a)–(j) in Definition 1.1.1 of a vector space hold for $\mathcal{S}$:

- Properties (a) and (f) hold by hypothesis.
- Properties (b), (c), (g), (h), (i), and (j) hold for *all* vectors in $\mathcal{V}$, so they certainly hold for all vectors in $\mathcal{S}$ too, since $\mathcal{S} \subseteq \mathcal{V}$.

> Every subspace, just like every vector space, must contain a zero vector.

- For property (d), we need to show that $\mathbf{0} \in \mathcal{S}$. If $\mathbf{v} \in \mathcal{S}$ then we know that $0\mathbf{v} \in \mathcal{S}$ too since $\mathcal{S}$ is closed under scalar multiplication. However, we know from Theorem 1.1.1(a) that $0\mathbf{v} = \mathbf{0}$, so we are done.
- For property (e), we need to show that if $\mathbf{v} \in \mathcal{S}$ then $-\mathbf{v} \in \mathcal{S}$ too. We know that $(-1)\mathbf{v} \in \mathcal{S}$ since $\mathcal{S}$ is closed under scalar multiplication, and we know from Theorem 1.1.1(b) that $(-1)\mathbf{v} = -\mathbf{v}$, so we are done.

∎

Example 1.1.6

The Set of Polynomials is a Subspace

Let $\mathcal{P}^p$ be the set of real-valued polynomials of degree at most p. Show that $\mathcal{P}^p$ a subspace of $\mathcal{F}$, the vector space of all real-valued functions.

Solution:

We just have to check the two properties described by Theorem 1.1.2. Hopefully it is somewhat clear that adding two polynomials of degree at most p results in another polynomial of degree at most p, and similarly multiplying a polynomial of degree at most p by a scalar results in another one, but we make this computation explicit.

> The **degree** of a polynomial is the largest exponent to which the variable is raised, so a polynomial of degree at most p looks like $f(x) = a_p x^p + \cdots + a_1 x + a_0$, where $a_p, \ldots, a_1, a_0 \in \mathbb{R}$. The degree of f is *exactly* p if $a_p \neq 0$. See Appendix A.2 for an introduction to polynomials.

Suppose $f(x) = a_p x^p + \cdots + a_1 x + a_0$ and $g(x) = b_p x^p + \cdots + b_1 x + b_0$ are polynomials of degree at most p (i.e., $f, g \in \mathcal{P}^p$) and $c \in \mathbb{R}$ is a scalar. Then:

a) We compute

$$(f+g)(x) = (a_p x^p + \cdots + a_1 x + a_0) + (b_p x^p + \cdots + b_1 x + b_0)$$
$$= (a_p + b_p)x^p + \cdots + (a_1 + b_1)x + (a_0 + b_0),$$

which is also a polynomial of degree at most p, so $f + g \in \mathcal{P}^p$.

b) Just like before, we compute

$$(cf)(x) = c(a_p x^p + \cdots + a_1 x + a_0)$$
$$= (ca_p)x^p + \cdots + (ca_1)x + (ca_0),$$

which is also a polynomial of degree at most p, so $cf \in \mathcal{P}^p$.

Slightly more generally, the set $\mathcal{P}^p(\mathbb{F})$ of polynomials with coefficients from a field $\mathbb{F}$ is also a subspace of the vector space $\mathcal{F}(\mathbb{F})$ of functions from $\mathbb{F}$ to itself, even when $\mathbb{F} \neq \mathbb{R}$. However, $\mathcal{P}^p(\mathbb{F})$ has some subtle properties that make it difficult to work with in general (see Exercise 1.2.34, for example), so we only consider it in the case when $\mathbb{F} = \mathbb{R}$ or $\mathbb{F} = \mathbb{C}$.

Recall that a function is called "differentiable" if it has a derivative.

Similarly, the set of continuous functions $\mathcal{C}$ is also a subspace of $\mathcal{F}$, as is the set of differentiable functions $\mathcal{D}$ (see Exercise 1.1.10). Since every polynomial is differentiable, and every differentiable function is continuous, it is even the case that $\mathcal{P}^p$ is a subspace of $\mathcal{D}$, which is a subspace of $\mathcal{C}$, which is a subspace of $\mathcal{F}$.

Example 1.1.7
The Set of Upper Triangular Matrices is a Subspace

Show that the set of $n \times n$ upper triangular matrices a subspace of $\mathcal{M}_n$.

Solution:
Again, we have to check the two properties described by Theorem 1.1.2, and it is again somewhat clear that both of these properties hold. For example, if we add two matrices with zeros below the diagonal, their sum will still have zeros below the diagonal.

Recall that a matrix A is called upper triangular if $a_{i,j} = 0$ whenever $i > j$. For example, a 2×2 upper triangular matrix has the form

$$A = \begin{bmatrix} a & b \\ 0 & c \end{bmatrix}.$$

We now formally prove that these properties hold. Suppose $A, B \in \mathcal{M}_n$ are upper triangular (i.e., $a_{i,j} = b_{i,j} = 0$ whenever $i > j$) and $c \in \mathbb{F}$ is a scalar.

a) The (i,j)-entry of $A + B$ is $a_{i,j} + b_{i,j} = 0 + 0 = 0$ whenever $i > j$, and

b) the (i,j)-entry of cA is $ca_{i,j} = 0c = 0$ whenever $i > j$.

Since both properties are satisfied, we conclude that the set of $n \times n$ upper triangular matrices is indeed a subspace of $\mathcal{M}_n$.

Similar to the previous example, the set $\mathcal{M}_n^S$ of $n \times n$ symmetric matrices is also a subspace of $\mathcal{M}_n$ (see Exercise 1.1.8). It is worth noting, however, that the set $\mathcal{M}_n^H$ of $n \times n$ *Hermitian* matrices is *not* a subspace of $\mathcal{M}_n(\mathbb{C})$ unless we regard $\mathcal{M}_n(\mathbb{C})$ as a vector space over $\mathbb{R}$ instead of $\mathbb{C}$ (which is possible, but quite non-standard).

Example 1.1.8
The Set of Non-Invertible Matrices is Not a Subspace

Show that the set of non-invertible 2×2 matrices is not a subspace of $\mathcal{M}_2$.

Solution:
This set is not a subspace because it is not closed under addition. For example, the following matrices are not invertible:

Similar examples can be used to show that, for all $n \geq 1$, the set of non-invertible $n \times n$ matrices is not a subspace of $\mathcal{M}_n$.

$$A = \begin{bmatrix} 1 & 0 \\ 0 & 0 \end{bmatrix}, \qquad B = \begin{bmatrix} 0 & 0 \\ 0 & 1 \end{bmatrix}.$$

However, their sum is

$$A + B = \begin{bmatrix} 1 & 0 \\ 0 & 1 \end{bmatrix},$$

which *is* invertible. It follows that property (a) of Theorem 1.1.2 does not hold, so this set is not a subspace of $\mathcal{M}_2$.

Example 1.1.9

The Set of Integer-Entry Vectors is Not a Subspace

Show that $\mathbb{Z}^3$, the set of 3-entry vectors with *integer* entries, is not a subspace of $\mathbb{R}^3$.

Solution:

This set is a subset of $\mathbb{R}^3$ and is closed under addition, but it is *not* a subspace because it is not closed under scalar multiplication. Because we are asking whether or not it is a subspace of $\mathbb{R}^3$, which uses $\mathbb{R}$ as its ground field, we must use the same scalars here as well. However, if $\mathbf{v} \in \mathbb{Z}^3$ and $c \in \mathbb{R}$ then $c\mathbf{v}$ may not be in $\mathbb{Z}^3$. For example, if

$$\mathbf{v} = (1,2,3) \in \mathbb{Z}^3 \quad \text{then} \quad \frac{1}{2}\mathbf{v} = (1/2,1,3/2) \notin \mathbb{Z}^3.$$

It follows that property (b) of Theorem 1.1.2 does not hold, so this set is not a subspace of $\mathbb{R}^3$.

Example 1.1.10

The Set of Eventually-Zero Sequences is a Subspace

In the notation c_{00}, the "c" refers to the sequences converging and the "00" refers to how they converge to 0 and eventually equal 0.

Show that the set $c_{00} \subset \mathbb{F}^{\mathbb{N}}$ of sequences with only finitely many non-zero entries is a subspace of the sequence space $\mathbb{F}^{\mathbb{N}}$.

Solution:

Once again, we have to check properties (a) and (b) of Theorem 1.1.2. That is, if $\mathbf{v}, \mathbf{w} \in c_{00}$ have only finitely many non-zero entries and $c \in \mathbb{F}$ is a scalar, then we have to show that (a) $\mathbf{v} + \mathbf{w}$ and (b) $c\mathbf{v}$ each have finitely many non-zero entries as well.

These properties are both straightforward to show—if $\mathbf{v}$ has m non-zero entries and $\mathbf{w}$ has n non-zero entries then $\mathbf{v} + \mathbf{w}$ has at most $m + n$ non-zero entries and $c\mathbf{v}$ has either m non-zero entries (if $c \neq 0$) or 0 non-zero entries (if $c = 0$).

1.1.2 Spans, Linear Combinations, and Independence

We now start re-introducing various aspects of linear algebra that we have already seen in an introductory course in the more concrete setting of subspaces of $\mathbb{R}^n$. All of these concepts (in particular, spans, linear combinations, and linear (in)dependence for now) behave almost exactly the same in general vector spaces as they do in $\mathbb{R}^n$ (and its subspaces), so our presentation of these topics is quite brief.

Definition 1.1.3

Linear Combina-
tions

Suppose $\mathcal{V}$ is a vector space over a field $\mathbb{F}$. A **linear combination** of the vectors $\mathbf{v}_1, \mathbf{v}_2, \ldots, \mathbf{v}_k \in \mathcal{V}$ is any vector of the form

$$c_1 \mathbf{v}_1 + c_2 \mathbf{v}_2 + \cdots + c_k \mathbf{v}_k,$$

It is important that the sum presented here is finite.

where $c_1, c_2, \ldots, c_k \in \mathbb{F}$.

For example, in the vector space $\mathcal{P}^2$ of polynomials with degree at most 2, the polynomial $p(x) = 3x^2 + x - 2$ is a linear combination of the polynomials x^2, x, and 1 (and in fact *every* polynomial in $\mathcal{P}^2$ is a linear combination of x^2, x, and 1). We now start looking at some less straightforward examples.

Example 1.1.11

Linear Combina-
tions of Polynomials

Suppose $f, g, h \in \mathcal{P}^2$ are given by the formulas

$$f(x) = x^2 - 3x - 4, \quad g(x) = x^2 - x + 2, \quad \text{and} \quad h(x) = 2x^2 - 3x + 1.$$

Determine whether or not f is a linear combination of g and h.

Solution:

We want to know whether or not there exist $c_1, c_2 \in \mathbb{R}$ such that

$$f(x) = c_1 g(x) + c_2 h(x).$$

Writing this equation out more explicitly gives

$$x^2 - 3x - 4 = c_1(x^2 - x + 2) + c_2(2x^2 - 3x + 1)$$
$$= (c_1 + 2c_2)x^2 + (-c_1 - 3c_2)x + (2c_1 + c_2).$$

We now use the fact that two polynomials are equal if and only if their coefficients are equal: we set the coefficients of x^2 on both sides of the equation equal to each other, the coefficients of x equal to each other, and the constant terms equal to each other. This gives us the linear system

$$1 = c_1 + 2c_2$$
$$-3 = -c_1 - 3c_2$$
$$-4 = 2c_1 + c_2.$$

This linear system can be solved using standard techniques (e.g., Gaussian elimination) to find the unique solution $c_1 = -3, c_2 = 2$. It follows that f *is* a linear combination of g and h: $f(x) = -3g(x) + 2h(x)$.

Example 1.1.12

Linear Combina-
tions of Matrices

Determine whether or not the identity matrix $I \in \mathcal{M}_2(\mathbb{C})$ is a linear combination of the three matrices

$$X = \begin{bmatrix} 0 & 1 \\ 1 & 0 \end{bmatrix}, \quad Y = \begin{bmatrix} 0 & -i \\ i & 0 \end{bmatrix}, \quad \text{and} \quad Z = \begin{bmatrix} 1 & 0 \\ 0 & -1 \end{bmatrix}.$$

*The four matrices $I, X, Y, Z \in \mathcal{M}_2(\mathbb{C})$ are sometimes called the **Pauli matrices**.*

Solution:

We want to know whether or not there exist $c_1, c_2, c_3 \in \mathbb{C}$ such that $I = c_1 X + c_2 Y + c_3 Z$. Writing this matrix equation out more explicitly gives

$$\begin{bmatrix} 1 & 0 \\ 0 & 1 \end{bmatrix} = c_1 \begin{bmatrix} 0 & 1 \\ 1 & 0 \end{bmatrix} + c_2 \begin{bmatrix} 0 & -i \\ i & 0 \end{bmatrix} + c_3 \begin{bmatrix} 1 & 0 \\ 0 & -1 \end{bmatrix}$$

$$= \begin{bmatrix} c_3 & c_1 - ic_2 \\ c_1 + ic_2 & -c_3 \end{bmatrix}.$$

The $(1,1)$-entry of the above matrix equation tells us that $c_3 = 1$, but the $(2,2)$-entry tells us that $c_3 = -1$, so this system of equations has no solution. It follows that I is *not* a linear combination of X, Y, and Z.

Linear combinations are useful for the way that they combine both of the vector space operations (vector addition and scalar multiplication)—instead of phrasing linear algebraic phenomena in terms of those two operations, we can often phrase them more elegantly in terms of linear combinations.

For example, we saw in Theorem 1.1.2 that a subspace is a subset of a vector space that is closed under vector addition and scalar multiplication. Equivalently, we can combine those two operations and just say that a subspace is a subset of a vector space that is closed under linear combinations (see Exercise 1.1.11). On the other hand, a subset B of a vector space that is *not* closed under linear combinations is necessarily not a subspace.

However, it is often useful to consider the smallest subspace containing B. We call this smallest subspace the **span** of B, and to construct it we just take all linear combinations of members of B:

Definition 1.1.4	Suppose $\mathcal{V}$ is a vector space and $B \subseteq \mathcal{V}$ is a set of vectors. The **span** of B, denoted by $\mathrm{span}(B)$, is the set of all (finite!) linear combinations of vectors from B:
Span	

$$\mathrm{span}(B) \stackrel{\text{def}}{=} \left\{ \sum_{j=1}^{k} c_j \mathbf{v}_j \ \middle|\ k \in \mathbb{N},\ c_j \in \mathbb{F} \text{ and } \mathbf{v}_j \in B \text{ for all } 1 \le j \le k \right\}.$$

Furthermore, if $\mathrm{span}(B) = \mathcal{V}$ then we say that $\mathcal{V}$ is **spanned** by B.

In $\mathbb{R}^n$, for example, the span of a single vector is the line through the origin in the direction of that vector, and the span of two non-parallel vectors is the plane through the origin containing those vectors (see Figure 1.1).

Two or more vectors can also span a line, if they are all parallel.

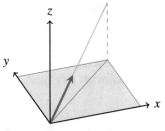

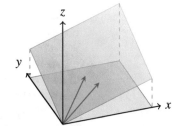

(a) A single vector spans a line.

(b) Two non-parallel vectors span a plane.

Figure 1.1: The span of a set of vectors is the smallest subspace that contains all of those vectors. In $\mathbb{R}^n$, this smallest subspace is a line, plane, or hyperplane containing all of the vectors in the set.

When we work in other vector spaces, we lose much of this geometric

We use span$(1,x,x^2)$ as slight shorthand for span$(\{1,x,x^2\})$— we sometimes omit the curly set braces.

interpretation, but algebraically spans still work much like they do in $\mathbb{R}^n$. For example, span$(1,x,x^2) = \mathcal{P}^2$ (the vector space of polynomials with degree at most 2) since every polynomial $f \in \mathcal{P}^2$ can be written in the form $f(x) = c_1 + c_2 x + c_3 x^2$ for some $c_1, c_2, c_3 \in \mathbb{R}$. Indeed, this is exactly what it means for a polynomial to have degree at most 2. More generally, span$(1,x,x^2,\ldots,x^p) = \mathcal{P}^p$.

However, it is important to keep in mind that linear combinations are always *finite*, even if B is not. To illustrate this point, consider the vector space $\mathcal{P} = $ span$(1,x,x^2,x^3,\ldots)$, which is the set of all polynomials (of any degree). If we recall from calculus that we can represent the function $f(x) = e^x$ in the form

This is called a **Taylor series** for e^x (see Appendix A.2.2).

$$e^x = \sum_{n=0}^{\infty} \frac{x^n}{n!} = 1 + x + \frac{x^2}{2} + \frac{x^3}{6} + \frac{x^4}{24} + \cdots,$$

we might expect that $e^x \in \mathcal{P}$, since we have written e^x as a sum of scalar multiples of 1, x, x^2, and so on. However, $e^x \notin \mathcal{P}$ since e^x can only be written as an *infinite* sum of polynomials, not a finite one (see Figure 1.2).

While all subspaces of $\mathbb{R}^n$ are "closed" (roughly speaking, they contain their edges / limits / boundaries), this example illustrates the fact that some vector spaces (like $\mathcal{P}$) are not.

To make this idea of a function being on the "boundary" of $\mathcal{P}$ precise, we need a way of measuring the distance between functions—we describe how to do this in Sections 1.3.4 and 1.D.

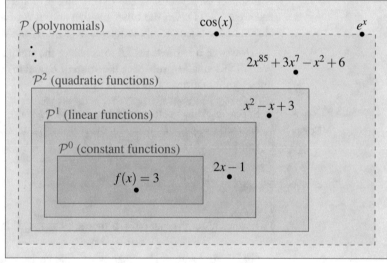

Figure 1.2: The vector spaces $\mathcal{P}^0 \subset \mathcal{P}^1 \subset \mathcal{P}^2 \subset \cdots \subset \mathcal{P}^p \subset \cdots \subset \mathcal{P} \subset \mathcal{F}$ are subspaces of each other in the manner indicated. The vector space $\mathcal{P}$ of all polynomials is interesting for the fact that it does not contain its boundary: functions like e^x and $\cos(x)$ can be approximated by polynomials and are thus on the boundary of $\mathcal{P}$, but are not polynomials themselves (i.e., they cannot be written as a *finite* linear combination of $1,x,x^2,\ldots$).

As we suggested earlier, our primary reason for being interested in the span of a set of vectors is that it is always a subspace. We now state and prove this fact rigorously.

Theorem 1.1.3

Spans are Subspaces

Let $\mathcal{V}$ be a vector space and let $B \subseteq \mathcal{V}$. Then span(B) is a subspace of $\mathcal{V}$.

Proof. We need to check that the two closure properties described by Theorem 1.1.2 are satisfied. We thus suppose that $\mathbf{v}, \mathbf{w} \in$ span(B) and $b \in \mathbb{F}$. Then (by the definition of span(B)) there exist scalars $c_1, c_2, \ldots, c_k, d_1, d_2, \ldots, d_\ell \in \mathbb{F}$

The sets $\{\mathbf{v}_1,\ldots,\mathbf{v}_k\}$ and $\{\mathbf{w}_1,\ldots,\mathbf{w}_\ell\}$ may be disjoint or they may overlap—it does not matter.

and vectors $\mathbf{v}_1, \mathbf{v}_2, \ldots, \mathbf{v}_k, \mathbf{w}_1, \mathbf{w}_2, \ldots, \mathbf{w}_\ell \in B$ such that

$$\mathbf{v} = c_1\mathbf{v}_1 + c_2\mathbf{v}_2 + \cdots + c_k\mathbf{v}_k \qquad \text{and} \qquad \mathbf{w} = d_1\mathbf{w}_1 + d_2\mathbf{w}_2 + \cdots + d_\ell\mathbf{w}_\ell.$$

To establish property (a) of Theorem 1.1.2, we note that

$$\mathbf{v} + \mathbf{w} = (c_1\mathbf{v}_1 + c_2\mathbf{v}_2 + \cdots + c_k\mathbf{v}_k) + (d_1\mathbf{w}_1 + d_2\mathbf{w}_2 + \cdots + d_\ell\mathbf{w}_\ell),$$

which is a linear combination of $\mathbf{v}_1, \mathbf{v}_2, \ldots, \mathbf{v}_k, \mathbf{w}_1, \mathbf{w}_2, \ldots, \mathbf{w}_\ell$, so we conclude that $\mathbf{v} + \mathbf{w} \in \text{span}(B)$.

Similarly, to show that property (b) holds, we check that

$$b\mathbf{v} = b(c_1\mathbf{v}_1 + c_2\mathbf{v}_2 + \cdots + c_k\mathbf{v}_k) = (bc_1)\mathbf{v}_1 + (bc_2)\mathbf{v}_2 + \cdots + (bc_k)\mathbf{v}_k,$$

which is a linear combination of $\mathbf{v}_1, \mathbf{v}_2, \ldots, \mathbf{v}_k$, so $b\mathbf{v} \in \text{span}(B)$. Since properties (a) and (b) are both satisfied, $\text{span}(B)$ is a subspace of $\mathcal{V}$. $\blacksquare$

Just like subspaces and spans, linear dependence and independence in general vector spaces are defined almost identically to how they are defined in $\mathbb{R}^n$. The biggest difference in this more general setting is that it is now useful to consider linear independence of sets containing infinitely many vectors (whereas any infinite subset of $\mathbb{R}^n$ is necessarily linearly dependent, so the definition of linear independence in $\mathbb{R}^n$ typically bakes finiteness right in).

Definition 1.1.5

Linear Dependence and Independence

Suppose $\mathcal{V}$ is a vector space over a field $\mathbb{F}$ and $B \subseteq \mathcal{V}$. We say that B is **linearly dependent** if there exist scalars $c_1, c_2, \ldots, c_k \in \mathbb{F}$, at least one of which is not zero, and vectors $\mathbf{v}_1, \mathbf{v}_2, \ldots, \mathbf{v}_k \in B$ such that

$$c_1\mathbf{v}_1 + c_2\mathbf{v}_2 + \cdots + c_k\mathbf{v}_k = \mathbf{0}.$$

If B is not linearly dependent then it is called **linearly independent**.

Notice that we choose finitely many vectors $\mathbf{v}_1, \mathbf{v}_2, \ldots, \mathbf{v}_k$ in this definition even if B is infinite.

Linear dependence and independence work very similarly to how they did in $\mathbb{R}^n$. The rough intuition for them is that linearly dependent sets are "redundant" in the sense that one of the vectors does not point in a "really different" direction than the other vectors—it can be obtained from the others via some linear combination. Geometrically, in $\mathbb{R}^n$ this corresponds to two or more vectors lying on a common line, three or more vectors lying on a common plane, and so on (see Figure 1.3).

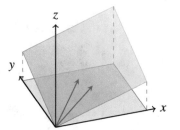

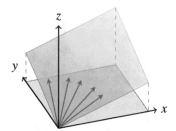

(a) Two non-parallel vectors are linearly independent.

(b) Three or more vectors on a common plane are linearly dependent.

Figure 1.3: A set of vectors is linearly independent if and only if each vector contributes a new direction or dimension.

Furthermore, just as was the case with linear (in)dependence in $\mathbb{R}^n$, we still have all of the following properties that make working with linear (in)dependence easier in certain special cases:

- A set of two vectors $B = \{\mathbf{v}, \mathbf{w}\}$ is linearly dependent if and only if $\mathbf{v}$ and $\mathbf{w}$ are scalar multiples of each other.
- A finite set of vectors $B = \{\mathbf{v}_1, \mathbf{v}_2, \ldots, \mathbf{v}_k\}$ is linearly independent if and only if the equation

$$c_1 \mathbf{v}_1 + c_2 \mathbf{v}_2 + \cdots + c_k \mathbf{v}_k = \mathbf{0}$$

has a unique solution: $c_1 = c_2 = \cdots = c_k = 0$.

- A (not necessarily finite) set B is linearly dependent if and only if there exists a vector $\mathbf{v} \in B$ that can be written as a linear combination of the other vectors from B.

Example 1.1.13

Linear Independence of Polynomials

Is the set $B = \{2x^2 + 1, x^2 - x + 1\}$ linearly independent in $\mathcal{P}^2$?

Solution:

Since this set contains just two polynomials, it suffices to just check whether or not they are scalar multiples of each other. Since they are *not* scalar multiples of each other, B is linearly independent.

Example 1.1.14

Linear Independence of Matrices

Determine whether or not the following set of matrices linearly independent in $\mathcal{M}_2(\mathbb{R})$:

$$B = \left\{ \begin{bmatrix} 1 & 1 \\ 1 & 1 \end{bmatrix}, \begin{bmatrix} 1 & -1 \\ -1 & 1 \end{bmatrix}, \begin{bmatrix} -1 & 2 \\ 2 & -1 \end{bmatrix} \right\}.$$

Solution:

Since this set is finite, we want to check whether the equation

$$c_1 \begin{bmatrix} 1 & 1 \\ 1 & 1 \end{bmatrix} + c_2 \begin{bmatrix} 1 & -1 \\ -1 & 1 \end{bmatrix} + c_3 \begin{bmatrix} -1 & 2 \\ 2 & -1 \end{bmatrix} = \begin{bmatrix} 0 & 0 \\ 0 & 0 \end{bmatrix}$$

has a unique solution (which would necessarily be $c_1 = c_2 = c_3 = 0$, corresponding to linear independence) or infinitely many solutions (corresponding to linear dependence). We can solve for c_1, c_2, and c_3 by comparing entries of the matrices on the left- and right-hand sides above to get the linear system

$$\begin{aligned} c_1 + c_2 - c_3 &= 0 \\ c_1 - c_2 + 2c_3 &= 0 \\ c_1 - c_2 + 2c_3 &= 0 \\ c_1 + c_2 - c_3 &= 0. \end{aligned}$$

In general, to find an explicit linear combination that demonstrates linear dependence, we can choose the free variable(s) to be any non-zero value(s) that we like.

Solving this linear system via our usual methods reveals that c_3 is a free variable (so there are infinitely many solutions) and $c_1 = -(1/2)c_3$, $c_2 = (3/2)c_3$. It follows that B is linearly dependent, and in particular, choosing $c_3 = 2$ gives $c_1 = -1$ and $c_2 = 3$, so

$$- \begin{bmatrix} 1 & 1 \\ 1 & 1 \end{bmatrix} + 3 \begin{bmatrix} 1 & -1 \\ -1 & 1 \end{bmatrix} + 2 \begin{bmatrix} -1 & 2 \\ 2 & -1 \end{bmatrix} = \begin{bmatrix} 0 & 0 \\ 0 & 0 \end{bmatrix}.$$

Example 1.1.15

Linear Independence of Many Polynomials

Is the set $B = \{1, x, x^2, \ldots, x^p\}$ linearly independent in $\mathcal{P}^p$?

Solution:

Since this set is finite, we want to check whether the equation

$$c_0 + c_1 x + c_2 x^2 + \cdots + c_p x^p = 0 \qquad (1.1.1)$$

has a unique solution (linear independence) or infinitely many solutions (linear dependence). By plugging $x = 0$ into that equation, we see that $c_0 = 0$.

Taking the derivative of both sides of Equation (1.1.1) then reveals that

$$c_1 + 2c_2 x + 3c_3 x^2 + \cdots + pc_p x^{p-1} = 0,$$

All we are doing here is showing that if $f \in \mathcal{P}^p$ satisfies $f(x) = 0$ for all x then all of its coefficients equal 0. This likely seems "obvious", but it is still good to pin down why it is true.

and plugging $x = 0$ into this equation gives $c_1 = 0$. By repeating this procedure (i.e., taking the derivative and then plugging in $x = 0$) we similarly see that $c_2 = c_3 = \cdots = c_p = 0$, so B is linearly independent.

Example 1.1.16

Linear Independence of an Infinite Set

Is the set $B = \{1, x, x^2, x^3, \ldots\}$ linearly independent in $\mathcal{P}$?

Solution:

This example is a bit trickier than the previous one since the set contains infinitely many polynomials (vectors), and we are *not* asking whether or not there exist scalars $c_0, c_1, c_2, \ldots$, not all equal to zero, such that

$$c_0 + c_1 x + c_2 x^2 + c_3 x^3 + \cdots = 0.$$

Instead, we are asking whether or not there exists some *finite* linear combination of $1, x, x^2, x^3, \ldots$ that adds to 0 (and does not have all coefficients equal to 0).

Let p be the largest power of x in such a linear combination—we want to know if there exists (not all zero) scalars $c_0, c_1, c_2, \ldots, c_p$ such that

Recall that in $\mathbb{R}^n$ it was not possible to have a linearly independent set consisting of infinitely many vectors (such sets had at most n vectors). This is one way in which general vector spaces can differ from $\mathbb{R}^n$.

$$c_0 + c_1 x + c_2 x^2 + \cdots + c_p x^p = 0.$$

It follows from the exact same argument used in Example 1.1.15 that this equation has unique solution $c_0 = c_1 = c_2 = \cdots = c_p = 0$, so B is a linearly independent set.

The previous examples could all be solved more or less just by "plugging and chugging"—we just set up the linear (in)dependence equation

$$c_1 \mathbf{v}_1 + c_2 \mathbf{v}_2 + \cdots + c_k \mathbf{v}_k = \mathbf{0}$$

and worked through the calculation to solve for $c_1, c_2, \ldots, c_k$. We now present some examples to show that, in some more exotic vector spaces like $\mathcal{F}$, checking linear independence is not always so straightforward, as there is no obvious or systematic way to solve that linear (in)dependence equation.

Example 1.1.17

Linear Independence of a Set of Functions

Is the set $B = \{\cos(x), \sin(x), \sin^2(x)\}$ linearly independent in $\mathcal{F}$?

Solution:

Since this set is finite, we want to determine whether or not there exist scalars $c_1, c_2, c_3 \in \mathbb{R}$ (not all equal to 0) such that

$$c_1 \cos(x) + c_2 \sin(x) + c_3 \sin^2(x) = 0.$$

We could also plug in other values of x to get other equations involving c_1, c_2, c_3.

Plugging in $x = 0$ tells us that $c_1 = 0$. Then plugging in $x = \pi/2$ tells us that $c_2 + c_3 = 0$, and plugging in $x = 3\pi/2$ tells us that $-c_2 + c_3 = 0$. Solving this system of equations involving c_2 and c_3 reveals that $c_2 = c_3 = 0$ as well, so B is a linearly independent set.

Example 1.1.18

Linear Dependence of a Set of Functions

Is the set $B = \{\sin^2(x), \cos^2(x), \cos(2x)\}$ linearly independent in $\mathcal{F}$?

Solution:

Since this set is finite, we want to determine whether or not there exist scalars $c_1, c_2, c_3 \in \mathbb{R}$ (not all equal to 0) such that

$$c_1 \sin^2(x) + c_2 \cos^2(x) + c_3 \cos(2x) = 0.$$

On the surface, this set *looks* linearly independent, and we could try proving it by plugging in specific x values to get some equations involving c_1, c_2, c_3 (just like in Example 1.1.17). However, it turns out that this won't work—the resulting system of linear equations will not have a unique solution, regardless of the x values we choose. To see why this is the case, recall the trigonometric identity

This identity follows from the angle-sum identity $\cos(\theta + \phi) = \cos(\theta)\cos(\phi) - \sin(\theta)\sin(\phi)$. In particular, plug in $\theta = x$ and $\phi = x$.

$$\cos(2x) = \cos^2(x) - \sin^2(x).$$

In particular, this tells us that if $c_1 = 1, c_2 = -1$ and $c_3 = 1$, then

$$\sin^2(x) - \cos^2(x) + \cos(2x) = 0,$$

so B is linearly dependent.

We introduce a somewhat more systematic method of proving linear independence of a set of functions in $\mathcal{F}$ in Exercise 1.1.21. However, checking linear (in)dependence in general can still be quite difficult.

1.1.3 Bases

In the final two examples of the previous subsection (i.e., Examples 1.1.17 and 1.1.18, which involve functions in $\mathcal{F}$), we had to fiddle around and "guess" an approach that would work to show linear (in)dependence. In Example 1.1.17, we stumbled upon a proof that the set is linearly independent by luckily picking a bunch of x values that gave us a linear system with a unique solution, and in Example 1.1.18 we were only able to prove linear dependence because we conveniently already knew a trigonometric identity that related the given functions to each other.

Another method of proving linear independence in $\mathcal{F}$ is explored in Exercise 1.1.21.

The reason that determining linear (in)dependence in $\mathcal{F}$ is so much more difficult than in $\mathcal{P}^p$ or $\mathcal{M}_{m,n}$ is that we do not have a nice basis for $\mathcal{F}$ that we can work with, whereas we do for the other vector spaces that we have introduced so far ($\mathbb{R}^n$, polynomials, and matrices). In fact, we have been working with those nice bases already without even realizing it, and without even knowing what a basis is in vector spaces other than $\mathbb{R}^n$.

With this in mind, we now introduce bases of arbitrary vector spaces and start discussing how they make vector spaces easier to work with. Not surprisingly, they are defined in almost exactly the same way as they are for subspaces of $\mathbb{R}^n$.

Definition 1.1.6	A **basis** of a vector space $\mathcal{V}$ is a set of vectors in $\mathcal{V}$ that
Bases	a) spans $\mathcal{V}$, and
	b) is linearly independent.

For example, in $\mathbb{R}^3$ there are three standard basis vectors: $\mathbf{e}_1 = (1,0,0)$, $\mathbf{e}_2 = (0,1,0)$, and $\mathbf{e}_3 = (0,0,1)$.

The prototypical example of a basis is the set consisting of the **standard basis vectors** $\mathbf{e}_j = (0,\ldots,0,1,0,\ldots,0) \in \mathbb{R}^n$ for $1 \leq j \leq n$, where there is a single 1 in the j-th position and 0s elsewhere. It is straightforward to show that the set $\{\mathbf{e}_1,\mathbf{e}_2,\ldots,\mathbf{e}_n\} \subset \mathbb{R}^n$ is a basis of $\mathbb{R}^n$, and in fact it is called the **standard basis** of $\mathbb{R}^n$.

As an example of a basis of a vector space other than $\mathbb{R}^n$, consider the set $B = \{1,x,x^2,\ldots,x^p\} \subset \mathcal{P}^p$. We showed in Example 1.1.15 that this set is linearly independent, and it spans $\mathcal{P}^p$ since every polynomial $f \in \mathcal{P}^p$ can be written as a linear combination of the members of B:

$$f(x) = c_0 + c_1 x + c_2 x^2 + \cdots + c_p x^p.$$

It follows that B is a basis of $\mathcal{P}^p$, and in fact it is called the **standard basis** of $\mathcal{P}^p$.

Keep in mind that, just like in $\mathbb{R}^n$, bases are very non-unique. If a vector space has one basis then it has many different bases.

Similarly, there is also a standard basis of $\mathcal{M}_{m,n}$. We define $E_{i,j} \in \mathcal{M}_{m,n}$ to be the matrix with a 1 in its (i,j)-entry and zeros elsewhere (these are called the **standard basis matrices**). For example, in $\mathcal{M}_{2,2}$ the standard basis matrices are

$$E_{1,1} = \begin{bmatrix} 1 & 0 \\ 0 & 0 \end{bmatrix}, \quad E_{1,2} = \begin{bmatrix} 0 & 1 \\ 0 & 0 \end{bmatrix}, \quad E_{2,1} = \begin{bmatrix} 0 & 0 \\ 1 & 0 \end{bmatrix}, \quad \text{and} \quad E_{2,2} = \begin{bmatrix} 0 & 0 \\ 0 & 1 \end{bmatrix}.$$

The set $\{E_{1,1}, E_{1,2}, \ldots, E_{m,n}\}$ is a basis (called the **standard basis**) of $\mathcal{M}_{m,n}$. This fact is hopefully believable enough, but it is proved explicitly in Exercise 1.1.13.

Example 1.1.19	Recall the matrices
Strange Basis of	
Matrices	$$X = \begin{bmatrix} 0 & 1 \\ 1 & 0 \end{bmatrix}, \quad Y = \begin{bmatrix} 0 & -i \\ i & 0 \end{bmatrix}, \quad \text{and} \quad Z = \begin{bmatrix} 1 & 0 \\ 0 & -1 \end{bmatrix},$$

which we introduced in Example 1.1.12. Is the set of matrices $B = \{I, X, Y, Z\}$ a basis of $\mathcal{M}_2(\mathbb{C})$?

Solution:

We start by checking whether or not $\text{span}(B) = \mathcal{M}_2(\mathbb{C})$. That is, we determine whether or not an arbitrary matrix $A \in \mathcal{M}_2(\mathbb{C})$ can be written as a linear combination of the members of B:

$$c_1 \begin{bmatrix} 1 & 0 \\ 0 & 1 \end{bmatrix} + c_2 \begin{bmatrix} 0 & 1 \\ 1 & 0 \end{bmatrix} + c_3 \begin{bmatrix} 0 & -i \\ i & 0 \end{bmatrix} + c_4 \begin{bmatrix} 1 & 0 \\ 0 & -1 \end{bmatrix} = \begin{bmatrix} a_{1,1} & a_{1,2} \\ a_{2,1} & a_{2,2} \end{bmatrix}.$$

By setting the entries of the matrices on the left- and right-hand-sides equal to each other, we arrive at the system of linear equations

$$c_1 + c_4 = a_{1,1}, \qquad c_2 - ic_3 = a_{1,2},$$
$$c_2 + ic_3 = a_{2,1}, \qquad c_1 - c_4 = a_{2,2}.$$

Keep in mind that $a_{1,1}, a_{1,2}, a_{2,1}$, and $a_{2,2}$ are *constants* in this linear system, whereas the variables are c_1, c_2, c_3, and c_4.

It is straightforward to check that, no matter what A is, this linear system has the unique solution

$$c_1 = \frac{1}{2}(a_{1,1} + a_{2,2}), \qquad c_2 = \frac{1}{2}(a_{1,2} + a_{2,1}),$$
$$c_3 = \frac{i}{2}(a_{1,2} - a_{2,1}), \qquad c_4 = \frac{1}{2}(a_{1,1} - a_{2,2}).$$

It follows that A is always a linear combination of the members of B, so $\text{span}(B) = \mathcal{M}_2(\mathbb{C})$. Similarly, since this linear combination is unique (and in particular, it is unique when $A = O$ is the zero matrix), we see that B is linearly independent and thus a basis of $\mathcal{M}_2(\mathbb{C})$.

Example 1.1.20

Strange Basis of Polynomials

Is the set of polynomials $B = \{1, x, 2x^2 - 1, 4x^3 - 3x\}$ a basis of $\mathcal{P}^3$?

Solution:
We start by checking whether or not $\text{span}(B) = \mathcal{P}^3$. That is, we determine whether or not an arbitrary polynomial $a_0 + a_1 x + a_2 x^2 + a_3 x^3 \in \mathcal{P}^3$ can be written as a linear combination of the members of B:

$$c_0 + c_1 x + c_2(2x^2 - 1) + c_3(4x^3 - 3x) = a_0 + a_1 x + a_2 x^2 + a_3 x^3.$$

By setting the coefficients of each power of x equal to each other (like we did in Example 1.1.11), we arrive at the system of linear equations

$$c_0 - c_2 = a_0, \qquad c_1 - 3c_3 = a_1,$$
$$2c_2 = a_2, \qquad 4c_3 = a_3.$$

It is straightforward to check that, no matter what a_0, a_1, a_2, and a_3 are, this linear system has the unique solution

$$c_0 = a_0 + \frac{1}{2}a_2, \quad c_1 = a_1 + \frac{3}{4}a_3, \quad c_2 = \frac{1}{2}a_2, \quad c_3 = \frac{1}{4}a_3.$$

It follows that $a_0 + a_1 x + a_2 x^2 + a_3 x^3$ is a linear combination of the members of B regardless of the values of a_0, a_1, a_2, and a_3, so $\text{span}(B) = \mathcal{P}^3$. Similarly, since the solution is unique (and in particular, it is unique when $a_0 + a_1 x + a_2 x^2 + a_3 x^3$ is the zero polynomial), we see that B is linearly independent and thus a basis of $\mathcal{P}^3$.

Example 1.1.21

Bases of Even and Odd Polynomials

Let $\mathcal{P}^E$ and $\mathcal{P}^O$ be the sets of even and odd polynomials, respectively:

$$\mathcal{P}^E \stackrel{\text{def}}{=} \{f \in \mathcal{P} : f(-x) = f(x)\} \quad \text{and} \quad \mathcal{P}^O \stackrel{\text{def}}{=} \{f \in \mathcal{P} : f(-x) = -f(x)\}.$$

Show that $\mathcal{P}^E$ and $\mathcal{P}^O$ are subspaces of $\mathcal{P}$, and find bases of them.

Solution:
We show that $\mathcal{P}^E$ is a subspace of $\mathcal{P}$ by checking the two closure

properties from Theorem 1.1.2.

a) If $f, g \in \mathcal{P}^E$ then

$$(f+g)(-x) = f(-x) + g(-x) = f(x) + g(x) = (f+g)(x),$$

so $f + g \in \mathcal{P}^E$ too.

b) If $f \in \mathcal{P}^E$ and $c \in \mathbb{R}$ then

$$(cf)(-x) = cf(-x) = cf(x) = (cf)(x),$$

so $cf \in \mathcal{P}^E$ too.

To find a basis of $\mathcal{P}^E$, we first notice that $\{1, x^2, x^4, \ldots\} \subset \mathcal{P}^E$. This set is linearly independent since it is a subset of the linearly independent set $\{1, x, x^2, x^3, \ldots\}$ from Example 1.1.16. To see that it spans $\mathcal{P}^E$, we notice that if

$$f(x) = a_0 + a_1 x + a_2 x^2 + a_3 x^3 + \cdots \in \mathcal{P}^E$$

then $f(x) + f(-x) = 2f(x)$, so

$$\begin{aligned} 2f(x) &= f(x) + f(-x) \\ &= (a_0 + a_1 x + a_2 x^2 + \cdots) + (a_0 - a_1 x + a_2 x^2 - \cdots) \\ &= 2(a_0 + a_2 x^2 + a_4 x^4 + \cdots). \end{aligned}$$

It follows that $f(x) = a_0 + a_2 x^2 + a_4 x^4 + \cdots \in \mathrm{span}(1, x^2, x^4, \ldots)$, so the set $\{1, x^2, x^4, \ldots\}$ spans $\mathcal{P}^E$ and is thus a basis of it.

A similar argument shows that $\mathcal{P}^O$ is a subspace of $\mathcal{P}$ with basis $\{x, x^3, x^5, \ldots\}$, but we leave the details to Exercise 1.1.14.

Just as was the case in $\mathbb{R}^n$, the reason why we require a basis to span a vector space $\mathcal{V}$ is so that we can write every vector $\mathbf{v} \in \mathcal{V}$ as a linear combination of those basis vectors, and the reason why we require a basis to be linearly independent is so that those linear combinations are unique. The following theorem pins this observation down, and it roughly says that linear independence is the property needed to remove redundancies in linear combinations.

Theorem 1.1.4 **Uniqueness of Linear Combinations**	Suppose B is a basis of a vector space $\mathcal{V}$. For every $\mathbf{v} \in \mathcal{V}$, there is exactly one way to write $\mathbf{v}$ as a linear combination of the vectors in B.

Proof. Since B spans $\mathcal{V}$, we know that every $\mathbf{v} \in \mathcal{V}$ can be written as a linear combination of the vectors in B, so all we have to do is use linear independence of B to show that this linear combination is unique.

Conversely, if B is a set with the property that every vector $\mathbf{v} \in \mathcal{V}$ can be written as a linear combination of the members of B in exactly one way, then B must be a basis of $\mathcal{V}$ (see Exercise 1.1.20).

Suppose that $\mathbf{v}$ could be written as a (finite) linear combination of vectors from B in two ways:

$$\begin{aligned} \mathbf{v} &= c_1 \mathbf{v}_1 + c_2 \mathbf{v}_2 + \cdots + c_k \mathbf{v}_k \quad \text{and} \\ \mathbf{v} &= d_1 \mathbf{w}_1 + d_2 \mathbf{w}_2 + \cdots + d_\ell \mathbf{w}_\ell. \end{aligned} \tag{1.1.2}$$

It might be the case that some of the $\mathbf{v}_i$ vectors equal some of the $\mathbf{w}_j$ vectors. Let m be the number of such pairs of equal vectors in these linear combinations, and order the vectors so that $\mathbf{v}_i = \mathbf{w}_i$ when $1 \leq i \leq m$, but $\mathbf{v}_i \neq \mathbf{w}_j$ when $i, j > m$.

Subtracting these linear combinations from each other then gives

$$\mathbf{0} = \mathbf{v} - \mathbf{v} = ((c_1 - d_1)\mathbf{v}_1 + \cdots + (c_m - d_m)\mathbf{v}_m)$$
$$+ (c_{m+1}\mathbf{v}_{m+1} + \cdots + c_k\mathbf{v}_k) - (d_{m+1}\mathbf{w}_{m+1} + \cdots + d_\ell\mathbf{w}_\ell).$$

Since B is a basis, and thus linearly independent, this linear combination tells us that $c_1 - d_1 = 0$, ..., $c_m - d_m = 0$, $c_{m+1} = \cdots = c_k = 0$, and $d_{m+1} = \cdots = d_\ell = 0$. It follows that the two linear combinations in Equation (1.1.2) are in fact the *same* linear combination, which proves uniqueness. ∎

Most properties of bases of subspaces of $\mathbb{R}^n$ carry over to bases of arbitrary vector spaces, as long as those bases are finite. However, we will see in the next section that vector spaces with infinite bases (like $\mathcal{P}$) can behave somewhat strangely, and for some vector spaces (like $\mathcal{F}$) it can be difficult to even say whether or not they *have* a basis.

Exercises

solutions to starred exercises on page 449

1.1.1 Determine which of the following sets are and are not subspaces of the indicated vector space.

*(a) The set $\{(x,y) \in \mathbb{R}^2 : x \geq 0, y \geq 0\}$ in $\mathbb{R}^2$.

(b) The set $\mathcal{M}_n^{sS}$ of skew-symmetric matrices (i.e., matrices B satisfying $B^T = -B$) in $\mathcal{M}_n$.

*(c) The set of invertible matrices in $\mathcal{M}_3(\mathbb{C})$.

(d) The set of polynomials f satisfying $f(4) = 0$ in $\mathcal{P}^3$.

*(e) The set of polynomials f satisfying $f(2) = 2$ in $\mathcal{P}^4$.

(f) The set of polynomials with even degree in $\mathcal{P}$.

*(g) The set of even functions (i.e., functions f satisfying $f(x) = f(-x)$ for all $x \in \mathbb{R}$) in $\mathcal{F}$.

(h) The set of functions

$$\{f \in \mathcal{F} : f'(x) + f(x) = 2\}$$

in $\mathcal{F}$ (here f' denotes the derivative of f).

*(i) The set of functions

$$\{f \in \mathcal{F} : f''(x) - 2f(x) = 0\}$$

in $\mathcal{F}$ (here f'' denotes the second derivative of f).

(j) The set of functions

$$\{f \in \mathcal{F} : \sin(x)f'(x) + e^x f(x) = 0\}$$

in $\mathcal{F}$.

*(k) The set of matrices with trace (i.e., sum of diagonal entries) equal to 0.

1.1.2 Determine which of the following sets are and are not linearly independent. If the set is linearly dependent, explicitly write one of the vectors as a linear combination of the other vectors.

*(a) $\left\{ \begin{bmatrix} 1 & 2 \\ 3 & 4 \end{bmatrix} \right\} \subset \mathcal{M}_2$

(b) $\{1+x, 1+x^2, x^2-x\} \subset \mathcal{P}^2$

*(c) $\{\sin^2(x), \cos^2(x), 1\} \subset \mathcal{F}$

(d) $\{\sin(x), \cos(x), 1\} \subset \mathcal{F}$

*(e) $\{\sin(x+1), \sin(x), \cos(x)\} \subset \mathcal{F}$

(f) $\{e^x, e^{-x}\} \subset \mathcal{F}$

**(g) $\{e^x, xe^x, x^2 e^x\} \subset \mathcal{F}$

1.1.3 Determine which of the following sets are and are not bases of the indicated vector space.

*(a) $\left\{ \begin{bmatrix} 1 & 0 \\ 0 & 1 \end{bmatrix}, \begin{bmatrix} 1 & 1 \\ 0 & 1 \end{bmatrix}, \begin{bmatrix} 1 & 0 \\ 1 & 1 \end{bmatrix} \right\} \subset \mathcal{M}_2$

(b) $\left\{ \begin{bmatrix} 1 & 0 \\ 0 & 1 \end{bmatrix}, \begin{bmatrix} 1 & 1 \\ 0 & 1 \end{bmatrix}, \begin{bmatrix} 1 & 0 \\ 1 & 1 \end{bmatrix}, \begin{bmatrix} 1 & 1 \\ 1 & 1 \end{bmatrix} \right\} \subset \mathcal{M}_2$

*(c) $\left\{ \begin{bmatrix} 1 & 0 \\ 0 & 1 \end{bmatrix}, \begin{bmatrix} 1 & 1 \\ 0 & 1 \end{bmatrix}, \begin{bmatrix} 1 & 0 \\ 1 & 1 \end{bmatrix}, \begin{bmatrix} 1 & 1 \\ 1 & 0 \end{bmatrix} \right\} \subset \mathcal{M}_2$

(d) $\{x+1, x-1\} \subset \mathcal{P}^1$

*(e) $\{x^2+1, x+1, x^2-x\} \subset \mathcal{P}^2$

(f) $\{x^2+1, x+1, x^2-x+1\} \subset \mathcal{P}^2$

*(g) $\{1, x-1, (x-1)^2, (x-1)^3, \ldots\} \subset \mathcal{P}$

(h) $\{e^x, e^{-x}\} \subset \mathcal{F}$

1.1.4 Determine which of the following statements are true and which are false.

(a) The empty set is a vector space.

*(b) Every vector space $\mathcal{V} \neq \{\mathbf{0}\}$ contains a subspace $\mathcal{W}$ such that $\mathcal{W} \neq \mathcal{V}$.

(c) If B and C are subsets of a vector space $\mathcal{V}$ with $B \subseteq C$ then $\text{span}(B) \subseteq \text{span}(C)$.

*(d) If B and C are subsets of a vector space $\mathcal{V}$ with $\text{span}(B) \subseteq \text{span}(C)$ then $B \subseteq C$.

(e) Linear combinations must contain only finitely many terms in the sum.

*(f) If B is a subset of a vector space $\mathcal{V}$ then the span of B is equal to the span of some finite subset of B.

(g) Bases must contain finitely many vectors.

*(h) A set containing a single vector must be linearly independent.

*1.1.5 Suppose $\mathcal{V}$ is a vector space over a field $\mathbb{F}$. Show that if $\mathbf{v} \in \mathcal{V}$ and $c \in \mathbb{F}$ then $(-c)\mathbf{v} = -(c\mathbf{v})$.

1.1.6 Consider the subset $\mathcal{S} = \{(a+bi, a-bi) : a, b \in \mathbb{R}\}$ of $\mathbb{C}^2$. Explain why $\mathcal{S}$ is a vector space over the field $\mathbb{R}$, but not over $\mathbb{C}$.

**1.1.7 Show that the sequence space $\mathbb{F}^\mathbb{N}$ is a vector space.

∗∗1.1.8 Show that the set $\mathcal{M}_n^S$ of $n \times n$ symmetric matrices (i.e., matrices A satisfying $A^T = A$) is a subspace of $\mathcal{M}_n$.

1.1.9 Let $\mathcal{P}_n$ denote the set of n-variable polynomials, $\mathcal{P}_n^p$ denote the set of n-variable polynomials of degree at most p, and $\mathcal{HP}_n^p$ denote the set of n-variable **homogeneous polynomials** (i.e., polynomials in which every term has the exact same degree) of degree p, together with the 0 function. For example,

$$x^3 y + xy^2 \in \mathcal{P}_2^4 \quad \text{and} \quad x^4 y^2 z + xy^3 z^3 \in \mathcal{HP}_3^7.$$

(a) Show that $\mathcal{P}_n$ is a vector space.
(b) Show that $\mathcal{P}_n^p$ is a subspace of $\mathcal{P}_n$.
(c) Show that $\mathcal{HP}_n^p$ is a subspace of $\mathcal{P}_n^p$.
   [Side note: We explore $\mathcal{HP}_n^p$ extensively in Section 3.B.]

∗∗1.1.10 Let $\mathcal{C}$ be the set of continuous real-valued functions, and let $\mathcal{D}$ be the set of differentiable real-valued functions.

(a) Briefly explain why $\mathcal{C}$ is a subspace of $\mathcal{F}$.
(b) Briefly explain why $\mathcal{D}$ is a subspace of $\mathcal{F}$.

∗∗1.1.11 Let $\mathcal{V}$ be a vector space over a field $\mathbb{F}$ and let $S \subseteq \mathcal{V}$ be non-empty. Show that S is a subspace of $\mathcal{V}$ if and only if it is closed under linear combinations:

$$c_1 \mathbf{v}_1 + c_2 \mathbf{v}_2 + \cdots + c_k \mathbf{v}_k \in S$$

for all $c_1, c_2, \ldots, c_k \in \mathbb{F}$ and $\mathbf{v}_1, \mathbf{v}_2, \ldots, \mathbf{v}_k \in S$.

1.1.12 Recall the geometric series

$$1 + x + x^2 + x^3 + \cdots = \frac{1}{1-x} \quad \text{whenever} \quad |x| < 1.$$

(a) Why does the above expression *not* imply that $1/(1-x)$ is a linear combination of $1, x, x^2, \ldots$?
(b) Show that the set $\left\{ 1/(1-x), 1, x, x^2, x^3, \ldots \right\}$ of functions from $(-1, 1)$ to $\mathbb{R}$ is linearly independent.

∗∗1.1.13 Prove that the standard basis

$$\{ E_{1,1}, E_{1,2}, \ldots, E_{m,n} \}$$

is indeed a basis of $\mathcal{M}_{m,n}$.

∗∗1.1.14 We showed in Example 1.1.21 that the set of even polynomials $\mathcal{P}^E$ is a subspace of $\mathcal{P}$ with $\{1, x^2, x^4, \ldots\}$ as a basis.

(a) Show that the set of odd polynomials $\mathcal{P}^O$ is also a subspace of $\mathcal{P}$.
(b) Show that $\{x, x^3, x^5, \ldots\}$ is a basis of $\mathcal{P}^O$.

1.1.15 We showed in Example 1.1.19 that the set of Pauli matrices $B = \{I, X, Y, Z\}$ is a basis of $\mathcal{M}_2(\mathbb{C})$. Show that it is also a basis of $\mathcal{M}_2^H$ (the vector space of 2×2 Hermitian matrices).

1.1.16 In $\mathbb{F}^{\mathbb{N}}$, let $\mathbf{e}_j = (0, \ldots, 0, 1, 0, \ldots)$ be the sequence with a single 1 in the j-th entry, and all other terms equal to 0.

(a) Show that $\{\mathbf{e}_1, \mathbf{e}_2, \mathbf{e}_3, \ldots\}$ is a basis of the subspace c_{00} from Example 1.1.10.
(b) Explain why $\{\mathbf{e}_1, \mathbf{e}_2, \mathbf{e}_3, \ldots\}$ is *not* a basis of $\mathbb{F}^{\mathbb{N}}$.

∗1.1.17 Let S_1 and S_2 be subspaces of a vector space $\mathcal{V}$.

(a) Show that $S_1 \cap S_2$ is also a subspace of $\mathcal{V}$.
(b) Provide an example to show that $S_1 \cup S_2$ might not be a subspace of $\mathcal{V}$.

1.1.18 Let B and C be subsets of a vector space $\mathcal{V}$.

(a) Show that $\text{span}(B \cap C) \subseteq \text{span}(B) \cap \text{span}(C)$.
(b) Provide an example for which $\text{span}(B \cap C) = \text{span}(B) \cap \text{span}(C)$ and another example for which $\text{span}(B \cap C) \subsetneq \text{span}(B) \cap \text{span}(C)$.

∗∗1.1.19 Let S_1 and S_2 be subspaces of a vector space $\mathcal{V}$. The **sum of S_1 and S_2** is defined by

$$S_1 + S_2 \overset{\text{def}}{=} \{\mathbf{v} + \mathbf{w} : \mathbf{v} \in S_1, \mathbf{w} \in S_2\}.$$

(a) If $\mathcal{V} = \mathbb{R}^3$, S_1 is the x-axis, and S_2 is the y-axis, what is $S_1 + S_2$?
(b) If $\mathcal{V} = \mathcal{M}_2$, $\mathcal{M}_2^S$ is the subspace of symmetric matrices, and $\mathcal{M}_2^{sS}$ is the subspace of skew-symmetric matrices (i.e., the matrices B satisfying $B^T = -B$), what is $\mathcal{M}_2^S + \mathcal{M}_2^{sS}$?
(c) Show that $S_1 + S_2$ is always a subspace of $\mathcal{V}$.

∗∗1.1.20 Show that the converse of Theorem 1.1.4 holds. That is, show that if $\mathcal{V}$ is a vector space and B is a set with the property that every vector $\mathbf{v} \in \mathcal{V}$ can be written as a linear combination of the members of B in exactly one way, then B must be a basis of $\mathcal{V}$.

∗∗1.1.21 In the space of functions $\mathcal{F}$, proving linear dependence or independence can be quite difficult. In this question, we derive one method that can help us if the functions are smooth enough (i.e., can be differentiated enough).

Given a set of n functions $f_1, f_2, \ldots, f_n : \mathbb{R} \to \mathbb{R}$ that are differentiable at least $n-1$ times, we define the following $n \times n$ matrix (note that each entry in this matrix is a function):

$$W(x) = \begin{bmatrix} f_1(x) & f_2(x) & \cdots & f_n(x) \\ f_1'(x) & f_2'(x) & \cdots & f_n'(x) \\ f_1''(x) & f_2''(x) & \cdots & f_n''(x) \\ \vdots & \vdots & \ddots & \vdots \\ f_1^{(n-1)}(x) & f_2^{(n-1)}(x) & \cdots & f_n^{(n-1)}(x) \end{bmatrix}.$$

The notation $f^{(n-1)}(x)$ above means the $(n-1)$-th derivative of f.

(a) Show that if $f_1, f_2, \ldots, f_n$ are linearly dependent, then $\det(W(x)) = 0$ for all $x \in \mathbb{R}$. [Hint: What can you say about the columns of $W(x)$?] [Side note: The determinant $\det(W(x))$ is called the **Wronskian** of $f_1, f_2, \ldots, f_n$.]
(b) Use the result of part (a) to show that the set $\{x, \ln(x), \sin(x)\}$ is linearly independent.
(c) Use the result of part (a) to show that the set $\{1, x, x^2, \ldots, x^n\}$ is linearly independent.
(d) This method *cannot* be used to prove linear dependence! For example, show that the functions $f_1(x) = x^2$ and $f_2(x) = x|x|$ are linearly independent, but $\det(W(x)) = 0$ for all x.

∗∗1.1.22 Let $a_1, a_2, \ldots, a_n$ be n distinct real numbers (i.e., $a_i \neq a_j$ whenever $i \neq j$). Show that the set of functions $\{e^{a_1 x}, e^{a_2 x}, \ldots, e^{a_n x}\}$ is linearly independent.

[Hint: Construct the Wronskian of this set of functions (see Exercise 1.1.21) and notice that it is the determinant of a Vandermonde matrix.]

1.2 Coordinates and Linear Transformations

We now start investigating what we can do with bases of vector spaces. It is perhaps not surprising that we can do many of the same things that we do with bases of subspaces of $\mathbb{R}^n$, like define the dimension of arbitrary vector spaces and construct coordinate vectors. However, bases really shine in this more general setting because we can use them to turn any question about a vector space $\mathcal{V}$ (as long as it has a finite basis) into one about $\mathbb{R}^n$ or $\mathbb{C}^n$ (or $\mathbb{F}^n$, where $\mathcal{V}$ is a vector space over the field $\mathbb{F}$).

In other words, we spend this section showing that we can think of general vector spaces (like polynomials or matrices) just as copies of $\mathbb{R}^n$ that have just been written in a different way, and we can think of linear transformations on these vector spaces just as matrices.

1.2.1 Dimension and Coordinate Vectors

Recall from Theorem 1.1.4 that every vector $\mathbf{v}$ in a vector space $\mathcal{V}$ can be written as a linear combination of the vectors from a basis B in exactly one way. We can interpret this fact as saying that the vectors in a basis each specify a different direction, and the (unique) coefficients in the linear combination specify how far $\mathbf{v}$ points in each of those directions.

In other words, $\mathbf{v} = (c_1, c_2)$.

For example, if $B = \{\mathbf{e}_1, \mathbf{e}_2\}$ is the standard basis of $\mathbb{R}^2$ then when we write $\mathbf{v} = c_1 \mathbf{e}_1 + c_2 \mathbf{e}_2$, the coefficient c_1 tells us how far $\mathbf{v}$ points in the direction of $\mathbf{e}_1$ (i.e., along the x-axis) and c_2 tells us how far $\mathbf{v}$ points in the direction of $\mathbf{e}_2$ (i.e., along the y-axis). A similar interpretation works with other bases of $\mathbb{R}^2$ (see Figure 1.4) and with bases of other vector spaces.

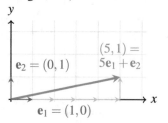

(a) The vector $(5, 1)$ points 5 times as far as $\mathbf{e}_1$ in its direction and 1 times as far as $\mathbf{e}_2$ in its direction.

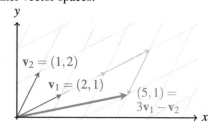

(b) The vector $(5, 1)$ points 3 times as far as $(2, 1)$ in its direction and exactly as far as $(1, 2)$ in the opposite of its direction.

Figure 1.4: When we write a vector as a linear combination of basis vectors, the coefficients of that linear combination tell us how far it points in the direction of those basis vectors.

We abuse notation a
bit when using bases.
Order does not
matter in sets like
$\{e_1, e_2\}$, but it matters
if that set is a basis.

It is often easier to just keep track of and work with these coefficients c_1, c_2, ..., c_n in a linear combination, rather than the original vector $\mathbf{v}$ itself. However, there is one technicality that we have to deal with before we can do this: the fact that bases are sets of vectors, and sets do not care about order. For example, $\{\mathbf{e}_1, \mathbf{e}_2\}$ and $\{\mathbf{e}_2, \mathbf{e}_1\}$ are both the same standard basis of $\mathbb{R}^2$. However, we want to be able to talk about things like the "first" vector in a basis and the "third" coefficient in a linear combination. For this reason, we typically consider bases to be **ordered**—even though they are written using the same notation as sets, their order really is meant as written.

Definition 1.2.1	Suppose $\mathcal{V}$ is a vector space over a field $\mathbb{F}$ with a finite (ordered) basis
Coordinate Vectors	$B = \{\mathbf{v}_1, \mathbf{v}_2, \ldots, \mathbf{v}_n\}$, and $\mathbf{v} \in \mathcal{V}$. Then the unique scalars $c_1, c_2, \ldots, c_n \in \mathbb{F}$ for which

$$\mathbf{v} = c_1 \mathbf{v}_1 + c_2 \mathbf{v}_2 + \cdots + c_n \mathbf{v}_n$$

are called the **coordinates** of $\mathbf{v}$ with respect to B, and the vector

$$[\mathbf{v}]_B \overset{\text{def}}{=} (c_1, c_2, \ldots, c_n)$$

is called the **coordinate vector** of $\mathbf{v}$ with respect to B.

It is worth noting that coordinate vectors in $\mathbb{R}^n$ (or $\mathbb{F}^n$ in general) with respect to the standard basis are equal to those vectors themselves: if $B = \{\mathbf{e}_1, \mathbf{e}_2, \ldots, \mathbf{e}_n\}$ then $[\mathbf{v}]_B = \mathbf{v}$ for all $\mathbf{v} \in \mathbb{R}^n$, since the first entry of $\mathbf{v}$ is the coefficient of $\mathbf{e}_1$, the second entry of $\mathbf{v}$ is the coefficient of $\mathbf{e}_2$, and so on. We can also construct coordinate vectors with respect to different bases and of vectors from other vector spaces though.

For example, the coordinate vector of $3 + 2x + x^2 \in \mathcal{P}^2$ with respect to the basis $B = \{1, x, x^2\}$ is $[3 + 2x + x^2]_B = (3, 2, 1)$: we just place the coefficients of the polynomial in a vector in $\mathbb{R}^3$, and this vector completely specifies the polynomial. More generally, the coordinate vector of the polynomial $a_0 + a_1 x + a_2 x^2 + \cdots + a_p x^p$ with respect to the standard basis $B = \{1, x, x^2, \ldots, x^p\}$ is

$$\left[a_0 + a_1 x + a_2 x^2 + \cdots + a_p x^p\right]_B = (a_0, a_1, a_2, \ldots, a_p).$$

The key idea behind coordinate vectors is that they let us use bases to treat every vector space with a finite basis just like we treat $\mathbb{F}^n$ (where $\mathbb{F}$ is the ground field and n is the number of vectors in the basis). In particular, coordinate vectors with respect to any basis B interact with vector addition and scalar multiplication how we would expect them to (see Exercise 1.2.22):

$$[\mathbf{v} + \mathbf{w}]_B = [\mathbf{v}]_B + [\mathbf{w}]_B \quad \text{and} \quad [c\mathbf{v}]_B = c[\mathbf{v}]_B \quad \text{for all} \quad \mathbf{v}, \mathbf{w} \in \mathcal{V}, c \in \mathbb{F}.$$

It follows that we can answer pretty much any linear algebraic question about vectors in an n-dimensional vector space $\mathcal{V}$ just by finding their coordinate vectors and then answering the corresponding question in $\mathbb{F}^n$ (which we can do via techniques from introductory linear algebra). For example, a set is linearly independent if and only if the set consisting of their coordinate vectors is linearly independent (see Exercise 1.2.23).

Example 1.2.1	Show that the following sets of vectors C are linearly independent in the
Checking Linear (In)Dependence via Coordinate Vectors	indicated vector space $\mathcal{V}$: a) $\mathcal{V} = \mathbb{R}^4$, $C = \{(0, 2, 0, 1), (0, -1, 1, 2), (3, -2, 1, 0)\}$,

b) $V = \mathcal{P}^3$, $C = \{2x+x^3,\ -x+x^2+2x^3,\ 3-2x+x^2\}$, and

c) $V = \mathcal{M}_2$, $C = \left\{ \begin{bmatrix} 0 & 2 \\ 0 & 1 \end{bmatrix}, \begin{bmatrix} 0 & -1 \\ 1 & 2 \end{bmatrix}, \begin{bmatrix} 3 & -2 \\ 1 & 0 \end{bmatrix} \right\}$.

Solutions:

a) Recall that C is linearly independent if and only if the linear system

$$c_1(0,2,0,1) + c_2(0,-1,1,2) + c_3(3,-2,1,0) = (0,0,0,0)$$

has a unique solution. Explicitly, this linear system has the form

$$
\begin{aligned}
3c_3 &= 0 \\
2c_1 - c_2 - 2c_3 &= 0 \\
c_2 + c_3 &= 0 \\
c_1 + 2c_2 &= 0
\end{aligned}
$$

Need a refresher on solving linear systems? See Appendix A.1.1.

which can be solved via Gaussian elimination to see that the unique solution is $c_1 = c_2 = c_3 = 0$, so C is indeed a linearly independent set.

b) We could check linear independence directly via the methods of Section 1.1.2, but we instead compute the coordinate vectors of the members of C with respect to the standard basis $B = \{1, x, x^2, x^3\}$ of $\mathcal{P}^3$:

$$
\begin{aligned}
[2x + x^3]_B &= (0,2,0,1), \\
[-x + x^2 + 2x^3]_B &= (0,-1,1,2), \quad \text{and} \\
[3 - 2x + x^2]_B &= (3,-2,1,0).
\end{aligned}
$$

These coordinate vectors are exactly the vectors from part (a), which we already showed are linearly independent in $\mathbb{R}^4$, so C is linearly independent in $\mathcal{P}^3$ as well.

c) Again, we start by computing the coordinate vectors of the members of C with respect to the standard basis $B = \{E_{1,1}, E_{1,2}, E_{2,1}, E_{2,2}\}$ of $\mathcal{M}_2$:

The notation here is quite unfortunate. The inner square brackets indicate that these are matrices, while the outer square brackets denote the coordinate vectors of these matrices.

$$\left[\begin{bmatrix} 0 & 2 \\ 0 & 1 \end{bmatrix} \right]_B = (0,2,0,1),$$

$$\left[\begin{bmatrix} 0 & -1 \\ 1 & 2 \end{bmatrix} \right]_B = (0,-1,1,2), \quad \text{and}$$

$$\left[\begin{bmatrix} 3 & -2 \\ 1 & 0 \end{bmatrix} \right]_B = (3,-2,1,0).$$

Again, these coordinate vectors are exactly the vectors from part (a), which we already showed are linearly independent in $\mathbb{R}^4$, so C is linearly independent in $\mathcal{M}_2$.

The above example highlights the fact that we can really think of $\mathbb{R}^4$, $\mathcal{P}^3$, and $\mathcal{M}_2$ as "essentially the same" vector spaces, just dressed up and displayed differently: we can work with the polynomial $a_0 + a_1 x + a_2 x^2 + a_3 x^3$ in the same way that we work with the vector $(a_0, a_1, a_2, a_3) \in \mathbb{R}^4$, which we can work with in the same way as the matrix

$$\begin{bmatrix} a_0 & a_1 \\ a_2 & a_3 \end{bmatrix} \in \mathcal{M}_2.$$

However, it is also often useful to represent vectors in bases other than the standard basis, so we briefly present an example that illustrates how to do this.

Example 1.2.2

Coordinate Vectors in Weird Bases

Find the coordinate vector of $2 + 7x + x^2 \in \mathcal{P}^2$ with respect to the basis $B = \{x + x^2, \ 1 + x^2, \ 1 + x\}$.

Solution:

We want to find scalars $c_1, c_2, c_3 \in \mathbb{R}$ such that

We did not explicitly prove that B is a basis of $\mathcal{P}^2$ here—try to convince yourself that it is.

$$2 + 7x + x^2 = c_1(x + x^2) + c_2(1 + x^2) + c_3(1 + x).$$

By matching up coefficients of powers of x on the left- and right-hand sides above, we arrive at the following system of linear equations:

$$\begin{aligned} c_2 + c_3 &= 2 \\ c_1 \quad\;\; + c_3 &= 7 \\ c_1 + c_2 \quad\;\; &= 1 \end{aligned}$$

To find c_1, c_2, c_3, just apply Gaussian elimination like usual.

This linear system has $c_1 = 3, c_2 = -2, c_3 = 4$ as its unique solution, so our desired coordinate vector is

$$\left[2 + 7x + x^2\right]_B = (c_1, c_2, c_3) = (3, -2, 4).$$

By making use of coordinate vectors, we can leech many of our results concerning bases of subspaces of $\mathbb{R}^n$ directly into this more general setting. For example, the following theorem tells us that we can roughly think of a spanning set as one that is "big" and a linearly independent set as one that is "small" (just like we did when we worked with these concepts for subspaces of $\mathbb{R}^n$), and bases are exactly the sweet spot in the middle where these two properties meet.

Theorem 1.2.1

Linearly Independent Sets Versus Spanning Sets

Suppose $n \geq 1$ is an integer and $\mathcal{V}$ is a vector space with a basis B consisting of n vectors.

a) Any set of more than n vectors in $\mathcal{V}$ must be linearly dependent, and
b) any set of fewer than n vectors cannot span $\mathcal{V}$.

Proof. For property (a), suppose that a set C has $m > n$ vectors, which we call $\mathbf{v}_1, \mathbf{v}_2, \ldots, \mathbf{v}_m$. To see that C is necessarily linearly dependent, we must show that there exist scalars $c_1, c_2, \ldots, c_m$, not all equal to zero, such that

$$c_1\mathbf{v}_1 + c_2\mathbf{v}_2 + \cdots + c_m\mathbf{v}_m = \mathbf{0}.$$

If we compute the coordinate vector of both sides of this equation with respect to B then we see that it is equivalent to

$$c_1[\mathbf{v}_1]_B + c_2[\mathbf{v}_2]_B + \cdots + c_m[\mathbf{v}_m]_B = \mathbf{0},$$

which is a homogeneous system of n linear equations in m variables. Since $m > n$, this is a "short and fat" linear system (i.e., its coefficient matrix has more columns than rows) and thus must have infinitely many solutions. In particular, it has at least one non-zero solution, from which it follows that C is linearly dependent.

Part (b) is proved in a similar way and thus left as Exercise 1.2.36. ∎

By recalling that a basis of a vector space $\mathcal{V}$ is a set that both spans $\mathcal{V}$ and is linearly independent, we immediately get the following corollary:

Corollary 1.2.2
Uniqueness of Size of Bases

Suppose $n \geq 1$ is an integer and $\mathcal{V}$ is a vector space with a basis consisting of n vectors. Then *every* basis of $\mathcal{V}$ has exactly n vectors.

For example, we saw in Example 1.1.19 that the set

$$\left\{ \begin{bmatrix} 1 & 0 \\ 0 & 1 \end{bmatrix}, \begin{bmatrix} 0 & 1 \\ 1 & 0 \end{bmatrix}, \begin{bmatrix} 0 & -i \\ i & 0 \end{bmatrix}, \begin{bmatrix} 1 & 0 \\ 0 & -1 \end{bmatrix} \right\}$$

is a basis of $\mathcal{M}_2(\mathbb{C})$, which is consistent with Corollary 1.2.2 since the standard basis $\{E_{1,1}, E_{1,2}, E_{2,1}, E_{2,2}\}$ also contains exactly 4 matrices. In a sense, this tells us that $\mathcal{M}_2(\mathbb{C})$ contains 4 "degrees of freedom" or requires 4 (complex) numbers to describe each matrix that it contains. This quantity (the number of vectors in any basis) gives a useful description of the "size" of the vector space that we are working with, so we give it a name:

Definition 1.2.2
Dimension of a Vector Space

A vector space $\mathcal{V}$ is called...

a) **finite-dimensional** if it has a finite basis. Its **dimension**, denoted by $\dim(\mathcal{V})$, is the number of vectors in any of its bases.

b) **infinite-dimensional** if it does not have a finite basis, and we write $\dim(\mathcal{V}) = \infty$.

For example, since $\mathbb{R}^n$ has $\{\mathbf{e}_1, \mathbf{e}_2, \ldots, \mathbf{e}_n\}$ as a basis, which has n vectors, we see that $\dim(\mathbb{R}^n) = n$. Similarly, the standard basis $\{E_{1,1}, E_{1,2}, \ldots, E_{m,n}\}$ of $\mathcal{M}_{m,n}$ contains mn vectors (matrices), so $\dim(\mathcal{M}_{m,n}) = mn$. We summarize the dimensions of some of the other commonly-occurring vector spaces that we have been investigating in Table 1.1.

Vector space $\mathcal{V}$	Standard basis	$\dim(\mathcal{V})$
$\mathbb{F}^n$	$\{\mathbf{e}_1, \mathbf{e}_2, \ldots, \mathbf{e}_n\}$	n
$\mathcal{M}_{m,n}$	$\{E_{1,1}, E_{1,2}, \ldots, E_{m,n}\}$	mn
$\mathcal{P}^p$	$\{1, x, x^2, \ldots, x^p\}$	$p+1$
$\mathcal{P}$	$\{1, x, x^2, \ldots\}$	∞
$\mathcal{F}$	$-$	∞

Table 1.1: The standard basis and dimension of some common vector spaces.

We refer to $\{\mathbf{0}\}$ as the **zero vector space**.

There is one special case that is worth special attention, and that is the vector space $\mathcal{V} = \{\mathbf{0}\}$. The only basis of this vector space is the empty set $\{\}$, not $\{\mathbf{0}\}$ as we might first guess, since any set containing $\mathbf{0}$ is necessarily linearly dependent. Since the empty set contains no vectors, we conclude that $\dim(\{\mathbf{0}\}) = 0$.

If we already know the dimension of the vector space that we are working with, it becomes much simpler to determine whether or not a given set is a basis of it. In particular, if a set contains the right number of vectors (i.e., its size coincides with the dimension of the vector space) then we can show that it is a basis just by showing that it is linearly independent *or* spanning—the other property comes for free (see Exercise 1.2.27).

Remark 1.2.1

Existence of Bases?

Notice that we have *not* proved a theorem saying that all vector spaces have bases (whereas this fact *is* true of subspaces of $\mathbb{R}^n$ and is typically proved in this case in introductory linear algebra textbooks). The reason for this omission is that constructing bases is much more complicated in arbitrary (potentially infinite-dimensional) vector spaces.

While it is true that all finite-dimensional vector spaces have bases (in fact, this is baked right into Definition 1.2.2(a)), it is much less clear what a basis of (for example) the vector space $\mathcal{F}$ of real-valued functions would look like. We could try sticking all of the "standard" functions that we know of into a set like

$$\{1, x, x^2, \ldots, |x|, \ln(x^2 + 1), \cos(x), \sin(x), 1/(3 + 2x^2), e^{6x}, \ldots\},$$

but there will always be lots of functions left over that are not linear combinations of these familiar functions.

Most (but not all!) mathematicians accept the axiom of choice, and thus would say that every vector space has a basis. However, the axiom of choice is non-constructive, so we still cannot actually write down explicit examples of bases in many infinite-dimensional vector spaces.

It turns out that the existence of bases depends on something called the "axiom of choice", which is a mathematical axiom that is independent of the other set-theoretic underpinnings of modern mathematics. In other words, we can neither prove that every vector space has a basis, nor can we construct a vector space that does not have one.

From a practical point of view, this means that it is simply not possible to write down a basis of many vector spaces like $\mathcal{F}$, even if they exist. Even in the space $\mathcal{C}$ of *continuous* real-valued functions, any basis necessarily contains some extremely hideous and pathological functions. Roughly speaking, the reason for this is that any finite linear combination of "nice" functions will still be "nice", but there are many "not nice" continuous functions out there.

Again, keep in mind that this does *not* mean that W is a linear combination of $\cos(2x)$, $\cos(4x)$, $\cos(8x)$, ..., since this sum has infinitely many terms in it.

For example, there are continuous functions that are nowhere differentiable (i.e., they do not have a derivative anywhere). This means that no matter how much we zoom in on the graph of the function, it never starts looking like a straight line, but rather looks jagged at all zoom levels. Every basis of $\mathcal{C}$ must contain strange functions like these, and an explicit example is the **Weierstrass function** W defined by

$$W(x) = \sum_{n=1}^{\infty} \frac{\cos(2^n x)}{2^n},$$

whose graph is displayed on the next page:

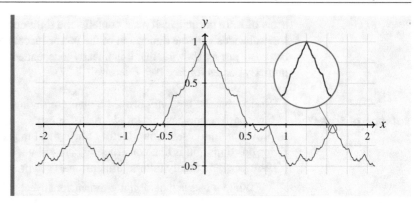

1.2.2   Change of Basis

When we change the basis B that we are working with, the resulting coordinate vectors change as well. For instance, we showed in Example 1.2.2 that, with respect to a certain basis B of $\mathcal{P}^2$, we have

$$\left[2 + 7x + x^2\right]_B = (3, -2, 4).$$

However, if $C = \{1, x, x^2\}$ is the standard basis of $\mathcal{P}^2$ then we have

$$\left[2 + 7x + x^2\right]_C = (2, 7, 1).$$

In order to convert coordinate vectors between two bases B and C, we could of course compute $\mathbf{v}$ from $[\mathbf{v}]_B$ and then compute $[\mathbf{v}]_C$ from $\mathbf{v}$. However, there is a "direct" way of doing this conversion that avoids the intermediate step of computing $\mathbf{v}$.

Definition 1.2.3	Suppose $\mathcal{V}$ is a vector space with bases $B = \{\mathbf{v}_1, \mathbf{v}_2, \ldots, \mathbf{v}_n\}$ and C. The
Change-of-Basis Matrix	**change-of-basis matrix** from B to C, denoted by $P_{C \leftarrow B}$, is the $n \times n$ matrix whose columns are the coordinate vectors $[\mathbf{v}_1]_C, [\mathbf{v}_2]_C, \ldots, [\mathbf{v}_n]_C$:

$$P_{C \leftarrow B} \overset{\text{def}}{=} \big[\, [\mathbf{v}_1]_C \mid [\mathbf{v}_2]_C \mid \cdots \mid [\mathbf{v}_n]_C \,\big].$$

It is worth emphasizing the fact that in the above definition, we place the coordinate vectors $[\mathbf{v}_1]_C, [\mathbf{v}_2]_C, \ldots, [\mathbf{v}_n]_C$ into the matrix $P_{C \leftarrow B}$ as its *columns*, not its rows. Nonetheless, when considering those vectors in isolation, we still write them using round parentheses and in a single row, like $[\mathbf{v}_2]_C = (1, 4, -3)$, just as we have been doing up until this point. The reason for this is that vectors (i.e., members of $\mathbb{F}^n$) do not have a shape—they are just lists of numbers, and we can arrange those lists however we like.

The change-of-basis matrix from B to C, as its name suggests, converts coordinate vectors with respect to B into coordinate vectors with respect to C. That is, we have the following theorem:

Theorem 1.2.3	Suppose B and C are bases of a finite-dimensional vector space $\mathcal{V}$, and let
Change-of-Basis Matrices	$P_{C \leftarrow B}$ be the change-of-basis matrix from B to C. Then

a) $P_{C \leftarrow B}[\mathbf{v}]_B = [\mathbf{v}]_C$ for all $\mathbf{v} \in \mathcal{V}$, and

b) $P_{C \leftarrow B}$ is invertible and $P_{C \leftarrow B}^{-1} = P_{B \leftarrow C}$.

Furthermore, $P_{C \leftarrow B}$ is the unique matrix with property (a).

Proof. Let $B = \{\mathbf{v}_1, \mathbf{v}_2, \ldots, \mathbf{v}_n\}$ so that we have names for the vectors in B.

For property (a), suppose $\mathbf{v} \in \mathcal{S}$ and write $\mathbf{v} = c_1\mathbf{v}_1 + c_2\mathbf{v}_2 + \cdots + c_n\mathbf{v}_n$, so that $[\mathbf{v}]_B = (c_1, c_2, \ldots, c_n)$. We can then directly compute

$$P_{C \leftarrow B}[\mathbf{v}]_B = \begin{bmatrix} [\mathbf{v}_1]_C & | & \cdots & | & [\mathbf{v}_n]_C \end{bmatrix} \begin{bmatrix} c_1 \\ \vdots \\ c_n \end{bmatrix} \qquad \text{(definition of } P_{C \leftarrow B})$$

$$= c_1[\mathbf{v}_1]_C + \cdots + c_n[\mathbf{v}_n]_C \qquad \text{(block matrix multiplication)}$$

$$= [c_1\mathbf{v}_1 + \cdots + c_n\mathbf{v}_n]_C \qquad \text{(by Exercise 1.2.22)}$$

$$= [\mathbf{v}]_C. \qquad (\mathbf{v} = c_1\mathbf{v}_1 + c_2\mathbf{v}_2 + \cdots c_n\mathbf{v}_n)$$

To see that property (b) holds, we just note that using property (a) twice tells us that $P_{B \leftarrow C}P_{C \leftarrow B}[\mathbf{v}]_B = [\mathbf{v}]_B$ for all $\mathbf{v} \in \mathcal{V}$, so $P_{B \leftarrow C}P_{C \leftarrow B} = I$, which implies $P_{C \leftarrow B}^{-1} = P_{B \leftarrow C}$.

Finally, to see that $P_{B \leftarrow C}$ the *unique* matrix satisfying property (a), suppose $P \in \mathcal{M}_n$ is any matrix for which $P[\mathbf{v}]_B = [\mathbf{v}]_C$ for all $\mathbf{v} \in \mathcal{V}$. For every $1 \leq j \leq n$, if $\mathbf{v} = \mathbf{v}_j$ then we see that $[\mathbf{v}]_B = [\mathbf{v}_j]_B = \mathbf{e}_j$ (the j-th standard basis vector in $\mathbb{F}^n$), so $P[\mathbf{v}]_B = P\mathbf{e}_j$ is the j-th column of P. On the other hand, it is also the case that $P[\mathbf{v}]_B = [\mathbf{v}]_C = [\mathbf{v}_j]_C$. The j-th column of P thus equals $[\mathbf{v}_j]_C$ for each $1 \leq j \leq n$, so $P = P_{C \leftarrow B}$. ∎

The two properties described by the above theorem are illustrated in Figure 1.5: the bases B and C provide two different ways of making the vector space $\mathcal{V}$ look like $\mathbb{F}^n$, and the change-of-basis matrices $P_{C \leftarrow B}$ and $P_{B \leftarrow C}$ convert these two different representations into each other.

One of the advantages of using change-of-basis matrices, rather than computing each coordinate vector directly, is that we can re-use change-of-basis matrices to change multiple vectors between bases.

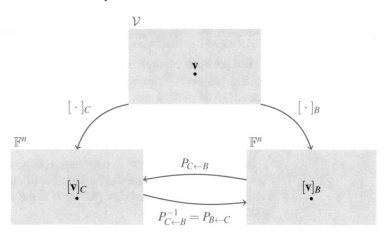

Figure 1.5: A visualization of the relationship between vectors, their coordinate vectors, and change-of-basis matrices. There are many different bases that let us think of $\mathcal{V}$ as $\mathbb{F}^n$, and change-of-basis matrices let us convert between them.

Example 1.2.3

Computing a Change-of-Basis Matrix

Find the change-of-basis matrices $P_{B \leftarrow C}$ and $P_{C \leftarrow B}$ for the bases

$$B = \{x + x^2, \ 1 + x^2, \ 1 + x\} \quad \text{and} \quad C = \{1, x, x^2\}$$

of $\mathcal{P}^2$. Then find the coordinate vector of $2 + 7x + x^2$ with respect to B.

Solution:

To start, we find the coordinate vectors of the members of B with respect to the basis C. Since C is the standard basis of $\mathcal{P}^2$, we can eyeball these coordinate vectors:

$$\left[x+x^2\right]_C = (0,1,1), \quad \left[1+x^2\right]_C = (1,0,1), \quad \text{and} \quad [1+x]_C = (1,1,0).$$

These vectors are the columns of the change-of-basis matrix $P_{C \leftarrow B}$:

$$P_{C \leftarrow B} = \begin{bmatrix} 0 & 1 & 1 \\ 1 & 0 & 1 \\ 1 & 1 & 0 \end{bmatrix}.$$

The simplest method for finding $P_{B \leftarrow C}$ from here is to compute the inverse of $P_{C \leftarrow B}$:

This inverse can be found by row-reducing $[\,A \mid I\,]$ to $[\,I \mid A^{-1}\,]$ (see Appendix A.1.3).

$$P_{B \leftarrow C} = P_{C \leftarrow B}^{-1} = \frac{1}{2} \begin{bmatrix} -1 & 1 & 1 \\ 1 & -1 & 1 \\ 1 & 1 & -1 \end{bmatrix}.$$

To then find $[2+7x+x^2]_B$, we just multiply $[2+7x+x^2]_C$ (which is easy to compute, since C is the standard basis) by $P_{B \leftarrow C}$:

$$\left[2+7x+x^2\right]_B = P_{B \leftarrow C}\left[2+7x+x^2\right]_C = \frac{1}{2} \begin{bmatrix} -1 & 1 & 1 \\ 1 & -1 & 1 \\ 1 & 1 & -1 \end{bmatrix} \begin{bmatrix} 2 \\ 7 \\ 1 \end{bmatrix} = \begin{bmatrix} 3 \\ -2 \\ 4 \end{bmatrix},$$

which is the same answer that we found in Example 1.2.2.

Generally, converting vectors into the standard basis is much easier than converting into other bases.

The previous example was not too difficult since C happened to be the standard basis of $\mathcal{P}^2$, so we could compute $P_{C \leftarrow B}$ just by "eyeballing" the standard basis coefficients, and we could then compute $P_{B \leftarrow C}$ just by taking the inverse of $P_{C \leftarrow B}$. However, if C *weren't* the standard basis, computing the columns of $P_{C \leftarrow B}$ would have been much more difficult—each column would require us to solve a linear system.

A quicker and easier way to compute a change-of-basis matrix is to change from the input basis B to the standard basis E (if one exists in the vector space being considered), and then change from E to the output basis C, as described by the following theorem:

Theorem 1.2.4

Computing Change-of-Basis Matrices (Method 1)

Let $\mathcal{V}$ be a finite-dimensional vector space with bases B, C, and E. Then

$$P_{C \leftarrow B} = P_{C \leftarrow E} P_{E \leftarrow B}.$$

Think of the "E"s in the middle of $P_{C \leftarrow B} = P_{C \leftarrow E} P_{E \leftarrow B}$ as "canceling out". The notation was designed specifically so that this works.

Proof. We just use the properties of change-of-basis matrices that we know from Theorem 1.2.3. If $\mathbf{v} \in \mathcal{V}$ is any vector then

$$(P_{C \leftarrow E} P_{E \leftarrow B})[\mathbf{v}]_B = P_{C \leftarrow E} (P_{E \leftarrow B}[\mathbf{v}]_B) = P_{C \leftarrow E}[\mathbf{v}]_E = [\mathbf{v}]_C.$$

It then follows from uniqueness of change-of-basis matrices that $P_{C \leftarrow E} P_{E \leftarrow B} = P_{C \leftarrow B}$. ∎

To actually make use of the above theorem, we note that if E is chosen to be the standard basis of $\mathcal{V}$ then $P_{E \leftarrow B}$ can be computed by eyeballing coefficients, and $P_{C \leftarrow E}$ can be computed by first eyeballing the entries of $P_{E \leftarrow C}$ and then inverting (like in Example 1.2.3).

Example 1.2.4

Computing a Change-of-Basis Matrix Between Ugly Bases

Use Theorem 1.2.4 to find the change-of-basis matrix $P_{C \leftarrow B}$, where

$$B = \left\{ \begin{bmatrix} 1 & 0 \\ 0 & 1 \end{bmatrix}, \begin{bmatrix} 0 & 1 \\ 1 & 0 \end{bmatrix}, \begin{bmatrix} 0 & -i \\ i & 0 \end{bmatrix}, \begin{bmatrix} 1 & 0 \\ 0 & -1 \end{bmatrix} \right\} \quad \text{and}$$

$$C = \left\{ \begin{bmatrix} 1 & 0 \\ 0 & 0 \end{bmatrix}, \begin{bmatrix} 1 & 1 \\ 0 & 0 \end{bmatrix}, \begin{bmatrix} 1 & 1 \\ 1 & 0 \end{bmatrix}, \begin{bmatrix} 1 & 1 \\ 1 & 1 \end{bmatrix} \right\}$$

We showed that the set B is indeed a basis of $\mathcal{M}_2(\mathbb{C})$ in Example 1.1.19. You should try to convince yourself that C is also a basis.

are bases of $\mathcal{M}_2(\mathbb{C})$. Then compute $[\mathbf{v}]_C$ if $[\mathbf{v}]_B = (1, 2, 3, 4)$.

Solution:

To start, we compute the change-of-basis matrices from B and C into the standard basis

$$E = \left\{ \begin{bmatrix} 1 & 0 \\ 0 & 0 \end{bmatrix}, \begin{bmatrix} 0 & 1 \\ 0 & 0 \end{bmatrix}, \begin{bmatrix} 0 & 0 \\ 1 & 0 \end{bmatrix}, \begin{bmatrix} 0 & 0 \\ 0 & 1 \end{bmatrix} \right\},$$

which can be done by inspection:

Coordinate vectors of matrices with respect to the standard matrix can be obtained just by writing the entries of the matrices, in order, row-by-row.

$$\left[\begin{bmatrix} 1 & 0 \\ 0 & 1 \end{bmatrix} \right]_E = (1, 0, 0, 1), \qquad \left[\begin{bmatrix} 0 & 1 \\ 1 & 0 \end{bmatrix} \right]_E = (0, 1, 1, 0),$$

$$\left[\begin{bmatrix} 0 & -i \\ i & 0 \end{bmatrix} \right]_E = (0, -i, i, 0), \quad \text{and} \quad \left[\begin{bmatrix} 1 & 0 \\ 0 & -1 \end{bmatrix} \right]_E = (1, 0, 0, -1).$$

We then get $P_{E \leftarrow B}$ by placing these vectors into a matrix as its columns:

$$P_{E \leftarrow B} = \begin{bmatrix} 1 & 0 & 0 & 1 \\ 0 & 1 & -i & 0 \\ 0 & 1 & i & 0 \\ 1 & 0 & 0 & -1 \end{bmatrix}.$$

A similar procedure can be used to compute $P_{E \leftarrow C}$, and then we find $P_{C \leftarrow E}$ by inverting:

$$P_{E \leftarrow C} = \begin{bmatrix} 1 & 1 & 1 & 1 \\ 0 & 1 & 1 & 1 \\ 0 & 0 & 1 & 1 \\ 0 & 0 & 0 & 1 \end{bmatrix}, \quad \text{so} \quad P_{C \leftarrow E} = P_{E \leftarrow C}^{-1} = \begin{bmatrix} 1 & -1 & 0 & 0 \\ 0 & 1 & -1 & 0 \\ 0 & 0 & 1 & -1 \\ 0 & 0 & 0 & 1 \end{bmatrix}.$$

To put this all together and compute $P_{C \leftarrow B}$, we multiply:

$$P_{C \leftarrow B} = P_{C \leftarrow E} P_{E \leftarrow B} = \begin{bmatrix} 1 & -1 & 0 & 0 \\ 0 & 1 & -1 & 0 \\ 0 & 0 & 1 & -1 \\ 0 & 0 & 0 & 1 \end{bmatrix} \begin{bmatrix} 1 & 0 & 0 & 1 \\ 0 & 1 & -i & 0 \\ 0 & 1 & i & 0 \\ 1 & 0 & 0 & -1 \end{bmatrix}$$

$$= \begin{bmatrix} 1 & -1 & i & 1 \\ 0 & 0 & -2i & 0 \\ -1 & 1 & i & 1 \\ 1 & 0 & 0 & -1 \end{bmatrix}.$$

Finally, since we were asked for $[\mathbf{v}]_C$ if $[\mathbf{v}]_B = (1,2,3,4)$, we compute

$$[\mathbf{v}]_C = P_{C \leftarrow B}[\mathbf{v}]_B = \begin{bmatrix} 1 & -1 & i & 1 \\ 0 & 0 & -2i & 0 \\ -1 & 1 & i & 1 \\ 1 & 0 & 0 & -1 \end{bmatrix} \begin{bmatrix} 1 \\ 2 \\ 3 \\ 4 \end{bmatrix} = \begin{bmatrix} 3+3i \\ -6i \\ 5+3i \\ -3 \end{bmatrix}.$$

The previous example perhaps seemed a bit long and involved. Indeed, to make use of Theorem 1.2.4, we have to invert $P_{E \leftarrow C}$ and then multiply two matrices together. The following theorem tells us that we can reduce the amount of work required slightly by instead row reducing a certain cleverly-chosen matrix based on $P_{E \leftarrow B}$ and $P_{E \leftarrow C}$. This takes about the same amount of time as inverting $P_{E \leftarrow C}$, and lets us avoid the matrix multiplication step that comes afterward.

Corollary 1.2.5

Computing Change-of-Basis Matrices (Method 2)

Let $\mathcal{V}$ be a finite-dimensional vector space with bases B, C, and E. Then the reduced row echelon form of the augmented matrix

$$[\, P_{E \leftarrow C} \mid P_{E \leftarrow B} \,] \quad \text{is} \quad [\, I \mid P_{C \leftarrow B} \,].$$

To help remember this corollary, notice that $[P_{E \leftarrow C} \mid P_{E \leftarrow B}]$ has C and B in the same order as $[I \mid P_{C \leftarrow B}]$, and we start with $[P_{E \leftarrow C} \mid P_{E \leftarrow B}]$ because changing into the standard basis E is easy.

Proof of Corollary 1.2.5. Recall that $P_{E \leftarrow C}$ is invertible, so its reduced row echelon form is I and thus the RREF of $[\, P_{E \leftarrow C} \mid P_{E \leftarrow B} \,]$ has the form $[\, I \mid X \,]$ for some matrix X.

To see that $X = P_{C \leftarrow B}$ (and thus complete the proof), recall that sequences of elementary row operations correspond to multiplication on the left by invertible matrices (see Appendix A.1.3), so there is an invertible matrix Q such that

$$Q[\, P_{E \leftarrow C} \mid P_{E \leftarrow B} \,] = [\, I \mid X \,].$$

On the other hand, block matrix multiplication shows that

$$Q[\, P_{E \leftarrow C} \mid P_{E \leftarrow B} \,] = [\, QP_{E \leftarrow C} \mid QP_{E \leftarrow B} \,].$$

Comparing the left blocks of these matrices shows that $QP_{E \leftarrow C} = I$, so $Q = P_{E \leftarrow C}^{-1} = P_{C \leftarrow E}$. Comparing the right blocks of these matrices then shows that $X = QP_{E \leftarrow B} = P_{C \leftarrow E}P_{E \leftarrow B} = P_{C \leftarrow B}$, as claimed. ∎

Example 1.2.5

Computing a Change-of-Basis Matrix via Row Operations

Use Corollary 1.2.5 to find the change-of-basis matrix $P_{C \leftarrow B}$, where

$$B = \left\{ \begin{bmatrix} 1 & 0 \\ 0 & 1 \end{bmatrix}, \begin{bmatrix} 0 & 1 \\ 1 & 0 \end{bmatrix}, \begin{bmatrix} 0 & -i \\ i & 0 \end{bmatrix}, \begin{bmatrix} 1 & 0 \\ 0 & -1 \end{bmatrix} \right\} \quad \text{and}$$

$$C = \left\{ \begin{bmatrix} 1 & 0 \\ 0 & 0 \end{bmatrix}, \begin{bmatrix} 1 & 1 \\ 0 & 0 \end{bmatrix}, \begin{bmatrix} 1 & 1 \\ 1 & 0 \end{bmatrix}, \begin{bmatrix} 1 & 1 \\ 1 & 1 \end{bmatrix} \right\}$$

are bases of $\mathcal{M}_2(\mathbb{C})$.

Solution:

Recall that we already computed $P_{E \leftarrow B}$ and $P_{E \leftarrow C}$ in Example 1.2.4:

$$P_{E \leftarrow B} = \begin{bmatrix} 1 & 0 & 0 & 1 \\ 0 & 1 & -i & 0 \\ 0 & 1 & i & 0 \\ 1 & 0 & 0 & -1 \end{bmatrix} \quad \text{and} \quad P_{E \leftarrow C} = \begin{bmatrix} 1 & 1 & 1 & 1 \\ 0 & 1 & 1 & 1 \\ 0 & 0 & 1 & 1 \\ 0 & 0 & 0 & 1 \end{bmatrix}.$$

To compute $P_{C \leftarrow B}$, we place these matrices side-by-side in an augmented matrix and then row reduce:

$$[\, P_{E \leftarrow C} \mid P_{E \leftarrow B} \,] = \left[\begin{array}{cccc|cccc} 1 & 1 & 1 & 1 & 1 & 0 & 0 & 1 \\ 0 & 1 & 1 & 1 & 0 & 1 & -i & 0 \\ 0 & 0 & 1 & 1 & 0 & 1 & i & 0 \\ 0 & 0 & 0 & 1 & 1 & 0 & 0 & -1 \end{array} \right]$$

$$\xrightarrow{R_1 - R_2} \left[\begin{array}{cccc|cccc} 1 & 0 & 0 & 0 & 1 & -1 & i & 1 \\ 0 & 1 & 1 & 1 & 0 & 1 & -i & 0 \\ 0 & 0 & 1 & 1 & 0 & 1 & i & 0 \\ 0 & 0 & 0 & 1 & 1 & 0 & 0 & -1 \end{array} \right]$$

$$\xrightarrow{R_2 - R_3} \left[\begin{array}{cccc|cccc} 1 & 0 & 0 & 0 & 1 & -1 & i & 1 \\ 0 & 1 & 0 & 0 & 0 & 0 & -2i & 0 \\ 0 & 0 & 1 & 1 & 0 & 1 & i & 0 \\ 0 & 0 & 0 & 1 & 1 & 0 & 0 & -1 \end{array} \right]$$

$$\xrightarrow{R_3 - R_4} \left[\begin{array}{cccc|cccc} 1 & 0 & 0 & 0 & 1 & -1 & i & 1 \\ 0 & 1 & 0 & 0 & 0 & 0 & -2i & 0 \\ 0 & 0 & 1 & 0 & -1 & 1 & i & 1 \\ 0 & 0 & 0 & 1 & 1 & 0 & 0 & -1 \end{array} \right]$$

$$= [\, I \mid P_{C \leftarrow B} \,].$$

Since the left block is now the identity matrix I, it follows from Corollary 1.2.5 that the right block is $P_{C \leftarrow B}$:

$$P_{C \leftarrow B} = \begin{bmatrix} 1 & -1 & i & 1 \\ 0 & 0 & -2i & 0 \\ -1 & 1 & i & 1 \\ 1 & 0 & 0 & -1 \end{bmatrix},$$

which agrees with the answer that we found in Example 1.2.4.

1.2.3   Linear Transformations

When working with vectors in $\mathbb{R}^n$, we often think of matrices as functions that move vectors around. That is, we fix a matrix $A \in \mathcal{M}_{m,n}$ and consider the function $T : \mathbb{R}^n \to \mathbb{R}^m$ defined by $T(\mathbf{v}) = A\mathbf{v}$. When we do this, many of the usual linear algebraic properties of the matrix A can be interpreted as properties of the function T: the rank of A is the dimension of the range of T, the absolute value of $\det(A)$ is a measure of how much T expands space, and so on.

Whenever we multiply a matrix by a vector, we interpret that vector as a column vector so that the matrix multiplication makes sense.

Any function T that is constructed in this way inherits some nice properties from matrix multiplication: $T(\mathbf{v} + \mathbf{w}) = A(\mathbf{v} + \mathbf{w}) = A\mathbf{v} + A\mathbf{w} = T(\mathbf{v}) + T(\mathbf{w})$ for all vectors $\mathbf{v}, \mathbf{w} \in \mathbb{R}^n$, and similarly $T(c\mathbf{v}) = A(c\mathbf{v}) = cA\mathbf{v} = cT(\mathbf{v})$ for all vectors $\mathbf{v} \in \mathbb{R}^n$ and all scalars $c \in \mathbb{R}$. We now explore functions between arbitrary vectors spaces that have these same two properties.

Definition 1.2.4

Linear Transformations

> Let $\mathcal{V}$ and $\mathcal{W}$ be vector spaces over the same field $\mathbb{F}$. A **linear transformation** is a function $T : \mathcal{V} \to \mathcal{W}$ that satisfies the following two properties:
>
> a) $T(\mathbf{v} + \mathbf{w}) = T(\mathbf{v}) + T(\mathbf{w})$ for all $\mathbf{v}, \mathbf{w} \in \mathcal{V}$, and
> b) $T(c\mathbf{v}) = cT(\mathbf{v})$ for all $\mathbf{v} \in \mathcal{V}$ and $c \in \mathbb{F}$.

All of our old examples of linear transformations (i.e., matrices) acting on $\mathcal{V} = \mathbb{R}^n$ satisfy this definition, and our intuition concerning how they move vectors around $\mathbb{R}^n$ still applies. However, there are linear transformations acting on other vector spaces that perhaps seem a bit surprising at first, so we spend some time looking at examples.

Example 1.2.6

The Transpose is a Linear Transformation

Show that the matrix transpose is a linear transformation. That is, show that the function $T : \mathcal{M}_{m,n} \to \mathcal{M}_{n,m}$ defined by $T(A) = A^T$ is a linear transformation.

Solution:

We need to show that the two properties of Definition 1.2.4 hold. That is, we need to show that $(A + B)^T = A^T + B^T$ and $(cA)^T = cA^T$ for all $A, B \in \mathcal{M}_{m,n}$ and $c \in \mathbb{F}$. Both of these properties follow almost immediately from the definition, so we do not dwell on them.

Example 1.2.7

The Trace is a Linear Transformation

The **trace** is the function $\operatorname{tr} : \mathcal{M}_n(\mathbb{F}) \to \mathbb{F}$ that adds up the diagonal entries of a matrix:

$$\operatorname{tr}(A) \stackrel{\text{def}}{=} a_{1,1} + a_{2,2} + \cdots + a_{n,n} \quad \text{for all} \quad A \in \mathcal{M}_n(\mathbb{F}).$$

Show that the trace is a linear transformation.

Solution:

We need to show that the two properties of Definition 1.2.4 hold:

The notation $[A + B]_{i,j}$ means the (i, j)-entry of $A + B$.

a) $\operatorname{tr}(A + B) = [A + B]_{1,1} + \cdots + [A + B]_{n,n} = a_{1,1} + b_{1,1} + \cdots + a_{n,n} + b_{n,n} = (a_{1,1} + \cdots + a_{n,n}) + (b_{1,1} + \cdots + b_{n,n}) = \operatorname{tr}(A) + \operatorname{tr}(B)$, and

b) $\operatorname{tr}(cA) = [cA]_{1,1} + \cdots + [cA]_{n,n} = ca_{1,1} + \cdots + ca_{n,n} = c\operatorname{tr}(A).$

It follows that the trace is a linear transformation.

Example 1.2.8

The Derivative is a Linear Transformation

Recall that $\mathcal{D}$ is the vector space of differentiable real-valued functions. Sorry for using two different "D"s in the same sentence to mean different things (curly $\mathcal{D}$ is a vector space, block D is the derivative linear transformation).

Show that the derivative is a linear transformation. That is, show that the function $D : \mathcal{D} \to \mathcal{F}$ defined by $D(f) = f'$ is a linear transformation.

Solution:

We need to show that the two properties of Definition 1.2.4 hold. That is, we need to show that $(f + g)' = f' + g'$ and $(cf)' = cf'$ for all $f, g \in \mathcal{D}$ and $c \in \mathbb{R}$. Both of these properties are typically presented in introductory calculus courses, so we do not prove them here.

There are also a couple of particularly commonly-occurring linear transformations that it is useful to give names to: the **zero transformation** $O : \mathcal{V} \to \mathcal{W}$ is the one defined by $O(\mathbf{v}) = \mathbf{0}$ for all $\mathbf{v} \in \mathcal{V}$, and the **identity transformation** $I : \mathcal{V} \to \mathcal{V}$ is the one defined by $I(\mathbf{v}) = \mathbf{v}$ for all $\mathbf{v} \in \mathcal{V}$. If we wish to clarify which spaces the identity and zero transformations are acting on, we use subscripts as in $I_\mathcal{V}$ and $O_{\mathcal{V},\mathcal{W}}$.

On the other hand, it is perhaps helpful to see an example of a linear algebraic function that is *not* a linear transformation.

Example 1.2.9

The Determinant is Not a Linear Transformation

Show that the determinant, $\det : \mathcal{M}_n(\mathbb{F}) \to \mathbb{F}$, is not a linear transformation when $n \geq 2$.

Solution:

To see that det is not a linear transformation, we need to show that at least one of the two defining properties from Definition 1.2.4 do not hold. Well, $\det(I) = 1$, so

$$\det(I + I) = \det(2I) = 2^n \det(I) = 2^n \neq 2 = \det(I) + \det(I),$$

so property (a) of that definition fails (property (b) also fails for a similar reason).

We now do for linear transformations what we did for vectors in the previous section: we give them coordinates so that we can explicitly write them down using numbers from the ground field $\mathbb{F}$. More specifically, just like every vector in a finite-dimensional vector space can be associated with a vector in $\mathbb{F}^n$, every linear transformation between vector spaces can be associated with a matrix in $\mathcal{M}_{m,n}(\mathbb{F})$:

Theorem 1.2.6

Standard Matrix of a Linear Transformation

Let $\mathcal{V}$ and $\mathcal{W}$ be vector spaces with bases B and D, respectively, where $B = \{\mathbf{v}_1, \mathbf{v}_2, \ldots, \mathbf{v}_n\}$ and $\mathcal{W}$ is m-dimensional. A function $T : \mathcal{V} \to \mathcal{W}$ is a linear transformation if and only if there exists a matrix $[T]_{D \leftarrow B} \in \mathcal{M}_{m,n}$ for which

$$[T(\mathbf{v})]_D = [T]_{D \leftarrow B}[\mathbf{v}]_B \quad \text{for all} \quad \mathbf{v} \in \mathcal{V}.$$

Furthermore, the unique matrix $[T]_{D \leftarrow B}$ with this property is called the **standard matrix** of T with respect to the bases B and D, and it is

$$[T]_{D \leftarrow B} \overset{\text{def}}{=} \big[\, [T(\mathbf{v}_1)]_D \mid [T(\mathbf{v}_2)]_D \mid \cdots \mid [T(\mathbf{v}_n)]_D \,\big].$$

In other words, this theorem tells us that instead of working with a vector $\mathbf{v} \in \mathcal{V}$, applying the linear transformation $T : \mathcal{V} \to \mathcal{W}$ to it, and *then* converting it into a coordinate vector with respect to the basis D, we can convert $\mathbf{v}$ to its coordinate vector $[\mathbf{v}]_B$ and then multiply by the matrix $[T]_{D \leftarrow B}$ (see Figure 1.6).

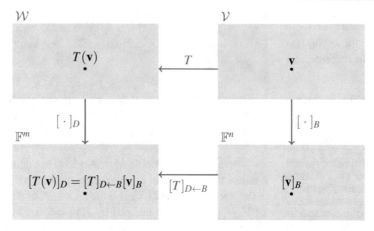

Read this figure as starting at the top-right corner and moving to the bottom-left.

Figure 1.6: A visualization of the relationship between linear transformations, their standard matrices, and coordinate vectors. Just like T sends $\mathbf{v}$ to $T(\mathbf{v})$, the standard matrix $[T]_{D\leftarrow B}$ sends the coordinate vector of $\mathbf{v}$ to the coordinate vector of $T(\mathbf{v})$.

Proof of Theorem 1.2.6. It is straightforward to show that a function that multiplies a coordinate vector by a matrix is a linear transformation, so we only prove that for every linear transformation $T : \mathcal{V} \to \mathcal{W}$, the matrix

$$[T]_{D\leftarrow B} = \big[\, [T(\mathbf{v}_1)]_D \mid [T(\mathbf{v}_2)]_D \mid \cdots \mid [T(\mathbf{v}_n)]_D \,\big]$$

satisfies $[T]_{D\leftarrow B}[\mathbf{v}]_B = [T(\mathbf{v})]_D$, and no other matrix has this property. To see that $[T]_{D\leftarrow B}[\mathbf{v}]_B = [T(\mathbf{v})]_D$, suppose that $[\mathbf{v}]_B = (c_1, c_2, \ldots, c_n)$ (i.e., $\mathbf{v} = c_1\mathbf{v}_1 + c_2\mathbf{v}_2 + \cdots + c_n\mathbf{v}_n$) and do block matrix multiplication:

This proof is almost identical to that of Theorem 1.2.3. The reason for this is that change-of-basis matrices are exactly the standard matrices of the identity transformation.

$$[T]_{D\leftarrow B}[\mathbf{v}]_B = \big[\, [T(\mathbf{v}_1)]_D \mid \cdots \mid [T(\mathbf{v}_n)]_D \,\big] \begin{bmatrix} c_1 \\ \vdots \\ c_n \end{bmatrix} \quad \text{(definition of } [T]_{D\leftarrow B})$$

$$= c_1[T(\mathbf{v}_1)]_D + \cdots + c_n[T(\mathbf{v}_n)]_D \quad \text{(block matrix mult.)}$$

$$= [c_1 T(\mathbf{v}_1) + \cdots + c_n T(\mathbf{v}_n)]_D \quad \text{(by Exercise 1.2.22)}$$

$$= [T(c_1\mathbf{v}_1 + \cdots + c_n\mathbf{v}_n)]_D \quad \text{(linearity of } T)$$

$$= [T(\mathbf{v})]_D. \quad (\mathbf{v} = c_1\mathbf{v}_1 + \cdots c_n\mathbf{v}_n)$$

The proof of uniqueness of $[T]_{D\leftarrow B}$ is almost identical to the proof of uniqueness of $P_{C\leftarrow B}$ from Theorem 1.2.3, so we leave it to Exercise 1.2.35. ∎

For simplicity of notation, in the special case when $\mathcal{V} = \mathcal{W}$ and $B = D$ we denote the standard matrix $[T]_{B\leftarrow B}$ simply by $[T]_B$. If $\mathcal{V}$ furthermore has a standard basis E then we sometimes denote $[T]_E$ simply by $[T]$.

Example 1.2.10

Standard Matrix of the Transposition Map

Find the standard matrix $[T]$ of the transposition map $T : \mathcal{M}_2 \to \mathcal{M}_2$ with respect to the standard basis $E = \{E_{1,1}, E_{1,2}, E_{2,1}, E_{2,2}\}$.

Solution:

We need to compute the coefficient vectors of $E_{1,1}^T$, $E_{1,2}^T$, $E_{2,1}^T$, and $E_{2,2}^T$, and place them (in that order) into the matrix $[T]$:

$$[E_{1,1}^T]_E = \begin{bmatrix} 1 & 0 \\ 0 & 0 \end{bmatrix}_E = (1,0,0,0), \quad [E_{1,2}^T]_E = \begin{bmatrix} 0 & 0 \\ 1 & 0 \end{bmatrix}_E = (0,0,1,0),$$

$$[E_{2,1}^T]_E = \begin{bmatrix} 0 & 1 \\ 0 & 0 \end{bmatrix}_E = (0,1,0,0), \quad [E_{2,2}^T]_E = \begin{bmatrix} 0 & 0 \\ 0 & 1 \end{bmatrix}_E = (0,0,0,1).$$

The standard matrix of T looks different depending on which bases B and D are used (just like coordinate vectors look different depending on the basis B).

It follows that

$$[T] = [T]_E = \left[\, [E_{1,1}^T]_E \,\middle|\, [E_{1,2}^T]_E \,\middle|\, [E_{2,1}^T]_E \,\middle|\, [E_{2,2}^T]_E \,\right] = \begin{bmatrix} 1 & 0 & 0 & 0 \\ 0 & 0 & 1 & 0 \\ 0 & 1 & 0 & 0 \\ 0 & 0 & 0 & 1 \end{bmatrix}.$$

We could be done at this point, but as a bit of a sanity check, it is perhaps useful to verify that $[T]_E[A]_E = [A^T]_E$ for all $A \in \mathcal{M}_2$. To this end, we notice that if

$$A = \begin{bmatrix} a & b \\ c & d \end{bmatrix} \quad \text{then} \quad [A]_E = (a,b,c,d),$$

We generalize this example to higher dimensions in Exercise 1.2.12.

so

$$[T]_E[A]_E = \begin{bmatrix} 1 & 0 & 0 & 0 \\ 0 & 0 & 1 & 0 \\ 0 & 1 & 0 & 0 \\ 0 & 0 & 0 & 1 \end{bmatrix}\begin{bmatrix} a \\ b \\ c \\ d \end{bmatrix} = \begin{bmatrix} a \\ c \\ b \\ d \end{bmatrix} = \begin{bmatrix} a & c \\ b & d \end{bmatrix}_E = [A^T]_E,$$

as desired.

Example 1.2.11

Standard Matrix of the Derivative

Find the standard matrix $[D]_{C \leftarrow B}$ of the derivative map $D : \mathcal{P}^3 \to \mathcal{P}^2$ with respect to the standard bases $B = \{1, x, x^2, x^3\} \subset \mathcal{P}^3$ and $C = \{1, x, x^2\} \subset \mathcal{P}^2$.

Solution:

We need to compute the coefficient vectors of $D(1) = 0$, $D(x) = 1$, $D(x^2) = 2x$, and $D(x^3) = 3x^2$, and place them (in that order) into the matrix $[D]_{C \leftarrow B}$:

$$[0]_C = (0,0,0), \quad [1]_C = (1,0,0), \quad [2x]_C = (0,2,0), \quad [3x^2]_C = (0,0,3).$$

It follows that

$$[D]_{C \leftarrow B} = \left[\, [0]_C \,\middle|\, [1]_C \,\middle|\, [2x]_C \,\middle|\, [3x^2]_C \,\right] = \begin{bmatrix} 0 & 1 & 0 & 0 \\ 0 & 0 & 2 & 0 \\ 0 & 0 & 0 & 3 \end{bmatrix}.$$

Once again, as a bit of a sanity check, it is perhaps useful to verify that $[D]_{C \leftarrow B}[f]_B = [f']_C$ for all $f \in \mathcal{P}^3$. If $f(x) = a_0 + a_1 x + a_2 x^2 + a_3 x^3$ then

$[f]_B = (a_0, a_1, a_2, a_3)$, so

$$[D]_{C \leftarrow B}[f]_B = \begin{bmatrix} 0 & 1 & 0 & 0 \\ 0 & 0 & 2 & 0 \\ 0 & 0 & 0 & 3 \end{bmatrix} \begin{bmatrix} a_0 \\ a_1 \\ a_2 \\ a_3 \end{bmatrix} = \begin{bmatrix} a_1 \\ 2a_2 \\ 3a_3 \end{bmatrix} = [a_1 + 2a_2x + 3a_3x^2]_C.$$

Since $a_1 + 2a_2x + 3a_3x^2 = (a_0 + a_1x + a_2x^2 + a_3x^3)' = f'(x)$, we thus see that we indeed have $[D]_{C \leftarrow B}[f]_B = [f']_C$, as expected.

Keep in mind that if we change the input and/or output vector spaces of the derivative map D, then its standard matrix can look quite a bit different (after all, even just changing the bases used on these vector spaces can change the standard matrix). For example, if we considered D as a linear transformation from $\mathcal{P}^3$ to $\mathcal{P}^3$, instead of from $\mathcal{P}^3$ to $\mathcal{P}^2$ as in the previous example, then its standard matrix would be 4×4 instead of 3×4 (it would have an extra zero row at its bottom).

The next example illustrates this observation with the derivative map on a slightly more exotic vector space.

Example 1.2.12

Standard Matrix of the Derivative (Again)

Let $B = \{e^x, xe^x, x^2e^x\}$ be a basis of the vector space $\mathcal{V} = \mathrm{span}(B)$. Find the standard matrix $[D]_B$ of the derivative map $D : \mathcal{V} \to \mathcal{V}$ with respect to B.

Solution:

We need to compute the coefficient vectors of $D(e^x) = e^x$, $D(xe^x) = e^x + xe^x$, and $D(x^2e^x) = 2xe^x + x^2e^x$, and place them (in that order) as columns into the matrix $[D]_B$:

Recall that spans are always subspaces. B was shown to be linearly independent in Exercise 1.1.2(g), so it is indeed a basis of $\mathcal{V}$.

$$[e^x]_B = (1,0,0), \quad [e^x + xe^x]_B = (1,1,0), \quad [2xe^x + x^2e^x]_B = (0,2,1).$$

It follows that

$$[D]_B = \left[\; [e^x]_B \; \middle| \; [e^x + xe^x]_B \; \middle| \; [2xe^x + x^2e^x]_B \; \right] = \begin{bmatrix} 1 & 1 & 0 \\ 0 & 1 & 2 \\ 0 & 0 & 1 \end{bmatrix}.$$

It is often useful to consider the effect of applying two or more linear transformations to a vector, one after another. Rather than thinking of these linear transformations as separate objects that are applied in sequence, we can combine their effect into a single new function that is called their composition.

More specifically, suppose $\mathcal{V}, \mathcal{W}$, and $\mathcal{X}$ are vector spaces and $T : \mathcal{V} \to \mathcal{W}$ and $S : \mathcal{W} \to \mathcal{X}$ are linear transformations. We say that the **composition** of S and T, denoted by $S \circ T$, is the function defined by

Like most things in this section, the composition $S \circ T$ should be read right-to-left. The input vector **v** goes in on the right, then T acts on it, and then S acts on that.

$$(S \circ T)(\mathbf{v}) \overset{\text{def}}{=} S(T(\mathbf{v})) \quad \text{for all} \quad \mathbf{v} \in \mathcal{V}.$$

That is, the composition $S \circ T$ of two linear transformations is the function that we get if we apply T first and then S. In other words, while T sends $\mathcal{V}$ to $\mathcal{W}$ and S sends $\mathcal{W}$ to $\mathcal{X}$, the composition $S \circ T$ skips the intermediate step and sends $\mathcal{V}$ directly to $\mathcal{X}$, as illustrated in Figure 1.7.

Importantly, $S \circ T$ is also a linear transformation, and its standard matrix can be obtained simply by multiplying together the standard matrices of S and T.

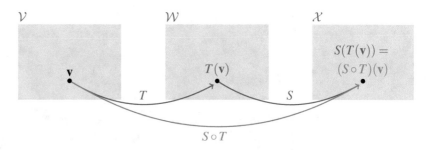

Figure 1.7: The composition of S and T, denotes by $S \circ T$, is the function that sends $\mathbf{v} \in \mathcal{V}$ to $S(T(\mathbf{v})) \in \mathcal{X}$.

In fact, this is the primary reason that matrix multiplication is defined in the seemingly bizarre way that it is—we want it to capture the idea of applying one matrix (linear transformation) after another to a vector.

Theorem 1.2.7

Composition of Linear Transformations

Suppose $\mathcal{V}$, $\mathcal{W}$, and $\mathcal{X}$ are finite-dimensional vector spaces with bases B, C, and D, respectively. If $T : \mathcal{V} \to \mathcal{W}$ and $S : \mathcal{W} \to \mathcal{X}$ are linear transformations then $S \circ T : \mathcal{V} \to \mathcal{X}$ is a linear transformation, and its standard matrix is

$$[S \circ T]_{D \leftarrow B} = [S]_{D \leftarrow C}[T]_{C \leftarrow B}.$$

Notice that the middle C subscripts match, while the left D subscripts and the right B subscripts also match.

Proof. Thanks to the uniqueness condition of Theorem 1.2.6, we just need to show that $[(S \circ T)(\mathbf{v})]_D = [S]_{D \leftarrow C}[T]_{C \leftarrow B}[\mathbf{v}]_B$ for all $\mathbf{v} \in \mathcal{V}$. To this end, we compute $[(S \circ T)(\mathbf{v})]_D$ by using Theorem 1.2.6 applied to each of S and T individually:

$$[(S \circ T)(\mathbf{v})]_D = [S(T(\mathbf{v}))]_D = [S]_{D \leftarrow C}[T(\mathbf{v})]_C = [S]_{D \leftarrow C}[T]_{C \leftarrow B}[\mathbf{v}]_B.$$

It follows that $S \circ T$ is a linear transformation, and its standard matrix is $[S]_{D \leftarrow C}[T]_{C \leftarrow B}$, as claimed. ∎

In the special case when the linear transformations that we are composing are equal to each other, we denote their composition by $T^2 \stackrel{\text{def}}{=} T \circ T$. More generally, we can define powers of a linear transformation $T : \mathcal{V} \to \mathcal{V}$ via

$$T^k \stackrel{\text{def}}{=} \underbrace{T \circ T \circ \cdots \circ T}_{k \text{ copies}}.$$

Example 1.2.13

Iterated Derivatives

Use standard matrices to compute the fourth derivative of $x^2 e^x + 2xe^x$.

Solution:
We let $B = \{e^x, xe^x, x^2 e^x\}$ and $\mathcal{V} = \text{span}(B)$ so that we can make use of the standard matrix

$$[D]_B = \begin{bmatrix} 1 & 1 & 0 \\ 0 & 1 & 2 \\ 0 & 0 & 1 \end{bmatrix}$$

of the derivative map D that we computed in Example 1.2.12. We know from Theorem 1.2.7 that the standard matrix of $D^2 = D \circ D$ (i.e., the linear map that takes the derivative twice) is

$$[D^2]_B = [D]_B^2 = \begin{bmatrix} 1 & 1 & 0 \\ 0 & 1 & 2 \\ 0 & 0 & 1 \end{bmatrix}^2 = \begin{bmatrix} 1 & 2 & 2 \\ 0 & 1 & 4 \\ 0 & 0 & 1 \end{bmatrix}.$$

Similar reasoning tells us that the standard matrix of $D^4 = D \circ D \circ D \circ D$ (i.e., the linear map that takes the derivative four times) is

$$[D^4]_B = [D^2]_B^2 = \begin{bmatrix} 1 & 2 & 2 \\ 0 & 1 & 4 \\ 0 & 0 & 1 \end{bmatrix}^2 = \begin{bmatrix} 1 & 4 & 12 \\ 0 & 1 & 8 \\ 0 & 0 & 1 \end{bmatrix}.$$

With this standard matrix in hand, one way to find the fourth derivative of $x^2 e^x + 2x e^x$ is to multiply its coordinate vector $(0,2,1)$ by $[D^4]_B$:

$$[(x^2 e^x + 2x e^x)'''']_B = [D^4]_B [x^2 e^x + 2x e^x]_B$$

$$= \begin{bmatrix} 1 & 4 & 12 \\ 0 & 1 & 8 \\ 0 & 0 & 1 \end{bmatrix} \begin{bmatrix} 0 \\ 2 \\ 1 \end{bmatrix} = \begin{bmatrix} 20 \\ 10 \\ 1 \end{bmatrix}.$$

It follows that $(x^2 e^x + 2x e^x)'''' = 20 e^x + 10x e^x + x^2 e^x$.

Recall from earlier that we learned how to convert a coordinate vector from the basis B to another one C by using a change-of-basis matrix $P_{C \leftarrow B}$. We now learn how to do the same thing for linear transformations: there is a reasonably direct method of converting a standard matrix with respect to bases B and D to a standard matrix with respect to bases C and E.

> The basis E here does *not* have to be the standard basis. There just are not enough letters in the alphabet.

Theorem 1.2.8

Change of Basis for Linear Transformations

Suppose $\mathcal{V}$ is a finite-dimensional vector space with bases B and C, $\mathcal{W}$ is a finite-dimensional vector space with bases D and E, and $T : \mathcal{V} \to \mathcal{W}$ is a linear transformation. Then

$$[T]_{E \leftarrow B} = P_{E \leftarrow D}[T]_{D \leftarrow C} P_{C \leftarrow B}.$$

> As always, notice in this theorem that adjacent subscripts always match (e.g., the two "D"s are next to each other, as are the two "C"s).

Proof. We simply multiply the matrix $P_{E \leftarrow D}[T]_{D \leftarrow C} P_{C \leftarrow B}$ on the right by an arbitrary coordinate vector $[\mathbf{v}]_B$, where $\mathbf{v} \in \mathcal{V}$. Well,

$$\begin{aligned} P_{E \leftarrow D}[T]_{D \leftarrow C} P_{C \leftarrow B}[\mathbf{v}]_B &= P_{E \leftarrow D}[T]_{D \leftarrow C}[\mathbf{v}]_C && \text{(since } P_{C \leftarrow B}[\mathbf{v}]_B = [\mathbf{v}]_C) \\ &= P_{E \leftarrow D}[T(\mathbf{v})]_D && \text{(by Theorem 1.2.6)} \\ &= [T(\mathbf{v})]_E. \end{aligned}$$

However, we know from Theorem 1.2.6 that $[T]_{E \leftarrow B}$ is the *unique* matrix for which $[T]_{E \leftarrow B}[\mathbf{v}]_B = [T(\mathbf{v})]_E$ for all $\mathbf{v} \in \mathcal{V}$, so it follows that $P_{E \leftarrow D}[T]_{D \leftarrow C} P_{C \leftarrow B} = [T]_{E \leftarrow B}$, as claimed. ∎

A schematic that illustrates the statement of the above theorem is provided by Figure 1.8. All it says is that there are two different (but equivalent) ways of converting $[\mathbf{v}]_B$ into $[T(\mathbf{v})]_E$: we could multiply $[\mathbf{v}]_B$ by $[T]_{E \leftarrow B}$, or we could convert $[\mathbf{v}]_B$ into basis C, then multiply by $[T]_{D \leftarrow C}$, and then convert into basis E.

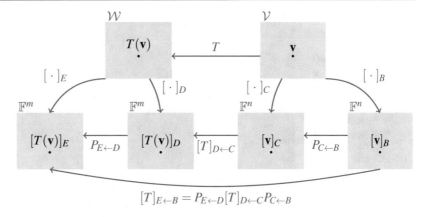

Again, read this figure as starting at the top-right corner and moving to the bottom-left.

$$[T]_{E \leftarrow B} = P_{E \leftarrow D}[T]_{D \leftarrow C}P_{C \leftarrow B}$$

Figure 1.8: A visualization of the relationship between linear transformations, standard matrices, change-of-basis matrices, and coordinate vectors. In particular, the bottom row illustrates Theorem 1.2.8, which says that we can construct $[T]_{E \leftarrow B}$ from $[T]_{D \leftarrow C}$ by multiplying on the right and left by the appropriate change-of-basis matrices. This image is basically just a combination of Figures 1.5 and 1.6.

Example 1.2.14

Representing the Transpose in Weird Bases

Compute the standard matrix $[T]_{C \leftarrow B}$ of the transpose map $T : \mathcal{M}_2(\mathbb{C}) \to \mathcal{M}_2(\mathbb{C})$, where

$$B = \left\{ \begin{bmatrix} 1 & 0 \\ 0 & 1 \end{bmatrix}, \begin{bmatrix} 0 & 1 \\ 1 & 0 \end{bmatrix}, \begin{bmatrix} 0 & -i \\ i & 0 \end{bmatrix}, \begin{bmatrix} 1 & 0 \\ 0 & -1 \end{bmatrix} \right\} \quad \text{and}$$

$$C = \left\{ \begin{bmatrix} 1 & 0 \\ 0 & 0 \end{bmatrix}, \begin{bmatrix} 1 & 1 \\ 0 & 0 \end{bmatrix}, \begin{bmatrix} 1 & 1 \\ 1 & 0 \end{bmatrix}, \begin{bmatrix} 1 & 1 \\ 1 & 1 \end{bmatrix} \right\}$$

are bases of $\mathcal{M}_2(\mathbb{C})$.

Solution:

Recall that if $E = \{E_{1,1}, E_{1,2}, E_{2,1}, E_{2,2}\}$ is the standard basis of $\mathcal{M}_2(\mathbb{C})$ then we already computed $P_{E \leftarrow B}$ and $P_{C \leftarrow E}$ in Example 1.2.4:

$$P_{E \leftarrow B} = \begin{bmatrix} 1 & 0 & 0 & 1 \\ 0 & 1 & -i & 0 \\ 0 & 1 & i & 0 \\ 1 & 0 & 0 & -1 \end{bmatrix} \quad \text{and} \quad P_{C \leftarrow E} = \begin{bmatrix} 1 & -1 & 0 & 0 \\ 0 & 1 & -1 & 0 \\ 0 & 0 & 1 & -1 \\ 0 & 0 & 0 & 1 \end{bmatrix},$$

as well as $[T]_E$ in Example 1.2.10:

$$[T]_E = \begin{bmatrix} 1 & 0 & 0 & 0 \\ 0 & 0 & 1 & 0 \\ 0 & 1 & 0 & 0 \\ 0 & 0 & 0 & 1 \end{bmatrix}.$$

Theorem 1.2.8 tells us that we can compute $[T]_{C \leftarrow B}$ just by multiplying these matrices together in the appropriate order:

We could also
compute $[T]_{C \leftarrow B}$
directly from the
definition, but that is
a lot more work as it
requires the
computation of
numerous ugly
coordinate vectors.

$$[T]_{C \leftarrow B} = P_{C \leftarrow E}[T]_E P_{E \leftarrow B}$$

$$= \begin{bmatrix} 1 & -1 & 0 & 0 \\ 0 & 1 & -1 & 0 \\ 0 & 0 & 1 & -1 \\ 0 & 0 & 0 & 1 \end{bmatrix} \begin{bmatrix} 1 & 0 & 0 & 0 \\ 0 & 0 & 1 & 0 \\ 0 & 1 & 0 & 0 \\ 0 & 0 & 0 & 1 \end{bmatrix} \begin{bmatrix} 1 & 0 & 0 & 1 \\ 0 & 1 & -i & 0 \\ 0 & 1 & i & 0 \\ 1 & 0 & 0 & -1 \end{bmatrix}$$

$$= \begin{bmatrix} 1 & -1 & 0 & 0 \\ 0 & 1 & -1 & 0 \\ 0 & 0 & 1 & -1 \\ 0 & 0 & 0 & 1 \end{bmatrix} \begin{bmatrix} 1 & 0 & 0 & 1 \\ 0 & 1 & i & 0 \\ 0 & 1 & -i & 0 \\ 1 & 0 & 0 & -1 \end{bmatrix}$$

$$= \begin{bmatrix} 1 & -1 & -i & 1 \\ 0 & 0 & 2i & 0 \\ -1 & 1 & -i & 1 \\ 1 & 0 & 0 & -1 \end{bmatrix}.$$

1.2.4 Properties of Linear Transformations

Now that we know how to represent linear transformations as matrices, we can introduce many properties of linear transformations that are analogous to properties of their standard matrices. Because of the close relationship between these two concepts, all of the techniques that we know for dealing with matrices carry over immediately to this more general setting, as long as the underlying vector spaces are finite-dimensional.

For example, we say that a linear transformation $T : V \to W$ is **invertible** if there exists a linear transformation $T^{-1} : W \to V$ such that

$$T^{-1}(T(\mathbf{v})) = \mathbf{v} \quad \text{for all} \quad \mathbf{v} \in V \quad \text{and}$$
$$T(T^{-1}(\mathbf{w})) = \mathbf{w} \quad \text{for all} \quad \mathbf{w} \in W.$$

In other words, $T^{-1} \circ T = I_V$ and $T \circ T^{-1} = I_W$. Our intuitive understanding of the inverse of a linear transformation is identical to that of matrices (T^{-1} is the linear transformation that "undoes" what T "does"), and the following theorem says that invertibility of T can in fact be determined from its standard matrix exactly how we might hope:

Theorem 1.2.9	Suppose V and W are finite-dimensional vector spaces with bases B and D, respectively, and $\dim(V) = \dim(W)$. Then a linear transformation $T : V \to W$ is invertible if and only if its standard matrix $[T]_{D \leftarrow B}$ is invertible, and $$[T^{-1}]_{B \leftarrow D} = ([T]_{D \leftarrow B})^{-1}.$$
Invertibility of Linear Transformations	

Proof. We make use of the fact from Exercise 1.2.30 that the standard matrix of the identity transformation is the identity matrix: $[T]_B = I$ if and only if $T = I_V$.

For the "only if" direction, note that if T is invertible then we have

$$I = [I_V]_B = [T^{-1} \circ T]_B = [T^{-1}]_{B \leftarrow D}[T]_{D \leftarrow B}.$$

Since $[T^{-1}]_{B \leftarrow D}$ and $[T]_{D \leftarrow B}$ multiply to the identity matrix, it follows that they are inverses of each other.

We show in Exercise 1.2.21 that if $\dim(\mathcal{V}) \neq \dim(\mathcal{W})$ then T cannot possibly be invertible.

For the "if" direction, suppose that $[T]_{D \leftarrow B}$ is invertible, with inverse matrix A. Then there is some linear transformation $S : \mathcal{W} \to \mathcal{V}$ such that $A = [S]_{B \leftarrow D}$, so for all $\mathbf{v} \in \mathcal{V}$ we have

$$[\mathbf{v}]_B = A[T]_{D \leftarrow B}[\mathbf{v}]_B = [S]_{D \leftarrow B}[T]_{D \leftarrow B}[\mathbf{v}]_B = [(S \circ T)(\mathbf{v})]_B.$$

This implies $[S \circ T]_B = I$, so $S \circ T = I_{\mathcal{V}}$, and a similar argument shows that $T \circ S = I_{\mathcal{W}}$. It follows that T is invertible, and its inverse is S. ∎

In a sense, the above theorem tells us that when considering the invertibility of linear transformations, we do not actually need to do anything new—everything just carries over from the invertibility of matrices straightforwardly. The important difference here is our perspective—we can now use everything we know about invertible matrices in many other situations where we want to invert an operation. The following example illustrates this observation by inverting the derivative (i.e., integrating).

Example 1.2.15

Indefinite Integrals via Standard Matrices

Use standard matrices to compute

$$\int (3x^2 e^x - xe^x)\, dx.$$

Solution:

The "standard" way to compute this indefinite integral directly would be to use integration by parts twice.

We let $B = \{e^x, xe^x, x^2 e^x\}$ and $\mathcal{V} = \operatorname{span}(B)$ so that we can make use of the standard matrix

$$[D]_B = \begin{bmatrix} 1 & 1 & 0 \\ 0 & 1 & 2 \\ 0 & 0 & 1 \end{bmatrix}$$

of the derivative map D that we computed in Example 1.2.12. Since the indefinite integral and derivative are inverse operations of one another on $\mathcal{V}$, D^{-1} is the integration linear transformation. It then follows from Theorem 1.2.9 that the standard matrix of D^{-1} is

See Appendix A.1.3 if you need a refresher on how to compute the inverse of a matrix. Recall that applying Gaussian elimination to $[A \mid I]$ produces $[I \mid A^{-1}]$.

$$[D^{-1}]_B = [D]_B^{-1} = \begin{bmatrix} 1 & -1 & 2 \\ 0 & 1 & -2 \\ 0 & 0 & 1 \end{bmatrix}.$$

It follows that, to find the coefficient vector of the indefinite integral of $3x^2 e^x - xe^x$, we compute

$$[D^{-1}]_B [3x^2 e^x - xe^x]_B = \begin{bmatrix} 1 & -1 & 2 \\ 0 & 1 & -2 \\ 0 & 0 & 1 \end{bmatrix} \begin{bmatrix} 0 \\ -1 \\ 3 \end{bmatrix} = \begin{bmatrix} 7 \\ -7 \\ 3 \end{bmatrix}.$$

We thus conclude that

$$\int (3x^2 e^x - xe^x)\, dx = 7e^x - 7xe^x + 3x^2 e^x + C.$$

It is worth pointing out that the method of integration presented in Example 1.2.15 does not work if the derivative map D is not invertible on the vector space in question. For example, if we tried to compute

$$\int x^3 \, dx$$

via this method, we could run into trouble since $D : \mathcal{P}^3 \to \mathcal{P}^3$ is not invertible: direct computation reveals that its standard matrix is 4×4, but has rank 3. After all, it has a 1-dimensional null space consisting of the coordinate vectors of the constant functions, which are all mapped to 0. One way to get around this problem is to use the *pseudo*inverse, which we introduce in Section 2.C.1.

Remark 1.2.2 **Invertibility in Infinite-Dimensional Vector Spaces** A "one-sided inverse" of a linear transformation T is a linear transformation T^{-1} such that $T^{-1} \circ T = I_\mathcal{V}$ or $T \circ T^{-1} = I_\mathcal{W}$ (but not necessarily both). Recall that $\mathbb{F}^\mathbb{N}$ is the vector space of infinite sequences of scalars from $\mathbb{F}$.	Because we can think of linear transformations acting on finite-dimensional vector spaces as matrices, almost any property or theorem involving matrices can be carried over to this new setting without much trouble. However, if the vector spaces involved are infinite-dimensional, then many of the nice properties that matrices have can break down. For example, a standard result of introductory linear algebra says that every one-sided inverse of a matrix, and thus every one-sided inverse of a linear transformation between finite-dimensional vector spaces, is necessarily a two-sided inverse as well. However, this property does not hold in infinite-dimensional vector spaces. For example, if $R : \mathbb{F}^\mathbb{N} \to \mathbb{F}^\mathbb{N}$ is the "right shift" map defined by $$R(c_1, c_2, c_3, \ldots) = (0, c_1, c_2, c_3, \ldots),$$ then it is straightforward to verify that R is linear (see Exercise 1.2.32). Furthermore, the "left shift" map $L : \mathbb{F}^\mathbb{N} \to \mathbb{F}^\mathbb{N}$ defined by $$L(c_1, c_2, c_3, \ldots) = (c_2, c_3, c_4, \ldots)$$ is a one-sided inverse of R, since $L \circ R = I_{\mathbb{F}^\mathbb{N}}$. However, L is *not* a two-sided inverse of R, since $$(R \circ L)(c_1, c_2, c_3, \ldots) = (0, c_2, c_3, \ldots),$$ so $R \circ L \neq I_{\mathbb{F}^\mathbb{N}}$.

We can also introduce many other properties and quantities associated with linear transformations, such as their range, null space, rank, and eigenvalues. In all of these cases, the definitions are almost identical to what they were for matrices, and in the finite-dimensional case they can all be handled simply by appealing to what we already know about matrices.

In particular, we now define the following concepts concerning a linear transformation $T : \mathcal{V} \to \mathcal{W}$:

We show that range(T) is a subspace of $\mathcal{W}$ and null(T) is a subspace of $\mathcal{V}$ in Exercise 1.2.24.

- range$(T) \stackrel{\text{def}}{=} \{T(\mathbf{x}) : \mathbf{x} \in \mathcal{V}\}$,
- null$(T) \stackrel{\text{def}}{=} \{\mathbf{x} \in \mathcal{V} : T(\mathbf{x}) = \mathbf{0}\}$,
- rank$(T) \stackrel{\text{def}}{=} \dim(\text{range}(T))$, and
- nullity$(T) \stackrel{\text{def}}{=} \dim(\text{null}(T))$.

In all of these cases, we can compute these quantities by converting everything to standard matrices and coordinate vectors, doing our computations on matrices using the techniques that we already know, and then converting back to linear transformations and vectors in $\mathcal{V}$. We now illustrate this technique with some examples, but keep in mind that if $\mathcal{V}$ and/or $\mathcal{W}$ are infinite-dimensional, then these concepts become quite a bit more delicate and are beyond the scope of this book.

Example 1.2.16

Range and Null Space of the Derivative

Determine the range, null space, rank, and nullity of the derivative map $D : \mathcal{P}^3 \to \mathcal{P}^3$.

Solution:

We first compute these objects directly from the definitions.

- The range of D is the set of all polynomials of the form $D(p)$, where $p \in \mathcal{P}^3$. Since the derivative of a degree-3 polynomial has degree-2 (and conversely, every degree-2 polynomial is the derivative of some degree-3 polynomial), we conclude that range$(D) = \mathcal{P}^2$.
- The null space of D is the set of all polynomials p for which $D(p) = 0$. Since $D(p) = 0$ if and only if p is constant, we conclude that null$(D) = \mathcal{P}^0$ (the constant functions).
- The rank of D is the dimension of its range, which is $\dim(\mathcal{P}^2) = 3$.
- The nullity of D is the dimension of its null space: $\dim(\mathcal{P}^0) = 1$.

Alternatively, we could have arrived at these answers by working with the standard matrix of D. Using an argument analogous to that of Example 1.2.11, we see that the standard matrix of D with respect to the standard basis $B = \{1, x, x^2, x^3\}$ is

In Example 1.2.11, D was a linear transformation into $\mathcal{P}^2$ (instead of $\mathcal{P}^3$) so its standard matrix there was 3×4 instead of 4×4.

$$[D]_B = \begin{bmatrix} 0 & 1 & 0 & 0 \\ 0 & 0 & 2 & 0 \\ 0 & 0 & 0 & 3 \\ 0 & 0 & 0 & 0 \end{bmatrix}.$$

Straightforward computation then shows the following:

- range$([D]_B) = \mathrm{span}\big((1,0,0,0),(0,1,0,0),(0,0,1,0)\big)$. These three vectors are $[1]_B$, $[x]_B$, and $[x^2]_B$, so range$(D) = \mathrm{span}(1,x,x^2) = \mathcal{P}^2$, as we saw earlier.
- null$([D]_B) = \mathrm{span}\big((1,0,0,0)\big)$. Since $(1,0,0,0) = [1]_B$, we conclude that null$(D) = \mathrm{span}(1) = \mathcal{P}^0$, as we saw earlier.
- rank$([D]_B) = 3$, so rank$(D) = 3$ as well.
- nullity$([D]_B) = 1$, so nullity$(D) = 1$ as well.

In the above example, we were able to learn about the range, null space, rank, and nullity of a linear transformation by considering the corresponding properties of its standard matrix (with respect to any basis). These facts are hopefully intuitive enough (after all, the entire reason we introduced standard matrices is because they act on $\mathbb{F}^n$ in the same way that the linear transformation acts on $\mathcal{V}$), but they are proved explicitly in Exercise 1.2.25.

If $T : \mathcal{V} \to \mathcal{W}$ with $\mathcal{V} \neq \mathcal{W}$ then $T(\mathbf{v})$ and $\lambda \mathbf{v}$ live in different vector spaces, so it does not make sense to talk about them being equal.

We can also define eigenvalues and eigenvectors of linear transformations in almost the exact same way that is done for matrices. If $\mathcal{V}$ is a vector space over a field $\mathbb{F}$ then a non-zero vector $\mathbf{v} \in \mathcal{V}$ is an **eigenvector** of a linear transformation $T : \mathcal{V} \to \mathcal{V}$ with corresponding **eigenvalue** $\lambda \in \mathbb{F}$ if $T(\mathbf{v}) = \lambda \mathbf{v}$. We furthermore say that the **eigenspace** corresponding to a particular eigenvalue is the set consisting of all eigenvectors corresponding to that eigenvalue, together with $\mathbf{0}$.

Just like the range and null space, eigenvalues and eigenvectors can be computed either straight from the definition or via the corresponding properties of a standard matrix.

Example 1.2.17

Eigenvalues and Eigenvectors of the Transpose

Compute the eigenvalues and corresponding eigenspaces of the transpose map $T : \mathcal{M}_2 \to \mathcal{M}_2$.

Solution:

We compute the eigenvalues and eigenspaces of T by making use of its standard matrix with respect to the standard basis $E = \{E_{1,1}, E_{1,2}, E_{2,1}, E_{2,2}\}$ that we computed in Example 1.2.10:

$$[T] = \begin{bmatrix} 1 & 0 & 0 & 0 \\ 0 & 0 & 1 & 0 \\ 0 & 1 & 0 & 0 \\ 0 & 0 & 0 & 1 \end{bmatrix}.$$

Then the characteristic polynomial $p_{[T]}$ of $[T]$ is

Computing determinants of 4×4 matrices is normally not particularly fun, but it's doing so here is not too bad since $[T]$ has so many zero entries.

$$p_{[T]}(\lambda) = \det\left(\begin{bmatrix} 1-\lambda & 0 & 0 & 0 \\ 0 & -\lambda & 1 & 0 \\ 0 & 1 & -\lambda & 0 \\ 0 & 0 & 0 & 1-\lambda \end{bmatrix}\right) = \lambda^2(1-\lambda)^2 - (1-\lambda)^2$$
$$= (\lambda^2 - 1)(1-\lambda)^2$$
$$= (\lambda-1)^3(\lambda+1).$$

It follows that the eigenvalues of $[T]$ are 1, with algebraic multiplicity 3, and -1, with algebraic multiplicity 1. We now find bases of its eigenspaces.

$\lambda = 1$: This eigenspace equals the null space of

$$A - \lambda I = A - I = \begin{bmatrix} 0 & 0 & 0 & 0 \\ 0 & -1 & 1 & 0 \\ 0 & 1 & -1 & 0 \\ 0 & 0 & 0 & 0 \end{bmatrix},$$

which has $B = \{(1,0,0,0), (0,1,1,0), (0,0,0,1)\}$ as a basis. These three basis vectors are the coordinate vectors of the matrices

$$\begin{bmatrix} 1 & 0 \\ 0 & 0 \end{bmatrix}, \quad \begin{bmatrix} 0 & 1 \\ 1 & 0 \end{bmatrix}, \quad \text{and} \quad \begin{bmatrix} 0 & 0 \\ 0 & 1 \end{bmatrix},$$

When we say "symmetric" here, we really mean symmetric: $A^T = A$. We do not mean "Hermitian", even if the underlying field is $\mathbb{C}$.

which thus make up a basis of the $\lambda = 1$ eigenspace of T. This makes sense since these matrices span the set $\mathcal{M}_2^S$ of symmetric 2×2 matrices. In other words, we have just restated the trivial fact that $A^T = A$ means that A is an eigenvector of the transpose map with eigenvalue 1 (i.e., $A^T = \lambda A$ with $\lambda = 1$).

$\lambda = -1$: Similarly, this eigenspace equals the null space of

$$A - \lambda I = A + I = \begin{bmatrix} 2 & 0 & 0 & 0 \\ 0 & 1 & 1 & 0 \\ 0 & 1 & 1 & 0 \\ 0 & 0 & 0 & 2 \end{bmatrix},$$

which has basis $B = \{(0,1,-1,0)\}$. This vector is the coordinate vector of the matrix

$$\begin{bmatrix} 0 & 1 \\ -1 & 0 \end{bmatrix},$$

which thus makes up a basis of the $\lambda = -1$ eigenspace of T. This makes sense since this matrix spans the set $\mathcal{M}_2^{sS}$ of skew-symmetric 2×2 matrices. Similar to before, we have just restated the trivial fact that $A^T = -A$ means that A is an eigenvector of the transpose map with eigenvalue -1 (i.e., $A^T = \lambda A$ with $\lambda = -1$).

The above example generalizes straightforwardly to higher dimensions: for any integer $n \geq 2$, the transpose map $T : \mathcal{M}_n \to \mathcal{M}_n$ only has eigenvalues ± 1, and the corresponding eigenspaces are exactly the spaces $\mathcal{M}_n^S$ and $\mathcal{M}_n^{sS}$ of symmetric and skew-symmetric matrices, respectively (see Exercise 1.2.13).

Because we can talk about eigenvalues and eigenvectors of linear transformations via their standard matrices, all of our results based on diagonalization automatically apply in this new setting with pretty much no extra work (as long as everything is finite-dimensional). We can thus diagonalize linear transformations, take arbitrary (non-integer) powers of linear transformations, and even apply strange functions like the exponential to linear transformations.

Refer to Appendix A.1.7 if you need a refresher on diagonalization.

Example 1.2.18

Square Root of the Transpose

Find a square root of the transpose map $T : \mathcal{M}_2(\mathbb{C}) \to \mathcal{M}_2(\mathbb{C})$. That is, find a linear transformation $S : \mathcal{M}_2(\mathbb{C}) \to \mathcal{M}_2(\mathbb{C})$ with the property that $S^2 = T$.

Solution:

Since we already know how to solve problems like this for matrices, we just do the corresponding matrix computation on the standard matrix $[T]$ rather than trying to solve it "directly" on T itself. That is, we find a matrix square root of $[T]$ via diagonalization.

Throughout this example, $E = \{E_{1,1}, E_{1,2}, E_{2,1}, E_{2,2}\}$ is the standard basis of $\mathcal{M}_2$.

First, we diagonalize $[T]$. We learned in Example 1.2.17 that $[T]$ has eigenvalues 1 and -1, with corresponding eigenspace bases equal to

$$\{(1,0,0,0),(0,1,1,0),(0,0,0,1)\} \quad \text{and} \quad \{(0,1,-1,0)\},$$

respectively. It follows that one way to diagonalize $[T]$ as $[T] = PDP^{-1}$ is

Recall that to diagonalize a matrix, we place the eigenvalues along the diagonal of D and bases of the corresponding eigenspaces as columns of P in the same order.

to choose

$$P = \begin{bmatrix} 1 & 0 & 0 & 0 \\ 0 & 1 & 0 & 1 \\ 0 & 1 & 0 & -1 \\ 0 & 0 & 1 & 0 \end{bmatrix}, \quad P^{-1} = \frac{1}{2}\begin{bmatrix} 2 & 0 & 0 & 0 \\ 0 & 1 & 1 & 0 \\ 0 & 0 & 0 & 2 \\ 0 & 1 & -1 & 0 \end{bmatrix}, \quad \text{and}$$

$$D = \begin{bmatrix} 1 & 0 & 0 & 0 \\ 0 & 1 & 0 & 0 \\ 0 & 0 & 1 & 0 \\ 0 & 0 & 0 & -1 \end{bmatrix}.$$

To find a square root of $[T]$, we just take a square root of each diagonal entry in D and then multiply these matrices back together. It follows that one square root of $[T]$ is given by

The $1/2$ in front of this diagonalization comes from P^{-1}.

$$[T]^{1/2} = PD^{1/2}P^{-1}$$

$$= \frac{1}{2}\begin{bmatrix} 1 & 0 & 0 & 0 \\ 0 & 1 & 0 & 1 \\ 0 & 1 & 0 & -1 \\ 0 & 0 & 1 & 0 \end{bmatrix}\begin{bmatrix} 1 & 0 & 0 & 0 \\ 0 & 1 & 0 & 0 \\ 0 & 0 & 1 & 0 \\ 0 & 0 & 0 & i \end{bmatrix}\begin{bmatrix} 2 & 0 & 0 & 0 \\ 0 & 1 & 1 & 0 \\ 0 & 0 & 0 & 2 \\ 0 & 1 & -1 & 0 \end{bmatrix}$$

$$= \frac{1}{2}\begin{bmatrix} 1 & 0 & 0 & 0 \\ 0 & 1 & 0 & 1 \\ 0 & 1 & 0 & -1 \\ 0 & 0 & 1 & 0 \end{bmatrix}\begin{bmatrix} 2 & 0 & 0 & 0 \\ 0 & 1 & 1 & 0 \\ 0 & 0 & 0 & 2 \\ 0 & i & -i & 0 \end{bmatrix}$$

$$= \frac{1}{2}\begin{bmatrix} 2 & 0 & 0 & 0 \\ 0 & 1+i & 1-i & 0 \\ 0 & 1-i & 1+i & 0 \\ 0 & 0 & 0 & 2 \end{bmatrix}.$$

To unravel $[T]^{1/2}$ back into our desired linear transformation S, we just look at how it acts on coordinate vectors:

$$\frac{1}{2}\begin{bmatrix} 2 & 0 & 0 & 0 \\ 0 & 1+i & 1-i & 0 \\ 0 & 1-i & 1+i & 0 \\ 0 & 0 & 0 & 2 \end{bmatrix}\begin{bmatrix} a \\ b \\ c \\ d \end{bmatrix} = \frac{1}{2}\begin{bmatrix} 2a \\ (1+i)b+(1-i)c \\ (1-i)b+(1+i)c \\ 2d \end{bmatrix},$$

so we conclude that the linear transformation $S : \mathcal{M}_2(\mathbb{C}) \to \mathcal{M}_2(\mathbb{C})$ defined by

To double-check our work, it is perhaps a good idea to apply S to a matrix twice and see that we indeed end up with the transpose of that matrix.

$$S\left(\begin{bmatrix} a & b \\ c & d \end{bmatrix}\right) = \frac{1}{2}\begin{bmatrix} 2a & (1+i)b+(1-i)c \\ (1-i)b+(1+i)c & 2d \end{bmatrix}$$

is a square root of T.

As perhaps an even weirder application of this method, we can also take non-integer powers of the derivative map. That is, we can talk about things like the "half derivative" of a function (just like we regularly talk about the second or third derivative of a function).

Example 1.2.19

How to Take Half of a Derivative

Let $B = \{\sin(x), \cos(x)\}$ and $V = \text{span}(B)$. Find a square root $D^{1/2}$ of the derivative map $D : V \to V$ and use it to compute $D^{1/2}(\sin(x))$ and $D^{1/2}(\cos(x))$.

Solution:

We think of $D^{1/2}$ as the "half derivative", just like D^2 is the second derivative.

Just like in the previous example, instead of working with D itself, we work with its standard matrix, which we can compute straightforwardly to be

$$[D]_B = \begin{bmatrix} 0 & -1 \\ 1 & 0 \end{bmatrix}.$$

Recall that the standard matrix of a counter-clockwise rotation by angle θ is

$$\begin{bmatrix} \cos(\theta) & -\sin(\theta) \\ \sin(\theta) & \cos(\theta) \end{bmatrix}.$$

Plugging in $\theta = \pi/2$ gives exactly $[D]_B$.

While we could compute $[D]_B^{1/2}$ by diagonalizing $[D]_B$ (we do exactly this in Exercise 1.2.18), it is simpler to recognize $[D]_B$ as a counter-clockwise rotation by $\pi/2$ radians:

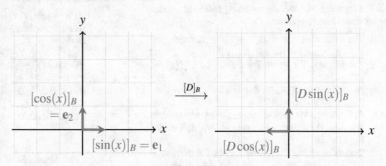

It is then trivial to find a square root of $[D]_B$: we just construct the standard matrix of the linear transformation that rotates counter-clockwise by $\pi/4$ radians:

$$[D]_B^{1/2} = \begin{bmatrix} \cos(\pi/4) & -\sin(\pi/4) \\ \sin(\pi/4) & \cos(\pi/4) \end{bmatrix} = \frac{1}{\sqrt{2}} \begin{bmatrix} 1 & -1 \\ 1 & 1 \end{bmatrix}.$$

Unraveling this standard matrix back into the linear transformation $D^{1/2}$ shows that the half derivatives of $\sin(x)$ and $\cos(x)$ are

To double-check our work, we could apply $D^{1/2}$ to $\sin(x)$ twice and see that we get $\cos(x)$.

$$D^{1/2}(\sin(x)) = \frac{1}{\sqrt{2}}(\sin(x) + \cos(x)) \quad \text{and}$$

$$D^{1/2}(\cos(x)) = \frac{1}{\sqrt{2}}(\cos(x) - \sin(x)).$$

We return to diagonalization and the idea of applying strange functions like the square root to matrices later, in Section 2.4.3.

Exercises

solutions to starred exercises on page 452

1.2.1 Find the coordinate vector of the given vector $\mathbf{v}$ with respect to the indicated basis B.

*(a) $\mathbf{v} = 3x - 2$, $B = \{x + 1, x - 1\} \subset \mathcal{P}^1$
(b) $\mathbf{v} = 2x^2 + 3$, $B = \{x^2 + 1, x + 1, x^2 - x + 1\} \subset \mathcal{P}^2$

*(c) $\mathbf{v} = \begin{bmatrix} 1 & 2 \\ 3 & 4 \end{bmatrix}$, $B = \{E_{1,1}, E_{2,2}, E_{1,2}, E_{2,1}\} \subset \mathcal{M}_2$

(d) $\mathbf{v} = \begin{bmatrix} 1 & 2 \\ 3 & 4 \end{bmatrix}$,

$$B = \left\{ \begin{bmatrix} 0 & 1 \\ 1 & 1 \end{bmatrix}, \begin{bmatrix} 1 & 0 \\ 1 & 1 \end{bmatrix}, \begin{bmatrix} 1 & 1 \\ 0 & 1 \end{bmatrix}, \begin{bmatrix} 1 & 1 \\ 1 & 0 \end{bmatrix} \right\}$$

1.2.2 Find a basis of the indicated vector space $\mathcal{V}$ and then determine its dimension.

(a) $\mathcal{V} = \mathcal{M}_2^S$ (the set of 2×2 symmetric matrices)

**(b) $\mathcal{V} = \mathcal{M}_n^S$

(c) $\mathcal{V} = \mathcal{M}_2^{sS} = \{A \in \mathcal{M}_2 : A^T = -A\}$
(the set of 2×2 skew-symmetric matrices)

**(d) $\mathcal{V} = \mathcal{M}_n^{sS} = \{A \in \mathcal{M}_n : A^T = -A\}$

(e) $\mathcal{V} = \mathcal{M}_2^H$ (the set of 2×2 Hermitian matrices)

*(f) $\mathcal{V} = \mathcal{M}_n^H$

(g) $\mathcal{V} = \{f \in \mathcal{P}^2 : f(0) = 0\}$

*(h) $\mathcal{V} = \{f \in \mathcal{P}^3 : f(3) = 0\}$

(i) $\mathcal{V} = \{f \in \mathcal{P}^4 : f(0) = f(1) = 0\}$

*(j) $\mathcal{V} = \{f \in \mathcal{P}^3 : f(x) - xf'(x) = 0\}$

(k) $\mathcal{V} = \text{span}\{(0,1,1),(1,2,-1),(-1,-1,2)\}$

*(l) $\mathcal{V} = \text{span}\{e^x, e^{2x}, e^{3x}\}$

1.2.3 Determine which of the following functions are and are not linear transformations.

(a) The function $T : \mathbb{R}^n \to \mathbb{R}^n$ defined by $T(v_1, v_2, \ldots, v_n) = (v_n, v_{n-1}, \ldots, v_1)$.

*(b) The function $T : \mathcal{P}^p \to \mathcal{P}^p$ defined by $T(f(x)) = f(2x-1)$.

(c) Matrix inversion (i.e., the function $\text{Inv} : \mathcal{M}_n \to \mathcal{M}_n$ defined by $\text{Inv}(A) = A^{-1}$).

*(d) One-sided matrix multiplication: given a fixed matrix $B \in \mathcal{M}_n$, the function $R_B : \mathcal{M}_n \to \mathcal{M}_n$ defined by $R_B(A) = AB$.

(e) Two-sided matrix conjugation: given a fixed matrix $B \in \mathcal{M}_{m,n}$, the function $T_B : \mathcal{M}_n \to \mathcal{M}_m$ defined by $T_B(A) = BAB^*$.

*(f) Conjugate transposition (i.e., the function $T : \mathcal{M}_n(\mathbb{C}) \to \mathcal{M}_n(\mathbb{C})$ defined by $T(A) = A^*$).

(g) The function $T : \mathcal{P}^2 \to \mathcal{P}^2$ defined by $T(f) = f(0) + f(1)x + f(2)x^2$.

1.2.4 Determine which of the following statements are true and which are false.

*(a) Every finite-dimensional vector space has a basis.

(b) The vector space $\mathcal{M}_{3,4}$ is 12-dimensional.

*(c) The vector space $\mathcal{P}^3$ is 3-dimensional.

(d) If 4 vectors span a particular vector space $\mathcal{V}$, then every set of 6 vectors in $\mathcal{V}$ is linearly dependent.

*(e) The zero vector space $\mathcal{V} = \{\mathbf{0}\}$ has dimension 1.

(f) If $\mathcal{V}$ is a vector space and $\mathbf{v}_1, \ldots, \mathbf{v}_n \in \mathcal{V}$ are such that $\text{span}(\mathbf{v}_1, \ldots, \mathbf{v}_n) = \mathcal{V}$, then $\{\mathbf{v}_1, \ldots, \mathbf{v}_n\}$ is a basis of $\mathcal{V}$.

*(g) The set $\{1, x, x^2, x^3, \ldots\}$ is a basis of $\mathcal{C}$ (the vector space of continuous functions).

(h) For all $A \in \mathcal{M}_n$ it is true that $\text{tr}(A) = \text{tr}(A^T)$.

*(i) The transposition map $T : \mathcal{M}_{m,n} \to \mathcal{M}_{n,m}$ is invertible.

(j) The derivative map $D : \mathcal{P}^5 \to \mathcal{P}^5$ is invertible.

1.2.5 Find the change-of-basis matrix $P_{C \leftarrow B}$ between the given bases B and C of their span.

(a) $B = \{1 + 2x, 2 - x^2, 1 + x + 2x^2\}$, $C = \{1, x, x^2\}$

*(b) $B = \{1, x, x^2\}$, $C = \{1 + 2x, 2 - x^2, 1 + x + 2x^2\}$

(c) $B = \{1 - x, 2 + x + x^2, 1 - x^2\}$,
$C = \{x + 3x^2, x, 1 - x + x^2\}$

*(d) $B = \left\{ \begin{bmatrix} 1 & 0 \\ 0 & 1 \end{bmatrix}, \begin{bmatrix} 1 & 1 \\ 0 & 1 \end{bmatrix}, \begin{bmatrix} 1 & 0 \\ 1 & 1 \end{bmatrix}, \begin{bmatrix} 1 & 1 \\ 1 & 0 \end{bmatrix} \right\}$,
$C = \{E_{1,1}, E_{1,2}, E_{2,1}, E_{2,2}\}$

(e) $B = \left\{ \begin{bmatrix} 1 & 0 \\ 0 & 1 \end{bmatrix}, \begin{bmatrix} 1 & 1 \\ 0 & 1 \end{bmatrix}, \begin{bmatrix} 1 & 0 \\ 1 & 1 \end{bmatrix}, \begin{bmatrix} 1 & 1 \\ 1 & 0 \end{bmatrix} \right\}$,
$C = \left\{ \begin{bmatrix} 1 & 0 \\ 0 & 1 \end{bmatrix}, \begin{bmatrix} 0 & 1 \\ 1 & 0 \end{bmatrix}, \begin{bmatrix} 0 & -i \\ i & 0 \end{bmatrix}, \begin{bmatrix} 1 & 0 \\ 0 & -1 \end{bmatrix} \right\}$

1.2.6 Find the standard matrix $[T]_{D \leftarrow B}$ of the linear transformation T with respect to the given bases B and D.

(a) $T : \mathbb{R}^3 \to \mathbb{R}^2$, $T(\mathbf{v}) = (v_1 - v_2 + 2v_3, 2v_1 - 3v_2 + v_3)$, B and D are the standard bases of $\mathbb{R}^3$ and $\mathbb{R}^2$, respectively.

*(b) $T : \mathcal{P}^2 \to \mathcal{P}^2$, $T(f(x)) = f(3x+1)$, $B = D = \{x^2, x, 1\}$.

(c) $T : \mathcal{P}^2 \to \mathcal{P}^3$, $T(f(x)) = \int_1^x f(t)\, dt$, $B = \{1, x, x^2\}$, $D = \{1, x, x^2, x^3\}$.

*(d) $T : \mathcal{M}_2 \to \mathcal{M}_2$, $T(X) = AX - XA^T$, $B = D$ is the standard basis of $\mathcal{M}_2$, and $A = \begin{bmatrix} 1 & 2 \\ 3 & 4 \end{bmatrix}$.

1.2.7 The dimension of a vector space depends on its ground field. What is the dimension of $\mathbb{C}^n$ as a vector space over $\mathbb{C}$? What is its dimension as a vector space over $\mathbb{R}$?

1.2.8 Consider the symmetric matrices

$$S_{i,i} = \mathbf{e}_i \mathbf{e}_i^T \quad \text{for} \quad 1 \le i \le n, \quad \text{and}$$

$$S_{i,j} = (\mathbf{e}_i + \mathbf{e}_j)(\mathbf{e}_i + \mathbf{e}_j)^T \quad \text{for} \quad 1 \le i < j \le n.$$

Show that $B = \{S_{i,j} : 1 \le i \le j \le n\}$ is a basis of $\mathcal{M}_n^S$.

[Side note: This basis is useful because every member of it has rank 1.]

1.2.9 Let $C = \{1 + x, 1 + x + x^2, x + x^2\}$ be a basis of $\mathcal{P}^2$. Find a basis B of $\mathcal{P}^2$ for which

$$P_{C \leftarrow B} = \begin{bmatrix} 1 & 1 & 2 \\ 1 & 2 & 1 \\ 2 & 1 & 1 \end{bmatrix}.$$

1.2.10 Use the method of Example 1.2.15 to compute the given indefinite integral.

(a) $\int (e^{-x} + 2xe^{-x})\, dx$

*(b) $\int e^x \sin(2x)\, dx$

(c) $\int x \sin(x)\, dx$

1.2.11 Recall that we computed the standard matrix $[T]$ of the transpose map on $\mathcal{M}_2$ in Example 1.2.10.

(a) Compute $[T]^2$. [Hint: Your answer should be a well-known named matrix.]

(b) Explain how we could have gotten the answer to part (a) without actually computing $[T]$.

∗∗1.2.12 Show that the standard matrix of the transpose map $T : \mathcal{M}_{m,n} \to \mathcal{M}_{n,m}$ with respect to the standard basis $E = \{E_{1,1}, E_{1,2}, \ldots, E_{m,n}\}$ is the $mn \times mn$ block matrix that, for all i and j, has $E_{j,i}$ in its (i,j)-block:

$$[T] = \begin{bmatrix} E_{1,1} & E_{2,1} & \cdots & E_{m,1} \\ E_{1,2} & E_{2,2} & \cdots & E_{m,2} \\ \vdots & \vdots & \ddots & \vdots \\ E_{1,n} & E_{2,n} & \cdots & E_{m,n} \end{bmatrix}.$$

∗∗1.2.13 Suppose $T : \mathcal{M}_n \to \mathcal{M}_n$ is the transpose map. Show that 1 and -1 are the only eigenvalues of T, and the corresponding eigenspaces are the spaces $\mathcal{M}_n^{\mathrm{S}}$ and $\mathcal{M}_n^{\mathrm{sS}}$ of symmetric and skew-symmetric matrices, respectively.

[Hint: It is probably easier to work directly with T rather than using its standard matrix like in Example 1.2.17.]

1.2.14 Suppose $T : \mathcal{M}_{2,3} \to \mathcal{M}_{3,2}$ is the transpose map. Explain why T does not have any eigenvalues or eigenvectors even though its standard matrix does.

∗1.2.15 Let $T : \mathcal{P}^2 \to \mathcal{P}^2$ be the linear transformation defined by $T(f(x)) = f(x+1)$.

(a) Find the range and null space of T.
(b) Find all eigenvalues and corresponding eigenspaces of T.
(c) Find a square root of T.

1.2.16 In Example 1.2.19, we came up with a formula for the half derivatives $D^{1/2}(\sin(x))$ and $D^{1/2}(\cos(x))$. Generalize this by coming up with a formula for $D^r(\sin(x))$ and $D^r(\cos(x))$ where $r \in \mathbb{R}$ is arbitrary.

1.2.17 Let $D : \mathcal{D} \to \mathcal{D}$ be the derivative map.

(a) Find an eigenvector of D with corresponding eigenvalue $\lambda = 1$.
(b) Show that every real number $\lambda \in \mathbb{R}$ is an eigenvalue of D. [Hint: You can tweak the eigenvector that you found in part (a) slightly to change its eigenvalue.]

∗∗1.2.18 In Example 1.2.19, we computed a square root of the matrix

$$[D]_B = \begin{bmatrix} 0 & -1 \\ 1 & 0 \end{bmatrix}$$

by considering how it acts on $\mathbb{R}^2$ geometrically. Now find one of its square roots by diagonalizing it (over $\mathbb{C}$).

1.2.19 Suppose $\mathcal{V}$ and $\mathcal{W}$ are vector spaces and let $B = \{\mathbf{v}_1, \mathbf{v}_2, \ldots, \mathbf{v}_n\} \subset \mathcal{V}$ and $C = \{\mathbf{w}_1, \mathbf{w}_2, \ldots, \mathbf{w}_n\} \subset \mathcal{W}$ be sets of vectors.

(a) Show that if B is a basis of $\mathcal{V}$ then the function $T : \mathcal{V} \to \mathcal{W}$ defined by

$$T(c_1\mathbf{v}_1 + \cdots + c_n\mathbf{v}_n) = c_1\mathbf{w}_1 + \cdots + c_n\mathbf{w}_n$$

is a linear transformation.
(b) Show that if C is also a basis of $\mathcal{W}$ then T is invertible.
(c) This definition of T may not make sense if B is not a basis of $\mathcal{V}$—explain why.

1.2.20 Suppose $\mathcal{V}$ and $\mathcal{W}$ are vector spaces and $T : \mathcal{V} \to \mathcal{W}$ is an invertible linear transformation. Let $B = \{\mathbf{v}_1, \mathbf{v}_2, \ldots, \mathbf{v}_n\} \subseteq \mathcal{V}$ be a set of vectors.

(a) Show that if B is linearly independent then so is $T(B) = \{T(\mathbf{v}_1), T(\mathbf{v}_2), \ldots, T(\mathbf{v}_n)\}$.
(b) Show that if B spans $\mathcal{V}$ then $T(B)$ spans $\mathcal{W}$.
(c) Show that if B is a basis of $\mathcal{V}$ then $T(B)$ is a basis of $\mathcal{W}$.
(d) Provide an example to show that none of the results of parts (a), (b), or (c) hold if T is not invertible.

∗∗1.2.21 Suppose $\mathcal{V}, \mathcal{W}$ are vector spaces and $T : \mathcal{V} \to \mathcal{W}$ is a linear transformation. Show that if T is invertible then $\dim(\mathcal{V}) = \dim(\mathcal{W})$.

∗∗1.2.22 Let $\mathcal{V}$ be a finite-dimensional vector space over a field $\mathbb{F}$ and suppose that B is a basis of $\mathcal{V}$.

(a) Show that $[\mathbf{v} + \mathbf{w}]_B = [\mathbf{v}]_B + [\mathbf{w}]_B$ for all $\mathbf{v}, \mathbf{w} \in \mathcal{V}$.
(b) Show that $[c\mathbf{v}]_B = c[\mathbf{v}]_B$ for all $\mathbf{v} \in \mathcal{V}$ and $c \in \mathbb{F}$.
(c) Suppose $\mathbf{v}, \mathbf{w} \in \mathcal{V}$. Show that $[\mathbf{v}]_B = [\mathbf{w}]_B$ if and only if $\mathbf{v} = \mathbf{w}$.

[Side note: This means that the function $T : \mathcal{V} \to \mathbb{F}^n$ defined by $T(\mathbf{v}) = [\mathbf{v}]_B$ is an invertible linear transformation.]

∗∗1.2.23 Let $\mathcal{V}$ be an n-dimensional vector space over a field $\mathbb{F}$ and suppose that B is a basis of $\mathcal{V}$.

(a) Show that a set $\{\mathbf{v}_1, \mathbf{v}_2, \ldots, \mathbf{v}_m\} \subset \mathcal{V}$ is linearly independent if and only if $\{[\mathbf{v}_1]_B, [\mathbf{v}_2]_B, \ldots, [\mathbf{v}_m]_B\} \subset \mathbb{F}^n$ is linearly independent.
(b) Show that $\mathrm{span}(\mathbf{v}_1, \mathbf{v}_2, \ldots, \mathbf{v}_m) = \mathcal{V}$ if and only if $\mathrm{span}\left([\mathbf{v}_1]_B, [\mathbf{v}_2]_B, \ldots, [\mathbf{v}_m]_B\right) = \mathbb{F}^n$.
(c) Show that a set $\{\mathbf{v}_1, \mathbf{v}_2, \ldots, \mathbf{v}_n\}$ is a basis of $\mathcal{V}$ if and only if $\{[\mathbf{v}_1]_B, [\mathbf{v}_2]_B, \ldots, [\mathbf{v}_n]_B\}$ is a basis of $\mathbb{F}^n$.

∗∗1.2.24 Suppose $\mathcal{V}$ and $\mathcal{W}$ are vector spaces and $T : \mathcal{V} \to \mathcal{W}$ is a linear transformation.

(a) Show that $\mathrm{range}(T)$ is a subspace of $\mathcal{W}$.
(b) Show that $\mathrm{null}(T)$ is a subspace of $\mathcal{V}$.

∗∗1.2.25 Suppose $\mathcal{V}$ and $\mathcal{W}$ are finite-dimensional vector spaces with bases B and D, respectively, and $T : \mathcal{V} \to \mathcal{W}$ is a linear transformation.

(a) Show that

$$\mathrm{range}(T) = \left\{\mathbf{w} \in \mathcal{W} : [\mathbf{w}]_D \in \mathrm{range}\left([T]_{D \leftarrow B}\right)\right\}.$$

(b) Show that

$$\mathrm{null}(T) = \left\{\mathbf{v} \in \mathcal{V} : [\mathbf{v}]_B \in \mathrm{null}\left([T]_{D \leftarrow B}\right)\right\}.$$

(c) Show that $\mathrm{rank}(T) = \mathrm{rank}\left([T]_{D \leftarrow B}\right)$.
(d) Show that $\mathrm{nullity}(T) = \mathrm{nullity}\left([T]_{D \leftarrow B}\right)$.

1.2.26 Suppose B is a subset of a finite-dimensional vector space $\mathcal{V}$.

(a) Show that if B is linearly independent then there is a basis C of $\mathcal{V}$ with $B \subseteq C \subseteq \mathcal{V}$.
(b) Show that if B spans $\mathcal{V}$ then there is a basis C of $\mathcal{V}$ with $C \subseteq B \subseteq \mathcal{V}$.

∗∗1.2.27 Suppose B is a subset of a finite-dimensional vector space $\mathcal{V}$ consisting of $\dim(\mathcal{V})$ vectors.

(a) Show that B is a basis of $\mathcal{V}$ if and only if it is linearly independent.
(b) Show that B is a basis of $\mathcal{V}$ if and only if it spans $\mathcal{V}$.

1.2.28 Suppose $\mathcal{V}$ and $\mathcal{W}$ are vector spaces, and let $\mathcal{L}(\mathcal{V}, \mathcal{W})$ be the set of linear transformations $T : \mathcal{V} \to \mathcal{W}$.

(a) Show that $\mathcal{L}(\mathcal{V}, \mathcal{W})$ is a vector space.
(b) If $\dim(\mathcal{V}) = n$ and $\dim(\mathcal{W}) = m$, what is $\dim(\mathcal{L}(\mathcal{V}, \mathcal{W}))$?

1.2.29 Let B and C be bases of a finite-dimensional vector space $\mathcal{V}$. Show that $P_{C \leftarrow B} = I$ if and only if $B = C$.

1.2.30 Let B be a basis of an n-dimensional vector space $\mathcal{V}$ and let $T : \mathcal{V} \to \mathcal{V}$ be a linear transformation. Show that $[T]_B = I_n$ if and only if $T = I_{\mathcal{V}}$.

1.2.31 Suppose that $\mathcal{V}$ is a vector space and $\mathcal{W} \subseteq \mathcal{V}$ is a subspace with $\dim(\mathcal{W}) = \dim(\mathcal{V})$.

(a) Show that if $\mathcal{V}$ is finite-dimensional then $\mathcal{W} = \mathcal{V}$.
(b) Provide an example to show that the conclusion of part (a) does not necessarily hold if $\mathcal{V}$ is infinite-dimensional.

1.2.32 Show that the "right shift" map R from Remark 1.2.2 is indeed a linear transformation.

1.2.33 Show that the "right shift" map R from Remark 1.2.2 has no eigenvalues or eigenvectors, regardless of the field $\mathbb{F}$.

[Side note: Recall that every square matrix over $\mathbb{C}$ has an eigenvalue and eigenvector, so the same is true of every linear transformation acting on a finite-dimensional complex vector space. This exercise shows that this claim no longer holds in infinite dimensions.]

1.2.34 Let $\mathcal{P}^p(\mathbb{F})$ denote the vector space of polynomials of degree $\leq p$ acting on the field $\mathbb{F}$ (and with coefficients from $\mathbb{F}$). We noted earlier that $\dim(\mathcal{P}^p(\mathbb{F})) = p + 1$ when $\mathbb{F} = \mathbb{R}$ or $\mathbb{F} = \mathbb{C}$. Show that $\dim(\mathcal{P}^2(\mathbb{Z}_2)) = 2$ (not 3), where $\mathbb{Z}_2 = \{0, 1\}$ is the finite field with 2 elements (see Appendix A.4).

1.2.35 Complete the proof of Theorem 1.2.6 by showing that the standard matrix $[T]_{D \leftarrow B}$ is unique.

1.2.36 Prove part (b) of Theorem 1.2.1. That is, show that if a vector space $\mathcal{V}$ has a basis B consisting of n vectors, then any set C with fewer than n vectors cannot span $\mathcal{V}$.

1.3 Isomorphisms and Linear Forms

Now that we are familiar with how most of the linear algebraic objects from $\mathbb{R}^n$ (e.g., subspaces, linear independence, bases, linear transformations, and eigenvalues) generalize to vector spaces in general, we take a bit of a detour to discuss some ideas that it did not really make sense to talk about when $\mathbb{R}^n$ was the only vector space in sight.

1.3.1 Isomorphisms

Recall that every finite-dimensional vector space $\mathcal{V}$ has a basis B, and we can use that basis to represent a vector $\mathbf{v} \in \mathcal{V}$ as a coordinate vector $[\mathbf{v}]_B \in \mathbb{F}^n$, where $\mathbb{F}$ is the ground field. We used this correspondence between $\mathcal{V}$ and $\mathbb{F}^n$ to motivate the idea that these vector spaces are "the same" in the sense that, in order to do a linear algebraic calculation in $\mathcal{V}$, we can instead do the corresponding calculation on coordinate vectors in $\mathbb{F}^n$.

We now make this idea of vector spaces being "the same" a bit more precise and clarify under exactly which conditions this "sameness" happens.

Definition 1.3.1 Isomorphisms	Suppose $\mathcal{V}$ and $\mathcal{W}$ are vector spaces over the same field. We say that $\mathcal{V}$ and $\mathcal{W}$ are **isomorphic**, denoted by $\mathcal{V} \cong \mathcal{W}$, if there exists an invertible linear transformation $T : \mathcal{V} \to \mathcal{W}$ (called an **isomorphism** from $\mathcal{V}$ to $\mathcal{W}$).

We can think of isomorphic vector spaces as having the same structure and the same vectors as each other, but different labels on those vectors. This is perhaps easiest to illustrate by considering the vector spaces $\mathcal{M}_{1,2}$ and $\mathcal{M}_{2,1}$

The expression "the map is not the territory" seems relevant here—we typically only care about the underlying vectors, not the form we use to write them down.

of row and column vectors, respectively. Vectors (i.e., matrices) in these vector spaces have the forms

$$[a \quad b] \in \mathcal{M}_{1,2} \quad \text{and} \quad \begin{bmatrix} a \\ b \end{bmatrix} \in \mathcal{M}_{2,1}.$$

The fact that we write the entries of vectors in $\mathcal{M}_{1,2}$ in a row whereas we write those from $\mathcal{M}_{2,1}$ in a column is often just as irrelevant as if we used a different font when writing the entries of the vectors in one of these vector spaces. Indeed, vector addition and scalar multiplication in these spaces are both performed entrywise, so it does not matter how we arrange or order those entries.

To formally see that $\mathcal{M}_{1,2}$ and $\mathcal{M}_{2,1}$ are isomorphic, we just construct the "obvious" isomorphism between them: the transpose map $T : \mathcal{M}_{1,2} \to \mathcal{M}_{2,1}$ satisfies

$$T([a \quad b]) = \begin{bmatrix} a \\ b \end{bmatrix}.$$

Furthermore, we already noted in Example 1.2.6 that the transpose is a linear transformation, and it is clearly invertible (it is its own inverse, since transposing twice gets us back to where we started), so it is indeed an isomorphism. The same argument works to show that each of $\mathbb{F}^n$, $\mathcal{M}_{n,1}$, and $\mathcal{M}_{1,n}$ are isomorphic.

Remark 1.3.1

Column Vectors and Row Vectors

The fact that $\mathbb{F}^n$, $\mathcal{M}_{n,1}$, and $\mathcal{M}_{1,n}$ are isomorphic justifies something that is typically done right from the beginning in linear algebra—treating members of $\mathbb{F}^n$ (vectors), members of $\mathcal{M}_{n,1}$ (column vectors), and members of $\mathcal{M}_{1,n}$ (row vectors) as the same thing.

When we walk about isomorphisms and vector spaces being "the same", we only mean with respect to the 10 defining properties of vector spaces (i.e., properties based on vector addition and scalar multiplication). We can add column vectors in the exact same way that we add row vectors, and similarly scalar multiplication works the exact same for those two types of vectors. However, other operations like matrix multiplication may behave differently on these two sets (e.g., if $A \in \mathcal{M}_n$ then $A\mathbf{x}$ makes sense when $\mathbf{x}$ is a column vector, but not when it is a row vector).

A "morphism" is a function and the prefix "iso" means "identical". The word "isomorphism" thus means a function that keeps things "the same".

As an even simpler example of an isomorphism, we have implicitly been using one when we say things like $\mathbf{v}^T \mathbf{w} = \mathbf{v} \cdot \mathbf{w}$ for all $\mathbf{v}, \mathbf{w} \in \mathbb{R}^n$. Indeed, the quantity $\mathbf{v} \cdot \mathbf{w}$ is a scalar in $\mathbb{R}$, whereas $\mathbf{v}^T \mathbf{w}$ is actually a 1×1 matrix (after all, it is obtained by multiplying a $1 \times n$ matrix by an $n \times 1$ matrix), so it does not quite make sense to say that they are "equal" to each other. However, the spaces $\mathbb{R}$ and $\mathcal{M}_{1,1}(\mathbb{R})$ are trivially isomorphic, so we typically sweep this technicality under the rug.

Before proceeding, it is worthwhile to specifically point out some basic properties of isomorphisms that follow almost immediately from facts that we already know about (invertible) linear transformations in general:

We prove these two properties in Exercise 1.3.6.

- If $T : \mathcal{V} \to \mathcal{W}$ is an isomorphism then so is $T^{-1} : \mathcal{W} \to \mathcal{V}$.
- If $T : \mathcal{V} \to \mathcal{W}$ and $S : \mathcal{W} \to \mathcal{X}$ are isomorphisms then so is $S \circ T : \mathcal{V} \to \mathcal{X}$. In particular, if $\mathcal{V} \cong \mathcal{W}$ and $\mathcal{W} \cong \mathcal{X}$ then $\mathcal{V} \cong \mathcal{X}$.

For example, we essentially showed in Example 1.2.1 that $\mathcal{M}_2 \cong \mathbb{R}^4$ and $\mathbb{R}^4 \cong \mathcal{P}^3$, so it follows that $\mathcal{M}_2 \cong \mathcal{P}^3$ as well. The fact that these vector spaces are isomorphic can be demonstrated by noting that the functions $T : \mathcal{M}_2 \to \mathbb{R}^4$ and $S : \mathbb{R}^4 \to \mathcal{P}^3$ defined by

$$T\left(\begin{bmatrix} a & b \\ c & d \end{bmatrix}\right) = (a,b,c,d) \quad \text{and} \quad S(a,b,c,d) = a + bx + cx^2 + dx^3$$

are clearly isomorphisms (i.e., invertible linear transformations).

Example 1.3.1

Isomorphism of a Space of Functions

Our method of coming up with this map T is very naïve—just send a basis of $\mathbb{R}^3$ to a basis of $\mathcal{V}$. This technique works fairly generally and is a good way of coming up with "obvious" isomorphisms.

Show that the vector spaces $\mathcal{V} = \text{span}(e^x, xe^x, x^2e^x)$ and $\mathbb{R}^3$ are isomorphic.

Solution:

The standard way to show that two spaces are isomorphic is to construct an isomorphism between them. To this end, consider the linear transformation $T : \mathbb{R}^3 \to \mathcal{V}$ defined by

$$T(a,b,c) = ae^x + bxe^x + cx^2e^x.$$

It is straightforward to show that this function is a linear transformation, so we just need to convince ourselves that it is invertible. To this end, we recall from Exercise 1.1.2(g) that $B = \{e^x, xe^x, x^2e^x\}$ is linearly independent and thus a basis of $\mathcal{V}$, so we can construct the standard matrix $[T]_{B \leftarrow E}$, where $E = \{\mathbf{e}_1, \mathbf{e}_2, \mathbf{e}_3\}$ is the standard basis of $\mathbb{R}^3$:

$$[T]_{B \leftarrow E} = \left[\; [T(1,0,0)]_B \; \big| \; [T(0,1,0)]_B \; \big| \; [T(0,0,1)]_B \; \right]$$

$$= \left[\; [e^x]_B \; \big| \; [xe^x]_B \; \big| \; [x^2e^x]_B \; \right] = \begin{bmatrix} 1 & 0 & 0 \\ 0 & 1 & 0 \\ 0 & 0 & 1 \end{bmatrix}.$$

Since $[T]_{B \leftarrow E}$ is clearly invertible (the identity matrix is its own inverse), T is invertible too and is thus an isomorphism.

Example 1.3.2

Polynomials are Isomorphic to Eventually-Zero Sequences

Show that the vector space of polynomials $\mathcal{P}$ and the vector space of eventually-zero sequences c_{00} from Example 1.1.10 are isomorphic.

Solution:

As always, our method of showing that these two spaces are isomorphic is to explicitly construct an isomorphism between them. As with the previous examples, there is an "obvious" choice of isomorphism $T : \mathcal{P} \to c_{00}$, and it is defined by

$$T(a_0 + a_1x + a_2x^2 + \cdots + a_px^p) = (a_0, a_1, a_2, \ldots, a_p, 0, 0, \ldots).$$

If we worked with $\mathbb{F}^{\mathbb{N}}$ here instead of c_{00}, this inverse would not work since the sum $a_0 + a_1x + a_2x^2 + \cdots$ might have infinitely many terms and thus not be a polynomial.

It is straightforward to show that this function is a linear transformation, so we just need to convince ourselves that it is invertible. To this end, we just explicitly construct its inverse $T^{-1} : c_{00} \to \mathcal{P}$:

$$T^{-1}(a_0, a_1, a_2, a_3, \ldots) = a_0 + a_1x + a_2x^2 + a_3x^3 + \cdots$$

At first glance, it might seem like the sum on the right is infinite and thus not a polynomial, but recall that every sequence in c_{00} has only finitely many non-zero entries, so there is indeed some final non-zero term

$a_p x^p$ in the sum. Furthermore, it is straightforward to check that

$$(T^{-1} \circ T)(a_0 + a_1 x + \cdots + a_p x^p) = a_0 + a_1 x + \cdots + a_p x^p \quad \text{and}$$
$$(T \circ T^{-1})(a_0, a_1, a_2, a_3, \ldots) = (a_0, a_1, a_2, a_3, \ldots)$$

for all $a_0 + a_1 x + \cdots + a_p x^p \in \mathcal{P}$ and all $(a_0, a_1, a_2, a_3, \ldots) \in c_{00}$, so T is indeed invertible and thus an isomorphism.

As a generalization of Example 1.3.1 and many of the earlier observations that we made, we now note that coordinate vectors let us immediately conclude that *every* n-dimensional vector space over a field $\mathbb{F}$ is isomorphic to $\mathbb{F}^n$.

Theorem 1.3.1 **Isomorphisms of Finite-Dimensional Vector Spaces**	Suppose $\mathcal{V}$ is an n-dimensional vector space over a field $\mathbb{F}$. Then $\mathcal{V} \cong \mathbb{F}^n$.

Proof. We just recall from Exercise 1.2.22 that if B is any basis of $\mathcal{V}$ then the function $T : \mathcal{V} \to \mathbb{F}^n$ defined by $T(\mathbf{v}) = [\mathbf{v}]_B$ is an invertible linear transformation (i.e., an isomorphism). ∎

In particular, the above theorem tells us that any two vector spaces of the same (finite) dimension over the same field are necessarily isomorphic, since they are both isomorphic to $\mathbb{F}^n$ and thus to each other. The following corollary states this observation precisely and also establishes its converse.

Corollary 1.3.2 **Finite-Dimensional Vector Spaces are Isomorphic**	Suppose $\mathcal{V}$ and $\mathcal{W}$ are vector spaces over the same field and $\mathcal{V}$ is finite-dimensional. Then $\mathcal{V} \cong \mathcal{W}$ if and only if $\dim(\mathcal{V}) = \dim(\mathcal{W})$.

Proof. We already explained how Theorem 1.3.1 gives us the "if" direction, so we now prove the "only if" direction. To this end, we just note that if $\mathcal{V} \cong \mathcal{W}$ then there is an invertible linear transformation $T : \mathcal{V} \to \mathcal{W}$, so Exercise 1.2.21 tells us that $\dim(\mathcal{V}) = \dim(\mathcal{W})$. ∎

Remark 1.3.2
Why Isomorphisms?

In a sense, we did not actually do anything new in this subsection—we already knew about linear transformations and invertibility, so it seems natural to wonder why we would bother adding the "isomorphism" layer of terminology on top of it.

An isomorphism $T : \mathcal{V} \to \mathcal{V}$ (i.e., from a vector space back to *itself*) is called an **automorphism**.

While it's true that there's nothing really "mathematically" new about isomorphisms, the important thing is the new perspective that it gives us. It is very useful to be able to think of vector spaces as being the same as each other, as it can provide us with new intuition or cut down the amount of work that we have to do.

For example, instead of having to do a computation or think about vector space properties in $\mathcal{P}^3$ or $\mathcal{M}_2$, we can do all of our work in $\mathbb{R}^4$, which is likely a fair bit more intuitive. Similarly, we can always work with whichever of c_{00} (the space of eventually-zero sequences) or $\mathcal{P}$ (the space of polynomials) we prefer, since an answer to any linear algebraic question in one of those spaces can be straightforwardly converted into an answer to the corresponding question in the other space via the isomorphism that we constructed in Example 1.3.2.

If it is necessary to clarify which type of isomorphism we are talking about, we call an isomorphism in the linear algebra sense a **vector space isomorphism**.

More generally, isomorphisms are used throughout all of mathematics, not just in linear algebra. In general, they are defined to be invertible maps that preserve whatever the relevant structures or operations are. In our setting, the relevant operations are scalar multiplication and vector addition, and those operations being preserved is exactly equivalent to the invertible map being a linear transformation.

1.3.2 Linear Forms

One of the simplest types of linear transformations are those that send vectors to scalars. For example, it is straightforward to check that the functions $f_1, f_2, \ldots, f_n : \mathbb{R}^n \to \mathbb{R}$ defined by

$$f_1(\mathbf{v}) = v_1, \quad f_2(\mathbf{v}) = v_2, \quad \ldots, \quad f_n(\mathbf{v}) = v_n,$$

where $\mathbf{v} = (v_1, v_2, \ldots, v_n)$, are linear transformations. We now give type of linear transformation a name to make it easier to discuss.

Definition 1.3.2 **Linear Forms**	Suppose $\mathcal{V}$ is a vector space over a field $\mathbb{F}$. Then a linear transformation $f : \mathcal{V} \to \mathbb{F}$ is called a **linear form**.

*Linear forms are sometimes instead called **linear functionals**.*

Linear forms can be thought of as giving us snapshots of vectors—knowing the value of $f(\mathbf{v})$ tells us what $\mathbf{v}$ looks like from one particular direction or angle (just like having a photograph tells us what an object looks like from one side), but not necessarily what it looks like as a whole. For example, the linear transformations f_1, f_2, $\ldots$, f_n described above each give us one of $\mathbf{v}$'s coordinates (i.e., they tell us what $\mathbf{v}$ looks like in the direction of one of the standard basis vectors), but tell us nothing about its other coordinates.

Alternatively, linear forms can be thought of as the building blocks that make up more general linear transformations. For example, consider the linear transformation $T : \mathbb{R}^2 \to \mathbb{R}^2$ (which is *not* a linear form) defined by

$$T(x,y) = (x + 2y, 3x - 4y) \quad \text{for all} \quad (x,y) \in \mathbb{R}^2.$$

If we define $f(x,y) = x + 2y$ and $g(x,y) = 3x - 4y$ then it is straightforward to check that f and g are each linear forms, and $T(x,y) = (f(x,y), g(x,y))$. That is, T just outputs the value of two linear forms. Similarly, every linear transformation into an n-dimensional vector space can be thought of as being made up of n linear forms (one for each of the n output dimensions).

Example 1.3.3 **(Half of)** **the Dot Product** **is a Linear Form**	Suppose $\mathbf{v} \in \mathbb{F}^n$ is a fixed vector. Show that the function $f_{\mathbf{v}} : \mathbb{F}^n \to \mathbb{F}$ defined by $$f_{\mathbf{v}}(\mathbf{w}) = v_1 w_1 + v_2 w_2 + \cdots + v_n w_n \quad \text{for all} \quad \mathbf{w} \in \mathbb{F}^n$$ is a linear form. **Solution:** 　　This follows immediately from the more general fact that multiplication by a matrix is a linear transformation. Indeed, if we let $A = \mathbf{v} \in \mathcal{M}_{1,n}$ be $\mathbf{v}$ as a row vector, then $f_{\mathbf{v}}(\mathbf{w}) = A\mathbf{w}$ for all column vectors $\mathbf{w}$.

Recall that the dot product on $\mathbb{R}^n$ is defined by
$$\mathbf{v} \cdot \mathbf{w} = v_1 w_1 + \cdots + v_n w_n.$$

In particular, if $\mathbb{F} = \mathbb{R}$ then the previous example tells us that

$$f_{\mathbf{v}}(\mathbf{w}) = \mathbf{v} \cdot \mathbf{w} \quad \text{for all} \quad \mathbf{w} \in \mathbb{R}^n$$

is a linear form. This is actually the "standard" example of a linear form, and the one that we should keep in mind as our intuition builder. We will see shortly that every linear form on a finite-dimensional vector space can be written in this way (in the exact same sense that every linear transformation can be written as a matrix).

Example 1.3.4

The Trace is a Linear Form

Show that the trace $\mathrm{tr} : \mathcal{M}_n(\mathbb{F}) \to \mathbb{F}$ is a linear form.

Solution:

We already showed in Example 1.2.7 that the trace is a linear transformation. Since it outputs scalars, it is necessarily a linear form.

Example 1.3.5

Evaluation is a Linear Form

Show that the function $E_2 : \mathcal{P}^3 \to \mathbb{R}$ defined by $E_2(f) = f(2)$ is a linear form.

Solution:

We just need to show that E_2 is a linear transformation, since it is clear that its output is (by definition) always a scalar. We thus check the two properties of Definition 1.2.4:

a) $E_2(f+g) = (f+g)(2) = f(2)+g(2) = E_2(f)+E_2(g)$.
b) $E_2(cf) = (cf)(2) = cf(2) = cE_2(f)$.

E_2 is called the **evaluation map** at $x = 2$. More generally, the function $E_x : \mathcal{P} \to \mathbb{R}$ defined by $E_x(f) = f(x)$ is also a linear form, regardless of the value of $x \in \mathbb{R}$.

The steps used in Example 1.3.5 might seem somewhat confusing at first, since we are applying a function (E_2) to functions (f and g). It is very important to be careful when working through this type of problem to make sure that the correct type of object is being fed into each function (e.g., $E_2(f)$ makes sense, but $E_2(4)$ does not, since E_2 takes a polynomial as its input).

Also, that example highlights our observation that linear forms give us one linear "piece of information" about vectors. In this case, knowing the value of $E_2(f) = f(2)$ tells us a little bit about the polynomial (i.e., the vector) f. If we also knew the value of three other linear forms on $f \in \mathcal{P}^3$ (e.g., $f(1)$, $f(3)$, and $f(4)$), we could use polynomial interpolation to reconstruct f itself.

Example 1.3.6

Integration is a Linear Form

Let $\mathcal{C}[a,b]$ be the vector space of continuous real-valued functions on the interval $[a,b]$. Show that the function $I : \mathcal{C}[a,b] \to \mathbb{R}$ defined by

$$I(f) = \int_a^b f(x)\, dx$$

is a linear form.

Recall that every continuous function is integrable, so this linear form makes sense.

Solution:

We just need to show that I is a linear transformation, since it is (yet again) clear that its output is always a scalar. We thus check the two properties of Definition 1.2.4:

a) By properties of integrals that are typically covered in calculus courses, we know that for all $f, g \in C[a,b]$ we have

$$I(f+g) = \int_a^b (f+g)(x)\, dx$$

$$= \int_a^b f(x)\, dx + \int_a^b g(x)\, dx = I(f) + I(g).$$

b) We similarly know that we can pull scalars in and out of integrals:

$$I(cf) = \int_a^b (cf)(x)\, dx = c \int_a^b f(x)\, dx = cI(f)$$

for all $f \in C[a,b]$ and $c \in \mathbb{R}$.

We now pin down the claim that we made earlier that every linear form on a finite-dimensional vector space looks like one half of the dot product. In particular, to make this work we just do what we always do when we want to make abstract vector space concepts more concrete—we represent vectors as coordinate vectors with respect to some basis.

Theorem 1.3.3 **The Form of** **Linear Forms**	Let B be a basis of a finite-dimensional vector space $\mathcal{V}$ over a field $\mathbb{F}$, and let $f : \mathcal{V} \to \mathbb{F}$ be a linear form. Then there exists a unique vector $\mathbf{v} \in \mathcal{V}$ such that $$f(\mathbf{w}) = [\mathbf{v}]_B^T [\mathbf{w}]_B \quad \text{for all} \quad \mathbf{w} \in \mathcal{V},$$ where we are treating $[\mathbf{v}]_B$ and $[\mathbf{w}]_B$ as column vectors.

Proof. Since f is a linear transformation, Theorem 1.2.6 tells us that it has a standard matrix—a matrix A such that $f(\mathbf{w}) = A[\mathbf{w}]_B$ for all $\mathbf{w} \in \mathcal{V}$. Since f maps into $\mathbb{F}$, which is 1-dimensional, the standard matrix A is $1 \times n$, where $n = \dim(\mathcal{V})$. It follows that A is a row vector, and since every vector in $\mathbb{F}^n$ is the coordinate vector of *some* vector in $\mathcal{V}$, we can find some $\mathbf{v} \in \mathcal{V}$ such that $A = [\mathbf{v}]_B^T$, so that $f(\mathbf{w}) = [\mathbf{v}]_B^T [\mathbf{w}]_B$.

Uniqueness of $\mathbf{v}$ follows immediately from uniqueness of standard matrices and of coordinate vectors. ∎

In the special case when $\mathbb{F} = \mathbb{R}$ or $\mathbb{F} = \mathbb{C}$, it makes sense to talk about the dot product of the coordinate vectors $[\mathbf{v}]_B$ and $[\mathbf{w}]_B$, and the above theorem can be rephrased as saying that there exists a unique vector $\mathbf{v} \in \mathcal{V}$ such that

> Recall that the dot product on $\mathbb{C}^n$ is defined by
> $\mathbf{v} \cdot \mathbf{w} = \overline{v_1} w_1 + \cdots + \overline{v_n} w_n.$

$$f(\mathbf{w}) = [\mathbf{v}]_B \cdot [\mathbf{w}]_B \quad \text{for all} \quad \mathbf{w} \in \mathcal{V}.$$

The only thing to be slightly careful of here is that if $\mathbb{F} = \mathbb{C}$ then $[\mathbf{v}]_B \cdot [\mathbf{w}]_B = [\mathbf{v}]_B^* [\mathbf{w}]_B$ (not $[\mathbf{v}]_B \cdot [\mathbf{w}]_B = [\mathbf{v}]_B^T [\mathbf{w}]_B$), so we have to absorb a complex conjugate into the vector $\mathbf{v}$ to make this reformulation work.

Example 1.3.7 **The Evaluation Map** **as a Dot Product** This example just illustrates how Theorem 1.3.3 works out for the linear form E_2.	Let $E_2 : \mathcal{P}^3 \to \mathbb{R}$ be the evaluation map from Example 1.3.5, defined by $E_2(f) = f(2)$, and let $E = \{1, x, x^2, x^3\}$ be the standard basis of $\mathcal{P}^3$. Find a polynomial $g \in \mathcal{P}^3$ such that $E_2(f) = [g]_E \cdot [f]_E$ for all $f \in \mathcal{P}^3$. **Solution:** If we write $f(x) = a + bx + cx^2 + dx^3$ (i.e., $[f]_E = (a, b, c, d)$) then $$E_2(f) = f(2) = a + 2b + 4c + 8d = (1, 2, 4, 8) \cdot (a, b, c, d).$$

It follows that we want to choose $g \in \mathcal{P}^3$ so that $[g]_E = (1, 2, 4, 8)$. In other words, we want $g(y) = 1 + 2y + 4y^2 + 8y^3$.

Slightly more generally, for all $x \in \mathbb{R}$, the evaluation map $E_x(f) = f(x)$ can be represented in this sense via the polynomial $g_x(y) = 1 + xy + x^2 y^2 + x^3 y^3$ (see Exercise 1.3.21).

The Dual Space

Theorem 1.3.3 tells us that every linear form on $\mathcal{V}$ corresponds to a particular vector in $\mathcal{V}$, at least when $\mathcal{V}$ is finite-dimensional, so it seems like there is an isomorphism lurking in the background here. We need to make one more definition before we can discuss this isomorphism properly.

Definition 1.3.3

Dual of a Vector Space

Let $\mathcal{V}$ be a vector space over a field $\mathbb{F}$. Then the **dual** of $\mathcal{V}$, denoted by $\mathcal{V}^*$, is the vector space consisting of all linear forms on $\mathcal{V}$.

The fact that $\mathcal{V}^*$ is indeed a vector space is established by Exercise 1.2.28(a), and part (b) of that same exercise even tells us that $\dim(\mathcal{V}^*) = \dim(\mathcal{V})$ when $\mathcal{V}$ is finite-dimensional, so $\mathcal{V}$ and $\mathcal{V}^*$ are isomorphic by Corollary 1.3.2. In fact, one simple isomorphism from $\mathcal{V}^*$ to $\mathcal{V}$ is exactly the one that sends a linear form f to its corresponding vector $\mathbf{v}$ from Theorem 1.3.3. However, this isomorphism between $\mathcal{V}$ and $\mathcal{V}^*$ is somewhat strange, as it depends on the particular choice of basis that we make on $\mathcal{V}$—if we change the basis B in Theorem 1.3.3 then the vector $\mathbf{v}$ corresponding to a linear form f changes as well.

The fact that the isomorphism between $\mathcal{V}$ and $\mathcal{V}^*$ is basis-dependent suggests that something somewhat unnatural is going on, as many (even finite-dimensional) vector spaces do not have a "natural" or "standard" choice of basis. However, if we go one step further and consider the **double-dual** space $\mathcal{V}^{**}$ consisting of linear forms acting on $\mathcal{V}^*$ then things become a bit more well-behaved, so we now briefly explore this double-dual space.

Yes, it is pretty awkward to think about what the members of $\mathcal{V}^{**}$ are. They are linear forms acting on linear forms (i.e., functions of functions).

Using the exact same ideas as earlier, if $\mathcal{V}$ is finite-dimensional then we still have $\dim(\mathcal{V}^{**}) = \dim(\mathcal{V}^*) = \dim(\mathcal{V})$, so all three of these vector spaces are isomorphic. However, $\mathcal{V}$ and $\mathcal{V}^{**}$ are isomorphic in a much more natural way than $\mathcal{V}$ and $\mathcal{V}^*$, since there is a basis-independent choice of isomorphism between them. To see what it is, notice that for every vector $\mathbf{v} \in \mathcal{V}$ we can define a linear form $\phi_\mathbf{v} \in \mathcal{V}^{**}$ by

This spot right here is exactly as abstract as this book gets. It's been a slow climb to this point, and from here on it's downhill back into more concrete things.

$$\phi_\mathbf{v}(f) = f(\mathbf{v}) \quad \text{for all} \quad f \in \mathcal{V}^*. \tag{1.3.1}$$

Showing that $\phi_\mathbf{v}$ is linear form does not require any clever insight—we just have to check the two defining properties from Definition 1.2.4, and each of these properties follows almost immediately from the relevant definitions. The hard part of this verification is keeping the notation straight and making sure that the correct type of object goes into and comes out of each function at every step:

a) For all $f, g \in \mathcal{V}^*$ we have

$$\begin{aligned}
\phi_\mathbf{v}(f + g) &= (f + g)(\mathbf{v}) &&\text{(definition of } \phi_\mathbf{v}) \\
&= f(\mathbf{v}) + g(\mathbf{v}) &&\text{(definition of "} + \text{" in } \mathcal{V}^*) \\
&= \phi_\mathbf{v}(f) + \phi_\mathbf{v}(g) &&\text{(definition of } \phi_\mathbf{v})
\end{aligned}$$

b) Similarly, for all $c \in \mathbb{F}$ and $f \in \mathcal{V}^*$ we have

$$\phi_{\mathbf{v}}(cf) = (cf)(\mathbf{v}) = c\big(f(\mathbf{v})\big) = c\phi_{\mathbf{v}}(f).$$

Example 1.3.8

The Double-Dual of Polynomials

Show that for each $\phi \in (\mathcal{P}^3)^{**}$ there exists $f \in \mathcal{P}^3$ such that

$$\phi(E_x) = f(x) \quad \text{for all} \quad x \in \mathbb{R},$$

where $E_x \in (\mathcal{P}^3)^*$ is the evaluation map at $x \in \mathbb{R}$ (see Example 1.3.5).

Solution:
We know from Exercise 1.3.22 that $B = \{E_1, E_2, E_3, E_4\}$ is a basis of $(\mathcal{P}^3)^*$, so for every $x \in \mathbb{R}$ there exist scalars $c_{1,x}, \ldots, c_{4,x}$ such that

$$E_x = c_{1,x}E_1 + c_{2,x}E_2 + c_{3,x}E_3 + c_{4,x}E_4.$$

We then expand the quantities

$$\begin{aligned}\phi(E_x) &= \phi(c_{1,x}E_1 + c_{2,x}E_2 + c_{3,x}E_3 + c_{4,x}E_4) \\ &= c_{1,x}\phi(E_1) + c_{2,x}\phi(E_2) + c_{3,x}\phi(E_3) + c_{4,x}\phi(E_4)\end{aligned}$$

and

$$\begin{aligned}f(x) = E_x(f) &= \big(c_{1,x}E_1 + c_{2,x}E_2 + c_{3,x}E_3 + c_{4,x}E_4\big)(f) \\ &= c_{1,x}E_1(f) + c_{2,x}E_2(f) + c_{3,x}E_3(f) + c_{4,x}E_4(f) \\ &= c_{1,x}f(1) + c_{2,x}f(2) + c_{3,x}f(3) + c_{4,x}f(4)\end{aligned}$$

for all $f \in \mathcal{P}^3$ and $x \in \mathbb{R}$.

In fact, f is uniquely determined by ϕ.

If we choose f so that $f(1) = \phi(E_1)$, $f(2) = \phi(E_2)$, $f(3) = \phi(E_3)$, and $f(4) = \phi(E_4)$ (which can be done by polynomial interpolation) then we find that $\phi(E_x) = f(x)$ for all $x \in \mathbb{R}$, as desired.

The above example is suggestive of a natural isomorphism between $\mathcal{V}$ and $\mathcal{V}^{**}$: if $\mathcal{V} = \mathcal{P}^3$ then every $\phi \in (\mathcal{P}^3)^{**}$ looks the exact same as a polynomial in $\mathcal{P}^3$; we just have to relabel the evaluation map E_x as x itself. The following theorem pins down how this "natural" isomorphism between $\mathcal{V}$ and $\mathcal{V}^{**}$ works for other vector spaces.

Theorem 1.3.4

Canonical Double-Dual Isomorphism

The function $T : \mathcal{V} \to \mathcal{V}^{**}$ defined by $T(\mathbf{v}) = \phi_{\mathbf{v}}$ is an isomorphism, where $\phi_{\mathbf{v}} \in \mathcal{V}^{**}$ is as defined in Equation (1.3.1).

Proof. We must show that T is linear and invertible, and we again do not have to be clever to do so. Rather, as long as we keep track of what space all of these objects live in and parse the notation carefully, then linearity and invertibility of T follow almost immediately from the relevant definitions.

Yes, T is a function that sends vectors to functions of vectors. If you recall that $\mathcal{V}$ itself might be a vector space made up of functions then your head might explode.

Before showing that T is linear, we first make a brief note on notation. Since T maps into $\mathcal{V}^{**}$, we know that $T(\mathbf{v})$ is a function acting on $\mathcal{V}^*$. We thus use the (admittedly unfortunate) notation $T(\mathbf{v})(f)$ to refer to the scalar value that results from applying the function $T(\mathbf{v})$ to $f \in \mathcal{V}^*$. Once this notational nightmare is understood, the proof of linearity is straightforward:

a) For all $f \in \mathcal{V}^*$ we have

$$
\begin{aligned}
T(\mathbf{v} + \mathbf{w})(f) &= \phi_{\mathbf{v}+\mathbf{w}}(f) && \text{(definition of } T) \\
&= f(\mathbf{v} + \mathbf{w}) && \text{(definition of } \phi_{\mathbf{v}+\mathbf{w}}) \\
&= f(\mathbf{v}) + f(\mathbf{w}) && \text{(linearity of each } f \in \mathcal{V}^*) \\
&= \phi_{\mathbf{v}}(f) + \phi_{\mathbf{w}}(f) && \text{(definition of } \phi_{\mathbf{v}} \text{ and } \phi_{\mathbf{w}}) \\
&= T(\mathbf{v})(f) + T(\mathbf{w})(f). && \text{(definition of } T)
\end{aligned}
$$

Since $T(\mathbf{v} + \mathbf{w})(f) = (T(\mathbf{v}) + T(\mathbf{w}))(f)$ for all $f \in \mathcal{V}^*$, we conclude that $T(\mathbf{v} + \mathbf{w}) = T(\mathbf{v}) + T(\mathbf{w})$.

b) Similarly,

$$
T(c\mathbf{v})(f) = \phi_{c\mathbf{v}}(f) = f(c\mathbf{v}) = cf(\mathbf{v}) = c\phi_{\mathbf{v}}(f) = cT(\mathbf{v})(f)
$$

for all $f \in \mathcal{V}^*$, so $T(c\mathbf{v}) = cT(\mathbf{v})$ for all $c \in \mathbb{F}$.

> If $\dim(\mathcal{V}) = \infty$, invertibility of T fails. Even if $\mathcal{V}$ has a basis, it is never the case that any of $\mathcal{V}, \mathcal{V}^*$, and $\mathcal{V}^{**}$ are isomorphic ($\mathcal{V}^{**}$ is "bigger" than $\mathcal{V}^*$, which is "bigger" than $\mathcal{V}$).

For invertibility, we claim that if $B = \{\mathbf{v}_1, \mathbf{v}_2, \ldots, \mathbf{v}_n\}$ is linearly independent then so is $C = \{T(\mathbf{v}_1), T(\mathbf{v}_2), \ldots, T(\mathbf{v}_n)\}$ (this claim is pinned down in Exercise 1.3.24). Since C contains $n = \dim(\mathcal{V}) = \dim(\mathcal{V}^{**})$ vectors, it must be a basis of $\mathcal{V}^{**}$ by Exercise 1.2.27(a). It follows that $[T]_{C \leftarrow B} = I$, which is invertible, so T is invertible as well. ∎

This double-dual space $\mathcal{V}^{**}$ and its correspondence with $\mathcal{V}$ likely still seems quite abstract, so it is useful to think about what it means when $\mathcal{V} = \mathbb{F}^n$, which we typically think of as consisting of column vectors. Theorem 1.3.3 tells us that each $f \in (\mathbb{F}^n)^*$ corresponds to some *row* vector $\mathbf{v}^T$ (in the sense that $f(\mathbf{w}) = \mathbf{v}^T\mathbf{w}$ for all $\mathbf{w} \in \mathbb{F}^n$). Theorem 1.3.4 says that if we go one step further, then each $\phi \in (\mathbb{F}^n)^{**}$ corresponds to some *column* vector $\mathbf{w} \in \mathbb{F}^n$ in the sense that $\phi(f) = f(\mathbf{w}) = \mathbf{v}^T\mathbf{w}$ for all $f \in \mathcal{V}^*$ (i.e., for all $\mathbf{v}^T$).

> The close relationship between $\mathcal{V}$ and $\mathcal{V}^{**}$ is why we use the term "dual space" in the first place—duality refers to an operation or concept that, when applied a second time, gets us back to where we started.

For this reason, it is convenient (and for the most part, acceptable) to think of $\mathcal{V}$ as consisting of column vectors and $\mathcal{V}^*$ as consisting of the corresponding row vectors. In fact, this is exactly why we use the notation $\mathcal{V}^*$ for the dual space in the first place—it is completely analogous to taking the (conjugate) transpose of the vector space $\mathcal{V}$. The fact that $\mathcal{V}^*$ is isomorphic to $\mathcal{V}$, but in a way that depends on the particular basis chosen, is analogous to the fact that if $\mathbf{v} \in \mathbb{F}^n$ is a column vector then $\mathbf{v}$ and $\mathbf{v}^T$ have the same size (and entries) but not shape, and the fact that $\mathcal{V}^{**}$ is so naturally isomorphic to $\mathcal{V}$ is analogous to the fact that $(\mathbf{v}^T)^T$ and $\mathbf{v}$ have the same size *and* shape (and are equal).

Remark 1.3.3

Linear Forms Versus Vector Pairings

While $\mathcal{V}^*$ is defined as a set of linear forms on $\mathcal{V}$, this can be cumbersome to think about once we start considering vector spaces like $\mathcal{V}^{**}$ (and, heaven forbid, $\mathcal{V}^{***}$), as it is somewhat difficult to make sense of what a function of a function (of a function...) "looks like".

Instead, notice that if $\mathbf{w} \in \mathcal{V}$ and $f \in \mathcal{V}^*$ then the expression $f(\mathbf{w})$ is linear in each of $\mathbf{w}$ and f, so we can think of it just as *combining* members of two vector spaces together in a linear way, rather than as members of one vector space *acting* on the members of another. One way of making this observation precise is via Theorem 1.3.3, which says that applying a linear form $f \in \mathcal{V}^*$ to $\mathbf{w} \in \mathcal{V}$ is the same as taking the dot product of two vectors $[\mathbf{v}]_B$ and $[\mathbf{w}]_B$ (at least in the finite-dimensional case).

1.3.3 Bilinearity and Beyond

As suggested by Remark 1.3.3, we are often interested not just in applying linear functions to vectors, but also in combining vectors from different vector spaces together in a linear way. We now introduce a way of doing exactly this.

Definition 1.3.4

Bilinear Forms

Suppose V and W are vector spaces over the same field $\mathbb{F}$. Then a function $f : V \times W \to \mathbb{F}$ is called a **bilinear form** if it satisfies the following properties:

  a) It is linear in its first argument:
     i) $f(\mathbf{v}_1 + \mathbf{v}_2, \mathbf{w}) = f(\mathbf{v}_1, \mathbf{w}) + f(\mathbf{v}_2, \mathbf{w})$ and
     ii) $f(c\mathbf{v}_1, \mathbf{w}) = cf(\mathbf{v}_1, \mathbf{w})$ for all $c \in \mathbb{F}$, $\mathbf{v}_1, \mathbf{v}_2 \in V$, and $\mathbf{w} \in W$.

  b) It is linear in its second argument:
     i) $f(\mathbf{v}, \mathbf{w}_1 + \mathbf{w}_2) = f(\mathbf{v}, \mathbf{w}_1) + f(\mathbf{v}, \mathbf{w}_2)$ and
     ii) $f(\mathbf{v}, c\mathbf{w}_1) = cf(\mathbf{v}, \mathbf{w}_1)$ for all $c \in \mathbb{F}$, $\mathbf{v} \in V$, and $\mathbf{w}_1, \mathbf{w}_2 \in W$.

> Recall that the notation $f : V \times W \to \mathbb{F}$ means that f takes two vectors as input—one from V and one from W—and provides a scalar from $\mathbb{F}$ as output.

While the above definition might seem like a mouthful, it simply says that f is a bilinear form exactly if it becomes a linear form when one of its inputs is held constant. That is, for every fixed vector $\mathbf{w} \in W$ the function $g_{\mathbf{w}} : V \to \mathbb{F}$ defined by $g_{\mathbf{w}}(\mathbf{v}) = f(\mathbf{v}, \mathbf{w})$ is a linear form, and similarly for every fixed vector $\mathbf{v} \in V$ the function $h_{\mathbf{v}} : W \to \mathbb{F}$ defined by $h_{\mathbf{v}}(\mathbf{w}) = f(\mathbf{v}, \mathbf{w})$ is a linear form. Yet again, we look at some examples to try to get a feeling for what bilinear forms look like.

Example 1.3.9

The Real Dot Product is a Bilinear Form

Show that the function $f : \mathbb{R}^n \times \mathbb{R}^n \to \mathbb{R}$ defined by

$$f(\mathbf{v}, \mathbf{w}) = \mathbf{v} \cdot \mathbf{w} \quad \text{for all} \quad \mathbf{v}, \mathbf{w} \in \mathbb{R}^n$$

is a bilinear form.

Solution:
    We could work through the four defining properties of bilinear forms, but an easier way to solve this problem is to recall from Example 1.3.3 that the dot product is a linear form if we keep the first vector $\mathbf{v}$ fixed, which establishes property (b) in Definition 1.3.4.

    Since $\mathbf{v} \cdot \mathbf{w} = \mathbf{w} \cdot \mathbf{v}$, it is also the case that the dot product is a linear form if we keep the second vector fixed, which in turn establishes property (a) in Definition 1.3.4. It follows that the dot product is indeed a bilinear form.

> The function $f(x, y) = xy$ is a bilinear form but not a linear transformation. Linear transformations must be linear "as a whole", whereas bilinear forms just need to be linear with respect to each variable independently.

The real dot product is the prototypical example of a bilinear form, so keep it in mind when working with bilinear forms abstractly to help make them seem a bit more concrete. Perhaps even more simply, notice that multiplication (of real numbers) is a bilinear form. That is, if we define a function $f : \mathbb{R} \times \mathbb{R} \to \mathbb{R}$ simply via $f(x, y) = xy$, then f is a bilinear form. This of course makes sense since multiplication of real numbers is just the one-dimensional dot product.

In order to simplify proofs that certain functions are bilinear forms, we can check linearity in the first argument by showing that $f(\mathbf{v}_1 + c\mathbf{v}_2, \mathbf{w}) = f(\mathbf{v}_1, \mathbf{w}) + cf(\mathbf{v}_2, \mathbf{w})$ for all $c \in \mathbb{F}$, $\mathbf{v}_1, \mathbf{v}_2 \in V$, and $\mathbf{w} \in W$, rather than checking vector addition and scalar multiplication separately as in conditions (a)(i) and (a)(ii) of Definition 1.3.4 (and we similarly check linearity in the second argument).

Example 1.3.10

The Dual Pairing is a Bilinear Form

Let $\mathcal{V}$ be a vector space over a field $\mathbb{F}$. Show that the function $g : \mathcal{V}^* \times \mathcal{V} \to \mathbb{F}$ defined by

$$g(f, \mathbf{v}) = f(\mathbf{v}) \quad \text{for all} \quad f \in \mathcal{V}^*, \ \mathbf{v} \in \mathcal{V}$$

is a bilinear form.

Solution:

We just notice that g is linear in each of its input arguments individually. For the first input argument, we have

$$g(f_1 + cf_2, \mathbf{v}) = (f_1 + cf_2)(\mathbf{v}) = f_1(\mathbf{v}) + cf_2(\mathbf{v}) = g(f_1, \mathbf{v}) + cg(f_2, \mathbf{v}),$$

for all $f_1, f_2 \in \mathcal{V}^*$, $\mathbf{v} \in \mathcal{V}$, and $c \in \mathbb{F}$ simply from the definition of addition and scalar multiplication of functions. Similarly, for the second input argument we have

$$g(f, \mathbf{v}_1 + c\mathbf{v}_2) = f(\mathbf{v}_1 + c\mathbf{v}_2) = f(\mathbf{v}_1) + cf(\mathbf{v}_2) = g(f, \mathbf{v}_1) + cg(f, \mathbf{v}_2),$$

for all $f \in \mathcal{V}^*$, $\mathbf{v}_1, \mathbf{v}_2 \in \mathcal{V}$, and $c \in \mathbb{F}$ since each $f \in \mathcal{V}^*$ is (by definition) linear.

Example 1.3.11

Matrices are Bilinear Forms

Let $A \in \mathcal{M}_{m,n}(\mathbb{F})$ be a matrix. Show that the function $f : \mathbb{F}^m \times \mathbb{F}^n \to \mathbb{F}$ defined by

$$f(\mathbf{v}, \mathbf{w}) = \mathbf{v}^T A \mathbf{w} \quad \text{for all} \quad \mathbf{v} \in \mathbb{F}^m, \ \mathbf{w} \in \mathbb{F}^n$$

is a bilinear form.

Solution:

Once again, we just check the defining properties from Definition 1.3.4, all of which follow straightforwardly from the corresponding properties of matrix multiplication:

In this example (and as always, unless specified otherwise), $\mathbb{F}$ refers to an arbitrary field.

a) For all $\mathbf{v}_1, \mathbf{v}_2 \in \mathbb{F}^m$, $\mathbf{w} \in \mathbb{F}^n$, and $c \in \mathbb{F}$ we have

$$f(\mathbf{v}_1 + c\mathbf{v}_2, \mathbf{w}) = (\mathbf{v}_1 + c\mathbf{v}_2)^T A \mathbf{w}$$
$$= \mathbf{v}_1^T A \mathbf{w} + c\mathbf{v}_2^T A \mathbf{w} = f(\mathbf{v}_1, \mathbf{w}) + cf(\mathbf{v}_2, \mathbf{w}).$$

b) Similarly, for all $\mathbf{v} \in \mathbb{F}^m$, $\mathbf{w}_1, \mathbf{w}_2 \in \mathbb{F}^n$, and $c \in \mathbb{F}$ we have

$$f(\mathbf{v}, \mathbf{w}_1 + c\mathbf{w}_2) = \mathbf{v}^T A(\mathbf{w}_1 + c\mathbf{w}_2)$$
$$= \mathbf{v}^T A \mathbf{w}_1 + c\mathbf{v}^T A \mathbf{w}_2 = f(\mathbf{v}, \mathbf{w}_1) + cf(\mathbf{v}, \mathbf{w}_2).$$

In fact, if $\mathbb{F} = \mathbb{R}$ and $A = I$ then the bilinear form f in Example 1.3.11 simplifies to $f(\mathbf{v}, \mathbf{w}) = \mathbf{v}^T \mathbf{w} = \mathbf{v} \cdot \mathbf{w}$, so we recover the fact that the real dot product is bilinear. This example also provides us with a fairly quick way of showing that certain functions are bilinear forms—if we can write them in terms of a matrix in this way then bilinearity follows immediately.

Example 1.3.12

A Numerical Bilinear Form

Show that the function $f : \mathbb{R}^2 \times \mathbb{R}^2 \to \mathbb{R}$ defined by

$$f(\mathbf{v}, \mathbf{w}) = 3v_1w_1 - 4v_1w_2 + 5v_2w_1 + v_2w_2 \quad \text{for all} \quad \mathbf{v}, \mathbf{w} \in \mathbb{R}^2$$

is a bilinear form.

Solution:

We could check the defining properties from Definition 1.3.4, but an easier way to show that f is bilinear is to notice that we can group its coefficients into a matrix as follows:

> Notice that the coefficient of v_iw_j goes in the (i, j)-entry of the matrix. This always happens.

$$f(\mathbf{v}, \mathbf{w}) = \begin{bmatrix} v_1 & v_2 \end{bmatrix} \begin{bmatrix} 3 & -4 \\ 5 & 1 \end{bmatrix} \begin{bmatrix} w_1 \\ w_2 \end{bmatrix}.$$

It follows that f is a bilinear form, since we showed in Example 1.3.11 that any function of this form is.

In fact, we now show that *every* bilinear form acting on finite-dimensional vector spaces can be written in this way. Just like matrices can be used to represent linear transformations, they can also be used to represent bilinear forms.

Theorem 1.3.5

The Form of Bilinear Forms

Let B and C be bases of m- and n-dimensional vector spaces $\mathcal{V}$ and $\mathcal{W}$, respectively, over a field $\mathbb{F}$, and let $f : \mathcal{V} \times \mathcal{W} \to \mathbb{F}$ be a bilinear form. There exists a unique matrix $A \in \mathcal{M}_{m,n}(\mathbb{F})$ such that

$$f(\mathbf{v}, \mathbf{w}) = [\mathbf{v}]_B^T A [\mathbf{w}]_C \quad \text{for all} \quad \mathbf{v} \in \mathcal{V}, \ \mathbf{w} \in \mathcal{W},$$

where we are treating $[\mathbf{v}]_B$ and $[\mathbf{w}]_C$ as column vectors.

Proof. We just use the fact that bilinear forms are linear when we keep one of their inputs constant, and we then leech off of the representation of linear forms that we already know from Theorem 1.3.3.

Specifically, if we denote the vectors in the basis B by $B = \{\mathbf{v}_1, \mathbf{v}_2, \ldots, \mathbf{v}_m\}$ then $[\mathbf{v}_j]_B = \mathbf{e}_j$ for all $1 \leq j \leq m$ and Theorem 1.3.3 tells us that the linear form $g_j : \mathcal{W} \to \mathbb{F}$ defined by $g_j(\mathbf{w}) = f(\mathbf{v}_j, \mathbf{w})$ can be written as $g_j(\mathbf{w}) = \mathbf{a}_j^T [\mathbf{w}]_C$ for some fixed (column) vector $\mathbf{a}_j \in \mathbb{F}^n$. If we let A be the matrix with rows $\mathbf{a}_1^T, \ldots \mathbf{a}_m^T$ (i.e., $A^T = [\ \mathbf{a}_1 \mid \cdots \mid \mathbf{a}_m \ ]$) then

> Here we use the fact that $\mathbf{e}_j^T A$ equals the j-th row of A (i.e., $\mathbf{a}_j^T$).

$$\begin{aligned} f(\mathbf{v}_j, \mathbf{w}) = g_j(\mathbf{w}) &= \mathbf{a}_j^T [\mathbf{w}]_C \\ &= \mathbf{e}_j^T A [\mathbf{w}]_C = [\mathbf{v}_j]_B^T A [\mathbf{w}]_C \quad \text{for all} \quad 1 \leq j \leq m, \ \mathbf{w} \in \mathcal{W}. \end{aligned}$$

To see that this same equation holds when we replace $\mathbf{v}_j$ by an arbitrary $\mathbf{v} \in \mathcal{V}$, we just use linearity in the first argument of f and the fact that every $\mathbf{v} \in \mathcal{V}$ can be written as a linear combination of the basis vectors from B (i.e.,

$\mathbf{v} = c_1\mathbf{v}_1 + \cdots + c_m\mathbf{v}_m$ for some $c_1, \ldots, c_m \in \mathbb{F}$):

$$f(\mathbf{v}, \mathbf{w}) = f\left(\sum_{j=1}^{m} c_j\mathbf{v}_j, \mathbf{w}\right)$$

$$= \sum_{j=1}^{m} c_j f(\mathbf{v}_j, \mathbf{w}) = \sum_{j=1}^{m} c_j [\mathbf{v}_j]_B^T A[\mathbf{w}]_C$$

$$= \left(\sum_{j=1}^{m} c_j\mathbf{e}_j\right)^T A[\mathbf{w}]_C = [\mathbf{v}]_B^T A[\mathbf{w}]_C$$

for all $\mathbf{v} \in \mathcal{V}$ and $\mathbf{w} \in \mathcal{W}$.

Finally, to see that A is unique we just note that if $C = \{\mathbf{w}_1, \mathbf{w}_2, \ldots, \mathbf{w}_n\}$ then $f(\mathbf{v}_i, \mathbf{w}_j) = [\mathbf{v}_i]_B^T A[\mathbf{w}_j]_C = \mathbf{e}_i^T A\mathbf{e}_j = a_{i,j}$ for all $1 \leq i \leq m$ and $1 \leq j \leq n$, so the entries of A are completely determined by f. ∎

As one particularly interesting example of how the above theorem works, we recall that the determinant is multilinear in the columns of the matrix it acts on, and thus in particular it can be interpreted as a bilinear form on 2×2 matrices. More specifically, if we define a function $f : \mathbb{F}^2 \times \mathbb{F}^2 \to \mathbb{F}$ via

$$f(\mathbf{v}, \mathbf{w}) = \det\left(\begin{bmatrix} \mathbf{v} \mid \mathbf{w} \end{bmatrix}\right)$$

then f is a bilinear form and thus can be represented by a 2×2 matrix.

Example 1.3.13
The 2 × 2 Determinant as a Matrix

Find the matrix $A \in \mathcal{M}_2(\mathbb{F})$ with the property that

$$\det\left(\begin{bmatrix} \mathbf{v} \mid \mathbf{w} \end{bmatrix}\right) = \mathbf{v}^T A\mathbf{w} \quad \text{for all} \quad \mathbf{v}, \mathbf{w} \in \mathbb{F}^2.$$

Solution:
Recall that $\det\left(\begin{bmatrix} \mathbf{v} \mid \mathbf{w} \end{bmatrix}\right) = v_1 w_2 - v_2 w_1$, while direct calculation shows that

$$\mathbf{v}^T A\mathbf{w} = a_{1,1} v_1 w_1 + a_{1,2} v_1 w_2 + a_{2,1} v_2 w_1 + a_{2,2} v_2 w_2.$$

By simply comparing these two expressions, we see that the unique matrix A that makes them equal to each other has entries $a_{1,1} = 0$, $a_{1,2} = 1$, $a_{2,1} = -1$, and $a_{2,2} = 0$. That is,

$$\det\left(\begin{bmatrix} \mathbf{v} \mid \mathbf{w} \end{bmatrix}\right) = \mathbf{v}^T A\mathbf{w} \quad \text{if and only if} \quad A = \begin{bmatrix} 0 & 1 \\ -1 & 0 \end{bmatrix}.$$

The fact that A is skew-symmetric (i.e., $A^T = -A$) corresponds to the fact that swapping two columns of a matrix multiplies its determinant by -1 (i.e., $\det\left(\begin{bmatrix} \mathbf{v} \mid \mathbf{w} \end{bmatrix}\right) = -\det\left(\begin{bmatrix} \mathbf{w} \mid \mathbf{v} \end{bmatrix}\right)$).

Multilinear Forms

In light of Example 1.3.13, it might be tempting to think that the determinant of a 3×3 matrix can be represented via a single fixed 3×3 matrix, but this is not the case—the determinant of a 3×3 matrix is not a bilinear form, but rather it is linear in the *three* columns of the input matrix. More generally, the determinant of a $p \times p$ matrix is *multi*linear—linear in each of its p columns. This generalization of bilinearity is captured by the following definition, which requires that the function being considered is a linear form when all except for one of its inputs are held constant.

A multilinear form
with p input
arguments is
sometimes called
p-linear.

Suppose $\mathcal{V}_1, \mathcal{V}_2, \ldots, \mathcal{V}_p$ are vector spaces over the same field $\mathbb{F}$. A function $f : \mathcal{V}_1 \times \mathcal{V}_2 \times \cdots \times \mathcal{V}_p \to \mathbb{F}$ is called a **multilinear form** if, for each $1 \leq j \leq p$ and each $\mathbf{v}_1 \in \mathcal{V}_1, \mathbf{v}_2 \in \mathcal{V}_2, \ldots, \mathbf{v}_p \in \mathcal{V}_p$, it is the case that the function $g : \mathcal{V}_j \to \mathbb{F}$ defined by

$$g(\mathbf{v}) = f(\mathbf{v}_1, \ldots, \mathbf{v}_{j-1}, \mathbf{v}, \mathbf{v}_{j+1}, \ldots, \mathbf{v}_p) \quad \text{for all} \quad \mathbf{v} \in \mathcal{V}_j$$

is a linear form.

When $p = 1$ or $p = 2$, this definition gives us exactly linear and bilinear forms, respectively. Just like linear forms can be represented by vectors (1-dimensional lists of numbers) and bilinear forms can be represented by matrices (2-dimensional arrays of numbers), multilinear forms in general can be represented by p-dimensional arrays of numbers.

We investigate
arrays and
multilinearity in much
more depth in
Chapter 3.

This characterization of multilinear forms is provided by the upcoming Theorem 1.3.6. To prepare ourselves for this theorem (since it looks quite ugly at first, so it helps to be prepared), recall that the entries of vectors are indexed by a single subscript (as in v_j for the j-th entry of $\mathbf{v}$) and the entries of matrices are indexed by two subscripts (as in $a_{i,j}$ for the (i, j)-entry of A). Similarly, the entries of a p-dimensional array are indexed by p subscripts (as in $a_{j_1, j_2, \ldots, j_p}$ for the $(j_1, j_2, \ldots, j_p)$-entry of the p-dimensional array A).

Just like a matrix can be thought of as made up of several column vectors, a 3-dimensional array can be thought of as made up of several matrices (see Figure 1.9), a 4-dimensional array can be thought of as made up of several 3-dimensional arrays, and so on (though this quickly becomes difficult to visualize).

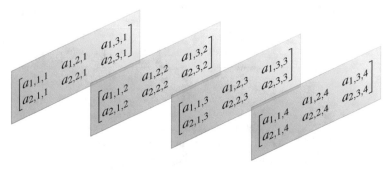

Figure 1.9: A visualization of a 3-dimensional array with 2 rows, 3 columns, and 4 "layers". We use $a_{i,j,k}$ to denote the entry of this array in the i-th row, j-th column, and k-th layer.

Suppose $\mathcal{V}_1, \ldots, \mathcal{V}_p$ are finite-dimensional vector spaces over a field $\mathbb{F}$ and $f : \mathcal{V}_1 \times \cdots \times \mathcal{V}_p \to \mathbb{F}$ is a multilinear form. For each $1 \leq i \leq p$ and $\mathbf{v}_i \in \mathcal{V}_i$, let $v_{i,j}$ denote the j-th coordinate of $\mathbf{v}_i$ with respect to some basis of $\mathcal{V}_i$. Then there exists a unique p-dimensional array (with entries $\{a_{j_1, \ldots, j_p}\}$), called the **standard array** of f, such that

$$f(\mathbf{v}_1, \ldots, \mathbf{v}_p) = \sum_{j_1, \ldots, j_p} a_{j_1, \ldots, j_p} v_{1, j_1} \cdots v_{p, j_p} \quad \text{for all} \quad \mathbf{v}_1 \in \mathcal{V}_1, \ldots, \mathbf{v}_p \in \mathcal{V}_p.$$

Proof. We proceed just as we did in the proof of the characterization of bilinear forms (Theorem 1.3.5), using induction on the number p of vector spaces.

We already know from Theorems 1.3.3 and 1.3.5 that this result holds when $p = 1$ or $p = 2$, which establishes the base case of the induction. For the inductive step, suppose that the result is true for all $(p-1)$-linear forms acting on $\mathcal{V}_2 \times \cdots \times \mathcal{V}_p$. If we let $B = \{\mathbf{w}_1, \mathbf{w}_2, \ldots, \mathbf{w}_m\}$ be a basis of $\mathcal{V}_1$ then the inductive hypothesis tells us that the $(p-1)$-linear forms $g_{j_1} : \mathcal{V}_2 \times \cdots \times \mathcal{V}_n \to \mathbb{F}$ defined by

> We use j_1 here instead of just j since it will be convenient later on.

$$g_{j_1}(\mathbf{v}_2, \ldots, \mathbf{v}_p) = f(\mathbf{w}_{j_1}, \mathbf{v}_2, \ldots, \mathbf{v}_p)$$

can be written as

> The scalar $a_{j_1, j_2, \ldots, j_p}$ here depends on j_1 (not just $j_2, \ldots, j_n$) since each choice of $\mathbf{w}_{j_1}$ gives a different $(p-1)$-linear form g_{j_1}.

$$f(\mathbf{w}_{j_1}, \mathbf{v}_2, \ldots, \mathbf{v}_p) = g_{j_1}(\mathbf{v}_2, \ldots, \mathbf{v}_p) = \sum_{j_2, \ldots, j_p} a_{j_1, j_2, \ldots, j_p} v_{2, j_2} \cdots v_{p, j_p}$$

for some fixed family of scalars $\{a_{j_1, j_2, \ldots, j_p}\}$.

If we write an arbitrary vector $\mathbf{v}_1 \in \mathcal{V}_1$ as a linear combination of the basis vectors $\mathbf{w}_1, \mathbf{w}_2, \ldots, \mathbf{w}_m$ (i.e., $\mathbf{v}_1 = v_{1,1}\mathbf{w}_1 + v_{1,2}\mathbf{w}_2 + \cdots + v_{1,m}\mathbf{w}_m$), it then follows from linearity of the first argument of f that

$$f(\mathbf{v}_1, \ldots, \mathbf{v}_p) = f\left(\sum_{j_1=1}^{m} v_{1, j_1} \mathbf{w}_{j_1}, \mathbf{v}_2, \ldots, \mathbf{v}_p\right) \qquad \left(\mathbf{v}_1 = \sum_{j_1=1}^{m} v_{1, j_1} \mathbf{w}_{j_1}\right)$$

$$= \sum_{j_1=1}^{m} v_{1, j_1} f(\mathbf{w}_{j_1}, \mathbf{v}_2, \ldots, \mathbf{v}_p) \qquad \text{(multilinearity of } f)$$

$$= \sum_{j_1=1}^{m} v_{1, j_1} \sum_{j_2, \ldots, j_p} a_{j_1, j_2, \ldots, j_p} v_{2, j_2} \cdots v_{p, j_p} \qquad \text{(inductive hypothesis)}$$

$$= \sum_{j_1, \ldots, j_p} a_{j_1, \ldots, j_p} v_{1, j_1} \cdots v_{p, j_p} \qquad \text{(group sums together)}$$

for all $\mathbf{v}_1 \in \mathcal{V}_1, \ldots, \mathbf{v}_p \in \mathcal{V}_p$, which completes the inductive step and shows that the family of scalars $\{a_{j_1, \ldots, j_p}\}$ exists.

To see that the scalars $\{a_{j_1, \ldots, j_p}\}$ are unique, just note that if we choose $\mathbf{v}_1$ to be the j_1-th member of the basis of $\mathcal{V}_1$, $\mathbf{v}_2$ to be the j_2-th member of the basis of $\mathcal{V}_2$, and so on, then we get

$$f(\mathbf{v}_1, \ldots, \mathbf{v}_p) = a_{j_1, \ldots, j_p}.$$

In particular, these scalars are completely determined by f. $\blacksquare$

For example, the determinant of a 3×3 matrix is a 3-linear (trilinear?) form, so it can be represented by a single fixed $3 \times 3 \times 3$ array (or equivalently, a family of $3^3 = 27$ scalars), just like the determinant of a 2×2 matrix is a bilinear form and thus can be represented by a 2×2 matrix (i.e., a family of $2^2 = 4$ scalars). The following example makes this observation explicit.

Example 1.3.14

The 3 × 3 Determinant as a 3 × 3 × 3 Array

Find the $3 \times 3 \times 3$ array A with the property that

$$\det\left(\left[\,\mathbf{v} \mid \mathbf{w} \mid \mathbf{x}\,\right]\right) = \sum_{i,j,k=1}^{3} a_{i,j,k} v_i w_j x_k \quad \text{for all} \quad \mathbf{v}, \mathbf{w}, \mathbf{x} \in \mathbb{F}^3.$$

Solution:

We recall the following explicit formula for the determinant of a 3×3 matrix:

$$\det\left(\left[\,\mathbf{v} \mid \mathbf{w} \mid \mathbf{x}\,\right]\right) = v_1 w_2 x_3 + v_2 w_3 x_1 + v_3 w_1 x_2 - v_1 w_3 x_2 - v_2 w_1 x_3 - v_3 w_2 x_1.$$

This is exactly the representation of the determinant that we want, and we can read the coefficients of the array A from it directly:

$$a_{1,2,3} = a_{2,3,1} = a_{3,1,2} = 1 \quad \text{and} \quad a_{1,3,2} = a_{2,1,3} = a_{3,2,1} = -1,$$

and all other entries of A equal 0. We can visualize this 3-dimensional array as follows, where the first subscript i indexes rows, the second subscript j indexes columns, and the third subscript k indexes layers:

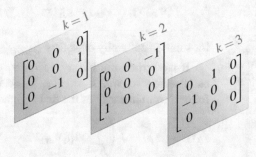

It is worth noting that the array that represents the 3×3 determinant in the above example is **antisymmetric**, which means that it is skew-symmetric no matter how we "slice" it. For example, the three layers shown are each skew-symmetric matrices, but the matrix consisting of each of their top rows is also skew-symmetric, as is the matrix consisting of each of their central columns, and so on. Remarkably, the array that corresponds to the determinant is, up to scaling, the unique array with this property (see Exercise 1.3.27), just like the matrix

$$A = \begin{bmatrix} 0 & 1 \\ -1 & 0 \end{bmatrix}$$

is, up to scaling, the unique 2×2 skew-symmetric matrix.

1.3.4 Inner Products

The dot product on $\mathbb{R}^n$ let us do several important geometrically-motivated things, like computing the lengths of vectors, the angles between them, and determining whether or not they are orthogonal. We now generalize the dot product to other vector spaces, which will let us carry out these same tasks in this new more general setting.

To this end, we make use of bilinear forms, for which we have already seen that the dot product serves as the prototypical example. However, bilinear forms are actually *too* general, since they do not satisfy all of the "nice" properties that the dot product satisfies. For example, bilinear forms do not typically mimic the commutativity property of the dot product (i.e., $\mathbf{v} \cdot \mathbf{w} = \mathbf{w} \cdot \mathbf{v}$, but most bilinear forms f have $f(\mathbf{v}, \mathbf{w}) \neq f(\mathbf{w}, \mathbf{v})$), nor are they typically positive definite (i.e., $\mathbf{v} \cdot \mathbf{v} \geq 0$, but most bilinear forms do not have $f(\mathbf{v}, \mathbf{v}) \geq 0$).

We now investigate functions that satisfy these two additional properties, with the caveat that the commutativity condition that we need looks slightly different than might be naïvely expected:

Definition 1.3.6

Inner Product

The notation $\langle \cdot, \cdot \rangle : \mathcal{V} \times \mathcal{V} \to \mathbb{F}$ means that $\langle \cdot, \cdot \rangle$ is a function that takes in two vectors from $\mathcal{V}$ and outputs a single number from $\mathbb{F}$.

Suppose that $\mathbb{F} = \mathbb{R}$ or $\mathbb{F} = \mathbb{C}$ and that $\mathcal{V}$ is a vector space over $\mathbb{F}$. Then an **inner product** on $\mathcal{V}$ is a function $\langle \cdot, \cdot \rangle : \mathcal{V} \times \mathcal{V} \to \mathbb{F}$ such that the following three properties hold for all $c \in \mathbb{F}$ and all $\mathbf{v}, \mathbf{w}, \mathbf{x} \in \mathcal{V}$:

a) $\langle \mathbf{v}, \mathbf{w} \rangle = \overline{\langle \mathbf{w}, \mathbf{v} \rangle}$ (conjugate symmetry)

b) $\langle \mathbf{v}, \mathbf{w} + c\mathbf{x} \rangle = \langle \mathbf{v}, \mathbf{w} \rangle + c\langle \mathbf{v}, \mathbf{x} \rangle$ (linearity)

c) $\langle \mathbf{v}, \mathbf{v} \rangle \geq 0$, with equality if and only if $\mathbf{v} = \mathbf{0}$.   (pos. definiteness)

"Sesquilinear" means "one-and-a-half linear".

If $\mathbb{F} = \mathbb{R}$ then inner products are indeed bilinear forms, since property (b) gives linearity in the second argument and then the symmetry condition (a) guarantees that it is also linear in its first argument. However, if $\mathbb{F} = \mathbb{C}$ then they are instead **sesquilinear forms**—they are linear in their second argument, but only *conjugate* linear in their first argument:

$$\langle \mathbf{v} + c\mathbf{x}, \mathbf{w} \rangle = \overline{\langle \mathbf{w}, \mathbf{v} + c\mathbf{x} \rangle} = \overline{\langle \mathbf{w}, \mathbf{v} \rangle} + \overline{c}\overline{\langle \mathbf{w}, \mathbf{x} \rangle} = \langle \mathbf{v}, \mathbf{w} \rangle + \overline{c}\langle \mathbf{x}, \mathbf{w} \rangle.$$

Remark 1.3.4

Why a Complex Conjugate?

Perhaps the only "weird" property in the definition of an inner product is the fact that we require $\langle \mathbf{v}, \mathbf{w} \rangle = \overline{\langle \mathbf{w}, \mathbf{v} \rangle}$ rather than the seemingly simpler $\langle \mathbf{v}, \mathbf{w} \rangle = \langle \mathbf{w}, \mathbf{v} \rangle$. The reason for this strange choice is that if $\mathbb{F} = \mathbb{C}$ then there does not actually exist any function satisfying $\langle \mathbf{v}, \mathbf{w} \rangle = \langle \mathbf{w}, \mathbf{v} \rangle$ as well as properties (b) and (c)—if there did, then for all $\mathbf{v} \neq \mathbf{0}$ we would have

$$0 < \langle i\mathbf{v}, i\mathbf{v} \rangle = i\langle i\mathbf{v}, \mathbf{v} \rangle = i\langle \mathbf{v}, i\mathbf{v} \rangle = i^2 \langle \mathbf{v}, \mathbf{v} \rangle = -\langle \mathbf{v}, \mathbf{v} \rangle < 0,$$

which makes no sense.

Some books instead define inner products to be linear in their *first* argument and conjugate linear in the *second*.

It is also worth noting that we restrict our attention to the fields $\mathbb{R}$ and $\mathbb{C}$ when discussing inner products, since otherwise it is not at all clear what the positive definiteness property $\langle \mathbf{v}, \mathbf{v} \rangle \geq 0$ even means. Many fields do not have a natural ordering on them, so it does not make sense to discuss whether or not their members are bigger than 0. In fact, there is also no natural ordering on the field $\mathbb{C}$, but that is okay because the conjugate symmetry condition $\langle \mathbf{v}, \mathbf{v} \rangle = \overline{\langle \mathbf{v}, \mathbf{v} \rangle}$ ensures that $\langle \mathbf{v}, \mathbf{v} \rangle$ is real.

To get a bit more comfortable with inner products, we now present several examples of standard inner products in the various vector spaces that we have been working with. As always though, keep the real dot product in mind as the canonical example of an inner product around which we build our intuition.

Example 1.3.15

Complex Dot Product

This inner product on $\mathbb{C}^n$ is also called the **dot product**.

Show that the function $\langle \cdot, \cdot \rangle : \mathbb{C}^n \times \mathbb{C}^n \to \mathbb{C}$ defined by

$$\langle \mathbf{v}, \mathbf{w} \rangle = \mathbf{v}^* \mathbf{w} = \sum_{i=1}^{n} \overline{v_i} w_i \quad \text{for all} \quad \mathbf{v}, \mathbf{w} \in \mathbb{C}^n$$

is an inner product on $\mathbb{C}^n$.

Solution:

We must check that the three properties described by Definition 1.3.6 hold. All of these properties follow very quickly from the corresponding properties of matrix multiplication and conjugate transposition:

Keep in mind that $\mathbf{w}^* \mathbf{v}$ is a *number*, so transposing it has no effect, so $(\mathbf{w}^* \mathbf{v})^* = \overline{\mathbf{w}^* \mathbf{v}}$.

a) $\langle \mathbf{v}, \mathbf{w} \rangle = \mathbf{v}^* \mathbf{w} = (\mathbf{w}^* \mathbf{v})^* = \overline{\mathbf{w}^* \mathbf{v}} = \overline{\langle \mathbf{w}, \mathbf{v} \rangle}$

b) $\langle \mathbf{v}, \mathbf{w} + c\mathbf{x} \rangle = \mathbf{v}^* (\mathbf{w} + c\mathbf{x}) = \mathbf{v}^* \mathbf{w} + c\mathbf{v}^* \mathbf{x} = \langle \mathbf{v}, \mathbf{w} \rangle + c\langle \mathbf{v}, \mathbf{x} \rangle$

c) $\langle \mathbf{v}, \mathbf{v} \rangle = \mathbf{v}^* \mathbf{v} = |v_1|^2 + \cdots + |v_n|^2$, which is non-negative, and it equals 0 if and only if $\mathbf{v} = \mathbf{0}$.

Just like in $\mathbb{R}^n$, we often denote the inner product (i.e., the dot product) from the above example by $\mathbf{v} \cdot \mathbf{w}$ instead of $\langle \mathbf{v}, \mathbf{w} \rangle$.

Example 1.3.16

The Frobenius Inner Product

Recall that $\mathrm{tr}(A^*B)$ is the trace of A^*B, which is the sum of its diagonal entries.

We call this inner product on $\mathcal{M}_{m,n}(\mathbb{F})$ the **Frobenius inner product**. Some books call it the **Hilbert–Schmidt inner product**.

Show that the function $\langle \cdot, \cdot \rangle : \mathcal{M}_{m,n}(\mathbb{F}) \times \mathcal{M}_{m,n}(\mathbb{F}) \to \mathbb{F}$ defined by

$$\langle A, B \rangle = \mathrm{tr}(A^*B) \quad \text{for all} \quad A, B \in \mathcal{M}_{m,n}(\mathbb{F}),$$

where $\mathbb{F} = \mathbb{R}$ or $\mathbb{F} = \mathbb{C}$, is an inner product on $\mathcal{M}_{m,n}(\mathbb{F})$.

Solution:

We could directly verify that the three properties described by Definition 1.3.6 hold, but it is perhaps more illuminating to compute

$$\langle A, B \rangle = \mathrm{tr}(A^*B) = \sum_{i=1}^{m} \sum_{j=1}^{n} \overline{a_{i,j}} b_{i,j}.$$

In other words, this inner product multiplies all of the entries of $\overline{A}$ by the corresponding entries of B and adds them up, just like the dot product on $\mathbb{F}^n$. For example, if $m = n = 2$ then

$$\left\langle \begin{bmatrix} a_{1,1} & a_{1,2} \\ a_{2,1} & a_{2,2} \end{bmatrix}, \begin{bmatrix} b_{1,1} & b_{1,2} \\ b_{2,1} & b_{2,2} \end{bmatrix} \right\rangle = \overline{a_{1,1}} b_{1,1} + \overline{a_{2,1}} b_{2,1} + \overline{a_{1,2}} b_{1,2} + \overline{a_{2,2}} b_{2,2}$$

$$= (a_{1,1}, a_{1,2}, a_{2,1}, a_{2,2}) \cdot (b_{1,1}, b_{1,2}, b_{2,1}, b_{2,2}).$$

More generally, if E is the standard basis of $\mathcal{M}_{m,n}(\mathbb{F})$ then we have

$$\langle A, B \rangle = [A]_E \cdot [B]_E \quad \text{for all} \quad A, B \in \mathcal{M}_{m,n}(\mathbb{F}).$$

In other words, this inner product is what we get if we forget about the shape of A and B and just take their dot product as if they were vectors in $\mathbb{F}^{mn}$. The fact that this is an inner product now follows directly from the fact that the dot product on $\mathbb{F}^n$ is an inner product.

Example 1.3.17

An Inner Product on Continuous Functions

Let $a < b$ be real numbers and let $\mathcal{C}[a,b]$ be the vector space of continuous functions on the real interval $[a,b]$. Show that the function $\langle \cdot, \cdot \rangle : \mathcal{C}[a,b] \times \mathcal{C}[a,b] \to \mathbb{R}$ defined by

$$\langle f, g \rangle = \int_a^b f(x) g(x) \, dx \quad \text{for all} \quad f, g \in \mathcal{C}[a,b]$$

is an inner product on $\mathcal{C}[a,b]$.

Solution:

We must check that the three properties described by Definition 1.3.6 hold. All of these properties follow quickly from the corresponding properties of definite integrals:

Since all numbers here are real, we do not need to worry about the complex conjugate in property (a).

a) We simply use commutativity of real number multiplication:

$$\langle f, g \rangle = \int_a^b f(x)g(x)\,dx = \int_a^b g(x)f(x)\,dx = \langle g, f \rangle.$$

b) This follows from linearity of integrals:

$$\langle f, g + ch \rangle = \int_a^b f(x)\big(g(x) + ch(x)\big)\,dx$$
$$= \int_a^b f(x)g(x)\,dx + c\int_a^b f(x)h(x)\,dx$$
$$= \langle f, g \rangle + c\langle f, h \rangle.$$

c) We just recall that the integral of a positive function is positive:

$$\langle f, f \rangle = \int_a^b f(x)^2\,dx \geq 0,$$

with equality if and only if $f(x) = 0$ for all $x \in [a, b]$ (i.e., f is the zero function).

We can make a bit more sense of the above inner product on $\mathcal{C}[a, b]$ if we think of definite integrals as "continuous sums". While the dot product $\mathbf{v} \cdot \mathbf{w}$ on $\mathbb{R}^n$ adds up all values of $v_j w_j$ for $1 \leq j \leq n$, this inner product $\langle f, g \rangle$ on $\mathcal{C}[a, b]$ "adds up" all values of $f(x)g(x)$ for $a \leq x \leq b$ (and weighs them appropriately so that the sum is finite).

All of the inner products that we have seen so far are called the **standard** inner products on spaces that they act on. That is, the dot product is the standard inner product on $\mathbb{R}^n$ or $\mathbb{C}^n$, the Frobenius inner product is the standard inner product on $\mathcal{M}_{m,n}$, and

$$\langle f, g \rangle = \int_a^b f(x)g(x)\,dx$$

is the standard inner product on $\mathcal{C}[a, b]$. We similarly use $\mathcal{P}[a, b]$ and $\mathcal{P}^p[a, b]$ to denote the spaces of polynomials (of degree at most p, respectively) acting on the real interval $[a, b]$, and we assume that the inner product acting on these spaces is this standard one unless we indicate otherwise.

Inner products can also look quite a bit different from the standard ones that we have seen so far, however. The following example illustrates how the same vector space can have multiple different inner products, and at first glance they might look quite different than the standard inner product.

Example 1.3.18
A Weird Inner Product

Show that the function $\langle \cdot, \cdot \rangle : \mathbb{R}^2 \times \mathbb{R}^2 \to \mathbb{R}$ defined by

$$\langle \mathbf{v}, \mathbf{w} \rangle = v_1 w_1 + 2v_1 w_2 + 2v_2 w_1 + 5v_2 w_2 \quad \text{for all} \quad \mathbf{v}, \mathbf{w} \in \mathbb{R}^2$$

is an inner product on $\mathbb{R}^2$.

Solution:
Properties (a) and (b) of Definition 1.3.6 follow fairly quickly from the definition of this function, but proving property (c) is somewhat trickier.

We will show in Theorem 1.4.3 that we can *always* rewrite inner products in a similar manner so as to make their positive definiteness "obvious".

To this end, it is helpful to rewrite this function in the form

$$\langle \mathbf{v}, \mathbf{w} \rangle = (v_1 + 2v_2)(w_1 + 2w_2) + v_2 w_2.$$

It follows that

$$\langle \mathbf{v}, \mathbf{v} \rangle = (v_1 + 2v_2)^2 + v_2^2 \geq 0,$$

with equality if and only if $v_1 + 2v_2 = v_2 = 0$, which happens if and only if $v_1 = v_2 = 0$ (i.e., $\mathbf{v} = \mathbf{0}$), as desired.

Recall that a vector space $\mathcal{V}$ is not just a set of vectors, but rather it also includes a particular addition and scalar multiplication operation as part of it. Similarly, if we have a particular inner product in mind then we typically group it together with $\mathcal{V}$ and call it an **inner product space**. If there is a possibility for confusion among different inner products (e.g., because there are multiple different inner product spaces $\mathcal{V}$ and $\mathcal{W}$ being used simultaneously) then we may write them using notation like $\langle \cdot, \cdot \rangle_{\mathcal{V}}$ or $\langle \cdot, \cdot \rangle_{\mathcal{W}}$.

The Norm Induced by the Inner Product

Now that we have inner products to work with, we can define the length of a vector in a manner that is completely analogous to how we did it with the dot product in $\mathbb{R}^n$. However, in this setting of general vector spaces, we are a bit beyond the point of being able to draw a geometric picture of what length means (for example, the "length" of a matrix does not quite make sense), so we change terminology slightly and instead call this function a "norm".

Definition 1.3.7

Norm Induced by the Inner Product

Suppose that $\mathcal{V}$ is an inner product space. Then the **norm induced by the inner product** is the function $\| \cdot \| : \mathcal{V} \to \mathbb{R}$ defined by

$$\|\mathbf{v}\| \stackrel{\text{def}}{=} \sqrt{\langle \mathbf{v}, \mathbf{v} \rangle} \quad \text{for all} \quad \mathbf{v} \in \mathcal{V}.$$

We use the notation $\| \cdot \|$ to refer to the standard length on $\mathbb{R}^n$ or $\mathbb{C}^n$, but also to refer to the norm induced by whichever inner product we are currently discussing. If there is ever a chance for confusion, we use subscripts to distinguish different norms.

When $\mathcal{V} = \mathbb{R}^n$ or $\mathcal{V} = \mathbb{C}^n$ and the inner product is just the usual dot product, the norm induced by the inner product is just the usual length of a vector, given by

$$\|\mathbf{v}\| = \sqrt{\mathbf{v} \cdot \mathbf{v}} = \sqrt{|v_1|^2 + |v_2|^2 + \cdots + |v_n|^2}.$$

However, if we change which inner product we are working with, then the norm induced by the inner product changes as well. For example, the norm on $\mathbb{R}^2$ induced by the weird inner product of Example 1.3.18 has the form

$$\|\mathbf{v}\|_* = \sqrt{\langle \mathbf{v}, \mathbf{v} \rangle} = \sqrt{(v_1 + 2v_2)^2 + v_2^2},$$

which is different from the norm that we are used to (we use the notation $\| \cdot \|_*$ just to differentiate this norm from the standard length $\| \cdot \|$).

The norm induced by the standard (Frobenius) inner product on $\mathcal{M}_{m,n}$ from Example 1.3.16 is given by

The Frobenius norm is also sometimes called the **Hilbert–Schmidt norm** and denoted by $\|A\|_{\text{HS}}$.

$$\|A\|_{\text{F}} = \sqrt{\text{tr}(A^*A)} = \sqrt{\sum_{i=1}^{m} \sum_{j=1}^{n} |a_{i,j}|^2}.$$

This norm on matrices is often called the **Frobenius norm**, and it is usually written as $\|A\|_{\text{F}}$ rather than just $\|A\|$ to avoid confusion with another matrix norm that we will see a bit later in this book.

Similarly, the norm on $\mathcal{C}[a,b]$ induced by the inner product of Example 1.3.17 has the form

$$\|f\| = \sqrt{\int_a^b f(x)^2 \, dx}.$$

Perhaps not surprisingly, the norm induced by an inner product satisfies the same basic properties as the length of a vector in $\mathbb{R}^n$. The next few theorems are devoted to establishing these properties.

Theorem 1.3.7

Properties of the Norm Induced by the Inner Product

Suppose that $\mathcal{V}$ is an inner product space, $\mathbf{v} \in \mathcal{V}$ is a vector, and $c \in \mathbb{F}$ is a scalar. Then the following properties of the norm induced by the inner product hold:

    a) $\|c\mathbf{v}\| = |c|\|\mathbf{v}\|$, and   (absolute homogeneity)
    b) $\|\mathbf{v}\| \geq 0$, with equality if and only if $\mathbf{v} = \mathbf{0}$.   (pos. definiteness)

Proof. Both of these properties follow fairly quickly from the definition. For property (a), we compute

$$\|c\mathbf{v}\| = \sqrt{\langle c\mathbf{v}, c\mathbf{v}\rangle} = \sqrt{c\langle \mathbf{v}, c\mathbf{v}\rangle} = \sqrt{c\overline{\langle \mathbf{v}, c\mathbf{v}\rangle}}$$
$$= \sqrt{c\overline{c}\overline{\langle \mathbf{v}, \mathbf{v}\rangle}} = \sqrt{|c|^2\|\mathbf{v}\|^2} = |c|\|\mathbf{v}\|.$$

Property (b) follows immediately from the property of inner products that says that $\langle \mathbf{v}, \mathbf{v}\rangle \geq 0$, with equality if and only if $\mathbf{v} = \mathbf{0}$. ∎

The above theorem tells us that it makes sense to break vectors into their "length" and "direction", just like we do with vectors in $\mathbb{R}^n$. Specifically, we can write every vector $\mathbf{v} \in \mathcal{V}$ in the form $\mathbf{v} = \|\mathbf{v}\|\mathbf{u}$, where $\mathbf{u} \in \mathcal{V}$ has $\|\mathbf{u}\| = 1$ (so we call $\mathbf{u}$ a **unit vector**). In particular, if $\mathbf{v} \neq \mathbf{0}$ then we can choose $\mathbf{u} = \mathbf{v}/\|\mathbf{v}\|$, which we think of as encoding the direction of $\mathbf{v}$.

The two other main properties that vector length in $\mathbb{R}^n$ satisfies are the Cauchy–Schwarz inequality and the triangle inequality. We now show that these same properties hold for the norm induced by any inner product.

Theorem 1.3.8

Cauchy–Schwarz Inequality for Inner Products

If $\mathcal{V}$ is an inner product space then

$$|\langle \mathbf{v}, \mathbf{w}\rangle| \leq \|\mathbf{v}\|\|\mathbf{w}\| \quad \text{for all} \quad \mathbf{v}, \mathbf{w} \in \mathcal{V}.$$

Furthermore, equality holds if and only if $\mathbf{v}$ and $\mathbf{w}$ are collinear (i.e., $\{\mathbf{v}, \mathbf{w}\}$ is a linearly dependent set).

Proof. We start by letting $c, d \in \mathbb{F}$ be arbitrary scalars and expanding the quantity $\|c\mathbf{v} + d\mathbf{w}\|^2$ in terms of the inner product:

$\text{Re}(z)$ is the real part of z. That is, if $z \in \mathbb{R}$ then $\text{Re}(z) = z$, and if $z = a + ib \in \mathbb{C}$ then $\text{Re}(z) = a$.

$$
\begin{aligned}
0 &\leq \|c\mathbf{v} + d\mathbf{w}\|^2 &&\text{(pos. definiteness)}\\
&= \langle c\mathbf{v} + d\mathbf{w}, c\mathbf{v} + d\mathbf{w}\rangle &&\text{(definition of } \|\cdot\|)\\
&= |c|^2\langle \mathbf{v}, \mathbf{v}\rangle + \overline{c}d\langle \mathbf{v}, \mathbf{w}\rangle + c\overline{d}\langle \mathbf{w}, \mathbf{v}\rangle + |d|^2\langle \mathbf{w}, \mathbf{w}\rangle &&\text{(sesquilinearity)}\\
&= |c|^2\|\mathbf{v}\|^2 + 2\text{Re}\big(\overline{c}d\langle \mathbf{v}, \mathbf{w}\rangle\big) + |d|^2\|\mathbf{w}\|^2. &&\text{(since } z + \overline{z} = 2\text{Re}(z))
\end{aligned}
$$

If $\mathbf{w} = \mathbf{0}$ then the Cauchy–Schwarz inequality holds trivially (it just says $0 \leq 0$), so we can assume that $\mathbf{w} \neq \mathbf{0}$. We can thus choose $c = \|\mathbf{w}\|$ and $d = -\langle \mathbf{w}, \mathbf{v} \rangle / \|\mathbf{w}\|$, which tells us that

To simplify, we use
the fact that

$\langle \mathbf{w}, \mathbf{v} \rangle \langle \mathbf{v}, \mathbf{w} \rangle = \langle \mathbf{w}, \mathbf{v} \rangle \overline{\langle \mathbf{w}, \mathbf{v} \rangle}$

$= |\langle \mathbf{w}, \mathbf{v} \rangle|^2.$

$$0 \leq \|\mathbf{v}\|^2 \|\mathbf{w}\|^2 - 2\mathrm{Re}\big(\|\mathbf{w}\| \langle \mathbf{w}, \mathbf{v} \rangle \langle \mathbf{v}, \mathbf{w} \rangle / \|\mathbf{w}\|\big) + |\langle \mathbf{w}, \mathbf{v} \rangle|^2 \|\mathbf{w}\|^2 / \|\mathbf{w}\|^2.$$
$$= \|\mathbf{v}\|^2 \|\mathbf{w}\|^2 - |\langle \mathbf{v}, \mathbf{w} \rangle|^2.$$

Rearranging and taking the square root of both sides gives us $|\langle \mathbf{v}, \mathbf{w} \rangle| \leq \|\mathbf{v}\| \|\mathbf{w}\|$, which is exactly the Cauchy–Schwarz inequality.

To see that equality holds if and only if $\{\mathbf{v}, \mathbf{w}\}$ is a linearly dependent set, suppose that $|\langle \mathbf{v}, \mathbf{w} \rangle| = \|\mathbf{v}\| \|\mathbf{w}\|$. We can then follow the above proof backward to see that $0 = \|c\mathbf{v} + d\mathbf{w}\|^2$, so $c\mathbf{v} + d\mathbf{w} = \mathbf{0}$ (where $c = \|\mathbf{w}\| \neq 0$), so $\{\mathbf{v}, \mathbf{w}\}$ is linearly dependent. In the opposite direction, if $\{\mathbf{v}, \mathbf{w}\}$ is linearly dependent then either $\mathbf{v} = \mathbf{0}$ (in which case equality clearly holds in the Cauchy–Schwarz inequality since both sides equal 0) or $\mathbf{w} = c\mathbf{v}$ for some $c \in \mathbb{F}$. Then

$$|\langle \mathbf{v}, \mathbf{w} \rangle| = |\langle \mathbf{v}, c\mathbf{v} \rangle| = |c| \|\mathbf{v}\|^2 = \|\mathbf{v}\| \|c\mathbf{v}\| = \|\mathbf{v}\| \|\mathbf{w}\|,$$

which completes the proof.   ∎

For example, if we apply the Cauchy–Schwarz inequality to the Frobenius inner product on $\mathcal{M}_{m,n}$, it tells us that

$$\big|\mathrm{tr}(A^*B)\big|^2 \leq \mathrm{tr}(A^*A)\mathrm{tr}(B^*B) \quad \text{for all} \quad A, B \in \mathcal{M}_{m,n},$$

and if we apply it to the standard inner product on $\mathcal{C}[a,b]$ then it says that

$$\left(\int_a^b f(x)g(x)\, dx \right)^2 \leq \left(\int_a^b f(x)^2\, dx \right) \left(\int_a^b g(x)^2\, dx \right) \quad \text{for all} \quad f, g \in \mathcal{C}[a,b].$$

These examples illustrate the utility of thinking abstractly about vector spaces. These matrix and integral inequalities are tricky to prove directly from properties of the trace and integrals, but follow straightforwardly when we forget about the fine details and only think about vector space properties.

Just as was the case in $\mathbb{R}^n$, the triangle inequality now follows very quickly from the Cauchy–Schwarz inequality.

Theorem 1.3.9

The Triangle Inequality for the Norm Induced by the Inner Product

If $\mathcal{V}$ is an inner product space then

$$\|\mathbf{v} + \mathbf{w}\| \leq \|\mathbf{v}\| + \|\mathbf{w}\| \quad \text{for all} \quad \mathbf{v}, \mathbf{w} \in \mathcal{V}.$$

Furthermore, equality holds if and only if $\mathbf{v}$ and $\mathbf{w}$ point in the same direction (i.e., $\mathbf{v} = \mathbf{0}$ or $\mathbf{w} = c\mathbf{v}$ for some $0 \leq c \in \mathbb{R}$).

Proof. We start by expanding $\|\mathbf{v} + \mathbf{w}\|^2$ in terms of the inner product:

For any $z = x + iy \in \mathbb{C}$,
we have $\mathrm{Re}(z) \leq |z|$
since $\mathrm{Re}(z) = x \leq$
$\sqrt{x^2 + y^2} = |x + iy| = |z|$.

$$
\begin{aligned}
\|\mathbf{v} + \mathbf{w}\|^2 &= \langle \mathbf{v} + \mathbf{w}, \mathbf{v} + \mathbf{w} \rangle \\
&= \langle \mathbf{v}, \mathbf{v} \rangle + \langle \mathbf{v}, \mathbf{w} \rangle + \langle \mathbf{w}, \mathbf{v} \rangle + \langle \mathbf{w}, \mathbf{w} \rangle && \text{(sesquilinearity)} \\
&= \|\mathbf{v}\|^2 + 2\mathrm{Re}(\langle \mathbf{v}, \mathbf{w} \rangle) + \|\mathbf{w}\|^2 && \text{(since } z + \bar{z} = 2\mathrm{Re}(z)) \\
&\leq \|\mathbf{v}\|^2 + 2|\langle \mathbf{v}, \mathbf{w} \rangle| + \|\mathbf{w}\|^2 && \text{(since } \mathrm{Re}(z) \leq |z|) \\
&\leq \|\mathbf{v}\|^2 + 2\|\mathbf{v}\|\|\mathbf{w}\| + \|\mathbf{w}\|^2 && \text{(by Cauchy–Schwarz)} \\
&= (\|\mathbf{v}\| + \|\mathbf{w}\|)^2.
\end{aligned}
$$

We can then take the square root of both sides of this inequality to see that $\|\mathbf{v} + \mathbf{w}\| \leq \|\mathbf{v}\| + \|\mathbf{w}\|$, as desired.

The above argument demonstrates that equality holds in the triangle inequality if and only if $\text{Re}(\langle \mathbf{v}, \mathbf{w} \rangle) = |\langle \mathbf{v}, \mathbf{w} \rangle| = \|\mathbf{v}\|\|\mathbf{w}\|$. We know from the Cauchy–Schwarz inequality that the second of these equalities holds if and only if $\{\mathbf{v}, \mathbf{w}\}$ is linearly dependent (i.e., $\mathbf{v} = \mathbf{0}$ or $\mathbf{w} = c\mathbf{v}$ for some $c \in \mathbb{F}$). In this case, the first equality holds if and only if $\mathbf{v} = \mathbf{0}$ or $\text{Re}(\langle \mathbf{v}, c\mathbf{v} \rangle) = |\langle \mathbf{v}, c\mathbf{v} \rangle|$. Well, $\text{Re}(\langle \mathbf{v}, c\mathbf{v} \rangle) = \text{Re}(c)\|\mathbf{v}\|^2$ and $|\langle \mathbf{v}, c\mathbf{v} \rangle| = |c|\|\mathbf{v}\|^2$, so we see that equality holds in the triangle inequality if and only if $\text{Re}(c) = |c|$ (i.e., $0 \leq c \in \mathbb{R}$). ∎

The fact that inner products cannot be expressed in terms of vector addition and scalar multiplication means that they give us some extra structure that vector spaces alone do not have.

It is worth noting that isomorphisms in general do not preserve inner products, since inner products cannot be derived from only scalar multiplication and vector addition (which isomorphisms *do* preserve). For example, the isomorphism $T : \mathbb{R}^2 \to \mathbb{R}^2$ defined by $T(\mathbf{v}) = (v_1 + 2v_2, v_2)$ has the property that

$$T(\mathbf{v}) \cdot T(\mathbf{w}) = (v_1 + 2v_2, v_2) \cdot (w_1 + 2w_2, w_2)$$
$$= v_1 w_1 + 2v_1 w_2 + 2v_2 w_1 + 5v_2 w_2.$$

In other words, T turns the usual dot product on $\mathbb{R}^2$ into the weird inner product from Example 1.3.18.

However, even though isomorphisms do not *preserve* inner products, they at least do always convert one inner product into another one. That is, if $\mathcal{V}$ and $\mathcal{W}$ are vector spaces, $\langle \cdot, \cdot \rangle_{\mathcal{W}}$ is an inner product on $\mathcal{W}$, and $T : \mathcal{V} \to \mathcal{W}$ is an isomorphism, then we can define an inner product on $\mathcal{V}$ via $\langle \mathbf{v}_1, \mathbf{v}_2 \rangle_{\mathcal{V}} = \langle T(\mathbf{v}_1), T(\mathbf{v}_2) \rangle_{\mathcal{W}}$ (see Exercise 1.3.25).

Exercises

solutions to starred exercises on page 454

1.3.1 Determine whether or not the given vector spaces $\mathcal{V}$ and $\mathcal{W}$ are isomorphic.

*(a) $\mathcal{V} = \mathbb{R}^6$, $\mathcal{W} = \mathcal{M}_{2,3}(\mathbb{R})$
(b) $\mathcal{V} = \mathcal{M}_2(\mathbb{C})$, $\mathcal{W} = \mathcal{M}_2(\mathbb{R})$
*(c) $\mathcal{V} = \mathcal{P}^8$, $\mathcal{W} = \mathcal{M}_3(\mathbb{R})$
(d) $\mathcal{V} = \mathbb{R}$, $\mathcal{W}$ is the vector space from Example 1.1.4
*(e) $\mathcal{V} = \mathcal{P}^8$, $\mathcal{W} = \mathcal{P}$
(f) $\mathcal{V} = \{A \in \mathcal{M}_2 : \text{tr}(A) = 0\}$, $\mathcal{W} = \mathcal{P}^2$

1.3.2 Determine which of the following functions are and are not linear forms.

*(a) The function $f : \mathbb{R}^n \to \mathbb{R}$ defined by $f(\mathbf{v}) = \|\mathbf{v}\|$.
(b) The function $f : \mathbb{F}^n \to \mathbb{F}$ defined by $f(\mathbf{v}) = v_1$.
*(c) The function $f : \mathcal{M}_2 \to \mathcal{M}_2$ defined by

$$f\left(\begin{bmatrix} a & b \\ c & d \end{bmatrix}\right) = \begin{bmatrix} c & a \\ d & b \end{bmatrix}.$$

(d) The determinant of a matrix (i.e., the function $\det : \mathcal{M}_n \to \mathbb{F}$).
*(e) The function $g : \mathcal{P} \to \mathbb{R}$ defined by $g(f) = f'(3)$, where f' is the derivative of f.
(f) The function $g : \mathcal{C} \to \mathbb{R}$ defined by $g(f) = \cos(f(0))$.

1.3.3 Determine which of the following functions are inner products.

*(a) On $\mathbb{R}^2$, the function

$$\langle \mathbf{v}, \mathbf{w} \rangle = v_1 w_1 + v_1 w_2 + v_2 w_2.$$

(b) On $\mathbb{R}^2$, the function

$$\langle \mathbf{v}, \mathbf{w} \rangle = v_1 w_1 + v_1 w_2 + v_2 w_1 + v_2 w_2.$$

(c) On $\mathbb{R}^2$, the function

$$\langle \mathbf{v}, \mathbf{w} \rangle = 3v_1 w_1 + v_1 w_2 + v_2 w_1 + 3v_2 w_2.$$

*(d) On $\mathcal{M}_n$, the function $\langle A, B \rangle = \text{tr}(A^* + B)$.
(e) On $\mathcal{P}^2$, the function

$$\langle ax^2 + bx + c, dx^2 + ex + f \rangle = ad + be + cf.$$

*(f) On $\mathcal{C}[-1, 1]$, the function

$$\langle f, g \rangle = \int_{-1}^{1} \frac{f(x)g(x)}{\sqrt{1 - x^2}} \, dx.$$

1.3.4 Determine which of the following statements are true and which are false.

*(a) If $T : \mathcal{V} \to \mathcal{W}$ is an isomorphism then so is $T^{-1} : \mathcal{W} \to \mathcal{V}$.
(b) $\mathbb{R}^n$ is isomorphic to $\mathbb{C}^n$.
*(c) Two vector spaces $\mathcal{V}$ and $\mathcal{W}$ over the same field are isomorphic if and only if $\dim(\mathcal{V}) = \dim(\mathcal{W})$.

(d) If $\mathbf{w} \in \mathbb{R}^n$ then the function $f_{\mathbf{w}} : \mathbb{R}^n \to \mathbb{R}$ defined by $f_{\mathbf{w}}(\mathbf{v}) = \mathbf{v} \cdot \mathbf{w}$ for all $\mathbf{v} \in \mathbb{R}^n$ is a linear form.

∗(e) If $\mathbf{w} \in \mathbb{C}^n$ then the function $f_{\mathbf{w}} : \mathbb{C}^n \to \mathbb{C}$ defined by $f_{\mathbf{w}}(\mathbf{v}) = \mathbf{v} \cdot \mathbf{w}$ for all $\mathbf{v} \in \mathbb{C}^n$ is a linear form.

(f) If $A \in \mathcal{M}_n(\mathbb{R})$ is invertible then so is the bilinear form $f(\mathbf{v}, \mathbf{w}) = \mathbf{v}^T A \mathbf{w}$.

∗(g) If $E_x : \mathcal{P}^2 \to \mathbb{R}$ is the evaluation map defined by $E_x(f) = f(x)$ then $E_1 + E_2 = E_3$.

1.3.5 Show that if we regard $\mathbb{C}^n$ as a vector space over $\mathbb{R}$ (instead of over $\mathbb{C}$ as usual) then it is isomorphic to $\mathbb{R}^{2n}$.

∗∗**1.3.6** Suppose $\mathcal{V}$, $\mathcal{W}$, and $\mathcal{X}$ are vector spaces and $T : \mathcal{V} \to \mathcal{W}$ and $S : \mathcal{W} \to \mathcal{X}$ are isomorphisms.

(a) Show that $T^{-1} : \mathcal{W} \to \mathcal{V}$ is an isomorphism.
(b) Show that $S \circ T : \mathcal{V} \to \mathcal{X}$ is an isomorphism.

∗∗**1.3.7** Show that for every linear form $f : \mathcal{M}_{m,n}(\mathbb{F}) \to \mathbb{F}$, there exists a matrix $A \in \mathcal{M}_{m,n}(\mathbb{F})$ such that $f(X) = \text{tr}(A^T X)$ for all $X \in \mathcal{M}_{m,n}(\mathbb{F})$.

∗∗**1.3.8** Show that the result of Exercise 1.3.7 still holds if we replace every instance of $\mathcal{M}_{m,n}(\mathbb{F})$ by $\mathcal{M}_n^S(\mathbb{F})$ or every instance of $\mathcal{M}_{m,n}(\mathbb{F})$ by $\mathcal{M}_n^H$ (and set $\mathbb{F} = \mathbb{R}$ in this latter case).

∗∗**1.3.9** Suppose $\mathcal{V}$ is an inner product space. Show that $\langle \mathbf{v}, \mathbf{0} \rangle = 0$ for all $\mathbf{v} \in \mathcal{V}$.

∗∗**1.3.10** Suppose $A \in \mathcal{M}_{m,n}(\mathbb{C})$. Show that $\|A\|_F = \|A^T\|_F = \|A^*\|_F$.

1.3.11 Suppose $a < b$ are real numbers and $f \in \mathcal{C}[a,b]$. Show that

$$\left(\frac{1}{b-a} \int_a^b f(x)\, dx \right)^2 \leq \frac{1}{b-a} \left(\int_a^b f(x)^2\, dx \right).$$

[Side note: In words, this says that the square of the average of a function is never larger than the average of its square.]

∗∗**1.3.12** Let $\mathcal{V}$ be an inner product space and let $\|\cdot\|$ be the norm induced by $\mathcal{V}$'s inner product. Show that if $\mathbf{v}, \mathbf{w} \in \mathcal{V}$ are such that $\langle \mathbf{v}, \mathbf{w} \rangle = 0$ then

$$\|\mathbf{v} + \mathbf{w}\|^2 = \|\mathbf{v}\|^2 + \|\mathbf{w}\|^2.$$

[Side note: This is called the **Pythagorean theorem for inner products**.]

∗∗**1.3.13** Let $\mathcal{V}$ be an inner product space and let $\|\cdot\|$ be the norm induced by $\mathcal{V}$'s inner product. Show that

$$\|\mathbf{v} + \mathbf{w}\|^2 + \|\mathbf{v} - \mathbf{w}\|^2 = 2\|\mathbf{v}\|^2 + 2\|\mathbf{w}\|^2$$

for all $\mathbf{v}, \mathbf{w} \in \mathcal{V}$.

[Side note: This is called the **parallelogram law**, since it relates the norms of the sides of a parallelogram to the norms of its diagonals.]

∗∗**1.3.14** Let $\mathcal{V}$ be an inner product space and let $\|\cdot\|$ be the norm induced by $\mathcal{V}$'s inner product.

(a) Show that if the ground field is $\mathbb{R}$ then

$$\langle \mathbf{v}, \mathbf{w} \rangle = \frac{1}{4} \left(\|\mathbf{v} + \mathbf{w}\|^2 - \|\mathbf{v} - \mathbf{w}\|^2 \right)$$

for all $\mathbf{v}, \mathbf{w} \in \mathcal{V}$.

(b) Show that if the ground field is $\mathbb{C}$ then

$$\langle \mathbf{v}, \mathbf{w} \rangle = \frac{1}{4} \sum_{k=0}^3 \frac{1}{i^k} \|\mathbf{v} + i^k \mathbf{w}\|^2$$

for all $\mathbf{v}, \mathbf{w} \in \mathcal{V}$.

[Side note: This is called the **polarization identity**.]

1.3.15 Let $\mathcal{V}$ be an inner product space and let $\|\cdot\|$ be the norm induced by $\mathcal{V}$'s inner product.

(a) Show that if the ground field is $\mathbb{R}$ then

$$\langle \mathbf{v}, \mathbf{w} \rangle = \|\mathbf{v}\|\|\mathbf{w}\| \left(1 - \frac{1}{2} \left\| \frac{\mathbf{v}}{\|\mathbf{v}\|} - \frac{\mathbf{w}}{\|\mathbf{w}\|} \right\|^2 \right)$$

for all $\mathbf{v}, \mathbf{w} \in \mathcal{V}$.
[Side note: This representation of the inner product implies the Cauchy–Schwarz inequality as an immediate corollary.]

(b) Show that if the ground field is $\mathbb{C}$ then

$$\langle \mathbf{v}, \mathbf{w} \rangle = \|\mathbf{v}\|\|\mathbf{w}\| \left(1 - \frac{1}{2} \left\| \frac{\mathbf{v}}{\|\mathbf{v}\|} - \frac{\mathbf{w}}{\|\mathbf{w}\|} \right\|^2 \right.$$
$$\left. + i - \frac{i}{2} \left\| \frac{\mathbf{v}}{\|\mathbf{v}\|} - \frac{i\mathbf{w}}{\|\mathbf{w}\|} \right\|^2 \right)$$

for all $\mathbf{v}, \mathbf{w} \in \mathcal{V}$.

∗**1.3.16** Suppose $\mathcal{V}$ is a vector space over a field $\mathbb{F}$ in which $1 + 1 \neq 0$, and $f : \mathcal{V} \times \mathcal{V} \to \mathbb{F}$ is a bilinear form.

(a) Show that $f(\mathbf{v}, \mathbf{w}) = -f(\mathbf{w}, \mathbf{v})$ for all $\mathbf{v}, \mathbf{w} \in \mathcal{V}$ (in which case f is called **skew-symmetric**) if and only if $f(\mathbf{v}, \mathbf{v}) = 0$ for all $\mathbf{v} \in \mathcal{V}$ (in which case f is called **alternating**).

(b) Explain what goes wrong in part (a) if $1 + 1 = 0$ (e.g., if $\mathbb{F} = \mathbb{Z}_2$ is the field of 2 elements described in Appendix A.4).

1.3.17 Suppose $\mathcal{V}$ is a vector space over a field $\mathbb{F}$ and $f : \mathcal{V} \times \mathcal{V} \to \mathbb{F}$ is a bilinear form.

(a) Show that $f(\mathbf{v}, \mathbf{w}) = f(\mathbf{w}, \mathbf{v})$ for all $\mathbf{v}, \mathbf{w} \in \mathcal{V}$ (in which case f is called **symmetric**) if and only if the matrix $A \in \mathcal{M}_n(\mathbb{F})$ from Theorem 1.3.5 is symmetric (i.e., satisfies $A^T = A$).

(b) Show that $f(\mathbf{v}, \mathbf{w}) = -f(\mathbf{w}, \mathbf{v})$ for all $\mathbf{v}, \mathbf{w} \in \mathcal{V}$ (in which case f is called **skew-symmetric**) if and only if the matrix $A \in \mathcal{M}_n(\mathbb{F})$ from Theorem 1.3.5 is skew-symmetric (i.e., satisfies $A^T = -A$).

∗∗**1.3.18** Suppose $\mathcal{V}$ and $\mathcal{W}$ are m- and n-dimensional vector spaces over $\mathbb{C}$, respectively, and let $f : \mathcal{V} \times \mathcal{W} \to \mathbb{C}$ be a sesquilinear form (i.e., a function that is linear in its second argument and conjugate linear in its first).

(a) Show that if B and C are bases of $\mathcal{V}$ and $\mathcal{W}$, respectively, then there exists a unique matrix $A \in \mathcal{M}_{m,n}(\mathbb{C})$ such that

$$f(\mathbf{v}, \mathbf{w}) = [\mathbf{v}]_B^* A [\mathbf{w}]_C \quad \text{for all} \quad \mathbf{v} \in \mathcal{V}, \ \mathbf{w} \in \mathcal{W}.$$

(b) Suppose $\mathcal{W} = \mathcal{V}$ and $C = B$. Show that f is **conjugate symmetric** (i.e., $f(\mathbf{v}, \mathbf{w}) = \overline{f(\mathbf{w}, \mathbf{v})}$ for all $\mathbf{v} \in \mathcal{V}$ and $\mathbf{w} \in \mathcal{W}$) if and only if the matrix A from part (a) is Hermitian.

(c) Suppose $\mathcal{W} = \mathcal{V}$ and $C = B$. Show that f is an inner product if and only if the matrix A from part (a) is Hermitian and satisfies $\mathbf{v}^* A \mathbf{v} \geq 0$ for all $\mathbf{v} \in \mathbb{C}^m$ with equality if and only if $\mathbf{v} = \mathbf{0}$.
[Side note: A matrix A with these properties is called **positive definite**, and we explore such matrices in Section 2.2.]

1.3.19 Suppose $\mathcal{V}$ is a vector space with basis $B = \{\mathbf{v}_1, \ldots, \mathbf{v}_n\}$. Define linear functionals $f_1, \ldots, f_n \in \mathcal{V}^*$ by

$$f_i(\mathbf{v}_j) = \begin{cases} 1 & \text{if } i = j, \\ 0 & \text{otherwise.} \end{cases}$$

Show that the set $B^* \stackrel{\text{def}}{=} \{f_1, \ldots, f_n\}$ is a basis of $\mathcal{V}^*$.

[Side note: B^* is called the **dual basis** of B.]

1.3.20 Let c_{00} be the vector space of eventually-zero real sequences from Example 1.1.10. Show that $c_{00}^* \cong \mathbb{R}^{\mathbb{N}}$.

[Side note: This example shows that the dual of an infinite-dimensional vector space can be much larger than the original vector space. For example, c_{00} has dimension equal to the cardinality of $\mathbb{N}$, but this exercise shows that c_{00}^* has dimension at least as large as the cardinality of $\mathbb{R}$. See also Exercise 1.3.23.]

∗∗ 1.3.21 Let $E_x : \mathcal{P}^p \to \mathbb{R}$ be the evaluation map defined by $E_x(f) = f(x)$ (see Example 1.3.5) and let $E = \{1, x, x^2, \ldots, x^p\}$ be the standard basis of $\mathcal{P}^p$. Find a polynomial $g_x \in \mathcal{P}^p$ such that $E_x(f) = [g_x]_E \cdot [f]_E$ for all $f \in \mathcal{P}^p$.

∗∗ 1.3.22 Let $E_x : \mathcal{P}^p \to \mathbb{R}$ be the evaluation map defined by $E_x(f) = f(x)$. Show that if $c_0, c_1, \ldots, c_p \in \mathbb{R}$ are distinct then $\{E_{c_0}, E_{c_1}, \ldots, E_{c_p}\}$ is a basis of $(\mathcal{P}^p)^*$.

1.3.23 Let $E_x : \mathcal{P} \to \mathbb{R}$ be the evaluation map defined by $E_x(f) = f(x)$. Show that the set

$$\{E_x : x \in \mathbb{R}\}$$

is linearly independent.

∗∗ 1.3.24 Let $T : \mathcal{V} \to \mathcal{V}^{**}$ be the canonical double-dual isomorphism described by Theorem 1.3.4. Complete the proof of that theorem by showing that if $B = \{\mathbf{v}_1, \mathbf{v}_2, \ldots, \mathbf{v}_n\} \subseteq \mathcal{V}$ is linearly independent then so is $C = \{T(\mathbf{v}_1), T(\mathbf{v}_2), \ldots, T(\mathbf{v}_n)\} \subseteq \mathcal{V}^{**}$.

∗∗ 1.3.25 Suppose that $\mathcal{V}$ and $\mathcal{W}$ are vector spaces over a field $\mathbb{F}$, $\langle \cdot, \cdot \rangle_{\mathcal{W}}$ is an inner product on $\mathcal{W}$, and $T : \mathcal{V} \to \mathcal{W}$ is an isomorphism. Show that the function $\langle \cdot, \cdot \rangle_{\mathcal{V}} : \mathcal{V} \times \mathcal{V} \to \mathbb{F}$ defined by $\langle \mathbf{v}_1, \mathbf{v}_2 \rangle_{\mathcal{V}} = \langle T(\mathbf{v}_1), T(\mathbf{v}_2) \rangle_{\mathcal{W}}$ is an inner product on $\mathcal{V}$.

1.3.26 Suppose $\mathcal{V}$ is a vector space over a field $\mathbb{F}$, $f : \mathcal{V} \times \cdots \times \mathcal{V} \to \mathbb{F}$ is a multilinear form, and $T_1, \ldots, T_n : \mathcal{V} \to \mathcal{V}$ are linear transformations. Show that the function $g : \mathcal{V} \times \cdots \times \mathcal{V} \to \mathbb{F}$ defined by

$$g(\mathbf{v}_1, \ldots, \mathbf{v}_n) = f(T_1(\mathbf{v}_1), \ldots, T_n(\mathbf{v}_n))$$

is a multilinear form.

∗∗ 1.3.27 An array A is called **antisymmetric** if swapping any two of its indices swaps the sign of its entries (i.e., $a_{j_1, \ldots, j_k, \ldots, j_\ell, \ldots, j_p} = -a_{j_1, \ldots, j_\ell, \ldots, j_k, \ldots, j_p}$ for all $k \neq \ell$). Show that for each $p \geq 2$ there is, up to scaling, only one $p \times p \times \cdots \times p$ (p times) antisymmetric array.

[Side note: This array corresponds to the determinant in the sense of Theorem 1.3.6.]

[Hint: Make use of permutations (see Appendix A.1.5) or properties of the determinant.]

1.4 Orthogonality and Adjoints

Now that we know how to generalize the dot product from $\mathbb{R}^n$ to other vector spaces (via inner products), it is worth revisiting the concept of orthogonality. Recall that two vectors $\mathbf{v}, \mathbf{w} \in \mathbb{R}^n$ are orthogonal when $\mathbf{v} \cdot \mathbf{w} = 0$. We define orthogonality in general inner product spaces completely analogously:

Definition 1.4.1
Orthogonality

Suppose $\mathcal{V}$ is an inner product space. Two vectors $\mathbf{v}, \mathbf{w} \in \mathcal{V}$ are called **orthogonal** if $\langle \mathbf{v}, \mathbf{w} \rangle = 0$.

In $\mathbb{R}^n$, we could think of "orthogonal" as a synonym for "perpendicular", since two vectors in $\mathbb{R}^n$ are orthogonal if and only if the angle between them is $\pi/2$. However, in general inner product spaces this geometric picture makes much less sense (for example, it does not quite make sense to say that the angle between two polynomials is $\pi/2$). For this reason, it is perhaps better to think of orthogonal vectors as ones that are "as linearly independent as possible" (see Figure 1.10 for a geometric justification of this interpretation).

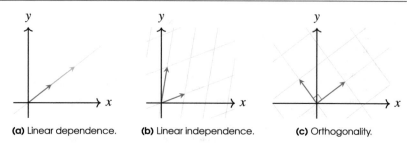

(a) Linear dependence.   **(b)** Linear independence.   **(c)** Orthogonality.

Figure 1.10: Orthogonality can be thought of as a stronger version of linear independence: not only do the vectors not point in the same direction, but in fact they point as far away from each other as possible.

In order to better justify this interpretation of orthogonality, we first need a notion of orthogonality for *sets* of vectors rather than just *pairs* of vectors. To make this leap, we simply say that a set of vectors B is **mutually orthogonal** if every two distinct vectors $\mathbf{v} \neq \mathbf{w} \in B$ are orthogonal. For example, a set of three vectors in $\mathbb{R}^3$ is mutually orthogonal if they all make right angles with each other (like the coordinate axes).

The following result pins down our claim that mutual orthogonality is a stronger property than linear independence.

Theorem 1.4.1 **Mutual Orthogonality** **Implies Linear** **Independence**	Suppose $\mathcal{V}$ is an inner product space. If $B \subset \mathcal{V}$ is a mutually orthogonal set of non-zero vectors then B is linearly independent.

Proof. We start by supposing that $\mathbf{v}_1, \mathbf{v}_2, \ldots, \mathbf{v}_k \in B$ and $c_1, c_2, \ldots, c_k \in \mathbb{F}$ (where $\mathbb{F}$ is the ground field) are such that

$$c_1 \mathbf{v}_1 + c_2 \mathbf{v}_2 + \cdots + c_k \mathbf{v}_k = \mathbf{0}. \tag{1.4.1}$$

Our goal is to show that $c_1 = c_2 = \cdots = c_k = 0$. To this end, we note that
$\langle \mathbf{v}_1, \mathbf{0} \rangle = 0$, but also

The fact that
$\langle \mathbf{v}_1, \mathbf{0} \rangle = 0$ is hopefully
intuitive enough. If
not, it was proved in
Exercise 1.3.9.

$$\begin{aligned}
\langle \mathbf{v}_1, \mathbf{0} \rangle &= \langle \mathbf{v}_1, c_1 \mathbf{v}_1 + c_2 \mathbf{v}_2 + \cdots + c_k \mathbf{v}_k \rangle && \text{(by Equation (1.4.1))} \\
&= c_1 \langle \mathbf{v}_1, \mathbf{v}_1 \rangle + c_2 \langle \mathbf{v}_1, \mathbf{v}_2 \rangle + \cdots + c_k \langle \mathbf{v}_1, \mathbf{v}_k \rangle && \text{(linearity of the i.p.)} \\
&= c_1 \langle \mathbf{v}_1, \mathbf{v}_1 \rangle + 0 + \cdots + 0 && \text{(B is mutually orthogonal)} \\
&= c_1 \|\mathbf{v}_1\|^2. && \text{(definition of norm)}
\end{aligned}$$

Since all of the vectors in B are non-zero we know that $\|\mathbf{v}_1\| \neq 0$, so this implies $c_1 = 0$.

A similar computation involving $\langle \mathbf{v}_2, \mathbf{0} \rangle$ shows that $c_2 = 0$, and so on up to $\langle \mathbf{v}_k, \mathbf{0} \rangle$ showing that $c_k = 0$, so we conclude $c_1 = c_2 = \cdots = c_k = 0$ and thus B is linearly independent. ∎

Example 1.4.1 **Checking Mutual** **Orthogonality of** **Polynomials** Recall that $\mathcal{P}^2[-1,1]$ is the vector space of polynomials of degree at most 2 acting on the interval $[-1,1]$.	Show that the set $B = \{1, x, 2x^2 - 1\} \subset \mathcal{P}^2[-1,1]$ is mutually orthogonal with respect to the inner product $$\langle f, g \rangle = \int_{-1}^{1} \frac{f(x)g(x)}{\sqrt{1 - x^2}} \, dx.$$ **Solution:** We explicitly compute all 3 possible inner products between these

3 polynomials, which requires some integration techniques that we may have not used in a while:

$$\langle 1, x \rangle = \int_{-1}^{1} \frac{x}{\sqrt{1-x^2}}\, dx = -\sqrt{1-x^2}\Big|_{-1}^{1} = 0 - 0 = 0,$$

All 3 of these integrals can be solved by making the substitution $x = \sin(u)$ and then integrating with respect to u.

$$\langle 1, 2x^2 - 1 \rangle = \int_{-1}^{1} \frac{2x^2 - 1}{\sqrt{1-x^2}}\, dx = -x\sqrt{1-x^2}\Big|_{-1}^{1} = 0 - 0 = 0, \quad \text{and}$$

$$\langle x, 2x^2 - 1 \rangle = \int_{-1}^{1} \frac{2x^3 - x}{\sqrt{1-x^2}}\, dx = \frac{-(2x^2 + 1)}{3}\sqrt{1-x^2}\Big|_{-1}^{1} = 0 - 0 = 0.$$

Since each pair of these polynomials is orthogonal with respect to this inner product, the set is mutually orthogonal.

When combined with Theorem 1.4.1, the above example shows that the set $\{1, x, 2x^2 - 1\}$ is linearly independent in $\mathcal{P}^2[-1, 1]$. It is worth observing that a set of vectors may be mutually orthogonal with respect to one inner product but not another—all that is needed to show linear independence in this way is that it is mutually orthogonal with respect to *at least one* inner product. Conversely, for every linearly independent set there is some inner product with respect to which it is mutually orthogonal, at least in the finite-dimensional case (see Exercise 1.4.25).

1.4.1   Orthonormal Bases

One of the most useful things that we could do with linear independence was introduce bases, which in turn let us give coordinates to vectors in arbitrary finite-dimensional vector spaces. We now spend some time investigating bases that are not just linearly independent, but are even mutually orthogonal and scaled so that all vectors in the basis have the same length. We will see that this additional structure makes these bases more well-behaved and much easier to work with than others.

Definition 1.4.2

Orthonormal Bases

Suppose $\mathcal{V}$ is an inner product space with basis $B \subset \mathcal{V}$. We say that B is an **orthonormal basis** of $\mathcal{V}$ if

  a) $\langle \mathbf{v}, \mathbf{w} \rangle = 0$ for all $\mathbf{v} \neq \mathbf{w} \in B$, and   (mutual orthogonality)
  b) $\|\mathbf{v}\| = 1$ for all $\mathbf{v} \in B$.   (normalization)

Before proceeding with examples, we note that determining whether or not a set is an orthonormal basis is, rather surprisingly, often easier than determining whether or not it is a (potentially non-orthonormal) basis. The reason for this is that checking mutual orthogonality (which just requires computing some inner products) is typically easier than checking linear independence (which requires solving a linear system), and the following theorem says that this is all that we have to check as long as the set we are working with has the "right" size (i.e., as many vectors as the dimension of the vector space).

Suppose V is a finite-dimensional inner product space and $B \subseteq V$ is a mutually orthogonal set consisting of unit vectors. Then B is an orthonormal basis of V if and only if $|B| = \dim(V)$.

The notation $|B|$ means the number of vectors in B.

Proof. Recall that Theorem 1.4.1 tells us that if B is mutually orthogonal and its members are unit vectors (and thus non-zero) then B is linearly independent. We then make use of Exercise 1.2.27(a), which tells us that a set with $\dim(V)$ vectors is a basis if and only if it is linearly independent. ∎

It is straightforward to check that the standard bases of each of $\mathbb{R}^n$, $\mathbb{C}^n$, and $\mathcal{M}_{m,n}$ are in fact orthonormal with respect to the standard inner products on these spaces (i.e., the dot product on $\mathbb{R}^n$ and $\mathbb{C}^n$ and the Frobenius inner product on $\mathcal{M}_{m,n}$). On the other hand, we have to be somewhat careful when working with $\mathcal{P}^p$, since the standard basis $\{1, x, x^2, \ldots, x^p\}$ is *not* orthonormal with respect to any of its standard inner products like

$$\langle f, g \rangle = \int_0^1 f(x)g(x)\,dx.$$

For example, performing the integration indicated above shows that $\langle 1, x \rangle = (1^2)/2 - (0^2)/2 = 1/2 \neq 0$, so the polynomials 1 and x are not orthogonal in this inner product.

Example 1.4.2

An Orthonormal Basis of Polynomials

Construct an orthonormal basis of $\mathcal{P}^2[-1,1]$ with respect to the inner product

$$\langle f, g \rangle = \int_{-1}^1 \frac{f(x)g(x)}{\sqrt{1-x^2}}\,dx.$$

Solution:
We already showed that the set $B = \{1, x, 2x^2 - 1\}$ is mutually orthogonal with respect to this inner product in Example 1.4.1. To turn this set into an orthonormal basis, we just normalize these polynomials (i.e., divide them by their norms):

Be careful here: we might guess that $\|1\| = 1$, but this is not true. We have to go through the computation with the indicated inner product.

$$\|1\| = \sqrt{\langle 1, 1 \rangle} = \sqrt{\int_{-1}^1 \frac{1}{\sqrt{1-x^2}}\,dx} = \sqrt{\pi},$$

$$\|x\| = \sqrt{\langle x, x \rangle} = \sqrt{\int_{-1}^1 \frac{x^2}{\sqrt{1-x^2}}\,dx} = \sqrt{\frac{\pi}{2}}, \quad \text{and}$$

These integrals can be evaluated by substituting $x = \sin(u)$.

$$\|2x^2 - 1\| = \sqrt{\langle 2x^2 - 1, 2x^2 - 1 \rangle} = \sqrt{\int_{-1}^1 \frac{(2x^2 - 1)^2}{\sqrt{1-x^2}}\,dx} = \sqrt{\frac{\pi}{2}}.$$

It follows that the set $C = \{1/\sqrt{\pi}, \sqrt{2}x/\sqrt{\pi}, \sqrt{2}(2x^2 - 1)/\sqrt{\pi}\}$ is a mutually orthogonal set of normalized vectors. Since C consists of $\dim(\mathcal{P}^2) = 3$ vectors, we know from Theorem 1.4.2 that it is an orthonormal basis of $\mathcal{P}^2[-1,1]$ (with respect to this inner product).

Example 1.4.3

**The Pauli Matrices
Form an
Orthonormal Basis**

These are, up to the
scalar factor $1/\sqrt{2}$,
the Pauli matrices
that we saw earlier
in Example 1.1.12.

$\overline{X}$ is the entrywise
complex conjugate
of X.

Show that the set

$$B = \left\{ \frac{1}{\sqrt{2}} \begin{bmatrix} 1 & 0 \\ 0 & 1 \end{bmatrix}, \frac{1}{\sqrt{2}} \begin{bmatrix} 0 & 1 \\ 1 & 0 \end{bmatrix}, \frac{1}{\sqrt{2}} \begin{bmatrix} 0 & -i \\ i & 0 \end{bmatrix}, \frac{1}{\sqrt{2}} \begin{bmatrix} 1 & 0 \\ 0 & -1 \end{bmatrix} \right\}$$

is an orthonormal basis of $\mathcal{M}_2(\mathbb{C})$ with the usual Frobenius inner product.

Solution:

Recall that the Frobenius inner product works just like the dot product on $\mathbb{C}^n$: $\langle X, Y \rangle$ is obtained by multiplying $\overline{X}$ and Y entrywise and adding up the results. We thus can see that most of these matrices are indeed orthogonal, since all of the terms being added up are 0. For example,

$$\left\langle \frac{1}{\sqrt{2}} \begin{bmatrix} 0 & 1 \\ 1 & 0 \end{bmatrix}, \frac{1}{\sqrt{2}} \begin{bmatrix} 1 & 0 \\ 0 & -1 \end{bmatrix} \right\rangle = \frac{1}{2}(0 \cdot 1 + 1 \cdot 0 + 1 \cdot 0 + 0 \cdot (-1))$$

$$= \frac{1}{2}(0 + 0 + 0 + 0) = 0.$$

For this reason, the only two other inner products that we explicitly check are the ones where the zeros do *not* match up in this way:

$$\left\langle \frac{1}{\sqrt{2}} \begin{bmatrix} 1 & 0 \\ 0 & 1 \end{bmatrix}, \frac{1}{\sqrt{2}} \begin{bmatrix} 1 & 0 \\ 0 & -1 \end{bmatrix} \right\rangle = \frac{1}{2}(1 + 0 + 0 + (-1)) = 0, \quad \text{and}$$

$$\left\langle \frac{1}{\sqrt{2}} \begin{bmatrix} 0 & 1 \\ 1 & 0 \end{bmatrix}, \frac{1}{\sqrt{2}} \begin{bmatrix} 0 & -i \\ i & 0 \end{bmatrix} \right\rangle = \frac{1}{2}(0 + (-i) + i + 0) = 0.$$

Since every pair of these matrices is orthogonal, B is mutually orthogonal.

To see that these matrices are properly normalized, we just note that it is straightforward to check that $\langle X, X \rangle = 1$ for all four of these matrices $X \in B$. It thus follows from Theorem 1.4.2 that, since B consists of 4 matrices and $\dim(\mathcal{M}_2(\mathbb{C})) = 4$, B is an orthonormal basis of $\mathcal{M}_2(\mathbb{C})$.

We already learned in Section 1.3.1 that all finite-dimensional vector spaces are isomorphic (i.e., "essentially the same") as $\mathbb{F}^n$. It thus seems natural to ask the corresponding question about inner products—do all inner products on a finite-dimensional inner product space $\mathcal{V}$ look like the usual dot product on $\mathbb{F}^n$ on some basis? It turns out that the answer to this question is "yes", and the bases for which this happens are exactly orthonormal bases.

Theorem 1.4.3

**All Inner Products
Look Like the Dot
Product**

If B is an orthonormal basis of a finite-dimensional inner product space $\mathcal{V}$ then

$$\langle \mathbf{v}, \mathbf{w} \rangle = [\mathbf{v}]_B \cdot [\mathbf{w}]_B \quad \text{for all} \quad \mathbf{v}, \mathbf{w} \in \mathcal{V}.$$

Proof. Suppose $B = \{\mathbf{u}_1, \mathbf{u}_2, \ldots \mathbf{u}_n\}$. Since B is a basis of $\mathcal{V}$, we can write $\mathbf{v} = c_1\mathbf{u}_1 + c_2\mathbf{u}_2 + \cdots + c_n\mathbf{u}_n$ and $\mathbf{w} = d_1\mathbf{u}_1 + d_2\mathbf{u}_2 + \cdots + d_n\mathbf{u}_n$. Using properties of inner products then reveals that

$$\langle \mathbf{v}, \mathbf{w} \rangle = \left\langle \sum_{i=1}^{n} c_i \mathbf{u}_i, \sum_{j=1}^{n} d_j \mathbf{u}_j \right\rangle$$

$$= \sum_{i,j=1}^{n} \overline{c_i} d_j \langle \mathbf{u}_i, \mathbf{u}_j \rangle \qquad \text{(sesquilinearity)}$$

$$= \sum_{j=1}^{n} \overline{c_j} d_j \qquad \text{(B is an orthonormal basis)}$$

$$= (c_1, c_2, \ldots, c_n) \cdot (d_1, d_2, \ldots, d_n) \quad \text{(definition of dot product)}$$

$$= [\mathbf{v}]_B \cdot [\mathbf{w}]_B, \qquad \text{(definition of coordinate vectors)}$$

An even more explicit characterization of inner products on $\mathcal{V} = \mathbb{R}^n$ and $\mathcal{V} = \mathbb{C}^n$ is presented in Exercise 1.4.24.

as desired. ∎

For example, we showed in Example 1.3.16 that if E is the standard basis of $\mathcal{M}_{m,n}$ then the Frobenius inner product on matrices satisfies

$$\langle A, B \rangle = [A]_E \cdot [B]_E \quad \text{for all} \quad A, B \in \mathcal{M}_{m,n}.$$

Theorem 1.4.3 says that the same is true if we replace E by *any* orthonormal basis of $\mathcal{M}_{m,n}$. However, for some inner products it is not quite so obvious how to find an orthonormal basis and thus represent it as a dot product.

Example 1.4.4

A Polynomial Inner Product as the Dot Product

Find a basis B of $\mathcal{P}^1[0,1]$ (with the standard inner product) with the property that $\langle f, g \rangle = [f]_B \cdot [g]_B$ for all $f, g \in \mathcal{P}^1[0,1]$.

Solution:

First, we recall that the standard inner product on $\mathcal{P}^1[0,1]$ is

$$\langle f, g \rangle = \int_0^1 f(x) g(x)\, dx.$$

It might be tempting to think that we should choose the standard basis $B = \{1, x\}$, but this does not work (for example, we noted earlier that $\langle 1, x \rangle = 1/2$, but $[1]_B \cdot [x]_B = 0$).

We could also choose h_1 to be any other non-zero vector, but $h_1(x) = 1$ seems like it will make the algebra work out nicely.

The problem with the standard basis is that it is not orthonormal in this inner product, so Theorem 1.4.3 does not apply to it. To construct an orthonormal basis of $\mathcal{P}^1[0,1]$ in this inner product, we start by finding any vector (function) $h_2 \in \mathcal{P}^1[0,1]$ that is orthogonal to $h_1(x) = 1$. To do so, we write $h_2(x) = ax + b$ for some scalars $a, b \in \mathbb{R}$ and solve $\langle h_1, h_2 \rangle = 0$:

$$0 = \langle h_1, h_2 \rangle = \langle 1, ax + b \rangle = \int_0^1 (ax + b)\, dx = \left(\frac{a}{2} x^2 + bx \right) \Big|_0^1 = \frac{a}{2} + b.$$

We can solve this equation by choosing $a = 2$ and $b = -1$ so that $h_2(x) = 2x - 1$.

Finally, we just need to rescale h_1 and h_2 so that they have norm 1. We can compute their norms as follows:

$$\|h_1\| = \|1\| = \sqrt{\langle 1,1\rangle} = \sqrt{\int_0^1 1\,dx} = 1 \quad \text{and}$$

$$\|h_2\| = \|2x-1\| = \sqrt{\langle 2x-1,2x-1\rangle} = \sqrt{\int_0^1 (2x-1)^2\,dx} = 1/\sqrt{3}.$$

> Be careful here: the fact that $\|1\| = 1$ is not quite as obvious as it seems, since $\|1\|$ must be computed via an integral (i.e., the inner product).

It follows that $B = \left\{1, \sqrt{3}(2x-1)\right\}$ is an orthonormal basis of $\mathcal{P}^1[0,1]$, so Theorem 1.4.3 tells us that $\langle f,g\rangle = [f]_B \cdot [g]_B$ for all $f,g \in \mathcal{P}^1[0,1]$.

The following corollary shows that we can similarly think of the norm induced by any inner product as "the same" as the usual vector length on $\mathbb{R}^n$ or $\mathbb{C}^n$:

Corollary 1.4.4

The Norm Induced by an Inner Product Looks Like Vector Length

If B is an orthonormal basis of a finite-dimensional inner product space V then

$$\|\mathbf{v}\| = \|[\mathbf{v}]_B\| \quad \text{for all} \quad \mathbf{v} \in V.$$

> Be slightly careful here—$\|\mathbf{v}\|$ refers to the norm induced by the inner product on V, whereas $\|[\mathbf{v}]_B\|$ refers to the length in $\mathbb{R}^n$ or $\mathbb{C}^n$.

Proof of Theorem 1.2.6. Just choose $\mathbf{w} = \mathbf{v}$ in Theorem 1.4.3. Then

$$\|\mathbf{v}\| = \sqrt{\langle \mathbf{v},\mathbf{v}\rangle} = \sqrt{[\mathbf{v}]_B \cdot [\mathbf{v}]_B} = \|[\mathbf{v}]_B\|. \qquad \blacksquare$$

In words, Corollary 1.4.4 says that the norm induced by an inner product just measures how long a vector's coordinate vector is when it is represented in an orthonormal basis. To give a bit of geometric intuition to what this means, consider the norm $\|\cdot\|_*$ on $\mathbb{R}^2$ induced by the weird inner product of Example 1.3.18, which has the form

$$\|\mathbf{v}\|_* = \sqrt{\langle \mathbf{v},\mathbf{v}\rangle} = \sqrt{(v_1 + 2v_2)^2 + v_2^2}.$$

We can think of this norm as measuring how long $\mathbf{v}$ is when it is represented in the basis $B = \{(1,0),(-2,1)\}$ instead of in the standard basis (see Figure 1.11). The reason that this works is simply that B is an orthonormal basis with respect to the weird inner product that induces this norm (we will see how this basis B was constructed shortly, in Example 1.4.6).

> B is not orthonormal with respect to the usual inner product (the dot product) on $\mathbb{R}^2$ though.

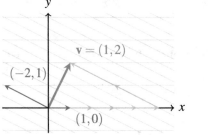

(a) Plotting using standard coordinates.

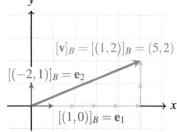

(b) Plotting using B coordinates.

Figure 1.11: The length of $\mathbf{v} = (1,2)$ is $\|\mathbf{v}\| = \sqrt{1^2 + 2^2} = \sqrt{5}$. On the other hand, $\|\mathbf{v}\|_*$ measures the length of $\mathbf{v}$ when it is represented in the basis $B = \{(1,0),(-2,1)\}$: $\|\mathbf{v}\|_* = \sqrt{(1+4)^2 + 2^2} = \sqrt{29}$ and $\|[\mathbf{v}]_B\| = \sqrt{5^2 + 2^2} = \sqrt{29}$.

We typically think of orthonormal bases are particularly "well-behaved" bases for which everything works out a bit more simply than it does for general bases. To give a bit of sense of what we mean by this, we now present a result that shows that finding coordinate vectors with respect to orthonormal bases is trivial. For example, recall that if $B = \{\mathbf{e}_1, \mathbf{e}_2, \ldots, \mathbf{e}_n\}$ is the standard basis of $\mathbb{R}^n$ then, for each $1 \leq j \leq n$, the j-th coordinate of a vector $\mathbf{v} \in \mathbb{R}^n$ is simply $\mathbf{e}_j \cdot \mathbf{v} = v_j$. The following theorem says that coordinates in *any* orthonormal basis can similarly be found simply by computing inner products (instead of solving a linear system, like we have to do to find coordinate vectors with respect to general bases).

Theorem 1.4.5

Coordinates with Respect to Orthonormal Bases

If $B = \{\mathbf{u}_1, \mathbf{u}_2, \ldots, \mathbf{u}_n\}$ is an orthonormal basis of a finite-dimensional inner product space $\mathcal{V}$ then

$$[\mathbf{v}]_B = \big(\langle \mathbf{u}_1, \mathbf{v} \rangle, \langle \mathbf{u}_2, \mathbf{v} \rangle, \ldots, \langle \mathbf{u}_n, \mathbf{v} \rangle\big) \quad \text{for all} \quad \mathbf{v} \in \mathcal{V}.$$

Proof. We simply make use of Theorem 1.4.3 to represent the inner product as the dot product of coordinate vectors. In particular, we recall that $[\mathbf{u}_j]_B = \mathbf{e}_j$ for all $1 \leq j \leq n$ and then compute

$[\mathbf{u}_j]_B = \mathbf{e}_j$ simply because $\mathbf{u}_j = 0\mathbf{u}_1 + \cdots + 1\mathbf{u}_j + \cdots + 0\mathbf{u}_n$, and sticking the coefficients of that linear combination in a vector gives $\mathbf{e}_j$.

$$\big(\langle \mathbf{u}_1, \mathbf{v} \rangle, \langle \mathbf{u}_2, \mathbf{v} \rangle, \ldots, \langle \mathbf{u}_n, \mathbf{v} \rangle\big) = \big([\mathbf{u}_1]_B \cdot [\mathbf{v}]_B, \ [\mathbf{u}_2]_B \cdot [\mathbf{v}]_B, \ \ldots, \ [\mathbf{u}_n]_B \cdot [\mathbf{v}]_B\big)$$
$$= \big(\mathbf{e}_1 \cdot [\mathbf{v}]_B, \ \mathbf{e}_2 \cdot [\mathbf{v}]_B, \ \ldots, \ \mathbf{e}_n \cdot [\mathbf{v}]_B\big),$$

which equals $[\mathbf{v}]_B$ since $\mathbf{e}_1 \cdot [\mathbf{v}]_B$ is the first entry of $[\mathbf{v}]_B$, $\mathbf{e}_2 \cdot [\mathbf{v}]_B$ is its second entry, and so on. ∎

The above theorem can be interpreted as telling us that the inner product between $\mathbf{v} \in \mathcal{V}$ and a unit vector $\mathbf{u} \in \mathcal{V}$ measures how far $\mathbf{v}$ points in the direction of $\mathbf{u}$ (see Figure 1.12).

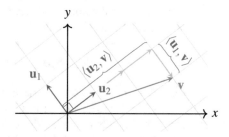

Figure 1.12: The coordinates of a vector $\mathbf{v}$ with respect to an orthonormal basis are simply the inner products of $\mathbf{v}$ with the basis vectors. In other words, the inner product of $\mathbf{v}$ with a unit vector tells us how far $\mathbf{v}$ extends in that direction.

Example 1.4.5

A Coordinate Vector with Respect to the Pauli Basis

Compute $[A]_B$ (the coordinate vector of $A \in \mathcal{M}_2(\mathbb{C})$ with respect to B) if

$$B = \left\{ \frac{1}{\sqrt{2}} \begin{bmatrix} 1 & 0 \\ 0 & 1 \end{bmatrix}, \frac{1}{\sqrt{2}} \begin{bmatrix} 0 & 1 \\ 1 & 0 \end{bmatrix}, \frac{1}{\sqrt{2}} \begin{bmatrix} 0 & -i \\ i & 0 \end{bmatrix}, \frac{1}{\sqrt{2}} \begin{bmatrix} 1 & 0 \\ 0 & -1 \end{bmatrix} \right\}$$

and $A = \begin{bmatrix} 5 & 2-3i \\ 2+3i & -3 \end{bmatrix}$.

Solution:

We already showed that B is an orthonormal basis of $\mathcal{M}_2(\mathbb{C})$ (with respect to the Frobenius inner product) in Example 1.4.3, so we can compute $[A]_B$ via Theorem 1.4.5. In particular,

$$\left\langle \frac{1}{\sqrt{2}} \begin{bmatrix} 1 & 0 \\ 0 & 1 \end{bmatrix}, A \right\rangle = \operatorname{tr}\left(\frac{1}{\sqrt{2}} \begin{bmatrix} 5 & 2-3i \\ 2+3i & -3 \end{bmatrix} \right) = \frac{2}{\sqrt{2}} = \sqrt{2},$$

$$\left\langle \frac{1}{\sqrt{2}} \begin{bmatrix} 0 & 1 \\ 1 & 0 \end{bmatrix}, A \right\rangle = \operatorname{tr}\left(\frac{1}{\sqrt{2}} \begin{bmatrix} 2+3i & -3 \\ 5 & 2-3i \end{bmatrix} \right) = \frac{4}{\sqrt{2}} = 2\sqrt{2},$$

$$\left\langle \frac{1}{\sqrt{2}} \begin{bmatrix} 0 & -i \\ i & 0 \end{bmatrix}, A \right\rangle = \operatorname{tr}\left(\frac{1}{\sqrt{2}} \begin{bmatrix} -2i+3 & 3i \\ 5i & 2i+3 \end{bmatrix} \right) = \frac{6}{\sqrt{2}} = 3\sqrt{2},$$

$$\left\langle \frac{1}{\sqrt{2}} \begin{bmatrix} 1 & 0 \\ 0 & -1 \end{bmatrix}, A \right\rangle = \operatorname{tr}\left(\frac{1}{\sqrt{2}} \begin{bmatrix} 5 & 2-3i \\ -2-3i & 3 \end{bmatrix} \right) = \frac{8}{\sqrt{2}} = 4\sqrt{2}.$$

> Recall that the Frobenius inner product is $\langle X, Y \rangle = \operatorname{tr}(X^*Y)$.

We thus conclude that $[A]_B = \sqrt{2}(1,2,3,4)$. We can verify our work by simply checking that it is indeed the case that

$$A = \begin{bmatrix} 5 & 2-3i \\ 2+3i & -3 \end{bmatrix} = \begin{bmatrix} 1 & 0 \\ 0 & 1 \end{bmatrix} + 2\begin{bmatrix} 0 & 1 \\ 1 & 0 \end{bmatrix} + 3\begin{bmatrix} 0 & -i \\ i & 0 \end{bmatrix} + 4\begin{bmatrix} 1 & 0 \\ 0 & -1 \end{bmatrix}.$$

> The fact that $[A]_B$ is real follows from A being Hermitian, since B is not just a basis of $\mathcal{M}_2(\mathbb{C})$ but also of the real vector space $\mathcal{M}_2^H$ of 2×2 Hermitian matrices.

Now that we have demonstrated how to determine whether or not a particular set *is* an orthonormal basis, and a little bit of what we can *do* with orthonormal bases, we turn to the question of how to *construct* an orthonormal basis. While this is reasonably intuitive in familiar inner product spaces like $\mathbb{R}^n$ or $\mathcal{M}_{m,n}$ (in both cases, we can just choose the standard basis), it becomes a bit more delicate when working in weirder vector spaces or with stranger inner products like the one on $\mathcal{P}^2[-1,1]$ from Example 1.4.1:

$$\langle f, g \rangle = \int_{-1}^{1} \frac{f(x)g(x)}{\sqrt{1-x^2}} \, dx.$$

Fortunately, there is indeed a standard method of turning any basis of a finite-dimensional vector space into an orthonormal one. Before stating the result in full generality, we illustrate how it works in $\mathbb{R}^2$.

Indeed, suppose that we have a (not necessarily orthonormal) basis $B = \{\mathbf{v}_1, \mathbf{v}_2\}$ of $\mathbb{R}^2$ that we want to turn into an orthonormal basis $C = \{\mathbf{u}_1, \mathbf{u}_2\}$. We start by simply defining $\mathbf{u}_1 = \mathbf{v}_1 / \|\mathbf{v}_1\|$, which will be the first vector in our orthonormal basis (after all, we want each vector in the basis to be normalized). To construct the next member of the orthonormal basis, we define

$$\mathbf{w}_2 = \mathbf{v}_2 - (\mathbf{u}_1 \cdot \mathbf{v}_2)\mathbf{u}_1, \quad \text{and then} \quad \mathbf{u}_2 = \mathbf{w}_2 / \|\mathbf{w}_2\|,$$

where we recall that $\mathbf{u}_1 \cdot \mathbf{v}_2$ measures how far $\mathbf{v}_2$ points in the direction $\mathbf{u}_1$. In words, $\mathbf{w}_2$ is the same as $\mathbf{v}_2$, but with the portion of $\mathbf{v}_2$ that points in the direction of $\mathbf{u}_1$ removed, leaving behind only the piece of it that is orthogonal to $\mathbf{u}_1$. The division by its length is just done so that the resulting vector $\mathbf{u}_2$ has length 1 (since we want an ortho*normal* basis, not just an orthogonal one). This construction of $\mathbf{u}_1$ and $\mathbf{u}_2$ from $\mathbf{v}_1$ and $\mathbf{v}_2$ is illustrated in Figure 1.13.

In higher dimensions, we would then continue in this way, adjusting each vector in the basis so that it is orthogonal to each of the previous vectors, and then normalizing it. The following theorem makes this precise and tells us that the result is indeed always an orthonormal basis, regardless of what vector space and inner product is being used.

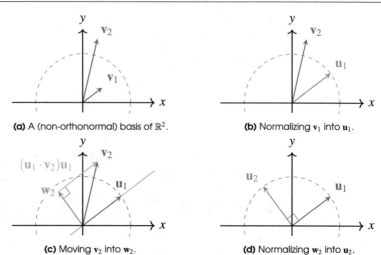

Figure 1.13: An illustration of our method for turning any basis of $\mathbb{R}^2$ into an orthonormal basis of $\mathbb{R}^2$. The process works by (b) normalizing one of the vectors, (c) moving the other vector so that they are orthogonal, and then (d) normalizing the second vector. In higher dimensions, the process continues in the same way by repositioning the vectors one at a time so that they are orthogonal to the rest, and then normalizing.

Theorem 1.4.6

The Gram–Schmidt Process

If $j = 1$ then the summation that defines $\mathbf{w}_j$ is empty and thus equals $\mathbf{0}$, so $\mathbf{w}_1 = \mathbf{v}_1$ and $\mathbf{u}_1 = \mathbf{v}_1/\|\mathbf{v}_1\|$.

Suppose $B = \{\mathbf{v}_1, \mathbf{v}_2, \ldots, \mathbf{v}_n\}$ is a linearly independent set in an inner product space $\mathcal{V}$. Define

$$\mathbf{w}_j = \mathbf{v}_j - \sum_{i=1}^{j-1} \langle \mathbf{u}_i, \mathbf{v}_j \rangle \mathbf{u}_i \quad \text{and} \quad \mathbf{u}_j = \frac{\mathbf{w}_j}{\|\mathbf{w}_j\|} \quad \text{for all} \quad j = 1, \ldots, n.$$

Then $C_j = \{\mathbf{u}_1, \mathbf{u}_2, \ldots, \mathbf{u}_j\}$ is an orthonormal basis of $\text{span}(\mathbf{v}_1, \mathbf{v}_2, \ldots, \mathbf{v}_j)$ for each $1 \leq j \leq n$. In particular, if B is a basis of $\mathcal{V}$ then C_n is an orthonormal basis of $\mathcal{V}$.

We emphasize that even though the formulas involved in the Gram–Schmidt process look ugly at first, each part of the formulas has a straightforward purpose. We want $\mathbf{w}_j$ to be orthogonal to $\mathbf{u}_i$ for each $1 \leq i < j$, so that is why we subtract off each $\langle \mathbf{u}_i, \mathbf{v}_j \rangle \mathbf{u}_i$. Similarly, we want each $\mathbf{u}_j$ to be a unit vector, which is why divide $\mathbf{w}_j$ by its norm.

Proof of Theorem 1.4.6. We prove this result by induction on j. For the base $j = 1$ case, we simply note that $\mathbf{u}_1$ is indeed a unit vector and $\text{span}(\mathbf{u}_1) = \text{span}(\mathbf{v}_1)$ since $\mathbf{u}_1$ and $\mathbf{v}_1$ are scalar multiples of each other.

For the inductive step, suppose that for some particular j we know that $\{\mathbf{u}_1, \mathbf{u}_2, \ldots, \mathbf{u}_j\}$ is a mutually orthogonal set of unit vectors and

$$\text{span}(\mathbf{u}_1, \mathbf{u}_2, \ldots, \mathbf{u}_j) = \text{span}(\mathbf{v}_1, \mathbf{v}_2, \ldots, \mathbf{v}_j). \tag{1.4.2}$$

We know that $\mathbf{v}_{j+1} \notin \text{span}(\mathbf{v}_1, \mathbf{v}_2, \ldots, \mathbf{v}_j)$, since B is linearly independent. It follows that $\mathbf{v}_{j+1} \notin \text{span}(\mathbf{u}_1, \mathbf{u}_2, \ldots, \mathbf{u}_j)$ as well, so the definition of $\mathbf{u}_{j+1}$ makes sense (i.e., $\mathbf{w}_{j+1} = \mathbf{v}_{j+1} - \sum_{i=1}^{j} \langle \mathbf{u}_i, \mathbf{v}_{j+1} \rangle \mathbf{u}_i \neq \mathbf{0}$, so we are not dividing by 0) and is a unit vector.

To see that $\mathbf{u}_{j+1}$ is orthogonal to each of $\mathbf{u}_1, \mathbf{u}_2, \ldots, \mathbf{u}_j$, suppose that

$1 \leq k \leq j$ and compute

$$\langle \mathbf{u}_k, \mathbf{u}_{j+1} \rangle = \left\langle \mathbf{u}_k, \frac{\mathbf{v}_{j+1} - \sum_{i=1}^{j} \langle \mathbf{u}_i, \mathbf{v}_{j+1} \rangle \mathbf{u}_i}{\left\| \mathbf{v}_{j+1} - \sum_{i=1}^{j} \langle \mathbf{u}_i, \mathbf{v}_{j+1} \rangle \mathbf{u}_i \right\|} \right\rangle \qquad \text{(definition of } \mathbf{u}_{j+1})$$

$$= \frac{\langle \mathbf{u}_k, \mathbf{v}_{j+1} \rangle - \sum_{i=1}^{j} \langle \mathbf{u}_i, \mathbf{v}_{j+1} \rangle \langle \mathbf{u}_k, \mathbf{u}_i \rangle}{\left\| \mathbf{v}_{j+1} - \sum_{i=1}^{j} \langle \mathbf{u}_i, \mathbf{v}_{j+1} \rangle \mathbf{u}_i \right\|} \qquad \text{(expand the inner product)}$$

$$= \frac{\langle \mathbf{u}_k, \mathbf{v}_{j+1} \rangle - \langle \mathbf{u}_k, \mathbf{v}_{j+1} \rangle}{\left\| \mathbf{v}_{j+1} - \sum_{i=1}^{j} \langle \mathbf{u}_i, \mathbf{v}_{j+1} \rangle \mathbf{u}_i \right\|} \qquad (k \leq j, \text{ so } \langle \mathbf{u}_k, \mathbf{u}_i \rangle = 0)$$

$$= 0. \qquad (\langle \mathbf{u}_k, \mathbf{v}_{j+1} \rangle - \langle \mathbf{u}_k, \mathbf{v}_{j+1} \rangle = 0)$$

All that remains is to show that

$$\text{span}(\mathbf{u}_1, \mathbf{u}_2, \ldots, \mathbf{u}_{j+1}) = \text{span}(\mathbf{v}_1, \mathbf{v}_2, \ldots, \mathbf{v}_{j+1}).$$

The fact that the only $(j+1)$-dimensional subspace of a $(j+1)$-dimensional vector space is that vector space itself is hopefully intuitive enough, but it was proved explicitly in Exercise 1.2.31.

By rearranging the definition of $\mathbf{u}_{j+1}$, we see that $\mathbf{v}_{j+1} \in \text{span}(\mathbf{u}_1, \mathbf{u}_2, \ldots, \mathbf{u}_{j+1})$. When we combine this fact with Equation (1.4.2), this implies

$$\text{span}(\mathbf{u}_1, \mathbf{u}_2, \ldots, \mathbf{u}_{j+1}) \supseteq \text{span}(\mathbf{v}_1, \mathbf{v}_2, \ldots, \mathbf{v}_{j+1}).$$

The $\mathbf{v}_i$'s are linearly independent, so the span on the right has dimension $j+1$. Similarly, the $\mathbf{u}_i$'s are linearly independent (they are mutually orthogonal, so linear independence follows from Theorem 1.4.1), so the span on the left also has dimension $j+1$, and thus the two spans must in fact be equal. ∎

Since finite-dimensional inner product spaces (by definition) have a basis consisting of finitely many vectors, and Theorem 1.4.6 tells us how to convert any such basis into one that is orthonormal, we now know that every finite-dimensional inner product space has an orthonormal basis:

Corollary 1.4.7

Existence of Orthonormal Bases

Every finite-dimensional inner product space has an orthonormal basis.

We now illustrate how to use the Gram–Schmidt process to find orthonormal bases of various inner product spaces.

Example 1.4.6

Finding an Orthonormal Basis with Respect to a Weird Inner Product

Use the Gram–Schmidt process to construct an orthonormal basis of $\mathbb{R}^2$ with respect to the weird inner product that we introduced back in Example 1.3.18:

$$\langle \mathbf{v}, \mathbf{w} \rangle = v_1 w_1 + 2v_1 w_2 + 2v_2 w_1 + 5v_2 w_2.$$

Solution:

We could also choose any other basis of $\mathbb{R}^2$ as our starting point.

The starting point of the Gram–Schmidt process is a basis of the given inner product space, so we start with the standard basis $B = \{\mathbf{e}_1, \mathbf{e}_2\}$ of $\mathbb{R}^2$. To turn this into an orthonormal (with respect to the weird inner product above) basis, we define

$$\mathbf{u}_1 = \frac{\mathbf{e}_1}{\|\mathbf{e}_1\|} = \mathbf{e}_1, \quad \text{since} \quad \|\mathbf{e}_1\| = \sqrt{\langle \mathbf{e}_1, \mathbf{e}_1 \rangle} = \sqrt{1+0+0+0} = 1.$$

Be careful when computing things like $\langle \mathbf{u}_1, \mathbf{e}_2 \rangle = \langle \mathbf{e}_1, \mathbf{e}_2 \rangle$. It is tempting to think that it equals $\mathbf{e}_1 \cdot \mathbf{e}_2 = 0$, but $\langle \mathbf{u}_1, \mathbf{e}_2 \rangle$ refers to the weird inner product, not the dot product.

Next, we let

$$\mathbf{w}_2 = \mathbf{e}_2 - \langle \mathbf{u}_1, \mathbf{e}_2 \rangle \mathbf{u}_1 = (0,1) - (0+2+0+0)(1,0) = (-2,1),$$

$$\mathbf{u}_2 = \mathbf{w}_2/\|\mathbf{w}_2\| = (-2,1), \quad \text{since} \quad \|\mathbf{w}_2\| = \sqrt{4-4-4+5} = 1.$$

It follows that $C = \{\mathbf{u}_1, \mathbf{u}_2\} = \{(1,0),(-2,1)\}$ is an orthonormal basis of $\mathbb{R}^2$ with respect to this weird inner product (in fact, this is exactly the orthonormal basis that we saw back in Figure 1.11).

Example 1.4.7

Finding an Orthonormal Basis of a Plane

Find an orthonormal basis (with respect to the usual dot product) of the plane $S \subset \mathbb{R}^3$ with equation $x - y - 2z = 0$.

Solution:

We start by picking any basis of S. Since S is 2-dimensional, a basis is made up of any two vectors in S that are not multiples of each other. By inspection, $\mathbf{v}_1 = (2,0,1)$ and $\mathbf{v}_2 = (3,1,1)$ are vectors that work, so we choose $B = \{\mathbf{v}_1, \mathbf{v}_2\}$.

$\mathbf{v}_1$ and $\mathbf{v}_2$ can be found by choosing x and y arbitrarily and using the equation $x - y - 2z = 0$ to solve for z.

To create an orthonormal basis from B, we apply the Gram–Schmidt process—we define

$$\mathbf{u}_1 = \frac{\mathbf{v}_1}{\|\mathbf{v}_1\|} = \frac{1}{\sqrt{5}}(2,0,1)$$

and

$$\mathbf{w}_2 = \mathbf{v}_2 - (\mathbf{u}_1 \cdot \mathbf{v}_2)\mathbf{u}_1 = (3,1,1) - \tfrac{7}{5}(2,0,1) = \frac{1}{5}(1,5,-2),$$

$$\mathbf{u}_2 = \frac{\mathbf{w}_2}{\|\mathbf{w}_2\|} = \frac{1}{\sqrt{30}}(1,5,-2).$$

Just like other bases, orthonormal bases are very non-unique. There are many other orthonormal bases of S.

It follows that $C = \{\mathbf{u}_1, \mathbf{u}_2\} = \left\{ \frac{1}{\sqrt{5}}(2,0,1), \frac{1}{\sqrt{30}}(1,5,-2) \right\}$ is an orthonormal basis of V, as displayed below:

$$B = \{\mathbf{v}_1, \mathbf{v}_2\} \qquad\qquad C = \{\mathbf{u}_1, \mathbf{u}_2\}$$
$$= \{(2,0,1),(3,1,1)\} \qquad\qquad = \left\{ \frac{1}{\sqrt{5}}(2,0,1), \frac{1}{\sqrt{30}}(1,5,-2) \right\}$$

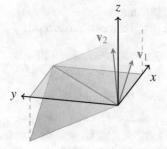
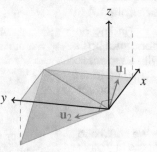

Example 1.4.8

Finding an Orthonormal Basis of Polynomials

Recall that the standard basis is *not* orthonormal in this inner product since, for example, $\langle 1, x \rangle = 1/2$.

Notice that $\{h_1, h_2\}$ is exactly the orthonormal basis of $\mathcal{P}^1[0,1]$ that we constructed back in Example 1.4.4. We were doing the Gram–Schmidt process back there without realizing it.

However, we should not expect to be able to directly "see" whether or not a basis of $\mathcal{P}^2[0,1]$ (or any of its variants) is orthonormal.

Find an orthonormal basis of $\mathcal{P}^2[0,1]$ with respect to the inner product

$$\langle f, g \rangle = \int_0^1 f(x)g(x)\, dx.$$

Solution:

Once again, we apply the Gram–Schmidt process to the standard basis $B = \{1, x, x^2\}$ to create an orthonormal basis $C = \{h_1, h_2, h_3\}$. To start, we define $h_1(x) = 1/\|1\| = 1$. The next member of the orthonormal basis is computed via

$$g_2(x) = x - \langle h_1, x \rangle h_1(x) = x - \langle 1, x \rangle 1 = x - 1/2,$$

$$h_2(x) = g_2(x)/\|g_2\| = (x - 1/2)\Big/\sqrt{\int_0^1 (x - 1/2)^2\, dx} = \sqrt{3}(2x - 1).$$

The last member of the orthonormal basis C is similarly computed via

$$\begin{aligned}
g_3(x) &= x^2 - \langle h_1, x^2 \rangle h_1(x) - \langle h_2, x^2 \rangle h_2(x) \\
&= x^2 - \langle 1, x^2 \rangle 1 - 12 \langle x - 1/2, x^2 \rangle (x - 1/2) \\
&= x^2 - 1/3 - (x - 1/2) = x^2 - x + 1/6, \quad \text{and}
\end{aligned}$$

$$\begin{aligned}
h_3(x) &= g_3(x)/\|g_3\| \\
&= (x^2 - x + 1/6)\Big/\sqrt{\int_0^1 (x^2 - x + 1/6)^2\, dx} \\
&= \sqrt{5}(6x^2 - 6x + 1).
\end{aligned}$$

It follows that $C = \big\{1, \sqrt{3}(2x - 1), \sqrt{5}(6x^2 - 6x + 1)\big\}$ is an orthonormal basis of $\mathcal{P}^2[0,1]$. While this basis looks a fair bit uglier than the standard basis $\{1, x, x^2\}$ algebraically, its members are more symmetric about the midpoint $x = 1/2$ and more evenly distributed across the interval $[0,1]$ geometrically, as shown below:

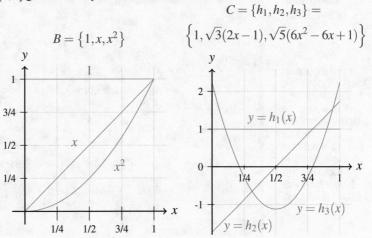

$$B = \{1, x, x^2\} \qquad\qquad C = \{h_1, h_2, h_3\} = \big\{1, \sqrt{3}(2x - 1), \sqrt{5}(6x^2 - 6x + 1)\big\}$$

1.4.2   Adjoint Transformations

There is one final operation that is used extensively in introductory linear algebra that we have not yet generalized to vector spaces beyond $\mathbb{F}^n$, and that is the (conjugate) transpose of a matrix. We now fill in this gap by introducing the adjoint of a linear transformation, which can be thought of as finally answering the question of why the transpose is an important operation (after all, why would we expect that swapping the rows and columns of a matrix should tell us anything useful?).

Definition 1.4.3	Suppose that $\mathcal{V}$ and $\mathcal{W}$ are inner product spaces and $T : \mathcal{V} \to \mathcal{W}$ is a linear transformation. Then a linear transformation $T^* : \mathcal{W} \to \mathcal{V}$ is called the **adjoint** of T if
Adjoint	
Transformation	$$\langle T(\mathbf{v}), \mathbf{w} \rangle = \langle \mathbf{v}, T^*(\mathbf{w}) \rangle \quad \text{for all} \quad \mathbf{v} \in \mathcal{V}, \ \mathbf{w} \in \mathcal{W}.$$

In fact, this finally explains why the *conjugate* transpose is typically the "right" version of the transpose for complex matrices.

For example, it is straightforward (but slightly tedious and unenlightening, so we leave it to Exercise 1.4.17) to show that real matrices $A \in \mathcal{M}_{m,n}(\mathbb{R})$ satisfy

$$(A\mathbf{v}) \cdot \mathbf{w} = \mathbf{v} \cdot (A^T \mathbf{w}) \quad \text{for all} \quad \mathbf{v} \in \mathbb{R}^n, \ \mathbf{w} \in \mathbb{R}^m,$$

so the transposed matrix A^T is the adjoint of A. Similarly, for every complex matrix $A \in \mathcal{M}_{m,n}(\mathbb{C})$ we have

$$(A\mathbf{v}) \cdot \mathbf{w} = \mathbf{v} \cdot (A^* \mathbf{w}) \quad \text{for all} \quad \mathbf{v} \in \mathbb{C}^n, \ \mathbf{w} \in \mathbb{C}^m,$$

The ground field must be $\mathbb{R}$ or $\mathbb{C}$ in order for inner products, and thus adjoints, to make sense.

so the conjugate transpose matrix A^* is the adjoint of A. However, it also makes sense to talk about the adjoint of linear transformations between more exotic vector spaces, as we now demonstrate with the trace (which we recall from Example 1.2.7 is the linear transformation $\text{tr} : \mathcal{M}_n(\mathbb{F}) \to \mathbb{F}$ that adds up the diagonal entries of a matrix).

Example 1.4.9	Show that the adjoint of the trace $\text{tr} : \mathcal{M}_n(\mathbb{F}) \to \mathbb{F}$ with respect to the standard Frobenius inner product is given by
The Adjoint	
of the Trace	$$\text{tr}^*(c) = cI \quad \text{for all} \quad c \in \mathbb{F}.$$

Solution:

In the equation $\langle c, \text{tr}(A) \rangle = \langle cI, A \rangle$, the left inner product is on $\mathbb{F}$ (i.e., it is the 1-dimensional dot product $\langle c, \text{tr}(A) \rangle = \overline{c}\,\text{tr}(A)$) and the right inner product is the Frobenius inner product on $\mathcal{M}_n(\mathbb{F})$.

Our goal is to show that $\langle c, \text{tr}(A) \rangle = \langle cI, A \rangle$ for all $A \in \mathcal{M}_n(\mathbb{F})$ and all $c \in \mathbb{F}$. Recall that the Frobenius inner product is defined by $\langle A, B \rangle = \text{tr}(A^* B)$, so this condition is equivalent to

$$\overline{c}\,\text{tr}(A) = \text{tr}\big((cI)^* A\big) \quad \text{for all} \quad A \in \mathcal{M}_n(\mathbb{F}), \ c \in \mathbb{F}.$$

This equation holds simply by virtue of linearity of the trace:

$$\text{tr}\big((cI)^* A\big) = \text{tr}\big(\overline{c}I^* A\big) = \overline{c}\,\text{tr}(IA) = \overline{c}\,\text{tr}(A),$$

as desired.

The previous example is somewhat unsatisfying for two reasons. First, it does not illustrate how to actually find the adjoint of a linear transformation, but rather it only shows how to verify that two linear transformations are adjoints of each other. How could have we *found* the adjoint $\text{tr}^*(c) = cI$ if it were not given to us? Second, how do we know that there is not another linear transformation that is also an adjoint of the trace? That is, how do we know that $\text{tr}^*(c) = cI$ is *the* adjoint of the trace rather than just *an* adjoint of it?

The following theorem answers both of these questions by showing that, in finite dimensions, every linear transformation has exactly one adjoint, and it can be computed by making use of orthonormal bases of the two vector spaces V and W.

Theorem 1.4.8

Existence and Uniqueness of the Adjoint

Suppose that V and W are finite-dimensional inner product spaces with orthonormal bases B and C, respectively. If $T : V \to W$ is a linear transformation then there exists a unique adjoint transformation $T^* : W \to V$, and its standard matrix satisfies

$$\left[T^*\right]_{B \leftarrow C} = [T]^*_{C \leftarrow B}.$$

In infinite dimensions, some linear transformations fail to have an adjoint (see Remark 1.4.1). However, if it exists then it is still unique (see Exercise 1.4.23).

Proof. To prove uniqueness of T^*, suppose that T^* exists, let $\mathbf{v} \in V$ and $\mathbf{w} \in W$, and compute $\langle T(\mathbf{v}), \mathbf{w} \rangle$ in two different ways:

$$
\begin{aligned}
\langle T(\mathbf{v}), \mathbf{w} \rangle &= \langle \mathbf{v}, T^*(\mathbf{w}) \rangle && \text{(definition of } T^*\text{)} \\
&= [\mathbf{v}]_B \cdot \left[T^*(\mathbf{w})\right]_B && \text{(Theorem 1.4.3)} \\
&= [\mathbf{v}]_B \cdot \left(\left[T^*\right]_{B \leftarrow C}[\mathbf{w}]_C\right) && \text{(definition of standard matrix)} \\
&= [\mathbf{v}]_B^* \left[T^*\right]_{B \leftarrow C}[\mathbf{w}]_C. && \text{(definition of dot product)}
\end{aligned}
$$

Similarly,

$$
\begin{aligned}
\langle T(\mathbf{v}), \mathbf{w} \rangle &= [T(\mathbf{v})]_C \cdot [\mathbf{w}]_C && \text{(Theorem 1.4.3)} \\
&= \left([T]_{C \leftarrow B}[\mathbf{v}]_B\right) \cdot [\mathbf{w}]_C && \text{(definition of standard matrix)} \\
&= [\mathbf{v}]_B^* [T]^*_{C \leftarrow B}[\mathbf{w}]_C. && \text{(definition of dot product)}
\end{aligned}
$$

It follows that $[\mathbf{v}]_B^* \left[T^*\right]_{B \leftarrow C}[\mathbf{w}]_C = [\mathbf{v}]_B^* [T]^*_{C \leftarrow B}[\mathbf{w}]_C$ for all $[\mathbf{v}]_B \in \mathbb{F}^n$ and all $[\mathbf{w}]_C \in \mathbb{F}^m$ (where $\mathbb{F}$ is the ground field). If we choose $\mathbf{v}$ to be the i-th vector in the basis B and $\mathbf{w}$ to be the j-th vector in C, then $[\mathbf{v}]_B = \mathbf{e}_i$ and $[\mathbf{w}]_C = \mathbf{e}_j$, so

$$[\mathbf{v}]_B^* \left[T^*\right]_{B \leftarrow C}[\mathbf{w}]_C = \mathbf{e}_i^T \left[T^*\right]_{B \leftarrow C}\mathbf{e}_j \quad \text{is the } (i,j)\text{-entry of } \left[T^*\right]_{B \leftarrow C}, \quad \text{and}$$

$$[\mathbf{v}]_B^* [T]^*_{C \leftarrow B}[\mathbf{w}]_C = \mathbf{e}_i^T [T]^*_{C \leftarrow B}\mathbf{e}_j \quad \text{is the } (i,j)\text{-entry of } [T]^*_{C \leftarrow B}.$$

Since these quantities are equal for all i and j, it follows that $\left[T^*\right]_{B \leftarrow C} = [T]^*_{C \leftarrow B}$. Uniqueness of T^* now follows immediately from uniqueness of standard matrices.

Existence of T^* follows from that fact that we can choose T^* to be the linear transformation with standard matrix $[T]^*_{C \leftarrow B}$ and then follow the above argument backward to verify that $\langle T(\mathbf{v}), \mathbf{w} \rangle = \langle \mathbf{v}, T^*(\mathbf{w}) \rangle$ for all $\mathbf{v} \in V$ and $\mathbf{w} \in W$. ∎

It is worth emphasizing that the final claim of Theorem 1.4.8 does *not* necessarily hold if B or C are not orthonormal. For example, the standard matrix of the differentiation map $D : \mathcal{P}^2 \to \mathcal{P}^2$ with respect to the standard basis $E = \{1, x, x^2\}$ is

$$[D]_E = \begin{bmatrix} 0 & 1 & 0 \\ 0 & 0 & 2 \\ 0 & 0 & 0 \end{bmatrix}.$$

We thus might expect that if we equip $\mathcal{P}^2$ with the standard inner product

$$\langle f, g \rangle = \int_0^1 f(x)g(x)\, dx$$

then we would have

$$[D^*]_E = [D]_E^* = \begin{bmatrix} 0 & 0 & 0 \\ 1 & 0 & 0 \\ 0 & 2 & 0 \end{bmatrix}.$$

This formula for D^* is also incorrect even if we use pretty much any other natural inner product on $\mathcal{P}^2$.

which would give $D^*(ax^2 + bx + c) = cx + 2bx^2$. However, it is straightforward to check that formula for D^* is incorrect since, for example,

$$\langle 1, D(x) \rangle = \int_0^1 D(x)\, dx = \int_0^1 1\, dx = 1, \quad \text{but}$$

$$\langle D^*(1), x \rangle = \int_0^1 D^*(1)x\, dx = \int_0^1 x^2\, dx = 1/3.$$

To fix the above problem and find the *actual* adjoint of D, we must mimic the above calculation with an orthonormal basis of $\mathcal{P}^2$ rather than the standard basis $E = \{1, x, x^2\}$.

Example 1.4.10

The Adjoint of the Derivative

Compute the adjoint of the differentiation map $D : \mathcal{P}^2[0,1] \to \mathcal{P}^2[0,1]$ (with respect to the standard inner product).

Solution:

Fortunately, we already computed an orthonormal basis of $\mathcal{P}^2[0,1]$ back in Example 1.4.8, and it is $C = \{h_1, h_2, h_3\}$, where

$$h_1(x) = 1, \quad h_2(x) = \sqrt{3}(2x - 1), \quad \text{and} \quad h_3(x) = \sqrt{5}(6x^2 - 6x + 1).$$

To find the standard matrix of D with respect to this basis, we compute

$$D(h_1(x)) = 0,$$
$$D(h_2(x)) = 2\sqrt{3} = 2\sqrt{3}h_1(x), \quad \text{and}$$
$$D(h_3(x)) = 12\sqrt{5}x - 6\sqrt{5} = 2\sqrt{15}h_2(x).$$

Then the standard matrices of D and D^* are given by

$$[D]_C = \begin{bmatrix} 0 & 2\sqrt{3} & 0 \\ 0 & 0 & 2\sqrt{15} \\ 0 & 0 & 0 \end{bmatrix} \quad \text{and} \quad [D^*]_C = \begin{bmatrix} 0 & 0 & 0 \\ 2\sqrt{3} & 0 & 0 \\ 0 & 2\sqrt{15} & 0 \end{bmatrix},$$

so D^* is the linear transformation with the property that $D^*(h_1) = 2\sqrt{3}h_2$, $D^*(h_2) = 2\sqrt{15}h_3$, and $D^*(h_3) = 0$.

This is a fine and dandy answer already, but it can perhaps be made a bit more intuitive if we instead describe D^* in terms of what it does to 1,

x, and x^2, rather than h_1, h_2, and h_3. That is, we compute

$$D^*(1) = D^*(h_1(x)) = 2\sqrt{3}h_2(x) = 6(2x-1),$$
$$D^*(x) = D^*\left(\frac{1}{2\sqrt{3}}h_2(x) + \frac{1}{2}h_1(x)\right)$$
$$= \sqrt{5}h_3(x) + \sqrt{3}h_2(x)$$
$$= 2(15x^2 - 12x + 1), \quad \text{and}$$
$$D^*(x^2) = D^*\left(\frac{1}{6\sqrt{5}}h_3(x) + \frac{1}{2\sqrt{3}}h_2(x) + \frac{1}{3}h_1(x)\right)$$
$$= \sqrt{5}h_3(x) + \frac{2}{\sqrt{3}}h_2(x)$$
$$= 6x^2 - 2x - 1.$$

Putting this all together shows that

$$D^*(ax^2 + bx + c) = 6(a+5b)x^2 - 2(a+12b-6c)x - (a-2b+6c).$$

Example 1.4.11

The Adjoint of the Transpose

Show that the adjoint of the transposition map $T : \mathcal{M}_{m,n} \to \mathcal{M}_{n,m}$, with respect to the Frobenius inner product, is also the transposition map.

Solution:

Our goal is to show that $\langle A^T, B \rangle = \langle A, B^T \rangle$ for all $A \in \mathcal{M}_{m,n}$ and $B \in \mathcal{M}_{n,m}$. Recall that the Frobenius inner product is defined by $\langle A, B \rangle = \operatorname{tr}(A^*B)$, so this is equivalent to

$$\operatorname{tr}(\overline{A}B) = \operatorname{tr}(A^*B^T) \quad \text{for all} \quad A \in \mathcal{M}_{m,n},\ B \in \mathcal{M}_{n,m}.$$

These two quantities can be shown to be equal by brute-force calculation of the traces and matrix multiplications in terms of the entries of A and B, but a more elegant way is to use properties of the trace and transpose:

$$\begin{aligned}
\operatorname{tr}(\overline{A}B) &= \operatorname{tr}\left((\overline{A}B)^T\right) && \text{(transpose does not change trace)} \\
&= \operatorname{tr}(B^T A^*) && \text{(transpose of a product)} \\
&= \operatorname{tr}(A^* B^T). && \text{(cyclic commutativity of trace)}
\end{aligned}$$

The situation presented in Example 1.4.11, where a linear transformation is its own adjoint, is important enough that we give it a name:

Definition 1.4.4

Self-Adjoint Transformations

If $\mathcal{V}$ is an inner product space then a linear transformation $T : \mathcal{V} \to \mathcal{V}$ is called **self-adjoint** if $T^* = T$.

For example, a matrix in $\mathcal{M}_n(\mathbb{R})$ is symmetric (i.e., $A = A^T$) if and only if it is a self-adjoint linear transformation on $\mathbb{R}^n$, and a matrix in $\mathcal{M}_n(\mathbb{C})$ is Hermitian (i.e., $A = A^*$) if and only if it is a self-adjoint linear transformation on $\mathbb{C}^n$. Slightly more generally, Theorem 1.4.8 tells us that a linear transformation on a finite-dimensional inner product space $\mathcal{V}$ is self-adjoint if and only if its standard matrix (with respect to some orthonormal basis of $\mathcal{V}$) is symmetric or Hermitian (depending on whether the underlying field is $\mathbb{R}$ or $\mathbb{C}$).

The fact that the transposition map on $\mathcal{M}_n$ is self-adjoint tells us that its standard matrix (with respect to an orthonormal basis, like the standard basis) is symmetric. In light of this, it is perhaps worthwhile looking back at

Example 1.2.10, where we explicitly computed its standard matrix in the $n = 2$ case (which is indeed symmetric).

<table>
<tr><td>

Remark 1.4.1

The Adjoint in Infinite Dimensions

</td><td>

The reason that Theorem 1.4.8 specifies that the vector spaces must be finite-dimensional is that some linear transformations acting on infinite-dimensional vector spaces do not have an adjoint. To demonstrate this phenomenon, consider the vector space c_{00} of all eventually-zero sequences of real numbers (which we first introduced in Example 1.1.10), together with inner product

</td></tr>
</table>

$$\langle (v_1, v_2, v_3, \ldots), (w_1, w_2, w_3, \ldots) \rangle = \sum_{i=1}^{\infty} v_i w_i.$$

Then consider the linear transformation $T : c_{00} \to c_{00}$ defined by

Since all of the sequences here are eventually zero, all of the sums considered here only have finitely many non-zero terms, so we do not need to worry about limits or convergence.

$$T\big((v_1, v_2, v_3, \ldots)\big) = \left(\sum_{i=1}^{\infty} v_i, \sum_{i=2}^{\infty} v_i, \sum_{i=3}^{\infty} v_i, \ldots \right).$$

A straightforward calculation reveals that, if the adjoint $T^* : c_{00} \to c_{00}$ exists, it must have the form

$$T^*(w_1, w_2, w_3, \ldots) = \left(\sum_{i=1}^{1} w_i, \sum_{i=1}^{2} w_i, \sum_{i=1}^{3} w_i, \ldots \right).$$

However, this is not actually a linear transformation on c_{00} since, for example, it would give

$$T^*(1, 0, 0, \ldots) = (1, 1, 1, \ldots),$$

which is not in c_{00} since its entries are not eventually 0.

1.4.3 Unitary Matrices

Properties of invertible matrices like this one are summarized in Theorem A.1.1.

Recall that invertible matrices are exactly the matrices whose columns form a basis of $\mathbb{R}^n$ (or $\mathbb{F}^n$ more generally). Now that we understand orthonormal bases and think of them as the "nicest" bases out there, it seems natural to ask what additional properties invertible matrices have if their columns form an orthonormal basis, rather than just any old basis. We now give a name to these matrices.

<table>
<tr><td>

Definition 1.4.5

Unitary Matrix

</td><td>

If $\mathbb{F} = \mathbb{R}$ or $\mathbb{F} = \mathbb{C}$ then a matrix $U \in \mathcal{M}_n(\mathbb{F})$ is called a **unitary matrix** if its columns form an orthonormal basis of $\mathbb{F}^n$ (with respect to the usual dot product).

</td></tr>
</table>

For example, the identity matrix is unitary since its columns are the standard basis vectors $\mathbf{e}_1, \mathbf{e}_2, \ldots, \mathbf{e}_n$, which form the standard basis of $\mathbb{F}^n$, which is orthonormal. As a slightly less trivial example, consider the matrix

$$U = \frac{1}{\sqrt{2}} \begin{bmatrix} 1 & -1 \\ 1 & 1 \end{bmatrix},$$

which we can show is unitary simply by noting that $\{(1,1)/\sqrt{2}, (1,-1)/\sqrt{2}\}$ (i.e., the set consisting of the columns of U) is an orthonormal basis of $\mathbb{R}^2$.

For a refresher on how to think of matrices geometrically as linear transformations, see Appendix A.1.2.

We can make geometric sense of unitary matrices if we recall that the columns of a matrix tell us where that matrix sends the standard basis vectors $\mathbf{e}_1, \mathbf{e}_2, \ldots, \mathbf{e}_n$. Thus, just like invertible matrices are those that send the unit square grid to a parallelogram grid (without squishing it down to a smaller dimension), unitary matrices are those that send the unit square grid to a (potentially rotated or reflected) unit square grid, as in Figure 1.14.

We will show shortly, in Examples 1.4.12 and 1.4.13, that all rotation matrices and all reflection matrices are indeed unitary.

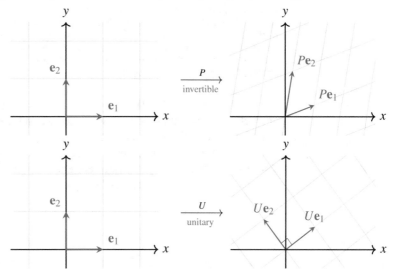

Figure 1.14: A non-zero matrix $P \in \mathcal{M}_2$ is invertible if and only if it sends the unit square grid to a parallelogram grid (whereas it is non-invertible if and only if it sends that grid to a line). A matrix $U \in \mathcal{M}_2$ is unitary if and only if it sends the unit square grid to a unit square grid that is potentially rotated and/or reflected, but not skewed.

For this reason, we often think of unitary matrices as the most "rigid" or "well-behaved" invertible matrices that exist—they preserve not just the dimension of $\mathbb{F}^n$, but also its shape (but maybe not its orientation). The following theorem provides several additional characterizations of unitary matrices that can help us understand them in other ways and perhaps make them a bit more intuitive.

Theorem 1.4.9

Characterization of Unitary Matrices

Suppose $\mathbb{F} = \mathbb{R}$ or $\mathbb{F} = \mathbb{C}$ and $U \in \mathcal{M}_n(\mathbb{F})$. The following are equivalent:

a) U is unitary,
b) U^* is unitary,
c) $UU^* = I$,
d) $U^*U = I$,
e) $(U\mathbf{v}) \cdot (U\mathbf{w}) = \mathbf{v} \cdot \mathbf{w}$ for all $\mathbf{v}, \mathbf{w} \in \mathbb{F}^n$, and
f) $\|U\mathbf{v}\| = \|\mathbf{v}\|$ for all $\mathbf{v} \in \mathbb{F}^n$.

In this theorem, $\|\cdot\|$ refers to the standard length in $\mathbb{F}^n$. We discuss how to generalize unitary matrices to other inner products, norms, and vector spaces in Section 1.D.3.

Before proving this result, it is worthwhile to think about what some of its characterizations really mean. Conditions (c) and (d) tell us that unitary matrices are not only invertible, but they are the matrices U for which their inverse equals their adjoint (i.e., $U^{-1} = U^*$). Algebraically, this is extremely convenient since it is trivial to compute the adjoint (i.e., conjugate transpose) of a matrix, so it is thus trivial to compute the inverse of a unitary matrix. The other properties of Theorem 1.4.9 can also be thought of as stronger versions of properties of invertible matrices, as summarized in Table 1.2.

Property of invertible P	Property of unitary U
P^{-1} exists	$U^{-1} = U^*$
$\|P\mathbf{v}\| \neq 0$ whenever $\|\mathbf{v}\| \neq 0$	$\|U\mathbf{v}\| = \|\mathbf{v}\|$ for all $\mathbf{v}$
columns of P are a basis	columns of U are an orthonormal basis
vec. space automorphism on $\mathbb{F}^n$	inner prod. space automorphism on $\mathbb{F}^n$

Table 1.2: A comparison of the properties of invertible matrices and the corresponding stronger properties of unitary matrices. The final properties (that invertible matrices are vector space automorphisms while unitary matrices are inner product space automorphisms) means that invertible matrices preserve linear combinations, whereas unitary matrices preserve linear combinations as well as the dot product (property (e) of Theorem 1.4.9).

The final two properties of Theorem 1.4.9 provide us with another natural geometric interpretation of unitary matrices. Condition (f) tells us that unitary matrices are exactly those that preserve the length of every vector. Similarly, since the dot product can be used to measure angles between vectors, condition (e) says that unitary matrices are exactly those that preserve the angle between every pair of vectors, as in Figure 1.15.

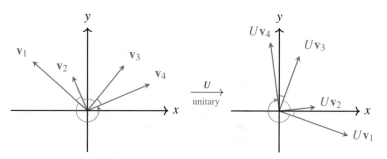

Figure 1.15: Unitary matrices are those that preserve the lengths of vectors as well as the angles between them.

To prove this theorem, we show that the 6 properties imply each other as follows:

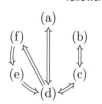

Proof of Theorem 1.4.9. We start by showing that conditions (a)–(d) are equivalent to each other. The equivalence of conditions (c) and (d) follows from the fact that, for square matrices, a one-sided inverse is necessarily a two-sided inverse.

To see that (a) is equivalent to (d), we write $U = \begin{bmatrix} \mathbf{u}_1 \mid \mathbf{u}_2 \mid \cdots \mid \mathbf{u}_n \end{bmatrix}$ and then use block matrix multiplication to multiply by U^*:

$$U^*U = \begin{bmatrix} \mathbf{u}_1^* \\ \hline \mathbf{u}_2^* \\ \hline \vdots \\ \hline \mathbf{u}_n^* \end{bmatrix} \begin{bmatrix} \mathbf{u}_1 \mid \mathbf{u}_2 \mid \cdots \mid \mathbf{u}_n \end{bmatrix} = \begin{bmatrix} \mathbf{u}_1 \cdot \mathbf{u}_1 & \mathbf{u}_1 \cdot \mathbf{u}_2 & \cdots & \mathbf{u}_1 \cdot \mathbf{u}_n \\ \mathbf{u}_2 \cdot \mathbf{u}_1 & \mathbf{u}_2 \cdot \mathbf{u}_2 & \cdots & \mathbf{u}_2 \cdot \mathbf{u}_n \\ \vdots & \vdots & \ddots & \vdots \\ \mathbf{u}_n \cdot \mathbf{u}_1 & \mathbf{u}_n \cdot \mathbf{u}_2 & \cdots & \mathbf{u}_n \cdot \mathbf{u}_n, \end{bmatrix}.$$

This product equals I if and only if its diagonal entries equal 1 and its off-diagonal entries equal 0. In other words, $U^*U = I$ if and only if $\mathbf{u}_i \cdot \mathbf{u}_i = 1$ for all i and $\mathbf{u}_i \cdot \mathbf{u}_j = 0$ whenever $i \neq j$. This says exactly that $\{\mathbf{u}_1, \mathbf{u}_2, \ldots, \mathbf{u}_n\}$ is a set of mutually orthogonal normalized vectors. Since it consists of exactly n

vectors, Theorem 1.4.2 tells us that this is equivalent to $\{\mathbf{u}_1, \mathbf{u}_2, \ldots, \mathbf{u}_n\}$ being an orthonormal basis of $\mathbb{F}^n$, which is exactly the definition of U being unitary.

The equivalence of (b) and (c) follows by applying the same argument as in the previous paragraph to U^* instead of U, so all that remains is to show that conditions (d), (e), and (f) are equivalent. We prove these equivalences by showing the chain of implications (d) $\implies$ (f) $\implies$ (e) $\implies$ (d).

To see that (d) implies (f), suppose $U^*U = I$. Then for all $\mathbf{v} \in \mathbb{F}^n$ we have

$$\|U\mathbf{v}\|^2 = (U\mathbf{v}) \cdot (U\mathbf{v}) = (U^*U\mathbf{v}) \cdot \mathbf{v} = \mathbf{v} \cdot \mathbf{v} = \|\mathbf{v}\|^2,$$

as desired.

The implication (f) $\implies$ (e) is the "tough one" of this proof.

For the implication (f) $\implies$ (e), note that if $\|U\mathbf{v}\|^2 = \|\mathbf{v}\|^2$ for all $\mathbf{v} \in \mathbb{F}^n$ then $(U\mathbf{v}) \cdot (U\mathbf{v}) = \mathbf{v} \cdot \mathbf{v}$ for all $\mathbf{v} \in \mathbb{F}^n$. If $\mathbf{x}, \mathbf{y} \in \mathbb{F}^n$ then this tells us (by choosing $\mathbf{v} = \mathbf{x} + \mathbf{y}$) that $\big(U(\mathbf{x}+\mathbf{y})\big) \cdot \big(U(\mathbf{x}+\mathbf{y})\big) = (\mathbf{x}+\mathbf{y}) \cdot (\mathbf{x}+\mathbf{y})$. Expanding this dot product on both the left and right then gives

Here we use the fact that $(U\mathbf{x}) \cdot (U\mathbf{y}) + (U\mathbf{y}) \cdot (U\mathbf{x}) = (U\mathbf{x}) \cdot (U\mathbf{y}) + \overline{(U\mathbf{x}) \cdot (U\mathbf{y})} = 2\mathrm{Re}\big((U\mathbf{x}) \cdot (U\mathbf{y})\big)$.

$$(U\mathbf{x}) \cdot (U\mathbf{x}) + 2\mathrm{Re}\big((U\mathbf{x}) \cdot (U\mathbf{y})\big) + (U\mathbf{y}) \cdot (U\mathbf{y}) = \mathbf{x} \cdot \mathbf{x} + 2\mathrm{Re}(\mathbf{x} \cdot \mathbf{y}) + \mathbf{y} \cdot \mathbf{y}.$$

By then using the facts that $(U\mathbf{x}) \cdot (U\mathbf{x}) = \mathbf{x} \cdot \mathbf{x}$ and $(U\mathbf{y}) \cdot (U\mathbf{y}) = \mathbf{y} \cdot \mathbf{y}$, we can simplify the above equation to the form

$$\mathrm{Re}\big((U\mathbf{x}) \cdot (U\mathbf{y})\big) = \mathrm{Re}(\mathbf{x} \cdot \mathbf{y}).$$

If $\mathbb{F} = \mathbb{R}$ then this implies $(U\mathbf{x}) \cdot (U\mathbf{y}) = \mathbf{x} \cdot \mathbf{y}$ for all $\mathbf{x}, \mathbf{y} \in \mathbb{F}^n$, as desired. If instead $\mathbb{F} = \mathbb{C}$ then we can repeat the above argument with $\mathbf{v} = \mathbf{x} + i\mathbf{y}$ to see that

$$\mathrm{Im}\big((U\mathbf{x}) \cdot (U\mathbf{y})\big) = \mathrm{Im}(\mathbf{x} \cdot \mathbf{y}),$$

so in this case we have $(U\mathbf{x}) \cdot (U\mathbf{y}) = \mathbf{x} \cdot \mathbf{y}$ for all $\mathbf{x}, \mathbf{y} \in \mathbb{F}^n$ too, establishing (e).

Finally, to see that (e) $\implies$ (d), note that if we rearrange $(U\mathbf{v}) \cdot (U\mathbf{w}) = \mathbf{v} \cdot \mathbf{w}$ slightly, we get

$$\big((U^*U - I)\mathbf{v}\big) \cdot \mathbf{w} = 0 \quad \text{for all} \quad \mathbf{v}, \mathbf{w} \in \mathbb{F}^n.$$

If we choose $\mathbf{w} = (U^*U - I)\mathbf{v}$ then this implies $\big\|(U^*U - I)\mathbf{v}\big\|^2 = 0$ for all $\mathbf{v} \in \mathbb{F}^n$, so $(U^*U - I)\mathbf{v} = 0$ for all $\mathbf{v} \in \mathbb{F}^n$. This in turn implies $U^*U - I = O$, so $U^*U = I$, which completes the proof. ∎

Checking whether or not a matrix is unitary is now quite simple, since we just have to check whether or not $U^*U = I$. For example, if we return to the matrix

The fact that this matrix is unitary makes sense geometrically if we notice that it rotates $\mathbb{R}^2$ counter-clockwise by $\pi/4$ (45°).

$$U = \frac{1}{\sqrt{2}} \begin{bmatrix} 1 & -1 \\ 1 & 1 \end{bmatrix}$$

from earlier, we can now check that it is unitary simply by computing

$$U^*U = \frac{1}{2} \begin{bmatrix} 1 & 1 \\ -1 & 1 \end{bmatrix} \begin{bmatrix} 1 & -1 \\ 1 & 1 \end{bmatrix} = \begin{bmatrix} 1 & 0 \\ 0 & 1 \end{bmatrix}.$$

Since $U^*U = I$, Theorem 1.4.9 tells us that U is unitary. The following example generalizes the above calculation.

Example 1.4.12

Rotation Matrices are Unitary

Recall from introductory linear algebra that the standard matrix of the linear transformation $R^\theta : \mathbb{R}^2 \to \mathbb{R}^2$ that rotates $\mathbb{R}^2$ counter-clockwise by an angle of θ is

$$\left[R^\theta\right] = \begin{bmatrix} \cos(\theta) & -\sin(\theta) \\ \sin(\theta) & \cos(\theta) \end{bmatrix}.$$

Show that $\left[R^\theta\right]$ is unitary.

Solution:

Since rotation matrices do not change the length of vectors, we know that they must be unitary. To verify this a bit more directly we compute $\left[R^\theta\right]^*\left[R^\theta\right]$:

$$\left[R^\theta\right]^*\left[R^\theta\right] = \begin{bmatrix} \cos(\theta) & \sin(\theta) \\ -\sin(\theta) & \cos(\theta) \end{bmatrix} \begin{bmatrix} \cos(\theta) & -\sin(\theta) \\ \sin(\theta) & \cos(\theta) \end{bmatrix}$$

$$= \begin{bmatrix} \cos^2(\theta)+\sin^2(\theta) & \sin(\theta)\cos(\theta)-\cos(\theta)\sin(\theta) \\ \cos(\theta)\sin(\theta)-\sin(\theta)\cos(\theta) & \sin^2(\theta)+\cos^2(\theta) \end{bmatrix}$$

Recall that $\sin^2(\theta)+\cos^2(\theta) = 1$ for all $\theta \in \mathbb{R}$.

$$= \begin{bmatrix} 1 & 0 \\ 0 & 1 \end{bmatrix}.$$

Since $\left[R^\theta\right]^*\left[R^\theta\right] = I$, we conclude that $\left[R^\theta\right]$ is unitary.

Example 1.4.13

Reflection Matrices are Unitary

Recall from introductory linear algebra that the standard matrix of the linear transformation $F_{\mathbf{u}} : \mathbb{R}^n \to \mathbb{R}^n$ that reflects $\mathbb{R}^n$ through the line in the direction of the unit vector $\mathbf{u} \in \mathbb{R}^n$ is

$$[F_{\mathbf{u}}] = 2\mathbf{u}\mathbf{u}^T - I.$$

Show that $[F_{\mathbf{u}}]$ is unitary.

Solution:

Again, reflection matrices do not change the length of vectors, so we know that they must be unitary. To see this a bit more directly, we compute $[F_{\mathbf{u}}]^*[F_{\mathbf{u}}]$:

$$[F_{\mathbf{u}}]^*[F_{\mathbf{u}}] = (2\mathbf{u}\mathbf{u}^T - I)^*(2\mathbf{u}\mathbf{u}^T - I) = 4\mathbf{u}(\mathbf{u}^T\mathbf{u})\mathbf{u}^T - 4\mathbf{u}\mathbf{u}^T + I$$

$$= 4\mathbf{u}\mathbf{u}^T - 4\mathbf{u}\mathbf{u}^T + I = I,$$

where the third equality comes from the fact that $\mathbf{u}$ is a unit vector, so $\mathbf{u}^T\mathbf{u} = \|\mathbf{u}\|^2 = 1$. Since $[F_{\mathbf{u}}]^*[F_{\mathbf{u}}] = I$, we conclude the $[F_{\mathbf{u}}]$ is unitary.

Again, the previous two examples provide exactly the intuition that we should have for unitary matrices—they are the ones that rotate and/or reflect $\mathbb{F}^n$, but do not stretch, shrink, or otherwise distort it. They can be thought of as very rigid linear transformations that leave the size and shape of $\mathbb{F}^n$ intact, but possibly change its orientation.

Remark 1.4.2

Orthogonal Matrices

Many sources refer to real unitary matrices as **orthogonal matrices** (but still refer to complex unitary matrices as unitary). However, we dislike this terminology for numerous reasons:

a) The columns of orthogonal matrices (i.e., real unitary matrices) are mutually ortho*normal*, not just mutually orthogonal.

b) Two orthogonal matrices (i.e., real unitary matrices) need not be orthogonal to each other in any particular inner product on $\mathcal{M}_n(\mathbb{R})$.

c) There just is no reason to use separate terminology depending on whether the matrix is real or complex.

We thus do not use the term "orthogonal matrix" again in this book, and we genuinely hope that it falls out of use.

1.4.4 Projections

As one final application of inner products and orthogonality, we now introduce projections, which we roughly think of as linear transformations that squish vectors down into some given subspace. For example, when discussing the Gram–Schmidt process, we implicitly used the fact that if $\mathbf{u} \in \mathcal{V}$ is a unit vector then the linear transformation $P_{\mathbf{u}} : \mathcal{V} \to \mathcal{V}$ defined by $P_{\mathbf{u}}(\mathbf{v}) = \langle \mathbf{u}, \mathbf{v} \rangle \mathbf{u}$ squishes $\mathbf{v}$ down onto span($\mathbf{u}$), as in Figure 1.16. Indeed, $P_{\mathbf{u}}$ is a projection onto span($\mathbf{u}$).

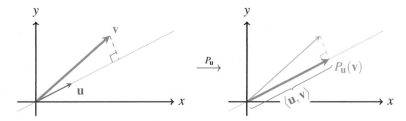

Figure 1.16: Given a unit vector $\mathbf{u}$, the linear transformation $P_{\mathbf{u}}(\mathbf{v}) = \langle \mathbf{u}, \mathbf{v} \rangle \mathbf{u}$ is a projection onto the line in the direction of $\mathbf{u}$.

Intuitively, projecting a vector onto a subspace is analogous to casting a vector's shadow onto a surface as in Figure 1.17, or looking at a 3D object from one side (and thus seeing a 2D image of it, like when we look at objects on computer screens). The simplest projections to work with mathematically are those that project at a right angle (as in Figures 1.16 and 1.17(b))—these are called **orthogonal projections**, whereas if they project at another angle (as in Figure 1.17(a)) then they are called **oblique projections**.

The key observation that lets us get our hands on projections mathematically is that a linear transformation $P : \mathcal{V} \to \mathcal{V}$ projects onto its range (i.e., it leaves every vector in its range alone) if and only if $P^2 = P$, since $P(\mathbf{v}) \in \text{range}(P)$ regardless of $\mathbf{v} \in \mathcal{V}$, so applying P again does not change it (see Figure 1.18(a)).

Recall that $P^2 = P \circ P$.

If we furthermore require that every vector is projected down at a right angle (that is, it is projected orthogonally) then we need $\mathcal{V}$ to be an inner product space and we want $\langle P(\mathbf{v}), \mathbf{v} - P(\mathbf{v}) \rangle = 0$ for all $\mathbf{v} \in \mathcal{V}$ (see Figure 1.18(b)). Remarkably, this property is equivalent to P being self-adjoint (i.e., $P = P^*$), but proving this fact is somewhat tedious, so we defer it to Exercises 1.4.29 and 2.1.23.

One of the greatest joys that this author gets is from glossing over ugly details and making the reader work them out.

With all of this in mind, we are finally able to define (orthogonal) projections in general:

When the sun is
directly above us in
the sky, our shadow
is our orthogonal
projection. At any
other time, our
shadow is just an
oblique projection.

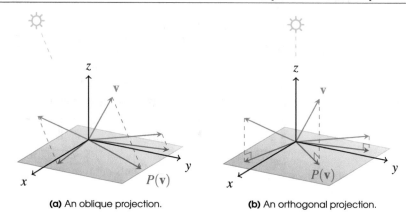

(a) An oblique projection. **(b)** An orthogonal projection.

Figure 1.17: A projection P can be thought of as a linear transformation that casts the shadow of a vector onto a subspace from a far-off light source. Here, $\mathbb{R}^3$ is projected onto the xy-plane.

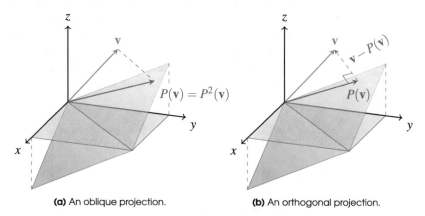

(a) An oblique projection. **(b)** An orthogonal projection.

Figure 1.18: A rank-2 projection $P : \mathbb{R}^3 \to \mathbb{R}^3$ projects onto a plane (its range). After it projects once, projecting again has no additional effect, so $P^2 = P$. An *orthogonal* projection is one for which, as in (b), $P(\mathbf{v})$ is always orthogonal to $\mathbf{v} - P(\mathbf{v})$.

Definition 1.4.6 **Projections**	Suppose that $\mathcal{V}$ is a vector space and $P : \mathcal{V} \to \mathcal{V}$ is a linear transformation. a) If $P^2 = P$ then P is called a **projection**. b) If $\mathcal{V}$ is an inner product space and $P^2 = P = P^*$ then P is called an **orthogonal projection**. We furthermore say that P **projects onto** range(P).

Example 1.4.14 **Determining if a Matrix is a Projection**	Determine which of the following matrices are projections. If they are projections, determine whether or not they are orthogonal and describe the subspace of $\mathbb{R}^n$ that they project onto. a) $P = \begin{bmatrix} 1 & -1 \\ 1 & 1 \end{bmatrix}$ b) $Q = \begin{bmatrix} 1 & 0 & 0 \\ 0 & 1 & 1/2 \\ 0 & 0 & 0 \end{bmatrix}$

c) $R = \begin{bmatrix} 5/6 & 1/6 & -1/3 \\ 1/6 & 5/6 & 1/3 \\ -1/3 & 1/3 & 1/3 \end{bmatrix}$

Solutions:

a) This matrix is not a projection, since direct computation shows that

$$P^2 = \begin{bmatrix} 0 & -2 \\ 2 & 0 \end{bmatrix} \neq \begin{bmatrix} 1 & -1 \\ 1 & 1 \end{bmatrix} = P.$$

b) Again, we just check whether or not $Q^2 = Q$:

$$Q^2 = \begin{bmatrix} 1 & 0 & 0 \\ 0 & 1 & 1/2 \\ 0 & 0 & 0 \end{bmatrix} \begin{bmatrix} 1 & 0 & 0 \\ 0 & 1 & 1/2 \\ 0 & 0 & 0 \end{bmatrix} = \begin{bmatrix} 1 & 0 & 0 \\ 0 & 1 & 1/2 \\ 0 & 0 & 0 \end{bmatrix} = Q,$$

In fact, Q is exactly the projection onto the xy-plane that was depicted in Figure 1.17(a).

so Q is a projection, but $Q^* \neq Q$ so it is not an *orthogonal* projection. Since its columns span the xy-plane, that is its range (i.e., the subspace that it projects onto).

c) This matrix is a projection, since

$$R^2 = \left(\frac{1}{6} \begin{bmatrix} 5 & 1 & -2 \\ 1 & 5 & 2 \\ -2 & 2 & 2 \end{bmatrix} \right)^2$$

$$= \frac{1}{36} \begin{bmatrix} 30 & 6 & -12 \\ 6 & 30 & 12 \\ -12 & 12 & 12 \end{bmatrix} = \frac{1}{6} \begin{bmatrix} 5 & 1 & -2 \\ 1 & 5 & 2 \\ -2 & 2 & 2 \end{bmatrix} = R.$$

R is the projection that was depicted in Figure 1.18(b).

Furthermore, R is an orthogonal projection since $R^* = R$. We can compute $\text{range}(R) = \text{span}\{(2,0,-1),(0,2,1)\}$ using techniques from introductory linear algebra, which is the plane with equation $x - y + 2z = 0$.

We return to oblique projections in Section 1.B.2.

Although projections in general have their uses, we are primarily interested in orthogonal projections, and will focus on them for the remainder of this section. One of the nicest features of orthogonal projections is that they are uniquely determined by the subspace that they project onto (i.e., there is only one orthogonal projection for each subspace), and they can be computed in a straightforward way from any orthonormal basis of that subspace, at least in finite dimensions.

In order to get more comfortable with constructing and making use of orthogonal projections, we start by describing what they look like in the concrete setting of matrices that project $\mathbb{F}^n$ down onto some subspace of it.

Theorem 1.4.10

Construction of Orthogonal Projections

Suppose $\mathbb{F} = \mathbb{R}$ or $\mathbb{F} = \mathbb{C}$ and let $\mathcal{S}$ be an m-dimensional subspace of $\mathbb{F}^n$. Then there is a unique orthogonal projection P onto $\mathcal{S}$, and it is given by

$$P = AA^*,$$

where $A \in \mathcal{M}_{n,m}(\mathbb{F})$ is a matrix with any orthonormal basis of $\mathcal{S}$ as its columns.

Recall that $(AB)^* = B^*A^*$. Plugging in $B = A^*$ gives $(AA^*)^* = AA^*$.

Proof. We start by showing that the matrix $P = AA^*$ is indeed an orthogonal projection onto $\mathcal{S}$. To verify this claim, we write $A = \begin{bmatrix} \mathbf{u}_1 \mid \mathbf{u}_2 \mid \cdots \mid \mathbf{u}_m \end{bmatrix}$. Then notice that $P^* = (AA^*)^* = AA^* = P$ and

$$P^2 = (AA^*)(AA^*) = A(A^*A)A^*$$

Our method of computing A^*A here is almost identical to the one from the proof of Theorem 1.4.9. Recall that $\{\mathbf{u}_1, \ldots, \mathbf{u}_m\}$ is an orthonormal basis of $\mathcal{S}$.

$$= A \begin{bmatrix} \mathbf{u}_1 \cdot \mathbf{u}_1 & \mathbf{u}_1 \cdot \mathbf{u}_2 & \cdots & \mathbf{u}_1 \cdot \mathbf{u}_m \\ \mathbf{u}_2 \cdot \mathbf{u}_1 & \mathbf{u}_2 \cdot \mathbf{u}_2 & \cdots & \mathbf{u}_2 \cdot \mathbf{u}_m \\ \vdots & \vdots & \ddots & \vdots \\ \mathbf{u}_m \cdot \mathbf{u}_1 & \mathbf{u}_m \cdot \mathbf{u}_2 & \cdots & \mathbf{u}_m \cdot \mathbf{u}_m \end{bmatrix} A^*$$

$$= AI_m A^* = AA^* = P.$$

It follows that P is an orthogonal projection. To see that $\text{range}(P) = \mathcal{S}$, we simply notice that $\text{range}(P) = \text{range}(A^*A) = \text{range}(A)$ by Theorem A.1.2, and $\text{range}(A)$ is the span of its columns (by that same theorem), which is $\mathcal{S}$ (since those columns were chosen specifically to be an orthonormal basis of $\mathcal{S}$).

Uniqueness of P lets us (in finite dimensions) talk about *the* orthogonal projection onto a given subspace, rather than just *an* orthogonal projection onto it.

To see that P is unique, suppose that Q is another orthogonal projection onto $\mathcal{S}$. If $\{\mathbf{u}_1, \mathbf{u}_2, \ldots, \mathbf{u}_m\}$ is an orthonormal basis of $\mathcal{S}$ then we can extend it to an orthonormal basis $\{\mathbf{u}_1, \mathbf{u}_2, \ldots, \mathbf{u}_n\}$ of all of $\mathbb{F}^n$ via Exercise 1.4.20. We then claim that

$$P\mathbf{u}_j = Q\mathbf{u}_j = \begin{cases} \mathbf{u}_j, & \text{if } 1 \leq j \leq m \\ \mathbf{0}, & \text{if } m < j \leq n \end{cases}.$$

To see why this is the case, we notice that $P\mathbf{u}_j = Q\mathbf{u}_j = \mathbf{u}_j$ for $1 \leq j \leq m$ because $\mathbf{u}_j \in \mathcal{S}$ and P, Q leave everything in $\mathcal{S}$ alone. The fact that $Q\mathbf{u}_j = \mathbf{0}$ for $j > m$ follows from the fact that $Q^2 = Q = Q^*$, so

$$\langle Q\mathbf{u}_j, Q\mathbf{u}_j \rangle = \langle Q^*Q\mathbf{u}_j, \mathbf{u}_j \rangle = \langle Q^2\mathbf{u}_j, \mathbf{u}_j \rangle = \langle Q\mathbf{u}_j, \mathbf{u}_j \rangle.$$

Since $\mathbf{u}_j$ is orthogonal to each of $\mathbf{u}_1, \mathbf{u}_2, \ldots, \mathbf{u}_m$ and thus everything in $\text{range}(Q) = \mathcal{S}$, we then have $\langle Q\mathbf{u}_j, \mathbf{u}_j \rangle = 0$, so $\|Q\mathbf{u}_j\|^2 = \langle Q\mathbf{u}_j, Q\mathbf{u}_j \rangle = 0$ too, and thus $Q\mathbf{u}_j = \mathbf{0}$. The proof that $P\mathbf{u}_j = \mathbf{0}$ when $j > m$ is identical.

Since a matrix (linear transformation) is completely determined by how it acts on a basis of $\mathbb{F}^n$, the fact that $P\mathbf{u}_j = Q\mathbf{u}_j$ for all $1 \leq j \leq n$ implies $P = Q$, so all orthogonal projections onto $\mathcal{S}$ are the same. ∎

In the special case when $\mathcal{S}$ is 1-dimensional (i.e., a line), the above result simply says that $P = \mathbf{u}\mathbf{u}^*$, where $\mathbf{u}$ is a unit vector pointing in the direction of that line. It follows that $P\mathbf{v} = (\mathbf{u}\mathbf{u}^*)\mathbf{v} = (\mathbf{u} \cdot \mathbf{v})\mathbf{u}$, which recovers the fact that we noted earlier about functions of this form (well, functions of the form $P(\mathbf{v}) = \langle \mathbf{u}, \mathbf{v} \rangle \mathbf{u}$) projecting down onto the line in the direction of $\mathbf{u}$.

More generally, if we expand out the product $P = AA^*$ using block matrix multiplication, we see that if $\{\mathbf{u}_1, \mathbf{u}_2, \ldots, \mathbf{u}_m\}$ is an orthonormal basis of $\mathcal{S}$ then

This is a special case of the rank-one sum decomposition from Theorem A.1.3.

$$P = \begin{bmatrix} \mathbf{u}_1 \mid \mathbf{u}_2 \mid \cdots \mid \mathbf{u}_m \end{bmatrix} \begin{bmatrix} \mathbf{u}_1^* \\ \hline \mathbf{u}_2^* \\ \hline \vdots \\ \hline \mathbf{u}_m^* \end{bmatrix} = \sum_{j=1}^m \mathbf{u}_j \mathbf{u}_j^*.$$

Example 1.4.15

Finding an Orthogonal Projection Onto a Plane

Construct the orthogonal projection P onto the plane $\mathcal{S} \subset \mathbb{R}^3$ with equation $x - y - 2z = 0$.

Solution:

Recall from Example 1.4.7 that one orthonormal basis of $\mathcal{S}$ is

Even though there are lots of orthonormal bases of $\mathcal{S}$, they all produce the same projection P.

$$C = \{\mathbf{u}_1, \mathbf{u}_2\} = \left\{ \tfrac{1}{\sqrt{5}}(2, 0, 1), \tfrac{1}{\sqrt{30}}(1, 5, -2) \right\}.$$

It follows from Theorem 1.4.10 that the (unique!) orthogonal projection onto $\mathcal{S}$ is

It is worth comparing P to the orthogonal projection onto the plane $x - y + 2z = 0$ from Example 1.4.14(c).

$$P = \mathbf{u}_1 \mathbf{u}_1^* + \mathbf{u}_2 \mathbf{u}_2^* = \frac{1}{5} \begin{bmatrix} 2 \\ 0 \\ 1 \end{bmatrix} \begin{bmatrix} 2 & 0 & 1 \end{bmatrix} + \frac{1}{30} \begin{bmatrix} 1 \\ 5 \\ -2 \end{bmatrix} \begin{bmatrix} 1 & 5 & -2 \end{bmatrix}$$

$$= \frac{1}{6} \begin{bmatrix} 5 & 1 & 2 \\ 1 & 5 & -2 \\ 2 & -2 & 2 \end{bmatrix}.$$

One of the most useful features of orthogonal projections is that they do not just project a vector $\mathbf{v}$ anywhere in their range, but rather they always project down to the closest vector in their range, as illustrated in Figure 1.19. This observation hopefully makes some intuitive sense, since the shortest path from us to the ceiling above us is along a line pointing straight up (i.e., the shortest path is orthogonal to the ceiling), but we make it precise in the following theorem.

The distance between two vectors $\mathbf{v}$ and $\mathbf{w}$ is $\|\mathbf{v} - \mathbf{w}\|$.

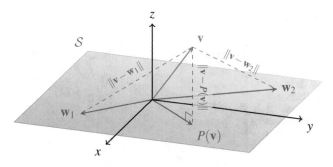

Figure 1.19: The fastest way to get from a vector to a nearby plane is to go "straight down" to it. In other words, the closest vector to $\mathbf{v}$ in a subspace $\mathcal{S}$ is $P(\mathbf{v})$, where P is the orthogonal projection of onto $\mathcal{S}$. That is, $\|\mathbf{v} - P(\mathbf{v})\| \leq \|\mathbf{v} - \mathbf{w}\|$ for all $\mathbf{w} \in \mathcal{S}$.

Theorem 1.4.11

Orthogonal Projections Find the Closest Point in a Subspace

Suppose P is an orthogonal projection onto a subspace S of an inner product space V. Then

$$\|\mathbf{v} - P(\mathbf{v})\| \leq \|\mathbf{v} - \mathbf{w}\| \quad \text{for all} \quad \mathbf{v} \in V, \, \mathbf{w} \in S.$$

Furthermore, equality holds if and only if $\mathbf{w} = P(\mathbf{v})$.

Proof. We first notice that $\mathbf{v} - P(\mathbf{v})$ is orthogonal to every vector $\mathbf{x} \in S$, since

$$\langle \mathbf{v} - P(\mathbf{v}), \mathbf{x} \rangle = \langle \mathbf{v} - P(\mathbf{v}), P(\mathbf{x}) \rangle = \langle P^*(\mathbf{v} - P(\mathbf{v})), \mathbf{x} \rangle$$
$$= \langle P(\mathbf{v} - P(\mathbf{v})), \mathbf{x} \rangle = \langle P(\mathbf{v}) - P(\mathbf{v}), \mathbf{x} \rangle = 0.$$

If we now let $\mathbf{w} \in S$ be arbitrary and choose $\mathbf{x} = P(\mathbf{v}) - \mathbf{w}$ (which is in S since $P(\mathbf{v})$ and $\mathbf{w}$ are), we see that $\mathbf{v} - P(\mathbf{v})$ is orthogonal to $P(\mathbf{v}) - \mathbf{w}$. It follows that

$$\|\mathbf{v} - \mathbf{w}\|^2 = \|(\mathbf{v} - P(\mathbf{v})) + (P(\mathbf{v}) - \mathbf{w})\|^2 = \|\mathbf{v} - P(\mathbf{v})\|^2 + \|P(\mathbf{v}) - \mathbf{w}\|^2,$$

where the final equality follows from the Pythagorean theorem for inner products (Exercise 1.3.12). Since $\|P(\mathbf{v}) - \mathbf{w}\|^2 \geq 0$, it follows immediately that $\|\mathbf{v} - \mathbf{w}\| \geq \|\mathbf{v} - P(\mathbf{v})\|$. Furthermore, equality holds if and only if $\|P(\mathbf{v}) - \mathbf{w}\| = 0$, which happens if and only if $P(\mathbf{v}) = \mathbf{w}$. ∎

Example 1.4.16

Finding the Closest Vector in a Plane

Find the closest vector to $\mathbf{v} = (3, -2, 2)$ in the plane $S \subset \mathbb{R}^3$ defined by the equation $x - 4y + z = 0$.

Solution:

Theorem 1.4.11 tells us that the closest vector to $\mathbf{v}$ in S is $P\mathbf{v}$, where P is the orthogonal projection onto S. To construct P, we first need an orthonormal basis of S so that we can use Theorem 1.4.10. To find an orthonormal basis of S, we proceed as we did in Example 1.4.7—we apply the Gram–Schmidt process to any set consisting of two linearly independent vectors in S, like $B = \{(1, 0, -1), (0, 1, 4)\}$.

To find these vectors $(1,0,-1)$ and $(0,1,4)$, choose x and y arbitrarily and then solve for z via $x - 4y + z = 0$.

Applying the Gram–Schmidt process to B gives the following orthonormal basis C of S:

$$C = \left\{ \tfrac{1}{\sqrt{2}}(1, 0, -1), \tfrac{1}{3}(2, 1, 2) \right\}.$$

Theorem 1.4.10 then tells us that the orthogonal projection onto S is

Even though there are lots of different orthonormal bases of S, they all produce this same orthogonal projection P.

$$P = AA^* = \frac{1}{18} \begin{bmatrix} 17 & 4 & -1 \\ 4 & 2 & 4 \\ -1 & 4 & 17 \end{bmatrix}, \quad \text{where} \quad A = \begin{bmatrix} 1/\sqrt{2} & 2/3 \\ 0 & 1/3 \\ -1/\sqrt{2} & 2/3 \end{bmatrix}.$$

It follows that the closest vector to $\mathbf{v} = (3, -2, 2)$ in S is

$$P\mathbf{v} = \frac{1}{18} \begin{bmatrix} 17 & 4 & -1 \\ 4 & 2 & 4 \\ -1 & 4 & 17 \end{bmatrix} \begin{bmatrix} 3 \\ -2 \\ 2 \end{bmatrix} = \frac{1}{18} \begin{bmatrix} 41 \\ 16 \\ 23 \end{bmatrix},$$

which is illustrated on the next page:

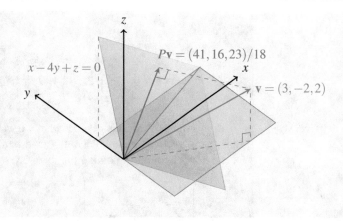

Example 1.4.17

Almost Solving a Linear System

Show that the linear system $A\mathbf{x} = \mathbf{b}$ is inconsistent, where

$$A = \begin{bmatrix} 1 & 2 & 3 \\ 4 & 5 & 6 \\ 7 & 8 & 9 \end{bmatrix} \quad \text{and} \quad \mathbf{b} = \begin{bmatrix} 2 \\ -1 \\ 0 \end{bmatrix}.$$

Recall that "inconsistent" means "has no solutions".

Then find the closest thing to a solution—that is, find a vector $\mathbf{x} \in \mathbb{R}^3$ that minimizes $\|A\mathbf{x} - \mathbf{b}\|$.

Solution:

To see that this linear system is inconsistent (i.e., has no solutions), we just row reduce the augmented matrix $[\, A \mid \mathbf{b}\,]$:

$$\begin{bmatrix} 1 & 2 & 3 & 2 \\ 4 & 5 & 6 & -1 \\ 7 & 8 & 9 & 0 \end{bmatrix} \xrightarrow{\text{row reduce}} \begin{bmatrix} 1 & 0 & -1 & 0 \\ 0 & 1 & 2 & 0 \\ 0 & 0 & 0 & 1 \end{bmatrix}.$$

The bottom row of this row echelon form tells us that the original linear system has no solutions, since it corresponds to the equation $0x_1 + 0x_2 + 0x_3 = 1$.

The fact that this linear system is inconsistent means exactly that $\mathbf{b} \notin \text{range}(A)$, and to find that "closest thing" to a solution (i.e., to minimize $\|A\mathbf{x} - \mathbf{b}\|$), we just orthogonally project $\mathbf{b}$ onto $\text{range}(A)$. We thus start by constructing an orthonormal basis C of $\text{range}(A)$:

This orthonormal basis can be found by applying the Gram–Schmidt process to the first two columns of A, which span its range (see Theorem A.1.2).

$$C = \left\{ \tfrac{1}{\sqrt{66}}(1,4,7), \tfrac{1}{\sqrt{11}}(-3,-1,1) \right\}.$$

The orthogonal projection onto $\text{range}(A)$ is then

We develop a more direct method of constructing the projection onto $\text{range}(A)$ in Exercise 1.4.30.

$$P = BB^* = \frac{1}{6} \begin{bmatrix} 5 & 2 & -1 \\ 2 & 2 & 2 \\ -1 & 2 & 5 \end{bmatrix}, \quad \text{where} \quad B = \begin{bmatrix} 1/\sqrt{66} & -3/\sqrt{11} \\ 4/\sqrt{66} & -1/\sqrt{11} \\ 7/\sqrt{66} & 1/\sqrt{11} \end{bmatrix},$$

which tells us (via Theorem 1.4.11) that the vector $A\mathbf{x}$ that minimizes

$\|A\mathbf{x} - \mathbf{b}\|$ is

$$A\mathbf{x} = P\mathbf{b} = \frac{1}{6}\begin{bmatrix} 5 & 2 & -1 \\ 2 & 2 & 2 \\ -1 & 2 & 5 \end{bmatrix}\begin{bmatrix} 2 \\ -1 \\ 0 \end{bmatrix} = \frac{1}{3}\begin{bmatrix} 4 \\ 1 \\ -2 \end{bmatrix}.$$

Unfortunately, this is not quite what we want—we have found $A\mathbf{x}$ (i.e., the closest vector to $\mathbf{b}$ in range(A)), but we want to find $\mathbf{x}$ itself. Fortunately, $\mathbf{x}$ can now be found by solving the linear system $A\mathbf{x} = (4, 1, -2)/3$:

$$\begin{bmatrix} 1 & 2 & 3 & | & 4/3 \\ 4 & 5 & 6 & | & 1/3 \\ 7 & 8 & 9 & | & -2/3 \end{bmatrix} \xrightarrow{\text{row reduce}} \begin{bmatrix} 1 & 0 & -1 & | & -2 \\ 0 & 1 & 2 & | & 5/3 \\ 0 & 0 & 0 & | & 0 \end{bmatrix}.$$

It follows that $\mathbf{x}$ can be chosen to be any vector of the form $\mathbf{x} = (-2, 5/3, 0) + x_3(1, -2, 1)$, where x_3 is a free variable. Choosing $x_3 = 0$ gives $\mathbf{x} = (-2, 5/3, 0)$, for example.

The method of Example 1.4.17 for using projections to find the "closest thing" to a solution of an unsolvable linear system is called **linear least squares**, and it is extremely widely-used in statistics. If we want to fit a model to a set of data points, we typically have far more data points (equations) than parameters of the model (variables), so our model will typically not exactly match all of the data. However, with linear least squares we can find the model that comes as close as possible to matching the data. We return to this method and investigate it in more depth and with some new machinery in Section 2.C.1.

While Theorem 1.4.10 only applies directly in the finite-dimensional case, we can extend them somewhat to the case where only the subspace being projected onto is finite-dimensional, but the source vector space $\mathcal{V}$ is potentially infinite-dimensional. We have to be slightly more careful in this situation, since it does not make sense to even talk about the standard matrix of the projection when $\mathcal{V}$ is infinite-dimensional:

Theorem 1.4.12

Another Construction of Orthogonal Projections

Uniqueness of P in this case is a bit trickier, and requires an additional assumption like the axiom of choice or $\mathcal{V}$ being finite-dimensional.

Suppose $\mathcal{V}$ is an inner product space and $\mathcal{S} \subseteq \mathcal{V}$ is an n-dimensional subspace with orthonormal basis $\{\mathbf{u}_1, \mathbf{u}_2, \ldots, \mathbf{u}_n\}$. Then the linear transformation $P : \mathcal{V} \to \mathcal{V}$ defined by

$$P(\mathbf{v}) = \langle \mathbf{u}_1, \mathbf{v}\rangle\mathbf{u}_1 + \langle \mathbf{u}_2, \mathbf{v}\rangle\mathbf{u}_2 + \cdots + \langle \mathbf{u}_n, \mathbf{v}\rangle\mathbf{u}_n \quad \text{for all} \quad \mathbf{v} \in \mathcal{V}$$

is an orthogonal projection onto $\mathcal{S}$.

Proof. To see that $P^2 = P$, we just use linearity of the second entry of the inner product to see that

$$P\big(P(\mathbf{v})\big) = \sum_{i=1}^{n} \langle \mathbf{u}_i, P(\mathbf{v}) \rangle \mathbf{u}_i \qquad \text{(definition of } P(\mathbf{v}))$$

It is worth comparing this result to Theorem 1.4.5—we are just giving coordinates to $P(\mathbf{v})$ with respect to $\{\mathbf{u}_1, \mathbf{u}_2, \ldots, \mathbf{u}_n\}$.

$$= \sum_{i=1}^{n} \left\langle \mathbf{u}_i, \sum_{j=1}^{n} \langle \mathbf{u}_j, \mathbf{v} \rangle \mathbf{u}_j \right\rangle \mathbf{u}_i \quad \text{(definition of } P(\mathbf{v}))$$

$$= \sum_{i,j=1}^{n} \langle \mathbf{u}_j, \mathbf{v} \rangle \langle \mathbf{u}_i, \mathbf{u}_j \rangle \mathbf{u}_i \qquad \text{(linearity of the inner product)}$$

$$= \sum_{i=1}^{n} \langle \mathbf{u}_i, \mathbf{v} \rangle \mathbf{u}_i = P(\mathbf{v}). \qquad (\{\mathbf{u}_1, \ldots, \mathbf{u}_n\} \text{ is an orthonormal basis)}$$

Similarly, to see that $P^* = P$ we use sesquilinearity of the inner product to see that

For the second-to-last equality here, recall that inner products are *conjugate* linear in their first argument, so $\langle \mathbf{u}_i, \mathbf{v} \rangle \langle \mathbf{w}, \mathbf{u}_i \rangle = \langle \langle \mathbf{w}, \mathbf{u}_i \rangle \mathbf{u}_i, \mathbf{v} \rangle = \langle \langle \mathbf{u}_i, \mathbf{w} \rangle \mathbf{u}_i, \mathbf{v} \rangle.$

$$\langle \mathbf{w}, P(\mathbf{v}) \rangle = \left\langle \mathbf{w}, \sum_{i=1}^{n} \langle \mathbf{u}_i, \mathbf{v} \rangle \mathbf{u}_i \right\rangle = \sum_{i=1}^{n} \langle \mathbf{u}_i, \mathbf{v} \rangle \langle \mathbf{w}, \mathbf{u}_i \rangle$$

$$= \left\langle \sum_{i=1}^{n} \langle \mathbf{u}_i, \mathbf{w} \rangle \mathbf{u}_i, \mathbf{v} \right\rangle = \langle P(\mathbf{w}), \mathbf{v} \rangle$$

for all $\mathbf{v}, \mathbf{w} \in \mathcal{V}$. The fact that $P^* = P$ follows from the definition of adjoint given in Definition 1.4.3. $\blacksquare$

While Theorems 1.4.10 and 1.4.12 perhaps look quite different on the surface, the latter simply reduces to the former when $\mathcal{V} = \mathbb{F}^n$. To see this, simply notice that in this case Theorem 1.4.12 says that

Keep in mind that $\mathbf{u}_j^* \mathbf{v}$ is a scalar, so there is no problem commuting it past $\mathbf{u}_j$ in the final equality here.

$$P(\mathbf{v}) = \sum_{j=1}^{n} \langle \mathbf{u}_j, \mathbf{v} \rangle \mathbf{u}_j = \sum_{j=1}^{n} (\mathbf{u}_j^* \mathbf{v}) \mathbf{u}_j = \sum_{j=1}^{n} \mathbf{u}_j (\mathbf{u}_j^* \mathbf{v}) = \left(\sum_{j=1}^{n} \mathbf{u}_j \mathbf{u}_j^* \right) \mathbf{v},$$

which agrees with the formula from Theorem 1.4.10.

Example 1.4.18

Finding the Closest Polynomial to a Function

Find the degree-2 polynomial f with the property that the integral

$$\int_{-1}^{1} \left| e^x - f(x) \right|^2 dx$$

is as small as possible.

Solution:

The important fact to identify here is that we are being asked to minimize $\|e^x - f(x)\|^2$ (which is equivalent to minimizing $\|e^x - f(x)\|$) as f ranges over the subspace $\mathcal{P}^2[-1,1]$ of $\mathcal{C}[-1,1]$. By Theorem 1.4.11, our goal is thus to construct $P(e^x)$, where P is the orthogonal projection from $\mathcal{C}[-1,1]$ onto $\mathcal{P}^2[-1,1]$, which is guaranteed to exist by Theorem 1.4.12 since $\mathcal{P}^2[-1,1]$ is finite-dimensional. To construct P, we need an orthonormal basis of $\mathcal{P}^2[-1,1]$, and one such basis is

This basis can be found by applying the Gram–Schmidt process to the standard basis $\{1, x, x^2\}$ much like we did in Example 1.4.8.

$$C = \big\{ p_1(x), p_2(x), p_3(x) \big\} = \Big\{ \tfrac{1}{\sqrt{2}}, \tfrac{\sqrt{3}}{\sqrt{2}} x, \tfrac{\sqrt{5}}{\sqrt{8}} (3x^2 - 1) \Big\}.$$

Next, we compute the inner product of e^x with each of these basis polynomials:

$$\langle p_1(x), e^x \rangle = \tfrac{1}{\sqrt{2}} \int_{-1}^{1} e^x \, dx = \tfrac{1}{\sqrt{2}} \left(e - \tfrac{1}{e} \right) \approx 1.6620,$$

$$\langle p_2(x), e^x \rangle = \tfrac{\sqrt{3}}{\sqrt{2}} \int_{-1}^{1} x e^x \, dx = \tfrac{\sqrt{6}}{e} \approx 0.9011, \quad \text{and}$$

$$\langle p_3(x), e^x \rangle = \tfrac{\sqrt{5}}{\sqrt{8}} \int_{-1}^{1} (3x^2 - 1) e^x \, dx = \tfrac{\sqrt{5}(e^2 - 7)}{\sqrt{2} e} \approx 0.2263.$$

Theorem 1.4.12 then tells us that

$$P(e^x) = \langle p_1(x), e^x \rangle p_1(x) + \langle p_2(x), e^x \rangle p_2(x) + \langle p_3(x), e^x \rangle p_3(x)$$
$$= \tfrac{1}{2} \left(e - \tfrac{1}{e} \right) + \tfrac{3}{e} x + \tfrac{5(e^2 - 7)}{4e} (3x^2 - 1)$$
$$\approx 0.5367 x^2 + 1.1036 x + 0.9963.$$

It is worth noting that the solution to the previous example is rather close to the degree-2 Taylor polynomial for e^x, which is $x^2/2 + x + 1$. The reason that these polynomials are close to each other, but not exactly the same as each other, is that the Taylor polynomial is the polynomial that best approximates e^x at $x = 0$, whereas the one that we constructed in the previous example is the polynomial that best approximates e^x on the whole interval $[-1,1]$ (see Figure 1.20). In fact, if we similarly use orthogonal projections to find polynomials that approximate e^x on the interval $[-c,c]$ for some scalar $c > 0$, those polynomials get closer and closer to the Taylor polynomial as c goes to 0 (see Exercise 1.4.32).

For a refresher on Taylor polynomials, see Appendix A.2.2.

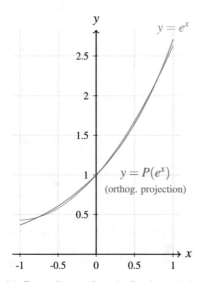

(a) The orthogonal projection $y = P(e^x) \approx 0.5367x^2 + 1.1036x + 0.9963$ approximates $y = e^x$ well on the interval $[-1,1]$.

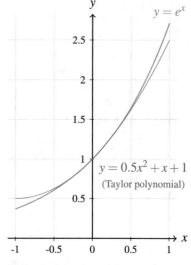

(b) The Taylor polynomial $y = 0.5x^2 + x + 1$ approximates $y = e^x$ better near $x = 0$, but worse near $x = \pm1$.

Figure 1.20: Orthogonal projections of e^x and its Taylor polynomials each approximate it well, but orthogonal projections provide a better approximation over an entire interval ($[-1,1]$ in this case) while Taylor polynomials provide a better approximation at a specific point ($x = 0$ in this case).

Remark 1.4.3

Projecting Onto Infinite-Dimensional Subspaces

When projecting onto infinite-dimensional subspaces, we must be much more careful, as neither of Theorem 1.4.10 nor 1.4.12 apply in this situation. In fact, there are infinite-dimensional subspaces that cannot be projected onto by any orthogonal projection.

For example, if there existed an orthogonal projection from $\mathcal{C}[-1,1]$ onto $\mathcal{P}[-1,1]$, then Theorem 1.4.11 (which *does* hold even in the infinite-dimensional case) would tell us that

$$\left\| e^x - P(e^x) \right\| \le \left\| e^x - f(x) \right\|$$

for all polynomials $f \in \mathcal{P}[-1,1]$. However, this is impossible since we can find polynomials $f(x)$ that make the norm on the right as close to 0 as we like. For example, if we define

$$T_p(x) = \sum_{n=1}^{p} \frac{x^n}{n!}$$

A proof that this limit equals 0 is outside of the scope of this book, but is typically covered in differential calculus courses. The idea is simply that as the degree of the Taylor polynomial increases, it approximates e^x better and better.

to be the degree-p Taylor polynomial for e^x centered at 0, then

$$\lim_{p \to \infty} \left\| e^x - T_p(x) \right\|^2 = \lim_{p \to \infty} \int_{-1}^{1} \left| e^x - T_p(x) \right|^2 dx = 0.$$

Since $\left\| e^x - P(e^x) \right\| \le \left\| e^x - T_p(x) \right\|$ for all p, this implies $\left\| e^x - P(e^x) \right\| = 0$, so $e^x = P(e^x)$. However, this is impossible since $e^x \notin \mathcal{P}[-1,1]$, so the orthogonal projection P does not actually exist.

We close this section with one final simple result that says that orthogonal projections never increase the norm of a vector. While this hopefully seems somewhat intuitive, it is important to keep in mind that it does not hold for oblique projections. For example, if the sun is straight overhead then shadows (i.e., projections) are shorter than the objects that cast them, but if the sun is low in the sky then our shadow may be longer than we are tall.

Theorem 1.4.13

Orthogonal Projections Never Increase the Norm

Suppose $\mathcal{V}$ is an inner product space and let $P : \mathcal{V} \to \mathcal{V}$ be an orthogonal projection onto some subspace of $\mathcal{V}$. Then

$$\|P(\mathbf{v})\| \le \|\mathbf{v}\| \quad \text{for all} \quad \mathbf{v} \in \mathcal{V}.$$

Furthermore, equality holds if and only if $P(\mathbf{v}) = \mathbf{v}$ (i.e., $\mathbf{v} \in \text{range}(P)$).

Proof. We just move things around in the inner product and use the Cauchy–Schwarz inequality:

Keep in mind that P is an orthogonal projection, so $P^* = P$ and $P^2 = P$ here.

$$\begin{aligned}
\|P(\mathbf{v})\|^2 = \langle P(\mathbf{v}), P(\mathbf{v}) \rangle &= \langle (P^* \circ P)(\mathbf{v}), \mathbf{v} \rangle \\
&= \langle P^2(\mathbf{v}), \mathbf{v} \rangle = \langle P(\mathbf{v}), \mathbf{v} \rangle \le \|P(\mathbf{v})\| \|\mathbf{v}\|.
\end{aligned} \tag{1.4.3}$$

If $\|P(\mathbf{v})\| = 0$ then the desired inequality is trivial, and if $\|P(\mathbf{v})\| \ne 0$ then we can divide Inequality (1.4.3) through by it to see that $\|P(\mathbf{v})\| \le \|\mathbf{v}\|$, as desired.

To verify the claim about the equality condition, suppose $\mathbf{v} \ne \mathbf{0}$ (otherwise the statement is trivial) and note that equality holds in the Cauchy–Schwarz inequality step of Inequality (1.4.3) if and only if $P(\mathbf{v})$ is a multiple of $\mathbf{v}$ (i.e., $\mathbf{v}$

is an eigenvector of P). Since Exercise 1.4.31 tells us that all eigenvalues of P are 0 or 1, we conclude that $P(\mathbf{v})=\mathbf{v}$ (if $P(\mathbf{v})=\mathbf{0}$ then $\|P(\mathbf{v})\|=0\neq\|\mathbf{v}\|$).   ∎

We show a bit later, in Exercise 2.3.25, that the above result completely characterizes orthogonal projections. That is, if a projection $P:\mathcal{V}\to\mathcal{V}$ is such that $\|P(\mathbf{v})\|\leq\|\mathbf{v}\|$ for all $\mathbf{v}\in\mathcal{V}$ then P must be an *orthogonal* projection.

Exercises

solutions to starred exercises on page 456

1.4.1 Determine which of the following sets are orthonormal bases of their span in $\mathbb{R}^n$ (with respect to the dot product).

* (a) $B=\{(0,1,1)\}$
* (b) $B=\{(1,1)/\sqrt{2},(1,-1)/\sqrt{2}\}$
* (c) $B=\{(1,1)/\sqrt{2},(1,0),(1,2)/\sqrt{5}\}$
* (d) $B=\{(1,0,2)/\sqrt{5},(0,1,0),(2,0,-1)/\sqrt{5}\}$
* (e) $B=\{(1,1,1,1)/2,(0,1,-1,0)/\sqrt{2},$
  $\qquad (1,0,0,-1)/\sqrt{2}\}$
* (f) $B=\{(1,2,2,1)/\sqrt{10},(1,0,0,-1)/\sqrt{2},$
  $\qquad (0,1,-1,0)/\sqrt{2},(2,1,-1,-2)/\sqrt{10}\}$

1.4.2 Find an orthonormal basis (with respect to the standard inner product on the indicated inner product space) for the span of the given set of vectors:

* (a) $\{(0,3,4)\}\subseteq\mathbb{R}^3$
* (b) $\{(3,4),(7,32)\}\subseteq\mathbb{R}^2$.
* (c) $\{(1,0,1),(2,1,3)\}\subseteq\mathbb{R}^3$
* (d) $\{\sin(x),\cos(x)\}\subseteq\mathcal{C}[-\pi,\pi]$.
* (e) $\{1,x,x^2\}\subseteq\mathcal{P}[0,1]$.
* (f) The following three matrices in $\mathcal{M}_2$:

$$\begin{bmatrix}1&1\\1&1\end{bmatrix},\begin{bmatrix}1&0\\0&1\end{bmatrix},\text{ and }\begin{bmatrix}0&2\\1&-1\end{bmatrix}.$$

1.4.3 For each of the following linear transformations T, find the adjoint T^*:

* (a) $T:\mathbb{R}^2\to\mathbb{R}^3$, defined by $T(\mathbf{v})=(v_1,v_1+v_2,v_2)$.
* (b) $T:\mathbb{R}^n\to\mathbb{R}$, defined by $T(\mathbf{v})=v_1$.
* (c) $T:\mathbb{R}^2\to\mathbb{R}^2$, defined by $T(\mathbf{v})=(v_1+v_2,v_1-v_2)$, where the inner product on $\mathbb{R}^2$ is $\langle\mathbf{v},\mathbf{w}\rangle=v_1w_1-v_1w_2-v_2w_1+2v_2w_2$.
* (d) $T:\mathcal{M}_2\to\mathcal{M}_2$, defined by

$$T\left(\begin{bmatrix}a&b\\c&d\end{bmatrix}\right)=\begin{bmatrix}d&-b\\-c&a\end{bmatrix}.$$

1.4.4 Construct the orthogonal projection onto the indicated subspace of $\mathbb{R}^n$.

* (a) The x-axis in $\mathbb{R}^2$.
* (b) The plane in $\mathbb{R}^3$ with equation $x+2y+3z=0$.
* (c) The range of the matrix $A=\begin{bmatrix}1&0&1\\1&1&2\\0&1&1\end{bmatrix}$.

1.4.5 Determine which of the following statements are true and which are false.

* (a) If B and C are orthonormal bases of finite-dimensional inner product spaces $\mathcal{V}$ and $\mathcal{W}$, respectively, and $T:\mathcal{V}\to\mathcal{W}$ is a linear transformation, then

$$[T^*]_{B\leftarrow C}=[T]^*_{C\leftarrow B}.$$

* (b) If $A,B\in\mathcal{M}_n(\mathbb{C})$ are Hermitian matrices then so is $A+B$.
* (c) If $A,B\in\mathcal{M}_n(\mathbb{R})$ are symmetric matrices then so is AB.
* (d) There exists a set of 6 mutually orthogonal non-zero vectors in $\mathbb{R}^4$.
* (e) If $U,V\in\mathcal{M}_n$ are unitary matrices, then so is $U+V$.
* (f) If $U\in\mathcal{M}_n$ is a unitary matrix, then U^{-1} exists and is also unitary.
* (g) The identity transformation is an orthogonal projection.
* (h) If $P:\mathcal{V}\to\mathcal{V}$ is a projection then so is $I_\mathcal{V}-P$.

1.4.6 Let $D:\mathcal{P}^2[-1,1]\to\mathcal{P}^2[-1,1]$ be the differentiation map, and recall that $\mathcal{P}^2[-1,1]$ is the vector space of polynomials with degree at most 2 with respect to the standard inner product

$$\langle f,g\rangle=\int_{-1}^{1}f(x)g(x)\,dx.$$

Compute $D^*(2x+1)$.

1.4.7 Find a vector $\mathbf{x}\in\mathbb{R}^3$ that minimizes $\|A\mathbf{x}-\mathbf{b}\|$, where

$$A=\begin{bmatrix}1&1&1\\1&-1&0\\2&0&1\end{bmatrix}\quad\text{and}\quad\mathbf{b}=\begin{bmatrix}1\\1\\1\end{bmatrix}.$$

** **1.4.8** Suppose $U,V\in\mathcal{M}_n$ are unitary matrices. Show that UV is also unitary.

1.4.9 Show that if $U\in\mathcal{M}_n$ is unitary then $\|U\|_\mathrm{F}=\sqrt{n}$ (recall that $\|U\|_\mathrm{F}$ is the Frobenius norm of U).

* **1.4.10** Show that the eigenvalues of unitary matrices lie on the unit circle in the complex plane. That is, show that if $U\in\mathcal{M}_n$ is a unitary matrix, and λ is an eigenvalue of U, then $|\lambda|=1$.

** **1.4.11** Show that if $U\in\mathcal{M}_n$ is a unitary matrix then $|\det(U)|=1$.

** **1.4.12** Show that if $U\in\mathcal{M}_n$ is unitary and upper triangular then it must be diagonal and its diagonal entries must have magnitude 1.

1.4.13 Let $\{\mathbf{u}_1,\mathbf{u}_2,\ldots,\mathbf{u}_n\}$ be any orthonormal basis of $\mathbb{F}^n$ (where $\mathbb{F}=\mathbb{R}$ or $\mathbb{F}=\mathbb{C}$). Show that

$$\sum_{j=1}^{n}\mathbf{u}_j\mathbf{u}_j^*=I.$$

1.4.14 Let $A\in\mathcal{M}_n$ have mutually orthogonal (but *not* necessarily normalized) columns.

(a) Show that A^*A is a diagonal matrix.
(b) Give an example to show that AA^* might not be a diagonal matrix.

∗ **1.4.15** Let $\omega = e^{2\pi i/n}$ (which is an n-th root of unity). Show that the **Fourier matrix** $F \in \mathcal{M}_n(\mathbb{C})$ defined by

$$F = \frac{1}{\sqrt{n}} \begin{bmatrix} 1 & 1 & 1 & \cdots & 1 \\ 1 & \omega & \omega^2 & \cdots & \omega^{n-1} \\ 1 & \omega^2 & \omega^4 & \cdots & \omega^{2n-2} \\ \vdots & \vdots & \vdots & \ddots & \vdots \\ 1 & \omega^{n-1} & \omega^{2n-2} & \cdots & \omega^{(n-1)(n-1)} \end{bmatrix}.$$

is unitary.

[Side note: F is, up to scaling, a Vandermonde matrix.]

[Hint: Try to convince yourself that $\sum_{k=0}^{n-1} \omega^k = 0$.]

1.4.16 Suppose that B and C are bases of a finite-dimensional inner product space $\mathcal{V}$.

(a) Show that if B and C are each orthonormal then $[\mathbf{v}]_B \cdot [\mathbf{w}]_B = [\mathbf{v}]_C \cdot [\mathbf{w}]_C$.
(b) Provide an example that shows that if B and C are *not* both orthonormal bases, then it might be the case that $[\mathbf{v}]_B \cdot [\mathbf{w}]_B \neq [\mathbf{v}]_C \cdot [\mathbf{w}]_C$.

∗∗**1.4.17** Suppose $A \in \mathcal{M}_{m,n}(\mathbb{F})$ and $B \in \mathcal{M}_{n,m}(\mathbb{F})$.

(a) Suppose $\mathbb{F} = \mathbb{R}$. Show that $(A\mathbf{x}) \cdot \mathbf{y} = \mathbf{x} \cdot (B\mathbf{y})$ for all $\mathbf{x} \in \mathbb{R}^n$ and $\mathbf{y} \in \mathbb{R}^m$ if and only if $B = A^T$.
(b) Suppose $\mathbb{F} = \mathbb{C}$. Show that $(A\mathbf{x}) \cdot \mathbf{y} = \mathbf{x} \cdot (B\mathbf{y})$ for all $\mathbf{x} \in \mathbb{C}^n$ and $\mathbf{y} \in \mathbb{C}^m$ if and only if $B = A^*$.

[Side note: In other words, the adjoint of A is either A^T or A^*, depending on the ground field.]

∗∗**1.4.18** Suppose $\mathbb{F} = \mathbb{R}$ or $\mathbb{F} = \mathbb{C}$, E is the standard basis of $\mathbb{F}^n$, and B is any basis of $\mathbb{F}^n$. Show that a change-of-basis matrix $P_{E \leftarrow B}$ is unitary if and only if B is an *orthonormal* basis of $\mathbb{F}^n$.

∗∗**1.4.19** Suppose $\mathbb{F} = \mathbb{R}$ or $\mathbb{F} = \mathbb{C}$ and $B, C \in \mathcal{M}_{m,n}(\mathbb{F})$. Show that the following are equivalent:

a) $B^*B = C^*C$,
b) $(B\mathbf{v}) \cdot (B\mathbf{w}) = (C\mathbf{v}) \cdot (C\mathbf{w})$ for all $\mathbf{v}, \mathbf{w} \in \mathbb{F}^n$, and
c) $\|B\mathbf{v}\| = \|C\mathbf{v}\|$ for all $\mathbf{v} \in \mathbb{F}^n$.

[Hint: If $C = I$ then we get some of the characterizations of unitary matrices from Theorem 1.4.9. Mimic that proof.]

∗∗**1.4.20** Suppose B is a mutually orthogonal set of unit vectors in a finite-dimensional inner product space $\mathcal{V}$. Show that there is an orthonormal basis C of $\mathcal{V}$ with $B \subseteq C \subseteq \mathcal{V}$.

[Side note: The analogous result for non-orthogonal bases was established in Exercise 1.2.26.]

1.4.21 Suppose $\mathcal{V}$ and $\mathcal{W}$ are inner product spaces and $T : \mathcal{V} \to \mathcal{W}$ is a linear transformation with adjoint T^*. Show that $(T^*)^* = T$.

∗∗**1.4.22** Suppose $\mathcal{V}$ and $\mathcal{W}$ are finite-dimensional inner product spaces and $T : \mathcal{V} \to \mathcal{W}$ is a linear transformation with adjoint T^*. Show that $\text{rank}(T^*) = \text{rank}(T)$.

∗∗**1.4.23** Show that every linear transformation $T : \mathcal{V} \to \mathcal{W}$ has at most one adjoint map, even when $\mathcal{V}$ and $\mathcal{W}$ are infinite-dimensional. [Hint: Use Exercise 1.4.27.]

∗∗**1.4.24** Suppose $\mathbb{F} = \mathbb{R}$ or $\mathbb{F} = \mathbb{C}$ and consider a function $\langle \cdot, \cdot \rangle : \mathbb{F}^n \times \mathbb{F}^n \to \mathbb{F}$.

(a) Show that $\langle \cdot, \cdot \rangle$ is an inner product if and only if there exists an invertible matrix $P \in \mathcal{M}_n(\mathbb{F})$ such that
$$\langle \mathbf{v}, \mathbf{w} \rangle = \mathbf{v}^*(P^*P)\mathbf{w} \quad \text{for all} \quad \mathbf{v}, \mathbf{w}, \in \mathbb{F}^n.$$
[Hint: Change a basis in Theorem 1.4.3.]
(b) Find a matrix P associated with the weird inner product from Example 1.3.18. That is, for that inner product find a matrix $P \in \mathcal{M}_2(\mathbb{R})$ such that
$$\langle \mathbf{v}, \mathbf{w} \rangle = \mathbf{v}^*(P^*P)\mathbf{w} \quad \text{for all} \quad \mathbf{v}, \mathbf{w}, \in \mathbb{R}^2.$$
(c) Explain why $\langle \cdot, \cdot \rangle$ is not an inner product if the matrix P from part (a) is not invertible.

∗∗**1.4.25** Suppose $\mathcal{V}$ is a finite-dimensional vector space and $B \subset \mathcal{V}$ is linearly independent. Show that there is an inner product on $\mathcal{V}$ with respect to which B is orthonormal.

[Hint: Use the method of Exercise 1.4.24 to construct an inner product.]

1.4.26 Find an inner product on $\mathbb{R}^2$ with respect to which the set $\{(1,0),(1,1)\}$ is an orthonormal basis.

∗∗**1.4.27** Suppose $\mathcal{V}$ and $\mathcal{W}$ are inner product spaces and $T : \mathcal{V} \to \mathcal{W}$ is a linear transformation. Show that
$$\langle T(\mathbf{v}), \mathbf{w} \rangle = 0 \quad \text{for all} \quad \mathbf{v} \in \mathcal{V} \text{ and } \mathbf{w} \in \mathcal{W}.$$
if and only if $T = O$.

∗∗**1.4.28** Suppose $\mathcal{V}$ is an inner product space and $T : \mathcal{V} \to \mathcal{V}$ is a linear transformation.

(a) Suppose the ground field is $\mathbb{C}$. Show that
$$\langle T(\mathbf{v}), \mathbf{v} \rangle = 0 \quad \text{for all} \quad \mathbf{v} \in \mathcal{V}$$
if and only if $T = O$.
[Hint: Mimic part of the proof of Theorem 1.4.9.]
(b) Suppose the ground field is $\mathbb{R}$. Show that
$$\langle T(\mathbf{v}), \mathbf{v} \rangle = 0 \quad \text{for all} \quad \mathbf{v} \in \mathcal{V}$$
if and only if $T^* = -T$.

∗∗**1.4.29** In this exercise, we show that if $\mathcal{V}$ is a finite-dimensional inner product space over $\mathbb{C}$ and $P : \mathcal{V} \to \mathcal{V}$ is a projection then P is orthogonal (i.e., $\langle P(\mathbf{v}), \mathbf{v} - P(\mathbf{v}) \rangle = 0$ for all $\mathbf{v} \in \mathcal{V}$) if and only if it is self-adjoint (i.e., $P = P^*$).

(a) Show that if P is self-adjoint then it is orthogonal.
(b) Use Exercise 1.4.28 to show that if P is orthogonal then it is self-adjoint.

[Side note: The result of this exercise is still true over $\mathbb{R}$, but it is more difficult to show—see Exercise 2.1.23.]

∗∗**1.4.30** Suppose $A \in \mathcal{M}_{m,n}$ has linearly independent columns.

(a) Show that A^*A is invertible.
(b) Show that $P = A(A^*A)^{-1}A^*$ is the orthogonal projection onto range(A).

[Side note: This exercise generalizes Theorem 1.4.10 to the case when the columns of A are just a basis of its range, but not necessarily an orthonormal one.]

1.4.31 Show that if $P \in \mathcal{M}_n$ is a projection then there exists an invertible matrix $Q \in \mathcal{M}_n$ such that

$$P = Q \begin{bmatrix} I_r & O \\ O & O \end{bmatrix} Q^{-1},$$

where $r = \text{rank}(P)$. In other words, every projection is diagonalizable and has all eigenvalues equal to 0 or 1.

[Hint: What eigen-properties do the vectors in the range and null space of P have?]

[Side note: We prove a stronger decomposition for *orthogonal* projections in Exercise 2.1.24.]

1.4.32 Let $0 < c \in \mathbb{R}$ be a scalar.

(a) Suppose $P_c : \mathcal{C}[-c,c] \to \mathcal{P}^2[-c,c]$ is an orthogonal projection. Compute $P_c(e_x)$. [Hint: We worked through the $c = 1$ case in Example 1.4.18.]
(b) Compute the polynomial $\lim_{c \to 0^+} P_c(e_x)$ and notice that it equals the degree-2 Taylor polynomial of e^x at $x = 0$. Provide a (not necessarily rigorous) explanation for why we would expect these two polynomials to coincide.

1.4.33 Much like we can use polynomials to approximate functions via orthogonal projections, we can also use trigonometric functions to approximate them. Doing so gives us something called the function's **Fourier series**.

(a) Show that, for each $n \geq 1$, the set

$$B_n = \{1, \sin(x), \sin(2x), \ldots, \sin(nx),$$
$$\cos(x), \cos(2x), \ldots, \cos(nx)\}$$

is mutually orthogonal in the usual inner product on $\mathcal{C}[-\pi, \pi]$.
(b) Rescale the members of B_n so that they have norm equal to 1.
(c) Orthogonally project the function $f(x) = x$ onto span(B_n).
[Side note: These are called the **Fourier approximations** of f, and letting $n \to \infty$ gives its Fourier series.]
(d) Use computer software to plot the function $f(x) = x$ from part (c), as well as its projection onto span(B_5), on the interval $[-\pi, \pi]$.

1.5   Summary and Review

In this chapter, we investigated how the various properties of vectors in $\mathbb{R}^n$ and matrices in $\mathcal{M}_{m,n}(\mathbb{R})$ from introductory linear algebra can be generalized to many other settings by working with vector spaces instead of $\mathbb{R}^n$, and linear transformations between them instead of $\mathcal{M}_{m,n}(\mathbb{R})$. In particular, any property or theorem about $\mathbb{R}^n$ that only depends on vector addition and scalar multiplication (i.e., linear combinations) carries over straightforwardly with essentially no changes to abstract vector spaces and the linear transformations acting on them. Examples of such properties include:

These concepts all work over any field $\mathbb{F}$.

- Subspaces, linear (in)dependence, and spanning sets.
- Coordinate vectors, change-of-basis matrices, and standard matrices.
- Invertibility, rank, range, nullity, and null space of linear transformations.
- Eigenvalues, eigenvectors, and diagonalization.

However, the dot product (and properties based on it, like the length of a vector) cannot be described solely in terms of scalar multiplication and vector addition, so we introduced inner products as their abstract generalization. With inner products in hand, we were able to extend the following ideas from $\mathbb{R}^n$ to more general inner product spaces:

These concepts require us to be working over one of the fields $\mathbb{R}$ or $\mathbb{C}$.

- The length of a vector (the norm induced by the inner product).
- Orthogonality.
- The transpose of a matrix (the adjoint of a linear transformation).
- Orthogonal projections.

We could have also used inner products to define angles in general vector spaces.

Having inner products to work with also let us introduce orthonormal bases and unitary matrices, which can be thought of as the "best-behaved" bases and invertible matrices that exist, respectively.

In the finite-dimensional case, none of the aforementioned topics change much when going from $\mathbb{R}^n$ to abstract vector spaces, since every such vector space is isomorphic to $\mathbb{R}^n$ (or $\mathbb{F}^n$, if the vector space is over a field $\mathbb{F}$). In particular, to check some purely linear-algebraic concept like linear independence or invertibility, we can simply convert abstract vectors and linear transformations into vectors in $\mathbb{F}^n$ and matrices in $\mathcal{M}_{m,n}(\mathbb{F})$, respectively, and check the corresponding property there. To check some property that also depends on an inner product like orthogonality or the length of a vector, we can similarly convert everything into vectors in $\mathbb{F}^n$ and matrices in $\mathcal{M}_{m,n}(\mathbb{F})$ as long as we are careful to represent the vectors and matrices in *orthonormal* bases.

For this reason, not much is lost in (finite-dimensional) linear algebra if we explicitly work with $\mathbb{F}^n$ instead of abstract vector spaces, and with matrices instead of linear transformations. We will often switch back and forth between these two perspectives, depending on which is more convenient for the topic at hand. For example, we spend most of Chapter 2 working specifically with matrices, though we will occasionally make a remark about what our results say for linear transformations.

Exercises

solutions to starred exercises on page 459

1.5.1 Determine which of the following statements are true and which are false.

*(a) If $B, C \subseteq \mathcal{V}$ satisfy $\text{span}(B) = \text{span}(C)$ then $B = C$.

(b) If $B, C \subseteq \mathcal{V}$ satisfy $B \subseteq C$ then $\text{span}(B) \subseteq \text{span}(C)$.

*(c) If B and C are bases of finite-dimensional inner product spaces $\mathcal{V}$ and $\mathcal{W}$, respectively, and $T : \mathcal{V} \to \mathcal{W}$ is an invertible linear transformation, then

$$\left[T^{-1}\right]_{B \leftarrow C} = [T]_{C \leftarrow B}^{-1}.$$

(d) If B and C are bases of finite-dimensional inner product spaces $\mathcal{V}$ and $\mathcal{W}$, respectively, and $T : \mathcal{V} \to \mathcal{W}$ is a linear transformation, then

$$[T^*]_{B \leftarrow C} = [T]_{C \leftarrow B}^*.$$

*(e) If $\mathcal{V}$ is a vector space over $\mathbb{C}$ then every linear transformation $T : \mathcal{V} \to \mathcal{V}$ has an eigenvalue.

(f) If $U, V \in \mathcal{M}_n$ are unitary matrices, then so is UV.

1.5.2 Let $\mathcal{S}_1$ and $\mathcal{S}_2$ be finite-dimensional subspaces of a vector space $\mathcal{V}$. Recall from Exercise 1.1.19 that sum of $\mathcal{S}_1$ and $\mathcal{S}_2$ is defined by

$$\mathcal{S}_1 + \mathcal{S}_2 \stackrel{\text{def}}{=} \{\mathbf{v} + \mathbf{w} : \mathbf{v} \in \mathcal{S}_1, \mathbf{w} \in \mathcal{S}_2\},$$

which is also a subspace of $\mathcal{V}$.

(a) Show that

$$\dim(\mathcal{S}_1 + \mathcal{S}_2) = \dim(\mathcal{S}_1) + \dim(\mathcal{S}_2) - \dim(\mathcal{S}_1 \cap \mathcal{S}_2).$$

(b) Show that $\dim(\mathcal{S}_1 \cap \mathcal{S}_2) \geq \dim(\mathcal{S}_1) + \dim(\mathcal{S}_2) - \dim(\mathcal{V})$.

1.5.3 Show that if f is a degree-3 polynomial then $\{f, f', f'', f'''\}$ is a basis of $\mathcal{P}^3$ (where f' is the first derivative of f, and so on).

*1.5.4 Let $\mathcal{P}_2^p$ denote the vector space of 2-variable polynomials of degree at most p. For example,

$$x^2 y^3 - 2x^4 y \in \mathcal{P}_2^5 \quad \text{and} \quad 3x^3 y - x^7 \in \mathcal{P}_2^7.$$

Construct a basis of $\mathcal{P}_2^p$ and compute $\dim(\mathcal{P}_2^p)$.

[Side note: We generalize this exercise to $\mathcal{P}_n^p$, the vector space of n-variable polynomials of degree at most p, in Exercise 3.B.4.]

1.5.5 Explain why the function

$$\langle f, g \rangle = \int_a^b f(x)g(x)\,dx$$

is an inner product on each of $\mathcal{P}[a, b]$, $\mathcal{P}$ (and their subspaces like $\mathcal{P}^p[a, b]$ and $\mathcal{P}^p$), and $\mathcal{C}[a, b]$ (see Example 1.3.17), but *not* on $\mathcal{C}$.

*1.5.6 Suppose $A, B \in \mathcal{M}_{m,n}$ and consider the linear transformation $T_{A,B} : \mathcal{M}_n \to \mathcal{M}_m$ defined by

$$T_{A,B}(X) = AXB^*.$$

Compute T^* (with respect to the standard Frobenius inner product).

1.5.7 How many diagonal unitary matrices are there in $\mathcal{M}_n(\mathbb{R})$? Your answer should be a function of n.

1.A   Extra Topic: More About the Trace

Recall that the trace of a square matrix $A \in \mathcal{M}_n$ is the sum of its diagonal entries:

$$\text{tr}(A) = a_{1,1} + a_{2,2} + \cdots + a_{n,n}.$$

Recall that a function f is **similarity-invariant** if $f(A) = f(PAP^{-1})$ for all invertible P.

While we are already familiar with some nice features of the trace (such as the fact that it is similarity-invariant, and the fact that $\text{tr}(A)$ also equals the sum of the eigenvalues of A), it still seems somewhat arbitrary—why does adding up the diagonal entries of a matrix tell us anything interesting?

In this section, we explore the trace in a bit more depth in an effort to explain where it "comes from", in the same sense that the determinant can be thought of as the answer to the question of how to measure how much a linear transformation expands or contracts space. In brief, the trace is the "most natural" or "most useful" linear form on $\mathcal{M}_n$—it can be thought of as the additive counterpart of the determinant, which in some sense is similarly the "most natural" multiplicative function on $\mathcal{M}_n$.

1.A.1   Algebraic Characterizations of the Trace

Algebraically, what makes the trace interesting is the fact that $\text{tr}(AB) = \text{tr}(BA)$ for all $A \in \mathcal{M}_{m,n}$ and $B \in \mathcal{M}_{n,m}$, since

$$\text{tr}(AB) = \sum_{i=1}^{m}[AB]_{i,i} = \sum_{i=1}^{m}\sum_{j=1}^{n} a_{i,j}b_{j,i}$$

$$= \sum_{j=1}^{n}\sum_{i=1}^{m} b_{j,i}a_{i,j} = \sum_{j=1}^{n}[BA]_{j,j} = \text{tr}(BA).$$

Even though matrix multiplication itself is not commutative, this property lets us treat it as if it were commutative in some situations. To illustrate what we mean by this, we note that the following example can be solved very quickly with the trace, but is quite difficult to solve otherwise.

Example 1.A.1

The Matrix $AB - BA$

Show that there do not exist matrices $A, B \in \mathcal{M}_n$ such that $AB - BA = I$.

Solution:
To see why such matrices cannot exist, simply take the trace of both sides of the equation:

The matrix $AB - BA$ is sometimes called the **commutator** of A and B, and is denoted by $[A, B] \overset{\text{def}}{=} AB - BA$.

$$\text{tr}(AB - BA) = \text{tr}(AB) - \text{tr}(BA) = \text{tr}(AB) - \text{tr}(AB) = 0,$$

but $\text{tr}(I) = n$. Since $n \neq 0$, no such matrices A and B can exist.

Remark 1.A.1

A Characterization of Trace-Zero Matrices

The previous example can actually be extended into a theorem. Using the exact same logic as in that example, we can see that the matrix equation $AB - BA = C$ can only ever have a solution when $\text{tr}(C) = 0$, since

$$\text{tr}(C) = \text{tr}(AB - BA) = \text{tr}(AB) - \text{tr}(BA) = \text{tr}(AB) - \text{tr}(AB) = 0.$$

This fact is proved in (AM57).

Remarkably, the converse of this observation is also true—for any matrix C with $\text{tr}(C) = 0$, there exist matrices A and B of the same size such that

> $AB - BA = C$. Proving this fact is quite technical and outside of the scope of this book, but we prove a slightly weaker result in Exercise 1.A.6.

One of the most remarkable facts about the trace is that, not only does it satisfy this commutativity property, but it is essentially the only linear form that does so. In particular, the following theorem says that the only linear forms for which $f(AB) = f(BA)$ are the trace and its scalar multiples:

Theorem 1.A.1

Commutativity Defines the Trace

> Let $f : \mathcal{M}_n \to \mathbb{F}$ be a linear form with the following properties:
>   a) $f(AB) = f(BA)$ for all $A, B \in \mathcal{M}_n$, and
>   b) $f(I) = n$.
> Then $f(A) = \text{tr}(A)$ for all $A \in \mathcal{M}_n$.

Proof. Every matrix $A \in \mathcal{M}_n(\mathbb{F})$ can be written as a linear combination of the standard basis matrices:

Recall that $E_{i,j}$ is the matrix with a 1 as its (i,j)-entry and all other entries equal to 0.

$$A = \sum_{i,j=1}^{n} a_{i,j} E_{i,j}.$$

Since f is linear, it follows that

$$f(A) = f\left(\sum_{i,j=1}^{n} a_{i,j} E_{i,j} \right) = \sum_{i,j=1}^{n} a_{i,j} f(E_{i,j}).$$

Our goal is thus to show that $f(E_{i,j}) = 0$ whenever $i \neq j$ and $f(E_{i,j}) = 1$ whenever $i = j$, since that would imply

$$f(A) = \sum_{i=1}^{n} a_{i,i} = \text{tr}(A).$$

To this end, we first recall that since f is linear it must be the case that $f(O) = 0$. Next, we notice that $E_{i,j} E_{j,j} = E_{i,j}$ but $E_{j,j} E_{i,j} = O$ whenever $i \neq j$, so property (a) of f given in the statement of the theorem implies

$$0 = f(O) = f\left(E_{j,j} E_{i,j} \right) = f\left(E_{i,j} E_{j,j} \right) = f(E_{i,j}) \quad \text{whenever} \quad i \neq j,$$

which is one of the two facts that we wanted to show.

To similarly prove the other fact (i.e., $f(E_{j,j}) = 1$ for all $1 \leq j \leq n$), we notice that $E_{1,j} E_{j,1} = E_{1,1}$, but $E_{j,1} E_{1,1} = E_{j,j}$ for all $1 \leq j \leq n$, so

$$f(E_{j,j}) = f\left(E_{j,1} E_{1,j} \right) = f\left(E_{1,j} E_{j,1} \right) = f(E_{1,1}) \quad \text{for all} \quad 1 \leq j \leq n.$$

However, since $I = E_{1,1} + E_{2,2} + \cdots + E_{n,n}$, it then follows that

$$\begin{aligned}
n = f(I) &= f(E_{1,1} + E_{2,2} + \cdots + E_{n,n}) \\
&= f(E_{1,1}) + f(E_{2,2}) + \cdots + f(E_{n,n}) = n f(E_{1,1}),
\end{aligned}$$

so $f(E_{1,1}) = 1$, and similarly $f(E_{j,j}) = 1$ for all $1 \leq j \leq n$, which completes the proof. ∎

There are also a few other ways of thinking of the trace as the unique function with certain properties. For example, there are numerous functions of matrices that are similarity-invariant (e.g., the rank, trace, and determinant), but the following corollary says that the trace is (up to scaling) the only one that is linear.

Corollary 1.A.2

Similarity-Invariance Defines the Trace

Suppose $\mathbb{F} = \mathbb{R}$ or $\mathbb{F} = \mathbb{C}$, and $f : \mathcal{M}_n(\mathbb{F}) \to \mathbb{F}$ is a linear form with the following properties:

    a) $f(A) = f(PAP^{-1})$ for all $A, P \in \mathcal{M}_n(\mathbb{F})$ with P invertible, and

    b) $f(I) = n$.

Then $f(A) = \text{tr}(A)$ for all $A \in \mathcal{M}_n(\mathbb{F})$.

The idea behind this corollary is that if A or B is invertible, then AB is similar to BA (see Exercise 1.A.5). It follows that if f is similarity-invariant then $f(AB) = f(BA)$ as long as at least one of A or B is invertible, which *almost* tells us, via Theorem 1.A.1, that f must be the trace (up to scaling).

However, making the jump from $f(AB) = f(BA)$ whenever at least one of A or B invertible, to $f(AB) = f(BA)$ for *all* A and B, is somewhat delicate. We thus return to this corollary in Section 2.D.3 (in particular, see Theorem 2.D.4), where we prove it via matrix analysis techniques. The idea is that, since every matrix is arbitrarily close to an invertible matrix, if this property holds for all invertible matrices then continuity of f tells us that it must in fact hold for non-invertible matrices too.

As one final way of characterizing the trace, notice that if $P \in \mathcal{M}_n$ is a (not necessarily orthogonal) projection (i.e., $P^2 = P$), then $\text{tr}(P) = \text{rank}(P)$ (see Exercise 1.A.4). In other words, $\text{tr}(P)$ is the dimension of the subspace projected onto by P. Our final result of this subsection says that this property characterizes the trace—it is the *only* linear form with this property. In a sense, this means that the trace can be thought of as a linearized version of the rank or dimension-counting function.

Theorem 1.A.3

Rank of Projections Defines the Trace

Let $f : \mathcal{M}_n \to \mathbb{F}$ be a linear form. The following are equivalent:

    a) $f(P) = \text{rank}(P)$ for all projections $P \in \mathcal{M}_n$.

    b) $f(A) = \text{tr}(A)$ for all $A \in \mathcal{M}_n$.

Proof. To see that (a) $\implies$ (b), we proceed much like in the proof of Theorem 1.A.1—our goal is to show that $f(E_{j,j}) = 1$ and $f(E_{i,j}) = 0$ for all $1 \leq i \neq j \leq n$.

> In fact, this shows that if f is a linear function for which $f(P) = 1$ for all *rank*-1 projections $P \in \mathcal{M}_n$ then $f = \text{tr}$.

The reason that $f(E_{j,j}) = 1$ is simply that $E_{j,j}$ is a rank-1 projection for each $1 \leq j \leq n$. To see that $f(E_{i,j}) = 0$ when $i \neq j$, notice that $E_{j,j} + E_{i,j}$ is also a rank-1 (oblique) projection, so $f(E_{j,j} + E_{i,j}) = 1$. However, since $f(E_{j,j}) = 1$, linearity of f tells us that $f(E_{i,j}) = 0$. It follows that for every matrix $A \in \mathcal{M}_n(\mathbb{F})$ we have

$$f(A) = f\left(\sum_{i,j=1}^{n} a_{i,j} E_{i,j} \right) = \sum_{i,j=1}^{n} a_{i,j} f(E_{i,j}) = \sum_{j=1}^{n} a_{j,j} = \text{tr}(A),$$

as desired.

The fact that (b) $\implies$ (a) is left to Exercise 1.A.4. $\blacksquare$

If we have an inner product to work with (and are thus working over the field $\mathbb{F} = \mathbb{R}$ or $\mathbb{F} = \mathbb{C}$), we can ask whether or not the above theorem can be strengthened to consider only *orthogonal* projections (i.e., projections P for which $P^* = P$). It turns out that if $\mathbb{F} = \mathbb{C}$ then this works—if $f(P) = \text{rank}(P)$ for all orthogonal projections $P \in \mathcal{M}_n(\mathbb{C})$ then $f(A) = \text{tr}(A)$ for all $A \in \mathcal{M}_n(\mathbb{C})$. However, this is not true if $\mathbb{F} = \mathbb{R}$ (see Exercise 1.A.7).

1.A.2   Geometric Interpretation of the Trace

Since the trace of a matrix is similarity-invariant, it should have some sort of geometric interpretation that depends only on the underlying linear transformation (after all, similarity-invariance means exactly that it does not depend on the basis that we use to represent that linear transformation). This is a bit more difficult to see than it was for the rank or determinant, but one such geometric interpretation is the best linear approximation of the determinant.

We return to similarity-invariance and its geometric interpretation in Section 2.4.

The above statement can be made more precise with derivatives. In particular, the following theorem says that if we start at the identity transformation (i.e., the linear transformation that does nothing) and move slightly in the direction of A, then space is expanded by an amount proportional to $\mathrm{tr}(A)$. More specifically, the directional derivative of the determinant in the direction of A is its trace:

Theorem 1.A.4
Directional Derivative of the Determinant

Suppose $\mathbb{F} = \mathbb{R}$ or $\mathbb{F} = \mathbb{C}$, let $A \in \mathcal{M}_n(\mathbb{F})$, and define a function $f_A : \mathbb{F} \to \mathbb{F}$ by $f_A(x) = \det(I + xA)$. Then $f_A'(0) = \mathrm{tr}(A)$.

Before proving this theorem, it is perhaps useful to try to picture it. If $A \in \mathcal{M}_n$ has columns $\mathbf{a}_1, \mathbf{a}_2, \ldots, \mathbf{a}_n$ then the linear transformation $I + xA$ adds $x\mathbf{a}_1, x\mathbf{a}_2, \ldots, x\mathbf{a}_n$ to the side vectors of the unit square (or cube, or hypercube...). The majority of the change in the determinant of $I + xA$ thus comes from how much $x\mathbf{a}_1$ points in the direction of $\mathbf{e}_1$ (i.e., $xa_{1,1}$) plus the amount that $x\mathbf{a}_2$ points in the direction of $\mathbf{e}_2$ (i.e., $xa_{2,2}$), and so on (see Figure 1.21), which equals $x\mathrm{tr}(A)$. In other words, $\mathrm{tr}(A)$ provides the first-order (or linear) approximation for how $\det(I + xA)$ changes when x is small.

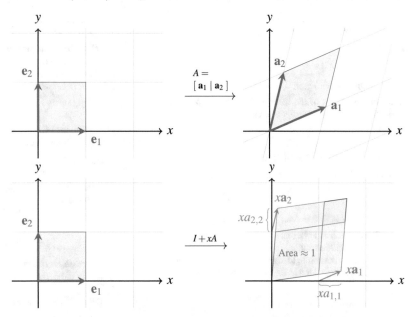

Figure 1.21: The determinant $\det(I + xA)$ can be split up into four pieces, as indicated here at the bottom-right. The blue region has area approximately equal to 1, the purple region has area proportional to x^2, and the orange region has area $xa_{1,1} + xa_{2,2} = x\mathrm{tr}(A)$. When x is close to 0, this orange region is much larger than the purple region and thus determines the growth rate of $\det(I + xA)$.

Proof of Theorem 1.A.4. Recall that the characteristic polynomial p_A of A has the form

$$p_A(\lambda) = \det(A - \lambda I) = (-1)^n \lambda^n + (-1)^{n-1}\mathrm{tr}(A)\lambda^{n-1} + \cdots + \det(A).$$

If we let $\lambda = -1/x$ and then multiply this characteristic polynomial through by x^n, we see that

$$\begin{aligned}
f_A(x) = \det(I + xA) &= x^n \det(A + (1/x)I) \\
&= x^n p_A(-1/x) \\
&= x^n\left((1/x)^n + \mathrm{tr}(A)(1/x)^{n-1} + \cdots + \det(A)\right) \\
&= 1 + x\mathrm{tr}(A) + \cdots + \det(A)x^n.
\end{aligned}$$

Recall that the derivative of x^n is nx^{n-1}.

It follows that

$$f_A'(x) = \mathrm{tr}(A) + \cdots + \det(A)nx^{n-1},$$

where each of the terms hidden in the "$\cdots$" has at least one power of x in it. We thus conclude that $f_A'(0) = \mathrm{tr}(A)$, as claimed. ∎

Exercises

solutions to starred exercises on page 459

1.A.1 Determine which of the following statements are true and which are false.

(a) If $A, B \in \mathcal{M}_n$ then $\mathrm{tr}(A + B) = \mathrm{tr}(A) + \mathrm{tr}(B)$.
*(b) If $A, B \in \mathcal{M}_n$ then $\mathrm{tr}(AB) = \mathrm{tr}(A)\mathrm{tr}(B)$.
(c) If $A \in \mathcal{M}_n$ is invertible, then $\mathrm{tr}(A^{-1}) = 1/\mathrm{tr}(A)$.
*(d) If $A \in \mathcal{M}_n$ then $\mathrm{tr}(A) = \mathrm{tr}(A^T)$.
(e) If $A \in \mathcal{M}_n$ has k-dimensional range then $\mathrm{tr}(A) = k$.

**1.A.2 Suppose $A, B \in \mathcal{M}_n$ are such that A is symmetric (i.e, $A = A^T$) and B is skew-symmetric (i.e., $B = -B^T$). Show that $\mathrm{tr}(AB) = 0$.

1.A.3 Suppose $A \in \mathcal{M}_{m,n}(\mathbb{C})$. Show that $\mathrm{tr}(A^*A) \geq 0$.

[Side note: Matrices of the form A^*A are called **positive semidefinite**, and we investigate them thoroughly in Section 2.2.]

**1.A.4 Show that if $P \in \mathcal{M}_n$ then $\mathrm{tr}(P) = \mathrm{rank}(P)$.

[Side note: This is the implication (b) $\implies$ (a) of Theorem 1.A.3.]

**1.A.5 Suppose $A, B \in \mathcal{M}_n$.

(a) Show that if at least one of A or B is invertible then AB and BA are similar.
(b) Provide an example to show that if A and B are not invertible then AB and BA may not be similar.

**1.A.6 Consider the two sets

$$\mathcal{Z} = \{C \in \mathcal{M}_n : \mathrm{tr}(C) = 0\} \quad \text{and}$$
$$\mathcal{W} = \mathrm{span}\{AB - BA : A, B \in \mathcal{M}_n\}.$$

In this exercise, we show that $\mathcal{W} = \mathcal{Z}$. [Side note: Refer back to Remark 1.A.1 for some context.]

(a) Show that $\mathcal{W}$ is a subspace of $\mathcal{Z}$, which is a subspace of $\mathcal{M}_n$.
(b) Compute $\dim(\mathcal{Z})$.

(c) Show that $\mathcal{W} = \mathcal{Z}$. [Hint: Find sufficiently many linearly independent matrices in $\mathcal{W}$ to show that its dimension coincides with that of $\mathcal{Z}$.]

**1.A.7 Suppose $\mathbb{F} = \mathbb{R}$ or $\mathbb{F} = \mathbb{C}$ and let $f : \mathcal{M}_n(\mathbb{F}) \to \mathbb{F}$ be a linear form with the property that $f(P) = \mathrm{rank}(P)$ for all orthogonal projections $P \in \mathcal{M}_n(\mathbb{F})$.

(a) Show that if $\mathbb{F} = \mathbb{C}$ then $f(A) = \mathrm{tr}(A)$ for all $A \in \mathcal{M}_n(\mathbb{C})$.
[Hint: Take inspiration from the proof of Theorem 1.A.3, but use some rank-2 projections too.]
(b) Provide an example to show that if $\mathbb{F} = \mathbb{R}$ then it is *not* necessarily the case that $f(A) = \mathrm{tr}(A)$ for all $A \in \mathcal{M}_n(\mathbb{R})$.

**1.A.8 Suppose $\mathbb{F} = \mathbb{R}$ or $\mathbb{F} = \mathbb{C}$ and let $A, B \in \mathcal{M}_n(\mathbb{F})$ be such that A is invertible.

(a) Show that if $f_{A,B}(x) = \det(A + xB)$ then $f_{A,B}'(0) = \det(A)\mathrm{tr}(A^{-1}B)$.
(b) Suppose that the entries of A depend in a differentiable way on a parameter $t \in \mathbb{F}$ (so we denote it by $A(t)$ from now on). Explain why

$$\frac{d}{dt}\det(A(t)) = f_{A(t),dA/dt}'(0),$$

where dA/dt refers to the matrix that is obtained by taking the derivative of each entry of A with respect to t.
[Hint: Taylor's theorem from Appendix A.2 might help.]
(c) Show that

$$\frac{d}{dt}\det(A(t)) = \det(A(t))\mathrm{tr}\left(A(t)^{-1}\frac{dA}{dt}\right).$$

[Side note: This is called **Jacobi's formula**, and we generalize it to non-invertible matrices in Theorem 2.D.6.]

1.B   Extra Topic: Direct Sum, Orthogonal Complement

It is often useful to break apart a large vector space into multiple subspaces that do not intersect each other (except at the zero vector, where intersection is unavoidable). For example, it is somewhat natural to think of $\mathbb{R}^2$ as being made up of two copies of $\mathbb{R}$, since every vector $(x,y) \in \mathbb{R}^2$ can be written in the form $(x,0) + (0,y)$, and the subspaces $\{(x,0) : x \in \mathbb{R}\}$ and $\{(0,y) : y \in \mathbb{R}\}$ are each isomorphic to $\mathbb{R}$ in a natural way.

Similarly, we can think of $\mathbb{R}^3$ as being made up of either three copies of $\mathbb{R}$, or a copy of $\mathbb{R}^2$ and a copy of $\mathbb{R}$, as illustrated in Figure 1.22. The direct sum provides a way of making this idea precise, and we explore it thoroughly in this section.

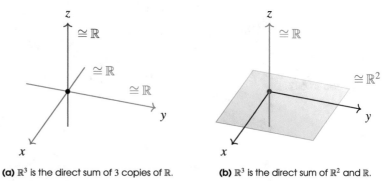

(a) $\mathbb{R}^3$ is the direct sum of 3 copies of $\mathbb{R}$.   **(b)** $\mathbb{R}^3$ is the direct sum of $\mathbb{R}^2$ and $\mathbb{R}$.

Figure 1.22: The direct sum lets us break down $\mathbb{R}^3$ (and other vector spaces) into smaller subspaces that do not intersect each other except at the origin.

1.B.1   The Internal Direct Sum

We now pin down exactly what we mean by the afore-mentioned idea of splitting up a vector space into two non-intersecting subspaces.

Definition 1.B.1

The Internal Direct Sum

> Let $\mathcal{V}$ be a vector space with subspaces $\mathcal{S}_1$ and $\mathcal{S}_2$. We say that $\mathcal{V}$ is the **(internal) direct sum** of $\mathcal{S}_1$ and $\mathcal{S}_2$, denoted by $\mathcal{V} = \mathcal{S}_1 \oplus \mathcal{S}_2$, if
>
> a) $\operatorname{span}(\mathcal{S}_1 \cup \mathcal{S}_2) = \mathcal{V}$, and
> b) $\mathcal{S}_1 \cap \mathcal{S}_2 = \{\mathbf{0}\}$.

For now, we just refer to this as a direct sum (without caring about the "internal" part of its name). We will distinguish between this and another type of direct sum later.

The two defining properties of the direct sum mimic very closely the two defining properties of bases (Definition 1.1.6). Just like bases must span the entire vector space, so too must subspaces in a direct sum, and just like bases must be "small enough" that they are linearly independent, subspaces in a direct sum must be "small enough" that they only contain the zero vector in common.

It is also worth noting that the defining property (a) of the direct sum is equivalent to saying that every vector $\mathbf{v} \in \mathcal{V}$ can be written in the form $\mathbf{v} = \mathbf{v}_1 + \mathbf{v}_2$ for some $\mathbf{v}_1 \in \mathcal{S}_1$ and $\mathbf{v}_2 \in \mathcal{S}_2$. The reason for this is simply that in any linear combination of vectors from $\mathcal{S}_1$ and $\mathcal{S}_2$, we can group the terms from S_1 into $\mathbf{v}_1$ and the terms from S_2 into $\mathbf{v}_2$. That is, if we write

$$\mathbf{v} = \underbrace{(c_1\mathbf{x}_1 + c_2\mathbf{x}_2 + \cdots + c_k\mathbf{x}_k)}_{\text{call this } \mathbf{v}_1} + \underbrace{(d_1\mathbf{y}_1 + d_2\mathbf{y}_2 + \cdots + d_m\mathbf{y}_m)}_{\text{call this } \mathbf{v}_2}, \qquad (1.B.1)$$

where $\mathbf{x}_1, \ldots, \mathbf{x}_k \in S_1$, $\mathbf{y}_1, \ldots, \mathbf{y}_m \in S_2$, and $c_1, \ldots, c_k, d_1, \ldots, d_m \in \mathbb{F}$, then we can just define $\mathbf{v}_1$ and $\mathbf{v}_2$ to be the parenthesized terms indicated above.

Example 1.B.1

Checking Whether or Not Subspaces Make a Direct Sum

Determine whether or not $\mathbb{R}^3 = S_1 \oplus S_2$, where S_1 and S_2 are the given subspaces.

a) S_1 is the x-axis and S_2 is the y-axis.
b) S_1 is xy-plane and S_2 is the yz-plane.
c) S_1 is the line through the origin in the direction of the vector $(0,1,1)$ and S_2 is the xy-plane.

Solutions:

a) To determine whether or not $V = S_1 \oplus S_2$, we must check the two defining properties of the direct sum from Definition 1.B.1. Indeed, property (a) does not hold since $\mathrm{span}(S_1 \cup S_2)$ is just the xy-plane, not all of $\mathbb{R}^3$, so $\mathbb{R}^3 \neq S_1 \oplus S_2$:

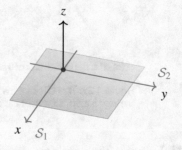

b) Again, we check the two defining properties of the direct sum. Property (a) holds since every vector in $\mathbb{R}^3$ can be written as a linear combination of vectors in the xy-plane and the yz-plane. However, property (b) does not hold since $S_1 \cap S_2$ is the y-axis (not just $\{\mathbf{0}\}$),

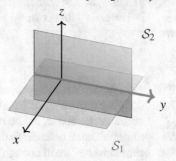

so $\mathbb{R}^3 \neq S_1 \oplus S_2$.

c) To see that $\mathrm{span}(S_1 \cup S_2) = \mathbb{R}^3$ we must show that we can write every vector $(x,y,z) \in \mathbb{R}^3$ as a linear combination of vectors from S_1 and S_2. One way to do this is to notice that

$$(x,y,z) = (0,z,z) + (x,y-z,0).$$

Since $(0, z, z) \in \mathcal{S}_1$ and $(x, y - z, 0) \in \mathcal{S}_2$, we have

$$(x, y, z) \in \text{span}(\mathcal{S}_1 \cup \mathcal{S}_2),$$

so $\text{span}(\mathcal{S}_1 \cup \mathcal{S}_2) = \mathbb{R}^3$.

To see that $\mathcal{S}_1 \cap \mathcal{S}_2 = \{\mathbf{0}\}$, suppose $(x, y, z) \in \mathcal{S}_1 \cap \mathcal{S}_2$. Since $(x, y, z) \in \mathcal{S}_1$, we know that $x = 0$ and $y = z$. Since $(x, y, z) \in \mathcal{S}_2$, we know that $z = 0$. It follows that $(x, y, z) = (0, 0, 0) = \mathbf{0}$, so $\mathcal{S}_1 \cap \mathcal{S}_2 = \{\mathbf{0}\}$, and

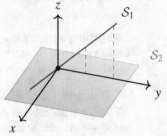

Notice that the line $\mathcal{S}_1$ is not orthogonal to the plane $\mathcal{S}_2$. The intuition here is the same as that of linear independence, not of orthogonality.

thus $\mathbb{R}^3 = \mathcal{S}_1 \oplus \mathcal{S}_2$.

The direct sum extends straightforwardly to three or more subspaces as well. In general, we say that $\mathcal{V}$ is the (internal) direct sum of subspaces $\mathcal{S}_1, \mathcal{S}_2, \ldots, \mathcal{S}_k$ if $\text{span}(\mathcal{S}_1 \cup \mathcal{S}_2 \cup \cdots \cup \mathcal{S}_k) = \mathcal{V}$ and

The notation with the big "$\cup$" union symbol here is analogous to big-Σ notation for sums.

$$\mathcal{S}_i \cap \text{span}\left(\bigcup_{j \neq i} \mathcal{S}_j \right) = \{\mathbf{0}\} \quad \text{for all} \quad 1 \leq i \leq k. \tag{1.B.2}$$

Indeed, Equation (1.B.2) looks somewhat complicated on the surface, but it just says that each subspace $\mathcal{S}_i$ ($1 \leq i \leq k$) has no non-zero intersection with (the span of) the rest of them. In this case we write either

$$\mathcal{V} = \mathcal{S}_1 \oplus \mathcal{S}_2 \oplus \cdots \oplus \mathcal{S}_k \quad \text{or} \quad \mathcal{V} = \bigoplus_{j=1}^{k} \mathcal{S}_j.$$

For example, if $\mathcal{S}_1$, $\mathcal{S}_2$, and $\mathcal{S}_3$ are the x-, y-, and z-axes in $\mathbb{R}^3$, then $\mathbb{R}^3 = \mathcal{S}_1 \oplus \mathcal{S}_2 \oplus \mathcal{S}_3$. More generally, $\mathbb{R}^n$ can be written as the direct sum of its n coordinate axes. Even more generally, given any basis $\{\mathbf{v}_1, \mathbf{v}_2, \ldots, \mathbf{v}_k\}$ of a finite-dimensional vector space $\mathcal{V}$, it is the case that

This claim is proved in Exercise 1.B.4.

$$\mathcal{V} = \text{span}(\mathbf{v}_1) \oplus \text{span}(\mathbf{v}_2) \oplus \cdots \oplus \text{span}(\mathbf{v}_k).$$

We thus think of the direct sum as a higher-dimensional generalization of bases: while bases break vector spaces down into 1-dimensional subspaces (i.e., the lines in the direction of the basis vectors) that only intersect at the zero vector, direct sums allow for subspaces of any dimension.

Given this close connection between bases and direct sums, it should not be surprising that many of our theorems concerning bases from Section 1.2 generalize in a straightforward way to direct sums. Our first such result is analogous to the fact that there is a unique way to write every vector in a vector space as a linear combination of basis vectors (Theorem 1.1.4).

Theorem 1.B.1

Uniqueness of Sums in Direct Sums

Suppose that $\mathcal{V}$ is a vector space with subspaces $\mathcal{S}_1, \mathcal{S}_2 \subseteq \mathcal{V}$ satisfying $\mathcal{V} = \mathcal{S}_1 \oplus \mathcal{S}_2$. For every $\mathbf{v} \in \mathcal{V}$, there is exactly one way to write $\mathbf{v}$ in the form

$$\mathbf{v} = \mathbf{v}_1 + \mathbf{v}_2 \quad \text{with} \quad \mathbf{v}_1 \in \mathcal{S}_1 \quad \text{and} \quad \mathbf{v}_2 \in \mathcal{S}_2.$$

Proof. We already noted that $\mathbf{v}$ can be written in this form back in Equation (1.B.1). To see uniqueness, suppose that there exist $\mathbf{v}_1, \mathbf{w}_1 \in \mathcal{S}_1$ and $\mathbf{v}_2, \mathbf{w}_2 \in \mathcal{S}_2$ such that

$$\mathbf{v} = \mathbf{v}_1 + \mathbf{v}_2 = \mathbf{w}_1 + \mathbf{w}_2.$$

Recall that $\mathcal{S}_1$ is a subspace, so if $\mathbf{v}_1, \mathbf{w}_1 \in \mathcal{S}_1$ then $\mathbf{v}_1 - \mathbf{w}_1 \in \mathcal{S}_1$ too (and similarly for $\mathcal{S}_2$).

Subtracting $\mathbf{v}_2 + \mathbf{w}_1$ from the above equation shows that $\mathbf{v}_1 - \mathbf{w}_1 = \mathbf{w}_2 - \mathbf{v}_2$. Since $\mathbf{v}_1 - \mathbf{w}_1 \in \mathcal{S}_1$ and $\mathbf{w}_2 - \mathbf{v}_2 \in \mathcal{S}_2$, and $\mathcal{S}_1 \cap \mathcal{S}_2 = \{\mathbf{0}\}$, this implies $\mathbf{v}_1 - \mathbf{w}_1 = \mathbf{w}_2 - \mathbf{v}_2 = \mathbf{0}$, so $\mathbf{v}_1 = \mathbf{w}_1$ and $\mathbf{v}_2 = \mathbf{w}_2$, as desired. ∎

Similarly, we now show that if we combine bases of the subspaces $\mathcal{S}_1$ and $\mathcal{S}_2$ then we get a basis of $\mathcal{V} = \mathcal{S}_1 \oplus \mathcal{S}_2$. This hopefully makes some intuitive sense—every vector in $\mathcal{V}$ can be represented uniquely as a sum of vectors in $\mathcal{S}_1$ and $\mathcal{S}_2$, and every vector in those subspaces can be represented uniquely as a linear combination of their basis vectors.

Theorem 1.B.2

Bases of Direct Sums

Suppose that $\mathcal{V}$ is a vector space and $\mathcal{S}_1, \mathcal{S}_2 \subseteq \mathcal{V}$ are subspaces with bases B and C, respectively. Then

a) $\text{span}(B \cup C) = \mathcal{V}$ if and only if $\text{span}(\mathcal{S}_1 \cup \mathcal{S}_2) = \mathcal{V}$, and

b) $B \cup C$ is linearly independent if and only if $\mathcal{S}_1 \cap \mathcal{S}_2 = \{\mathbf{0}\}$.

In particular, $B \cup C$ is a basis of $\mathcal{V}$ if and only if $\mathcal{V} = \mathcal{S}_1 \oplus \mathcal{S}_2$.

Here, $B \cup C$ refers to the union of B and C as a *multiset*, so that if there is a common vector in each of B and C then we immediately regard $B \cup C$ as linearly dependent.

Proof. For part (a), we note that if $\text{span}(B \cup C) = \mathcal{V}$ then it must be the case that $\text{span}(\mathcal{S}_1 \cup \mathcal{S}_2) = \mathcal{V}$ as well, since $\text{span}(B \cup C) \subset \text{span}(\mathcal{S}_1 \cup \mathcal{S}_2)$. In the other direction, if $\text{span}(\mathcal{S}_1 \cup \mathcal{S}_2) = \mathcal{V}$ then we can write every $\mathbf{v} \in \mathcal{V}$ in the form $\mathbf{v} = \mathbf{v}_1 + \mathbf{v}_2$, where $\mathbf{v}_1 \in \mathcal{S}_1$ and $\mathbf{v}_2 \in \mathcal{S}_2$. Since B and C are bases of $\mathcal{S}_1$ and $\mathcal{S}_2$, respectively, we can write $\mathbf{v}_1$ and $\mathbf{v}_2$ as linear combinations of vectors from those sets:

$$\mathbf{v} = \mathbf{v}_1 + \mathbf{v}_2 = \left(c_1 \mathbf{x}_1 + c_2 \mathbf{x}_2 + \cdots + c_k \mathbf{x}_k \right) + \left(d_1 \mathbf{y}_1 + d_2 \mathbf{y}_2 + \cdots + d_m \mathbf{y}_m \right),$$

where $\mathbf{x}_1, \mathbf{x}_2, \ldots, \mathbf{x}_k \in \mathcal{S}_1$, $\mathbf{y}_1, \mathbf{y}_2, \ldots, \mathbf{y}_m \in \mathcal{S}_2$, and $c_1, c_2, \ldots, c_k, d_1, d_2, \ldots, d_m \in \mathbb{F}$. We have thus written $\mathbf{v}$ as a linear combination of vectors from $B \cup C$, so $\text{span}(B \cup C) = \mathcal{V}$.

For part (b), suppose that $\mathcal{S}_1 \cap \mathcal{S}_2 = \{\mathbf{0}\}$ and consider some linear combination of vectors from $B \cup C$ that equals the zero vector:

$$\underbrace{\left(c_1 \mathbf{x}_1 + c_2 \mathbf{x}_2 + \cdots + c_k \mathbf{x}_k \right)}_{\text{call this } \mathbf{v}_1} + \underbrace{\left(d_1 \mathbf{y}_1 + d_2 \mathbf{y}_2 + \cdots + d_m \mathbf{y}_m \right)}_{\text{call this } \mathbf{v}_2} = \mathbf{0}, \qquad (1.B.3)$$

As was the case with proofs about bases, these proofs are largely uninspiring definition-chasing affairs. The theorems themselves should be somewhat intuitive though.

where the vectors and scalars come from the same spaces as they did in part (a). Our goal is to show that $c_1 = c_2 = \cdots = c_k = 0$ and $d_1 = d_2 = \cdots = d_m = 0$, which implies linear independence of $B \cup C$.

To this end, notice that Equation (1.B.3) says that $\mathbf{0} = \mathbf{v}_1 + \mathbf{v}_2$, where $\mathbf{v}_1 \in \mathcal{S}_1$ and $\mathbf{v}_2 \in \mathcal{S}_2$. Since $\mathbf{0} = \mathbf{0} + \mathbf{0}$ is another way of writing $\mathbf{0}$ as a sum of something from $\mathcal{S}_1$ and something from $\mathcal{S}_2$, Theorem 1.B.1 tells us that $\mathbf{v}_1 = \mathbf{0}$ and $\mathbf{v}_2 = \mathbf{0}$. It follows that

$$c_1 \mathbf{x}_1 + c_2 \mathbf{x}_2 + \cdots + c_k \mathbf{x}_k = \mathbf{0} \quad \text{and} \quad d_1 \mathbf{y}_1 + d_2 \mathbf{y}_2 + \cdots + d_m \mathbf{y}_m = \mathbf{0},$$

so linear independence of B implies $c_1 = c_2 = \cdots = c_k = 0$ and linear independence of C similarly implies $d_1 = d_2 = \cdots = d_m = 0$. We thus conclude that $B \cup C$ is linearly independent.

All that remains is to show that $B \cup C$ being linearly independent implies $\mathcal{S}_1 \cap \mathcal{S}_2 = \{\mathbf{0}\}$. This proof has gone on long enough already, so we leave this final implication to Exercise 1.B.5. ∎

For example, the above theorem implies that if we take any basis of a vector space $\mathcal{V}$ and partition that basis in any way, then the spans of those partitions form a direct sum decomposition of $\mathcal{V}$. For example, if $B = \{\mathbf{v}_1, \mathbf{v}_2, \mathbf{v}_3, \mathbf{v}_4\}$ is a basis of a (4-dimensional) vector space $\mathcal{V}$ then

$$\begin{aligned}
\mathcal{V} &= \operatorname{span}(\mathbf{v}_1) \oplus \operatorname{span}(\mathbf{v}_2) \oplus \operatorname{span}(\mathbf{v}_3) \oplus \operatorname{span}(\mathbf{v}_4) \\
&= \operatorname{span}(\mathbf{v}_1 \cup \mathbf{v}_2) \oplus \operatorname{span}(\mathbf{v}_3) \oplus \operatorname{span}(\mathbf{v}_4) \\
&= \operatorname{span}(\mathbf{v}_1) \oplus \operatorname{span}(\mathbf{v}_2 \cup \mathbf{v}_4) \oplus \operatorname{span}(\mathbf{v}_3) \\
&= \operatorname{span}(\mathbf{v}_1 \cup \mathbf{v}_3 \cup \mathbf{v}_4) \oplus \operatorname{span}(\mathbf{v}_2),
\end{aligned}$$

as well as many other possibilities.

Example 1.B.2 **Even and Odd Polynomials as a Direct Sum**	Let $\mathcal{P}^E$ and $\mathcal{P}^O$ be the subspaces of $\mathcal{P}$ consisting of the even and odd polynomials, respectively:

$$\mathcal{P}^E = \{f \in \mathcal{P} : f(-x) = f(x)\} \quad \text{and} \quad \mathcal{P}^O = \{f \in \mathcal{P} : f(-x) = -f(x)\}.$$

Show that $\mathcal{P} = \mathcal{P}^E \oplus \mathcal{P}^O$.

We introduced $\mathcal{P}^E$ and $\mathcal{P}^O$ in Example 1.1.21.

Solution:

We could directly show that $\operatorname{span}(\mathcal{P}^E \cup \mathcal{P}^O) = \mathcal{P}$ and $\mathcal{P}^E \cap \mathcal{P}^O = \{\mathbf{0}\}$, but perhaps an easier way is consider how the bases of these vector spaces relate to each other. Recall from Example 1.1.21 that $B = \{1, x^2, x^4, \ldots\}$ is a basis of $\mathcal{P}^E$ and $C = \{x, x^3, x^5, \ldots\}$ is a basis of $\mathcal{P}^O$. Since

$$B \cup C = \{1, x, x^2, x^3, \ldots\}$$

is a basis of $\mathcal{P}$, we conclude from Theorem 1.B.2 that $\mathcal{P} = \mathcal{P}^E \oplus \mathcal{P}^O$.

Another direct consequence of Theorem 1.B.2 is the (hopefully intuitive) fact that we can only write a vector space as a direct sum of subspaces if the dimensions of those subspaces sum up to the dimension of the original space. Note that this agrees with our intuition from Example 1.B.1 that subspaces in a direct sum must not be too big, but also must not be too small.

Corollary 1.B.3 **Dimension of (Internal) Direct Sums**	Suppose that $\mathcal{V}$ is a finite-dimensional vector space with subspaces $\mathcal{S}_1, \mathcal{S}_2 \subseteq \mathcal{V}$ satisfying $\mathcal{V} = \mathcal{S}_1 \oplus \mathcal{S}_2$. Then

$$\dim(\mathcal{V}) = \dim(\mathcal{S}_1) + \dim(\mathcal{S}_2).$$

Proof. We recall from Definition 1.2.2 that the dimension of a finite-dimensional vector space is the number of vectors in any of its bases. If B and C are bases of $\mathcal{S}_1$ and $\mathcal{S}_2$, respectively, then the fact that $\mathcal{S}_1 \cap \mathcal{S}_2 = \{\mathbf{0}\}$ implies $B \cap C = \{\}$, since $\mathbf{0}$ cannot be a member of a basis. It follows that $|B \cup C| = |B| + |C|$, so

Recall that $|B|$ means the number of vectors in B.

$$\dim(\mathcal{V}) = |B \cup C| = |B| + |C| = \dim(\mathcal{S}_1) + \dim(\mathcal{S}_2). \quad \blacksquare$$

For example, we can now see straight away that the subspaces S_1 and S_2 from Examples 1.B.1(a) and (b) do not form a direct sum decomposition of $\mathbb{R}^3$ since in part (a) we have $\dim(S_1) + \dim(S_2) = 1 + 1 = 2 \neq 3$, and in part (b) we have $\dim(S_1) + \dim(S_2) = 2 + 2 = 4 \neq 3$. It is perhaps worthwhile to work through a somewhat more exotic example in a vector space other than $\mathbb{R}^n$.

Example 1.B.3

The Cartesian Decomposition of Matrices

Let $\mathcal{M}_n^S$ and $\mathcal{M}_n^{sS}$ be the subspaces of $\mathcal{M}_n$ consisting of the symmetric and skew-symmetric matrices, respectively. Show that

$$\mathcal{M}_n = \mathcal{M}_n^S \oplus \mathcal{M}_n^{sS}.$$

Also compute the dimensions of each of these vector spaces.

Recall that $A \in \mathcal{M}_n$ is symmetric if $A^T = A$ and it is skew-symmetric if $A^T = -A$.

Solution:

To see that property (b) of Definition 1.B.1 holds, suppose that $A \in \mathcal{M}_n$ is both symmetric and skew-symmetric. Then

$$A = A^T = -A,$$

from which it follows that $A = O$, so $\mathcal{M}_n^S \cap \mathcal{M}_n^{sS} = \{O\}$.

This example only works if the ground field does not have $1 + 1 = 0$ (like in $\mathbb{Z}_2$, for example). Fields with this property are said to have "characteristic 2", and they often require special attention.

For property (a), we have to show that every matrix $A \in \mathcal{M}_n$ can be written in the form $A = B + C$, where $B^T = B$ and $C^T = -C$. We can check via direct computation that

$$A = \underbrace{\frac{1}{2}\left(A + A^T\right)}_{\text{symmetric}} + \underbrace{\frac{1}{2}\left(A - A^T\right)}_{\text{skew-symmetric}},$$

so we can choose $B = (A + A^T)/2$ and $C = (A - A^T)/2$ (and it is straightforward to check that $B^T = B$ and $C^T = -C$). We thus conclude that $\mathcal{M}_n = \mathcal{M}_n^S \oplus \mathcal{M}_n^{sS}$, as desired.

We computed the dimensions of these vector spaces in Exercise 1.2.2:

$$\dim(\mathcal{M}_n^S) = \frac{n(n+1)}{2}, \quad \dim(\mathcal{M}_n^{sS}) = \frac{n(n-1)}{2}, \quad \text{and} \quad \dim(\mathcal{M}_n) = n^2.$$

Note that these quantities are in agreement with Corollary 1.B.3, since

$$\dim(\mathcal{M}_n^S) + \dim(\mathcal{M}_n^{sS}) = \frac{n(n+1)}{2} + \frac{n(n-1)}{2} = n^2 = \dim(\mathcal{M}_n).$$

Remark 1.B.1

The Complex Cartesian Decomposition

While the decomposition of Example 1.B.3 works fine for complex matrices (as well as matrices over almost any other field), a slightly different decomposition that makes use of the *conjugate* transpose is typically used in that setting instead. Indeed, we can write every matrix $A \in \mathcal{M}_n(\mathbb{C})$ as a sum of a Hermitian and a skew-Hermitian matrix via

$$A = \underbrace{\frac{1}{2}\left(A + A^*\right)}_{\text{Hermitian}} + \underbrace{\frac{1}{2}\left(A - A^*\right)}_{\text{skew-Hermitian}}. \tag{1.B.4}$$

It is thus tempting to write $\mathcal{M}_n(\mathbb{C}) = \mathcal{M}_n^H \oplus \mathcal{M}_n^{sH}$ (where now $\mathcal{M}_n^H$

and $\mathcal{M}_n^{sH}$ denote the sets of Hermitian and skew-Hermitian matrices, respectively). However, this is only true if we are thinking of $\mathcal{M}_n(\mathbb{C})$ as a $2n^2$-dimensional vector space over $\mathbb{R}$ (*not* as an n^2-dimensional vector space over $\mathbb{C}$ like usual). The reason for this is simply that $\mathcal{M}_n^H$ and $\mathcal{M}_n^{sH}$ are not even subspaces of $\mathcal{M}_n(\mathbb{C})$ over the field $\mathbb{C}$ (refer back to Example 1.1.5—the point is that if A is Hermitian or skew-Hermitian then so is cA when $c \in \mathbb{R}$, but not necessarily when $c \in \mathbb{C}$).

> The matrix $(A + A^*)/2$ is sometimes called the **real part** or **Hermitian part** of A and denoted by $\mathrm{Re}(A)$, and similarly $(A - A^*)/(2i)$ is called its **imaginary part** or **skew-Hermitian part** and denoted by $\mathrm{Im}(A)$. Then
>
> $A = \mathrm{Re}(A) + i\mathrm{Im}(A).$

It is also worth noting that, in light of Theorem 1.B.1, Equation (1.B.4) provides the *unique* way of writing a matrix as a sum of a Hermitian and skew-Hermitian matrix. This is sometimes called the **Cartesian decomposition** of A (as is the closely-related decomposition of Example 1.B.3), and it is analogous to the fact that every complex number can be written as a sum of a real number and imaginary number:

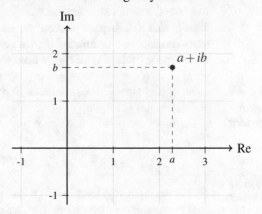

> We return to this idea of "matrix versions" of certain subsets of complex numbers in Figure 2.6.

Indeed, Hermitian matrices are often thought of as the "matrix version" of real numbers, and skew-Hermitian matrices as the "matrix version" of imaginary numbers.

Just like we can use dimensionality arguments to make it easier to determine whether or not a set is a basis of a vector space (via Theorem 1.2.1 or Exercise 1.2.27), we can also use dimensionality of subspaces to help us determine whether or not they form a direct sum decomposition of a vector space. In particular, if subspaces have the right size, we only need to check one of the two properties from Definition 1.B.1 that define the direct sum, not both.

> **Theorem 1.B.4**
>
> **Using Dimension to Check a Direct Sum Decomposition**

Suppose $\mathcal{V}$ is a finite-dimensional vector space with subspaces $\mathcal{S}_1, \mathcal{S}_2 \subseteq \mathcal{V}$.
  a) If $\dim(\mathcal{V}) \neq \dim(\mathcal{S}_1) + \dim(\mathcal{S}_2)$ then $\mathcal{V} \neq \mathcal{S}_1 \oplus \mathcal{S}_2$.
  b) If $\dim(\mathcal{V}) = \dim(\mathcal{S}_1) + \dim(\mathcal{S}_2)$ then the following are equivalent:
    i) $\mathrm{span}(\mathcal{S}_1 \cup \mathcal{S}_2) = \mathcal{V}$,
    ii) $\mathcal{S}_1 \cap \mathcal{S}_2 = \{\mathbf{0}\}$, and
    iii) $\mathcal{V} = \mathcal{S}_1 \oplus \mathcal{S}_2$.

Proof. Part (a) of the theorem follows immediately from Corollary 1.B.3, so we focus our attention on part (b). Also, condition (iii) immediately implies conditions (i) and (ii), since those two conditions define exactly what $\mathcal{V} = \mathcal{S}_1 \oplus \mathcal{S}_2$ means. We thus just need to show that condition (i) implies (iii) and that (ii) also implies (iii).

To see that condition (i) implies condition (iii), we first note that if B and C

are bases of $\mathcal{S}_1$ and $\mathcal{S}_2$, respectively, then

$$|B \cup C| \leq |B| + |C| = \dim(\mathcal{S}_1) + \dim(\mathcal{S}_2) = \dim(\mathcal{V}).$$

However, equality must actually be attained since Theorem 1.B.2 tells us that (since (i) holds) $\text{span}(B \cup C) = \mathcal{V}$. It then follows from Exercise 1.2.27 (since $B \cup C$ spans $\mathcal{V}$ and has $\dim(\mathcal{V})$ vectors) that $B \cup C$ is a basis of $\mathcal{V}$, so using Theorem 1.B.2 again tells us that $\mathcal{V} = \mathcal{S}_1 \oplus \mathcal{S}_2$.

The proof that condition (ii) implies condition (iii) is similar, and left as Exercise 1.B.6. $\blacksquare$

1.B.2 The Orthogonal Complement

Just like the direct sum can be thought of as a "subspace version" of bases, there is also a "subspace version" of *orthogonal* bases. As is usually the case when dealing with orthogonality, we need slightly more structure here than a vector space itself provides—we need an inner product as well.

Definition 1.B.2

Orthogonal Complement

Suppose $\mathcal{V}$ is an inner product space and $B \subseteq \mathcal{V}$ is a set of vectors. The **orthogonal complement** of B, denoted by $B^\perp$, is the subspace of $\mathcal{V}$ consisting of the vectors that are orthogonal to everything in B:

$$B^\perp \overset{\text{def}}{=} \left\{ \mathbf{v} \in \mathcal{V} : \langle \mathbf{v}, \mathbf{w} \rangle = 0 \text{ for all } \mathbf{w} \in B \right\}.$$

$B^\perp$ is read as "*B* perp", where "perp" is short for perpendicular.

For example, two lines in $\mathbb{R}^2$ that are perpendicular to each other are orthogonal complements of each other, as are a plane and a line in $\mathbb{R}^3$ that intersect at right angles (see Figure 1.23). The idea is that orthogonal complements break an inner product space down into subspaces that only intersect at the zero vector, much like (internal) direct sums do, but with the added restriction that those subspaces must be orthogonal to each other.

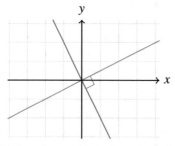

(a) The orthogonal complement of a line in $\mathbb{R}^2$ is the line perpendicular to it (through the origin).

(b) The orthogonal complement of the z-axis in $\mathbb{R}^3$ is the xy-plane (and vice-versa).

Figure 1.23: Orthogonal complements in $\mathbb{R}^2$ and $\mathbb{R}^3$.

It is straightforward to show that $B^\perp$ is always a subspace of $\mathcal{V}$ (even if B is not), so we leave the proof of this fact to Exercise 1.B.10. Furthermore, if $\mathcal{S}$ is a subspace of $\mathcal{V}$ then we will see shortly that $(\mathcal{S}^\perp)^\perp = \mathcal{S}$, at least in the finite-dimensional case, so orthogonal complement subspaces come in pairs. For now, we look at some examples.

Example 1.B.4

Orthogonal Complements in Euclidean Space

Describe the orthogonal complements of the following subsets of $\mathbb{R}^n$.

a) The line in $\mathbb{R}^2$ through the origin and the vector $(2,1)$.
b) The vector $(1,-1,2) \in \mathbb{R}^3$.

Solutions:

Actually, we want to find all vectors **v** orthogonal to *every* *multiple* of $(2,1)$. However, **v** is orthogonal to a non-zero multiple of $(2,1)$ if and only if it is orthogonal to $(2,1)$ itself.

a) We want to determine which vectors are orthogonal to $(2,1)$. Well, if $\mathbf{v} = (v_1,v_2)$ then we can rearrange the equation

$$(v_1,v_2) \cdot (2,1) = 2v_1 + v_2 = 0$$

to the form $v_2 = -2v_1$, or $\mathbf{v} = v_1(1,-2)$. It follows that the orthogonal complement is the set of scalar multiples of $(1,-2)$:

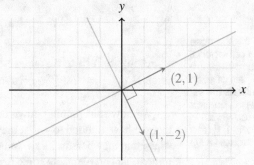

b) We want to determine which vectors are orthogonal to $(1,-1,2)$: the vectors $\mathbf{v} = (v_1,v_2,v_3)$ with $v_1 - v_2 + 2v_3 = 0$. This is a (quite degenerate) linear system with solutions of the form

$$\mathbf{v} = (v_1,v_2,v_3) = v_2(1,1,0) + v_3(-2,0,1),$$

where v_2 and v_3 are free variables. It follows that the orthogonal complement is the plane in $\mathbb{R}^3$ with basis $\{(1,1,0),(-2,0,1)\}$:

We replaced the basis vector $(1,1,0)$ with $(2,2,0)$ just to make this picture a bit prettier. The author is very superficial.

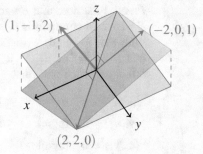

Example 1.B.5

Orthogonal Complements of Matrices

Describe the orthogonal complement of the set in $\mathcal{M}_n$ consisting of just the identity matrix: $B = \{I\}$.

Solution:

Recall from Example 1.3.16 that the standard (Frobenius) inner product on $\mathcal{M}_n$ is $\langle X,Y \rangle = \operatorname{tr}(X^*Y)$, so $X \in B^\perp$ if and only if $\langle X,I \rangle = \operatorname{tr}(X^*) = 0$. This is equivalent to simply requiring that $\operatorname{tr}(X) = 0$.

Example 1.B.6

Orthogonal Complements of Polynomials

Describe the orthogonal complement of the set $B = \{x, x^3\} \subseteq \mathcal{P}^3[-1, 1]$.

Solution:

Our goal is to find all polynomials $f(x) = ax^3 + bx^2 + cx + d$ with the property that

$$\langle f(x), x \rangle = \int_{-1}^{1} x f(x)\, dx = 0 \quad \text{and} \quad \langle f(x), x^3 \rangle = \int_{-1}^{1} x^3 f(x)\, dx = 0.$$

Straightforward calculation shows that

$$\langle f(x), x \rangle = \int_{-1}^{1} (ax^4 + bx^3 + cx^2 + dx)\, dx = \frac{2a}{5} + \frac{2c}{3} \quad \text{and}$$

$$\langle f(x), x^3 \rangle = \int_{-1}^{1} (ax^6 + bx^5 + cx^4 + dx^3)\, dx = \frac{2a}{7} + \frac{2c}{5}.$$

Similarly, it is the case that the orthogonal complement of the set of even polynomials is the set of odd polynomials (and vice-versa) in $\mathcal{P}[-1, 1]$ (see Exercise 1.B.17).

Setting both of these quantities equal to 0 and then solving for a and c gives $a = c = 0$, so $f \in B^\perp$ if and only if $f(x) = bx^2 + d$.

We now start pinning down the various details of orthogonal complements that we have alluded to—that they behave like an orthogonal version of direct sums, that they behave like a "subspace version" of orthonormal bases, and that orthogonal complements come in pairs. The following theorem does most of the heavy lifting in this direction, and it is completely analogous to Theorem 1.B.2 for (internal) direct sums.

Theorem 1.B.5

Orthonormal Bases of Orthogonal Complements

Suppose $\mathcal{V}$ is a finite-dimensional inner product space and $\mathcal{S} \subseteq \mathcal{V}$ is a subspace. If B is an orthonormal basis of $\mathcal{S}$ and C is an orthonormal basis of $\mathcal{S}^\perp$ then $B \cup C$ is an orthonormal basis of $\mathcal{V}$.

Proof. First note that B and C are disjoint since the only vector that $\mathcal{S}$ and $\mathcal{S}^\perp$ have in common is $\mathbf{0}$, since that is the only vector orthogonal to itself. With that in mind, write $B = \{\mathbf{u}_1, \mathbf{u}_2, \ldots, \mathbf{u}_m\}$ and $C = \{\mathbf{v}_1, \mathbf{v}_2, \ldots, \mathbf{v}_n\}$, so that $B \cup C = \{\mathbf{u}_1, \mathbf{u}_2, \ldots, \mathbf{u}_m, \mathbf{v}_1, \mathbf{v}_2, \ldots, \mathbf{v}_n\}$. To see that $B \cup C$ is a mutually orthogonal set, notice that

$$\langle \mathbf{u}_i, \mathbf{u}_j \rangle = 0 \quad \text{for all} \quad 1 \le i \ne j \le m, \quad \text{since } B \text{ is an orthonormal basis,}$$

$$\langle \mathbf{v}_i, \mathbf{v}_j \rangle = 0 \quad \text{for all} \quad 1 \le i \ne j \le n, \quad \text{since } C \text{ is an orthonormal basis, and}$$

$$\langle \mathbf{u}_i, \mathbf{v}_j \rangle = 0 \quad \text{for all} \quad 1 \le i \le m, 1 \le j \le n, \quad \text{since } \mathbf{u}_i \in \mathcal{S} \text{ and } \mathbf{v}_j \in \mathcal{S}^\perp.$$

We thus just need to show that $\text{span}(B \cup C) = \mathcal{V}$. To this end, recall from Exercise 1.4.20 that we can extend $B \cup C$ to an orthonormal basis of $\mathcal{V}$: we can find $k \ge 0$ unit vectors $\mathbf{w}_1, \mathbf{w}_2, \ldots \mathbf{w}_k$ such that the set

$$\{\mathbf{u}_1, \mathbf{u}_2, \ldots, \mathbf{u}_m, \mathbf{v}_1, \mathbf{v}_2, \ldots, \mathbf{v}_n, \mathbf{w}_1, \mathbf{w}_2, \ldots \mathbf{w}_k\}$$

is an orthonormal basis of $\mathcal{V}$. However, since this is an orthonormal basis, we know that $\langle \mathbf{w}_i, \mathbf{u}_j \rangle = 0$ for all $1 \le i \le k$ and $1 \le j \le m$, so in fact $\mathbf{w}_i \in \mathcal{S}^\perp$ for all i. This implies that $\{\mathbf{v}_1, \mathbf{v}_2, \ldots, \mathbf{v}_n, \mathbf{w}_1, \mathbf{w}_2, \ldots, \mathbf{w}_k\}$ is a mutually orthogonal subset of $\mathcal{S}^\perp$ consisting of $n + k$ vectors. However, since $\dim(\mathcal{S}^\perp) = |C| = n$ we conclude that the only possibility is that $k = 0$, so $B \cup C$ itself is a basis of $\mathcal{V}$. ∎

The above theorem has a few immediate (but very useful) corollaries. For example, we can now show that orthogonal complements really are a stronger version of direct sums:

Theorem 1.B.6

Orthogonal Complements are Direct Sums

Suppose $\mathcal{V}$ is a finite-dimensional inner product space with subspaces $\mathcal{S}_1, \mathcal{S}_2 \subseteq \mathcal{V}$. The following are equivalent:

a) $\mathcal{S}_2 = \mathcal{S}_1^\perp$.
b) $\mathcal{V} = \mathcal{S}_1 \oplus \mathcal{S}_2$ and $\langle \mathbf{v}, \mathbf{w} \rangle = 0$ for all $\mathbf{v} \in \mathcal{S}_1$ and $\mathbf{w} \in \mathcal{S}_2$.

In particular, this theorem tells us that if $\mathcal{V}$ is finite-dimensional then for every subspace $\mathcal{S} \subseteq \mathcal{V}$ we have $\mathcal{V} = \mathcal{S} \oplus \mathcal{S}^\perp$.

Proof. If property (a) holds (i.e., $\mathcal{S}_2 = \mathcal{S}_1^\perp$) then the fact that $\langle \mathbf{v}, \mathbf{w} \rangle = 0$ for all $\mathbf{v} \in \mathcal{S}_1$ and $\mathbf{w} \in \mathcal{S}_2$ is clear, so we just need to show that $\mathcal{V} = \mathcal{S}_1 \oplus \mathcal{S}_2$. Well, Theorem 1.B.5 tells us that if B and C are orthonormal bases of $\mathcal{S}_1$ and $\mathcal{S}_2$, respectively, then $B \cup C$ is an orthonormal basis of $\mathcal{V}$. Theorem 1.B.2 then tells us that $\mathcal{V} = \mathcal{S}_1 \oplus \mathcal{S}_2$.

In the other direction, property (b) immediately implies $\mathcal{S}_2 \subseteq \mathcal{S}_1^\perp$, so we just need to show the opposite inclusion. To that end, suppose $\mathbf{w} \in \mathcal{S}_1^\perp$ (i.e., $\langle \mathbf{v}, \mathbf{w} \rangle = 0$ for all $\mathbf{v} \in \mathcal{S}_1$). Then, since $\mathbf{w} \in \mathcal{V} = \mathcal{S}_1 \oplus \mathcal{S}_2$, we can write $\mathbf{w} = \mathbf{w}_1 + \mathbf{w}_2$ for some $\mathbf{w}_1 \in \mathcal{S}_1$ and $\mathbf{w}_2 \in \mathcal{S}_2$. It follows that

$$0 = \langle \mathbf{v}, \mathbf{w} \rangle = \langle \mathbf{v}, \mathbf{w}_1 \rangle + \langle \mathbf{v}, \mathbf{w}_2 \rangle = \langle \mathbf{v}, \mathbf{w}_1 \rangle$$

for all $\mathbf{v} \in \mathcal{S}_1$, where the final equality follows from the fact that $\mathbf{w}_2 \in \mathcal{S}_2$ and thus $\langle \mathbf{v}, \mathbf{w}_2 \rangle = 0$ by property (b). Choosing $\mathbf{v} = \mathbf{w}_1$ then gives $\langle \mathbf{w}_1, \mathbf{w}_1 \rangle = 0$, so $\mathbf{w}_1 = \mathbf{0}$, so $\mathbf{w} = \mathbf{w}_2 \in \mathcal{S}_2$, as desired. ∎

Example 1.B.7

The Orthogonal Complement of Symmetric Matrices

Suppose $\mathbb{F} = \mathbb{R}$ or $\mathbb{F} = \mathbb{C}$ and let $\mathcal{M}_n^S$ and $\mathcal{M}_n^{sS}$ be the subspaces of $\mathcal{M}_n(\mathbb{F})$ consisting of the symmetric and skew-symmetric matrices, respectively. Show that $\left(\mathcal{M}_n^S\right)^\perp = \mathcal{M}_n^{sS}$.

Solution:
Recall from Example 1.B.3 that $\mathcal{M}_n = \mathcal{M}_n^S \oplus \mathcal{M}_n^{sS}$. Furthermore, we can use basic properties of the trace to see that if $A \in \mathcal{M}_n^S$ and $B \in \mathcal{M}_n^{sS}$ then

Compare this orthogonality property with Exercise 1.A.2.

$$\langle A, B \rangle = \text{tr}(A^*B) = -\text{tr}(\overline{A}B^T) = -\text{tr}\left((BA^*)^T\right)$$
$$= -\text{tr}(BA^*) = -\text{tr}(A^*B) = -\langle A, B \rangle,$$

so $\langle A, B \rangle = 0$. It then follows from Theorem 1.B.6 that $\left(\mathcal{M}_n^S\right)^\perp = \mathcal{M}_n^{sS}$.

Theorem 1.B.6 tells us that for every subspace $\mathcal{S}$ of $\mathcal{V}$, we have $\mathcal{V} = \mathcal{S} \oplus \mathcal{S}^\perp$, so everything that we already know about direct sums also applies to orthogonal complements. For example:

Most of these connections between the orthogonal complement and direct sum break down in infinite dimensions—see Exercise 1.B.18.

- If $\mathcal{V}$ is a finite-dimensional inner product space with subspace $\mathcal{S} \subseteq \mathcal{V}$ and $\mathbf{v} \in \mathcal{V}$, then there exist unique vectors $\mathbf{v}_1 \in \mathcal{S}$ and $\mathbf{v}_2 \in \mathcal{S}^\perp$ such that $\mathbf{v} = \mathbf{v}_1 + \mathbf{v}_2$ (via Theorem 1.B.1).

- If $\mathcal{V}$ is a finite-dimensional inner product space with subspace $\mathcal{S} \subseteq \mathcal{V}$, then $\dim(\mathcal{S}) + \dim(\mathcal{S}^\perp) = \dim(\mathcal{V})$ (via Corollary 1.B.3).

These results also tell us that if $\mathcal{S}$ is a subspace of a finite-dimensional inner product space then $(\mathcal{S}^\perp)^\perp = \mathcal{S}$. Slightly more generally, we have the following fact:

(!) If B is any subset (not necessarily a subspace) of a finite-dimensional inner product space then $(B^\perp)^\perp = \text{span}(B)$.

This fact is proved in Exercise 1.B.14 and illustrated in Figure 1.24. Note in particular that after taking the orthogonal complement of any set B once, further orthogonal complements just bounce back and forth between $B^\perp$ and $\text{span}(B)$.

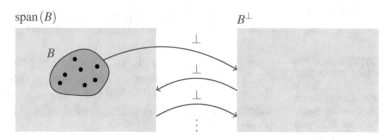

Figure 1.24: The orthogonal complement of B is $B^\perp$, and the orthogonal complement of $B^\perp$ is $\text{span}(B)$. After that point, taking the orthogonal complement of $\text{span}(B)$ results in $B^\perp$ again, and vice-versa.

Example 1.B.8

(Orthogonal) Projections and Direct Sums

If P is an orthogonal projection then $\mathcal{V}$ must actually be an inner product space, not just a vector space.

Suppose $\mathcal{V}$ is a vector space and $P : \mathcal{V} \to \mathcal{V}$ is a projection (i.e., $P^2 = P$).

a) Show that $\mathcal{V} = \text{range}(P) \oplus \text{null}(P)$.

b) Show that if P is an *orthogonal* projection (i.e., $P^* = P$) then $\text{range}(P)^\perp = \text{null}(P)$.

Solutions:

a) The fact that $\text{range}(P) \cap \text{null}(P) = \{\mathbf{0}\}$ is follows from noting that if $\mathbf{v} \in \text{range}(P)$ then $P(\mathbf{v}) = \mathbf{v}$, and if $\mathbf{v} \in \text{null}(P)$ then $P(\mathbf{v}) = \mathbf{0}$. Comparing these two equations shows that $\mathbf{v} = \mathbf{0}$.

To see that $\mathcal{V} = \text{span}(\text{range}(P) \cup \text{null}(P))$ (and thus $\mathcal{V} = \text{range}(P) \oplus \text{null}(P)$), notice that every vector $\mathbf{v} \in \mathcal{V}$ can be written in the form $\mathbf{v} = P(\mathbf{v}) + (\mathbf{v} - P(\mathbf{v}))$. The first vector $P(\mathbf{v})$ is in $\text{range}(P)$ and the second vector $\mathbf{v} - P(\mathbf{v})$ satisfies

$$P(\mathbf{v} - P(\mathbf{v})) = P(\mathbf{v}) - P^2(\mathbf{v}) = P(\mathbf{v}) - P(\mathbf{v}) = \mathbf{0},$$

so it is in $\text{null}(P)$. We have thus written $\mathbf{v}$ as a sum of vectors from $\text{range}(P)$ and $\text{null}(P)$, so we are done.

b) We just observe that $\mathbf{w} \in \text{range}(P)^\perp$ is equivalent to several other conditions:

$$\begin{aligned}
\mathbf{w} \in \text{range}(P)^\perp &\iff \langle P(\mathbf{v}), \mathbf{w} \rangle = 0 \quad \text{for all} \quad \mathbf{v} \in \mathcal{V} \\
&\iff \langle \mathbf{v}, P^*(\mathbf{w}) \rangle = 0 \quad \text{for all} \quad \mathbf{v} \in \mathcal{V} \\
&\iff \langle \mathbf{v}, P(\mathbf{w}) \rangle = 0 \quad \text{for all} \quad \mathbf{v} \in \mathcal{V} \\
&\iff P(\mathbf{w}) = \mathbf{0} \\
&\iff \mathbf{w} \in \text{null}(P).
\end{aligned}$$

It follows that $\text{range}(P)^\perp = \text{null}(P)$, as desired.

The above example tells us that every projection breaks space down into two pieces—its range and null space, respectively—one of which is projects *onto* and one of which it project *along*. Furthermore, the projection is orthogonal if and only if these two component subspaces are orthogonal to each other (see Figure 1.25).

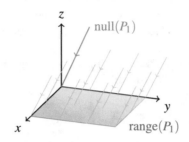

(a) An oblique projection onto the projection's range along its null space, which direct sum to all of $\mathbb{R}^3$.

(b) An orthogonal projection onto the projection's range along its orthogonal complement.

Figure 1.25: An (a) oblique projection P_1 and an (b) orthogonal projection P_2 projecting along their null spaces onto their ranges.

In fact, just like orthogonal projections are completely determined by their range (refer back to Theorem 1.4.10), oblique projections are uniquely determined by their range and null space (see Exercise 1.B.7). For this reason, just like we often talk about *the* orthogonal projection P onto a particular subspace (range(P)), we similarly talk about *the* (not necessarily orthogonal) projection P onto one subspace (range(P)) along another one (null(P)).

Orthogonality of the Fundamental Subspaces

The range and null space of a linear transformation acting on a finite-dimensional inner product space, as well as the range and null space of its adjoint, are sometimes collectively referred to as its **fundamental subspaces**. We saw above that the direct sum and orthogonal complement play an important role when dealing with the fundamental subspaces of projections. Somewhat surprisingly, they actually play an important role in the fundamental subspaces of every linear transformation.

For example, by using standard techniques from introductory linear algebra, we can see that the matrix

$$A = \begin{bmatrix} 1 & 0 & 1 & 0 & -1 \\ 1 & 1 & 0 & 0 & 1 \\ -1 & 0 & -1 & 1 & 4 \\ 2 & 1 & 1 & -1 & -3 \end{bmatrix} \tag{1.B.5}$$

has the following sets as bases of its four fundamental subspaces:

Here, $A^* = A^T$ since A is real.

subspaces of $\mathbb{R}^4$ $\begin{cases} \text{range}(A): & \{(1,1,-1,2),(0,1,0,1),(0,0,1,-1)\} \\ \text{null}(A^*): & \{(0,1,-1,-1)\} \end{cases}$

subspaces of $\mathbb{R}^5$ $\begin{cases} \text{range}(A^*): & \{(1,0,1,0,-1),(0,1,-1,0,2),(0,0,0,1,3)\} \\ \text{null}(A): & \{(-1,1,1,0,0),(1,-2,0,-3,1)\}. \end{cases}$

There is a lot of structure that is suggested by these bases—the dimensions of range(A) and null(A^*) add up to the dimension of the output space $\mathbb{R}^4$ that they live in, and similarly the dimensions of range(A^*) and null(A) add up to the dimension of the input space $\mathbb{R}^5$ that they live in. Furthermore, it is straightforward to check that the vector in this basis of null(A^*) is orthogonal to each of the basis vectors for range(A), and the vectors in this basis for null(A) are orthogonal to each of the basis vectors for range(A^*). All of these facts can be explained by observing that the fundamental subspaces of any linear transformation are in fact orthogonal complements of each other:

Theorem 1.B.7

Orthogonality of the Fundamental Subspaces

Suppose $\mathcal{V}$ and $\mathcal{W}$ are finite-dimensional inner product spaces and $T : \mathcal{V} \to \mathcal{W}$ is a linear transformation. Then

a) range$(T)^\perp$ = null(T^*), and
b) null$(T)^\perp$ = range(T^*).

Proof. The proof of this theorem is surprisingly straightforward. For part (a), we just argue as we did in Example 1.B.8(b)—we observe that $\mathbf{w} \in \text{range}(T)^\perp$ is equivalent to several other conditions:

To help remember this theorem, note that each equation in it contains exactly one T, one T^*, one range, one null space, and one orthogonal complement.

$$\mathbf{w} \in \text{range}(T)^\perp \iff \langle T(\mathbf{v}), \mathbf{w} \rangle = 0 \quad \text{for all} \quad \mathbf{v} \in \mathcal{V}$$
$$\iff \langle \mathbf{v}, T^*(\mathbf{w}) \rangle = 0 \quad \text{for all} \quad \mathbf{v} \in \mathcal{V}$$
$$\iff T^*(\mathbf{w}) = \mathbf{0}$$
$$\iff \mathbf{w} \in \text{null}(T^*).$$

It follows that range$(T)^\perp$ = null(T^*), as desired. Part (b) of the theorem can now be proved by making use of part (a), and is left to Exercise 1.B.15. ∎

The way to think of this theorem is as saying that, for every linear transformation $T : \mathcal{V} \to \mathcal{W}$, we can decompose the input space $\mathcal{V}$ into an orthogonal direct sum $\mathcal{V} = \text{range}(T^*) \oplus \text{null}(T)$ such that T acts like an invertible map on one space (range(T^*)) and acts like the zero map on the other (null(T)). Similarly, we can decompose the output space $\mathcal{W}$ into an orthogonal direct sum $\mathcal{W} = \text{range}(T) \oplus \text{null}(T^*)$ such that T maps all of $\mathcal{V}$ onto one space (range(T)) and maps nothing into the other (null(T^*)). These relationships between the four fundamental subspaces are illustrated in Figure 1.26.

Okay, T maps things to $\mathbf{0} \in \text{null}(T^*)$, but that's it! The rest of null(T^*) is untouched.

For example, the matrix A from Equation (1.B.5) acts as a rank 3 linear transformation that sends $\mathbb{R}^5$ to $\mathbb{R}^4$. This means that there is a 3-dimensional subspace range(A^*) $\subseteq \mathbb{R}^5$ on which A just "shuffles things around" to another 3-dimensional subspace range(A) $\subseteq \mathbb{R}^4$. The orthogonal complement of range(A^*) is null(A), which accounts for the other 2 dimensions of $\mathbb{R}^5$ that are "squashed away".

We return to the fundamental subspaces in Section 2.3.1.

If we recall from Exercise 1.4.22 that every linear transformation T acting on a finite-dimensional inner product space has rank(T) = rank(T^*), we immediately get the following corollary that tells us how large the fundamental subspaces are compared to each other.

Corollary 1.B.8

Dimensions of the Fundamental Subspaces

Suppose $\mathcal{V}$ and $\mathcal{W}$ are finite-dimensional inner product spaces and $T : \mathcal{V} \to \mathcal{W}$ is a linear transformation. Then

a) rank(T) + nullity(T) = dim($\mathcal{V}$), and
b) rank(T) + nullity(T^*) = dim($\mathcal{W}$).

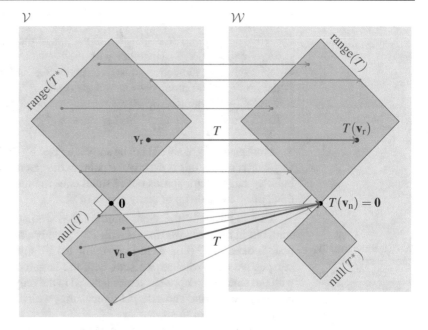

We make the idea that T acts "like" an invertible linear transformation on $\text{range}(T^*)$ precise in Exercise 1.B.16.

Figure 1.26: Given a linear transformation $T : \mathcal{V} \to \mathcal{W}$, $\text{range}(T^*)$ and $\text{null}(T)$ are orthogonal complements in $\mathcal{V}$, while $\text{range}(T)$ and $\text{null}(T^*)$ are orthogonal complements in $\mathcal{W}$. These particular orthogonal decompositions of $\mathcal{V}$ and $\mathcal{W}$ are useful because T acts like the zero map on $\text{null}(T)$ (i.e., $T(\mathbf{v}_n) = \mathbf{0}$ for each $\mathbf{v}_n \in \text{null}(T)$) and like an invertible linear transformation on $\text{range}(T^*)$ (i.e., for each $\mathbf{w} \in \text{range}(T)$ there exists a unique $\mathbf{v}_r \in \text{range}(T^*)$ such that $T(\mathbf{v}_r) = \mathbf{w}$).

1.B.3   The External Direct Sum

The internal direct sum that we saw in Section 1.B.1 works by first defining a vector space and then "breaking it apart" into two subspaces. We can also flip the direct sum around so as to start off with two vector spaces and then create their "external" direct sum, which is a larger vector space that more or less contains the original two vector spaces as subspaces. The following definition makes this idea precise.

Definition 1.B.3

The External Direct Sum

In other words, $\mathcal{V} \oplus \mathcal{W}$ is the Cartesian product of $\mathcal{V}$ and $\mathcal{W}$, together with the entry-wise addition and scalar multiplication operations.

Let $\mathcal{V}$ and $\mathcal{W}$ be vector spaces over the same field $\mathbb{F}$. Then the **external direct sum** of $\mathcal{V}$ and $\mathcal{W}$, denoted by $\mathcal{V} \oplus \mathcal{W}$, is the vector space with vectors and operations defined as follows:

Vectors: ordered pairs $(\mathbf{v}, \mathbf{w})$, where $\mathbf{v} \in \mathcal{V}$ and $\mathbf{w} \in \mathcal{W}$.

Vector addition: $(\mathbf{v}_1, \mathbf{w}_1) + (\mathbf{v}_2, \mathbf{w}_2) = (\mathbf{v}_1 + \mathbf{v}_2, \mathbf{w}_1 + \mathbf{w}_2)$ for all $\mathbf{v}_1, \mathbf{v}_2 \in \mathcal{V}$ and $\mathbf{w}_1, \mathbf{w}_2 \in \mathcal{W}$.

Scalar mult.: $c(\mathbf{v}, \mathbf{w}) = (c\mathbf{v}, c\mathbf{w})$ for all $c \in \mathbb{F}$, $\mathbf{v} \in \mathcal{V}$, and $\mathbf{w} \in \mathcal{W}$.

It is hopefully believable that the external direct sum $\mathcal{V} \oplus \mathcal{W}$ really is a vector space, so we leave the proof of that claim to Exercise 1.B.19. For now, we look at the canonical example that motivates the external direct sum.

Example 1.B.9

The Direct Sum of $\mathbb{F}^n$

Show that $\mathbb{F} \oplus \mathbb{F}^2$ (where "$\oplus$" here means the *external* direct sum) is isomorphic to $\mathbb{F}^3$.

Solution:
By definition, $\mathbb{F} \oplus \mathbb{F}^2$ consists of vectors of the form $(x, (y, z))$, where $x \in \mathbb{F}$ and $(y, z) \in \mathbb{F}^2$ (together with the "obvious" vector addition and scalar multiplication operations). It is straightforward to check that erasing the inner set of parentheses is an isomorphism (i.e., the linear map $T : \mathbb{F} \oplus \mathbb{F}^2 \to \mathbb{F}^3$ defined by $T(x, (y, z)) = (x, y, z)$ is an isomorphism), so $\mathbb{F} \oplus \mathbb{F}^2 \cong \mathbb{F}^3$.

More generally, it should not be surprising that $\mathbb{F}^m \oplus \mathbb{F}^n \cong \mathbb{F}^{m+n}$ in a natural way. Furthermore, this is exactly the point of the external direct sum—it lets us build up large vector spaces out of small ones in much the same way that we think of $\mathbb{F}^n$ as made up of many copies of $\mathbb{F}$.

This is basically the same intuition that we had for the *internal* direct sum, but in that case we *started* with $\mathbb{F}^{m+n}$ and broke it down into subspaces that "looked like" (i.e., were isomorphic to) $\mathbb{F}^m$ and $\mathbb{F}^n$. This difference in perspective (i.e., whether we start with the large container vector space or with the smaller component vector spaces) is the only appreciable difference between the internal and external direct sums, which is why we use the same notation for each of them.

> We will show that the internal and external direct sums are isomorphic shortly.

Theorem 1.B.9

Bases of the External Direct Sum

Suppose $\mathcal{V}$ and $\mathcal{W}$ are vector spaces with bases B and C, respectively, and define the following subsets of the external direct sum $\mathcal{V} \oplus \mathcal{W}$:

$$B' = \{(\mathbf{v}, \mathbf{0}) : \mathbf{v} \in B\} \quad \text{and} \quad C' = \{(\mathbf{0}, \mathbf{w}) : \mathbf{w} \in C\}.$$

Then $B' \cup C'$ is a basis of $\mathcal{V} \oplus \mathcal{W}$.

Proof. To see that $\text{span}(B' \cup C') = \mathcal{V} \oplus \mathcal{W}$, suppose $(\mathbf{x}, \mathbf{y}) \in \mathcal{V} \oplus \mathcal{W}$ (i.e., $\mathbf{x} \in \mathcal{V}$ and $\mathbf{y} \in \mathcal{W}$). Since B and C are bases of $\mathcal{V}$ and $\mathcal{W}$, respectively, we can find $\mathbf{v}_1, \mathbf{v}_2, \ldots, \mathbf{v}_k \in B$, $\mathbf{w}_1, \mathbf{w}_2, \ldots, \mathbf{w}_m \in C$, and scalars $c_1, c_2, \ldots, c_k$ and $d_1, d_2, \ldots, d_m$ such that

$$\mathbf{x} = \sum_{i=1}^{k} c_i \mathbf{v}_i \quad \text{and} \quad \mathbf{y} = \sum_{j=1}^{m} d_j \mathbf{w}_j.$$

It follows that

$$(\mathbf{x}, \mathbf{y}) = (\mathbf{x}, \mathbf{0}) + (\mathbf{0}, \mathbf{y}) = \left(\sum_{i=1}^{k} c_i \mathbf{v}_i, \mathbf{0} \right) + \left(\mathbf{0}, \sum_{j=1}^{m} d_j \mathbf{w}_j \right)$$

$$= \sum_{i=1}^{k} c_i (\mathbf{v}_i, \mathbf{0}) + \sum_{j=1}^{m} d_j (\mathbf{0}, \mathbf{w}_j),$$

which is a linear combination of vectors from $B' \cup C'$.

The proof of linear independence is similar, so we leave it to Exercise 1.B.20. ∎

In the finite-dimensional case, the above result immediately implies the following corollary, which helps clarify why we refer to the external direct sum as a "sum" in the first place.

Corollary 1.B.10	Suppose $\mathcal{V}$ and $\mathcal{W}$ are finite-dimensional vector spaces with external direct
Dimension of (External) Direct Sums	sum $\mathcal{V} \oplus \mathcal{W}$. Then $$\dim(\mathcal{V} \oplus \mathcal{W}) = \dim(\mathcal{V}) + \dim(\mathcal{W}).$$

Compare this corollary (and its proof) to Corollary 1.B.3.

Proof. Just observe that the sets B' and C' from Theorem 1.B.9 have empty intersection (keep in mind that they cannot even have $(\mathbf{0},\mathbf{0})$ in common, since B and C are bases so $\mathbf{0} \notin B,C$), so

$$\dim(\mathcal{V} \oplus \mathcal{W}) = |B' \cup C'| = |B'| + |C'| = |B| + |C| = \dim(\mathcal{V}) + \dim(\mathcal{W}),$$

as claimed. ∎

We close this section by noting that, in practice, people often just talk about the direct sum, without specifying whether they mean the internal or external one (much like we use the same notation for each of them). The reason for this is two-fold:

- First, it is always clear from context whether a direct sum is internal or external. If the components in the direct sum were first defined and then a larger vector space was constructed via their direct sum, it is external. On the other hand, if a single vector space was first defined and then it was broken down into two component subspaces, the direct sum is internal.

- Second, the internal and external direct sums are isomorphic in a natural way. If $\mathcal{V}$ and $\mathcal{W}$ are vector spaces with external direct sum $\mathcal{V} \oplus \mathcal{W}$, then we cannot quite say that $\mathcal{V} \oplus \mathcal{W}$ is the *internal* direct sum of $\mathcal{V}$ and $\mathcal{W}$, since they are not even subspaces of $\mathcal{V} \oplus \mathcal{W}$. However, $\mathcal{V} \oplus \mathcal{W}$ *is* the internal direct sum of its subspaces

$$\mathcal{V}' = \big\{(\mathbf{v},\mathbf{0}) : \mathbf{v} \in \mathcal{V}\big\} \quad \text{and} \quad \mathcal{W}' = \big\{(\mathbf{0},\mathbf{w}) : \mathbf{w} \in \mathcal{W}\big\},$$

which are pretty clearly isomorphic to $\mathcal{V}$ and $\mathcal{W}$, respectively (see Exercise 1.B.21).

Exercises

solutions to starred exercises on page 460

1.B.1 Find a basis for the orthogonal complement of each of the following sets in the indicated inner product space.

∗(a) $\{(3,2)\} \subset \mathbb{R}^2$
(b) $\{(3,2),(1,2)\} \subset \mathbb{R}^2$
∗(c) $\{(0,0,0)\} \subset \mathbb{R}^3$
(d) $\{(1,1,1),(2,1,0)\} \subset \mathbb{R}^3$
∗(e) $\{(1,2,3),(1,1,1),(3,2,1)\} \subset \mathbb{R}^3$
(f) $\{(1,1,1,1),(1,2,3,1)\} \subset \mathbb{R}^4$
∗(g) $\mathcal{M}_2^S \subset \mathcal{M}_2$ (the set of symmetric 2×2 matrices)
(h) $\mathcal{P}^1[-1,1] \subset \mathcal{P}^3[-1,1]$

1.B.2 Compute a basis of each of the four fundamental subspaces of the following matrices and verify that they satisfy the orthogonality relations of Theorem 1.B.7.

∗(a) $\begin{bmatrix} 1 & 2 \\ 3 & 6 \end{bmatrix}$

(b) $\begin{bmatrix} 2 & 1 & 3 & 1 \\ 4 & -1 & 2 & 3 \end{bmatrix}$

∗(c) $\begin{bmatrix} 1 & 2 & 3 \\ 4 & 5 & 7 \\ 7 & 8 & 9 \end{bmatrix}$

(d) $\begin{bmatrix} 1 & 3 & 4 & 1 \\ 2 & 1 & 2 & 2 \\ -1 & 2 & 2 & -1 \end{bmatrix}$

1.B.3 Determine which of the following statements are true and which are false.

   (a) If $S_1, S_2 \subseteq V$ are subspaces such that $V = S_1 \oplus S_2$ then $V = S_2 \oplus S_1$ too.
  *(b) If V is an inner product space then $V^\perp = \{\mathbf{0}\}$.
   (c) If $A \in \mathcal{M}_{m,n}$, $\mathbf{v} \in \text{range}(A)$, and $\mathbf{w} \in \text{null}(A)$ then $\mathbf{v} \cdot \mathbf{w} = 0$.
  *(d) If $A \in \mathcal{M}_{m,n}$, $\mathbf{v} \in \text{range}(A)$, and $\mathbf{w} \in \text{null}(A^*)$ then $\mathbf{v} \cdot \mathbf{w} = 0$.
   (e) If a vector space V has subspaces $S_1, S_2, \ldots, S_k$ satisfying $\text{span}(S_1 \cup S_2 \cup \cdots \cup S_k) = V$ and $S_i \cap S_j = \{\mathbf{0}\}$ for all $i \neq j$ then $V = S_1 \oplus \cdots \oplus S_k$.
  *(f) The set
$$\{(\mathbf{e}_1, \mathbf{e}_1), (\mathbf{e}_1, \mathbf{e}_2), (\mathbf{e}_1, \mathbf{e}_3), (\mathbf{e}_2, \mathbf{e}_1), (\mathbf{e}_2, \mathbf{e}_2), (\mathbf{e}_2, \mathbf{e}_3)\}$$
is a basis of the external direct sum $\mathbb{R}^2 \oplus \mathbb{R}^3$.
   (g) The external direct sum $\mathcal{P}^2 \oplus \mathcal{P}^3$ has dimension 6.

**1.B.4 Suppose V is a finite-dimensional vector space with basis $\{\mathbf{v}_1, \mathbf{v}_2, \ldots, \mathbf{v}_k\}$. Show that
$$V = \text{span}(\mathbf{v}_1) \oplus \text{span}(\mathbf{v}_2) \oplus \cdots \oplus \text{span}(\mathbf{v}_k).$$

** 1.B.5 Suppose V is a vector space with subspaces $S_1, S_2 \subseteq V$ that have bases B and C, respectively. Show that if $B \cup C$ (as a multiset) is linearly independent then $S_1 \cap S_2 = \{\mathbf{0}\}$.

[Side note: This completes the proof of Theorem 1.B.2.]

**1.B.6 Complete the proof of Theorem 1.B.4 by showing that condition (b)(ii) implies condition (b)(iii). That is, show that if V is a finite-dimensional vector space with subspaces $S_1, S_2 \subseteq V$ such that $\dim(V) = \dim(S_1) + \dim(S_2)$ and $S_1 \cap S_2 = \{\mathbf{0}\}$, then $V = S_1 \oplus S_2$.

**1.B.7 Suppose V is a finite-dimensional vector space with subspaces $S_1, S_2 \subseteq V$. Show that there is at most one projection $P : V \to V$ with $\text{range}(P) = S_1$ and $\text{null}(P) = S_2$.

[Side note: As long as $S_1 \oplus S_2 = V$, there is actually *exactly one* such projection, by Exercise 1.B.8.]

**1.B.8 Suppose $A, B \in \mathcal{M}_{m,n}$ have linearly independent columns and $\text{range}(A) \oplus \text{null}(B^*) = \mathbb{F}^m$ (where $\mathbb{F} = \mathbb{R}$ or $\mathbb{F} = \mathbb{C}$).

   (a) Show that B^*A is invertible.
   (b) Show that $P = A(B^*A)^{-1}B^*$ is the projection onto $\text{range}(A)$ along $\text{null}(B^*)$.

[Side note: Compare this exercise with Exercise 1.4.30, which covered orthogonal projections.]

1.B.9 Let S be a subspace of a finite-dimensional inner product space V. Show that if P is the orthogonal projection onto S then $I - P$ is the orthogonal projection onto $S^\perp$.

** 1.B.10 Let B be a set of vectors in an inner product space V. Show that $B^\perp$ is a subspace of V.

1.B.11 Suppose that B and C are sets of vectors in an inner product space V such that $B \subseteq C$. Show that $C^\perp \subseteq B^\perp$.

1.B.12 Suppose V is a finite-dimensional inner product space and $S, W_1, W_2 \subseteq V$ are subspaces for which $V = S \oplus W_1 = S \oplus W_2$.

   (a) Show that if $\langle \mathbf{v}, \mathbf{w}_1 \rangle = \langle \mathbf{v}, \mathbf{w}_2 \rangle = 0$ for all $\mathbf{v} \in S$, $\mathbf{w}_1 \in W_1$ and $\mathbf{w}_2 \in W_2$ then $W_1 = W_2$.
   (b) Provide an example to show that W_1 may not equal W_2 if we do not have the orthogonality requirement of part (a).

1.B.13 Show that if V is a finite-dimensional inner product space and $B \subseteq V$ then $B^\perp = \big(\text{span}(B)\big)^\perp$.

**1.B.14 Show that if V is a finite-dimensional inner product space and $B \subseteq V$ then $(B^\perp)^\perp = \text{span}(B)$.

[Hint: Make use of Exercises 1.B.12 and 1.B.13.]

**1.B.15 Prove part (b) of Theorem 1.B.7. That is, show that if V and W are finite-dimensional inner product spaces and $T : V \to W$ is a linear transformation then $\text{null}(T)^\perp = \text{range}(T^*)$.

** 1.B.16 Suppose V and W are finite-dimensional inner product spaces and $T : V \to W$ is a linear transformation. Show that the linear transformation $S : \text{range}(T^*) \to \text{range}(T)$ defined by $S(\mathbf{v}) = T(\mathbf{v})$ is invertible.

[Side note: S is called the **restriction** of T to $\text{range}(T^*)$.]

** 1.B.17 Let $\mathcal{P}^E[-1, 1]$ and $\mathcal{P}^O[-1, 1]$ denote the subspaces of even and odd polynomials, respectively, in $\mathcal{P}[-1, 1]$. Show that $(\mathcal{P}^E[-1, 1])^\perp = \mathcal{P}^O[-1, 1]$.

**1.B.18 Let $C[0, 1]$ be the inner product space of continuous functions on the real interval $[0, 1]$ and let $S \subset C[0, 1]$ be the subspace
$$S = \big\{ f \in C[0, 1] : f(0) = 0 \big\}.$$

   (a) Show that $S^\perp = \{\mathbf{0}\}$.
       [Hint: If $f \in S^\perp$, consider the function $g \in S$ defined by $g(x) = xf(x)$.]
   (b) Show that $(S^\perp)^\perp \neq S$.

[Side note: This result does not contradict Theorem 1.B.6 or Exercise 1.B.14 since $C[0, 1]$ is not finite-dimensional.]

**1.B.19 Show that if V and W are vector spaces over the same field then their external direct sum $V \oplus W$ is a vector space.

**1.B.20 Complete the proof of Theorem 1.B.9 by showing that if B and C are bases of vector spaces V and W, respectively, then the set $B' \cup C'$ (where B' and C' are as defined in the statement of that theorem) is linearly independent.

**1.B.21 In this exercise, we pin down the details that show that the internal and external direct sums are isomorphic. Let V and W be vector spaces over the same field.

   (a) Show that the sets
$$V' = \{(\mathbf{v}, \mathbf{0}) : \mathbf{v} \in V\} \quad \text{and} \quad W' = \{(\mathbf{0}, \mathbf{w}) : \mathbf{w} \in W\}$$
       are subspaces of the external direct sum $V \oplus W$.
   (b) Show that $V' \cong V$ and $W' \cong W$.
   (c) Show that $V \oplus W = V' \oplus W'$, where the direct sum on the left is external and the one on the right is internal.

1.C Extra Topic: The QR Decomposition

Many linear algebraic algorithms and procedures can be rephrased as a certain way of decomposing a matrix into a product of simpler matrices. For example, Gaussian elimination is the standard method that is used to solve systems of linear equations, and it is essentially equivalent to a matrix decomposition that the reader may be familiar with: the LU decomposition, which says that most matrices $A \in \mathcal{M}_{m,n}$ can be written in the form

$$A = LU,$$

where $L \in \mathcal{M}_m$ is lower triangular and $U \in \mathcal{M}_{m,n}$ is upper triangular. The rough idea is that L encodes the "forward elimination" portion of Gaussian elimination that gets A into row echelon form, U encodes the "backward substitution" step that solves for the variables from row echelon form (for details, see [Joh20], Section 2.D], for example).

Some matrices do not have an LU decomposition, but can be written in the form $A = PLU$, where $P \in \mathcal{M}_m$ is a permutation matrix that encodes the swap row operations used in Gaussian elimination.

In this section, we explore a matrix decomposition that is essentially equivalent to the Gram–Schmidt process (Theorem 1.4.6) in a very similar sense. Since the Gram–Schmidt process tells us how to convert any basis of $\mathbb{R}^n$ or $\mathbb{C}^n$ into an *orthonormal* basis of the same space, the corresponding matrix decomposition analogously provides us with a way of turning any invertible matrix (i.e., a matrix with columns that form a basis of $\mathbb{R}^n$ or $\mathbb{C}^n$) into a unitary matrix (i.e., a matrix with columns that form an *orthonormal* basis of $\mathbb{R}^n$ or $\mathbb{C}^n$).

1.C.1 Statement and Examples

We now state the main theorem of this section, which says that every matrix can be written as a product of a unitary matrix and an upper triangular matrix. As indicated earlier, our proof of this fact, as well as our method of actually computing this matrix decomposition, both come directly from the Gram–Schmidt process.

Theorem 1.C.1

QR Decomposition

Suppose $\mathbb{F} = \mathbb{R}$ or $\mathbb{F} = \mathbb{C}$, and $A \in \mathcal{M}_{m,n}(\mathbb{F})$. There exists a unitary matrix $U \in \mathcal{M}_m(\mathbb{F})$ and an upper triangular matrix $T \in \mathcal{M}_{m,n}(\mathbb{F})$ with non-negative real entries on its diagonal such that

$$A = UT.$$

We call such a decomposition of A a **QR decomposition**.

Before proving this theorem, we clarify that an upper triangular matrix T is one for which $t_{i,j} = 0$ whenever $i > j$, even if T is not square. For example,

$$\begin{bmatrix} 1 & 2 & 3 \\ 0 & 4 & 5 \\ 0 & 0 & 6 \end{bmatrix}, \quad \begin{bmatrix} 1 & 2 \\ 0 & 3 \\ 0 & 0 \end{bmatrix}, \quad \text{and} \quad \begin{bmatrix} 1 & 2 & 3 & 4 \\ 0 & 5 & 6 & 7 \\ 0 & 0 & 8 & 9 \end{bmatrix}$$

are all examples of upper triangular matrices.

In particular, this first
argument covers the
case when A is
square and
invertible (see
Theorem A.1.1).

Proof of Theorem 1.C.1. We start by proving this result in the special case when $m \geq n$ and A has linearly independent columns. We partition A as a block matrix according to its columns, which we denote by $\mathbf{v}_1, \mathbf{v}_2, \ldots, \mathbf{v}_n \in \mathbb{F}^m$:

$$A = \left[\; \mathbf{v}_1 \mid \mathbf{v}_2 \mid \cdots \mid \mathbf{v}_n \;\right].$$

To see that A has a QR decomposition, we recall that the Gram–Schmidt process (Theorem 1.4.6) tells us that there is a set of vectors $\{\mathbf{u}_1, \mathbf{u}_2, \ldots, \mathbf{u}_n\}$ with the property that, for each $1 \leq j \leq n$, $\{\mathbf{u}_1, \mathbf{u}_2, \ldots, \mathbf{u}_j\}$ is an orthonormal basis of $\mathrm{span}(\mathbf{v}_1, \mathbf{v}_2, \ldots, \mathbf{v}_j)$. Specifically, we can write $\mathbf{v}_j$ as a linear combination

$$\mathbf{v}_j = t_{1,j}\mathbf{u}_1 + t_{2,j}\mathbf{u}_2 + \cdots + t_{j,j}\mathbf{u}_j,$$

where

This formula for $t_{i,j}$
follows from
rearranging the
formula in
Theorem 1.4.6 so as
to solve for $\mathbf{v}_j$ (and
choosing the inner
product to be the
dot product). It also
follows from
Theorem 1.4.5, which
tells us that
$t_{j,j} = \mathbf{u}_j \cdot \mathbf{v}_j$ too.

$$t_{j,j} = \left\| \mathbf{v}_j - \sum_{i=1}^{j-1}(\mathbf{u}_i \cdot \mathbf{v}_j)\mathbf{u}_i \right\| \quad \text{and} \quad t_{i,j} = \begin{cases} \mathbf{u}_i \cdot \mathbf{v}_j & \text{if } i < j, \text{ and} \\ 0 & \text{if } i > j. \end{cases}$$

We then extend $\{\mathbf{u}_1, \mathbf{u}_2, \ldots, \mathbf{u}_n\}$ to an orthonormal basis $\{\mathbf{u}_1, \mathbf{u}_2, \ldots, \mathbf{u}_m\}$ of $\mathbb{F}^m$ and define $U = \left[\; \mathbf{u}_1 \mid \mathbf{u}_2 \mid \cdots \mid \mathbf{u}_m \;\right]$, noting that orthonormality of its columns implies that U is unitary. We also define $T \in \mathcal{M}_{m,n}$ to be the upper triangular matrix with $t_{i,j}$ as its (i,j)-entry for all $1 \leq i \leq m$ and $1 \leq j \leq n$ (noting that its diagonal entries $t_{j,j}$ are clearly real and non-negative, as required). Block matrix multiplication then shows that the j-th column of UT is

$$UT\mathbf{e}_j = \left[\; \mathbf{u}_1 \mid \mathbf{u}_2 \mid \cdots \mid \mathbf{u}_m \;\right] \begin{bmatrix} t_{1,j} \\ \vdots \\ t_{j,j} \\ 0 \\ \vdots \\ 0 \end{bmatrix} = t_{1,j}\mathbf{u}_1 + t_{2,j}\mathbf{u}_2 + \cdots + t_{j,j}\mathbf{u}_j = \mathbf{v}_j,$$

which is the j-th column of A, for all $1 \leq j \leq n$. It follows that $UT = A$, which completes the proof in the case when $m \geq n$ and A has linearly independent columns.

If $n > m$ then it is not possible for the columns of A to be linearly independent. However, if we write $A = \left[\; B \mid C \;\right]$ with $B \in \mathcal{M}_m$ having linearly independent columns then the previous argument shows that B has a QR decomposition $B = UT$. We can then write

In other words, B is
invertible.

$$A = U\left[\; T \mid U^*C \;\right],$$

which is a QR decomposition of A.

The name "QR"
decomposition is
mostly just a
historical
artifact—when it
was first introduced,
the unitary matrix U
was called Q all the
upper triangular
matrix T was called
R (for "right
triangular").

We defer the proof of the case when the leftmost $\min\{m, n\}$ columns of A do not form a linearly independent set to Section 2.D.3 (see Theorem 2.D.5 in particular). The rough idea is that we can approximate a QR decomposition of any matrix A as well as we like via QR decompositions of nearby matrices that have their leftmost $\min\{m, n\}$ columns being linearly independent. ∎

Before computing some explicit QR decompositions of matrices, we make some brief observations about it:

- The proof above shows that if $A \in \mathcal{M}_n$ is invertible then the diagonal entries of T are in fact strictly positive, not just non-negative (since $\mathbf{v}_j \neq \sum_{i=1}^{j-1}(\mathbf{u}_i \cdot \mathbf{v}_j)\mathbf{u}_i$). However, if A is not invertible then some diagonal entries of T will in fact equal 0.

- If $A \in \mathcal{M}_n$ is invertible then its QR decomposition is unique (see Exercise 1.C.5). However, uniqueness fails for non-invertible matrices.

- If $A \in \mathcal{M}_{m,n}(\mathbb{R})$ then we can choose the matrices U and T in the QR decomposition to be real as well.

Example 1.C.1

Computing a QR Decomposition

Compute the QR decomposition of the matrix $A = \begin{bmatrix} 1 & 3 & 3 \\ 2 & 2 & -2 \\ -2 & 2 & 1 \end{bmatrix}$.

Solution:

As suggested by the proof of Theorem 1.C.1, we can find the QR decomposition of A by applying the Gram–Schmidt process to the columns $\mathbf{v}_1$, $\mathbf{v}_2$, and $\mathbf{v}_3$ of A to recursively construct mutually orthogonal vectors $\mathbf{w}_j = \mathbf{v}_j - \sum_{i=1}^{j-1}(\mathbf{u}_i \cdot \mathbf{v}_j)\mathbf{u}_i$ and their normalizations $\mathbf{u}_j = \mathbf{w}_j/\|\mathbf{w}_j\|$ for $j = 1, 2, 3$:

The "lower-triangular" inner products like $\mathbf{u}_2 \cdot \mathbf{v}_1$ exist, but are irrelevant for the Gram–Schmidt process and QR decomposition. Also, the "diagonal" inner products come for free since $\mathbf{u}_j \cdot \mathbf{v}_j = \|\mathbf{w}_j\|$, and we already computed this norm when computing $\mathbf{u}_j = \mathbf{w}_j/\|\mathbf{w}_j\|$.

j	$\mathbf{w}_j$	$\mathbf{u}_j$	$\mathbf{u}_j \cdot \mathbf{v}_1$	$\mathbf{u}_j \cdot \mathbf{v}_2$	$\mathbf{u}_j \cdot \mathbf{v}_3$
1	$(1,2,-2)$	$(1,2,-2)/3$	3	1	-1
2	$(8,4,8)/3$	$(2,1,2)/3$	–	4	2
3	$(2,-2,-1)$	$(2,-2,-1)/3$	–	–	3

It follows that A has QR decomposition $A = UT$, where

$$U = \begin{bmatrix} \mathbf{u}_1 & | & \mathbf{u}_2 & | & \mathbf{u}_3 \end{bmatrix} = \frac{1}{3}\begin{bmatrix} 1 & 2 & 2 \\ 2 & 1 & -2 \\ -2 & 2 & -1 \end{bmatrix} \quad \text{and}$$

$$T = \begin{bmatrix} \mathbf{u}_1 \cdot \mathbf{v}_1 & \mathbf{u}_1 \cdot \mathbf{v}_2 & \mathbf{u}_1 \cdot \mathbf{v}_3 \\ 0 & \mathbf{u}_2 \cdot \mathbf{v}_2 & \mathbf{u}_2 \cdot \mathbf{v}_3 \\ 0 & 0 & \mathbf{u}_3 \cdot \mathbf{v}_3 \end{bmatrix} = \begin{bmatrix} 3 & 1 & -1 \\ 0 & 4 & 2 \\ 0 & 0 & 3 \end{bmatrix}.$$

Example 1.C.2

Computing a Rectangular QR Decomposition

Compute a QR decomposition of the matrix $A = \begin{bmatrix} 3 & 0 & 1 & 2 \\ -2 & -1 & -3 & 2 \\ -6 & -2 & -2 & 5 \end{bmatrix}$.

Solution:

Since A has more columns than rows, its columns cannot possibly form a linearly independent set. We thus just apply the Gram–Schmidt process to its leftmost 3 columns $\mathbf{v}_1$, $\mathbf{v}_2$, and $\mathbf{v}_3$, while just computing dot products with its 4th column $\mathbf{v}_4$ for later use:

We showed in the proof of Theorem 1.C.1 that the 4th column of T is $U^*\mathbf{v}_4$, whose entries are exactly the dot products in the final column here: $\mathbf{u}_j \cdot \mathbf{v}_4$ for $j = 1, 2, 3$.

j	$\mathbf{w}_j$	$\mathbf{u}_j$	$\mathbf{u}_j \cdot \mathbf{v}_1$	$\mathbf{u}_j \cdot \mathbf{v}_2$	$\mathbf{u}_j \cdot \mathbf{v}_3$	$\mathbf{u}_j \cdot \mathbf{v}_4$
1	$(3,-2,-6)$	$(3,-2,-6)/7$	7	2	3	-4
2	$(-6,-3,-2)/7$	$(-6,-3,-2)/7$	–	1	1	-4
3	$(4,-12,6)/7$	$(2,-6,3)/7$	–	–	2	1

It follows that A has QR decomposition $A = UT$, where

$$U = \begin{bmatrix} \mathbf{u}_1 \mid \mathbf{u}_2 \mid \mathbf{u}_3 \end{bmatrix} = \frac{1}{7} \begin{bmatrix} 3 & -6 & 2 \\ -2 & -3 & -6 \\ -6 & -2 & 3 \end{bmatrix} \quad \text{and}$$

$$T = \begin{bmatrix} \mathbf{u}_1 \cdot \mathbf{v}_1 & \mathbf{u}_1 \cdot \mathbf{v}_2 & \mathbf{u}_1 \cdot \mathbf{v}_3 & \mathbf{u}_1 \cdot \mathbf{v}_4 \\ 0 & \mathbf{u}_2 \cdot \mathbf{v}_2 & \mathbf{u}_2 \cdot \mathbf{v}_3 & \mathbf{u}_2 \cdot \mathbf{v}_4 \\ 0 & 0 & \mathbf{u}_3 \cdot \mathbf{v}_3 & \mathbf{u}_3 \cdot \mathbf{v}_4 \end{bmatrix} = \begin{bmatrix} 7 & 2 & 3 & -4 \\ 0 & 1 & 1 & -4 \\ 0 & 0 & 2 & 1 \end{bmatrix}.$$

Example 1.C.3

Computing a Tall/Thin QR Decomposition

Compute a QR decomposition of the matrix $A = \begin{bmatrix} -1 & -3 & 1 \\ -2 & 2 & -2 \\ 2 & -2 & 0 \\ 4 & 0 & -2 \end{bmatrix}$.

Solution:

Since A has more rows than columns, we can start by applying the Gram–Schmidt process to its columns, but this will only get us the leftmost 3 columns of the unitary matrix U:

j	$\mathbf{w}_j$	$\mathbf{u}_j$	$\mathbf{u}_j \cdot \mathbf{v}_1$	$\mathbf{u}_j \cdot \mathbf{v}_2$	$\mathbf{u}_j \cdot \mathbf{v}_3$
1	$(-1,-2,2,4)$	$(-1,-2,2,4)/5$	5	-1	-1
2	$(-16,8,-8,4)/5$	$(-4,2,-2,1)/5$	$-$	4	-2
3	$(-4,-8,-2,-4)/5$	$(-2,-4,-1,-2)/5$	$-$	$-$	2

We showed that every mutually orthogonal set of unit vectors can be extended to an orthonormal basis in Exercise 1.4.20. To do so, just add vector not in the span of the current set, apply Gram–Schmidt, and repeat.

To find its 4th column, we just extend its first 3 columns $\{\mathbf{u}_1, \mathbf{u}_2, \mathbf{u}_3\}$ to an orthonormal basis of $\mathbb{R}^4$. Up to sign, the unique unit vector $\mathbf{u}_4$ that works as the 4th column of U is $\mathbf{u}_4 = (2,-1,-4,2)/5$, so it follows that A has QR decomposition $A = UT$, where

$$U = \begin{bmatrix} \mathbf{u}_1 \mid \mathbf{u}_2 \mid \mathbf{u}_3 \mid \mathbf{u}_4 \end{bmatrix} = \frac{1}{5} \begin{bmatrix} -1 & -4 & -2 & 2 \\ -2 & 2 & -4 & -1 \\ 2 & -2 & -1 & -4 \\ 4 & 1 & -2 & 2 \end{bmatrix} \quad \text{and}$$

$$T = \begin{bmatrix} \mathbf{u}_1 \cdot \mathbf{v}_1 & \mathbf{u}_1 \cdot \mathbf{v}_2 & \mathbf{u}_1 \cdot \mathbf{v}_3 \\ 0 & \mathbf{u}_2 \cdot \mathbf{v}_2 & \mathbf{u}_2 \cdot \mathbf{v}_3 \\ 0 & 0 & \mathbf{u}_3 \cdot \mathbf{v}_3 \\ 0 & 0 & 0 \end{bmatrix} = \begin{bmatrix} 5 & -1 & -1 \\ 0 & 4 & -2 \\ 0 & 0 & 2 \\ 0 & 0 & 0 \end{bmatrix}.$$

Remark 1.C.1

Computing QR Decompositions

The method of computing the QR decomposition that we presented here, based on the Gram–Schmidt process, is typically not actually used in practice. The reason for this is that the Gram–Schmidt process is numerically unstable. If a set of vectors is "close" to linearly dependent then changing those vectors even slightly can drastically change the resulting orthonormal basis, and thus small errors in the entries of A can lead to a wildly incorrect QR decomposition.

Numerically stable methods for computing the QR decomposition (and numerical methods for linear algebraic tasks in general) are outside of the scope of this book, so the interested reader is directed to a book like [TB97] for their treatment.

1.C.2   Consequences and Applications

One of the primary uses of the QR decomposition is as a method for solving systems of linear equations more quickly than we otherwise could. To see how this works, suppose we have already computed a QR decomposition $A = UT$ of the coefficient matrix of the linear system $A\mathbf{x} = \mathbf{b}$. Then $UT\mathbf{x} = \mathbf{b}$, which is a linear system that we can solve via the following two-step procedure:

- First, solve the linear system $U\mathbf{y} = \mathbf{b}$ for the vector $\mathbf{y}$ by setting $\mathbf{y} = U^*\mathbf{b}$.
- Next, solve the linear system $T\mathbf{x} = \mathbf{y}$ for the vector $\mathbf{x}$. This linear system is straightforward to solve via backward elimination due to the triangular shape of T.

Once we have obtained the vector $\mathbf{x}$ via this procedure, it is the case that

$$A\mathbf{x} = UT\mathbf{x} = U(T\mathbf{x}) = U\mathbf{y} = \mathbf{b},$$

so $\mathbf{x}$ is indeed a solution of the original linear system, as desired.

Example 1.C.4

Solving Linear Systems via a QR Decomposition

Use the QR decomposition to find all solutions of the linear system

$$\begin{bmatrix} 3 & 0 & 1 & 2 \\ -2 & -1 & -3 & 2 \\ -6 & -2 & -2 & 5 \end{bmatrix} \begin{bmatrix} w \\ x \\ y \\ z \end{bmatrix} = \begin{bmatrix} 1 \\ 0 \\ 4 \end{bmatrix}.$$

Solution:

We constructed the following QR decomposition $A = UT$ of the coefficient matrix A in Example 1.C.2:

$$U = \frac{1}{7} \begin{bmatrix} 3 & -6 & 2 \\ -2 & -3 & -6 \\ -6 & -2 & 3 \end{bmatrix} \quad \text{and} \quad T = \begin{bmatrix} 7 & 2 & 3 & -4 \\ 0 & 1 & 1 & -4 \\ 0 & 0 & 2 & 1 \end{bmatrix}.$$

If $\mathbf{b} = (1,0,4)$ then setting $\mathbf{y} = U^*\mathbf{b}$ gives

$$\mathbf{y} = \frac{1}{7} \begin{bmatrix} 3 & -2 & -6 \\ -6 & -3 & -2 \\ 2 & -6 & 3 \end{bmatrix} \begin{bmatrix} 1 \\ 0 \\ 4 \end{bmatrix} = \begin{bmatrix} -3 \\ -2 \\ 2 \end{bmatrix}.$$

See Appendix A.1.1 if
you need a refresher
on linear systems.

Next, we solve the upper triangular system $T\mathbf{x} = \mathbf{y}$:

$$
\left[\begin{array}{ccc|c}
7 & 2 & 3 & -4 \\
0 & 1 & 1 & -4 \\
0 & 0 & 2 & 1
\end{array}\begin{array}{|c} -3 \\ -2 \\ 2 \end{array}\right]
\begin{array}{c} R_1/7 \\ R_3/2 \\ \longrightarrow \end{array}
\left[\begin{array}{cccc|c}
1 & 2/7 & 3/7 & -4/7 & -3/7 \\
0 & 1 & 1 & -4 & -2 \\
0 & 0 & 1 & 1/2 & 1
\end{array}\right]
$$

$$
\begin{array}{c} R_1 - \frac{3}{7}R_3 \\ R_2 - R_3 \\ \longrightarrow \end{array}
\left[\begin{array}{cccc|c}
1 & 2/7 & 0 & -11/14 & -6/7 \\
0 & 1 & 0 & -9/2 & -3 \\
0 & 0 & 1 & 1/2 & 1
\end{array}\right]
$$

$$
\begin{array}{c} R_1 - \frac{2}{7}R_2 \\ \longrightarrow \end{array}
\left[\begin{array}{cccc|c}
1 & 0 & 0 & 1/2 & 0 \\
0 & 1 & 0 & -9/2 & -3 \\
0 & 0 & 1 & 1/2 & 1
\end{array}\right].
$$

It follows that z is a free variable, and w, x, and y are leading variables with $w = -z/2$, $x = -3 + 9z/2$, and $y = 1 - z/2$. It follows that the solutions of this linear system (as well as the original linear system) are the vectors of the form $(w, x, y, z) = (0, -3, 1, 0) + z(-1, 9, -1, 2)/2$.

Remark 1.C.2

Multiple Methods for Solving Repeated Linear Systems

While solving a linear system via the QR decomposition is simpler than solving it directly via Gaussian elimination, actually computing the QR decomposition in the first place takes just as long as solving the linear system directly. For this reason, the QR decomposition is typically only used in this context to solve *multiple* linear systems, each of which has the same coefficient matrix (which we can pre-compute a QR decomposition of) but different right-hand-side vectors.

There are two other standard methods for solving repeated linear systems of the form $A\mathbf{x}_j = \mathbf{b}_j$ ($j = 1, 2, 3, \ldots$) that it is worth comparing to the QR decomposition:

- We could pre-compute A^{-1} and then set $\mathbf{x}_j = A^{-1}\mathbf{b}_j$ for each j. This method has the advantage of being conceptually simple, but it is slower and less numerically stable than the QR decomposition.

If $A \in \mathcal{M}_n$ then all three of these methods take $O(n^3)$ operations to do the pre-computation and then $O(n^2)$ operations to solve a linear system, compared to the $O(n^3)$ operations needed to solve a linear system directly via Gaussian elimination.

- We could pre-compute an LU decomposition $A = LU$, where L is lower triangular and U is upper triangular, and then solve the pair of triangular linear systems $L\mathbf{y}_j = \mathbf{b}_j$ and $U\mathbf{x}_j = \mathbf{y}_j$ for each j. This method is roughly twice as quick as the QR decomposition, but its numerical stability lies somewhere between that of the QR decomposition and the method based on A^{-1} described above.

Again, justification of the above claims is outside of the scope of this book, so we direct the interested reader to a book on numerical linear algebra like [TB97].

Once we have the QR decomposition of a (square) matrix, we can also use it to quickly compute the absolute value of its determinant, since determinants of unitary and triangular matrices are both easy to deal with.

Theorem 1.C.2

Determinant via QR Decomposition

If $A \in \mathcal{M}_n$ has QR decomposition $A = UT$ with $U \in \mathcal{M}_n$ unitary and $T \in \mathcal{M}_n$ upper triangular, then

$$|\det(A)| = t_{1,1} \cdot t_{2,2} \cdots t_{n,n}.$$

Proof of Theorem 1.C.1. We just string together three facts about the determinant that we already know:

We review these
properties of
determinants in
Appendix A.1.5.

- The determinant is multiplicative, so $|\det(A)| = |\det(U)||\det(T)|$,
- U is unitary, so Exercise 1.4.11 tells us that $|\det(U)| = 1$, and
- T is upper triangular, so $|\det(T)| = |t_{1,1} \cdot t_{2,2} \cdots t_{n,n}| = t_{1,1} \cdot t_{2,2} \cdots t_{n,n}$, with the final equality following from the fact that the diagonal entries of T are non-negative. ∎

For example, we saw in Example 1.C.1 that the matrix

$$A = \begin{bmatrix} 1 & 3 & 3 \\ 2 & 2 & -2 \\ -2 & 2 & 1 \end{bmatrix}$$

has QR decomposition $A = UT$ with

$$U = \frac{1}{3} \begin{bmatrix} 1 & 2 & 2 \\ 2 & 1 & -2 \\ -2 & 2 & -1 \end{bmatrix} \quad \text{and} \quad T = \begin{bmatrix} 3 & 1 & -1 \\ 0 & 4 & 2 \\ 0 & 0 & 3 \end{bmatrix}.$$

It follows that $|\det(A)|$ is the product of the diagonal entries of T: $|\det(A)| = 3 \cdot 4 \cdot 3 = 36$.

In fact, $\det(A) = 36$.

Exercises

solutions to starred exercises on page 462

1.C.1 Compute the QR Decomposition of each of the following matrices.

*(a) $\begin{bmatrix} 3 & 4 \\ 4 & 2 \end{bmatrix}$

(b) $\begin{bmatrix} 15 & -17 & -1 \\ 8 & -6 & 4 \end{bmatrix}$

*(c) $\begin{bmatrix} 6 & 3 \\ 3 & 2 \\ -6 & -2 \end{bmatrix}$

(d) $\begin{bmatrix} 0 & 0 & -1 \\ 4 & -1 & -2 \\ -3 & -3 & -1 \end{bmatrix}$

*(e) $\begin{bmatrix} 2 & 1 & 1 & 4 \\ 1 & 1 & 0 & 1 \\ 4 & 2 & 1 & -2 \\ 2 & 2 & 2 & 2 \end{bmatrix}$

(f) $\begin{bmatrix} -11 & 4 & -2 & -1 \\ 10 & -5 & -3 & 0 \\ -2 & -2 & -4 & -2 \end{bmatrix}$

1.C.2 Solve each of the following linear systems $A\mathbf{x} = \mathbf{b}$ by making use of the provided QR decomposition $A = UT$:

*(a) $\mathbf{b} = \begin{bmatrix} 1 \\ 2 \end{bmatrix}$, $U = \frac{1}{\sqrt{2}} \begin{bmatrix} 1 & 1 \\ -1 & 1 \end{bmatrix}$, $T = \begin{bmatrix} 2 & 1 \\ 0 & 3 \end{bmatrix}$

(b) $\mathbf{b} = \begin{bmatrix} 3 \\ -1 \end{bmatrix}$, $U = \frac{1}{5} \begin{bmatrix} 3 & -4 \\ 4 & 3 \end{bmatrix}$, $T = \begin{bmatrix} 2 & 3 & 1 \\ 0 & 1 & -1 \end{bmatrix}$

*(c) $\mathbf{b} = \begin{bmatrix} 1 \\ 2 \\ 0 \end{bmatrix}$, $U = \frac{1}{7} \begin{bmatrix} -3 & 6 & 2 \\ -2 & -3 & 6 \\ 6 & 2 & 3 \end{bmatrix}$

$T = \begin{bmatrix} 3 & -1 & 2 \\ 0 & 2 & 1 \\ 0 & 0 & 1 \end{bmatrix}$

(d) $\mathbf{b} = \begin{bmatrix} 1 \\ 1 \\ -2 \end{bmatrix}$, $U = \frac{1}{35} \begin{bmatrix} -15 & 30 & 10 \\ 18 & -1 & 30 \\ 26 & 18 & -15 \end{bmatrix}$

$T = \begin{bmatrix} 2 & 1 & 0 & 1 \\ 0 & 1 & 3 & -2 \\ 0 & 0 & 2 & 2 \end{bmatrix}$

1.C.3 Determine which of the following statements are true and which are false.

*(a) Every matrix $A \in \mathcal{M}_{m,n}(\mathbb{R})$ has a QR decomposition $A = UT$ with U and T real.

(b) If $A \in \mathcal{M}_n$ has QR decomposition $A = UT$ then $\det(A) = t_{1,1} \cdot t_{2,2} \cdots t_{n,n}$.

*(c) If $A \in \mathcal{M}_{m,n}$ has QR decomposition $A = UT$ then $\text{range}(A) = \text{range}(T)$.

(d) If $A \in \mathcal{M}_n$ is invertible and has QR decomposition $A = UT$ then, for each $1 \le j \le n$, the span of the leftmost j columns of A equals the span of the leftmost j columns of U.

1.C.4 Suppose that $A \in \mathcal{M}_n$ is an upper triangular matrix.

(a) Show that A is invertible if and only if all of its diagonal entries are non-zero.

(b) Show that if A is invertible then A^{-1} is also upper triangular. [Hint: First show that if $\mathbf{b}$ has its last k entries equal to 0 (for some k) then the solution $\mathbf{x}$ to $A\mathbf{x} = \mathbf{b}$ also has its last k entries equal to 0.]

(c) Show that if A is invertible then the diagonal entries of A^{-1} are the reciprocals of the diagonal entries of A, in the same order.

****1.C.5** Show that if $A \in \mathcal{M}_n$ is invertible then its QR decomposition is unique.

[Hint: Exercise 1.4.12 might help.]

1.C.6 Suppose that $A \in \mathcal{M}_2$ is the (non-invertible) matrix with QR decomposition $A = UT$, where

$$U = \frac{1}{5} \begin{bmatrix} 3 & 4 \\ 4 & -3 \end{bmatrix} \quad \text{and} \quad T = \begin{bmatrix} 0 & 0 \\ 0 & 1 \end{bmatrix}.$$

Find another QR decomposition of A.

[Side note: Contrast this with the fact from Exercise 1.C.5 that QR decompositions of invertible matrices are unique.]

***1.C.7** In this exercise, we investigate when the QR decomposition of a rectangular matrix $A \in \mathcal{M}_{m,n}$ is unique.

(a) Show that if $n \ge m$ and the left $m \times m$ block of A is invertible then its QR decomposition is unique.

(b) Provide an example that shows that if $n < m$ then the QR decomposition of A may not be unique, even if its top $n \times n$ block is invertible.

1.C.8 Suppose $A \in \mathcal{M}_{m,n}$. In this exercise, we demonstrate the existence of several variants of the QR decomposition.

(a) Show that there exists a unitary matrix U and a *lower* triangular matrix S such that $A = US$. [Hint: What happens to a matrix's QR decomposition if we swap its rows and/or columns?]

(b) Show that there exists a unitary matrix U and an upper triangular matrix T such that $A = TU$.

(c) Show that there exists a unitary matrix U and a lower triangular matrix S such that $A = SU$.

1.D   Extra Topic: Norms and Isometries

We now investigate how we can measure the length of vectors in arbitrary vector spaces. We already know how to do this in $\mathbb{R}^n$—the length of a vector $\mathbf{v} \in \mathbb{R}^n$ is

$$\|\mathbf{v}\| = \sqrt{\mathbf{v} \cdot \mathbf{v}} = \sqrt{v_1^2 + v_2^2 + \cdots + v_n^2}.$$

Slightly more generally, we saw in Section 1.3.4 that we can use the norm induced by the inner product to measure length in any inner product space:

$$\|\mathbf{v}\| = \sqrt{\langle \mathbf{v}, \mathbf{v} \rangle}.$$

However, it is sometimes preferable to use a measure of size that does not rely on us first defining an underlying inner product. We refer to such functions as **norms**, and we simply define them to be the functions that satisfy the usual properties that the length on $\mathbb{R}^n$ or the norm induced by an inner product satisfy.

Definition 1.D.1

Norm of a Vector

Suppose that $\mathbb{F} = \mathbb{R}$ or $\mathbb{F} = \mathbb{C}$ and that $\mathcal{V}$ is a vector space over $\mathbb{F}$. Then a **norm** on $\mathcal{V}$ is a function $\|\cdot\| : \mathcal{V} \to \mathbb{R}$ such that the following three properties hold for all $c \in \mathbb{F}$ and all $\mathbf{v}, \mathbf{w} \in \mathcal{V}$:

a) $\|c\mathbf{v}\| = |c| \|\mathbf{v}\|$ (absolute homogeneity)

b) $\|\mathbf{v} + \mathbf{w}\| \le \|\mathbf{v}\| + \|\mathbf{w}\|$ (triangle inequality)

c) $\|\mathbf{v}\| \ge 0$, with equality if and only if $\mathbf{v} = \mathbf{0}$ (positive definiteness)

The motivation for why each of the above defining properties should hold for any reasonable measure of "length" or "size" is hopefully somewhat clear— (a) if we multiply a vector by a scalar then its length scaled by the same amount, (b) the shortest path between two points is the straight line connecting them, and (c) we do not want lengths to be negative.

Every norm induced by an inner product is indeed a norm, as was established by Theorems 1.3.7 (which established properties (a) and (c)) and 1.3.9 (which established property (b)). However, there are also many different norms out there that are not induced by any inner product, both on $\mathbb{R}^n$ and on other vector spaces. We now present several examples.

Example 1.D.1

The 1-Norm (or "Taxicab" Norm)

The **1-norm** on $\mathbb{C}^n$ is the function defined by

$$\|\mathbf{v}\|_1 \overset{\text{def}}{=} |v_1| + |v_2| + \cdots + |v_n| \quad \text{for all} \quad \mathbf{v} \in \mathbb{C}^n.$$

Show that the 1-norm is indeed a norm.

Solution:
We have to check the three defining properties of norms (Definition 1.D.1). If $\mathbf{v}, \mathbf{w} \in \mathbb{C}^n$ and $c \in \mathbb{C}$ then:

a) $\|c\mathbf{v}\|_1 = |cv_1| + \cdots + |cv_n| = |c|(|v_1| + \cdots + |v_n|) = |c|\|\mathbf{v}\|_1.$

b) First, we note that $|v + w| \le |v| + |w|$ for all $v, w \in \mathbb{C}$ (this statement is equivalent to the usual triangle inequality on $\mathbb{R}^2$). We then have

$$\begin{aligned}
\|\mathbf{v} + \mathbf{w}\|_1 &= |v_1 + w_1| + \cdots + |v_n + w_n| \\
&\le (|v_1| + |w_1|) + \cdots + (|v_n| + |w_n|) \\
&= (|v_1| + \cdots + |v_n|) + (|w_1| + \cdots + |w_n|) \\
&= \|\mathbf{v}\|_1 + \|\mathbf{w}\|_1.
\end{aligned}$$

c) The fact that $\|\mathbf{v}\|_1 = |v_1| + \cdots + |v_n| \ge 0$ is clear. To see that $\|\mathbf{v}\|_1 = 0$ if and only if $\mathbf{v} = \mathbf{0}$, we similarly just notice that $|v_1| + \cdots + |v_n| = 0$ if and only if $v_1 = \cdots = v_n = 0$ (indeed, if any v_j were non-zero then $|v_j| > 0$, so $|v_1| + \cdots + |v_n| > 0$ too).

It is worth observing that the 1-norm measures the amount of distance that must be traveled to go from a vector's tail to its head when moving in the direction of the standard basis vectors. For this reason, it is sometimes called the **taxicab norm**: we imagine a square grid on $\mathbb{R}^2$ as the streets along which a taxi can travel to get from the tail to the head of a vector $\mathbf{v}$, and this norm measures how far the taxi must drive, as illustrated below for the vector $\mathbf{v} = (4,3)$:

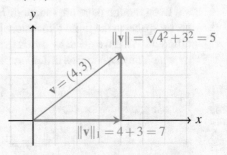

In analogy with the 1-norm from the above example, we refer to the usual vector length on $\mathbb{C}^n$, defined by

$$\|\mathbf{v}\|_2 \stackrel{\text{def}}{=} \sqrt{|v_1|^2 + |v_2|^2 + \cdots + |v_n|^2},$$

as the **2-norm** in this section, in reference to the exponent that appears in the terms being summed. We sometimes denote it by $\|\cdot\|_2$ instead of $\|\cdot\|$ to avoid confusion with the notation used for norms in general.

<table>
<tr><td>

Example 1.D.2

The ∞-Norm (or "Max" Norm)

</td><td>

The ∞-**norm** (or **max norm**) on $\mathbb{C}^n$ is the function defined by

$$\|\mathbf{v}\|_\infty \stackrel{\text{def}}{=} \max_{1 \le j \le n} \{|v_j|\} \quad \text{for all} \quad \mathbf{v} \in \mathbb{C}^n.$$

Show that the ∞-norm is indeed a norm.

Solution:
   Again, we have to check the three properties that define norms. If $\mathbf{v}, \mathbf{w} \in \mathbb{C}^n$ and $c \in \mathbb{C}$ then:

a) $\|c\mathbf{v}\|_\infty = \max_{1 \le j \le n}\{|cv_j|\} = |c|\max_{1 \le j \le n}\{|v_j|\} = |c|\|\mathbf{v}\|_\infty$.

b) By again making use of the fact that $|v + w| \le |v| + |w|$ for all $v, w \in \mathbb{C}$, we see that

$$\begin{aligned}
\|\mathbf{v} + \mathbf{w}\|_\infty &= \max_{1 \le j \le n}\{|v_j + w_j|\} \\
&\le \max_{1 \le j \le n}\{|v_j| + |w_j|\} \\
&\le \max_{1 \le j \le n}\{|v_j|\} + \max_{1 \le j \le n}\{|w_j|\} \\
&= \|\mathbf{v}\|_\infty + \|\mathbf{w}\|_\infty.
\end{aligned}$$

</td></tr>
</table>

The second inequality comes from two maximizations allowing more freedom (and thus a higher maximum) than one maximization. Equality holds if both maximums are attained at the same subscript j.

c) The fact that $\|\mathbf{v}\|_\infty \ge 0$ is clear. Similarly, $\|\mathbf{v}\|_\infty = 0$ if and only if the largest entry of $\mathbf{v}$ equals zero, if and only if $|v_j| = 0$ for all j, if and only if $\mathbf{v} = \mathbf{0}$.

One useful way of visualizing norms on $\mathbb{R}^n$ is to draw their **unit ball**—the set of vectors $\mathbf{v} \in \mathbb{R}^n$ satisfying $\|\mathbf{v}\| \le 1$. For the 2-norm, the unit ball is exactly the circle (or sphere, or hypersphere...) with radius 1 centered at the origin, together with its interior. For the ∞-norm, the unit ball is the set of vectors with the property that $|v_j| \le 1$ for all j, which is exactly the square (or cube, or hypercube...) that circumscribes the unit circle. Similarly, the unit ball of the 1-norm is the diamond inscribed within that unit circle. These unit balls are illustrated in $\mathbb{R}^2$ in Figure 1.27.

Even though there are lots of different norms that we can construct on any vector space, in the finite-dimensional case it turns out that they cannot be "too" different. What we mean by this is that there is at most a multiplicative constant by which any two norms differ—a property called **equivalence** of norms.

<table>
<tr><td>

Theorem 1.D.1

Equivalence of Norms

It does not matter which of $\|\cdot\|_a$ or $\|\cdot\|_b$ is in the middle in this theorem. The given inequality is equivalent to

</td><td>

Let $\|\cdot\|_a$ and $\|\cdot\|_b$ be norms on a finite-dimensional vector space $\mathcal{V}$. There exist real scalars $c, C > 0$ such that

$$c\|\mathbf{v}\|_a \le \|\mathbf{v}\|_b \le C\|\mathbf{v}\|_a \quad \text{for all} \quad \mathbf{v} \in \mathcal{V}.$$

</td></tr>
</table>

The proof of the above theorem is rather technical and involved, so we defer it to Appendix B.1. However, for certain choices of norms it is straightforward

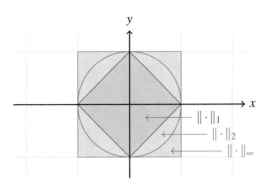

Figure 1.27: The unit balls for the 1-, 2-, and ∞-norms on $\mathbb{R}^2$.

to explicitly find constants c and C that work. For example, we can directly check that for any vector $\mathbf{v} \in \mathbb{C}^n$ we have

$$\|\mathbf{v}\|_\infty = \max_{1 \leq j \leq n} \left\{ |v_j| \right\} \leq |v_1| + |v_2| + \cdots + |v_n| = \|\mathbf{v}\|_1,$$

where the inequality holds because there is some particular index i with $|v_i| = \max_{1 \leq j \leq n} \left\{ |v_j| \right\}$, and the sum on the right contains this $|v_i|$ plus other non-negative terms.

To establish a bound between $\|\mathbf{v}\|_\infty$ and $\|\mathbf{v}\|_1$ that goes in the other direction, we similarly compute

The middle inequality just says that each $|v_i|$ is no larger than the largest $|v_i|$.

$$\|\mathbf{v}\|_1 = \sum_{i=1}^{n} |v_i| \leq \sum_{i=1}^{n} \left(\max_{1 \leq j \leq n} |v_j| \right) = \sum_{i=1}^{n} \|\mathbf{v}\|_\infty = n\|\mathbf{v}\|_\infty.$$

so $\|\mathbf{v}\|_1 \leq n\|\mathbf{v}\|_\infty$ (i.e., in the notation of Theorem 1.D.1, if we have $\|\cdot\|_a = \|\cdot\|_\infty$ and $\|\cdot\|_b = \|\cdot\|_1$, then we can choose $c = 1$ and $C = n$).

Geometrically, the fact that all norms on a finite-dimensional vector space are equivalent just means that their unit balls can be stretched or shrunk to contain each other, and the scalars c and C from Theorem 1.D.1 tell us what factor they must be stretched by to do so. For example, the equivalence of the 1- and ∞-norms is illustrated in Figure 1.28.

However, it is worth observing that Theorem 1.D.1 does not hold in infinite-dimensional vector spaces. That is, in infinite-dimensional vector spaces, we can construct norms $\|\cdot\|_a$ and $\|\cdot\|_b$ with the property that the ratio $\|\mathbf{v}\|_b/\|\mathbf{v}\|_a$ can be made as large as we like by suitably choosing $\mathbf{v}$ (see Exercise 1.D.26), so there does not exist a constant C for which $\|\mathbf{v}\|_b \leq C\|\mathbf{v}\|_a$. We call such norms **inequivalent**.

1.D.1 The p-Norms

We now investigate a family of norms that generalize the 1-, 2-, and ∞-norms on $\mathbb{C}^n$ in a fairly straightforward way:

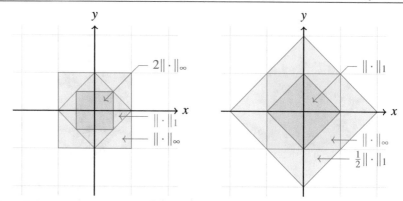

(a) From smallest to largest, the unit balls of the norms $2\|\cdot\|_\infty, \|\cdot\|_1$, and $\|\cdot\|_\infty$.

(b) From smallest to largest, the unit balls of the norms $\|\cdot\|_1, \|\cdot\|_\infty$, and $\frac{1}{2}\|\cdot\|_1$.

Figure 1.28: An illustration in $\mathbb{R}^2$ of the fact that (a) $\|\mathbf{v}\|_\infty \leq \|\mathbf{v}\|_1 \leq 2\|\mathbf{v}\|_\infty$ and (b) $\frac{1}{2}\|\mathbf{v}\|_1 \leq \|\mathbf{v}\|_\infty \leq \|\mathbf{v}\|_1$. In particular, one norm is lower bounded by another one if and only if the unit ball of the former norm is contained within the unit ball of the latter.

Definition 1.D.2

**The p-Norm
of a Vector**

If $p \geq 1$ is a real number, then the **p-norm** on $\mathbb{C}^n$ is defined by

$$\|\mathbf{v}\|_p \stackrel{\text{def}}{=} \left(\sum_{j=1}^n |v_j|^p \right)^{1/p} \quad \text{for all} \quad \mathbf{v} \in \mathbb{C}^n.$$

It should be reasonably clear that if $p = 1$ then the p-norm is exactly the "1-norm" from Example 1.D.1 (which explains why we referred to it as the 1-norm in the first place), and if $p = 2$ then it is the standard vector length. Furthermore, it is also the case that

$$\lim_{p \to \infty} \|\mathbf{v}\|_p = \max_{1 \leq j \leq n} \{|v_j|\} \quad \text{for all} \quad \mathbf{v} \in \mathbb{C}^n,$$

which is exactly the ∞-norm of $\mathbf{v}$ (and thus explains why we called it the ∞-norm in the first place). To informally see why this limit holds, we just notice that increasing the exponent p places more and more importance on the largest component of $\mathbf{v}$ compared to the others (this is proved more precisely in Exercise 1.D.8).

To verify that the p-norm is indeed a norm, we have to check the three properties of norms from Definition 1.D.1. Properties (a) and (c) (absolute homogeneity and positive definiteness) are straightforward enough:

a) If $\mathbf{v} \in \mathbb{C}^n$ and $c \in \mathbb{C}$ then

$$\|c\mathbf{v}\|_p = \left(|cv_1|^p + \cdots + |cv_n|^p\right)^{1/p} = |c|\left(|v_1|^p + \cdots + |v_n|^p\right)^{1/p} = |c|\|\mathbf{v}\|_p.$$

c) The fact that $\|\mathbf{v}\|_p \geq 0$ is clear. To see that $\|\mathbf{v}\|_p = 0$ if and only if $\mathbf{v} = \mathbf{0}$, we just notice that $\|\mathbf{v}\|_p = 0$ if and only if $|v_1|^p + \cdots + |v_n|^p = 0$, if and only if $v_1 = \cdots = v_n = 0$, if and only if $\mathbf{v} = \mathbf{0}$.

Proving that the triangle inequality holds for the p-norm is much more involved, so we state it separately as a theorem. This inequality is important enough and useful enough that it is given its own name—it is called **Minkowski's inequality**.

Theorem 1.D.2

Minkowski's Inequality

To see that f is convex, we notice that its second derivative is $f''(x) = p(p-1)x^{p-2}$, which is non-negative as long as $p \geq 1$ and $x \geq 0$ (see Theorem A.5.2).

For an introduction to convex functions, see Appendix A.5.2.

If $1 \leq p \leq \infty$ then $\|\mathbf{v} + \mathbf{w}\|_p \leq \|\mathbf{v}\|_p + \|\mathbf{w}\|_p$ for all $\mathbf{v}, \mathbf{w} \in \mathbb{C}^n$.

Proof. First, consider the function $f : [0, \infty) \to \mathbb{R}$ defined by $f(x) = x^p$. Standard calculus techniques show that f is convex (sometimes called "concave up" in introductory calculus courses) whenever $x \geq 0$—any line connecting two points on the graph of f lies above its graph (see Figure 1.29).

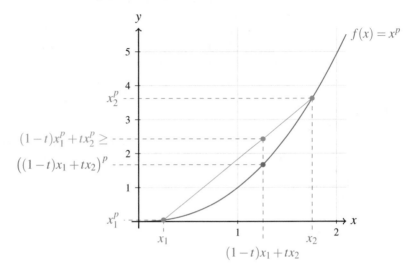

Figure 1.29: The function $f(x) = x^p$ (pictured here with $p = 2.3$) is convex on the interval $[0, \infty)$, so any line segment between two points on its graph lies above the graph itself.

Algebraically, this means that

$$\big((1-t)x_1 + tx_2\big)^p \leq (1-t)x_1^p + tx_2^p \quad \text{for all } x_1, x_2 \geq 0,\ 0 \leq t \leq 1. \quad (1.D.1)$$

Now suppose $\mathbf{v}, \mathbf{w} \in \mathbb{C}^n$ are non-zero vectors (if either one of them is the zero vector, Minkowski's inequality is trivial). Then we can write $\mathbf{v} = \|\mathbf{v}\|_p \mathbf{x}$ and $\mathbf{w} = \|\mathbf{w}\|_p \mathbf{y}$, where $\mathbf{x}$ and $\mathbf{y}$ are scaled so that $\|\mathbf{x}\|_p = \|\mathbf{y}\|_p = 1$. Then for any $0 \leq t \leq 1$, we have

Writing vectors as scaled unit vectors is a common technique when proving inequalities involving norms.

$$\big\|(1-t)\mathbf{x} + t\mathbf{y}\big\|_p^p = \sum_{j=1}^n |(1-t)x_j + ty_j|^p \qquad \text{(definition of } \|\cdot\|_p)$$

$$\leq \sum_{j=1}^n \big((1-t)|x_j| + t|y_j|\big)^p \qquad \text{(triangle inequality on } \mathbb{C})$$

$$\leq \sum_{j=1}^n \big((1-t)|x_j|^p + t|y_j|^p\big) \qquad \text{(Equation (1.D.1))}$$

$$= (1-t)\|\mathbf{x}\|_p^p + t\|\mathbf{y}\|_p^p \qquad \text{(definition of } \|\cdot\|_p)$$

$$= 1. \qquad (\|\mathbf{x}\|_p = \|\mathbf{y}\|_p = 1)$$

In particular, if we choose $t = \|\mathbf{w}\|_p / (\|\mathbf{v}\|_p + \|\mathbf{w}\|_p)$ then $1 - t = \|\mathbf{v}\|_p / (\|\mathbf{v}\|_p + \|\mathbf{w}\|_p)$, and the quantity above simplifies to

$$\big\|(1-t)\mathbf{x} + t\mathbf{y}\big\|_p^p = \frac{\big\|\|\mathbf{v}\|_p \mathbf{x} + \|\mathbf{w}\|_p \mathbf{y}\big\|_p^p}{(\|\mathbf{v}\|_p + \|\mathbf{w}\|_p)^p} = \frac{\|\mathbf{v} + \mathbf{w}\|_p^p}{(\|\mathbf{v}\|_p + \|\mathbf{w}\|_p)^p}.$$

Since this quantity is no larger than 1, multiplying through by $(\|\mathbf{v}\|_p + \|\mathbf{w}\|_p)^p$ tells us that

$$\|\mathbf{v} + \mathbf{w}\|_p^p \leq \big(\|\mathbf{v}\|_p + \|\mathbf{w}\|_p\big)^p.$$

Taking the p-th root of both sides of this inequality gives us exactly Minkowski's inequality and thus completes the proof. ∎

Equivalence and Hölder's Inequality

Now that we know that p-norms are indeed norms, it is instructive to try to draw their unit balls, much like we did in Figure 1.27 for the 1-, 2-, and ∞-norms. The following theorem shows that the p-norm can only decrease as p increases, which will help us draw the unit balls shortly.

Theorem 1.D.3

Monotonicity of p-Norms

> If $1 \leq p \leq q \leq \infty$ then $\|\mathbf{v}\|_q \leq \|\mathbf{v}\|_p$ for all $\mathbf{v} \in \mathbb{C}^n$.

Again, notice that proving this theorem was made easier by first rescaling the vector to have length 1.

Proof. If $\mathbf{v} = \mathbf{0}$ then this inequality is trivial, so suppose $\mathbf{v} \neq \mathbf{0}$ and consider the vector $\mathbf{w} = \mathbf{v}/\|\mathbf{v}\|_q$. Then $\|\mathbf{w}\|_q = 1$, so $|w_j| \leq 1$ for all j. It follows that $|w_j|^p \geq |w_j|^q$ for all j as well, so

$$\|\mathbf{w}\|_p = \left(\sum_{j=1}^{n} |w_j|^p\right)^{1/p} \geq \left(\sum_{j=1}^{n} |w_j|^q\right)^{1/p} = \|\mathbf{w}\|_q^{q/p} = 1.$$

Since $\mathbf{w} = \mathbf{v}/\|\mathbf{v}\|_q$, this implies that $\|\mathbf{w}\|_p = \|\mathbf{v}\|_p/\|\mathbf{v}\|_q \geq 1$, so $\|\mathbf{v}\|_p \geq \|\mathbf{v}\|_q$, as desired.

The argument above covers the case when both p and q are finite. The case when $q = \infty$ is proved in Exercise 1.D.7. ∎

The reason for the unit ball inclusion is that $\|\mathbf{v}\|_p \leq 1$ implies $\|\mathbf{v}\|_q \leq \|\mathbf{v}\|_p = 1$ too. Generally (even beyond just p-norms), larger norms have smaller unit balls.

By recalling that larger norms have smaller unit balls, we can interpret Theorem 1.D.3 as saying that the unit ball of the p-norm is contained within the unit ball of the q-norm whenever $1 \leq p \leq q \leq \infty$. When we combine this with the fact that we already know what the unit balls look like when $p = 1, 2$, or ∞ (refer back to Figure 1.27), we get a pretty good idea of what they look like for all values of p (see Figure 1.30). In particular, the sides of the unit ball gradually "bulge out" as p increases from 1 to ∞.

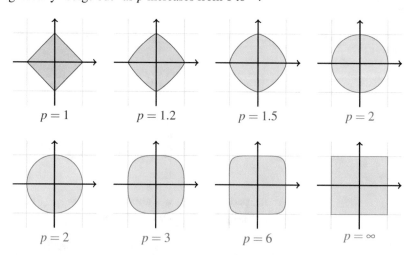

Figure 1.30: The unit ball of the p-norm on $\mathbb{R}^2$ for several values of $1 \leq p \leq \infty$.

The inequalities of Theorem 1.D.3 can be thought of as one half of the equivalence (refer back to Theorem 1.D.1) between the norms $\|\cdot\|_p$ and $\|\cdot\|_q$. The other inequality in this norm equivalence is trickier to pin down, so we start by solving this problem just for the 1-, 2-, and ∞-norms.

Theorem 1.D.4

1-, 2-, and ∞-Norm Inequalities

> If $\mathbf{v} \in \mathbb{C}^n$ then $\|\mathbf{v}\|_1 \leq \sqrt{n}\|\mathbf{v}\|_2 \leq n\|\mathbf{v}\|_\infty$.

Proof of Theorem 1.C.1. For the left inequality, we start by defining two vectors $\mathbf{x}, \mathbf{y} \in \mathbb{R}^n$:

$$\mathbf{x} = (1, 1, \ldots, 1) \quad \text{and} \quad \mathbf{y} = (|v_1|, |v_2|, \ldots, |v_n|).$$

By the Cauchy–Schwarz inequality, we know that $|\mathbf{x} \cdot \mathbf{y}| \leq \|\mathbf{x}\|_2 \|\mathbf{y}\|_2$. However, plugging everything into the relevant definitions then shows that

$$|\mathbf{x} \cdot \mathbf{y}| = \sum_{j=1}^n 1 \cdot |v_j| = \|\mathbf{v}\|_1,$$

and

$$\|\mathbf{x}\|_2 = \sqrt{\sum_{j=1}^n 1^2} = \sqrt{n} \quad \text{and} \quad \|\mathbf{y}\|_2 = \sqrt{\sum_{j=1}^n |v_j|^2} = \|\mathbf{v}\|_2.$$

It follows that $\|\mathbf{v}\|_1 = |\mathbf{x} \cdot \mathbf{y}| \leq \|\mathbf{x}\|_2 \|\mathbf{y}\|_2 = \sqrt{n}\|\mathbf{v}\|_2$, as claimed.

To prove the second inequality, we notice that

Again, the inequality here follows simply because each $|v_i|$ is no larger than the largest $|v_i|$.

$$\|\mathbf{v}\|_2 = \sqrt{\sum_{i=1}^n |v_i|^2} \leq \sqrt{\sum_{i=1}^n \left(\max_{1 \leq j \leq n} |v_j| \right)^2} = \sqrt{\sum_{i=1}^n \|\mathbf{v}\|_\infty^2} = \sqrt{n}\|\mathbf{v}\|_\infty. \quad \blacksquare$$

Before we can prove an analogous result for arbitrary p- and q-norms, we first need one more technical helper theorem. Just like Minkowski's inequality (Theorem 1.D.2) generalizes the triangle inequality from the 2-norm to arbitrary p-norms, the following theorem generalizes the Cauchy–Schwarz inequality from the 2-norm to arbitrary p-norms.

Theorem 1.D.5

Hölder's Inequality

> Let $1 \leq p, q \leq \infty$ be such that $1/p + 1/q = 1$. Then
>
> $$|\mathbf{v} \cdot \mathbf{w}| \leq \|\mathbf{v}\|_p \|\mathbf{w}\|_q \quad \text{for all} \quad \mathbf{v}, \mathbf{w} \in \mathbb{C}^n.$$

Before proving this theorem, it is worth focusing a bit on the somewhat strange relationship between p and q that it requires. First, notice that if $p = q = 2$ then $1/p + 1/q = 1$, so this is how we get the Cauchy–Schwarz inequality as a special case of Hölder's inequality. On the other hand, if one of p or q is smaller than 2 then the other one must be correspondingly larger than 2 in order for $1/p + 1/q = 1$ to hold—as p decreases from 2 to 1, q increases from 2 to ∞ (see Figure 1.31).

We could explicitly solve for q in terms of p and get $q = p/(p-1)$, but doing so makes the symmetry between p and q less clear (e.g., it is also the case that $p = q/(q-1)$), so we do not usually do so. It is also worth noting that if $p = \infty$ or $q = \infty$ then we take the convention that $1/\infty = 0$ in this setting, so that if $p = \infty$ then $q = 1$ and if $q = \infty$ then $p = 1$.

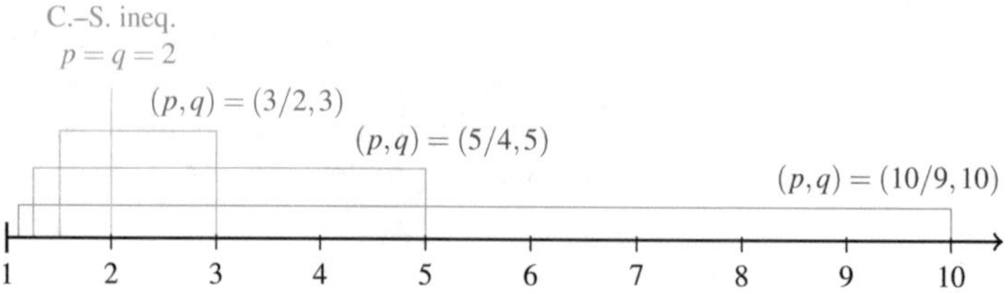

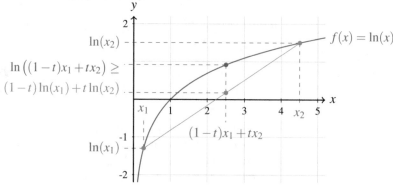

We call a pair of numbers p and q **conjugate** if $1/p + 1/q = 1$. We also call the pairs $(p,q) = (1,\infty)$ and $(p,q) = (\infty,1)$ conjugate.

Figure 1.31: Some pairs of numbers (p,q) for which $1/p + 1/q = 1$ and thus Hölder's inequality (Theorem 1.D.5) applies to them.

Inequality (1.D.2) is called **Young's inequality**, and it depends on the fact that $1/p + 1/q = 1$.

Proof of Theorem 1.D.5. Before proving the desired inequality involving vectors, we first prove the following inequality involving real numbers $x, y \geq 0$:

$$xy \leq \frac{x^p}{p} + \frac{y^q}{q}. \tag{1.D.2}$$

To see why this inequality holds, first notice that it is trivial if $x = 0$ or $y = 0$, so we can assume from now on that $x, y > 0$. Consider the function $f : (0,\infty) \to \mathbb{R}$ defined by $f(x) = \ln(x)$. Standard calculus techniques show that f is concave (sometimes called "concave down" in introductory calculus courses) whenever $x > 0$—any line connecting two points on the graph of f lies below its graph (see Figure 1.32).

$\ln(x)$ refers to the natural logarithm (i.e., the base-e logarithm) of x.

To see that f is concave, we notice that its second derivative is $f''(x) = -1/x^2$, which is negative as long as $x > 0$ (see Theorem A.5.2).

Figure 1.32: The function $f(x) = \ln(x)$ is concave on the interval $(0,\infty)$, so any line segment between two points on its graph lies below the graph itself.

Algebraically, this tells us that

$$\ln\big((1-t)x_1 + tx_2\big) \geq (1-t)\ln(x_1) + t\ln(x_2) \quad \text{for all} \quad x_1, x_2 > 0,\ 0 \leq t \leq 1.$$

In particular, if we choose $x_1 = x^p$, $x_2 = y^q$, and $t = 1/q$ then $1 - t = 1 - 1/q = 1/p$ and the above inequality becomes

Recall that $\ln(x^p) = p\ln(x)$, so $\ln(x^p)/p = \ln(x)$.

$$\ln\left(\frac{x^p}{p} + \frac{y^q}{q}\right) \geq \frac{\ln(x^p)}{p} + \frac{\ln(y^q)}{q} = \ln(x) + \ln(y) = \ln(xy).$$

Exponentiating both sides of this inequality (i.e., raising e to the power of the left- and right-hand-sides) then gives us exactly Inequality (1.D.2), which we were trying to prove.

To now prove Hölder's inequality, we first observe that it suffices to prove the case when $\|\mathbf{v}\|_p = \|\mathbf{w}\|_q = 1$, since multiplying either $\mathbf{v}$ or $\mathbf{w}$ by a scalar has no effect on whether or not the inequality $|\mathbf{v} \cdot \mathbf{w}| \leq \|\mathbf{v}\|_p \|\mathbf{w}\|_q$ holds (both sides of it just get multiplied by that same scalar). To see that $|\mathbf{v} \cdot \mathbf{w}| \leq 1$ in this case, and thus complete the proof, we compute

$$|\mathbf{v}\cdot\mathbf{w}| = \left|\sum_{j=1}^{n} v_j w_j\right| \leq \sum_{j=1}^{n} |v_j w_j| \qquad \text{(triangle inequality on } \mathbb{C}\text{)}$$

$$\leq \sum_{j=1}^{n} \frac{|v_j|^p}{p} + \sum_{j=1}^{n} \frac{|w_j|^q}{q} \qquad \text{(Inequality (1.D.2))}$$

$$= \frac{\|\mathbf{v}\|_p^p}{p} + \frac{\|\mathbf{w}\|_q^q}{q} \qquad \text{(definition of } p-, q\text{-norm)}$$

$$= \frac{1}{p} + \frac{1}{q} \qquad \text{(since } \|\mathbf{v}\|_p = \|\mathbf{w}\|_q = 1\text{)}$$

$$= 1.$$

As one final note, we have only explicitly proved this theorem in the case when both p and q are finite. The case when $p = \infty$ or $q = \infty$ is actually much simpler, and thus left to Exercise 1.D.9. ∎

With Hölder's inequality now at our disposal, we can finally find the multiplicative constants that can be used to bound different p-norms in terms of each other (i.e., we can quantify the equivalence of the p-norms that was guaranteed by Theorem 1.D.1).

Theorem 1.D.6

Equivalence of p-Norms

If $1 \leq p \leq q \leq \infty$ then

$$\|\mathbf{v}\|_q \leq \|\mathbf{v}\|_p \leq n^{1/p-1/q}\|\mathbf{v}\|_q \quad \text{for all} \quad \mathbf{v} \in \mathbb{C}^n.$$

Proof. We already proved the left inequality in Theorem 1.D.3, so we just need to prove the right inequality. To this end, consider the vectors $\mathbf{x} = (|v_1|^p, |v_2|^p, \ldots, |v_n|^p)$ and $\mathbf{y} = (1, 1, \ldots, 1) \in \mathbb{C}^n$ and define $\tilde{p} = q/p$ and $\tilde{q} = q/(q-p)$. It is straightforward to check that $1/\tilde{p} + 1/\tilde{q} = 1$, so Hölder's inequality then tells us that

We introduce $\tilde{p}$ and $\tilde{q}$ and apply Hölder's inequality to them instead of p and q since it might not even be the case that $1/p + 1/q = 1$.

$$\|\mathbf{v}\|_p^p = \sum_{j=1}^{n} |v_i|^p \qquad \text{(definition of } \|\mathbf{v}\|_p\text{)}$$

$$= |\mathbf{x}\cdot\mathbf{y}| \leq \|\mathbf{x}\|_{\tilde{p}}\|\mathbf{y}\|_{\tilde{q}} \qquad \text{(Hölder's inequality)}$$

$$= \left(\sum_{j=1}^{n} \left(|v_j|^p\right)^{\tilde{p}}\right)^{1/\tilde{p}} \left(\sum_{j=1}^{n} 1^{\tilde{q}}\right)^{1/\tilde{q}} \qquad \text{(definition of } \|\mathbf{x}\|_{\tilde{p}} \text{ and } \|\mathbf{y}\|_{\tilde{q}}\text{)}$$

$$= \left(\sum_{j=1}^{n} |v_j|^q\right)^{p/q} \left(n^{(q-p)/q}\right) \qquad (p\tilde{p} = p(q/p) = q, \text{ etc.})$$

$$= \left(n^{(q-p)/q}\right)\|\mathbf{v}\|_q^p. \qquad \text{(definition of } \|\mathbf{v}\|_q\text{)}$$

Taking the p-th root of both sides of this inequality then shows us that

$$\|\mathbf{v}\|_p \leq n^{(q-p)/(pq)}\|\mathbf{v}\|_q = n^{1/p-1/q}\|\mathbf{v}\|_q,$$

as desired. ∎

If p and q are some combination of 1, 2, or ∞ then the above result recovers earlier inequalities like those of Theorem 1.D.4. For example, if $(p,q) = (1,2)$ then this theorem says that

$$\|\mathbf{v}\|_2 \leq \|\mathbf{v}\|_1 \leq n^{1-1/2}\|\mathbf{v}\|_2 = \sqrt{n}\|\mathbf{v}\|_2,$$

which we already knew. Furthermore, we note that the constants ($c = 1$ and $C = n^{1/p-1/q}$) given in Theorem 1.D.6 are the best possible—for each inequality, there are particular choices of vectors $\mathbf{v} \in \mathbb{C}^n$ that attain equality (see Exercise 1.D.12).

The p-Norms for Functions

$\mathcal{C}[a,b]$ denotes the vector space of continuous functions on the interval $[a,b]$.

Before we move onto investigating other properties of norms, it is worth noting that the p-norms can also be defined for continuous functions in a manner completely analogous to how we defined them on $\mathbb{C}^n$ above.

Definition 1.D.3

The p-Norm of a Function

If $p \geq 1$ is a real number, then the **p-norm** on $\mathcal{C}[a,b]$ is defined by

$$\|f\|_p \overset{\text{def}}{=} \left(\int_a^b |f(x)|^p \, dx \right)^{1/p} \quad \text{for all} \quad f \in \mathcal{C}[a,b].$$

The Extreme Value Theorem guarantees that this maximum is attained, since f is continuous.

Similarly, for the $p = \infty$ case we simply define $\|f\|_p$ to be the maximum value that f attains on the interval $[a,b]$:

$$\|f\|_\infty \overset{\text{def}}{=} \max_{a \leq x \leq b} \left\{ |f(x)| \right\}.$$

Defining norms on functions in this way perhaps seems a bit more natural if we notice that we can think of a vector $\mathbf{v} \in \mathbb{C}^n$ as a function from $\{1, 2, \ldots, n\}$ to $\mathbb{C}^n$ (in particular, this function sends j to v_j). Definition 1.D.3 can then be thought of as simply extending the p-norms from functions on discrete sets of numbers like $\{1, 2, \ldots, n\}$ to functions on continuous intervals like $[a,b]$.

Example 1.D.3

Computing the p-Norm of a Function

Compute the p-norms of the function $f(x) = x+1$ in $\mathcal{C}[0,1]$.

Solution:

For finite values of p, we just directly compute the value of the desired integral:

We do not have to take the absolute value of f since $f(x) \geq 0$ on the interval $[0,1]$.

$$\|x+1\|_p = \left(\int_0^1 (x+1)^p \, dx \right)^{1/p}$$

$$= \left(\frac{(x+1)^{p+1}}{p+1} \bigg|_{x=0}^1 \right)^{1/p} = \left(\frac{2^{p+1}-1}{p+1} \right)^{1/p}.$$

For the $p = \infty$ case, we just notice that the maximum value of $f(x)$ on the interval $[0,1]$ is

$$\|x+1\|_\infty = \max_{0 \leq x \leq 1} \{x+1\} = 1+1 = 2.$$

It is worth noting that $\lim_{x \to \infty} \|x+1\|_p = 2$ as well, as we might hope.

One property of p-norms that does not carryover is the rightmost inequality in Theorem 1.D.6.

Furthermore, most of the previous theorems that we saw concerning p-norms carry through with minimal changes to this new setting too, simply by replacing vectors by functions and sums by integrals. To illustrate this fact, we now prove Minkowski's inequality for the p-norm of a function. However, we leave numerous other properties of the p-norm of a function to the exercises.

| Theorem 1.D.7 | If $1 \leq p \leq \infty$ then $\|f+g\|_p \leq \|f\|_p + \|g\|_p$ for all $f, g \in \mathcal{C}[a,b]$. |

Minkowski's Inequality for Functions

Proof. Just like in our original proof of Minkowski's inequality, notice that the function that sends x to x^p is convex, so

$$((1-t)x_1 + tx_2)^p \leq (1-t)x_1^p + tx_2^p \quad \text{for all} \quad x_1, x_2 \geq 0, \ 0 \leq t \leq 1.$$

This proof is almost identical to the proof of Theorem 1.D.2, but with sums replaced by integrals, vectors replaced by functions, and coordinates of vectors (like x_j) replaced by function values (like $\hat{f}(x)$).

Now suppose that $f, g \in \mathcal{C}[a,b]$ are non-zero functions (if either one of them is the zero function, Minkowski's inequality is trivial). Then we can write $f = \|f\|_p \hat{f}$ and $g = \|g\|_p \hat{g}$, where $\hat{f}$ and $\hat{g}$ are unit vectors in the p-norm (i.e., $\|\hat{f}\|_p = \|\hat{g}\|_p = 1$). Then for every $0 \leq t \leq 1$, we have

$$
\begin{aligned}
\left\|(1-t)\hat{f} + t\hat{g}\right\|_p^p &= \int_a^b \left|(1-t)\hat{f}(x) + t\hat{g}(x)\right|^p dx && \text{(definition of } \|\cdot\|_p) \\
&\leq \int_a^b \left((1-t)|\hat{f}(x)| + t|\hat{g}(x)|\right)^p dx && \text{(triangle ineq. on } \mathbb{C}) \\
&\leq \int_a^b \left((1-t)|\hat{f}(x)|^p + t|\hat{g}(x)|^p\right) dx && \text{(since } x^p \text{ is convex)} \\
&= (1-t)\|\hat{f}\|_p^p + t\|\hat{g}\|_p^p && \text{(definition of } \|\cdot\|_p) \\
&= 1. && (\|\hat{f}\|_p = \|\hat{g}\|_p = 1)
\end{aligned}
$$

We apologize for having norms within norms here. Ugh.

In particular, if we choose $t = \|g\|_p / (\|f\|_p + \|g\|_p)$ then $1 - t = \|f\|_p / (\|f\|_p + \|g\|_p)$, and the quantity above simplifies to

$$\left\|(1-t)\hat{f} + t\hat{g}\right\|_p^p = \frac{\left\|\|f\|_p\hat{f} + \|g\|_p\hat{g}\right\|_p^p}{(\|f\|_p + \|g\|_p)^p} = \frac{\|f+g\|_p^p}{(\|f\|_p + \|g\|_p)^p}.$$

Since this quantity is no larger than 1, multiplying through by $\left(\|f\|_p + \|g\|_p\right)^p$ tells us that

$$\|f+g\|_p^p \leq \left(\|f\|_p + \|g\|_p\right)^p.$$

Taking the p-th root of both sides of this inequality gives us exactly Minkowski's inequality. ∎

We prove Hölder's inequality for functions in Exercise 1.D.14.

Minkowski's inequality establishes the triangle inequality of the p-norm of a function, just like it did for the p-norm of vectors in $\mathbb{C}^n$. The other two defining properties of norms (absolute homogeneity and positive definiteness) are much easier to prove, and are left as Exercise 1.D.13.

1.D.2 From Norms Back to Inner Products

We have now seen that there are many different norms out there, but the norms induced by inner products serve a particularly important role and are a bit easier to work with (for example, they satisfy the Cauchy–Schwarz inequality). It thus seems natural to ask how we can determine whether or not a given norm is induced by an inner product, and if so how we can recover that inner product. For example, although the 1-norm on $\mathbb{R}^2$ is not induced by the standard inner product (i.e., the dot product), it does not seem obvious whether or not it is induced by some more exotic inner product.

To answer this question for the 1-norm, suppose for a moment that there were an inner product $\langle \cdot, \cdot \rangle$ on $\mathbb{R}^2$ that induces the 1-norm : $\|\mathbf{v}\|_1 = \langle \mathbf{v}, \mathbf{v} \rangle$ for all $\mathbf{v} \in \mathbb{R}^2$. By using the fact that $\|\mathbf{e}_1\|_1 = \|\mathbf{e}_2\|_1 = 1$ and $\|\mathbf{e}_1 + \mathbf{e}_2\|_1 = 2$, we see that

$$
\begin{aligned}
4 = \|\mathbf{e}_1 + \mathbf{e}_2\|_1^2 &= \langle \mathbf{e}_1 + \mathbf{e}_2, \mathbf{e}_1 + \mathbf{e}_2 \rangle \\
&= \|\mathbf{e}_1\|_1^2 + 2\mathrm{Re}(\langle \mathbf{e}_1, \mathbf{e}_2 \rangle) + \|\mathbf{e}_2\|_1^2 = 2 + 2\mathrm{Re}(\langle \mathbf{e}_1, \mathbf{e}_2 \rangle).
\end{aligned}
$$

> The upcoming theorem tells us how to turn a norm back into an inner product, so one way of interpreting it is as a converse of Definition 1.3.7, which gave us a method of turning an inner product into a norm.

By rearranging and simplifying, we thus see that $\mathrm{Re}(\langle \mathbf{e}_1, \mathbf{e}_2 \rangle) = 1$, so $|\langle \mathbf{e}_1, \mathbf{e}_2 \rangle| \geq 1$. However, the Cauchy–Schwarz inequality tells us that $|\langle \mathbf{e}_1, \mathbf{e}_2 \rangle| \leq 1$ and in fact $|\langle \mathbf{e}_1, \mathbf{e}_2 \rangle| < 1$ since $\mathbf{e}_1$ and $\mathbf{e}_2$ are not scalar multiples of each other. We have thus arrived at a contradiction that shows that no such inner product exists—the 1-norm is not induced by any inner product.

The above argument was very specific to the 1-norm, however, and it is not immediately clear whether or not a similar argument can be applied to the ∞-norm, 7-norm, or other even more exotic norms. Before introducing a theorem that solves this problem, we introduce one additional minor piece of terminology and notation. We say that a vector space $\mathcal{V}$, together with a particular norm, is a **normed vector space**, and we denote the norm by $\| \cdot \|_{\mathcal{V}}$ rather than just $\| \cdot \|$ in order to avoid any potential confusion with other norms.

> **Theorem 1.D.8**
>
> **Jordan–von Neumann Theorem**

Let $\mathcal{V}$ be a normed vector space. Then $\| \cdot \|_{\mathcal{V}}$ is induced by an inner product if and only if

$$
2\|\mathbf{v}\|_{\mathcal{V}}^2 + 2\|\mathbf{w}\|_{\mathcal{V}}^2 = \|\mathbf{v} + \mathbf{w}\|_{\mathcal{V}}^2 + \|\mathbf{v} - \mathbf{w}\|_{\mathcal{V}}^2 \quad \text{for all} \quad \mathbf{v}, \mathbf{w} \in \mathcal{V}.
$$

Before proving this theorem, we note that the equation $2\|\mathbf{v}\|_{\mathcal{V}}^2 + 2\|\mathbf{w}\|_{\mathcal{V}}^2 = \|\mathbf{v} + \mathbf{w}\|_{\mathcal{V}}^2 + \|\mathbf{v} - \mathbf{w}\|_{\mathcal{V}}^2$ is sometimes called the **parallelogram law**, since it relates the lengths of the sides of a parallelogram to the lengths of its diagonals, as illustrated in Figure 1.33. In particular, it says that the sum of squares of the side lengths of a parallelogram always equals the sum of squares of diagonal lengths of that parallelogram, as long as the way that we measure "length" comes from an inner product.

> The parallelogram law is a generalization of the Pythagorean theorem, which is what we get if we apply the parallelogram law to rectangles.

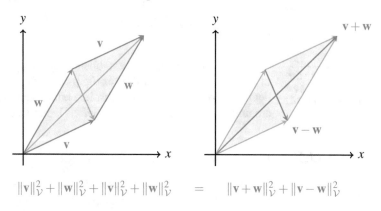

$$
\|\mathbf{v}\|_{\mathcal{V}}^2 + \|\mathbf{w}\|_{\mathcal{V}}^2 + \|\mathbf{v}\|_{\mathcal{V}}^2 + \|\mathbf{w}\|_{\mathcal{V}}^2 \quad = \quad \|\mathbf{v} + \mathbf{w}\|_{\mathcal{V}}^2 + \|\mathbf{v} - \mathbf{w}\|_{\mathcal{V}}^2
$$

Figure 1.33: The Jordan–von Neumann theorem says that a norm is induced by an inner product if and only if the sum of squares of norms of diagonals of a parallelogram equals the sum of squares of norms of its sides (i.e., if and only if the parallelogram law holds for that norm).

> We originally proved the "only if" direction back in Exercise 1.3.13.

Proof of Theorem 1.D.8. To see the "only if" claim, we note that if $\| \cdot \|_{\mathcal{V}}$ is induced by an inner product $\langle \cdot, \cdot \rangle$ then some algebra shows that

$$\begin{aligned}
\|\mathbf{v} + \mathbf{w}\|_{\mathcal{V}}^2 + \|\mathbf{v} - \mathbf{w}\|_{\mathcal{V}}^2 &= \langle \mathbf{v} + \mathbf{w}, \mathbf{v} + \mathbf{w} \rangle + \langle \mathbf{v} - \mathbf{w}, \mathbf{v} - \mathbf{w} \rangle \\
&= \big(\langle \mathbf{v}, \mathbf{v} \rangle + \langle \mathbf{v}, \mathbf{w} \rangle + \langle \mathbf{w}, \mathbf{v} \rangle + \langle \mathbf{w}, \mathbf{w} \rangle \big) \\
&\quad + \big(\langle \mathbf{v}, \mathbf{v} \rangle - \langle \mathbf{v}, \mathbf{w} \rangle - \langle \mathbf{w}, \mathbf{v} \rangle + \langle \mathbf{w}, \mathbf{w} \rangle \big) \\
&= 2\langle \mathbf{v}, \mathbf{v} \rangle + 2\langle \mathbf{w}, \mathbf{w} \rangle \\
&= 2\|\mathbf{v}\|_{\mathcal{V}}^2 + 2\|\mathbf{w}\|_{\mathcal{V}}^2.
\end{aligned}$$

The "if" direction of the proof is much more involved, and we only prove the case when the underlying field is $\mathbb{F} = \mathbb{R}$ (we leave the $\mathbb{F} = \mathbb{C}$ case to Exercise 1.D.16). To this end, suppose that the parallelogram law holds and define the following function $\langle \cdot, \cdot \rangle : \mathcal{V} \times \mathcal{V} \to \mathbb{R}$ (which we will show is an inner product):

*This formula for an inner product in terms of its induced norm is sometimes called the **polarization identity**. We first encountered this identity back in Exercise 1.3.14.*

$$\langle \mathbf{v}, \mathbf{w} \rangle = \frac{1}{4} \big(\|\mathbf{v} + \mathbf{w}\|_{\mathcal{V}}^2 - \|\mathbf{v} - \mathbf{w}\|_{\mathcal{V}}^2 \big).$$

To see that this function really is an inner product, we have to check the three defining properties of inner products from Definition 1.3.6. Property (a) is straightforward, since

$$\langle \mathbf{v}, \mathbf{w} \rangle = \frac{1}{4} \big(\|\mathbf{v} + \mathbf{w}\|_{\mathcal{V}}^2 - \|\mathbf{v} - \mathbf{w}\|_{\mathcal{V}}^2 \big) = \frac{1}{4} \big(\|\mathbf{w} + \mathbf{v}\|_{\mathcal{V}}^2 - \|\mathbf{w} - \mathbf{v}\|_{\mathcal{V}}^2 \big) = \langle \mathbf{w}, \mathbf{v} \rangle.$$

Similarly, property (c) follows fairly quickly from the definition, since

$$\langle \mathbf{v}, \mathbf{v} \rangle = \frac{1}{4} \big(\|\mathbf{v} + \mathbf{v}\|_{\mathcal{V}}^2 - \|\mathbf{v} - \mathbf{v}\|_{\mathcal{V}}^2 \big) = \frac{1}{4} \|2\mathbf{v}\|_{\mathcal{V}}^2 = \|\mathbf{v}\|_{\mathcal{V}}^2,$$

which is non-negative and equals 0 if and only if $\mathbf{v} = \mathbf{0}$. In fact, this furthermore shows that the norm induced by this inner product is indeed $\|\mathbf{v}\|_{\mathcal{V}}$.

Properties (a) and (c) do not actually depend on the parallelogram law holding—that is only required for property (b).

All that remains is to prove property (b) of inner products holds (i.e., $\langle \mathbf{v}, \mathbf{w} + c\mathbf{x} \rangle = \langle \mathbf{v}, \mathbf{w} \rangle + c\langle \mathbf{v}, \mathbf{x} \rangle$ for all $\mathbf{v}, \mathbf{w}, \mathbf{x} \in \mathcal{V}$ and all $c \in \mathbb{R}$). This task is significantly more involved than proving properties (a) and (c) was, so we split it up into four steps:

i) First, we show that $\langle \mathbf{v}, \mathbf{w} + \mathbf{x} \rangle = \langle \mathbf{v}, \mathbf{w} \rangle + \langle \mathbf{v}, \mathbf{x} \rangle$ for all $\mathbf{v}, \mathbf{w}, \mathbf{x} \in \mathcal{V}$. To this end we use the parallelogram law (i.e., the hypothesis of this theorem) to see that

$$2\|\mathbf{v} + \mathbf{x}\|_{\mathcal{V}}^2 + 2\|\mathbf{w}\|_{\mathcal{V}}^2 = \|\mathbf{v} + \mathbf{w} + \mathbf{x}\|_{\mathcal{V}}^2 + \|\mathbf{v} - \mathbf{w} + \mathbf{x}\|_{\mathcal{V}}^2.$$

Rearranging slightly gives

$$\|\mathbf{v} + \mathbf{w} + \mathbf{x}\|_{\mathcal{V}}^2 = 2\|\mathbf{v} + \mathbf{x}\|_{\mathcal{V}}^2 + 2\|\mathbf{w}\|_{\mathcal{V}}^2 - \|\mathbf{v} - \mathbf{w} + \mathbf{x}\|_{\mathcal{V}}^2.$$

By repeating this argument with the roles of $\mathbf{w}$ and $\mathbf{x}$ swapped, we similarly see that

$$\|\mathbf{v} + \mathbf{w} + \mathbf{x}\|_{\mathcal{V}}^2 = 2\|\mathbf{v} + \mathbf{w}\|_{\mathcal{V}}^2 + 2\|\mathbf{x}\|_{\mathcal{V}}^2 - \|\mathbf{v} + \mathbf{w} - \mathbf{x}\|_{\mathcal{V}}^2.$$

Averaging these two equations gives

$$\begin{aligned}
\|\mathbf{v} + \mathbf{w} + \mathbf{x}\|_{\mathcal{V}}^2 = {}&\|\mathbf{w}\|_{\mathcal{V}}^2 + \|\mathbf{x}\|_{\mathcal{V}}^2 + \|\mathbf{v} + \mathbf{w}\|_{\mathcal{V}}^2 + \|\mathbf{v} + \mathbf{x}\|_{\mathcal{V}}^2 \\
&- \tfrac{1}{2}\|\mathbf{v} - \mathbf{w} + \mathbf{x}\|_{\mathcal{V}}^2 - \tfrac{1}{2}\|\mathbf{v} + \mathbf{w} - \mathbf{x}\|_{\mathcal{V}}^2.
\end{aligned} \qquad \text{(1.D.3)}$$

Inner products
provide more
structure than just a
norm does, so this
theorem can be
thought of as telling
us exactly when that
extra structure is
present.

By repeating this entire argument with $\mathbf{w}$ and $\mathbf{x}$ replaced by $-\mathbf{w}$ and $-\mathbf{x}$, respectively, we similarly see that

$$\|\mathbf{v} - \mathbf{w} - \mathbf{x}\|_{\mathcal{V}}^2 = \|\mathbf{w}\|_{\mathcal{V}}^2 + \|\mathbf{x}\|_{\mathcal{V}}^2 + \|\mathbf{v} - \mathbf{w}\|_{\mathcal{V}}^2 + \|\mathbf{v} - \mathbf{x}\|_{\mathcal{V}}^2$$
$$- \tfrac{1}{2}\|\mathbf{v} + \mathbf{w} - \mathbf{x}\|_{\mathcal{V}}^2 - \tfrac{1}{2}\|\mathbf{v} - \mathbf{w} + \mathbf{x}\|_{\mathcal{V}}^2. \qquad (1.D.4)$$

Finally, we can use Equations (1.D.3) and (1.D.4) to get what we want:

$$\langle \mathbf{v}, \mathbf{w} + \mathbf{x} \rangle = \frac{1}{4}\left(\|\mathbf{v} + \mathbf{w} + \mathbf{x}\|_{\mathcal{V}}^2 - \|\mathbf{v} - \mathbf{w} - \mathbf{x}\|_{\mathcal{V}}^2\right)$$
$$= \frac{1}{4}\left(\|\mathbf{v} + \mathbf{w}\|_{\mathcal{V}}^2 + \|\mathbf{v} + \mathbf{x}\|_{\mathcal{V}}^2 - \|\mathbf{v} - \mathbf{w}\|_{\mathcal{V}}^2 - \|\mathbf{v} - \mathbf{x}\|_{\mathcal{V}}^2\right)$$
$$= \frac{1}{4}\left(\|\mathbf{v} + \mathbf{w}\|_{\mathcal{V}}^2 - \|\mathbf{v} - \mathbf{w}\|_{\mathcal{V}}^2\right) + \frac{1}{4}\left(\|\mathbf{v} + \mathbf{x}\|_{\mathcal{V}}^2 - \|\mathbf{v} - \mathbf{x}\|_{\mathcal{V}}^2\right)$$
$$= \langle \mathbf{v}, \mathbf{w} \rangle + \langle \mathbf{v}, \mathbf{x} \rangle,$$

as desired. Furthermore, by replacing $\mathbf{x}$ by $c\mathbf{x}$ above, we see that $\langle \mathbf{v}, \mathbf{w} + c\mathbf{x} \rangle = \langle \mathbf{v}, \mathbf{w} \rangle + \langle \mathbf{v}, c\mathbf{x} \rangle$ for all $c \in \mathbb{R}$. Thus, all that remains is to show that $\langle \mathbf{v}, c\mathbf{x} \rangle = c\langle \mathbf{v}, \mathbf{x} \rangle$ for all $c \in \mathbb{R}$, which is what the remaining three steps of this proof are devoted to demonstrating.

ii) We now show that $\langle \mathbf{v}, c\mathbf{x} \rangle = c\langle \mathbf{v}, \mathbf{x} \rangle$ for all *integers* c. If $c > 0$ is an integer, then this fact follows from (i) and induction (for example, (i) tells us that $\langle \mathbf{v}, 2\mathbf{x} \rangle = \langle \mathbf{v}, \mathbf{x} + \mathbf{x} \rangle = \langle \mathbf{v}, \mathbf{x} \rangle + \langle \mathbf{v}, \mathbf{x} \rangle = 2\langle \mathbf{v}, \mathbf{x} \rangle$).
If $c = 0$ then this fact is trivial, since it just says that

$$\langle \mathbf{v}, 0\mathbf{x} \rangle = \frac{1}{4}\left(\|\mathbf{v} + 0\mathbf{x}\|_{\mathcal{V}}^2 - \|\mathbf{v} - 0\mathbf{x}\|_{\mathcal{V}}^2\right) = \frac{1}{4}\left(\|\mathbf{v}\|_{\mathcal{V}}^2 - \|\mathbf{v}\|_{\mathcal{V}}^2\right) = 0 = 0\langle \mathbf{v}, \mathbf{x} \rangle.$$

If $c = -1$, then

$$\langle \mathbf{v}, -\mathbf{x} \rangle = \frac{1}{4}\left(\|\mathbf{v} - \mathbf{x}\|_{\mathcal{V}}^2 - \|\mathbf{v} + \mathbf{x}\|_{\mathcal{V}}^2\right)$$
$$= \frac{-1}{4}\left(\|\mathbf{v} + \mathbf{x}\|_{\mathcal{V}}^2 - \|\mathbf{v} - \mathbf{x}\|_{\mathcal{V}}^2\right) = -\langle \mathbf{v}, \mathbf{x} \rangle.$$

Finally, if $c < -1$ is an integer, then the result follows by combining the fact that it holds when $c = -1$ and when c is a positive integer.

We denote the set of
rational numbers
by $\mathbb{Q}$.

iii) We now show that $\langle \mathbf{v}, c\mathbf{x} \rangle = c\langle \mathbf{v}, \mathbf{x} \rangle$ for all *rational* c. If c is rational then we can write it as $c = p/q$, where p and q are integers and $q \neq 0$. Then

$$q\langle \mathbf{v}, c\mathbf{x} \rangle = q\langle \mathbf{v}, p(\mathbf{x}/q) \rangle = qp\langle \mathbf{v}, \mathbf{x}/q \rangle = p\langle \mathbf{v}, q\mathbf{x}/q \rangle = p\langle \mathbf{v}, \mathbf{x} \rangle,$$

where we used (ii) and the fact that p and q are integers in the second and third equalities. Dividing the above equation through by q then gives $\langle \mathbf{v}, c\mathbf{x} \rangle = (p/q)\langle \mathbf{v}, \mathbf{x} \rangle = c\langle \mathbf{v}, \mathbf{x} \rangle$, as desired.

Yes, this proof is *still*
going on.

iv) Finally, to see that $\langle \mathbf{v}, c\mathbf{x} \rangle = c\langle \mathbf{v}, \mathbf{x} \rangle$ for *all* $c \in \mathbb{R}$, we use the fact that, for each fixed $\mathbf{v}, \mathbf{w} \in \mathcal{V}$, the function $f : \mathbb{R} \to \mathbb{R}$ defined by

$$f(c) = \frac{1}{c}\langle \mathbf{v}, c\mathbf{x} \rangle = \frac{1}{4c}\left(\|\mathbf{v} + c\mathbf{w}\|_{\mathcal{V}}^2 - \|\mathbf{v} - c\mathbf{w}\|_{\mathcal{V}}^2\right)$$

is continuous (this follows from the fact that all norms are continuous when restricted to a finite-dimensional subspace like $\operatorname{span}(\mathbf{v}, \mathbf{w})$—see Section 2.D and Appendix B.1 for more discussion of this fact), and compositions and sums/differences of continuous functions are continuous).

We expand on this type of argument considerably in Section 2.D.

Well, we just showed in (iii) that $f(c) = \langle \mathbf{v}, \mathbf{w} \rangle$ for all $c \in \mathbb{Q}$. When combined with continuity of f, this means that $f(c) = \langle \mathbf{v}, \mathbf{w} \rangle$ for all $c \in \mathbb{R}$. This type of fact is typically covered in analysis courses and texts, but hopefully it is intuitive enough even if you have not taken such a course—the rational numbers are dense in $\mathbb{R}$, so if there were a real number c with $f(c) \neq \langle \mathbf{v}, \mathbf{w} \rangle$ then the graph of f would have to "jump" at c (see Figure 1.34) and thus f would not be continuous.

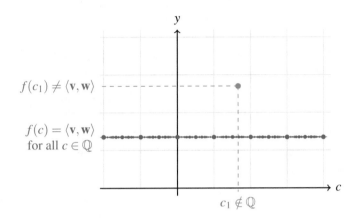

Figure 1.34: If a continuous function $f(c)$ is constant on the rationals, then it must be constant on all of $\mathbb{R}$—otherwise, it would have a discontinuity.

We have thus shown that $\langle \cdot, \cdot \rangle$ is indeed an inner product, and that it induces $\| \cdot \|_V$, so the proof is complete. ∎

Example 1.D.4

Which p-Norms are Induced by an Inner Product?

If we are being super technical, this argument only works when $n \geq 2$. The p-norms are all the same (and thus are all induced by an inner product) when $n = 1$.

Determine which of the p-norms ($1 \leq p \leq \infty$) on $\mathbb{C}^n$ are induced by an inner product.

Solution:

We already know that the 2-norm is induced by the standard inner product (the dot product). To see that none of the other p-norms are induced by an inner product, we can check that the parallelogram law does not hold when $\mathbf{v} = \mathbf{e}_1$ and $\mathbf{w} = \mathbf{e}_2$:

$$2\|\mathbf{e}_1\|_p^2 + 2\|\mathbf{e}_2\|_p^2 = 2 + 2 = 4 \quad \text{and}$$

$$\|\mathbf{e}_1 + \mathbf{e}_2\|_p^2 + \|\mathbf{e}_1 - \mathbf{e}_2\|_p^2 = \begin{cases} 1 + 1 = 2 & \text{if } p = \infty, \\ 2^{2/p} + 2^{2/p} = 2^{1+2/p} & \text{otherwise.} \end{cases}$$

The parallelogram law thus does not hold whenever $4 \neq 2^{1+2/p}$ (i.e., whenever $p \neq 2$). It follows from Theorem 1.D.8 that the p-norms are not induced by any inner product when $p \neq 2$.

In fact, the only norms that we have investigated so far that are induced by an inner product are the "obvious" ones: the 2-norm on $\mathbb{F}^n$ and $\mathcal{C}[a,b]$, the Frobenius norm on $\mathcal{M}_{m,n}(\mathbb{F})$, and the norms that we can construct from the inner products described by Corollary 1.4.4.

1.D.3   Isometries

Recall from Section 1.4.3 that if $\mathbb{F} = \mathbb{R}$ or $\mathbb{F} = \mathbb{C}$ then a unitary matrix $U \in \mathcal{M}_n(\mathbb{F})$ is one that preserves the usual 2-norm of all vectors: $\|U\mathbf{v}\|_2 = \|\mathbf{v}\|_2$ for all $\mathbf{v} \in \mathbb{F}^n$. Unitary matrices are extraordinary for the fact that they have so many different, but equivalent, characterizations (see Theorem 1.4.9). In this section, we generalize this idea and look at what we can say about linear transformations that preserve *any* particular norm, including ones that are not induced by any inner product.

Definition 1.D.4

Isometries

Let $\mathcal{V}$ and $\mathcal{W}$ be normed vector spaces and let $T : \mathcal{V} \to \mathcal{W}$ be a linear transformation. We say that T is an **isometry** if

$$\|T(\mathbf{v})\|_{\mathcal{W}} = \|\mathbf{v}\|_{\mathcal{V}} \quad \text{for all} \quad \mathbf{v} \in \mathcal{V}.$$

In other words, a unitary matrix is an isometry of the 2-norm on $\mathbb{F}^n$.

For example, the linear transformation $T : \mathbb{R}^2 \to \mathbb{R}^3$ (where we use the usual 2-norm on $\mathbb{R}^2$ and $\mathbb{R}^3$) defined by $T(v_1, v_2) = (v_1, v_2, 0)$ is an isometry since

$$\|T(\mathbf{v})\|_2 = \sqrt{v_1^2 + v_2^2 + 0^2} = \sqrt{v_1^2 + v_2^2} = \|\mathbf{v}\|_2 \quad \text{for all} \quad \mathbf{v} \in \mathcal{V}.$$

Whether or not a linear transformation is an isometry depends heavily on the norms that are being used on $\mathcal{V}$ and $\mathcal{W}$, so they need to be specified unless they are clear from context.

Example 1.D.5

An Isometry in the 2-Norm But Not the 1-Norm

Consider the matrix $U = \dfrac{1}{\sqrt{2}} \begin{bmatrix} 1 & -1 \\ 1 & 1 \end{bmatrix}$ that acts on $\mathbb{R}^2$.

a) Show that U is an isometry of the 2-norm.
b) Show that U is not an isometry of the 1-norm.

Solutions:

a) It is straightforward to check that U is unitary and thus an isometry of the 2-norm:

$$U^*U = \frac{1}{2} \begin{bmatrix} 1 & 1 \\ -1 & 1 \end{bmatrix} \begin{bmatrix} 1 & -1 \\ 1 & 1 \end{bmatrix} = \frac{1}{2} \begin{bmatrix} 2 & 0 \\ 0 & 2 \end{bmatrix} = I_2.$$

b) We simply have to find any vector $\mathbf{v} \in \mathbb{R}^2$ with the property that $\|U\mathbf{v}\|_1 \neq \|\mathbf{v}\|_1$. Well, $\mathbf{v} = \mathbf{e}_1$ works, since

$$\|U\mathbf{e}_1\|_1 = \left\| \tfrac{1}{\sqrt{2}}(1,1) \right\|_1 = \sqrt{2} \quad \text{but} \quad \|\mathbf{e}_1\|_1 = 1.$$

Recall that a rotation matrix (by an angle θ counter-clockwise) is a matrix of the form

$$\begin{bmatrix} \cos(\theta) & -\sin(\theta) \\ \sin(\theta) & \cos(\theta) \end{bmatrix}.$$

Geometrically, part (a) makes sense because U acts as a rotation counter-clockwise by $\pi/4$:

$$U = \begin{bmatrix} \cos(\pi/4) & -\sin(\pi/4) \\ \sin(\pi/4) & \cos(\pi/4) \end{bmatrix},$$

and rotating vectors does not change their 2-norm (see Example 1.4.12). However, rotating vectors can indeed change their 1-norm, as demonstrated in part (b):

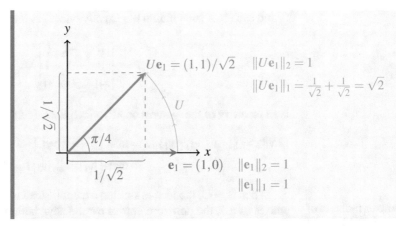

In situations where we have an inner product to work with (like in part (a) of the previous example), we can characterize isometries in much the same way that we characterized unitary matrices back in Section 1.4.3. In fact, the proof of the upcoming theorem is so similar to that of the equivalence of conditions (d)–(f) in Theorem 1.4.9 that we defer it to Exercise 1.D.22.

Theorem 1.D.9

Characterization of Isometries

Let V and W be inner product spaces and let $T : V \to W$ be a linear transformation. Then the following are equivalent:

    a) T is an isometry (with respect to the norms induced by the inner products on V and W),

    b) $T^* \circ T = I_V$, and

    c) $\langle T(\mathbf{x}), T(\mathbf{y}) \rangle_W = \langle \mathbf{x}, \mathbf{y} \rangle_V$ for all $\mathbf{x}, \mathbf{y} \in V$.

If we take the standard matrix of both sides of part (b) of this theorem with respect to an orthonormal basis B of V, then we see that T is an isometry if and only if

> The first equality here follows from Theorem 1.4.8 and relies on B being an *orthonormal* basis.

$$[T]_B^*[T]_B = [T^*]_B[T]_B = [T^* \circ T]_B = [I_V]_B = I.$$

In the special case when $V = W$, this observation can be phrased as follows:

> (!) If B is an orthonormal basis of an inner product space V, then $T : V \to V$ is an isometry of the norm induced by the inner product on V if and only if $[T]_B$ is unitary.

On the other hand, if a norm is *not* induced by an inner product then it can be quite a bit more difficult to get our hands on what its isometries are. The following theorem provides the answer for the ∞-norm:

Theorem 1.D.10

Isometries of the ∞-Norm

Suppose $\mathbb{F} = \mathbb{R}$ or $\mathbb{F} = \mathbb{C}$, and $P \in \mathcal{M}_n(\mathbb{F})$. Then

$$\|P\mathbf{v}\|_\infty = \|\mathbf{v}\|_\infty \quad \text{for all} \quad \mathbf{v} \in \mathbb{F}^n$$

if and only if every row and column of P has a single non-zero entry, and that entry has magnitude (i.e., absolute value) equal to 1.

> Real matrices of this type are sometimes called **signed permutation matrices**.

Before proving this theorem, it is worth illustrating exactly what it means. When $\mathbb{F} = \mathbb{R}$, it says that isometries of the ∞-norm only have entries 1, -1, and

0, and each row and column has exactly one non-zero entry. For example,

$$P = \begin{bmatrix} 0 & -1 & 0 \\ 0 & 0 & 1 \\ -1 & 0 & 0 \end{bmatrix}$$

is an isometry of the ∞-norm on $\mathbb{R}^3$, which is straightforward to verify directly:

$$\|P\mathbf{v}\|_\infty = \|(-v_2, v_3, -v_1)\|_\infty = \max\{|-v_2|, |v_3|, |-v_1|\}$$
$$= \max\{|v_1|, |v_2|, |v_3|\} = \|\mathbf{v}\|_\infty \quad \text{for all} \quad \mathbf{v} \in \mathbb{R}^3.$$

Complex matrices of this type are sometimes called **complex permutation matrices**.

When $\mathbb{F} = \mathbb{C}$, the idea is similar, except instead of just allowing non-zero entries of ± 1, the non-zero entries can lie anywhere on the unit circle in the complex plane.

Proof of Theorem 1.D.10. The "if" direction is not too difficult to prove. We just note that any P with the specified form just permutes the entries of $\mathbf{v}$ and possibly multiplies them by a number with absolute value 1, and such an operation leaves $\|\mathbf{v}\|_\infty = \max_j\{|v_j|\}$ unchanged.

For the "only if" direction, we only prove the $\mathbb{F} = \mathbb{R}$ case and leave the $\mathbb{F} = \mathbb{C}$ case to Exercise 1.D.25. Suppose $P = \begin{bmatrix} \mathbf{p}_1 \mid \mathbf{p}_2 \mid \cdots \mid \mathbf{p}_n \end{bmatrix}$ is an isometry of the ∞-norm. Then $P\mathbf{e}_j = \mathbf{p}_j$ for all j, so

$$\max_{1 \leq i \leq n}\{|p_{i,j}|\} = \|\mathbf{p}_j\|_\infty = \|P\mathbf{e}_j\|_\infty = \|\mathbf{e}_j\|_\infty = 1 \quad \text{for all} \quad 1 \leq j \leq n.$$

In particular, this implies that every column of P has at least one entry with absolute value 1. Similarly, $P(\mathbf{e}_j \pm \mathbf{e}_k) = \mathbf{p}_j \pm \mathbf{p}_k$ for all j, k, so

It turns out that the 1-norm has the exact same isometries (see Exercise 1.D.24). In fact, all p-norms other than the 2-norm have these same isometries (CL92, LS94), but proving this is beyond the scope of this book.

$$\max_{1 \leq i \leq n}\{|p_{i,j} \pm p_{i,k}|\} = \|\mathbf{p}_j \pm \mathbf{p}_k\|_\infty = \|P(\mathbf{e}_j \pm \mathbf{e}_k)\|_\infty$$
$$= \|\mathbf{e}_j \pm \mathbf{e}_k\|_\infty = 1 \quad \text{for all} \quad 1 \leq j \neq k \leq n.$$

It follows that if $|p_{i,j}| = 1$ then $p_{i,k} = 0$ for all $k \neq j$ (i.e., the rest of the i-th row of P equals zero). Since each column $\mathbf{p}_j$ contains an entry with $|p_{i,j}| = 1$, it follows that every row and column contains exactly one non-zero entry—the one with absolute value 1. ∎

In particular, it is worth noting that the above theorem says that all isometries of the ∞-norm are also isometries of the 2-norm, since $P^*P = I$ for any matrix of the form described by Theorem 1.D.10.

Exercises

solutions to starred exercises on page 463

1.D.1 Determine which of the following functions $\|\cdot\|$ are norms on the specified vector space $\mathcal{V}$.

*(a) $\mathcal{V} = \mathbb{R}^2$, $\|\mathbf{v}\| = |v_1| + 3|v_2|$.

(b) $\mathcal{V} = \mathbb{R}^2$, $\|\mathbf{v}\| = 2|v_1| - |v_2|$.

*(c) $\mathcal{V} = \mathbb{R}^2$, $\|\mathbf{v}\| = |v_1^3 + v_1^2 v_2 + v_1 v_2^2 + v_2^3|^{1/3}$.

(d) $\mathcal{V} = \mathbb{R}^2$, $\|\mathbf{v}\| = (|v_1^3 + 3v_1^2 v_2 + 3v_1 v_2^2 + v_2^3| + |v_2|^3)^{1/3}$.

*(e) $\mathcal{V} = \mathcal{P}$ (the vector space of polynomials), $\|p\|$ is the degree of p (i.e., the highest power of x appearing in $p(x)$).

(f) $\mathcal{V} = \mathcal{P}[0, 1]$,

$$\|p\| = \max_{0 \leq x \leq 1}\{|p(x)|\} + \max_{0 \leq x \leq 1}\{|p'(x)|\},$$

where p' is the derivative of p.

1.D.2 Determine whether or not each of the following norms $\|\cdot\|$ are induced by an inner product on the indicated vector space $\mathcal{V}$.

(a) $\mathcal{V} = \mathbb{R}^2$, $\|\mathbf{v}\| = \sqrt{v_1^2 + 2v_1 v_2 + 3v_2^2}$.

*(b) $\mathcal{V} = \mathbb{C}^2$, $\|\mathbf{v}\| = (|v_1|^3 + 3|v_2|^3)^{1/3}$.

(c) $\mathcal{V} = \mathcal{C}[0,1]$, $\|f\| = \max\limits_{0 \le x \le 1} \{|f(x)|\}$.

1.D.3   Determine which of the following statements are true and which are false.

*(a) If $\|\cdot\|$ is a norm on a vector space $\mathcal{V}$ and $0 < c \in \mathbb{R}$ then $c\|\cdot\|$ is also a norm on $\mathcal{V}$.
  (b) Every norm $\|\cdot\|$ on $\mathcal{M}_n(\mathbb{R})$ arises from some inner product on $\mathcal{M}_n(\mathbb{R})$ via $\|A\| = \sqrt{\langle A, A \rangle}$.
*(c) If $S, T : \mathcal{V} \to \mathcal{W}$ are isometries then so is $S + T$.
  (d) If $S, T : \mathcal{V} \to \mathcal{V}$ are isometries of the same norm then so is $S \circ T$.
*(e) If $T : \mathcal{V} \to \mathcal{W}$ is an isometry then T^{-1} exists and is also an isometry.

1.D.4   Which of the three defining properties of norms fails if we define the p-norm on $\mathbb{F}^n$ as usual in Definition 1.D.2, but with $0 < p < 1$?

1.D.5   Show that norms are never linear transformations. That is, show that if $\mathcal{V}$ is a non-zero vector space and $\|\cdot\| : \mathcal{V} \to \mathbb{R}$ is a norm, then $\|\cdot\|$ is not a linear transformation.

1.D.6   Suppose $\mathcal{V}$ is a normed vector space and let $\mathbf{v}, \mathbf{w} \in \mathcal{V}$. Show that $\|\mathbf{v} - \mathbf{w}\| \ge \|\mathbf{v}\| - \|\mathbf{w}\|$. [Side note: This is called the **reverse triangle inequality**.]

1.D.7   Suppose $\mathbf{v} \in \mathbb{C}^n$ and $1 \le p \le \infty$. Show that $\|\mathbf{v}\|_\infty \le \|\mathbf{v}\|_p$.

1.D.8   In this exercise, we show that
$$\lim_{p \to \infty} \|\mathbf{v}\|_p = \|\mathbf{v}\|_\infty \quad \text{for all} \quad \mathbf{v} \in \mathbb{C}^n.$$

(a) Briefly explain why $\lim_{p \to \infty} |v|^p = 0$ whenever $v \in \mathbb{C}$ satisfies $|v| < 1$.
(b) Show that if $\mathbf{v} \in \mathbb{C}^n$ is such that $\|\mathbf{v}\|_\infty = 1$ then
$$\lim_{p \to \infty} \|\mathbf{v}\|_p = 1.$$
[Hint: Be slightly careful—there might be multiple values of j for which $|v_j| = 1$.]
(c) Use absolute homogeneity of the p-norms and the ∞-norm to show that
$$\lim_{p \to \infty} \|\mathbf{v}\|_p = \|\mathbf{v}\|_\infty \quad \text{for all} \quad \mathbf{v} \in \mathbb{C}^n.$$

1.D.9   Prove the $p = 1, q = \infty$ case of Hölder's inequality. That is, show that
$$|\mathbf{v} \cdot \mathbf{w}| \le \|\mathbf{v}\|_1 \|\mathbf{w}\|_\infty \quad \text{for all} \quad \mathbf{v}, \mathbf{w} \in \mathbb{C}^n.$$

1.D.10   In this exercise, we determine when equality holds in Minkowski's inequality (Theorem 1.D.2) on $\mathbb{C}^n$.

(a) Show that if $p > 1$ or $p = \infty$ then $\|\mathbf{v} + \mathbf{w}\|_p = \|\mathbf{v}\|_p + \|\mathbf{w}\|_p$ if and only if either $\mathbf{w} = \mathbf{0}$ or $\mathbf{v} = c\mathbf{w}$ for some $0 \le c \in \mathbb{R}$.
(b) Show that if $p = 1$ then $\|\mathbf{v} + \mathbf{w}\|_p = \|\mathbf{v}\|_p + \|\mathbf{w}\|_p$ if and only if there exist non-negative scalars $\{c_j\} \subseteq \mathbb{R}$ such that, for each $1 \le j \le n$, either $w_j = 0$ or $v_j = c_j w_j$.

1.D.11   In this exercise, we determine when equality holds in Hölder's inequality (Theorem 1.D.5) on $\mathbb{C}^n$.

(a) Show that if $p, q > 1$ are such that $1/p + 1/q$ then $|\mathbf{v} \cdot \mathbf{w}| = \|\mathbf{v}\|_p \|\mathbf{w}\|_q$ if and only if either $\mathbf{w} = \mathbf{0}$ or there exists a scalar $0 \le c \in \mathbb{R}$ such that $|v_j|^p = c|w_j|^q$ for all $1 \le j \le n$.
(b) Show that if $p = \infty$ and $q = 1$ then $|\mathbf{v} \cdot \mathbf{w}| = \|\mathbf{v}\|_p \|\mathbf{w}\|_q$ if and only if $\mathbf{v} = c(w_1/|w_1|, w_2/|w_2|, \ldots, w_n/|w_n|)$ for some $0 \le c \in \mathbb{R}$ (and if $w_j = 0$ for some j then v_j can be chosen so that $|v_j| \le 1$ arbitrarily).

1.D.12   In this exercise, we show that the bounds of Theorem 1.D.6 are tight (i.e., the constants in it are as good as possible). Suppose $1 \le p \le q \le \infty$.

(a) Find a vector $\mathbf{v} \in \mathbb{C}^n$ such that $\|\mathbf{v}\|_p = \|\mathbf{v}\|_q$.
(b) Find a vector $\mathbf{v} \in \mathbb{C}^n$ such that
$$\|\mathbf{v}\|_p = \left(n^{\frac{1}{p} - \frac{1}{q}} \right) \|\mathbf{v}\|_q.$$

1.D.13   Show that the p-norm of a function (see Definition 1.D.3) is indeed a norm.

1.D.14   Let $1 \le p, q \le \infty$ be such that $1/p + 1/q = 1$. Show that
$$\int_a^b |f(x)g(x)| \, dx \le \|f\|_p \|g\|_q \quad \text{for all} \quad f, g \in \mathcal{C}[a,b].$$
[Side note: This is Hölder's inequality for functions.]

1.D.15   Show that the Jordan–von Neumann theorem (Theorem 1.D.8) still holds if we replace the parallelogram law by the inequality
$$2\|\mathbf{v}\|_\mathcal{V}^2 + 2\|\mathbf{w}\|_\mathcal{V}^2 \le \|\mathbf{v} + \mathbf{w}\|_\mathcal{V}^2 + \|\mathbf{v} - \mathbf{w}\|_\mathcal{V}^2$$
for all $\mathbf{v}, \mathbf{w} \in \mathcal{V}$. [Hint: Consider vectors of the form $\mathbf{v} = \mathbf{x} + \mathbf{y}$ and $\mathbf{w} = \mathbf{x} - \mathbf{y}$.]

1.D.16   Prove the "if" direction of the Jordan–von Neumann theorem when the field is $\mathbb{F} = \mathbb{C}$. In particular, show that if the parallelogram law holds on a complex normed vector space $\mathcal{V}$ then the function defined by
$$\langle \mathbf{v}, \mathbf{w} \rangle = \frac{1}{4} \sum_{k=0}^{3} \frac{1}{i^k} \|\mathbf{v} + i^k \mathbf{w}\|_\mathcal{V}^2$$
is an inner product on $\mathcal{V}$.

[Hint: Show that $\langle \mathbf{v}, i\mathbf{w} \rangle = i\langle \mathbf{v}, \mathbf{w} \rangle$ and apply the theorem from the real case to the real and imaginary parts of $\langle \mathbf{v}, \mathbf{w} \rangle$.]

1.D.17   Suppose $A \in \mathcal{M}_{m,n}(\mathbb{C})$, and $1 \le p \le \infty$.

(a) Show that the following quantity $\|\cdot\|_p$ is a norm on $\mathcal{M}_{m,n}(\mathbb{C})$:
$$\|A\|_p \overset{\text{def}}{=} \max_{\mathbf{v} \in \mathbb{C}^n} \left\{ \frac{\|A\mathbf{v}\|_p}{\|\mathbf{v}\|_p} : \mathbf{v} \ne \mathbf{0} \right\}.$$
[Side note: This is called the **induced p-norm** of A. In the $p = 2$ case it is also called the **operator norm** of A, and we explore it in Section 2.3.3.]
(b) Show that $\|A\mathbf{v}\|_p \le \|A\|_p \|\mathbf{v}\|_p$ for all $\mathbf{v} \in \mathbb{C}^n$.
(c) Show that $\|A^*\|_q = \|A\|_p$, where $1 \le q \le \infty$ is such that $1/p + 1/q = 1$.
[Hint: Make use of Exercise 1.D.11.]
(d) Show that this norm is not induced by an inner product, regardless of the value of p (as long as $m, n \ge 2$).

1.D.18 Suppose $A \in \mathcal{M}_{m,n}(\mathbb{C})$, and consider the induced p-norms introduced in Exercise 1.D.17.

(a) Show that the induced 1-norm of A is its maximal column sum:
$$\|A\|_1 = \max_{1 \le j \le n} \left\{ \sum_{i=1}^{m} |a_{i,j}| \right\}.$$

(b) Show that the induced ∞-norm of A is its maximal row sum:
$$\|A\|_\infty = \max_{1 \le i \le m} \left\{ \sum_{j=1}^{n} |a_{i,j}| \right\}.$$

1.D.19 Find all isometries of the norm on $\mathbb{R}^2$ defined by $\|\mathbf{v}\| = \max\{|v_1|, 2|v_2|\}$.

∗**1.D.20** Suppose that $\mathcal{V}$ and $\mathcal{W}$ are finite-dimensional normed vector spaces. Show that if $T : \mathcal{V} \to \mathcal{W}$ is an isometry then $\dim(\mathcal{V}) \le \dim(\mathcal{W})$.

1.D.21 Suppose that $\mathcal{V}$ is a normed vector space and $T : \mathcal{V} \to \mathcal{V}$ is an isometry.

(a) Show that if $\mathcal{V}$ is finite-dimensional then T^{-1} exists and is an isometry.

(b) Provide an example that demonstrates that if $\mathcal{V}$ is infinite-dimensional then T might not be invertible.

∗∗**1.D.22** Prove Theorem 1.D.9.

1.D.23 Suppose $U \in \mathcal{M}_{m,n}(\mathbb{C})$. Show that U is an isometry of the 2-norm if and only if its columns are mutually orthogonal and all have length 1.

[Side note: If $m = n$ then this recovers the usual definition of unitary matrices.]

∗∗**1.D.24** Show that $P \in \mathcal{M}_n(\mathbb{C})$ is an isometry of the 1-norm if and only if every row and column of P has a single non-zero entry, and that entry has absolute value 1.

[Hint: Try mimicking the proof of Theorem 1.D.10. It might be helpful to use the result of Exercise 1.D.10(b).]

∗∗**1.D.25** In the proof of Theorem 1.D.10, we only proved the "only if" direction in the case when $\mathbb{F} = \mathbb{R}$. Explain what changes/additions need to be made to the proof to handle the $\mathbb{F} = \mathbb{C}$ case.

∗∗**1.D.26** Show that the 1-norm and the ∞-norm on $\mathcal{P}[0,1]$ are inequivalent. That is, show that there do not exist real scalars $c, C > 0$ such that
$$c\|p\|_1 \le \|p\|_\infty \le C\|p\|_1 \quad \text{for all} \quad p \in \mathcal{P}[0,1].$$

[Hint: Consider the polynomials of the form $p_n(x) = x^n$ for some fixed $n \ge 1$.]

∗∗**1.D.27** Suppose that $\|\cdot\|_a$, $\|\cdot\|_b$, and $\|\cdot\|_c$ are any three norms on a vector space $\mathcal{V}$ for which $\|\cdot\|_a$ and $\|\cdot\|_b$ are equivalent, and so are $\|\cdot\|_b$ and $\|\cdot\|_c$. Show that $\|\cdot\|_a$ and $\|\cdot\|_c$ are equivalent as well (in other words, show that equivalence of norms is **transitive**).

∗∗**1.D.28** Suppose $\mathbb{F} = \mathbb{R}$ or $\mathbb{F} = \mathbb{C}$ and let $\mathcal{V}$ be a finite-dimensional vector space over $\mathbb{F}$ with basis B. Show that a function $\|\cdot\|_\mathcal{V} : \mathcal{V} \to \mathbb{F}$ is a norm if and only if there exists a norm $\|\cdot\|_{\mathbb{F}^n} : \mathbb{F}^n \to \mathbb{F}$ such that
$$\|\mathbf{v}\|_\mathcal{V} = \left\| [\mathbf{v}]_B \right\|_{\mathbb{F}^n} \quad \text{for all} \quad \mathbf{v} \in \mathcal{V}.$$

In words, this means that every norm on a finite-dimensional vector space looks like a norm on $\mathbb{F}^n$ (compare with Corollary 1.4.4).

2. Matrix Decompositions

> We think basis-free, we write basis-free, but when the chips are down we close the office door and compute with matrices like fury.
>
> Irving Kaplansky

We also explored the QR decomposition in Section 1.C1.C, which says that every matrix can be written as a product of a unitary matrix and an upper triangular matrix.

Many of the most useful linear algebraic tools are those that involve breaking matrices down into the product of two or more simpler matrices (or equivalently, breaking linear transformations down into the composition of two or more simpler linear transformations). The standard example of this technique from introductory linear algebra is diagonalization, which says that it is often the case that we can write a matrix $A \in \mathcal{M}_n$ in the form $A = PDP^{-1}$, where P is invertible and D is diagonal.

More explicitly, some variant of the following theorem is typically proved in introductory linear algebra courses:

Theorem 2.0.1

Characterization of Diagonalizability

Suppose $A \in \mathcal{M}_n(\mathbb{F})$. The following are equivalent:

a) A is diagonalizable over $\mathbb{F}$.

b) There exists a basis of $\mathbb{F}^n$ consisting of eigenvectors of A.

Furthermore, in any diagonalization $A = PDP^{-1}$, the eigenvalues of A are the diagonal entries of D and the corresponding eigenvectors are the columns of P in the same order.

By "diagonalizable over $\mathbb{F}$", we mean that we can choose $D, P \in \mathcal{M}_n(\mathbb{F})$.

One of the standard examples of why diagonalization is so useful is that we can use it to compute large powers of matrices quickly. In particular, if $A = PDP^{-1}$ is a diagonalization of $A \in \mathcal{M}_n$, then for every integer $k \geq 1$ we have

$$A^k = \underbrace{\left(PDP^{-1}\right)\overbrace{\left(PDP^{-1}\right)}^{P^{-1}P=I}\overbrace{\left(PDP^{-1}\right)}^{P^{-1}P=I}\cdots\left(PDP^{-1}\right)}_{k \text{ times}} = PD^kP^{-1},$$

Another brief review of diagonalization is provided in Appendix A.1.7.

and D^k is trivial to compute (for diagonal matrices, matrix multiplication is the same as entrywise multiplication). It follows that after diagonalizing a matrix, we can compute any power of it via just two matrix multiplications—pre-multiplication of D^k by P, and post-multiplication of D^k by P^{-1}. In a sense, we have off-loaded the difficulty of computing matrix powers into the difficulty of diagonalizing a matrix.

© Springer Nature Switzerland AG 2021
N. Johnston, *Advanced Linear and Matrix Algebra*,
https://doi.org/10.1007/978-3-030-52815-7_2

If we think of the matrix A as a linear transformation acting on $\mathbb{F}^n$ then diagonalizing A is equivalent to finding a basis B of $\mathbb{F}^n$ in which the standard matrix of that linear transformation is diagonal. That is, if we define $T : \mathbb{F}^n \to \mathbb{F}^n$ by $T(\mathbf{v}) = A\mathbf{v}$ then Theorem 1.2.8 tells us that $[T]_B = P_{B \leftarrow E} A P_{E \leftarrow B}$, where E is the standard basis of $\mathbb{F}^n$. Since $P_{B \leftarrow E} = P_{E \leftarrow B}^{-1}$ (by Theorem 1.2.3), a slight rearrangement shows that $A = P_{E \leftarrow B}[T]_B P_{E \leftarrow B}^{-1}$, so A is diagonalizable if and only if there is a basis B in which $[T]_B$ is diagonal (see Figure 2.1).

> Here we use the fact that $[T]_E = A$. For a refresher on change-of-basis matrices and standard matrices, refer back to Section 1.2.2.

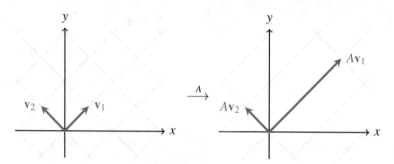

Figure 2.1: The matrix $A = \begin{bmatrix} 2 & 1 \\ 1 & 2 \end{bmatrix}$ "looks diagonal" (i.e., just stretches space, but does not skew or rotate it) when viewed in the basis $B = \{\mathbf{v}_1, \mathbf{v}_2\} = \{(1,1),(-1,1)\}$

> With this terminology in hand, a matrix is diagonalizable if and only if it is similar to a diagonal matrix.

We say that two matrices $A, B \in \mathcal{M}_n(\mathbb{F})$ are **similar** if there exists an invertible matrix $P \in \mathcal{M}_n(\mathbb{F})$ such that $A = PBP^{-1}$ (and we call such a transformation of A into B a **similarity transformation**). The argument that we just provided shows that two matrices are similar if and only if they represent the same linear transformation on $\mathbb{F}^n$, just represented in different bases. The main goal of this chapter is to better understand similarity transformations (and thus change of bases). More specifically, we investigate the following questions:

1) What if A is not diagonalizable—how "close" to diagonal can we make $D = P^{-1}AP$ when P is invertible? That is, how close to diagonal can we make A via a similarity transformation?

2) How simple can we make A if we multiply by something other than P and P^{-1} on the left and right? For example, how simple can we make A via a **unitary similarity transformation**—a similarity transformation in which the invertible matrix P is unitary? Since a change-of-basis matrix $P_{E \leftarrow B}$ is unitary if and only if B is an *orthonormal* basis of $\mathbb{F}^n$ (see Exercise 1.4.18), this is equivalent to asking how simple we can make the standard matrix of a linear transformation if we represent it in an *orthonormal* basis.

2.1 The Schur and Spectral Decompositions

We start this chapter by probing question (2) above—how simple can we make a matrix via a unitary similarity? That is, given a matrix $A \in \mathcal{M}_n(\mathbb{C})$, how simple can U^*AU be if U is a unitary matrix? Note that we need the field here to be $\mathbb{R}$ or $\mathbb{C}$ so that it even makes sense to talk about unitary matrices, and for now we restrict our attention to $\mathbb{C}$ since the answer in the real case will follow as a corollary of the answer in the complex case.

2.1.1   Schur Triangularization

Recall that an upper triangular matrix is one with zeros in every position below the main diagonal.

We know that we cannot hope in general to get a diagonal matrix via unitary similarity, since not every matrix is even diagonalizable via *any* similarity. However, the following theorem, which is the main workhorse for the rest of this section, says that we can get partway there and always get an upper triangular matrix.

Theorem 2.1.1

Schur Triangularization

Suppose $A \in M_n(\mathbb{C})$. There exists a unitary matrix $U \in M_n(\mathbb{C})$ and an upper triangular matrix $T \in M_n(\mathbb{C})$ such that

$$A = UTU^*.$$

Proof. We prove the result by induction on n (the size of A). For the base case, we simply notice that the result is trivial if $n = 1$, since every 1×1 matrix is upper triangular.

For the inductive step, suppose that every $(n-1) \times (n-1)$ matrix can be upper triangularized (i.e., can be written in the form described by the theorem). Since A is complex, the fundamental theorem of algebra (Theorem A.3.1) tells us that its characteristic polynomial p_A has a root, so A has an eigenvalue. If we denote one of its corresponding eigenvectors by $\mathbf{v} \in \mathbb{C}^n$, then we recall that eigenvectors are non-zero by definition, so we can scale it so that $\|\mathbf{v}\| = 1$.

This step is why we need to work over $\mathbb{C}$, not $\mathbb{R}$. Not all polynomials factor over $\mathbb{R}$, so not all real matrices have real eigenvalues.

We can then extend $\mathbf{v}$ to an orthonormal basis of $\mathbb{C}^n$ via Exercise 1.4.20. In other words, we can find a unitary matrix $V \in M_n(\mathbb{C})$ with $\mathbf{v}$ as its first column:

$$V = \begin{bmatrix} \mathbf{v} \mid V_2 \end{bmatrix},$$

To extend $\mathbf{v}$ to an orthonormal basis, pick any vector not in its span, apply Gram–Schmidt, and repeat.

where $V_2 \in M_{n,n-1}(\mathbb{C})$ satisfies $V_2^* \mathbf{v} = \mathbf{0}$ (since V is unitary so $\mathbf{v}$ is orthogonal to every column of V_2). Direct computation then shows that

The final step in this computation follows because $A\mathbf{v} = \lambda \mathbf{v}$, $\mathbf{v}^* \mathbf{v} = \|\mathbf{v}\|^2 = 1$, and $V_2^* \mathbf{v} = \mathbf{0}$.

$$V^*AV = \begin{bmatrix} \mathbf{v} \mid V_2 \end{bmatrix}^* A \begin{bmatrix} \mathbf{v} \mid V_2 \end{bmatrix} = \begin{bmatrix} \mathbf{v}^* \\ V_2^* \end{bmatrix} \begin{bmatrix} A\mathbf{v} \mid AV_2 \end{bmatrix}$$

$$= \begin{bmatrix} \mathbf{v}^*A\mathbf{v} & \mathbf{v}^*AV_2 \\ \lambda V_2^*\mathbf{v} & V_2^*AV_2 \end{bmatrix} = \begin{bmatrix} \lambda & \mathbf{v}^*AV_2 \\ \mathbf{0} & V_2^*AV_2 \end{bmatrix}.$$

We now apply the inductive hypothesis—since $V_2^*AV_2$ is an $(n-1) \times (n-1)$ matrix, there exists a unitary matrix $U_2 \in M_{n-1}(\mathbb{C})$ and an upper triangular $T_2 \in M_{n-1}(\mathbb{C})$ such that $V_2^*AV_2 = U_2T_2U_2^*$. It follows that

$$V^*AV = \begin{bmatrix} \lambda & \mathbf{v}^*AV_2 \\ \mathbf{0} & U_2T_2U_2^* \end{bmatrix} = \begin{bmatrix} 1 & \mathbf{0}^T \\ \mathbf{0} & U_2 \end{bmatrix} \begin{bmatrix} \lambda & \mathbf{v}^*AV_2 \\ \mathbf{0} & T_2 \end{bmatrix} \begin{bmatrix} 1 & \mathbf{0}^T \\ \mathbf{0} & U_2^* \end{bmatrix}.$$

By multiplying on the left by V and on the right by V^*, we see that $A = UTU^*$, where

Recall from Exercise 1.4.8 that the product of unitary matrices is unitary.

$$U = V \begin{bmatrix} 1 & \mathbf{0}^T \\ \mathbf{0} & U_2 \end{bmatrix} \quad \text{and} \quad T = \begin{bmatrix} \lambda & \mathbf{v}^*AV_2 \\ \mathbf{0} & T_2 \end{bmatrix}.$$

Since U is unitary and T is upper triangular, this completes the inductive step and the proof. ∎

In a Schur triangularization $A = UTU^*$, the diagonal entries of T are necessarily the same as the eigenvalues of A. To see why this is the case, recall that (a) A and T must have the same eigenvalues since they are similar, and

(b) the eigenvalues of a triangular matrix are its diagonal entries. However, the other pieces of Schur triangularization are highly non-unique—the unitary matrix U and the entries in the strictly upper triangular portion of T can vary wildly.

Example 2.1.1

Computing Schur Triangularizations

Compute a Schur triangularization of the following matrices:

a) $A = \begin{bmatrix} 1 & 2 \\ 5 & 4 \end{bmatrix}$

b) $B = \begin{bmatrix} 1 & 2 & 2 \\ 2 & 1 & 2 \\ 3 & -3 & 4 \end{bmatrix}$

Solutions:

a) We construct the triangularization by mimicking the proof of Theorem 2.1.1, so we start by constructing a unitary matrix whose first column is an eigenvector of A. It is straightforward to show that its eigenvalues are -1 and 6, with corresponding eigenvectors $(1, -1)$ and $(2, 5)$, respectively.

> We computed these eigenvalues and eigenvectors in Example A.1.1. See Appendix A.1.6 if you need a refresher on eigenvalues and eigenvectors.

If we (arbitrarily) choose the eigenvalue/eigenvector pair $\lambda = -1$ and $\mathbf{v} = (1, -1)/\sqrt{2}$, then we can extend $\mathbf{v}$ to the orthonormal basis

$$\left\{ \tfrac{1}{\sqrt{2}}(1, -1), \tfrac{1}{\sqrt{2}}(1, 1) \right\}$$

of $\mathbb{C}^2$ simply by inspection. Placing these vectors as columns into a unitary matrix then gives us a Schur triangularization $A = UTU^*$ as follows:

$$U = \frac{1}{\sqrt{2}} \begin{bmatrix} 1 & 1 \\ -1 & 1 \end{bmatrix} \quad \text{and} \quad T = U^*AU = \begin{bmatrix} -1 & -3 \\ 0 & 6 \end{bmatrix}.$$

b) Again, we mimic the proof of Theorem 2.1.1 and thus start by constructing a unitary matrix whose first column is an eigenvector of B. Its eigenvalues are -1, 3, and 4, with corresponding eigenvectors $(-4, 1, 3)$, $(1, 1, 0)$, and $(2, 2, 1)$, respectively. We (arbitrarily) choose the pair $\lambda = 4, \mathbf{v} = (2, 2, 1)/3$, and then we compute an orthonormal basis of $\mathbb{C}^3$ that contains $\mathbf{v}$ (notice that we normalized $\mathbf{v}$ so that this is possible).

> If we chose a different eigenvalue/eigenvector pair, we would still get a valid Schur triangularization of B, but it would look completely different than the one we compute here.

To this end, we note that the unit vector $\mathbf{w} = (1, -1, 0)/\sqrt{2}$ is clearly orthogonal to $\mathbf{v}$, so we just need to find one more unit vector orthogonal to both $\mathbf{v}$ and $\mathbf{w}$. We can either solve a linear system, or use the Gram–Schmidt process, or use the cross product to find $\mathbf{x} = (-1, -1, 4)/\sqrt{18}$, which does the job. It follows that the set

$$\{\mathbf{v}, \mathbf{w}, \mathbf{x}\} = \left\{ \tfrac{1}{3}(2, 2, 1), \tfrac{1}{\sqrt{2}}(1, -1, 0), \tfrac{1}{\sqrt{18}}(-1, -1, 4) \right\}$$

is an orthonormal basis of $\mathbb{C}^3$ containing $\mathbf{v}$, and if we stick these vectors as columns into a unitary matrix V_1 then we have

$$V_1 = \begin{bmatrix} 2/3 & 1/\sqrt{2} & -1/\sqrt{18} \\ 2/3 & -1/\sqrt{2} & -1/\sqrt{18} \\ 1/3 & 0 & 4/\sqrt{18} \end{bmatrix}, \quad V_1^*BV_1 = \begin{bmatrix} 4 & \sqrt{2} & 2\sqrt{2} \\ 0 & -1 & 0 \\ 0 & 4 & 3 \end{bmatrix}.$$

For larger matrices, we have to iterate even more. For example, to compute a Schur triangularization of a 4×4 matrix, we must compute an eigenvector of a 4×4 matrix, then a 3×3 matrix, and then a 2×2 matrix.

We now iterate—we repeat this entire procedure with the bottom-right 2×2 block of $V_1^* B V_1$. Well, the eigenvalues of

$$\begin{bmatrix} -1 & 0 \\ 4 & 3 \end{bmatrix}$$

are -1 and 3, with corresponding eigenvectors $(1, -1)$ and $(0, 1)$, respectively. The orthonormal basis $\{(1, -1)/\sqrt{2}, (1, 1)/\sqrt{2}\}$ contains one of its eigenvectors, so we define V_2 to be the unitary matrix with these vectors as its columns. Then B has Schur triangularization $B = UTU^*$ via

In these examples, the eigenvectors were simple enough to extend to orthonormal bases by inspection. In more complicated situations, use the Gram–Schmidt process.

$$U = V_1 \begin{bmatrix} 1 & \mathbf{0}^T \\ \hline \mathbf{0} & V_2 \end{bmatrix}$$

$$= \begin{bmatrix} 2/3 & 1/\sqrt{2} & -1/\sqrt{18} \\ 2/3 & -1/\sqrt{2} & -1/\sqrt{18} \\ 1/3 & 0 & 4/\sqrt{18} \end{bmatrix} \begin{bmatrix} 1 & 0 & 0 \\ \hline 0 & 1/\sqrt{2} & 1/\sqrt{2} \\ 0 & -1/\sqrt{2} & 1/\sqrt{2} \end{bmatrix}$$

$$= \frac{1}{3} \begin{bmatrix} 2 & 2 & 1 \\ 2 & -1 & -2 \\ 1 & -2 & 2 \end{bmatrix}$$

and

$$T = U^* B U = \frac{1}{9} \begin{bmatrix} 2 & 2 & 1 \\ 2 & -1 & -2 \\ 1 & -2 & 2 \end{bmatrix} \begin{bmatrix} 1 & 2 & 2 \\ 2 & 1 & 2 \\ 3 & -3 & 4 \end{bmatrix} \begin{bmatrix} 2 & 2 & 1 \\ 2 & -1 & -2 \\ 1 & -2 & 2 \end{bmatrix}$$

$$= \begin{bmatrix} 4 & -1 & 3 \\ 0 & -1 & -4 \\ 0 & 0 & 3 \end{bmatrix}.$$

As demonstrated by the previous example, computing a Schur triangularization by hand is quite tedious. If A is $n \times n$ then constructing a Schur triangularization requires us to find an eigenvalue and eigenvector of an $n \times n$ matrix, extend it to an orthonormal basis, finding an eigenvector of an $(n-1) \times (n-1)$ matrix, extend *it* it to an orthonormal basis, and so on down to a 2×2 matrix.

While there are somewhat faster numerical algorithms that can be used to compute Schur triangularizations in practice, we do not present them here. Rather, we are interested in Schur triangularizations for their theoretical properties and the fact that they will help us establish other, more practical, matrix decompositions. In other words, we frequently make use of the fact that Schur triangularizations *exist*, but we rarely need to actually compute them.

Remark 2.1.1

Schur Triangularizations are Complex

Schur triangularizations are one of the few things that really do require us to work with complex matrices, as opposed to real matrices (or even matrices over some more exotic field). The reason for this is that the first step in finding a Schur triangularization is finding an eigenvector, and some real matrices do not have any real eigenvectors. For example, the matrix

$$A = \begin{bmatrix} 1 & -2 \\ 1 & -1 \end{bmatrix}$$

has no real eigenvalues and thus no real Schur triangularization (since the diagonal entries of its triangularization T necessarily have the same eigenvalues as A). However, it does have a complex Schur triangularization: $A = UTU^*$, where

$$U = \frac{1}{\sqrt{6}} \begin{bmatrix} \sqrt{2}(1+i) & 1+i \\ \sqrt{2} & -2 \end{bmatrix} \quad \text{and} \quad T = \frac{1}{\sqrt{2}} \begin{bmatrix} i\sqrt{2} & 3-i \\ 0 & -i\sqrt{2} \end{bmatrix}.$$

Recall that the trace of a matrix (denoted by tr($\cdot$)) is the sum of its diagonal entries.

While it is not quite as powerful as diagonalization, the beauty of Schur triangularization is that it applies to *every* square matrix, which makes it an extremely useful tool when trying to prove theorems about (potentially non-diagonalizable) matrices. For example, we can use it to provide a short and simple proof of the following relationship between the determinant, trace, and eigenvalues of a matrix (the reader may have already seen this theorem for diagonalizable matrices, or may have already proved it in general by painstakingly analyzing coefficients of the characteristic polynomial).

Theorem 2.1.2

Determinant and Trace in Terms of Eigenvalues

Let $A \in \mathcal{M}_n(\mathbb{C})$ have eigenvalues $\lambda_1, \lambda_2, \ldots, \lambda_n$ (listed according to algebraic multiplicity). Then

$$\det(A) = \lambda_1 \lambda_2 \cdots \lambda_n \quad \text{and} \quad \text{tr}(A) = \lambda_1 + \lambda_2 + \cdots + \lambda_n.$$

Proof. Both formulas are proved by using Schur triangularization to write $A = UTU^*$ with U unitary and T upper triangular, and recalling that the diagonal entries of T are its eigenvalues, which (since T and A are similar) are also the eigenvalues of A. In particular,

$$
\begin{aligned}
\det(A) &= \det(UTU^*) && \text{(Schur triangularization)} \\
&= \det(U)\det(T)\det(U^*) && \text{(multiplicativity of determinant)} \\
&= \det(UU^*)\det(T) && \text{(multiplicativity of determinant)} \\
&= \det(T) && (\det(UU^*) = \det(I) = 1) \\
&= \lambda_1 \lambda_2 \cdots \lambda_n. && \text{(determinant of a triangular matrix)}
\end{aligned}
$$

For a refresher on properties of the determinant like the ones we use here, see Appendix A.1.5.

The corresponding statement about the trace is proved analogously:

$$
\begin{aligned}
\text{tr}(A) &= \text{tr}(UTU^*) && \text{(Schur triangularization)} \\
&= \text{tr}(U^*UT) && \text{(cyclic commutativity of trace)} \\
&= \text{tr}(T) && (U^*U = I) \\
&= \lambda_1 + \lambda_2 + \cdots + \lambda_n. && \text{(definition of trace)} \quad \blacksquare
\end{aligned}
$$

As one more immediate application of Schur triangularization, we prove an important result called the Cayley–Hamilton theorem, which says that every matrix satisfies its own characteristic polynomial.

Theorem 2.1.3

Cayley–Hamilton

Suppose $A \in \mathcal{M}_n(\mathbb{C})$ has characteristic polynomial $p_A(\lambda) = \det(A - \lambda I)$. Then $p_A(A) = O$.

In fact, $a_n = (-1)^n$.

Before proving this result, we clarify exactly what we mean by it. Since the characteristic polynomial p_A really is a polynomial, there are coefficients $c_0, c_1, \ldots, c_n \in \mathbb{C}$ such that

$$p_A(\lambda) = a_n \lambda^n + \cdots + a_2 \lambda^2 + a_1 \lambda + a_0.$$

What we mean by $p_A(A)$ is that we plug A into this polynomial in the naïve way—we just replace each power of λ by the corresponding power of A:

Recall that, for every matrix $A \in \mathcal{M}_n$, we have $A^0 = I$.

$$p_A(A) = a_n A^n + \cdots + a_2 A^2 + a_1 A + a_0 I.$$

For example, if $A = \begin{bmatrix} 1 & 2 \\ 3 & 4 \end{bmatrix}$ then

$$p_A(\lambda) = \det(A - \lambda I) = \det\left(\begin{bmatrix} 1 - \lambda & 2 \\ 3 & 4 - \lambda \end{bmatrix}\right) = (1 - \lambda)(4 - \lambda) - 6$$
$$= \lambda^2 - 5\lambda - 2,$$

Be careful with the constant term: -2 becomes $-2I$ when plugging a matrix into the polynomial.

so $p_A(A) = A^2 - 5A - 2I$. The Cayley–Hamilton theorem says that this equals the zero matrix, which we can verify directly:

$$A^2 - 5A - 2I = \begin{bmatrix} 7 & 10 \\ 15 & 22 \end{bmatrix} - 5\begin{bmatrix} 1 & 2 \\ 3 & 4 \end{bmatrix} - 2\begin{bmatrix} 1 & 0 \\ 0 & 1 \end{bmatrix} = \begin{bmatrix} 0 & 0 \\ 0 & 0 \end{bmatrix}.$$

We now prove this theorem in full generality.

Proof of Theorem 2.1.3. Because we are working over $\mathbb{C}$, the fundamental theorem of algebra (Theorem A.3.1) says that the characteristic polynomial of A has a root and can thus be factored:

$$p_A(\lambda) = (\lambda_1 - \lambda)(\lambda_2 - \lambda) \cdots (\lambda_n - \lambda), \quad \text{so}$$
$$p_A(A) = (\lambda_1 I - A)(\lambda_2 I - A) \cdots (\lambda_n I - A).$$

To simplify things, we use Schur triangularization to write $A = UTU^*$, where $U \in \mathcal{M}_n(\mathbb{C})$ is unitary and $T \in \mathcal{M}_n(\mathbb{C})$ is upper triangular. Then by factoring somewhat cleverly, we see that

This is a fairly standard technique that Schur triangularization gives us—it lets us just prove certain statements for triangular matrices without losing any generality.

$$\begin{aligned} p_A(A) &= (\lambda_1 I - UTU^*) \cdots (\lambda_n I - UTU^*) & \text{(plug in } A = UTU^*) \\ &= (\lambda_1 UU^* - UTU^*) \cdots (\lambda_n UU^* - UTU^*) & \text{(since } I = UU^*) \\ &= \big(U(\lambda_1 I - T)U^*\big) \cdots \big(U(\lambda_n I - T)U^*\big) & \text{(factor out } U \text{ and } U^*) \\ &= U(\lambda_1 I - T) \cdots (\lambda_n I - T)U^* & \text{(since } U^*U = I) \\ &= U p_A(T) U^*. \end{aligned}$$

It follows that we just need to prove that $p_A(T) = O$.

To this end, we assume for simplicity that the diagonal entries of T are in the same order that we have been using for the eigenvalues of A (i.e., the first diagonal entry of T is λ_1, the second is λ_2, and so on). Then for all $1 \leq j \leq n$, $\lambda_j I - T$ is an upper triangular matrix with diagonal entries $\lambda_1 - \lambda_j$, $\lambda_2 - \lambda_j$, ..., $\lambda_n - \lambda_j$. That is,

We use asterisks $(*)$ in the place of matrix entries whose values we do not care about.

$$\lambda_1 I - T = \begin{bmatrix} 0 & * & \cdots & * \\ 0 & \lambda_1 - \lambda_2 & \cdots & * \\ \vdots & \vdots & \ddots & \vdots \\ 0 & 0 & \cdots & \lambda_1 - \lambda_n \end{bmatrix},$$

and similarly for $\lambda_2 I - T$, ..., $\lambda_n I - T$. We now claim that the leftmost k columns of $(\lambda_1 I - T)(\lambda_2 I - T) \cdots (\lambda_k I - T)$ consist entirely of zeros whenever $1 \le k \le n$, and thus $p_A(T) = (\lambda_1 I - T)(\lambda_2 I - T) \cdots (\lambda_n I - T) = O$. Proving this claim is a straightforward (albeit rather ugly) matrix multiplication exercise, which we solve via induction.

Another proof of the Cayley–Hamilton theorem is provided in Section 2.D.4.

The base case $k = 1$ is straightforward, as we wrote down the matrix $\lambda_1 I - T$ above, and its leftmost column is indeed the zero vector. For the inductive step, define $T_k = (\lambda_1 I - T)(\lambda_2 I - T) \cdots (\lambda_k I - T)$ and suppose that the leftmost k columns of T_k consist entirely of zeros. We know that each T_k is upper triangular (since the product of upper triangular matrices is again upper triangular), so some rather unpleasant block matrix multiplication shows that

$$T_{k+1} = T_k\big(\lambda_{k+1}I - T\big)$$

Recall that O_k is the $k \times k$ zero matrix. O (without a subscript) of whatever size makes sense in that portion of the block matrix ($(n-k) \times k$ in this case).

$$= \left[\begin{array}{c|c} O_k & * \\ \hline & \begin{array}{cccc} * & * & \cdots & * \\ 0 & * & \cdots & * \\ \vdots & \vdots & \ddots & \vdots \\ 0 & 0 & \cdots & * \end{array} \\ O & \end{array}\right] \left[\begin{array}{c|c} * & \begin{array}{cccc} * & & & * \end{array} \\ \hline & \begin{array}{cccc} 0 & * & \cdots & * \\ 0 & \lambda_{k+1} - \lambda_{k+2} & \cdots & * \\ \vdots & \vdots & \ddots & \vdots \\ 0 & 0 & \cdots & \lambda_{k+1} - \lambda_n \end{array} \\ O & \end{array}\right]$$

$$= \left[\begin{array}{c|cc} O_k & \mathbf{0} & * \\ \hline & \begin{array}{ccc} 0 & * \cdots & * \\ 0 & * \cdots & * \\ \vdots & \vdots \ddots & \vdots \\ 0 & 0 \cdots & * \end{array} \\ O & \end{array}\right],$$

which has its leftmost $k + 1$ columns equal to the zero vector. This completes the inductive step and the proof. ∎

More generally, if $A \in \mathcal{M}_n(\mathbb{C})$ then the constant coefficient of p_A is $\det(A)$ and its coefficient of λ^{n-1} is $(-1)^{n-1}\mathrm{tr}(A)$.

For example, because the characteristic polynomial of a 2×2 matrix A is $p_A(\lambda) = \lambda^2 - \mathrm{tr}(A)\lambda + \det(A)$, in this case the Cayley–Hamilton theorem says that every 2×2 matrix satisfies the equation $A^2 = \mathrm{tr}(A)A - \det(A)I$, which can be verified directly by giving names to the 4 entries of A and computing all of the indicated quantities:

$$\mathrm{tr}(A)A - \det(A)I = (a+d)\begin{bmatrix} a & b \\ c & d \end{bmatrix} - (ad-bc)\begin{bmatrix} 1 & 0 \\ 0 & 1 \end{bmatrix}$$

$$= \begin{bmatrix} a^2 + bc & ab + bd \\ ac + cd & bc + d^2 \end{bmatrix} = A^2.$$

One useful feature of the Cayley–Hamilton theorem is that if $A \in \mathcal{M}_n(\mathbb{C})$ then it lets us write every power of A as a linear combination of $I, A, A^2, \ldots, A^{n-1}$. In other words, it tells us that the powers of A are all contained within some (at most) n-dimensional subspace of the n^2-dimensional vector space $\mathcal{M}_n(\mathbb{C})$.

Example 2.1.2

Matrix Powers via Cayley–Hamilton

Suppose

$$A = \begin{bmatrix} 1 & 2 \\ 3 & 4 \end{bmatrix}.$$

Use the Cayley–Hamilton theorem to come up with a formula for A^4 as a linear combination of A and I.

Solution:

As noted earlier, the characteristic polynomial of A is $p_A(\lambda) = \lambda^2 - 5\lambda - 2$, so $A^2 - 5A - 2I = O$. Rearranging somewhat gives $A^2 = 5A + 2I$. To get higher powers of A as linear combinations of A and I, just multiply this equation through by A:

$$A^3 = 5A^2 + 2A = 5(5A + 2I) + 2A = 25A + 10I + 2A = 27A + 10I, \text{ and}$$
$$A^4 = 27A^2 + 10A = 27(5A + 2I) + 10A = 135A + 54I + 10A$$
$$= 145A + 54I.$$

> Another way to solve this would be to square both sides of the equation $A^2 = 5A + 2I$, which gives $A^4 = 25A^2 + 20A + 4I$. Substituting in $A^2 = 5A + 2I$ then gives the answer.

Example 2.1.3

Matrix Inverses via Cayley–Hamilton

Use the Cayley–Hamilton theorem to find the inverse of the matrix

$$A = \begin{bmatrix} 1 & 2 \\ 3 & 4 \end{bmatrix}.$$

Solution:

As before, we know from the Cayley–Hamilton theorem that $A^2 - 5A - 2I = O$. If we solve this equation for I, we get $I = (A^2 - 5A)/2$. Factoring this equation gives $I = A(A - 5I)/2$, from which it follows that A is invertible, with inverse

> Alternatively, first check that $\det(A) \neq 0$ so that we know A^{-1} exists, and then find it by multiplying both sides of $A^2 - 5A - 2I = O$ by A^{-1}.

$$A^{-1} = (A - 5I)/2 = \frac{1}{2}\left(\begin{bmatrix} 1 & 2 \\ 3 & 4 \end{bmatrix} - \begin{bmatrix} 5 & 0 \\ 0 & 5 \end{bmatrix} \right) = \frac{1}{2} \begin{bmatrix} -4 & 2 \\ 3 & -1 \end{bmatrix}.$$

Example 2.1.4

Large Matrix Powers via Cayley–Hamilton

Suppose

$$A = \begin{bmatrix} 0 & 2 & 0 \\ 1 & 1 & -1 \\ -1 & 1 & 1 \end{bmatrix}.$$

Use the Cayley–Hamilton theorem to compute A^{314}.

Solution:

The characteristic polynomial of A is

$$p_A(\lambda) = \det(A - \lambda I) = (-\lambda)(1-\lambda)^2 + 2 + 0 - \lambda - 2(1-\lambda) - 0$$
$$= -\lambda^3 + 2\lambda^2.$$

The Cayley–Hamilton theorem then says that $A^3 = 2A^2$. Multiplying that equation by A repeatedly gives us $A^4 = 2A^3 = 4A^2, A^5 = 2A^4 = 4A^3 = 8A^2$, and in general, $A^n = 2^{n-2}A^2$. It follows that

$$A^{314} = 2^{312}A^2 = 2^{312} \begin{bmatrix} 2 & 2 & -2 \\ 2 & 2 & -2 \\ 0 & 0 & 0 \end{bmatrix} = 2^{313} \begin{bmatrix} 1 & 1 & -1 \\ 1 & 1 & -1 \\ 0 & 0 & 0 \end{bmatrix}.$$

It is maybe worth noting that the final example above is somewhat contrived, since it only works out so cleanly due to the given matrix having a very simple characteristic polynomial. For more complicated matrices, large powers are typically still best computed via diagonalization.

2.1.2   Normal Matrices and the Complex Spectral Decomposition

The reason that Schur triangularization is such a useful theoretical tool is that it applies to every matrix in $\mathcal{M}_n(\mathbb{C})$ (unlike diagonalization, for example, which has the annoying restriction of only applying to matrices with a basis of eigenvectors). However, upper triangular matrices can still be somewhat challenging to work with, as evidenced by how technical the proof of the Cayley–Hamilton theorem (Theorem 2.1.3) was. We thus now start looking at when Schur triangularization actually results in a *diagonal* matrix, rather than just an upper triangular one.

It turns out that the answer is related to a new class of matrices that we have not yet encountered, so we start with a definition:

Definition 2.1.1
Normal Matrix

A matrix $A \in \mathcal{M}_n(\mathbb{C})$ is called **normal** if $A^*A = AA^*$.

Many of the important families of matrices that we are already familiar with are normal. For example, every Hermitian matrix is normal, since if $A^* = A$ then $A^*A = A^2 = AA^*$, and a similar argument shows that skew-Hermitian matrices are normal as well. Similarly, every unitary matrix U is normal, since $U^*U = I = UU^*$ in this case, but there are also normal matrices that are neither Hermitian, nor skew-Hermitian, nor unitary.

Recall that $A \in \mathcal{M}_n(\mathbb{C})$ is skew-Hermitian if $A^* = -A$.

Example 2.1.5
A Normal Matrix

Show that the matrix
$$A = \begin{bmatrix} 1 & 1 & 0 \\ 0 & 1 & 1 \\ 1 & 0 & 1 \end{bmatrix}$$

is normal but not Hermitian, skew-Hermitian, or unitary.

Solution:
To see that A is normal, we directly compute

$$A^*A = \begin{bmatrix} 1 & 0 & 1 \\ 1 & 1 & 0 \\ 0 & 1 & 1 \end{bmatrix} \begin{bmatrix} 1 & 1 & 0 \\ 0 & 1 & 1 \\ 1 & 0 & 1 \end{bmatrix} = \begin{bmatrix} 2 & 1 & 1 \\ 1 & 2 & 1 \\ 1 & 1 & 2 \end{bmatrix} = AA^*.$$

The fact that A is not unitary follows from the fact that A^*A (as computed above) does not equal the identity matrix, and the fact that A is neither Hermitian nor skew-Hermitian can be seen just by inspecting the entries of the matrix.

Our primary interest in normal matrices comes from the following theorem, which says that normal matrices are exactly those that can be diagonalized by a unitary matrix. Equivalently, they are exactly the matrices whose Schur triangularizations are actually diagonal.

Theorem 2.1.4
Spectral Decomposition (Complex Version)

Suppose $A \in \mathcal{M}_n(\mathbb{C})$. Then there exists a unitary matrix $U \in \mathcal{M}_n(\mathbb{C})$ and diagonal matrix $D \in \mathcal{M}_n(\mathbb{C})$ such that

$$A = UDU^*$$

if and only if A is normal (i.e., $A^*A = AA^*$).

Proof. To see the "only if" direction, we just compute

$$A^*A = (UDU^*)^*(UDU^*) = UD^*U^*UDU^* = UD^*DU^*, \quad \text{and}$$
$$AA^* = (UDU^*)(UDU^*)^* = UDU^*UD^*U^* = UDD^*U^*.$$

Since D and D^* are diagonal, they commute, so $UD^*DU^* = UDD^*U^*$. We thus conclude that $A^*A = AA^*$, so A is normal.

For the "if" direction, use Schur triangularization to write $A = UTU^*$, where U is unitary and T is upper triangular. Since A is normal, we know that $A^*A = AA^*$. It follows that

$$UTT^*U^* = (UTU^*)(UTU^*)^* = AA^*$$
$$= A^*A = (UTU^*)^*(UTU^*) = UT^*TU^*.$$

Multiplying on the left by U^* and on the right by U then shows that $T^*T = TT^*$ (i.e., T is also normal). Our goal now is to show that, since T is both upper triangular and normal, it must in fact be diagonal.

To this end, we compute the diagonal entries of T^*T and TT^*, starting with $[T^*T]_{1,1} = [TT^*]_{1,1}$. It is perhaps useful to write out the matrices T and T^* to highlight what this equality tells us:

> Recall that the notation $[T^*T]_{i,j}$ means the (i,j)-entry of the matrix T^*T.

$$[T^*T]_{1,1} = \begin{bmatrix} \overline{t_{1,1}} & 0 & \cdots & 0 \\ \overline{t_{1,2}} & \overline{t_{2,2}} & \cdots & 0 \\ \vdots & \vdots & \ddots & \vdots \\ \overline{t_{1,n}} & \overline{t_{2,n}} & \cdots & \overline{t_{n,n}} \end{bmatrix} \begin{bmatrix} t_{1,1} & t_{1,2} & \cdots & t_{1,n} \\ 0 & t_{2,2} & \cdots & t_{2,n} \\ \vdots & \vdots & \ddots & \vdots \\ 0 & 0 & \cdots & t_{n,n} \end{bmatrix}_{1,1}$$

$$= |t_{1,1}|^2,$$

and

$$[TT^*]_{1,1} = \begin{bmatrix} t_{1,1} & t_{1,2} & \cdots & t_{1,n} \\ 0 & t_{2,2} & \cdots & t_{2,n} \\ \vdots & \vdots & \ddots & \vdots \\ 0 & 0 & \cdots & t_{n,n} \end{bmatrix} \begin{bmatrix} \overline{t_{1,1}} & 0 & \cdots & 0 \\ \overline{t_{1,2}} & \overline{t_{2,2}} & \cdots & 0 \\ \vdots & \vdots & \ddots & \vdots \\ \overline{t_{1,n}} & \overline{t_{2,n}} & \cdots & \overline{t_{n,n}} \end{bmatrix}_{1,1}$$

$$= |t_{1,1}|^2 + |t_{1,2}|^2 + \cdots + |t_{1,n}|^2.$$

We thus see that $[T^*T]_{1,1} = [TT^*]_{1,1}$ implies $|t_{1,1}|^2 = |t_{1,1}|^2 + |t_{1,2}|^2 + \cdots + |t_{1,n}|^2$, and since each term in that sum is non-negative it follows that $|t_{1,2}|^2 = \cdots = |t_{1,n}|^2 = 0$. In other words, the only non-zero entry in the first row of T is its $(1,1)$-entry.

We now repeat this argument with $[T^*T]_{2,2} = [TT^*]_{2,2}$. Again, direct computation shows that

> More of the entries of T and T^* equal zero this time because we now know that the off-diagonal entries of the first row of T are all zero.

$$[T^*T]_{2,2} = \begin{bmatrix} \overline{t_{1,1}} & 0 & \cdots & 0 \\ 0 & \overline{t_{2,2}} & \cdots & 0 \\ \vdots & \vdots & \ddots & \vdots \\ 0 & \overline{t_{2,n}} & \cdots & \overline{t_{n,n}} \end{bmatrix} \begin{bmatrix} t_{1,1} & 0 & \cdots & 0 \\ 0 & t_{2,2} & \cdots & t_{2,n} \\ \vdots & \vdots & \ddots & \vdots \\ 0 & 0 & \cdots & t_{n,n} \end{bmatrix}_{2,2}$$

$$= |t_{2,2}|^2,$$

and

$$[TT^*]_{2,2} = \begin{bmatrix} \begin{bmatrix} t_{1,1} & 0 & \cdots & 0 \\ 0 & t_{2,2} & \cdots & t_{2,n} \\ \vdots & \vdots & \ddots & \vdots \\ 0 & 0 & \cdots & t_{n,n} \end{bmatrix} \begin{bmatrix} \overline{t_{1,1}} & 0 & \cdots & 0 \\ 0 & \overline{t_{2,2}} & \cdots & 0 \\ \vdots & \vdots & \ddots & \vdots \\ 0 & \overline{t_{2,n}} & \cdots & \overline{t_{n,n}} \end{bmatrix} \end{bmatrix}_{2,2}$$

$$= |t_{2,2}|^2 + |t_{2,3}|^2 + \cdots + |t_{2,n}|^2,$$

which implies $|t_{2,2}|^2 = |t_{2,2}|^2 + |t_{2,3}|^2 + \cdots + |t_{2,n}|^2$. Since each term in this sum is non-negative, it follows that $|t_{2,3}|^2 = \cdots = |t_{2,n}|^2 = 0$, so the only non-zero entry in the second row of T is its $(2,2)$-entry.

By repeating this argument for $[T^*T]_{k,k} = [TT^*]_{k,k}$ for each $3 \le k \le n$, we similarly see that all of the off-diagonal entries in T equal 0, so T is diagonal. We can thus simply choose $D = T$, and the proof is complete. ∎

The spectral decomposition is one of the most important theorems in all of linear algebra, and it can be interpreted in at least three different ways, so it is worth spending some time thinking about how to think about this theorem.

Interpretation 1. A matrix is normal if and only if its Schur triangularizations are actually diagonalizations. To be completely clear, the following statements about a matrix $A \in \mathcal{M}_n(\mathbb{C})$ are all equivalent:

- *A* is normal.
- *At least one* Schur triangularization of *A* is diagonal.
- *Every* Schur triangularization of *A* is diagonal.

The equivalence of the final two points above might seem somewhat strange given how non-unique Schur triangularizations are, but most of that non-uniqueness comes from the strictly upper triangular portion of the triangular matrix T in $A = UTU^*$. Setting all of the strictly upper triangular entries of T to 0 gets rid of most of this non-uniqueness.

Interpretation 2. A matrix is normal if and only if it is diagonalizable (in the usual sense of Theorem 2.0.1) via a unitary matrix. In particular, recall the columns of the matrix P in a diagonalization $A = PDP^{-1}$ necessarily form a basis (of $\mathbb{C}^n$) of eigenvectors of A. It follows that we can compute spectral decompositions simply via the usual diagonalization procedure, but making sure to choose the eigenvectors to be an *orthonormal* basis of $\mathbb{C}^n$. This method is much quicker and easier to use than the method suggested by Interpretation 1, since Schur triangularizations are awful to compute.

The "usual diagonalization procedure" is illustrated in Appendix A.1.7.

Example 2.1.6

A Small Spectral Decomposition

Find a spectral decomposition of the matrix $A = \begin{bmatrix} 1 & 2 \\ 2 & 1 \end{bmatrix}$.

Solution:

We start by finding its eigenvalues:

$$\det(A - \lambda I) = \det\left(\begin{bmatrix} 1-\lambda & 2 \\ 2 & 1-\lambda \end{bmatrix} \right) = (1-\lambda)^2 - 4 = \lambda^2 - 2\lambda - 3.$$

Setting this polynomial equal to 0 and solving (either via factoring or the quadratic equation) gives $\lambda = -1$ and $\lambda = 3$ as the eigenvalues of A. To find the corresponding eigenvectors, we solve the equations $(A - \lambda I)\mathbf{v} = \mathbf{0}$ for each of $\lambda = -1$ and $\lambda = 3$.

$\lambda = -1$: The system of equations $(A + I)\mathbf{v} = \mathbf{0}$ can be solved as follows:

$$\begin{bmatrix} 2 & 2 & | & 0 \\ 2 & 2 & | & 0 \end{bmatrix} \xrightarrow{R_2 - R_1} \begin{bmatrix} 2 & 2 & | & 0 \\ 0 & 0 & | & 0 \end{bmatrix} \xrightarrow{\frac{1}{2}R_1} \begin{bmatrix} 1 & 1 & | & 0 \\ 0 & 0 & | & 0 \end{bmatrix}.$$

It follows that $v_2 = -v_1$, so any vector of the form $(v_1, -v_1)$ (with $v_1 \neq 0$) is an eigenvector to which this eigenvalue corresponds. However, we want to choose it to have length 1, since we want the eigenvectors to form an orthonormal basis of $\mathbb{C}^n$. We thus choose $v_1 = 1/\sqrt{2}$ so that the eigenvector is $(1, -1)/\sqrt{2}$.

We could have also chosen $v_1 = -1/\sqrt{2}$ (or even something like $v_1 = i/\sqrt{2}$ if we are feeling funky).

$\lambda = 3$: The system of equations $(A - 3I)\mathbf{v} = \mathbf{0}$ can be solved as follows:

$$\begin{bmatrix} -2 & 2 & | & 0 \\ 2 & -2 & | & 0 \end{bmatrix} \xrightarrow{R_2 + R_1} \begin{bmatrix} -2 & 2 & | & 0 \\ 0 & 0 & | & 0 \end{bmatrix}$$

$$\xrightarrow{\frac{-1}{2}R_1} \begin{bmatrix} 1 & -1 & | & 0 \\ 0 & 0 & | & 0 \end{bmatrix}.$$

This is just the usual diagonalization procedure, with some minor care being taken to choose eigenvectors to have length 1 (so that the matrix we place them in will be unitary).

We conclude that $v_2 = v_1$, so we choose $(1, 1)/\sqrt{2}$ to be our (unit length) eigenvector.

To construct a spectral decomposition of A, we then place the eigenvalues as the diagonal entries in a diagonal matrix D, and we place the eigenvectors to which they correspond as columns (in the same order) in a matrix U:

$$D = \begin{bmatrix} -1 & 0 \\ 0 & 3 \end{bmatrix} \quad \text{and} \quad U = \frac{1}{\sqrt{2}} \begin{bmatrix} 1 & 1 \\ -1 & 1 \end{bmatrix}.$$

It is then straightforward to verify that U is indeed unitary, and $A = UDU^*$, so we have found a spectral decomposition of A.

Interpretation 3. A matrix is normal if and only if it represents a linear transformation that stretches (but does not rotate or skew) some orthonormal basis of $\mathbb{C}^n$ (i.e., some unit square grid in $\mathbb{C}^n$, as in Figure 2.1). The reason for this is that if $A = UDU^*$ is a spectral decomposition then the (mutually orthogonal) columns $\mathbf{u}_1, \mathbf{u}_2, \ldots, \mathbf{u}_n$ of U are eigenvectors of A:

To see that $U^*\mathbf{u}_j = \mathbf{e}_j$, recall that $U\mathbf{e}_j = \mathbf{u}_j$ and multiply on the left by U^*.

$$A\mathbf{u}_j = UDU^*\mathbf{u}_j = UD\mathbf{e}_j = U(d_j\mathbf{e}_j) = d_j\mathbf{u}_j \quad \text{for all} \quad 1 \leq j \leq n.$$

Equivalently, normal matrices are the linear transformations that look diagonal in some *orthonormal* basis B of $\mathbb{C}^n$ (whereas diagonalizable matrices are the linear transformations that look diagonal in some not-necessarily-orthonormal basis).

In Example 2.1.6, we chose the eigenvectors so as to make the diagonalizing matrix unitary, but the fact that they were orthogonal to each other came for free as a consequence of A being normal. This always happens (for normal matrices) with eigenvectors corresponding to different eigenvalues—a fact that we now state formally as a corollary of the spectral decomposition.

Corollary 2.1.5

Normal Matrices have Orthogonal Eigenspaces

Suppose $A \in \mathcal{M}_n(\mathbb{C})$ is normal. If $\mathbf{v}, \mathbf{w} \in \mathbb{C}^n$ are eigenvectors of A corresponding to different eigenvalues then $\mathbf{v} \cdot \mathbf{w} = 0$.

Proof. The idea is simply that in a spectral decomposition $A = UDU^*$, the columns of U can be partitioned so as to span the eigenspaces of A, and since those columns are mutually orthogonal, so are all vectors in the eigenspaces.

More precisely, suppose $A\mathbf{v} = \lambda\mathbf{v}$ and $A\mathbf{w} = \mu\mathbf{w}$ for some $\lambda \neq \mu$ (i.e., these are the two different eigenvalues of A described by the theorem). By the spectral decomposition (Theorem 2.1.4), we can write $A = UDU^*$, where U is a unitary matrix and D is diagonal. Furthermore, we can permute the diagonal entries of D (and the columns of U correspondingly) so as to group all of the occurrences of λ and μ on the diagonal of D together:

> Here, k and ℓ are the multiplicities of the eigenvalues λ and μ of A (i.e., they are the dimensions of the corresponding eigenspaces).

$$D = \begin{bmatrix} \lambda I_k & O & O \\ O & \mu I_\ell & O \\ O & O & D_2 \end{bmatrix},$$

where D_2 is diagonal and does not contain either λ or μ as any of its diagonal entries. If we write $U = \begin{bmatrix} \mathbf{u}_1 & | & \mathbf{u}_2 & | & \cdots & | & \mathbf{u}_n \end{bmatrix}$ then $\{\mathbf{u}_1, \ldots, \mathbf{u}_k\}$ and $\{\mathbf{u}_{k+1}, \ldots, \mathbf{u}_{k+\ell}\}$ are orthonormal bases of the eigenspaces of A corresponding to the eigenvalues λ and μ, respectively. We can thus write $\mathbf{v}$ and $\mathbf{w}$ as linear combinations of these vectors:

> Be slightly careful—the spectral decomposition only says that there *exists* an orthonormal basis of $\mathbb{C}^n$ consisting of eigenvectors (i.e., the columns of U), not that *every* basis of eigenvectors is orthonormal.

$$\mathbf{v} = c_1\mathbf{u}_1 + \cdots + c_k\mathbf{u}_k \quad \text{and} \quad \mathbf{w} = d_1\mathbf{u}_{k+1} + \cdots + d_\ell\mathbf{u}_{k+\ell}.$$

We then simply compute

$$\mathbf{v} \cdot \mathbf{w} = \left(\sum_{i=1}^{k} c_i\mathbf{u}_i\right) \cdot \left(\sum_{j=1}^{\ell} d_j\mathbf{u}_{k+j}\right) = \sum_{i=1}^{k}\sum_{j=1}^{\ell} c_i d_j(\mathbf{u}_i \cdot \mathbf{u}_{k+j}) = 0,$$

since U is unitary so its columns are mutually orthogonal. ∎

The above corollary tells us that, when constructing a spectral decomposition of a normal matrix, eigenvectors corresponding to different eigenvalues will be orthogonal "for free", so we just need to make sure to scale them to have length 1. On the other hand, we have to be slightly careful when constructing multiple eigenvectors corresponding to the *same* eigenvalue—we have to *choose* them to be orthogonal to each other. This is a non-issue when the eigenvalues of the matrix are distinct (and thus its eigenspaces are 1-dimensional), but it requires some attention when some eigenvalues have multiplicity greater than 1.

> For diagonalizable matrices, algebraic and geometric multiplicities coincide, so here we just call them both "multiplicity".

Example 2.1.7

A Larger Spectral Decomposition

Find a spectral decomposition of the matrix $A = \begin{bmatrix} 1 & 1 & -1 & -1 \\ 1 & 1 & -1 & -1 \\ 1 & 1 & 1 & 1 \\ 1 & 1 & 1 & 1 \end{bmatrix}$.

Solution:

It is straightforward to check that $A^*A = AA^*$, so this matrix is normal and thus has a spectral decomposition. Its characteristic polynomial is $\lambda^2(\lambda^2 - 4\lambda + 8)$, so its eigenvalues are $\lambda = 0$, with algebraic multiplicity 2, and $\lambda = 2 \pm 2i$, each with algebraic multiplicity 1.

> No, this characteristic polynomial is not fun to come up with. Just compute $\det(A - \lambda I)$ like usual though.

For the eigenvalues $\lambda = 2 \pm 2i$, we find corresponding normalized

Recall that eigenvalues and eigenvectors of real matrices come in complex conjugate pairs.

eigenvectors to be $\mathbf{v} = (1 \pm i, 1 \pm i, 1 \mp i, 1 \mp i)/(2\sqrt{2})$, respectively. What happens when $\lambda = 0$ is a bit more interesting, so we now explicitly demonstrate the computation of those corresponding eigenvectors by solving the equation $(A - \lambda I)\mathbf{v} = \mathbf{0}$ when $\lambda = 0$:

$\lambda = 0$: The system of equations $(A - 0I)\mathbf{v} = \mathbf{0}$ can be solved as follows:

$$\left[\begin{array}{cccc|c} 1 & 1 & -1 & -1 & 0 \\ 1 & 1 & -1 & -1 & 0 \\ 1 & 1 & 1 & 1 & 0 \\ 1 & 1 & 1 & 1 & 0 \end{array}\right] \begin{array}{c} \\ R_2 - R_1 \\ R_3 - R_1 \\ R_4 - R_1 \\ \xrightarrow{} \end{array} \left[\begin{array}{cccc|c} 1 & 1 & -1 & -1 & 0 \\ 0 & 0 & 0 & 0 & 0 \\ 0 & 0 & 2 & 2 & 0 \\ 0 & 0 & 2 & 2 & 0 \end{array}\right]$$

These row operations are performed sequentially (i.e., one after another), not simultaneously.

$$\begin{array}{c} \frac{1}{2}R_3 \\ R_1 + R_3 \\ R_4 - 2R_3 \\ R_2 \leftrightarrow R_3 \\ \xrightarrow{} \end{array} \left[\begin{array}{cccc|c} 1 & 1 & 0 & 0 & 0 \\ 0 & 0 & 1 & 1 & 0 \\ 0 & 0 & 0 & 0 & 0 \\ 0 & 0 & 0 & 0 & 0 \end{array}\right].$$

The first row of the RREF above tells us that the eigenvectors $\mathbf{v} = (v_1, v_2, v_3, v_4)$ satisfy $v_1 = -v_2$, and its second row tells us that $v_3 = -v_4$, so the eigenvectors of A corresponding to the eigenvalue $\lambda = 0$ are exactly those of the form $\mathbf{v} = v_2(-1, 1, 0, 0) + v_4(0, 0, -1, 1)$.

We want to choose two vectors of this form (since $\lambda = 0$ has multiplicity 2), but we have to be slightly careful to choose them so that not only are they normalized, but they are also orthogonal to each other. In this case, there is an obvious choice: $(-1, 1, 0, 0)/\sqrt{2}$ and $(0, 0, -1, 1)/\sqrt{2}$ are orthogonal, so we choose them.

We could have also chosen $(1, -1, 0, 0)/\sqrt{2}$ and $(0, 0, 1, -1)/\sqrt{2}$, or $(1, -1, 1, -1)/2$ and $(1, -1, -1, 1)/2$, but *not* $(1, -1, 0, 0)/\sqrt{2}$ and $(1, -1, 1, -1)/2$, for example.

To construct a spectral decomposition of A, we then place the eigenvalues as the diagonal entries in a diagonal matrix D, and we place the normalized and orthogonal eigenvectors to which they correspond as columns (in the same order) in a matrix U:

$$D = \begin{bmatrix} 0 & 0 & 0 & 0 \\ 0 & 0 & 0 & 0 \\ 0 & 0 & 2+2i & 0 \\ 0 & 0 & 0 & 2-2i \end{bmatrix}, \quad U = \frac{1}{2\sqrt{2}}\begin{bmatrix} -2 & 0 & 1+i & 1-i \\ 2 & 0 & 1+i & 1-i \\ 0 & -2 & 1-i & 1+i \\ 0 & 2 & 1-i & 1+i \end{bmatrix}.$$

I'm glad that we did not do this for a 5×5 matrix. Maybe we'll save that for the exercises?

To double-check our work, we could verify that U is indeed unitary, and $A = UDU^*$, so we have indeed found a spectral decomposition of A.

In the previous example, we were able to find an orthonormal basis of the 2-dimensional eigenspace just by inspection. However, it might not always be so easy—if we cannot just "see" an orthonormal basis of an eigenspace, then we can construct one by applying the Gram–Schmidt process (Theorem 1.4.6) to any basis of the space.

Now that we know that every normal matrix is diagonalizable, it's worth taking a moment to remind ourselves of the relationships between the various families of matrices that we have introduced so far. See Figure 2.2 for a visual representation of these relationships and a reminder of which matrix families contain each other.

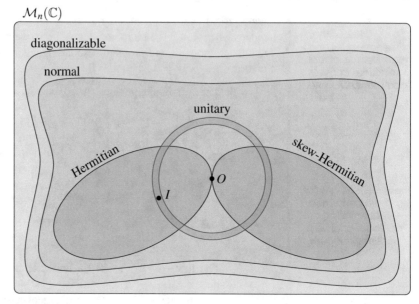

Figure 2.2: A visualization of the containments of several important families of matrices within each other. For example, every unitary matrix is normal, every normal matrix is diagonalizable, and the only matrix that is both Hermitian and skew-Hermitian is the zero matrix O.

The sizes of the sets shown here are slightly misleading. For example, the set of diagonalizable matrices is "dense" in $\mathcal{M}_n(\mathbb{C})$ (see Section 2.D) and thus quite large, whereas the set of normal matrices is tiny.

2.1.3 The Real Spectral Decomposition

Since the spectral decomposition applies to all (square) complex matrices, it automatically applies to all real matrices in the sense that, if $A \in \mathcal{M}_n(\mathbb{R})$ is normal, then we can find a unitary $U \in \mathcal{M}_n(\mathbb{C})$ and a diagonal $D \in \mathcal{M}_n(\mathbb{C})$ such that $A = UDU^*$. However, U and D might have complex entries, even if A is real. For example, the eigenvalues of

$$A = \begin{bmatrix} 0 & 1 \\ -1 & 0 \end{bmatrix}$$

We saw another example of a real matrix with a complex spectral decomposition in Example 2.1.7.

are $\pm i$, with corresponding eigenvectors $(1, \pm i)/\sqrt{2}$. It follows that a diagonal matrix D and unitary matrix U providing a spectral decomposition of A are

$$D = \begin{bmatrix} i & 0 \\ 0 & -i \end{bmatrix} \quad \text{and} \quad U = \frac{1}{\sqrt{2}} \begin{bmatrix} 1 & 1 \\ i & -i \end{bmatrix},$$

and furthermore there is no spectral decomposition of A that makes use of a real D and real U.

This observation raises a natural question: which real matrices have a *real* spectral decomposition (i.e., a spectral decomposition in which both the diagonal matrix D and unitary matrix U are real)? Certainly any such matrix must be symmetric since if $A = UDU^T$ for some diagonal $D \in \mathcal{M}_n(\mathbb{R})$ and unitary $U \in \mathcal{M}_n(\mathbb{R})$ then

Keep in mind that if U is real then $U^T = U^*$, so UDU^T really is a spectral decomposition.

$$A^T = (UDU^T)^T = (U^T)^T D^T U^T = UDU^T = A.$$

Remarkably, the converse of the above observation is also true—every real symmetric matrix has a spectral decomposition consisting of real matrices:

Theorem 2.1.6	Suppose $A \in \mathcal{M}_n(\mathbb{R})$. Then there exists a unitary matrix $U \in \mathcal{M}_n(\mathbb{R})$ and diagonal matrix $D \in \mathcal{M}_n(\mathbb{R})$ such that

Spectral Decomposition (Real Version)

$$A = UDU^T$$

if and only if A is symmetric (i.e., $A = A^T$).

Proof. We already proved the "only if" direction of the theorem, so we jump straight to proving the "if" direction. To this end, recall that the complex spectral decomposition (Theorem 2.1.4) says that we can find a *complex* unitary matrix $U \in \mathcal{M}_n(\mathbb{C})$ and diagonal matrix $D \in \mathcal{M}_n(\mathbb{C})$ such that $A = UDU^*$, and the columns of U are eigenvectors of A corresponding to the eigenvalues along the diagonal of D.

To see that D must in fact be real, we observe that since A is real and symmetric, we have

> In fact, this argument also shows that every (potentially complex) *Hermitian* matrix has only real eigenvalues.

$$\lambda \|\mathbf{v}\|^2 = \lambda \mathbf{v}^* \mathbf{v} = \mathbf{v}^* A \mathbf{v} = \mathbf{v}^* A^* \mathbf{v} = (\mathbf{v}^* A \mathbf{v})^* = (\lambda \mathbf{v}^* \mathbf{v})^* = \overline{\lambda} \|\mathbf{v}\|^2,$$

which implies $\lambda = \overline{\lambda}$ (since every eigenvector $\mathbf{v}$ is, by definition, non-zero), so λ is real.

To see that we can similarly choose U to be real, we must construct an orthonormal basis of $\mathbb{R}^n$ consisting of eigenvectors of A. To do so, we first recall from Corollary 2.1.5 that eigenvectors $\mathbf{v}$ and $\mathbf{w}$ of A corresponding to different eigenvalues $\lambda \neq \mu$, respectively, are *necessarily* orthogonal to each other, since real symmetric matrices are normal.

It thus suffices to show that we can find a real orthonormal basis of each eigenspace of A, and since we can use the Gram–Schmidt process (Theorem 1.4.6) to find an orthonormal basis of any subspace, it suffices to just show that each eigenspace of A is the *span* of a set of real vectors. The key observation that demonstrates this fact is that if a vector $\mathbf{v} \in \mathbb{C}^n$ is in the eigenspace of A corresponding to an eigenvalue λ then so is $\overline{\mathbf{v}}$, since taking the complex conjugate of the equation $A\mathbf{v} = \lambda\mathbf{v}$ gives

> Recall that $\overline{\mathbf{v}}$ is the complex conjugate of $\mathbf{v}$ (i.e., if $\mathbf{v} = \mathbf{x} + i\mathbf{y}$ for some $\mathbf{x}, \mathbf{y} \in \mathbb{R}^n$ then $\overline{\mathbf{v}} = \mathbf{x} - i\mathbf{y}$). Also, $\mathrm{Re}(\mathbf{v}) = (\mathbf{v} + \overline{\mathbf{v}})/2$ and $\mathrm{Im}(\mathbf{v}) = (\mathbf{v} - \overline{\mathbf{v}})/(2i)$ are called the **real** and **imaginary parts** of $\mathbf{v}$. Refer to Appendix A.3 for a refresher on complex numbers.

$$A\overline{\mathbf{v}} = \overline{A\mathbf{v}} = \overline{\lambda\mathbf{v}} = \lambda\overline{\mathbf{v}}.$$

In particular, this implies that if $\{\mathbf{v}_1, \mathbf{v}_2, \ldots, \mathbf{v}_k\}$ is any basis of the eigenspace, then the set $\{\mathrm{Re}(\mathbf{v}_1), \mathrm{Im}(\mathbf{v}_1), \mathrm{Re}(\mathbf{v}_2), \mathrm{Im}(\mathbf{v}_2), \ldots, \mathrm{Re}(\mathbf{v}_k), \mathrm{Im}(\mathbf{v}_k)\}$ has the same span. Since each vector in this set is real, the proof is now complete. ∎

Another way of phrasing the real spectral decomposition is as saying that if $\mathcal{V}$ is a real inner product space then a linear transformation $T : \mathcal{V} \to \mathcal{V}$ is self-adjoint if and only if T looks diagonal in some orthonormal basis of $\mathcal{V}$. Geometrically, this means that T is self-adjoint if and only if it looks like a rotation and/or reflection (i.e., a unitary matrix U), composed with a diagonal scaling, composed with a rotation and/or reflection back (i.e., the inverse unitary matrix U^*).

> The real spectral decomposition can also be proved "directly", without making use of its complex version. See Exercise 2.1.20.

This gives us the exact same geometric interpretation of the real spectral decomposition that we had for its complex counterpart—symmetric matrices are exactly those that stretch (but do not rotate or skew) some orthonormal basis. However, the important distinction here from the case of normal matrices is that the eigenvalues and eigenvalues of symmetric matrices are real, so we can really visualize this all happening in $\mathbb{R}^n$ rather than in $\mathbb{C}^n$, as in Figure 2.3.

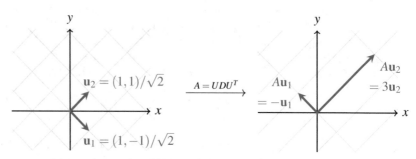

(a) A sends a square grid to a rectangular one with the same orientation.

Recall that $[\mathbf{v}]_B$ refers to the coordinate vector of $\mathbf{v}$ in the basis B. That is, $[\mathbf{v}]_B = U^*\mathbf{v}$ and U is the change-of-basis matrix from the orthonormal eigenbasis B to the standard basis $E = \{\mathbf{e}_1, \mathbf{e}_2\}$: $U = P_{E \leftarrow B}$.

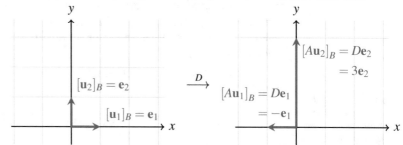

(b) A looks diagonal if we just rotate the plane (i.e., apply a unitary to it).

Figure 2.3: The matrix A from Example 2.1.6 looks diagonal if we view it in the orthonormal basis $B = \{(1,1)/\sqrt{2}, (1,-1)/\sqrt{2}\}$. There exists an orthonormal basis with this property precisely because A is symmetric.

In order to actually *compute* a real spectral decomposition of a symmetric matrix, we just do what we always do—we find the eigenvalues and eigenvectors of the matrix and use them to diagonalize it, taking the minor care necessary to ensure that the eigenvectors are chosen to be real, mutually orthogonal, and of length 1.

Example 2.1.8

A Real Spectral Decomposition

Find a real spectral decomposition of the matrix $A = \begin{bmatrix} 1 & 2 & 2 \\ 2 & 1 & 2 \\ 2 & 2 & 1 \end{bmatrix}$.

Solution:

Since this matrix is symmetric, we know that it has a real spectral decomposition. To compute it, we first find its eigenvalues:

$$\det(A - \lambda I) = \det\left(\begin{bmatrix} 1-\lambda & 2 & 2 \\ 2 & 1-\lambda & 2 \\ 2 & 2 & 1-\lambda \end{bmatrix}\right)$$
$$= (1-\lambda)^3 + 8 + 8 - 4(1-\lambda) - 4(1-\lambda) - 4(1-\lambda)$$
$$= (1+\lambda)^2(5-\lambda).$$

Setting the above characteristic polynomial equal to 0 gives $\lambda = -1$ (with multiplicity 2) or $\lambda = 5$ (with multiplicity 1). To find the corresponding eigenvectors, we solve the linear systems $(A - \lambda I)\mathbf{v} = \mathbf{0}$ for each of $\lambda = 5$ and $\lambda = -1$.

$\lambda = 5$: The system of equations $(A - 5I)\mathbf{v} = \mathbf{0}$ can be solved as follows:

$$\begin{bmatrix} -4 & 2 & 2 & | & 0 \\ 2 & -4 & 2 & | & 0 \\ 2 & 2 & -4 & | & 0 \end{bmatrix} \xrightarrow[\substack{R_2 + \frac{1}{2}R_1 \\ R_3 + \frac{1}{2}R_1}]{} \begin{bmatrix} -4 & 2 & 2 & | & 0 \\ 0 & -3 & 3 & | & 0 \\ 0 & 3 & -3 & | & 0 \end{bmatrix}$$

$$\xrightarrow[R_3 + R_2]{} \begin{bmatrix} -4 & 2 & 2 & | & 0 \\ 0 & -3 & 3 & | & 0 \\ 0 & 0 & 0 & | & 0 \end{bmatrix}.$$

From here we can use back substitution to see that $v_1 = v_2 = v_3$, so any vector of the form (v_1, v_1, v_1) (with $v_1 \neq 0$) is an eigenvector to which this eigenvalue corresponds. However, we want to choose it to have length 1, so we choose $v_1 = 1/\sqrt{3}$, so that the eigenvector is $(1, 1, 1)/\sqrt{3}$.

We could have also chosen $v_1 = -1$ so that the eigenvector would be $(-1, -1, -1)/\sqrt{3}$.

$\lambda = -1$: The system of equations $(A + I)\mathbf{v} = \mathbf{0}$ can be solved as follows:

$$\begin{bmatrix} 2 & 2 & 2 & | & 0 \\ 2 & 2 & 2 & | & 0 \\ 2 & 2 & 2 & | & 0 \end{bmatrix} \xrightarrow[\substack{R_2 - R_1 \\ R_3 - R_1}]{} \begin{bmatrix} 2 & 2 & 2 & | & 0 \\ 0 & 0 & 0 & | & 0 \\ 0 & 0 & 0 & | & 0 \end{bmatrix}.$$

We conclude that $v_1 = -v_2 - v_3$, so the eigenvectors to which this eigenvalue corresponds have the form $v_2(-1, 1, 0) + v_3(-1, 0, 1)$. Since this eigenspace has dimension 2, we have to choose two eigenvectors, and we furthermore have to be careful to choose them to form an orthonormal basis of their span (i.e., they must be orthogonal to each other, and we must scale them each to have length 1).

Alternatively, we could construct an orthonormal basis of this eigenspace by applying Gram–Schmidt to $\{(-1, 1, 0), (-1, 0, 1)\}$.

One of the easiest pairs of orthogonal eigenvectors to "eyeball" is $\{(-2, 1, 1), (0, 1, -1)\}$ (which correspond to choosing $(v_2, v_3) = (1, 1)$ and $(v_2, v_3) = (1, -1)$, respectively). After normalizing these eigenvectors, we get

$$\left\{ \tfrac{1}{\sqrt{6}}(-2, 1, 1), \tfrac{1}{\sqrt{2}}(0, 1, -1) \right\}$$

as the orthonormal basis of this eigenspace.

To construct a spectral decomposition of A, we then place the eigenvalues as the diagonal entries in a diagonal matrix D, and we place the eigenvectors to which they correspond as columns (in the same order) in a matrix U:

Note that we pulled a common factor of $1/\sqrt{6}$ outside of U, which makes its columns look slightly different from what we saw when computing the eigenvectors.

$$D = \begin{bmatrix} 5 & 0 & 0 \\ 0 & -1 & 0 \\ 0 & 0 & -1 \end{bmatrix} \quad \text{and} \quad U = \frac{1}{\sqrt{6}} \begin{bmatrix} \sqrt{2} & -2 & 0 \\ \sqrt{2} & 1 & \sqrt{3} \\ \sqrt{2} & 1 & -\sqrt{3} \end{bmatrix}.$$

To double-check our work, we could verify that U is indeed unitary, and $A = UDU^T$, so we have indeed found a real spectral decomposition of A.

Before moving on, it is worth having a brief look at Table 2.1, which summarizes the relationship between the real and complex spectral decompositions, and what they say about normal, Hermitian, and symmetric matrices.

Matrices that are *complex* symmetric (i.e., $A \in \mathcal{M}_n(\mathbb{C})$ and $A^T = A$) are not necessarily normal and thus may not have a spectral decomposition. However, they have a related decomposition that is presented in Exercise 2.3.26.

Matrix A	Decomp. $A = UDU^*$	Eigenvalues, eigenvectors
Normal $A^*A = AA^*$	$D \in \mathcal{M}_n(\mathbb{C})$ $U \in \mathcal{M}_n(\mathbb{C})$	Eigenvalues: complex Eigenvectors: complex
Hermitian $A^* = A$	$D \in \mathcal{M}_n(\mathbb{R})$ $U \in \mathcal{M}_n(\mathbb{C})$	Eigenvalues: real Eigenvectors: complex
Real symmetric $A^T = A$	$D \in \mathcal{M}_n(\mathbb{R})$ $U \in \mathcal{M}_n(\mathbb{R})$	Eigenvalues: real Eigenvectors: real

Table 2.1: A summary of which parts of the spectral decomposition are real and complex for different types of matrices.

Exercises

solutions to starred exercises on page 465

2.1.1 For each of the following matrices, say whether it is (i) unitary, (ii) Hermitian, (iii) skew-Hermitian, (iv) normal. It may have multiple properties or even none of the listed properties.

* (a) $\begin{bmatrix} 2 & 2 \\ -2 & 2 \end{bmatrix}$

(b) $\begin{bmatrix} 1 & 2 \\ 3 & 4 \end{bmatrix}$

* (c) $\dfrac{1}{\sqrt{5}} \begin{bmatrix} 1 & 2i \\ 2i & 1 \end{bmatrix}$

(d) $\begin{bmatrix} 1 & 0 \\ 0 & 1 \end{bmatrix}$

* (e) $\begin{bmatrix} 0 & 0 \\ 0 & 0 \end{bmatrix}$

(f) $\begin{bmatrix} 1 & 1+i \\ 1+i & 1 \end{bmatrix}$

* (g) $\begin{bmatrix} 0 & -i \\ i & 0 \end{bmatrix}$

(h) $\begin{bmatrix} i & 1 \\ -1 & 2i \end{bmatrix}$

2.1.2 Determine which of the following matrices are normal.

* (a) $\begin{bmatrix} 2 & -1 \\ 1 & 3 \end{bmatrix}$

(b) $\begin{bmatrix} 1 & 1 & 1 \\ 1 & 1 & 1 \end{bmatrix}$

* (c) $\begin{bmatrix} 1 & 1 \\ -1 & 1 \end{bmatrix}$

(d) $\begin{bmatrix} 1 & 2 & 3 \\ 3 & 1 & 2 \\ 2 & 3 & 1 \end{bmatrix}$

* (e) $\begin{bmatrix} 1 & 2 & -3i \\ 2 & 2 & 2 \\ 3i & 2 & 4 \end{bmatrix}$

(f) $\begin{bmatrix} 2+3i & 0 & 0 \\ 0 & 7i & 0 \\ 0 & 0 & 18 \end{bmatrix}$

* (g) $\begin{bmatrix} 1 & 2 & 0 \\ 3 & 4 & 5 \\ 0 & 6 & 7 \end{bmatrix}$

(h) $\begin{bmatrix} \sqrt{2} & \sqrt{2} & i \\ 1 & 1 & 1 \\ \sqrt{2} & i & 1 \end{bmatrix}$

2.1.3 Compute a Schur triangularization of the following matrices.

* (a) $\begin{bmatrix} 6 & -3 \\ 2 & 1 \end{bmatrix}$

(b) $\begin{bmatrix} 7 & -5 \\ -1 & 3 \end{bmatrix}$

2.1.4 Find a spectral decomposition of each of the following normal matrices.

* (a) $\begin{bmatrix} 3 & 2 \\ 2 & 3 \end{bmatrix}$

(b) $\begin{bmatrix} 1 & 1 \\ -1 & 1 \end{bmatrix}$

* (c) $\begin{bmatrix} 0 & -i \\ i & 0 \end{bmatrix}$

(d) $\begin{bmatrix} 1 & 0 & 1 \\ 0 & 1 & 0 \\ 1 & 0 & 1 \end{bmatrix}$

* (e) $\begin{bmatrix} 1 & 1 & 0 \\ 0 & 1 & 1 \\ 1 & 0 & 1 \end{bmatrix}$

(f) $\begin{bmatrix} 2i & 0 & 0 \\ 0 & 1+i & -1+i \\ 0 & -1+i & 1+i \end{bmatrix}$

2.1.5 Determine which of the following statements are true and which are false.

(a) If $A = UTU^*$ is a Schur triangularization of A then the eigenvalues of A are along the diagonal of T.
* (b) If $A, B \in \mathcal{M}_n(\mathbb{C})$ are normal then so is $A + B$.
(c) If $A, B \in \mathcal{M}_n(\mathbb{C})$ are normal then so is AB.
* (d) The set of normal matrices is a subspace of $\mathcal{M}_n(\mathbb{C})$.
(e) If $A, B \in \mathcal{M}_n(\mathbb{C})$ are similar and A is normal, then B is normal too.
* (f) If $A \in \mathcal{M}_n(\mathbb{R})$ is normal then there exists a unitary matrix $U \in \mathcal{M}_n(\mathbb{R})$ and a diagonal matrix $D \in \mathcal{M}_n(\mathbb{R})$ such that $A = UDU^T$.
(g) If $A = UTU^*$ is a Schur triangularization of A then $A^2 = UT^2U^*$ is a Schur triangularization of A^2.
* (h) If all of the eigenvalues of $A \in \mathcal{M}_n(\mathbb{C})$ are real, it must be Hermitian.
(i) If $A \in \mathcal{M}_3(\mathbb{C})$ has eigenvalues 1, 1, and 0 (counted via algebraic multiplicity), then $A^3 = 2A^2 - A$.

2.1.6 Suppose $A \in \mathcal{M}_n(\mathbb{C})$. Show that there exists a unitary matrix $U \in \mathcal{M}_n(\mathbb{C})$ and a *lower* triangular matrix $L \in \mathcal{M}_n(\mathbb{C})$ such that

$$A = ULU^*.$$

*2.1.7 Suppose $A \in \mathcal{M}_2(\mathbb{C})$. Use the Cayley–Hamilton theorem to find an explicit formula for A^{-1} in terms of the entries of A (assuming that A is invertible).

2.1.8 Compute A^{2718} if $A = \begin{bmatrix} 3 & \sqrt{2} & -3 \\ \sqrt{2} & 0 & \sqrt{2} \\ -3 & -\sqrt{2} & 3 \end{bmatrix}$.

[Hint: The Cayley–Hamilton theorem might help.]

*2.1.9** Suppose $A = \begin{bmatrix} 4 & 0 & 0 \\ 3 & 1 & -3 \\ 3 & -3 & 1 \end{bmatrix}$.

  (a) Use the Cayley–Hamilton theorem to write A^{-1} as a linear combination of I, A, and A^2.

  (b) Write A^{-1} as a linear combination of A^2, A^3, and A^4.

2.1.10 Suppose $A \in \mathcal{M}_n$ has characteristic polynomial $p_A(\lambda) = \det(A - \lambda I)$. Explain the problem with the following one-line "proof" of the Cayley–Hamilton theorem:

$$p_A(A) = \det(A - AI) = \det(O) = 0.$$

2.1.11 A matrix $A \in \mathcal{M}_n(\mathbb{C})$ is called **nilpotent** if there exists a positive integer k such that $A^k = O$.

  (a) Show that A is nilpotent if and only if all of its eigenvalues equal 0.

  (b) Show that if A is nilpotent then $A^k = O$ for some $k \leq n$.

** **2.1.12** Suppose that $A \in \mathcal{M}_n(\mathbb{C})$ has eigenvalues $\lambda_1, \lambda_2, \ldots, \lambda_n$ (listed according to algebraic multiplicity). Show that A is normal if and only if

$$\|A\|_F = \sqrt{\sum_{j=1}^{n} |\lambda_j|^2},$$

where $\|A\|_F$ is the Frobenius norm of A.

** **2.1.13** Show that $A \in \mathcal{M}_n(\mathbb{C})$ is normal if and only if $\|A\mathbf{v}\| = \|A^*\mathbf{v}\|$ for all $\mathbf{v} \in \mathbb{C}^n$.

[Hint: Make use of Exercise 1.4.19.]

** **2.1.14** Show that $A \in \mathcal{M}_n(\mathbb{C})$ is normal if and only if $A^* \in \text{span}\{I, A, A^2, A^3, \ldots\}$.

[Hint: Apply the spectral decomposition to A and think about interpolating polynomials.]

2.1.15 Suppose $A, B \in \mathcal{M}_n(\mathbb{C})$ are unitarily similar (i.e., there is a unitary $U \in \mathcal{M}_n(\mathbb{C})$ such that $B = UAU^*$). Show that if A is normal then so is B.

2.1.16 Suppose $T \in \mathcal{M}_n(\mathbb{C})$ is upper triangular. Show that T is diagonal if and only if it is normal.

[Hint: We actually showed this fact somewhere in this section—just explain where.]

[Side note: This exercise generalizes Exercise 1.4.12.]

** **2.1.17** Suppose $A \in \mathcal{M}_n(\mathbb{C})$ is normal.

  (a) Show that A is Hermitian if and only if its eigenvalues are real.

  (b) Show that A is skew-Hermitian if and only if its eigenvalues are imaginary.

  (c) Show that A is unitary if and only if its eigenvalues lie on the unit circle in the complex plane.

  (d) Explain why the "if" direction of each of the above statements fails if A is not normal.

2.1.18 Show that if $A \in \mathcal{M}_n(\mathbb{C})$ is Hermitian, then e^{iA} is unitary.

[Side note: Recall that e^{iA} is the matrix obtained via exponentiating the diagonal part of iA's diagonalization (which in this case is a spectral decomposition).]

** **2.1.19** Suppose $A \in \mathcal{M}_n$.

  (a) Show that if A is normal then the number of non-zero eigenvalues of A (counting algebraic multiplicity) equals rank(A).

  (b) Provide an example to show that the result of part (a) is not necessarily true if A is not normal.

** **2.1.20** In this exercise, we show how to prove the real spectral decomposition (Theorem 2.1.6) "directly", without relying on the complex spectral decomposition (Theorem 2.1.4). Suppose $A \in \mathcal{M}_n(\mathbb{R})$ is symmetric.

  (a) Show that all of the eigenvalues of A are real.

  (b) Show that A has a real eigenvector.

  (c) Complete the proof of the real spectral decomposition by mimicking the proof of Schur triangularization (Theorem 2.1.1).

2.1.21 Suppose $A \in \mathcal{M}_n(\mathbb{C})$ is skew-Hermitian (i.e., $A^* = -A$).

  (a) Show that $I + A$ is invertible.

  (b) Show that $U_A = (I + A)^{-1}(I - A)$ is unitary. [Side note: U_A is called the **Cayley transform** of A.]

  (c) Show that if A is real then $\det(U_A) = 1$.

2.1.22 Suppose $U \in \mathcal{M}_n(\mathbb{C})$ is a skew-symmetric unitary matrix (i.e., $U^T = -U$ and $U^*U = I$).

  (a) Show that n must be even (i.e., show that no such matrix exists when n is odd).

  (b) Show that the eigenvalues of U are $\pm i$, each with multiplicity equal to $n/2$.

  (c) Show that there is a unitary matrix $V \in \mathcal{M}_n(\mathbb{C})$ such that

$$U = VBV^*, \quad \text{where } Y = \begin{bmatrix} 0 & 1 \\ -1 & 0 \end{bmatrix} \text{ and }$$

$$B = \begin{bmatrix} Y & O & \cdots & O \\ O & Y & \cdots & O \\ \vdots & \vdots & \ddots & \vdots \\ O & O & \cdots & Y \end{bmatrix}.$$

[Hint: Find a complex spectral decomposition of Y.]

** **2.1.23** In this exercise, we finally show that if $\mathcal{V}$ is a finite-dimensional inner product space over $\mathbb{R}$ and $P : \mathcal{V} \to \mathcal{V}$ is a projection then P is orthogonal (i.e., $\langle P(\mathbf{v}), \mathbf{v} - P(\mathbf{v}) \rangle = 0$ for all $\mathbf{v} \in \mathcal{V}$) if and only if it is self-adjoint (i.e., $P = P^*$). Recall that we proved this when the ground field is $\mathbb{C}$ in Exercise 1.4.29.

  (a) Show that if P is self-adjoint then it is orthogonal.

(b) Use Exercise 1.4.28 to show that if P is orthogonal then the linear transformation $T = P - P^* \circ P$ satisfies $T^* = -T$.

(c) Use part (b) to show that if P is orthogonal then it is self-adjoint. [Hint: Represent T in an orthonormal basis, take its trace, and use Exercise 2.1.12.]

2.1.24 Show that if $P \in \mathcal{M}_n(\mathbb{C})$ is an orthogonal projection (i.e., $P = P^* = P^2$) then there exists a unitary matrix $U \in \mathcal{M}_n(\mathbb{C})$ such that

$$P = U \begin{bmatrix} I_r & O \\ O & O \end{bmatrix} U^*,$$

where $r = \operatorname{rank}(P)$.

2.1.25 A **circulant matrix** is a matrix $C \in \mathcal{M}_n(\mathbb{C})$ of the form

$$C = \begin{bmatrix} c_0 & c_1 & c_2 & \cdots & c_{n-1} \\ c_{n-1} & c_0 & c_1 & \cdots & c_{n-2} \\ c_{n-2} & c_{n-1} & c_0 & \cdots & c_{n-3} \\ \vdots & \vdots & \vdots & \ddots & \vdots \\ c_1 & c_2 & c_3 & \cdots & c_0 \end{bmatrix},$$

where $c_0, c_1, \ldots, c_{n-1}$ are scalars. Show that C can be diagonalized by the Fourier matrix F from Exercise 1.4.15. That is, show that F^*CF is diagonal.

[Hint: It suffices to show that the columns of F are eigenvectors of C.]

∗**2.1.26** Suppose $A, B \in \mathcal{M}_n(\mathbb{C})$ commute (i.e., $AB = BA$). Show that there is a vector $\mathbf{v} \in \mathbb{C}^n$ that is an eigenvector of each of A and B.

[Hint: This is *hard*. Maybe just prove it in the case when A has distinct eigenvalues, which is much easier. The general case can be proved using techniques like those used in the proof of Schur triangularization.]

2.1.27 Suppose $A, B \in \mathcal{M}_n(\mathbb{C})$ commute (i.e., $AB = BA$). Show that there is a *common* unitary matrix $U \in \mathcal{M}_n(\mathbb{C})$ that triangularizes them both:

$$A = UT_1U^* \quad \text{and} \quad B = UT_2U^*$$

for some upper triangular $T_1, T_2 \in \mathcal{M}_n(\mathbb{C})$.

[Hint: Use Exercise 2.1.26 and mimic the proof of Schur triangularization.]

2.1.28 Suppose $A, B \in \mathcal{M}_n(\mathbb{C})$ are normal. Show that A and B commute (i.e., $AB = BA$) if and only if there is a *common* unitary matrix $U \in \mathcal{M}_n(\mathbb{C})$ that diagonalizes them both:

$$A = UD_1U^* \quad \text{and} \quad B = UD_2U^*$$

for some diagonal $D_1, D_2 \in \mathcal{M}_n(\mathbb{C})$.

[Hint: Leech off of Exercise 2.1.27.]

[Side note: This result is still true if we replace "normal" by "real symmetric" and $\mathbb{C}$ by $\mathbb{R}$ throughout the exercise.]

2.1.29 Suppose $A, B \in \mathcal{M}_n(\mathbb{C})$ are diagonalizable. Show that A and B commute (i.e., $AB = BA$) if and only if there is a *common* invertible matrix $P \in \mathcal{M}_n(\mathbb{C})$ that diagonalizes them both:

$$A = PD_1P^{-1} \quad \text{and} \quad B = PD_2P^{-1}$$

for some diagonal $D_1, D_2 \in \mathcal{M}_n(\mathbb{C})$.

[Hint: When does a diagonal matrix D_1 commute with another matrix? Try proving the claim when the eigenvalues of A are distinct first, since that case is much easier. For the general case, induction might help you sidestep some of the ugly details.]

2.2 Positive Semidefiniteness

We have now seen that normal matrices play a particularly important role in linear algebra, especially when decomposing matrices. There is one particularly important sub-family of normal matrices that we now focus our attention on that play perhaps an even more important role.

Definition 2.2.1
Positive (Semi)Definite Matrices

Suppose $A \in \mathcal{M}_n(\mathbb{F})$ is self-adjoint. Then A is called

a) **positive semidefinite (PSD)** if $\mathbf{v}^*A\mathbf{v} \geq 0$ for all $\mathbf{v} \in \mathbb{F}^n$, and

b) **positive definite (PD)** if $\mathbf{v}^*A\mathbf{v} > 0$ for all $\mathbf{v} \neq \mathbf{0}$.

Since A is self-adjoint, $\mathbb{F} = \mathbb{R}$ or $\mathbb{F} = \mathbb{C}$ throughout this section. Adjoints only make sense in inner product spaces.

Positive (semi)definiteness is somewhat difficult to eyeball from the entries of a matrix, and we should emphasize that it does *not* mean that the entries of the matrix need to be positive. For example, if

$$A = \begin{bmatrix} 1 & -1 \\ -1 & 1 \end{bmatrix} \quad \text{and} \quad B = \begin{bmatrix} 1 & 2 \\ 2 & 1 \end{bmatrix}, \tag{2.2.1}$$

then A is positive semidefinite since

A 1×1 matrix (i.e., a scalar) is PSD if and only if it is non-negative. We often think of PSD matrices as the "matrix version" of the non-negative real numbers.

$$\mathbf{v}^*A\mathbf{v} = \begin{bmatrix} \overline{v_1} & \overline{v_2} \end{bmatrix} \begin{bmatrix} 1 & -1 \\ -1 & 1 \end{bmatrix} \begin{bmatrix} v_1 \\ v_2 \end{bmatrix}$$

$$= |v_1|^2 - \overline{v_1}v_2 - \overline{v_2}v_1 + |v_2|^2 = |v_1 - v_2|^2 \geq 0$$

for all $\mathbf{v} \in \mathbb{C}^2$. On the other hand, B is *not* positive semidefinite since if $\mathbf{v} = (1, -1)$ then $\mathbf{v}^*B\mathbf{v} = -2$.

While the defining property of positive definiteness seems quite strange at first, it actually has a very natural geometric interpretation. If we recall that $\mathbf{v}^*A\mathbf{v} = \mathbf{v} \cdot (A\mathbf{v})$ and that the angle θ between $\mathbf{v}$ and $A\mathbf{v}$ can be computed in terms of the dot product via

$$\theta = \arccos\left(\frac{\mathbf{v} \cdot (A\mathbf{v})}{\|\mathbf{v}\|\|A\mathbf{v}\|} \right),$$

For positive *semidefinite* matrices, the angle between $\mathbf{v}$ and $A\mathbf{v}$ is always acute or perpendicular (i.e., $0 \leq \theta \leq \pi/2$).

then we see that the positive definiteness property that $\mathbf{v}^*A\mathbf{v} > 0$ for all $\mathbf{v}$ simply says that the angle between $\mathbf{v}$ and $A\mathbf{v}$ is always acute (i.e., $0 \leq \theta < \pi/2$). We can thus think of positive (semi)definite matrices as the Hermitian matrices that do not rotate vectors "too much". In particular, $A\mathbf{v}$ is always in the same half-space as $\mathbf{v}$, as depicted in Figure 2.4.

For another geometric interpretation of positive (semi)definiteness, see Section 2.A.1.

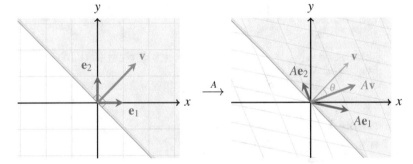

Figure 2.4: As a linear transformation, a positive definite matrix is one that keeps vectors pointing "mostly" in the same direction. In particular, the angle θ between $\mathbf{v}$ and $A\mathbf{v}$ never exceeds $\pi/2$, so $A\mathbf{v}$ is always in the same half-space as $\mathbf{v}$. Positive *semi*definiteness allows for $\mathbf{v}$ and $A\mathbf{v}$ to be perpendicular (i.e., orthogonal), so $A\mathbf{v}$ can be on the boundary of the half-space defined by $\mathbf{v}$.

2.2.1 Characterizing Positive (Semi)Definite Matrices

The definition of positive semidefinite matrices that was provided in Definition 2.2.1 perhaps looks a bit odd at first glance, and showing that a matrix is (or is not) positive semidefinite seems quite cumbersome so far. The following theorem characterizes these matrices in several other equivalent ways, some of which are a bit more illuminating and easier to work with. If we prefer, we can think of any one of these other characterizations of PSD matrices as their definition.

Theorem 2.2.1

Characterization of Positive Semidefinite Matrices

Recall that, since A is Hermitian, its eigenvalues are all real.

Suppose $A \in \mathcal{M}_n(\mathbb{F})$ is self-adjoint. The following are equivalent:

a) A is positive semidefinite,

b) All of the eigenvalues of A are non-negative,

c) There is a matrix $B \in \mathcal{M}_n(\mathbb{F})$ such that $A = B^*B$, and

d) There is a diagonal matrix $D \in \mathcal{M}_n(\mathbb{R})$ with non-negative diagonal entries and a unitary matrix $U \in \mathcal{M}_n(\mathbb{F})$ such that $A = UDU^*$.

Proof. We prove the theorem by showing that (a) $\implies$ (b) $\implies$ (d) $\implies$ (c) $\implies$ (a).

To see that (a) $\implies$ (b), let $\mathbf{v}$ be an eigenvector of A with corresponding eigenvalue λ. Then $A\mathbf{v} = \lambda\mathbf{v}$, and multiplying this equation on the left by $\mathbf{v}^*$ shows that $\mathbf{v}^*A\mathbf{v} = \lambda\mathbf{v}^*\mathbf{v} = \lambda\|\mathbf{v}\|^2$. Since A is positive semidefinite, we know that $\mathbf{v}^*A\mathbf{v} \geq 0$, so it follows that $\lambda \geq 0$ too.

To see that (b) $\implies$ (d), we just apply the spectral decomposition theorem (either the complex Theorem 2.1.4 or the real Theorem 2.1.6, as appropriate) to A.

To see why (d) $\implies$ (c), let $\sqrt{D}$ be the diagonal matrix that is obtained by taking the (non-negative) square root of the diagonal entries of D, and define $B = \sqrt{D}U^*$. Then $B^*B = (\sqrt{D}U^*)^*(\sqrt{D}U^*) = U\sqrt{D}^*\sqrt{D}U^* = UDU^* = A$.

Another characterization of positive semidefiniteness is provided in Exercise 2.2.25.

Finally, to see that (c) $\implies$ (a), we let $\mathbf{v} \in \mathbb{F}^n$ be any vector and we note that

$$\mathbf{v}^*A\mathbf{v} = \mathbf{v}^*B^*B\mathbf{v} = (B\mathbf{v})^*(B\mathbf{v}) = \|B\mathbf{v}\|^2 \geq 0.$$

It follows that A is positive semidefinite, so the proof is complete. ∎

Example 2.2.1

Demonstrating Positive Semidefiniteness

Explicitly show that all four properties of Theorem 2.2.1 hold for the matrix

$$A = \begin{bmatrix} 1 & -1 \\ -1 & 1 \end{bmatrix}.$$

Solution:

We already showed that $\mathbf{v}^*A\mathbf{v} = |v_1 - v_2|^2 \geq 0$ for all $\mathbf{v} \in \mathbb{C}^2$ at the start of this section, which shows that A is PSD. We now verify that properties (b)–(d) of Theorem 2.2.1 are all satisfied as well.

For property (b), we can explicitly compute the eigenvalues of A:

In practice, checking non-negativity of its eigenvalues is the simplest of these methods of checking positive semidefiniteness.

$$\det\left(\begin{bmatrix} 1-\lambda & -1 \\ -1 & 1-\lambda \end{bmatrix}\right) = (1-\lambda)^2 - 1 = \lambda^2 - 2\lambda = \lambda(\lambda - 2) = 0,$$

so the eigenvalues of A are 0 and 2, which are indeed non-negative.

For property (d), we want to find a unitary matrix U such that $A = UDU^*$, where D has 2 and 0 (the eigenvalues of A) along its diagonal. We know from the spectral decomposition that we can construct U by placing the normalized eigenvectors of A into U as columns. Eigenvectors corresponding to the eigenvalues 2 and 0 are $\mathbf{v} = (1, -1)$ and $\mathbf{v} = (1, 1)$,

We check property (d) before (c) so that we can mimic the proof of Theorem 2.2.1.

respectively, so

$$U = \frac{1}{\sqrt{2}} \begin{bmatrix} 1 & 1 \\ -1 & 1 \end{bmatrix} \quad \text{and} \quad D = \begin{bmatrix} 2 & 0 \\ 0 & 0 \end{bmatrix}.$$

It is then straightforward to verify that $A = UDU^*$, as desired.

For property (c), we let

$$B = \sqrt{D}U^* = \frac{1}{\sqrt{2}} \begin{bmatrix} \sqrt{2} & 0 \\ 0 & 0 \end{bmatrix} \begin{bmatrix} 1 & -1 \\ 1 & 1 \end{bmatrix} = \begin{bmatrix} 1 & -1 \\ 0 & 0 \end{bmatrix}.$$

Direct computation verifies that it is indeed the case that $A = B^*B$.

The characterization of positive semidefinite matrices provided by Theorem 2.2.1 can be tweaked slightly into a characterization of positive *definite* matrices by just making the relevant matrices invertible in each statement. In particular, we get the following theorem, which just has one or two words changed in each statement (we have italicized the words that changed from the previous theorem to make them easier to compare).

Theorem 2.2.2

Characterization of Positive Definite Matrices

Suppose $A \in \mathcal{M}_n(\mathbb{F})$ is self-adjoint. The following are equivalent:

a) A is positive *definite*,

b) All of the eigenvalues of A are *strictly positive*,

c) There is an *invertible* matrix $B \in \mathcal{M}_n(\mathbb{F})$ such that $A = B^*B$, and

d) There is a diagonal matrix $D \in \mathcal{M}_n(\mathbb{R})$ with *strictly positive* diagonal entries and a unitary matrix $U \in \mathcal{M}_n(\mathbb{F})$ such that $A = UDU^*$.

The proof of the above theorem is almost identical to that of Theorem 2.2.1, so we leave it to Exercise 2.2.23. Instead, we note that there are two additional characterizations of positive definite matrices that we would like to present, but we first need some "helper" theorems that make it a bit easier to work with positive (semi)definite matrices. The first of these results tells us how we can manipulate positive semidefinite matrices without destroying positive semidefiniteness.

Theorem 2.2.3

Modifying Positive (Semi)Definite Matrices

Suppose $A, B \in \mathcal{M}_n$ are positive (semi)definite, $P \in \mathcal{M}_{n,m}$ is any matrix, and $c > 0$ is a real scalar. Then

a) $A + B$ is positive (semi)definite,

b) cA is positive (semi)definite,

c) A^T is positive (semi)definite, and

d) P^*AP is positive semidefinite. Furthermore, if A is positive definite then P^*AP is positive definite if and only if $\text{rank}(P) = m$.

We return to this problem of asking what operations transform PSD matrices into PSD matrices in Section 3.A.

Proof. These properties all follow fairly quickly from the definition of positive semidefiniteness, so we leave the proof of (a), (b), and (c) to Exercise 2.2.24.

To show that property (d) holds, observe that for all $\mathbf{v} \in \mathbb{F}^n$ we have

$$0 \leq (P\mathbf{v})^*A(P\mathbf{v}) = \mathbf{v}^*(P^*AP)\mathbf{v},$$

so P^*AP is positive semidefinite as well. If A is positive definite then positive definiteness of P^*AP is equivalent to the requirement that $P\mathbf{v} \neq \mathbf{0}$ whenever

$\mathbf{v} \neq \mathbf{0}$, which is equivalent in turn to $\text{null}(P) = \{\mathbf{0}\}$ (i.e., $\text{nullity}(P) = 0$). By the rank-nullity theorem (see Theorem A.1.2), this is equivalent to $\text{rank}(P) = m$, which completes the proof. ∎

As a bit of motivation for our next "helper" result, notice that the diagonal entries of a positive semidefinite matrix A must be non-negative (or strictly positive, if A is positive *definite*), since

> Recall that $\mathbf{e}_j$ is the j-th standard basis vector.

$$0 \le \mathbf{e}_j^* A \mathbf{e}_j = a_{j,j} \quad \text{for all} \quad 1 \le j \le n.$$

The following theorem provides a natural generalization of this fact to block matrices.

Theorem 2.2.4

Diagonal Blocks of PSD Block Matrices

Suppose that the self-adjoint block matrix

$$A = \begin{bmatrix} A_{1,1} & A_{1,2} & \cdots & A_{1,n} \\ A_{1,2}^* & A_{2,2} & \cdots & A_{2,n} \\ \vdots & \vdots & \ddots & \vdots \\ A_{1,n}^* & A_{2,n}^* & \cdots & A_{n,n} \end{bmatrix}$$

> The diagonal blocks here must be square for this block matrix to make sense (e.g., $A_{1,1}$ is square since it has $A_{1,2}$ to its right and $A_{1,2}^*$ below it).

is positive (semi)definite. Then the diagonal blocks $A_{1,1}, A_{2,2}, \ldots, A_{n,n}$ must be positive (semi)definite.

Proof. We use property (d) of Theorem 2.2.3. In particular, consider the block matrices

$$P_1 = \begin{bmatrix} I \\ O \\ \vdots \\ O \end{bmatrix}, \quad P_2 = \begin{bmatrix} O \\ I \\ \vdots \\ O \end{bmatrix}, \quad \ldots, \quad P_n = \begin{bmatrix} O \\ O \\ \vdots \\ I \end{bmatrix},$$

where the sizes of the O and I blocks are such that the matrix multiplication AP_j makes sense for all $1 \le j \le n$. It is then straightforward to verify that $P_j^* A P_j = A_{j,j}$ for all $1 \le j \le n$, so $A_{j,j}$ must be positive semidefinite by Theorem 2.2.3(d). Furthermore, each P_j has rank equal to its number of columns, so $A_{j,j}$ is positive *definite* whenever A is. ∎

Example 2.2.2

Showing Matrices are Not Positive Semidefinite

Show that the following matrices are not positive semidefinite.

a) $\begin{bmatrix} 3 & 2 & 1 \\ 2 & 4 & 2 \\ 1 & 2 & -1 \end{bmatrix}$

b) $\begin{bmatrix} 3 & -1 & 1 & -1 \\ -1 & 1 & 2 & 1 \\ 1 & 2 & 1 & 2 \\ -1 & 1 & 2 & 3 \end{bmatrix}$

> Techniques like these ones for showing that a matrix is (not) PSD are useful since we often do not want to compute eigenvalues of 4×4 (or larger) matrices.

Solutions:

a) This matrix is not positive semidefinite since its third diagonal entry is -1, and positive semidefinite matrices cannot have negative diagonal entries.

b) This matrix is not positive semidefinite since its central 2×2 diagonal block is

$$\begin{bmatrix} 1 & 2 \\ 2 & 1 \end{bmatrix},$$

which is exactly the matrix B from Equation (2.2.1) that we showed is not positive semidefinite earlier.

We are now in a position to introduce the two additional characterizations of positive definite matrices that we are actually interested in, and that we introduced the above "helper" theorems for. It is worth noting that both of these upcoming characterizations really are specific to positive *definite* matrices. While they *can* be extended to positive *semi*definite matrices, they are significantly easier to use and interpret in the definite (i.e., invertible) case.

The first of these results says that positive definite matrices exactly characterize inner products on $\mathbb{F}^n$:

Theorem 2.2.5

Positive Definite Matrices Make Inner Products

A function $\langle \cdot, \cdot \rangle : \mathbb{F}^n \times \mathbb{F}^n \to \mathbb{F}$ is an inner product if and only if there exists a positive definite matrix $A \in \mathcal{M}_n(\mathbb{F})$ such that

$$\langle \mathbf{v}, \mathbf{w} \rangle = \mathbf{v}^* A \mathbf{w} \quad \text{for all} \quad \mathbf{v}, \mathbf{w} \in \mathbb{F}^n.$$

> If you need a refresher on inner products, refer back to Section 1.4.

Proof. We start by showing that if $\langle \cdot, \cdot \rangle$ is an inner product then such a positive definite matrix A must exist. To this end, recall from Theorem 1.4.3 that there exists a basis B of $\mathbb{F}^n$ such that

$$\langle \mathbf{v}, \mathbf{w} \rangle = [\mathbf{v}]_B \cdot [\mathbf{w}]_B \quad \text{for all} \quad \mathbf{v}, \mathbf{w} \in \mathbb{F}^n.$$

Well, let $P_{B \leftarrow E}$ be the change-of-basis matrix from the standard basis E to B. Then $[\mathbf{v}]_B = P_{B \leftarrow E} \mathbf{v}$ and $[\mathbf{w}]_B = P_{B \leftarrow E} \mathbf{w}$, so

$$\langle \mathbf{v}, \mathbf{w} \rangle = [\mathbf{v}]_B \cdot [\mathbf{w}]_B = (P_{B \leftarrow E} \mathbf{v}) \cdot (P_{B \leftarrow E} \mathbf{w}) = \mathbf{v}^* (P_{B \leftarrow E}^* P_{B \leftarrow E}) \mathbf{w}$$

for all $\mathbf{v}, \mathbf{w} \in \mathbb{F}^n$. Since change-of-basis matrices are invertible, it follows from Theorem 2.2.2(c) that $A = P_{B \leftarrow E}^* P_{B \leftarrow E}$ is positive definite, which is what we wanted.

In the other direction, we must show that every function $\langle \cdot, \cdot \rangle$ of the form $\langle \mathbf{v}, \mathbf{w} \rangle = \mathbf{v}^* A \mathbf{w}$ is necessarily an inner product when A is positive definite. We thus have to show that all three defining properties of inner products from Definition 1.3.6 hold.

For property (a), we note that $A = A^*$, so

> Here we used the fact that if $c \in \mathbb{C}$ is a scalar, then $\bar{c} = c^*$. If $\mathbb{F} = \mathbb{R}$ then these complex conjugations just vanish.

$$\overline{\langle \mathbf{w}, \mathbf{v} \rangle} = \overline{\mathbf{w}^* A \mathbf{v}} = (\mathbf{w}^* A \mathbf{v})^* = \mathbf{v}^* A^* \mathbf{w} = \mathbf{v}^* A \mathbf{w} = \langle \mathbf{v}, \mathbf{w} \rangle \quad \text{for all} \quad \mathbf{v}, \mathbf{w} \in \mathbb{F}^n.$$

For property (b), we check that

$$\langle \mathbf{v}, \mathbf{w} + c\mathbf{x} \rangle = \mathbf{v}^* A (\mathbf{w} + c\mathbf{x}) = \mathbf{v}^* A \mathbf{w} + c(\mathbf{v}^* A \mathbf{x}) = \langle \mathbf{v}, \mathbf{w} \rangle + c \langle \mathbf{v}, \mathbf{x} \rangle$$

for all $\mathbf{v}, \mathbf{w}, \mathbf{x} \in \mathbb{F}^n$ and all $c \in \mathbb{F}$.

> Notice that we called inner products "positive definite" way back when we first introduced them in Definition 1.3.6.

Finally, for property (c) we note that A is positive definite, so

$$\langle \mathbf{v}, \mathbf{v} \rangle = \mathbf{v}^* A \mathbf{v} \geq 0 \quad \text{for all} \quad \mathbf{v} \in \mathbb{F}^n,$$

with equality if and only if $\mathbf{v} = \mathbf{0}$. It follows that $\langle \cdot, \cdot \rangle$ is indeed an inner product, which completes the proof. ∎

Example 2.2.3

A Weird Inner Product (Again)

Show that the function $\langle \cdot, \cdot \rangle : \mathbb{R}^2 \times \mathbb{R}^2 \to \mathbb{R}$ defined by

$$\langle \mathbf{v}, \mathbf{w} \rangle = v_1 w_1 + 2 v_1 w_2 + 2 v_2 w_1 + 5 v_2 w_2 \quad \text{for all} \quad \mathbf{v}, \mathbf{w} \in \mathbb{R}^2$$

is an inner product on $\mathbb{R}^2$.

Solution:

We already showed this function is an inner product back in Example 1.3.18 in a rather brute-force manner. Now that we understand inner products better, we can be much more slick—we just notice that we can rewrite this function in the form

> To construct this matrix A, just notice that multiplying out $\mathbf{v}^T A \mathbf{w}$ gives $\sum_{i,j} a_{i,j} v_i w_j$, so we just let $a_{i,j}$ be the coefficient of $v_i w_j$.

$$\langle \mathbf{v}, \mathbf{w} \rangle = \mathbf{v}^T A \mathbf{w}, \quad \text{where} \quad A = \begin{bmatrix} 1 & 2 \\ 2 & 5 \end{bmatrix}.$$

It is straightforward to check that A is positive definite (its eigenvalues are $3 \pm 2\sqrt{2} > 0$, for example), so Theorem 2.2.5 tells us that this function is an inner product.

There is one final characterization of positive definite matrices that we now present. This theorem is useful because it gives us another way of checking positive definiteness that is sometimes easier than computing eigenvalues.

Theorem 2.2.6

Sylvester's Criterion

Suppose $A \in \mathcal{M}_n$ is self-adjoint. Then A is positive definite if and only if, for all $1 \leq k \leq n$, the determinant of the top-left $k \times k$ block of A is strictly positive.

Before proving this theorem, it is perhaps useful to present an example that demonstrates exactly what it says and how to use it.

Example 2.2.4

Applying Sylvester's Criterion

Use Sylvester's criterion to show that the following matrix is positive definite:

$$A = \begin{bmatrix} 2 & -1 & i \\ -1 & 2 & 1 \\ -i & 1 & 2 \end{bmatrix}.$$

Solution:

We have to check that the top-left 1×1, 2×2, and 3×3 blocks of A all have positive determinants:

> For 2×2 matrices, recall that
> $$\det\left(\begin{bmatrix} a & b \\ c & d \end{bmatrix} \right)$$
> $$= ad - bc.$$
> For larger matrices, we can use Theorem A.1.4 (or many other methods) to compute determinants.

$$\det([2]) = 2 > 0$$

$$\det\left(\begin{bmatrix} 2 & -1 \\ -1 & 2 \end{bmatrix} \right) = 4 - 1 = 3 > 0, \quad \text{and}$$

$$\det\left(\begin{bmatrix} 2 & -1 & i \\ -1 & 2 & 1 \\ -i & 1 & 2 \end{bmatrix} \right) = 8 + i - i - 2 - 2 - 2 = 2 > 0.$$

It follows from Sylvester's criterion that A is positive definite.

Proof of Sylvester's criterion. For the "only if" direction of the proof, recall from Theorem 2.2.4 that if A is positive definite then so is the top-left $k \times k$ block of A, which we will call A_k for the remainder of this proof. Since A_k is positive definite, its eigenvalues are positive, so $\det(A_k)$ (i.e., the product of those eigenvalues) is positive too, as desired.

The "if" direction is somewhat trickier to pin down, and we prove it by induction on n (the size of A). For the base case, if $n = 1$ then it is clear that $\det(A) > 0$ implies that A is positive definite since the determinant of a scalar just equals that scalar itself.

For the inductive step, assume that the theorem holds for $(n-1) \times (n-1)$ matrices. To see that it must then hold for $n \times n$ matrices, notice that if $A \in \mathcal{M}_n(\mathbb{F})$ is as in the statement of the theorem and $\det(A_k) > 0$ for all $1 \le k \le n$, then $\det(A) > 0$ and (by the inductive hypothesis) A_{n-1} is positive definite. Let λ_i and λ_j be any two eigenvalues of A with corresponding orthogonal eigenvectors $\mathbf{v}$ and $\mathbf{x}$, respectively. Then define $\mathbf{x} = w_n\mathbf{v} - v_n\mathbf{w}$ and notice that $\mathbf{x} \ne \mathbf{0}$ (since $\{\mathbf{v}, \mathbf{w}\}$ is linearly independent), but $x_n = w_n v_n - v_n w_n = 0$. Since $x_n = 0$ and A_{n-1} is positive definite, it follows that

> We can choose **v** and **x** to be orthogonal by the spectral decomposition.

$$0 < \mathbf{x}^* A \mathbf{x}$$
$$= (w_n\mathbf{v} - v_n\mathbf{w})^* A (w_n\mathbf{v} - v_n\mathbf{w})$$
$$= |w_n|^2 \mathbf{v}^* A\mathbf{v} - w_n\overline{v_n}\mathbf{w}^* A\mathbf{v} - v_n\overline{w_n}\mathbf{v}^* A\mathbf{w} + |v_n|^2 \mathbf{w}^* A\mathbf{w}$$
$$= \lambda_i|w_n|^2 \mathbf{v}^* \mathbf{v} - \lambda_i w_n\overline{v_n}\mathbf{w}^* \mathbf{v} - \lambda_j v_n\overline{w_n}\mathbf{v}^* \mathbf{w} + \lambda_j|v_n|^2 \mathbf{w}^* \mathbf{w}$$
$$= \lambda_i|w_n|^2 \|\mathbf{v}\|^2 - 0 - 0 + \lambda_j|v_n|^2 \|\mathbf{w}\|^2.$$

> We have to be careful if $v_n = w_n = 0$, since then $\mathbf{x} = \mathbf{0}$. In this case, we instead define $\mathbf{x} = \mathbf{v}$ to fix up the proof.

> Recall that **v** and **w** are orthogonal, so $\mathbf{v}^*\mathbf{w} = \mathbf{v} \cdot \mathbf{w} = 0$.

It is thus not possible that both $\lambda_i \le 0$ and $\lambda_j \le 0$. Since λ_i and λ_j were *arbitrary* eigenvalues of A, it follows that A must have at most one non-positive eigenvalue. However, if it had *exactly* one non-positive eigenvalue then it would be the case that $\det(A) = \lambda_1\lambda_2\cdots\lambda_n \le 0$, which we know is not the case. It follows that all of A's eigenvalues are strictly positive, so A is positive definite by Theorem 2.2.2, which completes the proof. ∎

It might be tempting to think that Theorem 2.2.6 can be extended to positive *semi*definite matrices by just requiring that the determinant of each square top-left block of A is non-negative, rather than strictly positive. However, this does not work, as demonstrated by the matrix

$$A = \begin{bmatrix} 0 & 0 \\ 0 & -1 \end{bmatrix}.$$

For this matrix, the top-left block clearly has determinant 0, and straightforward computation shows that $\det(A) = 0$ as well. However, A is not positive semidefinite, since it has -1 as an eigenvalue.

The following theorem shows that there is indeed some hope though, and Sylvester's criterion does apply to 2×2 matrices as long as we add in the requirement that the bottom-right entry is non-negative as well.

Theorem 2.2.7

Positive (Semi)Definiteness for 2 × 2 Matrices

Suppose $A \in \mathcal{M}_2$ is self-adjoint.

a) A is positive semidefinite if and only if $a_{1,1}, a_{2,2}, \det(A) \ge 0$, and

b) A is positive definite if and only if $a_{1,1} > 0$ and $\det(A) > 0$.

Proof. Claim (b) is exactly Sylvester's criterion (Theorem 2.2.6) for 2×2 matrices, so we only need to prove (a).

For the "only if" direction, the fact that $a_{1,1}, a_{2,2} \ge 0$ whenever A is positive semidefinite follows simply from the fact that PSD matrices have non-negative diagonal entries (Theorem 2.2.4). The fact that $\det(A) \geq 0$ follows from the fact that A has non-negative eigenvalues, and $\det(A)$ is the product of them.

For the "if" direction, recall that the characteristic polynomial of A is

$$p_A(\lambda) = \det(A - \lambda I) = \lambda^2 - \text{tr}(A)\lambda + \det(A).$$

It follows from the quadratic formula that the eigenvalues of A are

$$\lambda = \frac{1}{2}\left(\text{tr}(A) \pm \sqrt{\text{tr}(A)^2 - 4\det(A)}\right).$$

This proof shows that we can replace the inequalities $a_{1,1}, a_{2,2} \geq 0$ in the statement of this theorem with the single inequality $\text{tr}(A) \geq 0$.

To see that these eigenvalues are non-negative (and thus A is positive semidefinite), we just observe that if $\det(A) \geq 0$ then $\text{tr}(A)^2 - 4\det(A) < \text{tr}(A)^2$, so $\text{tr}(A) - \sqrt{\text{tr}(A)^2 - 4\det(A)} \geq 0$. ∎

For example, if we apply this theorem to the matrices

$$A = \begin{bmatrix} 1 & -1 \\ -1 & 1 \end{bmatrix} \quad \text{and} \quad B = \begin{bmatrix} 1 & 2 \\ 2 & 1 \end{bmatrix}$$

from Equation (2.2.1), we see immediately that A is positive semidefinite (but not positive definite) because $\det(A) = 1 - 1 = 0 \geq 0$, but B is not positive semidefinite because $\det(B) = 1 - 4 = -3 < 0$.

Remark 2.2.1

Sylvester's Criterion for Positive *Semi*Definite Matrices

To extend Sylvester's criterion (Theorem 2.2.6) to positive *semi*definite matrices, we first need a bit of extra terminology. A **principal minor** of a square matrix $A \in \mathcal{M}_n$ is the determinant of a submatrix of A that is obtained by deleting some (or none) of its rows as well as the corresponding columns. For example, the principal minors of a 2×2 Hermitian matrix

$$A = \begin{bmatrix} a & b \\ \overline{b} & d \end{bmatrix}$$

are

Notice that these three principal minors are exactly the quantities that determined positive semidefiniteness in Theorem 2.2.7(a).

$$\det\left(\begin{bmatrix} a \end{bmatrix}\right) = a, \quad \det\left(\begin{bmatrix} d \end{bmatrix}\right) = d, \quad \text{and} \quad \det\left(\begin{bmatrix} a & b \\ \overline{b} & d \end{bmatrix}\right) = ad - |b|^2,$$

and the principal minors of a 3×3 Hermitian matrix

$$B = \begin{bmatrix} a & b & c \\ \overline{b} & d & e \\ \overline{c} & \overline{e} & f \end{bmatrix}$$

are $a, d, f, \det(B)$ itself, as well as

For example,

$$\begin{bmatrix} a & c \\ \overline{c} & f \end{bmatrix}$$

is obtained by deleting the second row and column of B.

$$\det\left(\begin{bmatrix} a & b \\ \overline{b} & d \end{bmatrix}\right) = ad - |b|^2, \quad \det\left(\begin{bmatrix} a & c \\ \overline{c} & f \end{bmatrix}\right) = af - |c|^2, \quad \text{and}$$

$$\det\left(\begin{bmatrix} d & e \\ \overline{e} & f \end{bmatrix}\right) = df - |e|^2.$$

The correct generalization of Sylvester's criterion to positive semidefinite matrices is that a matrix is positive semidefinite if and only if *all* of its principal minors are non-negative (whereas in the positive definite version of Sylvester's criterion we only needed to check the principal minors coming from the top-left corners of the matrix).

2.2.2 Diagonal Dominance and Gershgorin Discs

Intuitively, Theorem 2.2.7 tells us that a 2×2 matrix is positive (semi)definite exactly when its diagonal entries are sufficiently large compared to its off-diagonal entries. After all, if we expand out the inequality $\det(A) \geq 0$ explicitly in terms of the entries of A, we see that it is equivalent to

$$a_{1,1}a_{2,2} \geq a_{1,2}a_{2,1}.$$

This same intuition is well-founded even for larger matrices. However, to clarify exactly what we mean, we first need the following result that helps us bound the eigenvalues of a matrix based on simple information about its entries.

Theorem 2.2.8 **Gershgorin Disc Theorem** A closed disc is just a filled circle.	Suppose $A \in \mathcal{M}_n$ and define the following objects for each $1 \leq i \leq n$: • $r_i = \sum_{j \neq i} \lvert a_{i,j} \rvert$ (the sum of off-diagonal entries of the i-th row of A), • $D(a_{i,i}, r_i)$ is the closed disc in the complex plane centered at $a_{i,i}$ with radius r_i (we call this disc the i-th **Gershgorin disc** of A). Then every eigenvalue of A is located in at least one of the $D(a_{i,i}, r_i)$.

This theorem can be thought of as an approximation theorem. For a diagonal matrix we have $r_i = 0$ for all $1 \leq i \leq n$, so its Gershgorin discs have radius 0 and thus its eigenvalues are exactly its diagonal entries, which we knew already (in fact, the same is true of triangular matrices). However, if we were to change the off-diagonal entries of that matrix then the radii of its Gershgorin discs would increase, allowing its eigenvalues to wiggle around a bit.

Before proving this result, we illustrate it with an example.

Example 2.2.5 **Gershgorin Discs** The radius of the second disc is 1 because $\lvert -i \rvert = 1$.	Draw the Gershgorin discs for the following matrix, and show that its eigenvalues are contained in these discs: $$A = \begin{bmatrix} -1 & 2 \\ -i & 1+i \end{bmatrix}.$$ **Solution:** The Gershgorin discs are $D(-1, 2)$ and $D(1+i, 1)$. Direct calculation shows that the eigenvalues of A are 1 and $-1+i$, which are indeed contained within these discs:

In this subsection, we focus on the $\mathbb{F} = \mathbb{C}$ case since Gershgorin discs live naturally in the complex plane. These same results apply if $\mathbb{F} = \mathbb{R}$ simply because real matrices are complex.

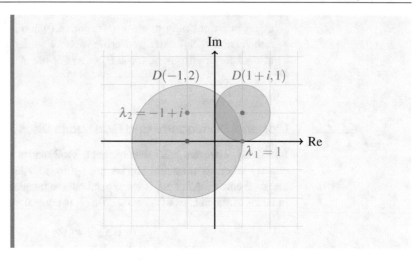

Proof of Theorem 2.2.8. Let λ be an eigenvalue of A corresponding to an eigenvector $\mathbf{v}$. Since $\mathbf{v} \neq \mathbf{0}$, we can scale it so that its largest entry is $v_i = 1$ and all other entries are no larger than 1 (i.e., $|v_j| \leq 1$ for all $j \neq i$). By looking at the i-th entry of the vector $A\mathbf{v} = \lambda\mathbf{v}$, we see that

In the notation of Section 1.D, we scale $\mathbf{v}$ so that $\|\mathbf{v}\|_\infty = 1$.

$$\sum_j a_{i,j} v_j = \lambda v_i = \lambda. \qquad \text{(since } v_i = 1\text{)}$$

Now notice that the sum on the left can be split up into the form

In fact, this proof shows that each eigenvalue lies in the Gershgorin disc corresponding to the largest entry of its corresponding eigenvectors.

$$\sum_j a_{i,j} v_j = \sum_{j \neq i} a_{i,j} v_j + a_{i,i}. \qquad \text{(again, since } v_i = 1\text{)}$$

By combining and slightly rearranging the two equations above, we get

$$\lambda - a_{i,i} = \sum_{j \neq i} a_{i,j} v_j.$$

Finally, taking the absolute value of both sides of this equation then shows that

We used the fact that $|wz| = |w||z|$ for all $w, z \in \mathbb{C}$ here.

$$\downarrow \text{ since } |v_j| \leq 1 \text{ for all } j \neq i$$

$$|\lambda - a_{i,i}| = \left| \sum_{j \neq i} a_{i,j} v_j \right| \leq \sum_{j \neq i} |a_{i,j}||v_j| \leq \sum_{j \neq i} |a_{i,j}| = r_i,$$

$$\uparrow \text{ triangle inequality}$$

which means exactly that $\lambda \in D(a_{i,i}, r_i)$. ∎

In order to improve the bounds provided by the Gershgorin disc theorem, recall that A and A^T have the same eigenvalues, which immediately implies that the eigenvalues of A also lie within the discs corresponding to the *columns* of A (instead of its rows). More specifically, if we define $c_i = \sum_{j \neq i} |a_{j,i}|$ (the sum of the off-diagonal entries of the i-th *column* of A) then it is also true that each eigenvalue of A is contained in at least one of the discs $D(a_{i,i}, c_i)$.

Example 2.2.6

Gershgorin Discs via Columns

Draw the Gershgorin discs based on the columns of the matrix from Example 2.2.5, and show that its eigenvalues are contained in those discs.

Solution:

The Gershgorin discs based on the columns of this matrix are $D(-1, 1)$ and $D(1 + i, 2)$, do indeed contain the eigenvalues 1 and $-1 + i$:

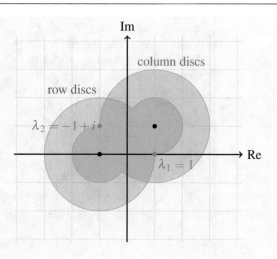

Remark 2.2.2

Counting Eigenvalues in Gershgorin Discs

Be careful when using the Gershgorin disc theorem. Since every complex matrix has exactly n eigenvalues (counting multiplicities) and n Gershgorin discs, it is tempting to conclude that *each* disc must contain an eigenvalue, but this is not necessarily true. For example, consider the matrix

$$A = \begin{bmatrix} 1 & 2 \\ -1 & -1 \end{bmatrix},$$

which has Gershgorin discs $D(1,2)$ and $D(-1,1)$. However, its eigenvalues are $\pm i$, both of which are contained in $D(1,2)$ and neither of which are contained in $D(-1,1)$:

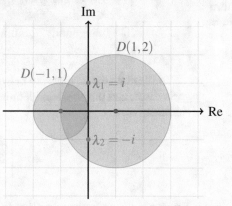

Proofs of the statements in this remark are above our pay grade, so we leave them to more specialized books like (HJ12)

However, in the case when the Gershgorin discs do not overlap, we can indeed conclude that each disc must contain exactly one eigenvalue. Slightly more generally, if we partition the Gershgorin discs into disjoint sets, then each set must contain exactly as many eigenvalues as Gershgorin discs. For example, the matrix

$$B = \begin{bmatrix} 4 & -1 & -1 \\ 1 & -1 & 1 \\ 1 & 0 & 5 \end{bmatrix}$$

has Gershgorin discs $D(4,2)$, $D(-1,2)$, and $D(5,1)$. Since $D(4,2)$ and $D(5,1)$ overlap, but $D(-1,2)$ is disjoint from them, we know that one of B's eigenvalues must be contained in $D(-1,2)$, and the other two must be contained in $D(4,2) \cup D(5,1)$. Indeed, the eigenvalues of B are approximately $\lambda_1 \approx -0.8345$ and $\lambda_{2,3} \approx 4.4172 \pm 0.9274i$:

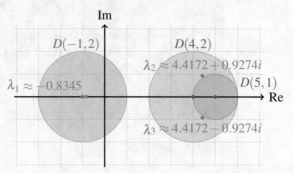

In fact, we could have even further bounded λ_1 to the real interval $[-2,0]$ by using the Gershgorin discs based on B's columns (one of these discs is $D(-1,1)$) and recalling that the eigenvalues of real matrices come in complex conjugate pairs.

Our primary purpose for introducing Gershgorin discs is that they will help us show that some matrices are positive (semi)definite shortly. First, we need to introduce one additional family of matrices:

Definition 2.2.2

Diagonally Dominant Matrices

A matrix $A \in \mathcal{M}_n$ is called

a) **diagonally dominant** if $|a_{i,i}| \geq \sum_{j \neq i} |a_{i,j}|$ for all $1 \leq i \leq n$, and

b) **strictly diagonally dominant** if $|a_{i,i}| > \sum_{j \neq i} |a_{i,j}|$ for all $1 \leq i \leq n$.

To illustrate the relationship between diagonally dominant matrices, Gershgorin discs, and positive semidefiniteness, consider the diagonally dominant Hermitian matrix

$$A = \begin{bmatrix} 2 & 0 & i \\ 0 & 7 & 1 \\ -i & 1 & 5 \end{bmatrix}. \tag{2.2.2}$$

This matrix has Gershgorin discs $D(2,1)$, $D(7,1)$, and $D(5,2)$. Furthermore, since A is Hermitian we know that its eigenvalues are real and are thus contained in the real interval $[1,8]$ (see Figure 2.5). In particular, this implies that its eigenvalues are strictly positive, so A is positive definite. This same type of argument works in general, and leads immediately to the following theorem:

Theorem 2.2.9

Diagonal Dominance Implies PSD

Suppose $A \in \mathcal{M}_n$ is self-adjoint with non-negative diagonal entries.

a) If A is diagonally dominant then it is positive semidefinite.

b) If A is strictly diagonally dominant then it is positive definite.

Proof. By the Gershgorin disc theorem, we know that the eigenvalues of A are contained in its Gershgorin discs in the complex plane. Since the diag-

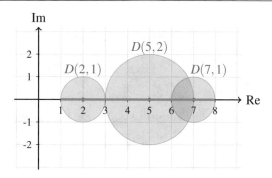

Figure 2.5: The Gershgorin discs of the Hermitian matrix from Equation (2.2.2) all lie in the right half of the complex plane, so it is positive definite.

The eigenvalues of this matrix are actually approximately 1.6806, 4.8767, and 7.4427.

onal entries of A are real and non-negative, the centers of these Gershgorin discs are located on the right half of the real axis. Furthermore, since A is diagonally dominant, the radii of these discs are smaller than the coordinates of their centers, so they do not cross the imaginary axis. It follows that every eigenvalue of A is non-negative (or strictly positive if A is strictly diagonally dominant), so A is positive semidefinite by Theorem 2.2.1 (or positive definite by Theorem 2.2.2). ∎

This is a one-way theorem: diagonally dominant matrices are PSD, but PSD matrices may not be diagonally dominant (see the matrix from Example 2.2.3, for example).

2.2.3 Unitary Freedom of PSD Decompositions

We saw in Theorem 2.2.1 that we can write every positive semidefinite matrix $A \in M_n$ in the form $A = B^*B$, where $B \in M_n$. However, this matrix B is not unique, since if $U \in M_n$ is any unitary matrix and we define $C = UB$ then $C^*C = (UB)^*(UB) = B^*U^*UB = B^*B = A$ as well. The following theorem says (among other things) that *all* such decompositions of A are related to each other in a similar way:

Theorem 2.2.10

Unitary Freedom of PSD Decompositions

Suppose $B, C \in M_{m,n}(\mathbb{F})$. The following are equivalent:

a) There exists a unitary matrix $U \in M_m(\mathbb{F})$ such that $C = UB$,
b) $B^*B = C^*C$,
c) $(B\mathbf{v}) \cdot (B\mathbf{w}) = (C\mathbf{v}) \cdot (C\mathbf{w})$ for all $\mathbf{v}, \mathbf{w} \in \mathbb{F}^n$, and
d) $\|B\mathbf{v}\| = \|C\mathbf{v}\|$ for all $\mathbf{v} \in \mathbb{F}^n$.

If $C = I$ then this theorem gives many of the characterizations of unitary matrices that we saw back in Theorem 1.4.9.

Proof. We showed above directly above the statement of the theorem that (a) implies (b), and we demonstrated the equivalence of conditions (b), (c), and (d) in Exercise 1.4.19. We thus only need to show that the conditions (b), (c), and (d) together imply condition (a).

To this end, note that since B^*B is positive semidefinite, it has a set of eigenvectors $\{\mathbf{v}_1, \ldots, \mathbf{v}_n\}$ (with corresponding eigenvalues $\lambda_1, \ldots, \lambda_n$, respectively) that form an orthonormal basis of $\mathbb{F}^n$. By Exercise 2.1.19, we know that $r = \text{rank}(B^*B)$ of these eigenvalues are non-zero, which we arrange so that $\lambda_1, \ldots, \lambda_r$ are the non-zero ones. We now prove some simple facts about these eigenvalues and eigenvectors:

i) $r \leq m$, since $\text{rank}(XY) \leq \min\{\text{rank}(X), \text{rank}(Y)\}$ in general, so $r = \text{rank}(B^*B) \leq \text{rank}(B) \leq \min\{m, n\}$.

ii) $B\mathbf{v}_1, \ldots, B\mathbf{v}_r$ are non-zero and $B\mathbf{v}_{r+1} = \cdots = B\mathbf{v}_n = \mathbf{0}$. These facts follow from noticing that, for each $1 \leq j \leq n$, we have

$$\|B\mathbf{v}_j\|^2 = (B\mathbf{v}_j) \cdot (B\mathbf{v}_j) = \mathbf{v}_j^*(B^*B\mathbf{v}_j) = \mathbf{v}_j^*(\lambda_j\mathbf{v}_j) = \lambda_j\|\mathbf{v}_j\|^2,$$

which equals zero if and only if $\lambda_j = 0$.

Recall that $\mathbf{v}_j$ is an eigenvector, and eigenvectors are (by definition) non-zero.

iii) The set $\{B\mathbf{v}_1, \ldots, B\mathbf{v}_r\}$ is mutually orthogonal, since if $i \neq j$ then

$$(B\mathbf{v}_i) \cdot (B\mathbf{v}_j) = \mathbf{v}_i^*(B^*B\mathbf{v}_j) = \mathbf{v}_i^*(\lambda_j\mathbf{v}_j) = \lambda_j\mathbf{v}_i^*\mathbf{v}_j = 0.$$

It follows that if we define vectors $\mathbf{w}_i = B\mathbf{v}_i/\|B\mathbf{v}_i\|$ for $1 \leq i \leq r$ then the set $\{\mathbf{w}_1, \ldots, \mathbf{w}_r\}$ is a mutually orthogonal set of unit vectors. We then know from Exercise 1.4.20 that we can extend this set to an orthonormal basis $\{\mathbf{w}_1, \ldots, \mathbf{w}_m\}$ of $\mathbb{F}^m$. By repeating this same argument with C in place of B, we similarly can construct an orthonormal basis $\{\mathbf{x}_1, \ldots, \mathbf{x}_m\}$ of $\mathbb{F}^m$ with the property that $\mathbf{x}_i = C\mathbf{v}_i/\|C\mathbf{v}_i\|$ for $1 \leq i \leq r$.

To extend $\{\mathbf{w}_1, \ldots, \mathbf{w}_r\}$ to an orthonormal basis, add any vector $\mathbf{z}$ not in its span, then apply the Gram–Schmidt process, and repeat.

Before we reach the home stretch of the proof, we need to establish one more property of this set of vectors:

iv) $B^*\mathbf{w}_j = C^*\mathbf{x}_j = \mathbf{0}$ whenever $j \geq r+1$. To see why this property holds, recall that $\mathbf{w}_j$ is orthogonal to each of $B\mathbf{v}_1, \ldots, B\mathbf{v}_n$. Since $\{\mathbf{v}_1, \ldots, \mathbf{v}_n\}$ is a basis of $\mathbb{F}^n$, it follows that $\mathbf{w}_j$ is orthogonal to everything in range(B). That is, $(B^*\mathbf{w}_j) \cdot \mathbf{v} = \mathbf{w}_j \cdot (B\mathbf{v}) = 0$ for all $\mathbf{v} \in \mathbb{F}^n$, so $B^*\mathbf{w}_j = \mathbf{0}$. The fact that $C^*\mathbf{x}_j = \mathbf{0}$ is proved similarly.

In the notation of Section 1.B.2, $\mathbf{w}_j \in \text{range}(B)^\perp$ and thus $\mathbf{w}_j \in \text{null}(B^*)$ (compare with Theorem 1.B.7).

With all of these preliminary properties out of the way, we now define the unitary matrices

$$U_1 = \begin{bmatrix} \mathbf{w}_1 \mid \cdots \mid \mathbf{w}_m \end{bmatrix} \quad \text{and} \quad U_2 = \begin{bmatrix} \mathbf{x}_1 \mid \cdots \mid \mathbf{x}_m \end{bmatrix}.$$

It then follows that

$$\begin{aligned}
B^*U_1 &= \begin{bmatrix} B^*\mathbf{w}_1 \mid \cdots \mid B^*\mathbf{w}_m \end{bmatrix} \\
&= \begin{bmatrix} \dfrac{B^*B\mathbf{v}_1}{\|B\mathbf{v}_1\|} \mid \cdots \mid \dfrac{B^*B\mathbf{v}_r}{\|B\mathbf{v}_r\|} \mid \mathbf{0} \mid \cdots \mid \mathbf{0} \end{bmatrix} & \text{(property (iv) above)} \\
&= \begin{bmatrix} \dfrac{C^*C\mathbf{v}_1}{\|C\mathbf{v}_1\|} \mid \cdots \mid \dfrac{C^*C\mathbf{v}_r}{\|C\mathbf{v}_r\|} \mid \mathbf{0} \mid \cdots \mid \mathbf{0} \end{bmatrix} & \text{(properties (b) and (d))} \\
&= \begin{bmatrix} C^*\mathbf{x}_1 \mid \cdots \mid C^*\mathbf{x}_m \end{bmatrix} & \text{(property (iv) again)} \\
&= C^*U_2.
\end{aligned}$$

By multiplying the equation $B^*U_1 = C^*U_2$ on the right by U_2^* and then taking the conjugate transpose of both sides, we see that $C = (U_2U_1^*)B$. Since $U_2U_1^*$ is unitary, we are done. $\blacksquare$

Recall from Theorem 1.4.9 that unitary matrices preserve the norm (induced by the usual dot product) and angles between vectors. The equivalence of conditions (a) and (b) in the above theorem can be thought of as the converse— if two sets of vectors have the same norm and pairwise angles as each other, then there must be a unitary matrix that transforms one set into the other. That is, if $\{\mathbf{v}_1, \ldots, \mathbf{v}_n\} \subseteq \mathbb{F}^m$ and $\{\mathbf{w}_1, \ldots, \mathbf{w}_n\} \subseteq \mathbb{F}^m$ are such that $\|\mathbf{v}_j\| = \|\mathbf{w}_j\|$ and $\mathbf{v}_i \cdot \mathbf{v}_j = \mathbf{w}_i \cdot \mathbf{w}_j$ for all i, j then there exists a unitary matrix $U \in \mathcal{M}_m(\mathbb{F})$ such that $\mathbf{w}_j = U\mathbf{v}_j$ for all j. After all, if $B = \begin{bmatrix} \mathbf{v}_1 \mid \mathbf{v}_2 \mid \cdots \mid \mathbf{v}_n \end{bmatrix}$ and $C = \begin{bmatrix} \mathbf{w}_1 \mid \mathbf{w}_2 \mid \cdots \mid \mathbf{w}_n \end{bmatrix}$ then $B^*B = C^*C$ if and only if these norms and dot products agree.

Now is a good time to have a look back at Figure 1.15.

Theorem 2.2.10 also raises the question of how simple we can make the matrix B in a positive semidefinite decomposition $A = B^*B$. The following theorem provides one possible answer: we can choose B so that it is also positive semidefinite.

Theorem 2.2.11

Principal Square Root of a Matrix

Suppose $A \in \mathcal{M}_n(\mathbb{F})$ is positive semidefinite. There exists a unique positive semidefinite matrix $P \in \mathcal{M}_n(\mathbb{F})$, called the **principal square root** of A, such that $A = P^2$.

Proof. To see that such a matrix P exists, we use the standard method of applying functions to diagonalizable matrices. Specifically, we use the spectral decomposition to write $A = UDU^*$, where $U \in \mathcal{M}_n(\mathbb{F})$ is unitary and $D \in \mathcal{M}_n(\mathbb{R})$ is diagonal with non-negative real numbers (the eigenvalues of A) as its diagonal entries. If we then define $P = U\sqrt{D}U^*$, where $\sqrt{D}$ is the diagonal matrix that contains the non-negative square roots of the entries of D, then

$$P^2 = (U\sqrt{D}U^*)(U\sqrt{D}U^*) = U\sqrt{D}^2U^* = UDU^* = A,$$

as desired.

To see that P is unique, suppose that $Q \in \mathcal{M}_n(\mathbb{F})$ is another positive semidefinite matrix for which $Q^2 = A$. We can use the spectral decomposition to write $Q = VFV^*$, where $V \in \mathcal{M}_n(\mathbb{F})$ is unitary and $F \in \mathcal{M}_n(\mathbb{R})$ is diagonal with non-negative real numbers as its diagonal entries. Since $VF^2V^* = Q^2 = A = UDU^*$, we conclude that the eigenvalues of Q^2 equal those of A, and thus the diagonal entries of F^2 equal those of D.

Be careful—it is tempting to try to show that $V = U$, but this is not true in general. For example, if $D = I$ then U and V can be anything.

Since we are free in the spectral decomposition to order the eigenvalues along the diagonals of F and D however we like, we can assume without loss of generality that $F = \sqrt{D}$. It then follows from the fact that $P^2 = A = Q^2$ that $VDV^* = UDU^*$, so $WD = DW$, where $W = U^*V$. Our goal now is to show that this implies $V\sqrt{D}V^* = U\sqrt{D}U^*$ (i.e., $Q = P$), which is equivalent to $W\sqrt{D} = \sqrt{D}W$.

Again, the spectral decomposition lets us order the diagonal entries of D (and thus of $\sqrt{D}$) however we like.

To this end, suppose that P has k distinct eigenvalues (i.e., $\sqrt{D}$ has k distinct diagonal entries), which we denote by $\lambda_1, \lambda_2, \ldots, \lambda_k$, and denote the multiplicity of each λ_j by m_j. We can then write $\sqrt{D}$ in block diagonal form as follows, where we have grouped repeated eigenvalues (if any) so as to be next to each other (i.e., in the same block):

$$\sqrt{D} = \begin{bmatrix} \lambda_1 I_{m_1} & O & \cdots & O \\ O & \lambda_2 I_{m_2} & \cdots & O \\ \vdots & \vdots & \ddots & \vdots \\ O & O & \cdots & \lambda_k I_{m_k} \end{bmatrix}.$$

If we partition W as a block matrix via blocks of the same size and shape, i.e.,

$$W = \begin{bmatrix} W_{1,1} & W_{1,2} & \cdots & W_{1,k} \\ W_{2,1} & W_{2,2} & \cdots & W_{2,k} \\ \vdots & \vdots & \ddots & \vdots \\ W_{k,1} & W_{k,2} & \cdots & W_{k,k} \end{bmatrix},$$

then block matrix multiplication shows that the equation $WD = DW$ is equivalent to $\lambda_i^2 W_{i,j} = \lambda_j^2 W_{i,j}$ for all $1 \le i \ne j \le k$. Since $\lambda_i^2 \ne \lambda_j^2$ when $i \ne j$, this implies that $W_{i,j} = O$ when $i \neq j$, so W is block diagonal. It then follows that

$$W\sqrt{D} = \begin{bmatrix} \lambda_1 W_{1,1} & O & \cdots & O \\ O & \lambda_2 W_{2,2} & \cdots & O \\ \vdots & \vdots & \ddots & \vdots \\ O & O & \cdots & \lambda_k W_{k,k} \end{bmatrix} = \sqrt{D}W$$

as well, which completes the proof. ∎

The principal square root P of a matrix A is typically denoted by $P = \sqrt{A}$, and it is directly analogous to the principal square root of a non-negative real number (indeed, for 1×1 matrices they are the exact same thing). This provides yet another reason why we typically think of positive semidefinite matrices as the "matrix version" of non-negative real numbers.

Example 2.2.7

Computing a Principal Square Root

Compute the principal square root of the matrix $A = \begin{bmatrix} 8 & 6 \\ 6 & 17 \end{bmatrix}$.

Solution:

We just compute a spectral decomposition of B and take the principal square root of the diagonal entries of its diagonal part. The eigenvalues of A are 5 and 20, and they have corresponding eigenvectors $(2, -1)$ and $(1, 2)$, respectively. It follows that A has spectral decomposition $A = UDU^*$, where

$$U = \frac{1}{\sqrt{5}} \begin{bmatrix} 2 & 1 \\ -1 & 2 \end{bmatrix} \quad \text{and} \quad D = \begin{bmatrix} 5 & 0 \\ 0 & 20 \end{bmatrix}.$$

> There are also three other square roots of A (obtained by taking some negative square roots in D), but the principal square root is the only one of them that is PSD.

The principal square root of A is thus

$$\sqrt{A} = U\sqrt{D}U^* = \frac{1}{5}\begin{bmatrix} 2 & 1 \\ -1 & 2 \end{bmatrix}\begin{bmatrix} \sqrt{5} & 0 \\ 0 & \sqrt{20} \end{bmatrix}\begin{bmatrix} 2 & -1 \\ 1 & 2 \end{bmatrix} = \frac{1}{\sqrt{5}}\begin{bmatrix} 6 & 2 \\ 2 & 9 \end{bmatrix}.$$

By combining our previous two theorems, we get a new matrix decomposition, which answers the question of how simple we can make a matrix by multiplying it on the left by a unitary matrix—we can always make it positive semidefinite.

Theorem 2.2.12

Polar Decomposition

> This is sometimes called the **right** polar decomposition of A. A **left** polar decomposition is $A = QV$, where Q is PSD and V is unitary. These two decompositions coincide (i.e., $Q = P$ and $V = U$) if and only if A is normal.

Suppose $A \in \mathcal{M}_n(\mathbb{F})$. There exists a unitary matrix $U \in \mathcal{M}_n(\mathbb{F})$ and a positive semidefinite matrix $P \in \mathcal{M}_n(\mathbb{F})$ such that

$$A = UP.$$

Furthermore, P is unique and is given by $P = \sqrt{A^*A}$, and U is unique if A is invertible.

Proof. Since A^*A is positive semidefinite, we know from Theorem 2.2.11 that we can define $P = \sqrt{A^*A}$ so that $A^*A = P^2 = P^*P$ and P is positive semidefinite. We then know from Theorem 2.2.10 that there exists a unitary matrix $U \in \mathcal{M}_n(\mathbb{F})$ such that $A = UP$.

To see uniqueness of P, suppose $A = U_1 P_1 = U_2 P_2$, where $U_1, U_2 \in \mathcal{M}_n(\mathbb{F})$

are unitary and $P_1, P_2 \in \mathcal{M}_n(\mathbb{F})$ are positive semidefinite. Then

$$A^*A = (U_1 P_1)^*(U_1 P_1) = P_1 U_1^* U_1 P_1 = P_1^2 \quad \text{and}$$
$$A^*A = (U_2 P_2)^*(U_2 P_2) = P_2 U_2^* U_2 P_2 = P_2^2.$$

Since $P_1^2 = P_2^2$ is positive semidefinite, is has a unique principal square root, so $P_1 = P_2$.

If A is invertible then uniqueness of U follows from the fact that we can rearrange the decomposition $A = UP$ to the form $U^* = PA^{-1}$, so $U = (PA^{-1})^*$. ∎

Similarly, the matrices $\sqrt{A^*A}$ and $\sqrt{AA^*}$ might be called the **left** and **right** absolute values of A, respectively, and they are equal if and only if A is normal.

The matrix $\sqrt{A^*A}$ in the polar decomposition can be thought of as the "matrix version" of the absolute value of a complex number $|z| = \sqrt{\bar{z}z}$. In fact, this matrix is sometimes even denoted by $|A| = \sqrt{A^*A}$. In a sense, this matrix provides a "regularization" of A that preserves many of its properties (such as rank and Frobenius norm—see Exercise 2.2.22), but also adds the desirable positive semidefiniteness property.

Example 2.2.8

Computing a Polar Decomposition

Compute the polar decomposition of the matrix $A = \begin{bmatrix} 2 & -1 \\ 2 & 4 \end{bmatrix}$.

Solution:

In order to find the polar decomposition $A = UP$, our first priority is to compute A^*A and then set $P = \sqrt{A^*A}$:

$$A^*A = \begin{bmatrix} 2 & 2 \\ -1 & 4 \end{bmatrix} \begin{bmatrix} 2 & -1 \\ 2 & 4 \end{bmatrix} = \begin{bmatrix} 8 & 6 \\ 6 & 17 \end{bmatrix},$$

which we computed the principal square root of in Example 2.2.7:

$$P = \sqrt{A^*A} = \frac{1}{\sqrt{5}} \begin{bmatrix} 6 & 2 \\ 2 & 9 \end{bmatrix}.$$

Next, as suggested by the proof of Theorem 2.2.12, we can find U by setting $U = (PA^{-1})^*$, since this rearranges to exactly $A = UP$:

The 10 in the denominator here comes from A^{-1}.

$$U = \left(\frac{1}{10\sqrt{5}} \begin{bmatrix} 6 & 2 \\ 2 & 9 \end{bmatrix} \begin{bmatrix} 4 & 1 \\ -2 & 2 \end{bmatrix} \right)^* = \frac{1}{\sqrt{5}} \begin{bmatrix} 2 & -1 \\ 1 & 2 \end{bmatrix}.$$

The polar decomposition is directly analogous to the fact that every complex number $z \in \mathbb{C}$ can be written in the form $z = re^{i\theta}$, where $r = |z|$ is non-negative and $e^{i\theta}$ lies on the unit circle in the complex plane. Indeed, we have already been thinking about positive semidefinite matrices as analogous to non-negative real numbers, and it similarly makes sense to think of unitary matrices as analogous to numbers on the complex unit circle (indeed, this is exactly what they are in the 1×1 case). After all, multiplication by $e^{i\theta}$ rotates numbers in the complex plane but does not change their absolute value, just like unitary matrices rotate and/or reflect vectors but do not change their length.

Have a look at Appendix A.3.3 if you need a refresher on the polar form of a complex number.

In fact, just like we think of PSD matrices as analogous to non-negative real numbers and unitary matrices as analogous to numbers on the complex unit circle, there are many other sets of matrices that it is useful to think of as analogous to subsets of the complex plane. For example, we think of the sets of Hermitian and skew-Hermitian matrices as analogous to the sets of real and imaginary numbers, respectively (see Figure 2.6). There are several ways in which these analogies can be justified:

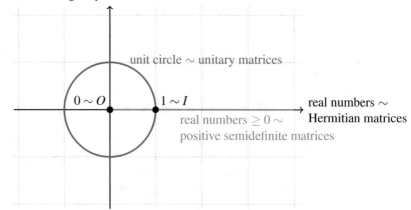

imaginary numbers $\sim$ skew-Hermitian matrices

Thinking about sets of matrices geometrically (even though it is a very high-dimensional space that we cannot *really* picture properly) is a very useful technique for building intuition.

unit circle $\sim$ unitary matrices

$0 \sim O$ $1 \sim I$

real numbers $\geq 0 \sim$
positive semidefinite matrices

real numbers $\sim$
Hermitian matrices

Figure 2.6: Several sets of normal matrices visualized on the complex plane as the sets of complex numbers to which they are analogous.

- Each of these sets of matrices have eigenvalues living in the corresponding set of complex numbers. For example, the eigenvalues of skew-Hermitian matrices are imaginary.

- Furthermore, the converse of the above point is true as long as we assume that the matrix is normal (see Exercise 2.1.17). For example, a normal matrix with all eigenvalues imaginary must be skew-Hermitian.

- We can write every matrix in the form $A = UP$, where U is unitary and P is PSD, just like we can write every complex number in the form $z = re^{i\theta}$, where $e^{i\theta}$ is on the unit circle and r is non-negative and real.

- We can write every matrix in the form $A = H + K$, where H is Hermitian and K is skew-Hermitian (see Remark 1.B.1), just like we can write every complex number in the form $z = h + k$, where h is real and k is imaginary.

- The only positive semidefinite unitary matrix is I (see Exercise 2.2.7), just like the only positive real number on the unit circle is 1.

- The only matrix that is both Hermitian and skew-Hermitian is O, just like the only complex number that is both real and imaginary is 0.

Remark 2.2.3

Unitary Multiplication and PSD Decompositions

Given a positive semidefinite matrix $A \in \mathcal{M}_n$, there are actually multiple different ways to simplify the matrix B in the PSD decomposition $A = B^*B$, and which one is best to use depends on what we want to do with it.

We showed in Theorem 2.2.11 that we can choose B to be positive semidefinite, and this led naturally to the polar decomposition of a matrix. However, there is another matrix decomposition (the Cholesky decomposition of Section 2.B.2) that says we can instead make B upper triangular with non-negative real numbers on the diagonal. This similarly leads to the QR decomposition of Section 1.C. The relationships between these four matrix decompositions are summarized here:

Decomposition of B	Decomposition of $A = B^*B$
polar decomposition: $B = UP$ P is positive semidefinite	principal square root: $A = P^2$
QR decomposition: $B = UT$ T is upper triangular with non-negative real numbers on the diagonal	Cholesky decomposition: $A = T^*T$

Exercises

solutions to starred exercises on page 467

2.2.1 Which of the following matrices are positive semidefinite? Positive definite?

* (a) $\begin{bmatrix} 1 & 0 \\ 0 & 1 \end{bmatrix}$

(b) $\begin{bmatrix} 4 & -2i \\ 2i & 1 \end{bmatrix}$

* (c) $\begin{bmatrix} 1 & 0 & 0 \\ 0 & 1 & 0 \end{bmatrix}$

(d) $\begin{bmatrix} 0 & 0 & 1 \\ 0 & 1 & 0 \\ 1 & 0 & 0 \end{bmatrix}$

* (e) $\begin{bmatrix} 1 & 0 & i \\ 0 & 1 & 0 \\ i & 0 & 1 \end{bmatrix}$

(f) $\begin{bmatrix} 1 & 1 & 1 \\ 1 & 2 & 1 \\ 1 & 1 & 2 \end{bmatrix}$

* (g) $\begin{bmatrix} 3 & 1 & 1 \\ 1 & 3 & 1 \\ 1 & 1 & 3 \end{bmatrix}$

(h) $\begin{bmatrix} 1 & 1 & 1 & 0 \\ 1 & 2 & 1 & 1 \\ 1 & 1 & 2 & i \\ 0 & 1 & -i & 2 \end{bmatrix}$

2.2.2 Compute the Gershgorin (row) discs of each of the following matrices.

* (a) $\begin{bmatrix} 1 & 2 \\ 3 & -1 \end{bmatrix}$

(b) $\begin{bmatrix} 2 & -2 \\ -i & 3 \end{bmatrix}$

* (c) $\begin{bmatrix} 1 & 0 & 0 \\ 0 & 2 & 0 \\ 0 & 0 & 3 \end{bmatrix}$

(d) $\begin{bmatrix} 0 & 0 & 1 \\ 0 & 3 & 0 \\ 2 & 0 & 0 \end{bmatrix}$

* (e) $\begin{bmatrix} 1 & 2 & i \\ -1 & 3 & 2 \\ i & -2 & 1 \end{bmatrix}$

(f) $\begin{bmatrix} -1 & 1 & 1 & 0 \\ 0 & 1 & 1 & i \\ 0 & 1 & 0 & 0 \\ 1 & 0 & -1 & i \end{bmatrix}$

2.2.3 Compute the principal square root of each of the following positive semidefinite matrices.

* (a) $\begin{bmatrix} 1 & -1 \\ -1 & 1 \end{bmatrix}$

(b) $\begin{bmatrix} 2 & 3 \\ 3 & 5 \end{bmatrix}$

* (c) $\begin{bmatrix} 1 & 0 & 0 \\ 0 & 2 & 0 \\ 0 & 0 & 3 \end{bmatrix}$

(d) $\begin{bmatrix} 2 & 1 & 1 \\ 1 & 2 & 1 \\ 1 & 1 & 2 \end{bmatrix}$

* (e) $\begin{bmatrix} 1 & 1 & 0 \\ 1 & 2 & 1 \\ 0 & 1 & 1 \end{bmatrix}$

(f) $\begin{bmatrix} 2 & 2 & 0 & 0 \\ 2 & 2 & 0 & 0 \\ 0 & 0 & 1 & \sqrt{3} \\ 0 & 0 & \sqrt{3} & 4 \end{bmatrix}$

2.2.4 Compute a polar decomposition of each of the following matrices.

* (a) $\begin{bmatrix} 1 & -1 \\ -1 & 1 \end{bmatrix}$

(b) $\begin{bmatrix} 1 & 0 \\ 3 & 2 \end{bmatrix}$

(c) $\begin{bmatrix} -1 & 0 & 0 \\ 0 & 1+i & 0 \\ 0 & 0 & 2i \end{bmatrix}$

* (d) $\begin{bmatrix} 0 & -1 & 0 \\ -1 & 1 & 1 \\ 2 & 3 & 1 \end{bmatrix}$

2.2.5 Determine which of the following statements are true and which are false.

* (a) If every entry of a matrix is real and non-negative then it is positive semidefinite.
  (b) The zero matrix $O \in \mathcal{M}_n$ is positive semidefinite.
* (c) The identity matrix $I \in \mathcal{M}_n$ is positive definite.
  (d) If $A, B \in \mathcal{M}_n$ are positive semidefinite matrices, then so is $A + B$.
* (e) If $A, B \in \mathcal{M}_n$ are positive semidefinite matrices, then so is AB.
  (f) If $A \in \mathcal{M}_n$ is a positive semidefinite matrix, then so is A^2.
* (g) Each Gershgorin disc of a matrix contains at least one of its eigenvalues.
  (h) The identity matrix is its own principal square root.
* (i) The only matrix $A \in \mathcal{M}_n$ such that $A^2 = I$ is $A = I$.
  (j) Every matrix $A \in \mathcal{M}_n$ has a unique polar decomposition.

2.2.6 For each of the following matrices, determine which values of $x \in \mathbb{R}$ result in the matrix being positive semidefinite.

* (a) $\begin{bmatrix} 1 & 0 \\ 0 & x \end{bmatrix}$

(b) $\begin{bmatrix} 1 & x \\ x & 9 \end{bmatrix}$

* (c) $\begin{bmatrix} 1 & x & 0 \\ x & 1 & x \\ 0 & x & 0 \end{bmatrix}$

(d) $\begin{bmatrix} 1 & x & 0 \\ x & 1 & x \\ 0 & x & 1 \end{bmatrix}$

2.2.7 Show that the only matrix that is both positive semidefinite and unitary is the identity matrix.

2.2.8 Show that if a matrix is positive definite then it is invertible, and its inverse is also positive definite.

2.2.9 Suppose $A \in \mathcal{M}_n$ is self-adjoint. Show that there exists a scalar $c \in \mathbb{R}$ such that $A + cI$ is positive definite.

2.2.10 Let $J \in \mathcal{M}_n$ be the $n \times n$ matrix with all entries equal to 1. Show that J is positive semidefinite.

∗∗2.2.11 Show that if $A \in \mathcal{M}_n$ is positive semidefinite and $a_{j,j} = 0$ for some $1 \leq j \leq n$ then the entire j-th row and j-th column of A must consist of zeros.

[Hint: What is $\mathbf{v}^*A\mathbf{v}$ if $\mathbf{v}$ has just two non-zero entries?]

2.2.12 Show that every (not necessarily Hermitian) strictly diagonally dominant matrix is invertible.

∗∗2.2.13 Show that the block diagonal matrix

$$A = \begin{bmatrix} A_1 & O & \cdots & O \\ O & A_2 & \cdots & O \\ \vdots & \vdots & \ddots & \vdots \\ O & O & \cdots & A_n \end{bmatrix}$$

is positive (semi)definite if and only if each of $A_1, A_2, \ldots, A_n$ are positive (semi)definite.

∗∗2.2.14 Show that $A \in \mathcal{M}_n$ is normal if and only if there exists a unitary matrix $U \in \mathcal{M}_n$ such that $A^* = UA$.

2.2.15 A matrix $A \in \mathcal{M}_n(\mathbb{R})$ with non-negative entries is called **row stochastic** if its rows each add up to 1 (i.e., $a_{i,1} + a_{i,2} + \cdots + a_{i,n} = 1$ for each $1 \leq i \leq n$). Show that each eigenvalue λ of a row stochastic matrix has $|\lambda| \leq 1$.

∗∗2.2.16 Suppose $\mathbb{F} = \mathbb{R}$ or $\mathbb{F} = \mathbb{C}$, and $A \in \mathcal{M}_n(\mathbb{F})$.

(a) Show that A is self-adjoint if and only if there exist positive semidefinite matrices $P, N \in \mathcal{M}_n(\mathbb{F})$ such that $A = P - N$. [Hint: Apply the spectral decomposition to A.]

(b) Show that if $\mathbb{F} = \mathbb{C}$ then A can be written as a linear combination of 4 or fewer positive semidefinite matrices, even if it is not Hermitian. [Hint: Have a look at Remark 1.B.1.]

(c) Explain why the result of part (b) does not hold if $\mathbb{F} = \mathbb{R}$.

2.2.17 Suppose $A \in \mathcal{M}_n$ is self-adjoint with p strictly positive eigenvalues. Show that the largest integer r with the property that P^*AP is positive definite for some $P \in \mathcal{M}_{n,r}$ is $r = p$.

2.2.18 Suppose $B \in \mathcal{M}_n$. Show that $\text{tr}\left(\sqrt{B^*B}\right) \geq |\text{tr}(B)|$.

[Side note: In other words, out of all possible PSD decompositions of a PSD matrix, its principal square root is the one with the largest trace.]

∗∗2.2.19 Suppose that $A, B, C \in \mathcal{M}_n$ are positive semidefinite.

(a) Show that $\text{tr}(A) \geq 0$.

(b) Show that $\text{tr}(AB) \geq 0$. [Hint: Decompose A and B.]

(c) Provide an example to show that it is *not* necessarily the case that $\text{tr}(ABC) \geq 0$. [Hint: Finding an example by hand might be tricky. If you have trouble, try writing a computer program that searches for an example by generating A, B, and C randomly.]

∗∗2.2.20 Suppose that $A \in \mathcal{M}_n$ is self-adjoint.

(a) Show that A is positive semidefinite if and only if $\text{tr}(AB) \geq 0$ for all positive semidefinite $B \in \mathcal{M}_n$. [Side note: This provides a converse to the statement of Exercise 2.2.19(b).]

(b) Show that A positive *definite* if and only if $\text{tr}(AB) > 0$ for all positive semidefinite $O \neq B \in \mathcal{M}_n$.

∗∗2.2.21 Suppose that $A \in \mathcal{M}_n$ is self-adjoint.

(a) Show that if there exists a scalar $c \in \mathbb{R}$ such that $\text{tr}(AB) \geq c$ for all positive semidefinite $B \in \mathcal{M}_n$ then A is positive semidefinite and $c \leq 0$.

(b) Show that if there exists a scalar $c \in \mathbb{R}$ such that $\text{tr}(AB) > c$ for all positive definite $B \in \mathcal{M}_n$ then A is positive semidefinite and $c \leq 0$.

∗∗2.2.22 Let $|A| = \sqrt{A^*A}$ be the **absolute value** of the matrix $A \in \mathcal{M}_n(\mathbb{F})$ that was discussed after Theorem 2.2.12.

(a) Show that $\text{rank}(|A|) = \text{rank}(A)$.

(b) Show that $\big\||A|\big\|_F = \|A\|_F$.

(c) Show that $\big\||A|\mathbf{v}\big\| = \|A\mathbf{v}\|$ for all $\mathbf{v} \in \mathbb{F}^n$.

∗∗2.2.23 Prove Theorem 2.2.2.

[Hint: Mimic the proof of Theorem 2.2.1 and just make minor changes where necessary.]

∗∗2.2.24 Recall Theorem 2.2.3, which described some ways in which we can combine PSD matrices to create new PSD matrices.

(a) Prove part (a) of the theorem.

(b) Prove part (b) of the theorem.

(c) Prove part (c) of the theorem.

∗∗2.2.25 Suppose $A \in \mathcal{M}_n(\mathbb{F})$ is self-adjoint.

(a) Show that A is positive semidefinite if and only if there exists a set of vectors $\{\mathbf{v}_1, \mathbf{v}_2, \ldots, \mathbf{v}_n\} \subset \mathbb{F}^n$ such that

$$a_{i,j} = \mathbf{v}_i \cdot \mathbf{v}_j \quad \text{for all} \quad 1 \leq i, j \leq n.$$

[Side note: A is sometimes called the **Gram matrix** of $\{\mathbf{v}_1, \mathbf{v}_2, \ldots, \mathbf{v}_n\}$.]

(b) Show that A is positive *definite* if and only if the set of vectors from part (a) is linearly independent.

2.2.26 Suppose that $A, B \in \mathcal{M}_n$ are positive definite.

(a) Show that all eigenvalues of AB are real and positive. [Hint: Multiply AB on the left by $\sqrt{A}^{-1}$ and on the right by $\sqrt{A}$.]

(b) Part (a) does *not* imply that AB is positive definite. Why not?

2.2.27 Let $A, B \in \mathcal{M}_n$ be positive definite matrices.

(a) Show that $\det(I + B) \geq 1 + \det(B)$.

(b) Show that $\det(A + B) \geq \det(A) + \det(B)$. [Hint: $\det(A + B) = \det(A) \det(I + \sqrt{A}^{-1} B \sqrt{A}^{-1})$.]

[Side note: The stronger inequality

$$(\det(A + B))^{1/n} \geq (\det(A))^{1/n} + (\det(B))^{1/n}$$

is also true for positive definite matrices (and is called **Minkowski's determinant theorem**), but proving it is quite difficult.]

∗∗2.2.28 In this exercise, we show that if $A \in \mathcal{M}_n(\mathbb{C})$ is written in terms of its columns as $A = [\, \mathbf{a}_1 \mid \mathbf{a}_2 \mid \cdots \mid \mathbf{a}_n \,]$, then

$$\big| \det(A) \big| \leq \|\mathbf{a}_1\| \|\mathbf{a}_2\| \cdots \|\mathbf{a}_n\|.$$

[Side note: This is called **Hadamard's inequality**.]

(a) Explain why it suffices to prove this inequality in the case when $\|\mathbf{a}_j\| = 1$ for all $1 \leq j \leq n$. Make this assumption throughout the rest of this question.

(b) Show that $\det(A^*A) \leq 1$. [Hint: Use the AM–GM inequality (Theorem A.5.3).]

(c) Conclude that $|\det(A)| \leq 1$ as well, thus completing the proof.

(d) Explain under which conditions equality is attained in Hadamard's inequality.

2.3 The Singular Value Decomposition

We are finally in a position to present what is possibly the most important theorem in all of linear algebra: the **singular value decomposition (SVD)**. On its surface, it can be thought of as a generalization of the spectral decomposition that applies to all matrices, rather than just normal matrices. However, we will also see that we can use this decomposition to do several unexpected things like compute the size of a matrix (in a way that typically makes more sense than the Frobenius norm), provide a new geometric interpretation of linear transformations, solve optimization problems, and construct an "almost inverse" for matrices that do not have an inverse.

Some of these applications of the SVD are deferred to Section 2.C.

Furthermore, many of the results from introductory linear algebra as well as earlier in this book can be seen as trivial consequences of the singular value decomposition. The spectral decomposition (Theorems 2.1.2 and 2.1.6), polar decomposition (Theorem 2.2.12), orthogonality of the fundamental subspaces (Theorem 1.B.7), and many more results can all be re-derived in a line or two from the SVD. In a sense, if we know the singular value decomposition then we know linear algebra.

Theorem 2.3.1

Singular Value Decomposition (SVD)

Suppose $\mathbb{F} = \mathbb{R}$ or $\mathbb{F} = \mathbb{C}$, and $A \in \mathcal{M}_{m,n}(\mathbb{F})$. There exist unitary matrices $U \in \mathcal{M}_m(\mathbb{F})$ and $V \in \mathcal{M}_n(\mathbb{F})$, and a diagonal matrix $\Sigma \in \mathcal{M}_{m,n}(\mathbb{R})$ with non-negative entries, such that

$$A = U\Sigma V^*.$$

Furthermore, in any such decomposition,

Note that Σ might not be square. When we say that it is a "diagonal matrix", we just mean that its (i, j)-entry is 0 whenever $i \neq j$.

- the diagonal entries of Σ (called the **singular values** of A) are the non-negative square roots of the eigenvalues of A^*A (or equivalently, of AA^*),

- the columns of U (called the **left singular vectors** of A) are eigenvectors of AA^*, and

- the columns of V (called the **right singular vectors** of A) are eigenvectors of A^*A.

Before proving the singular value decomposition, it's worth comparing the ways in which it is "better" and "worse" than the other matrix decompositions that we already know:

If $m \neq n$ then A^*A and AA^* have different sizes, but they still have essentially the same eigenvalues— whichever one is larger just has some extra 0 eigenvalues. The same is actually true of AB and BA for *any* A and B (see Exercise 2.B.11).

- **Better:** It applies to every single matrix (even rectangular ones). Every other matrix decomposition we have seen so far had at least *some* restrictions (e.g., diagonalization only applies to matrices with a basis of eigenvectors, the spectral decomposition only applies to normal matrices, and Schur triangularization only applies to square matrices).
- **Better:** The matrix Σ in the middle of the SVD is diagonal (and even has real non-negative entries). Schur triangularization only results in an upper triangular middle piece, and even diagonalization and the spectral decomposition do not guarantee an entrywise non-negative matrix.
- **Worse:** It requires *two* unitary matrices U and V, whereas all of our previous decompositions only required *one* unitary matrix or invertible matrix.

Proof of the singular value decomposition. To see that singular value decompositions exists, suppose that $m \geq n$ and construct a spectral decomposition of the positive semidefinite matrix $A^*A = VDV^*$ (if $m < n$ then we instead use the matrix AA^*, but otherwise the proof is almost identical). Since A^*A is positive semidefinite, its eigenvalues (i.e., the diagonal entries of D) are real and non-negative, so we can define $\Sigma \in \mathcal{M}_{m,n}$ by $[\Sigma]_{j,j} = \sqrt{d_{j,j}}$ for all j, and $[\Sigma]_{i,j} = 0$ if $i \neq j$.

In other words, Σ is the principal square root of D, but with extra zero rows so that it has the same size as A.

It follows that

$$A^*A = VDV^* = V\Sigma^*\Sigma V^* = (\Sigma V^*)^*(\Sigma V^*),$$

so the equivalence of conditions (a) and (b) in Theorem 2.2.10 tells us that there exists a unitary matrix $U \in \mathcal{M}_m(\mathbb{F})$ such that $A = U(\Sigma V^*)$, which is a singular value decomposition of A.

To check the "furthermore" claims we just note that if $A = U\Sigma V^*$ with U and V unitary and Σ diagonal then it must be the case that

We must write $\Sigma^*\Sigma$ here (instead of Σ^2) because Σ might not be square.

$$A^*A = (U\Sigma V^*)^*(U\Sigma V^*) = V(\Sigma^*\Sigma)V^*,$$

which is a diagonalization of A^*A. Since the only way to diagonalize a matrix is via its eigenvalues and eigenvectors (refer back to Theorem 2.0.1, for example), it follows that the columns of V are eigenvectors of A^*A and the diagonal entries of $\Sigma^*\Sigma$ (i.e., the squares of the diagonal entries of Σ) are the eigenvalues of A^*A. The statements about eigenvalues and eigenvectors of AA^* are proved in a similar manner. ∎

For a refresher on these facts about the rank of a matrix, see Exercise 2.1.19 and Theorem A.1.2, for example.

We typically denote singular values (i.e., the diagonal entries of Σ) by $\sigma_1, \sigma_2, \ldots, \sigma_{\min\{m,n\}}$, and we order them so that $\sigma_1 \geq \sigma_2 \geq \ldots \geq \sigma_{\min\{m,n\}}$. Note that exactly $\operatorname{rank}(A)$ of A's singular values are non-zero, since $\operatorname{rank}(U\Sigma V^*) = \operatorname{rank}(\Sigma)$, and the rank of a diagonal matrix is the number of non-zero diagonal entries. We often denote the rank of A simply by r, so we have $\sigma_1 \geq \cdots \geq \sigma_r > 0$ and $\sigma_{r+1} = \cdots = \sigma_{\min\{m,n\}} = 0$.

Example 2.3.1

Computing Singular Values

Compute the singular values of the matrix $A = \begin{bmatrix} 3 & 2 \\ -2 & 0 \end{bmatrix}$.

Solution:

The singular values are the square roots of the eigenvalues of A^*A.

Direct calculation shows that

$$A^*A = \begin{bmatrix} 13 & 6 \\ 6 & 4 \end{bmatrix},$$

which has characteristic polynomial

$$p_{A^*A}(\lambda) = \det(A^*A - \lambda I) = \det\left(\begin{bmatrix} 13 - \lambda & 6 \\ 6 & 4 - \lambda \end{bmatrix}\right)$$
$$= (13 - \lambda)(4 - \lambda) - 36 = \lambda^2 - 17\lambda + 16 = (\lambda - 1)(\lambda - 16).$$

We thus conclude that the eigenvalues of A^*A are $\lambda_1 = 16$ and $\lambda_2 = 1$, so the singular values of A are $\sigma_1 = \sqrt{16} = 4$ and $\sigma_2 = \sqrt{1} = 1$.

To compute a matrix's singular value decomposition (not just its singular values), we could construct spectral decompositions of each of A^*A (for the right singular vectors) and AA^* (for the left singular vectors). However, there is a simpler way of obtaining the left singular vectors that lets us avoid computing a second spectral decomposition. In particular, if we have already computed a spectral decomposition $A^*A = V(\Sigma^*\Sigma)V^*$, where $V = \begin{bmatrix} \mathbf{v}_1 \mid \mathbf{v}_2 \mid \cdots \mid \mathbf{v}_n \end{bmatrix}$ and Σ has non-zero diagonal entries $\sigma_1, \sigma_2, \ldots, \sigma_r$, then we can compute the remaining $U = \begin{bmatrix} \mathbf{u}_1 \mid \mathbf{u}_2 \mid \cdots \mid \mathbf{u}_m \end{bmatrix}$ piece of A's singular value decomposition by noting that $A = U\Sigma V^*$ implies

$$\begin{bmatrix} A\mathbf{v}_1 \mid A\mathbf{v}_2 \mid \cdots \mid A\mathbf{v}_n \end{bmatrix} = AV = U\Sigma = \begin{bmatrix} \sigma_1\mathbf{u}_1 \mid \cdots \mid \sigma_r\mathbf{u}_r \mid 0 \mid \cdots \mid 0 \end{bmatrix}.$$

The final $m - r$ columns of U do not really matter—they just need to "fill out" U to make it unitary (i.e., they must have length 1 and be mutually orthogonal).

It follows that the first $r = \text{rank}(A)$ columns of U can be obtained via

$$\mathbf{u}_j = \frac{1}{\sigma_j}A\mathbf{v}_j \qquad \text{for all} \quad 1 \le j \le r, \tag{2.3.1}$$

and the remaining $m - r$ columns can be obtained by extending $\{\mathbf{u}_i\}_{i=1}^r$ to an orthonormal basis of $\mathbb{F}^m$ (via the Gram–Schmidt process, for example).

Example 2.3.2

Computing a Singular Value Decomposition

Compute a singular value decomposition of the matrix $A = \begin{bmatrix} 1 & 2 & 3 \\ -1 & 0 & 1 \\ 3 & 2 & 1 \end{bmatrix}$.

Solution:

As discussed earlier, our first step is to find the "V" and "Σ" pieces of A's singular value decomposition, which we do by constructing a spectral decomposition of A^*A and taking the square roots of its eigenvalues. Well, direct calculation shows that

This is a good time to remind yourself of how to calculate eigenvalues and eigenvectors, in case you have gotten rusty. We do not go through all of the details here.

$$A^*A = \begin{bmatrix} 11 & 8 & 5 \\ 8 & 8 & 8 \\ 5 & 8 & 11 \end{bmatrix},$$

which has characteristic polynomial

$$p_{A^*A}(\lambda) = \det(A^*A - \lambda I) = -\lambda^3 + 30\lambda^2 - 144\lambda = -\lambda(\lambda - 6)(\lambda - 24).$$

It follows that the eigenvalues of A^*A are $\lambda_1 = 24$, $\lambda_2 = 6$, and $\lambda_3 = 0$,

In particular, A has rank 2 since 2 of its singular values are non-zero.

so the singular values of A are $\sigma_1 = \sqrt{24} = 2\sqrt{6}$, $\sigma_2 = \sqrt{6}$, and $\sigma_3 = 0$. The normalized eigenvectors corresponding to these eigenvalues are

$$\mathbf{v}_1 = \frac{1}{\sqrt{3}}\begin{bmatrix} 1 \\ 1 \\ 1 \end{bmatrix}, \qquad \mathbf{v}_2 = \frac{1}{\sqrt{2}}\begin{bmatrix} -1 \\ 0 \\ 1 \end{bmatrix}, \quad \text{and} \quad \mathbf{v}_3 = \frac{1}{\sqrt{6}}\begin{bmatrix} -1 \\ 2 \\ -1 \end{bmatrix},$$

respectively. We then place these eigenvectors into the matrix V as columns (in the same order as the corresponding eigenvalues/singular values) and obtain

We just pulled a $1/\sqrt{6}$ factor out of V to avoid having to write fractional entries—simplifying in another way (e.g., explicitly writing its top-left entry as $1/\sqrt{3}$) is fine too.

$$\Sigma = \begin{bmatrix} 2\sqrt{6} & 0 & 0 \\ 0 & \sqrt{6} & 0 \\ 0 & 0 & 0 \end{bmatrix} \quad \text{and} \quad V = \frac{1}{\sqrt{6}}\begin{bmatrix} \sqrt{2} & -\sqrt{3} & -1 \\ \sqrt{2} & 0 & 2 \\ \sqrt{2} & \sqrt{3} & -1 \end{bmatrix}.$$

Since 2 of the singular values are non-zero, we know that $\text{rank}(A) = 2$. We thus compute the first 2 columns of U via

$$\mathbf{u}_1 = \frac{1}{\sigma_1}A\mathbf{v}_1 = \frac{1}{2\sqrt{6}}\begin{bmatrix} 1 & 2 & 3 \\ -1 & 0 & 1 \\ 3 & 2 & 1 \end{bmatrix}\left(\frac{1}{\sqrt{3}}\begin{bmatrix} 1 \\ 1 \\ 1 \end{bmatrix}\right) = \frac{1}{\sqrt{2}}\begin{bmatrix} 1 \\ 0 \\ 1 \end{bmatrix},$$

$$\mathbf{u}_2 = \frac{1}{\sigma_2}A\mathbf{v}_2 = \frac{1}{\sqrt{6}}\begin{bmatrix} 1 & 2 & 3 \\ -1 & 0 & 1 \\ 3 & 2 & 1 \end{bmatrix}\left(\frac{1}{\sqrt{2}}\begin{bmatrix} -1 \\ 0 \\ 1 \end{bmatrix}\right) = \frac{1}{\sqrt{3}}\begin{bmatrix} 1 \\ 1 \\ -1 \end{bmatrix}.$$

The third column of U can be found by extending $\{\mathbf{u}_1, \mathbf{u}_2\}$ to an orthonormal basis of $\mathbb{R}^3$, which just means that we need to find a unit vector orthogonal to each of $\mathbf{u}_1$ and $\mathbf{u}_2$. We could do this by solving a linear system, using the Gram–Schmidt process, or via the cross product. Any of these methods quickly lead us to the vector $\mathbf{u}_3 = (1, -2, -1)/\sqrt{6}$, so the singular value decomposition $A = U\Sigma V^*$ is completed by choosing

Alternatively, $\mathbf{u}_3 = (-1, 2, 1)/\sqrt{6}$ would be fine too.

Again, we just pulled a $1/\sqrt{6}$ factor out of U to avoid fractions.

$$U = \frac{1}{\sqrt{6}}\begin{bmatrix} \sqrt{3} & \sqrt{2} & 1 \\ 0 & \sqrt{2} & -2 \\ \sqrt{3} & -\sqrt{2} & -1 \end{bmatrix}.$$

Before delving into what makes the singular value decomposition so useful, it is worth noting that if $A \in \mathcal{M}_{m,n}(\mathbb{F})$ has singular value decomposition $A = U\Sigma V^*$ then A^T and A^* have singular value decompositions

We use $\overline{V}$ to mean the entrywise complex conjugate of V. That is, $\overline{V} = (V^*)^T$.

$$A^T = \overline{V}\Sigma^T U^T \quad \text{and} \quad A^* = V\Sigma^* U^*.$$

In particular, Σ, Σ^T, and Σ^* all have the same diagonal entries, so A, A^T, and A^* all have the same singular values.

2.3.1   Geometric Interpretation and the Fundamental Subspaces

By recalling that unitary matrices correspond exactly to the linear transformations that rotate and/or reflect vectors in $\mathbb{F}^n$, but do not change their length, we see that the singular value decomposition has a very simple geometric interpretation. Specifically, it says that every matrix $A = U\Sigma V^* \in \mathcal{M}_{m,n}(\mathbb{F})$ acts as a linear transformation from $\mathbb{F}^n$ to $\mathbb{F}^m$ in the following way:

- First, it rotates and/or reflects $\mathbb{F}^n$ (i.e., V^* acts on $\mathbb{F}^n$).

- Then it stretches and/or shrinks $\mathbb{F}^n$ along the standard axes (i.e., the diagonal matrix Σ acts on $\mathbb{F}^n$) and then embeds it in $\mathbb{F}^m$ (i.e., either adds $m - n$ extra dimensions if $m > n$ or ignores $n - m$ of the dimensions if $m < n$).

- Finally, it rotates and/or reflects $\mathbb{F}^m$ (i.e., U acts on $\mathbb{F}^m$).

This geometric interpretation is illustrated in the $m = n = 2$ case in Figure 2.7. In particular, it is worth keeping track not only of how the linear transformation changes a unit square grid on $\mathbb{R}^2$ into a parallelogram grid, but also how it transforms the unit circle into an ellipse. Furthermore, the two radii of the ellipse are exactly the singular values σ_1 and σ_2 of the matrix, so we see that singular values provide another way of measuring how much a linear transformation expands space (much like eigenvalues and the determinant).

In fact, the product of a matrix's singular values equals the absolute value of its determinant (see Exercise 2.3.7).

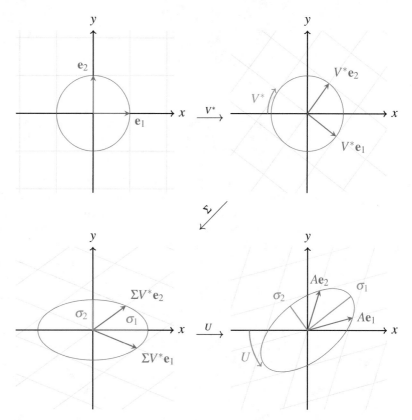

Figure 2.7: The singular value decomposition says that every linear transformation (i.e., multiplication by a matrix A) can be thought of as a rotation/reflection V^*, followed by a scaling along the standard axes Σ, followed by another rotation/reflection U.

This picture also extends naturally to higher dimensions. For example, a linear transformation acting on $\mathbb{R}^3$ sends the unit sphere to an ellipsoid whose radii are the 3 singular values of the standard matrix of that linear transformation. For example, we saw in Example 2.3.2 that the matrix

$$A = \begin{bmatrix} 1 & 2 & 3 \\ -1 & 0 & 1 \\ 3 & 2 & 1 \end{bmatrix}$$

has singular values $2\sqrt{6}, \sqrt{6}$, and 0, so we expect that A acts as a linear transformation that sends the unit sphere to an ellipsoid with radii $2\sqrt{6}, \sqrt{6}$, and 0. Since the third radius is 0, it is actually 2D ellipse living inside of $\mathbb{R}^3$—one of the dimensions is "squashed" by A, as illustrated in Figure 2.8.

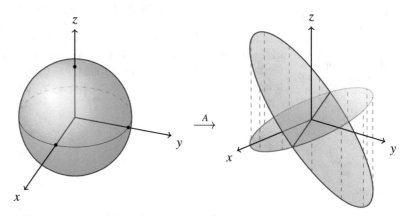

Figure 2.8: Every linear transformation on $\mathbb{R}^3$ sends the unit sphere to a (possibly degenerate) ellipsoid. The linear transformation displayed here is the one with the standard matrix A from Example 2.3.2.

The fact that the unit sphere is turned into a 2D ellipse by this matrix corresponds to the fact that it has rank 2, so its range is 2-dimensional. In fact, the first two left singular vectors (which point in the directions of the major and minor axes of the ellipse) form an orthonormal basis of the range. Similarly, the third right singular vector, $\mathbf{v}_3$ has the property that $A\mathbf{v}_3 = U\Sigma V^*\mathbf{v}_3 = U\Sigma\mathbf{e}_3 = \sigma_3 U\mathbf{e}_3 = \mathbf{0}$, since $\sigma_3 = 0$. It follows that $\mathbf{v}_3$ is in the null space of A.

This same type of argument works in general and leads to the following theorem, which shows that the singular value decomposition provides orthonormal bases for each of the four fundamental subspaces of a matrix:

Theorem 2.3.2

Bases of the Fundamental Subspaces

Let $A \in \mathcal{M}_{m,n}$ be a matrix with $\text{rank}(A) = r$ and singular value decomposition $A = U\Sigma V^*$, where

$$U = \begin{bmatrix} \mathbf{u}_1 \mid \mathbf{u}_2 \mid \cdots \mid \mathbf{u}_m \end{bmatrix} \quad \text{and} \quad V = \begin{bmatrix} \mathbf{v}_1 \mid \mathbf{v}_2 \mid \cdots \mid \mathbf{v}_n \end{bmatrix}.$$

Then

 a) $\{\mathbf{u}_1, \mathbf{u}_2, \ldots, \mathbf{u}_r\}$ is an orthonormal basis of $\text{range}(A)$,

 b) $\{\mathbf{u}_{r+1}, \mathbf{u}_{r+2}, \ldots, \mathbf{u}_m\}$ is an orthonormal basis of $\text{null}(A^*)$,

 c) $\{\mathbf{v}_1, \mathbf{v}_2, \ldots, \mathbf{v}_r\}$ is an orthonormal basis of $\text{range}(A^*)$, and

 d) $\{\mathbf{v}_{r+1}, \mathbf{v}_{r+2}, \ldots, \mathbf{v}_n\}$ is an orthonormal basis of $\text{null}(A)$.

Proof. First notice that we have

$$A\mathbf{v}_j = U\Sigma V^*\mathbf{v}_j = U\Sigma\mathbf{e}_j = \sigma_j U\mathbf{e}_j = \sigma_j\mathbf{u}_j \quad \text{for all} \quad 1 \le j \le r.$$

Dividing both sides by σ_j then shows that $\mathbf{u}_1, \mathbf{u}_2, \ldots, \mathbf{u}_r$ are all in the range of A. Since range(A) is (by definition) r-dimensional, and $\{\mathbf{u}_1, \mathbf{u}_2, \ldots, \mathbf{u}_r\}$ is a set of r mutually orthogonal unit vectors, it must be an orthonormal basis of range(A).

Similarly, $\sigma_j = 0$ whenever $j \geq r+1$, so

$$A\mathbf{v}_j = U\Sigma V^* \mathbf{v}_j = U\Sigma \mathbf{e}_j = \mathbf{0} \quad \text{for all} \quad j \geq r+1.$$

It follows that $\mathbf{v}_{r+1}, \mathbf{v}_{r+2}, \ldots, \mathbf{v}_n$ are all in the null space of A. Since the rank-nullity theorem (Theorem A.1.2(e)) tells us that null(A) has dimension $n - r$, and $\{\mathbf{v}_{r+1}, \mathbf{v}_{r+2}, \ldots, \mathbf{v}_n\}$ is a set of $n - r$ mutually orthogonal unit vectors, it must be an orthonormal basis of null(A).

The corresponding facts about range(A^*) and null(A^*) follow by applying these same arguments to A^* instead of A. ∎

Note that this theorem tells us immediately that everything in range(A) is orthogonal to everything in null(A^*), simply because the members of the set $\{\mathbf{u}_1, \mathbf{u}_2, \ldots, \mathbf{u}_r\}$ are all orthogonal to the members of $\{\mathbf{u}_{r+1}, \mathbf{u}_{r+2}, \ldots, \mathbf{u}_m\}$. A similar argument shows that everything in null(A) is orthogonal to everything range(A^*). In the terminology of Section 1.B.2, Theorem 2.3.2 shows that these fundamental subspaces are orthogonal complements of each other (i.e., this provides another proof of Theorem 1.B.7).

> Look back at Figure 1.26 for a geometric interpretation of these facts.

Example 2.3.3

Computing Bases of the Fundamental Subspaces via SVD

Compute a singular value decomposition of the matrix

$$A = \begin{bmatrix} 1 & 1 & 1 & -1 \\ 0 & 1 & 1 & 0 \\ -1 & 1 & 1 & 1 \end{bmatrix},$$

and use it to construct bases of the four fundamental subspaces of A.

Solution:

To compute the SVD of A, we could start by computing A^*A as in the previous example, but we can also construct an SVD from AA^* instead. Since A^*A is a 4×4 matrix and AA^* is a 3×3 matrix, working with AA^* will likely be easier, so that is what we do.

Direct calculation shows that

$$AA^* = \begin{bmatrix} 4 & 2 & 0 \\ 2 & 2 & 2 \\ 0 & 2 & 4 \end{bmatrix},$$

> In general, it is a good idea to compute the SVD of A from whichever of A^*A or AA^* is smaller.

which has eigenvalues $\lambda_1 = 6$, $\lambda_2 = 4$, and $\lambda_3 = 0$, so the singular values of A are $\sigma_1 = \sqrt{6}$, $\sigma_2 = \sqrt{4} = 2$, and $\sigma_3 = 0$. The normalized eigenvectors corresponding to these eigenvalues are

$$\mathbf{u}_1 = \frac{1}{\sqrt{3}} \begin{bmatrix} 1 \\ 1 \\ 1 \end{bmatrix}, \qquad \mathbf{u}_2 = \frac{1}{\sqrt{2}} \begin{bmatrix} -1 \\ 0 \\ 1 \end{bmatrix}, \quad \text{and} \quad \mathbf{u}_3 = \frac{1}{\sqrt{6}} \begin{bmatrix} -1 \\ 2 \\ -1 \end{bmatrix},$$

respectively. We then place these eigenvectors into the matrix U as columns and obtain

We put these as columns into U instead of V because we worked with AA^* instead of A^*A. Remember that U must be 3×3 and V must be 4×4.

$$\Sigma = \begin{bmatrix} \sqrt{6} & 0 & 0 & 0 \\ 0 & 2 & 0 & 0 \\ 0 & 0 & 0 & 0 \end{bmatrix} \quad \text{and} \quad U = \frac{1}{\sqrt{6}} \begin{bmatrix} \sqrt{2} & -\sqrt{3} & -1 \\ \sqrt{2} & 0 & 2 \\ \sqrt{2} & \sqrt{3} & -1 \end{bmatrix}.$$

Since 2 of the singular values are non-zero, we know that $\text{rank}(A) = 2$. We thus compute the first 2 columns of V via

Here, since we are working with AA^* instead of A^*A, we just swap the roles of U and V, and A and A^*, in Equation (2.3.1).

$$\mathbf{v}_1 = \frac{1}{\sigma_1} A^* \mathbf{u}_1 = \frac{1}{\sqrt{6}} \begin{bmatrix} 1 & 0 & -1 \\ 1 & 1 & 1 \\ 1 & 1 & 1 \\ -1 & 0 & 1 \end{bmatrix} \left(\frac{1}{\sqrt{3}} \begin{bmatrix} 1 \\ 1 \\ 1 \end{bmatrix} \right) = \frac{1}{\sqrt{2}} \begin{bmatrix} 0 \\ 1 \\ 1 \\ 0 \end{bmatrix},$$

$$\mathbf{v}_2 = \frac{1}{\sigma_2} A^* \mathbf{u}_2 = \frac{1}{2} \begin{bmatrix} 1 & 0 & -1 \\ 1 & 1 & 1 \\ 1 & 1 & 1 \\ -1 & 0 & 1 \end{bmatrix} \left(\frac{1}{\sqrt{2}} \begin{bmatrix} -1 \\ 0 \\ 1 \end{bmatrix} \right) = \frac{1}{\sqrt{2}} \begin{bmatrix} -1 \\ 0 \\ 0 \\ 1 \end{bmatrix}.$$

The third and fourth columns of V can be found by extending $\{\mathbf{v}_1, \mathbf{v}_2\}$ to an orthonormal basis of $\mathbb{R}^4$. We could do this via the Gram–Schmidt process, but in this case it is simple enough to "eyeball" vectors that work: we can choose $\mathbf{v}_3 = (0, 1, -1, 0)/\sqrt{2}$ and $\mathbf{v}_4 = (1, 0, 0, 1)/\sqrt{2}$. It follows that this singular value decomposition $A = U\Sigma V^*$ can be completed by choosing

$$V = \frac{1}{\sqrt{2}} \begin{bmatrix} 0 & -1 & 0 & 1 \\ 1 & 0 & 1 & 0 \\ 1 & 0 & -1 & 0 \\ 0 & 1 & 0 & 1 \end{bmatrix}.$$

We can then construct orthonormal bases of the four fundamental subspaces directly from Theorem 2.3.2 (recall that the rank of A is $r = 2$):

- range(A): $\{\mathbf{u}_1, \mathbf{u}_2\} = \{(1,1,1)/\sqrt{3}, (-1,0,1)/\sqrt{2}\}$,
- null(A^*): $\{\mathbf{u}_3\} = \{(-1,2,-1)/\sqrt{6}\}$,
- range(A^*): $\{\mathbf{v}_1, \mathbf{v}_2\} = \{(0,1,1,0)/\sqrt{2}, (-1,0,0,1)/\sqrt{2}\}$, and
- null(A): $\{\mathbf{v}_3, \mathbf{v}_4\} = \{(0,1,-1,0)/\sqrt{2}, (1,0,0,1)/\sqrt{2}\}$.

Remark 2.3.1

A Geometric Interpretation of the Adjoint

Keep in mind that A^* exists even if A is not square, in which case Σ^* has the same diagonal entries as Σ, but a different shape.

Up until now, the transpose of a matrix (and more generally, the adjoint of a linear transformation) has been one of the few linear algebraic concepts that we have not interpreted geometrically. The singular value decomposition lets us finally fill this gap.

Notice that if A has singular value decomposition $A = U\Sigma V^*$ then A^* and A^{-1} (if it exists) have singular value decompositions

$$A^* = V\Sigma^* U^* \quad \text{and} \quad A^{-1} = V\Sigma^{-1} U^*,$$

where Σ^{-1} is the diagonal matrix with $1/\sigma_1, \ldots, 1/\sigma_n$ on its diagonal.

In particular, each of A^* and A^{-1} act by undoing the rotations that A applies in the opposite order—they first apply U^*, which is the inverse of the unitary (rotation) U, then they apply a stretch, and then they apply V, which is the inverse of the unitary (rotation) V^*. The difference between A^* and A^{-1} is that A^{-1} also undoes the "stretch" Σ that A applies, whereas A^* simply stretches by the same amount.

We can thus think of A^* as rotating in the opposite direction as A, but stretching by the same amount (whereas A^{-1} rotates in the opposite direction as A and stretches by a reciprocal amount):

A might be more complicated than this (it might send the unit circle to an ellipse rather than to a circle), but it is a bit easier to visualize the relationships between these pictures when it sends circles to circles.

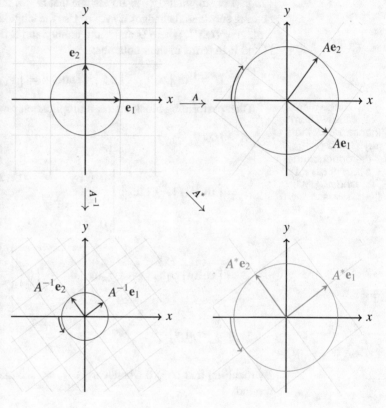

2.3.2   Relationship with Other Matrix Decompositions

The singular value decomposition also reduces to and clarifies several other matrix decompositions that we have seen, when we consider special cases of it. For example, recall from Theorem A.1.3 that if $A \in \mathcal{M}_{m,n}(\mathbb{F})$ has $\text{rank}(A) = r$ then we can find vectors $\{\mathbf{u}_j\}_{j=1}^r \subset \mathbb{F}^m$ and $\{\mathbf{v}_j\}_{j=1}^r \subset \mathbb{F}^n$ so that

This theorem is stated in Appendix A.1.4 just in the $\mathbb{F} = \mathbb{R}$ case, but it is true over any field.

$$A = \sum_{j=1}^r \mathbf{u}_j \mathbf{v}_j^T.$$

Furthermore, $\text{rank}(A)$ is the fewest number of terms possible in any such sum, and we can choose the sets $\{\mathbf{u}_j\}_{j=1}^r$ and $\{\mathbf{v}_j\}_{j=1}^r$ to be linearly independent.

One way of rephrasing the singular value decomposition is as saying that, if $\mathbb{F} = \mathbb{R}$ or $\mathbb{F} = \mathbb{C}$, then we can in fact choose the sets of vectors $\{\mathbf{u}_j\}_{j=1}^r$ and $\{\mathbf{v}_j\}_{j=1}^r$ to be mutually orthogonal (not just linearly independent):

Theorem 2.3.3

Orthogonal Rank-One Sum Decomposition

Suppose $\mathbb{F} = \mathbb{R}$ or $\mathbb{F} = \mathbb{C}$, and $A \in \mathcal{M}_{m,n}(\mathbb{F})$ has rank$(A) = r$. There exist orthonormal sets of vectors $\{\mathbf{u}_j\}_{j=1}^r \subset \mathbb{F}^m$ and $\{\mathbf{v}_j\}_{j=1}^r \subset \mathbb{F}^n$ such that

$$A = \sum_{j=1}^r \sigma_j \mathbf{u}_j \mathbf{v}_j^*,$$

where $\sigma_1 \geq \sigma_2 \geq \cdots \geq \sigma_r > 0$ are the non-zero singular values of A.

Proof. For simplicity, we again assume that $m \leq n$ throughout this proof, since nothing substantial changes if $m > n$. Use the singular value decomposition to write $A = U\Sigma V^*$, where U and V are unitary and Σ is diagonal, and then write U and V in terms of their columns:

$$U = \begin{bmatrix} \mathbf{u}_1 \mid \mathbf{u}_2 \mid \cdots \mid \mathbf{u}_m \end{bmatrix} \quad \text{and} \quad V = \begin{bmatrix} \mathbf{v}_1 \mid \mathbf{v}_2 \mid \cdots \mid \mathbf{v}_n \end{bmatrix}.$$

Furthermore, it follows from Theorem 2.3.2 that $\{\mathbf{u}_j\}_{j=1}^r$ and $\{\mathbf{v}_j\}_{j=1}^r$ are orthonormal bases of range(A) and range(A^*), respectively.

Then performing block matrix multiplication reveals that

$$A = U\Sigma V^*$$

$$= \begin{bmatrix} \mathbf{u}_1 \mid \mathbf{u}_2 \mid \cdots \mid \mathbf{u}_m \end{bmatrix} \begin{bmatrix} \sigma_1 & 0 & \cdots & 0 & 0 & \cdots & 0 \\ 0 & \sigma_2 & \cdots & 0 & 0 & \cdots & 0 \\ \vdots & \vdots & \ddots & \vdots & \vdots & \ddots & \vdots \\ 0 & 0 & \cdots & \sigma_m & 0 & \cdots & 0 \end{bmatrix} \begin{bmatrix} \mathbf{v}_1^* \\ \mathbf{v}_2^* \\ \vdots \\ \mathbf{v}_n^* \end{bmatrix}$$

$$= \begin{bmatrix} \sigma_1 \mathbf{u}_1 \mid \sigma_2 \mathbf{u}_2 \mid \cdots \mid \sigma_m \mathbf{u}_m \mid \mathbf{0} \mid \cdots \mid \mathbf{0} \end{bmatrix} \begin{bmatrix} \mathbf{v}_1^* \\ \mathbf{v}_2^* \\ \vdots \\ \mathbf{v}_n^* \end{bmatrix}$$

$$= \sum_{j=1}^m \sigma_j \mathbf{u}_j \mathbf{v}_j^*.$$

By recalling that $\sigma_j = 0$ whenever $j > r$, we see that this is exactly what we wanted. ∎

In fact, not only does the orthogonal rank-one sum decomposition follow from the singular value decomposition, but they are actually equivalent—we can essentially just follow the above proof backward to retrieve the singular value decomposition from the orthogonal rank-one sum decomposition. For this reason, this decomposition is sometimes just referred to as the singular value decomposition itself.

Example 2.3.4

Computing an Orthogonal Rank-One Sum Decomposition

Compute an orthogonal rank-one sum decomposition of the matrix

$$A = \begin{bmatrix} 1 & 1 & 1 & -1 \\ 0 & 1 & 1 & 0 \\ -1 & 1 & 1 & 1 \end{bmatrix}.$$

Solution:
This is the same matrix from Example 2.3.3, which has singular value

decomposition

$$U = \frac{1}{\sqrt{6}} \begin{bmatrix} \sqrt{2} & -\sqrt{3} & -1 \\ \sqrt{2} & 0 & 2 \\ \sqrt{2} & \sqrt{3} & -1 \end{bmatrix}, \quad \Sigma = \begin{bmatrix} \sqrt{6} & 0 & 0 & 0 \\ 0 & 2 & 0 & 0 \\ 0 & 0 & 0 & 0 \end{bmatrix} \quad \text{and}$$

$$V = \frac{1}{\sqrt{2}} \begin{bmatrix} 0 & -1 & 0 & 1 \\ 1 & 0 & 1 & 0 \\ 1 & 0 & -1 & 0 \\ 0 & 1 & 0 & 1 \end{bmatrix}.$$

Its orthogonal rank-one sum decomposition can be computed by adding up the outer products of the columns of U and V, scaled by the diagonal entries of Σ:

The term "outer product" means a product of a column vector and a row vector, like $\mathbf{u}\mathbf{v}^*$ (in contrast with an "inner product" like $\mathbf{u}^*\mathbf{v}$).

$$A = \sigma_1 \mathbf{u}_1 \mathbf{v}_1^* + \sigma_2 \mathbf{u}_2 \mathbf{v}_2^* = \begin{bmatrix} 1 \\ 1 \\ 1 \end{bmatrix} \begin{bmatrix} 0 & 1 & 1 & 0 \end{bmatrix} + \begin{bmatrix} -1 \\ 0 \\ 1 \end{bmatrix} \begin{bmatrix} -1 & 0 & 0 & 1 \end{bmatrix}.$$

If $A \in \mathcal{M}_n$ is positive semidefinite then the singular value decomposition coincides exactly with the spectral decomposition. Indeed, the spectral decomposition says that we can write $A = UDU^*$ with U unitary and D diagonal (with non-negative diagonal entries, thanks to A being positive semidefinite). This is also a singular value decomposition of A, with the added structure of having the two unitary matrices U and V being equal to each other. This immediately tells us the following important fact:

> ⚠ If $A \in \mathcal{M}_n$ is positive semidefinite then its singular values equal its eigenvalues.

Slightly more generally, there is also a close relationship between the singular value decomposition of a normal matrix and its spectral decomposition (and thus the singular values of a normal matrix and its eigenvalues):

Theorem 2.3.4

Singular Values of Normal Matrices

If $A \in \mathcal{M}_n$ is a normal matrix then its singular values are the absolute values of its eigenvalues.

Proof. Since A is normal, we can use the spectral decomposition to write $A = UDU^*$, where U is unitary and D is diagonal (with the not-necessarily-positive eigenvalues of A on its diagonal). If we write each λ_j in its polar form $\lambda_j = |\lambda_j| e^{i\theta_j}$ then

Alternatively, we could prove this theorem by recalling that the singular values of A are the square roots of A^*A. If A is normal then $A = UDU^*$, so $A^*A = UD^*DU^*$, which has eigenvalues $\overline{\lambda_j}\lambda_j = |\lambda_j|^2$.

$$D = \begin{bmatrix} \lambda_1 & 0 & \cdots & 0 \\ 0 & \lambda_2 & \cdots & 0 \\ \vdots & \vdots & \ddots & \vdots \\ 0 & 0 & \cdots & \lambda_n \end{bmatrix} = \underbrace{\begin{bmatrix} |\lambda_1| & 0 & \cdots & 0 \\ 0 & |\lambda_2| & \cdots & 0 \\ \vdots & \vdots & \ddots & \vdots \\ 0 & 0 & \cdots & |\lambda_n| \end{bmatrix}}_{\text{call this } \Sigma} \underbrace{\begin{bmatrix} e^{i\theta_1} & 0 & \cdots & 0 \\ 0 & e^{i\theta_2} & \cdots & 0 \\ \vdots & \vdots & \ddots & \vdots \\ 0 & 0 & \cdots & e^{i\theta_n} \end{bmatrix}}_{\text{call this } \Theta},$$

so that $A = UDU^* = U\Sigma\Theta U^* = U\Sigma(U\Theta^*)^*$. Since U and Θ are both unitary, so is $U\Theta^*$, so this is a singular value decomposition of A. It follows that the diagonal entries of Σ (i.e., $|\lambda_1|, |\lambda_2|, \ldots, |\lambda_n|$) are the singular values of A. ∎

To see that the above theorem does not hold for non-normal matrices, consider the matrix

$$A = \begin{bmatrix} 1 & 1 \\ 0 & 1 \end{bmatrix},$$

which has both eigenvalues equal to 1 (it is upper triangular, so its eigenvalues are its diagonal entries). However, its singular values are $(\sqrt{5} \pm 1)/2$. In fact, the conclusion of Theorem 2.3.4 does not hold for *any* non-normal matrix (see Exercise 2.3.20).

Finally, we note that the singular value decomposition is also "essentially equivalent" to the polar decomposition:

> (!) If $A \in \mathcal{M}_n$ has singular value decomposition $A = U\Sigma V^*$ then $A = (UV^*)(V\Sigma V^*)$ is a polar decomposition of A.

Keep in mind that this argument only works when A is square, but the SVD is more general and also applies to rectangular matrices.

The reason that the above fact holds is simply that UV^* is unitary whenever U and V are, and $V\Sigma V^*$ is positive semidefinite. This argument also works in reverse—if we start with a polar decomposition $A = UP$ and apply the spectral decomposition to $P = V\Sigma V^*$, we get $A = (UV)\Sigma V^*$ which is a singular value decomposition of A.

2.3.3 The Operator Norm

One of the primary uses of singular values is that they provide us with ways of measuring "how big" a matrix is. We have already seen two ways of doing this:

See Appendix A.1.5 for a refresher on this geometric interpretation of the determinant.

- The absolute value of the determinant of a matrix measures how much it expands space when acting as a linear transformation. That is, it is the area (or volume, or hypervolume, depending on the dimension) of the output of the unit square, cube, or hypercube after it is acted upon by the matrix.

- The Frobenius norm, which is simply the norm induced by the inner product in $\mathcal{M}_{m,n}$.

In fact, both of the above quantities come directly from singular values—we will see in Exercise 2.3.7 that if $A \in \mathcal{M}_n$ has singular values $\sigma_1, \sigma_2, \ldots, \sigma_n$ then $|\det(A)| = \sigma_1\sigma_2\cdots\sigma_n$, and we will see in Theorem 2.3.7 that $\|A\|_F^2 = \sigma_1^2 + \sigma_2^2 + \cdots + \sigma_n^2$.

However, the Frobenius norm is in some ways a very silly matrix norm. It is really "just" a vector norm—it only cares about the vector space structure of $\mathcal{M}_{m,n}$, not the fact that we can multiply matrices. There is thus no geometric interpretation of the Frobenius norm in terms of how the matrix acts as a linear transformation. For example, we can rearrange the entries of a matrix freely without changing its Frobenius norm, but doing so completely changes how it acts geometrically. In practice, the Frobenius norm is often just used for computational convenience—it is often not the "right" norm to work with, but it is so dirt easy to compute that we let it slide.

We now introduce another norm (i.e., way of measuring the "size" of a matrix) that really tells us something fundamental about how that matrix acts as a linear transformation. Specifically, it tells us the largest factor by which a matrix can stretch a vector. Also, as suggested earlier, we will see that this norm can also be phrased in terms of singular values (it simply equals the largest of them).

Definition 2.3.1 **Operator Norm**	Suppose $\mathbb{F} = \mathbb{R}$ or $\mathbb{F} = \mathbb{C}$, and $A \in \mathcal{M}_{m,n}(\mathbb{F})$. The **operator norm** of A, denoted by $\|A\|$, is any of the following (equivalent) quantities:

$$\|A\| \stackrel{\text{def}}{=} \max_{\mathbf{v} \in \mathbb{F}^n} \left\{ \frac{\|A\mathbf{v}\|}{\|\mathbf{v}\|} : \mathbf{v} \neq \mathbf{0} \right\}$$
$$= \max_{\mathbf{v} \in \mathbb{F}^n} \left\{ \|A\mathbf{v}\| : \|\mathbf{v}\| \leq 1 \right\}$$
$$= \max_{\mathbf{v} \in \mathbb{F}^n} \left\{ \|A\mathbf{v}\| : \|\mathbf{v}\| = 1 \right\}.$$

As its name suggests, the operator norm really is a norm in the sense of Section 1.D (see Exercise 2.3.16).

The fact that the three maximizations above really are equivalent to each other follows simply from rescaling the vectors $\mathbf{v}$ that are being maximized over. In particular, $\mathbf{v}$ is a vector that maximizes $\|A\mathbf{v}\|/\|\mathbf{v}\|$ (i.e., the first maximization) if and only if $\mathbf{v}/\|\mathbf{v}\|$ is a unit vector that maximizes $\|A\mathbf{v}\|$ (i.e., the second and third maximizations). The operator norm is typically considered the "default" matrix norm, so if we use the notation $\|A\|$ without any subscripts or other indicators, we typically mean the operator norm (just like $\|\mathbf{v}\|$ for a vector $\mathbf{v} \in \mathbb{F}^n$ typically refers to the norm induced by the dot product if no other context is provided).

Remark 2.3.2 **Induced Matrix Norms**	In the definition of the operator norm (Definition 2.3.1), the norm used on vectors in $\mathbb{F}^n$ is the usual norm induced by the dot product:

$$\|\mathbf{v}\| = \sqrt{\mathbf{v} \cdot \mathbf{v}} = \sqrt{|v_1|^2 + |v_2|^2 + \cdots + |v_n|^2}.$$

Most of the results from later in this section break down if we use a weird norm on the input and output vector space. For example, induced matrix norms are often very difficult to compute, but the operator norm is easy to compute.

However, it is also possible to define matrix norms (or more generally, norms of linear transformations between any two normed vector spaces— see Section 1.D) based on any norms on the input and output spaces. That is, given any normed vector spaces $\mathcal{V}$ and $\mathcal{W}$, we define the **induced norm** of a linear transformation $T : \mathcal{V} \to \mathcal{W}$ by

$$\|T\| \stackrel{\text{def}}{=} \max_{\mathbf{v} \in \mathcal{V}} \left\{ \|T(\mathbf{v})\|_{\mathcal{W}} : \|\mathbf{v}\|_{\mathcal{V}} = 1 \right\},$$

and the geometric interpretation of this norm is similar to that of the operator norm—$\|T\|$ measures how much T can stretch a vector, when "stretch" is measured in whatever norms we chose for $\mathcal{V}$ and $\mathcal{W}$.

Notice that a matrix cannot stretch any vector by more than a multiplicative factor of its operator norm. That is, if $A \in \mathcal{M}_{m,n}$ and $B \in \mathcal{M}_{n,p}$ then $\|A\mathbf{w}\| \leq \|A\|\|\mathbf{w}\|$ and $\|B\mathbf{v}\| \leq \|B\|\|\mathbf{v}\|$ for all $\mathbf{v} \in \mathbb{F}^p$ and $\mathbf{w} \in \mathbb{F}^n$. It follows that

Be careful: in expressions like $\|A\|\|\mathbf{w}\|$, the first norm is the operator norm (of a matrix) and the second norm is the norm (of a vector) induced by the dot product.

$$\|(AB)\mathbf{v}\| = \|A(B\mathbf{v})\| \leq \|A\|\|B\mathbf{v}\| \leq \|A\|\|B\|\|\mathbf{v}\| \quad \text{for all} \quad \mathbf{v} \in \mathbb{F}^p.$$

Dividing both sides by $\|\mathbf{v}\|$ shows that $\|(AB)\mathbf{v}\|/\|\mathbf{v}\| \leq \|A\|\|B\|$ for all $\mathbf{v} \neq \mathbf{0}$, so $\|AB\| \leq \|A\|\|B\|$. We thus say that the operator norm is **submultiplicative**. It turns out that the Frobenius norm is also submultiplicative, and we state these two results together as a theorem.

Let $A \in \mathcal{M}_{m,n}$ and $B \in \mathcal{M}_{n,p}$. Then

$$\|AB\| \leq \|A\|\|B\| \quad \text{and} \quad \|AB\|_{\mathrm{F}} \leq \|A\|_{\mathrm{F}}\|B\|_{\mathrm{F}}.$$

Proof. We already proved submultiplicativity for the operator norm, and we leave submultiplicativity of the Frobenius norm to Exercise 2.3.12. ∎

Unlike the Frobenius norm, it is not completely obvious how to actually compute the operator norm. It turns out that singular values save us here, and to see why, recall that every matrix sends the unit circle (or sphere, or hypersphere...) to an ellipse (or ellipsoid, or hyperellipsoid...). The operator norm asks for the norm of the longest vector on that ellipse, as illustrated in Figure 2.9. By comparing this visualization with the one from Figure 2.7, it should seem believable that the operator norm of a matrix just equals its largest singular value.

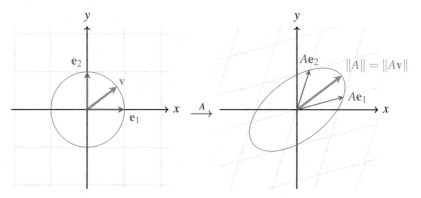

Figure 2.9: A visual representation of the operator norm. Matrices (linear transformations) transform the unit circle into an ellipse. The operator norm is the distance of the farthest point on that ellipse from the origin (i.e., the length of its semi-major axis).

To prove this relationship between the operator norm and singular values algebraically, we first need the following helper theorem that shows that multiplying a matrix on the left or right by a unitary matrix does not change its operator norm. For this reason, we say that the operator norm is **unitarily invariant**, and it turns out that the Frobenius norm also has this property:

Let $A \in \mathcal{M}_{m,n}$ and suppose $U \in \mathcal{M}_m$ and $V \in \mathcal{M}_n$ are unitary matrices. Then

$$\|UAV\| = \|A\| \quad \text{and} \quad \|UAV\|_{\mathrm{F}} = \|A\|_{\mathrm{F}}.$$

Proof. For the operator norm, we start by showing that every unitary matrix $U \in \mathcal{M}_m$ has $\|U\| = 1$. To this end, just recall from Theorem 1.4.9 that $\|U\mathbf{v}\| = \|\mathbf{v}\|$ for all $\mathbf{v} \in \mathbb{F}^m$, so $\|U\mathbf{v}\|/\|\mathbf{v}\| = 1$ for all $\mathbf{v} \neq \mathbf{0}$, which implies $\|U\| = 1$.

The fact that $\|U\| = 1$ whenever U is unitary is useful. Remember it!

We then know from submultiplicativity of the operator norm that

$$\|UAV\| \leq \|U\|\|AV\| \leq \|U\|\|A\|\|V\| = \|A\|.$$

However, by cleverly using the fact that $U^*U = I$ and $VV^* = I$, we can also deduce the opposite inequality:

$$\|A\| = \|(U^*U)A(VV^*)\| = \|U^*(UAV)V^*\| \leq \|U^*\|\|UAV\|\|V^*\| = \|UAV\|.$$

We thus conclude that $\|A\| = \|UAV\|$, as desired.

Unitary invariance for the Frobenius norm is somewhat more straightforward—we can directly compute

$$\begin{aligned}
\|UAV\|_{\mathrm{F}}^2 &= \mathrm{tr}\big((UAV)^*(UAV)\big) && \text{(definition of } \|\cdot\|_{\mathrm{F}}) \\
&= \mathrm{tr}(V^*A^*U^*UAV) && \text{(expand parentheses)} \\
&= \mathrm{tr}(V^*A^*AV) && (U \text{ is unitary}) \\
&= \mathrm{tr}(VV^*A^*A) && \text{(cyclic commutativity)} \\
&= \mathrm{tr}(A^*A) = \|A\|_{\mathrm{F}}^2. && (V \text{ is unitary}) \qquad\blacksquare
\end{aligned}$$

By combining unitary invariance with the singular value decomposition, we almost immediately confirm our observation that the operator norm should equal the matrix's largest singular value, and we also get a new formula for the Frobenius norm:

Theorem 2.3.7

Matrix Norms in Terms of Singular Values

Suppose $A \in \mathcal{M}_{m,n}$ has rank r and singular values $\sigma_1 \geq \sigma_2 \geq \cdots \geq \sigma_r > 0$. Then

$$\|A\| = \sigma_1 \quad \text{and} \quad \|A\|_{\mathrm{F}} = \sqrt{\sum_{j=1}^{r} \sigma_j^2}.$$

Proof. If we write A in its singular value decomposition $A = U\Sigma V^*$, then unitary invariance tells us that $\|A\| = \|\Sigma\|$ and $\|A\|_{\mathrm{F}} = \|\Sigma\|_{\mathrm{F}}$. The fact that $\|\Sigma\|_{\mathrm{F}} = \sqrt{\sum_{j=1}^{r} \sigma_j^2}$ then follows immediately from the fact that $\sigma_1, \sigma_2, \ldots, \sigma_r$ are the non-zero entries of Σ.

To see that $\|\Sigma\| = \sigma_1$, first note that direct matrix multiplication shows that

Recall that $\mathbf{e}_1 = (1, 0, \ldots, 0)$.

$$\|\Sigma \mathbf{e}_1\| = \|(\sigma_1, 0, \ldots, 0)\| = \sigma_1,$$

so $\|\Sigma\| \geq \sigma_1$. For the opposite inequality, note that for every $\mathbf{v} \in \mathbb{F}^n$ we have

$$\|\Sigma \mathbf{v}\| = \|(\sigma_1 v_1, \ldots, \sigma_r v_r, 0, \ldots, 0)\| = \sqrt{\sum_{i=1}^{r} \sigma_i^2 |v_i|^2} \leq \sqrt{\sigma_1^2 \sum_{i=1}^{r} |v_i|^2} \leq \sigma_1 \|\mathbf{v}\|.$$

since $\sigma_1 \geq \sigma_2 \geq \cdots \geq \sigma_r$ ↑

By dividing both sides of this inequality by $\|\mathbf{v}\|$, we see that $\|\Sigma \mathbf{v}\|/\|\mathbf{v}\| \leq \sigma_1$ whenever $\mathbf{v} \neq \mathbf{0}$, so $\|\Sigma\| \leq \sigma_1$. Since we already proved the opposite inequality, we conclude that $\|\Sigma\| = \sigma_1$, which completes the proof. $\blacksquare$

Example 2.3.5

Computing Matrix Norms

Compute the operator and Frobenius norms of $A = \begin{bmatrix} 1 & 2 & 3 \\ -1 & 0 & 1 \\ 3 & 2 & 1 \end{bmatrix}$.

Solution:

We saw in Example 2.3.2 that this matrix has non-zero singular values $\sigma_1 = 2\sqrt{6}$ and $\sigma_2 = \sqrt{6}$. By Theorem 2.3.7, it follows that

$$\|A\| = \sigma_1 = 2\sqrt{6} \quad \text{and}$$

$$\|A\|_{\mathrm{F}} = \sqrt{\sigma_1^2 + \sigma_2^2} = \sqrt{(2\sqrt{6})^2 + (\sqrt{6})^2} = \sqrt{24 + 6} = \sqrt{30}.$$

It is worth pointing out, however, that if we had not already precomputed the singular values of A, it would be quicker and easier to compute its Frobenius norm directly from the definition:

$$\|A\|_F = \sqrt{\sum_{i,j=1}^{3} |a_{i,j}|^2}$$

$$= \sqrt{|1|^2 + |2|^2 + |3|^2 + |-1|^2 + |0|^2 + |1|^2 + |3|^2 + |2|^2 + |1|^2}$$

$$= \sqrt{30}.$$

The characterization of the operator and Frobenius norms in terms of singular values are very useful for the fact that they provide us with several immediate corollaries that are not obvious from their definitions. For example, if $A \in \mathcal{M}_{m,n}$ then $\|A\| = \|A^T\| = \|A^*\|$ since singular values are unchanged by taking the (conjugate) transpose of a matrix (and similarly, $\|A\|_F = \|A^T\|_F = \|A^*\|_F$, but this can also be proved directly from the definition of the Frobenius norm).

> In fact, this property of the Frobenius norm was proved back in Exercise 1.3.10.

The following property is slightly less obvious, and provides us with one important condition under which equality is obtained in the submultiplicativity inequality $\|AB\| \leq \|A\|\|B\|$:

Corollary 2.3.8

The C^*-Property of the Operator Norm

If $A \in \mathcal{M}_{m,n}$ then $\|A^*A\| = \|A\|^2$.

Proof. Start by writing A in its singular value decomposition $A = U\Sigma V^*$, where Σ has largest diagonal entry σ_1 (the largest singular value of A). Then

$$\|A^*A\| = \|(U\Sigma V^*)^*(U\Sigma V^*)\| = \|V\Sigma^* U^* U\Sigma V^*\| = \|V\Sigma^*\Sigma V^*\| = \|\Sigma^*\Sigma\|,$$

with the final equality following from unitary invariance of the operator norm (Theorem 2.3.6).

> We use Theorem 2.3.7 twice here at the end: once to see that $\|\Sigma^*\Sigma\| = \sigma_1^2$ and once to see that $\sigma_1 = \|A\|$.

Since $\Sigma^*\Sigma$ is a diagonal matrix with largest entry σ_1^2, it follows that $\|\Sigma^*\Sigma\| = \sigma_1^2 = \|A\|^2$, which completes the proof. ∎

We close this section by noting that there are actually many matrix norms out there (just like we saw that there are many vector norms in Section 1.D), and many of the most useful ones come from singular values just like the operator and Frobenius norms. We explore another particularly important matrix norm of this type in Exercises 2.3.17–2.3.19.

Exercises

solutions to starred exercises on page 470

2.3.1 Compute a singular value decomposition of each of the following matrices.

* (a) $\begin{bmatrix} -1 & 3 \\ 3 & -1 \end{bmatrix}$

(b) $\begin{bmatrix} 1 & 1 & -1 \\ -1 & 1 & 1 \end{bmatrix}$

* (c) $\begin{bmatrix} 0 & 2 \\ 1 & 1 \\ -2 & 0 \end{bmatrix}$

(d) $\begin{bmatrix} 1 & 1 & -1 \\ 0 & 1 & 1 \\ 1 & 0 & 1 \end{bmatrix}$

* (e) $\begin{bmatrix} 2 & 2 & -2 \\ -4 & -1 & 4 \\ -4 & 2 & 4 \end{bmatrix}$

(f) $\begin{bmatrix} 1 & 1 & 1 & 1 \\ 2 & 2 & -2 & -2 \\ 3 & -3 & 3 & -3 \end{bmatrix}$

2.3.2 Compute the operator norm of each of the matrices from Exercise 2.3.1.

2.3.3 Compute orthonormal bases of the four fundamental subspaces of each of the matrices from Exercise 2.3.1.

2.3.4 Determine which of the following statements are true and which are false.

* (a) If λ is an eigenvalue of $A \in \mathcal{M}_n(\mathbb{C})$ then $|\lambda|$ is a singular value of A.
  (b) If σ is a singular value of $A \in \mathcal{M}_n(\mathbb{C})$ then σ^2 is a singular value of A^2.
* (c) If σ is a singular value of $A \in \mathcal{M}_{m,n}(\mathbb{C})$ then σ^2 is a singular value of A^*A.
  (d) If $A \in \mathcal{M}_{m,n}(\mathbb{C})$ then $\|A^*A\| = \|A\|^2$.
* (e) If $A \in \mathcal{M}_{m,n}(\mathbb{C})$ then $\|A^*A\|_F = \|A\|_F^2$.
  (f) If $A \in \mathcal{M}_n(\mathbb{C})$ is a diagonal matrix then its singular values are its diagonal entries.
* (g) If $A \in \mathcal{M}_n(\mathbb{C})$ has singular value decomposition $A = UDV^*$ then $A^2 = UD^2V^*$.
  (h) If $U \in \mathcal{M}_n$ is unitary then $\|U\| = 1$.
* (i) Every matrix has the same singular values as its transpose.

∗∗2.3.5 Show that $A \in \mathcal{M}_n$ is unitary if and only if all of its singular values equal 1.

∗∗2.3.6 Suppose $\mathbb{F} = \mathbb{R}$ or $\mathbb{F} = \mathbb{C}$, and $A \in \mathcal{M}_n(\mathbb{F})$. Show that $\operatorname{rank}(A) = r$ if and only if there exists a unitary matrix $U \in \mathcal{M}_n(\mathbb{F})$ and an invertible matrix $Q \in \mathcal{M}_n(\mathbb{F})$ such that

$$A = U \begin{bmatrix} I_r & O \\ O & O \end{bmatrix} Q.$$

∗∗2.3.7 Suppose $A \in \mathcal{M}_n$ has singular values $\sigma_1, \sigma_2, \ldots, \sigma_n$. Show that

$$|\det(A)| = \sigma_1 \sigma_2 \cdots \sigma_n.$$

2.3.8 Show that if λ is an eigenvalue of $A \in \mathcal{M}_n$ then $|\lambda| \leq \|A\|$.

2.3.9 Show that if $A \in \mathcal{M}_{m,n}$ then

$$\|A\| \leq \|A\|_F \leq \sqrt{\operatorname{rank}(A)} \|A\|.$$

∗2.3.10 Let $A \in \mathcal{M}_n$ be an invertible matrix. Show that $\|A^{-1}\| \geq 1/\|A\|$, and give an example where equality does not hold.

2.3.11 Suppose $a, b \in \mathbb{R}$ and

$$A = \begin{bmatrix} a & -b \\ b & a \end{bmatrix}.$$

Show that $\|A\| = \sqrt{a^2 + b^2}$.

[Side note: Recall from Remark A.3.2 that this matrix represents the complex number $a + bi$. This exercise shows that the operator norm gives exactly the magnitude of that complex number.]

∗∗2.3.12 Let $A \in \mathcal{M}_{m,n}$ and $B \in \mathcal{M}_{n,p}$.

(a) Show that $\|AB\|_F \leq \|A\|\|B\|_F$.
    [Hint: Apply the singular value decomposition to A.]
(b) Use part (a) to show that $\|AB\|_F \leq \|A\|_F\|B\|_F$.

∗2.3.13 Suppose $A \in \mathcal{M}_{m,n}$. Show that

$$\|A\| = \max_{\mathbf{v} \in \mathbb{F}^m, \mathbf{w} \in \mathbb{F}^n} \{|\mathbf{v}^*A\mathbf{w}| : \|\mathbf{v}\| = \|\mathbf{w}\| = 1\}.$$

2.3.14 Let $A \in \mathcal{M}_{m,n}$ have singular values $\sigma_1 \geq \sigma_2 \geq \cdots \geq 0$. Show that the block matrix

$$\begin{bmatrix} O & A \\ A^* & O \end{bmatrix}$$

has eigenvalues $\pm\sigma_1, \pm\sigma_2, \ldots$, together with $|m - n|$ extra 0 eigenvalues.

∗∗2.3.15 Suppose $A \in \mathcal{M}_{m,n}$ and $c \in \mathbb{R}$ is a scalar. Show that the block matrix

$$\begin{bmatrix} cI_m & A \\ A^* & cI_n \end{bmatrix}$$

is positive semidefinite if and only if $\|A\| \leq c$.

∗∗2.3.16 Show that the operator norm (Definition 2.3.1) is in fact a norm (i.e., satisfies the three properties of Definition 1.D.1).

∗∗2.3.17 The **trace norm** of a matrix $A \in \mathcal{M}_{m,n}$ is the sum of its singular values $\sigma_1, \sigma_2, \ldots, \sigma_r$:

$$\|A\|_{\operatorname{tr}} \stackrel{\text{def}}{=} \sigma_1 + \sigma_2 + \cdots + \sigma_r.$$

(a) Show that

$$\|A\|_{\operatorname{tr}} = \max_{B \in \mathcal{M}_{m,n}} \{|\langle A, B \rangle| : \|B\| \leq 1\}.$$

[Side note: For this reason, the trace norm and operator norm are sometimes said to be **dual** to each other.]
[Hint: To show the "$\geq$" inequality, cleverly pick B based on A's SVD. To show the (harder) "$\leq$" inequality, maybe also use Exercise 2.3.13.]

(b) Use part (a) to show that the trace norm is in fact a norm in the sense of Definition 1.D.1. That is, show that the trace norm satisfies the three properties of that definition.

2.3.18 Compute the trace norm (see Exercise 2.3.17) of each of the matrices from Exercise 2.3.1.

∗∗2.3.19 Show that the trace norm from Exercise 2.3.17 is unitarily-invariant. That is, show that if $A \in \mathcal{M}_{m,n}$, and $U \in \mathcal{M}_m$ and $V \in \mathcal{M}_n$ are unitary matrices, then $\|UAV\|_{\operatorname{tr}} = \|A\|_{\operatorname{tr}}$.

∗∗2.3.20 Prove the converse of Theorem 2.3.4. That is, suppose that $A \in \mathcal{M}_n(\mathbb{C})$ has eigenvalues $\lambda_1, \ldots, \lambda_n$ with $|\lambda_1| \geq \cdots \geq |\lambda_n|$ and singular values $\sigma_1 \geq \cdots \geq \sigma_n$. Show that if A is not normal then $\sigma_j \neq |\lambda_j|$ for some $1 \leq j \leq n$.

2.3.21 In this exercise, we show that every square matrix is a linear combination of two unitary matrices. Suppose $A \in \mathcal{M}_n(\mathbb{C})$ has singular value decomposition $A = U\Sigma V^*$.

(a) Show that if $\|A\| \leq 1$ then $I - \Sigma^2$ is positive semidefinite.
(b) Show that if $\|A\| \leq 1$ then $U(\Sigma \pm i\sqrt{I - \Sigma^2})V^*$ are both unitary matrices.
(c) Use part (b) to show that A can be written as a linear combination of two unitary matrices (regardless of the value of $\|A\|$).

2.3.22　Suppose that the 2×2 block matrix

$$\begin{bmatrix} A & B \\ B^* & C \end{bmatrix}$$

is positive semidefinite. Show that range$(B) \subseteq$ range(A).

[Hint: Show instead that null$(A) \subseteq$ null(B^*).]

2.3.23　Use the Jordan–von Neumann theorem (Theorem 1.D.8) to determine whether or not the operator norm is induced by some inner product on $\mathcal{M}_{m,n}$.

2.3.24　Suppose $A \in \mathcal{M}_{m,n}$ has singular values σ_1, σ_2, ..., σ_p (where $p = \min\{m,n\}$) and QR decomposition (see Section 1.C) $A = UT$ with $U \in \mathcal{M}_m$ unitary and $T \in \mathcal{M}_{m,n}$ upper triangular. Show that the product of the singular values of A equals the product of the diagonal entries of T:

$$\sigma_1 \cdot \sigma_2 \cdots \sigma_p = t_{1,1} \cdot t_{2,2} \cdots t_{p,p}.$$

∗∗2.3.25　Suppose $P \in \mathcal{M}_n(\mathbb{C})$ is a non-zero projection (i.e., $P^2 = P$).

(a) Show that $\|P\| \geq 1$.
(b) Show that if $P^* = P$ (i.e., P is an *orthogonal* projection) then $\|P\| = 1$.
(c) Show that if $\|P\| = 1$ then $P^* = P$.
  [Hint: Schur triangularize P.]
  [Side note: This can be seen as the converse of Theorem 1.4.13.]

∗∗2.3.26　A matrix $A \in \mathcal{M}_n(\mathbb{C})$ is called **complex symmetric** if $A^T = A$. For example,

$$A = \begin{bmatrix} 1 & i \\ i & 2-i \end{bmatrix}$$

is complex symmetric. In this exercise, we show that the singular value decomposition of these matrices can be chosen to have a special form.

(a) Provide an example to show that a complex symmetric matrix may not be normal and thus may not have a spectral decomposition.
(b) Suppose $A \in \mathcal{M}_n(\mathbb{C})$ is complex symmetric. Show that there exists a unitary matrix $V \in \mathcal{M}_n(\mathbb{C})$ such that, if we define $B = V^T A V$, then B is complex symmetric and B^*B is real.
  [Hint: Apply the spectral decomposition to A^*A.]
(c) Let B be as in part (b) and define $B_R = (B + B^*)/2$ and $B_I = (B - B^*)/(2i)$. Show that B_R and B_I are real, symmetric, and commute.
  [Hint: $B = B_R + iB_I$ and B^*B is real.]
  [Side note: Here we are using the Cartesian decomposition of B introduced in Remark 1.B.1.]
(d) Let B be as in part (b). Show that there is a unitary matrix $W \in \mathcal{M}_n(\mathbb{R})$ such that $W^T B W$ is diagonal.
  [Hint: Use Exercise 2.1.28—which matrices have we found that commute?]
(e) Use the unitary matrices V and W from parts (b) and (d) of this exercise to conclude that there exists a unitary matrix $U \in \mathcal{M}_n(\mathbb{C})$ and a diagonal matrix $D \in \mathcal{M}_n(\mathbb{R})$ with non-negative entries such that

$$A = UDU^T.$$

[Side note: This is called the **Takagi factorization** of A. Be somewhat careful—the entries on the diagonal of D are the singular values of A, not its eigenvalues.]

2.4　The Jordan Decomposition

All of the decompositions that we have introduced so far in this chapter—Schur triangularization (Theorem 2.1.1), the spectral decomposition (Theorems 2.1.4 and 2.1.6), the polar decomposition (Theorem 2.2.12), and the singular value decomposition (Theorem 2.3.1)—have focused on how much simpler we can make a matrix by either multiplying it by a unitary or applying a unitary similarity transformation to it.

We now switch gears a bit and return to the setting of diagonalization, where we allow for *arbitrary* (not necessarily unitary) similarity transformations, and we investigate how "close" to diagonal we can make a non-diagonalizable matrix. Solving this problem and coming up with a way to "almost-diagonalize" arbitrary matrices is important for at least two reasons:

Recall that diagonalization itself was characterized in Theorem 2.0.1. See also Appendix A.1.7 if you need a refresher.

- A linear transformation can be represented by its standard matrix, but by changing the basis that we are working in, the entries of that standard matrix can change considerably. The decomposition in this section tells us how to answer the question of whether or not two matrices represent the same linear transformation in different bases.

- The decomposition that we see in this section will provide us with a method of applying functions like e^x and $\sin(x)$ to matrices, rather than just to numbers. Some readers may have already learned how to apply functions to *diagonalizable* matrices, but going beyond diagonalizable matrices requires some extra mathematical machinery.

To get us started toward the decomposition that solves these problems, we first need a definition that suggests the "almost diagonal" form that we will be transforming matrices into.

Definition 2.4.1 **Jordan Blocks**	Given a scalar $\lambda \in \mathbb{C}$ and an integer $k \geq 1$, the **Jordan block** of **order k** corresponding to λ is the matrix $J_k(\lambda) \in \mathcal{M}_k(\mathbb{C})$ of the form

We say that $J_k(\lambda)$ has λ along its diagonal and 1 along its "superdiagonal".

$$J_k(\lambda) = \begin{bmatrix} \lambda & 1 & 0 & \cdots & 0 & 0 \\ 0 & \lambda & 1 & \cdots & 0 & 0 \\ 0 & 0 & \lambda & \cdots & 0 & 0 \\ \vdots & \vdots & \vdots & \ddots & \vdots & \vdots \\ 0 & 0 & 0 & \cdots & \lambda & 1 \\ 0 & 0 & 0 & \cdots & 0 & \lambda \end{bmatrix}.$$

For example, the following matrices are all Jordan blocks:

$$\begin{bmatrix} 7 \end{bmatrix}, \quad \begin{bmatrix} 4 & 1 \\ 0 & 4 \end{bmatrix}, \quad \text{and} \quad \begin{bmatrix} -2 & 1 & 0 \\ 0 & -2 & 1 \\ 0 & 0 & -2 \end{bmatrix},$$

but the following matrices are not:

$$\begin{bmatrix} 4 & 0 \\ 0 & 4 \end{bmatrix}, \quad \begin{bmatrix} 2 & 1 \\ 0 & 3 \end{bmatrix}, \quad \text{and} \quad \begin{bmatrix} 3 & 1 & 0 \\ 0 & 3 & 0 \\ 0 & 0 & 3 \end{bmatrix}.$$

The main result of this section says that every matrix is similar to one that is block diagonal, and whose diagonal blocks are Jordan blocks:

Theorem 2.4.1 **Jordan** **Decomposition**	If $A \in \mathcal{M}_n(\mathbb{C})$ then there exists an invertible matrix $P \in \mathcal{M}_n(\mathbb{C})$ and Jordan blocks $J_{k_1}(\lambda_1), J_{k_2}(\lambda_2), \ldots, J_{k_m}(\lambda_m)$ such that

$$A = P \begin{bmatrix} J_{k_1}(\lambda_1) & O & \cdots & O \\ O & J_{k_2}(\lambda_2) & \cdots & O \\ \vdots & \vdots & \ddots & \vdots \\ O & O & \cdots & J_{k_m}(\lambda_m) \end{bmatrix} P^{-1}.$$

We will see shortly that the numbers $\lambda_1, \lambda_2, \ldots, \lambda_m$ are necessarily the eigenvalues of A listed according to geometric multiplicity. Also, we must have $k_1 + \cdots + k_m = n$.

Furthermore, this block diagonal matrix is called the **Jordan canonical form** of A, and it is unique up to re-ordering the diagonal blocks.

For example, the following matrices are in Jordan canonical form:

$$\begin{bmatrix} 4 & 0 & 0 & 0 \\ 0 & 2 & 1 & 0 \\ 0 & 0 & 2 & 1 \\ 0 & 0 & 0 & 2 \end{bmatrix}, \quad \begin{bmatrix} 2 & 1 & 0 & 0 \\ 0 & 2 & 0 & 0 \\ 0 & 0 & 4 & 0 \\ 0 & 0 & 0 & 5 \end{bmatrix}, \quad \text{and} \quad \begin{bmatrix} 3 & 1 & 0 & 0 \\ 0 & 3 & 0 & 0 \\ 0 & 0 & 3 & 1 \\ 0 & 0 & 0 & 3 \end{bmatrix}.$$

Note in particular that there is no need for the eigenvalues corresponding to different Jordan blocks to be distinct (e.g., the third matrix above has two Jordan blocks, both of which are 2×2 and correspond to $\lambda = 3$). However, the following matrices are *not* in Jordan canonical form:

$$\begin{bmatrix} 2 & 1 & 0 & 0 \\ 0 & 3 & 0 & 0 \\ 0 & 0 & 3 & 0 \\ 0 & 0 & 0 & 3 \end{bmatrix}, \quad \begin{bmatrix} 2 & 0 & 0 & 0 \\ 1 & 2 & 0 & 0 \\ 0 & 0 & 1 & 0 \\ 0 & 0 & 0 & 1 \end{bmatrix}, \quad \text{and} \quad \begin{bmatrix} 3 & 0 & 1 & 0 \\ 0 & 3 & 0 & 0 \\ 0 & 0 & 3 & 0 \\ 0 & 0 & 0 & 2 \end{bmatrix}.$$

Diagonal matrices are all in Jordan canonical form, and their Jordan blocks all have sizes 1×1, so the Jordan decomposition really is a generalization of diagonalization. We now start introducing the tools needed to prove that every matrix has such a decomposition, to show that it is unique, and to actually compute it.

2.4.1 Uniqueness and Similarity

Before demonstrating how to compute the entire Jordan decomposition of a matrix $A \in \mathcal{M}_n(\mathbb{C})$, we investigate the computation of its Jordan canonical form. That is, we start by showing how to compute the matrix J, but not P, in the Jordan decomposition $A = PJP^{-1}$, assuming that it exists. As a natural by-product of this method, we will see that a matrix's Jordan canonical form is indeed unique, as stated by Theorem 2.4.1.

Since eigenvalues are unchanged by similarity transformations, and the eigenvalues of triangular matrices are their diagonal entries, we know that if the Jordan canonical form of A is

$$J = \begin{bmatrix} J_{k_1}(\lambda_1) & O & \cdots & O \\ O & J_{k_2}(\lambda_2) & \cdots & O \\ \vdots & \vdots & \ddots & \vdots \\ O & O & \cdots & J_{k_m}(\lambda_m) \end{bmatrix}$$

then the diagonal entries $\lambda_1, \lambda_2, \ldots, \lambda_m$ of J must be the eigenvalues of A. It is not obvious how to compute the orders $k_1, k_2, \ldots, k_m$ of the Jordan blocks corresponding to these eigenvalues, so we first introduce a new "helper" quantity that will get us most of the way there.

Definition 2.4.2	Suppose λ is an eigenvalue of $A \in \mathcal{M}_n$ and $k \geq 0$ is an integer. We say
Geometric Multiplicity of Order k	that the **geometric multiplicity** of **order k** of λ is the quantity

$$\gamma_k = \text{nullity}\big((A - \lambda I)^k\big).$$

If $k = 0$ then $(A - \lambda I)^k = I$, which has nullity 0, so $\gamma_0 = 0$ for every eigenvalue of every matrix. More interestingly, if $k = 1$ then this definition agrees with the definition of geometric multiplicity that we are already familiar with (i.e., γ_1 is simply the usual geometric multiplicity of the eigenvalue λ). Furthermore, the chain of inclusions

$$\text{null}(A - \lambda I) \subseteq \text{null}\big((A - \lambda I)^2\big) \subseteq \text{null}\big((A - \lambda I)^3\big) \subseteq \cdots$$

shows that the geometric multiplicities satisfy $\gamma_1 \leq \gamma_2 \leq \gamma_3 \leq \ldots$. Before presenting a theorem that tells us explicitly how to use the geometric multiplicities to compute the orders of the Jordan blocks in a matrix's Jordan canonical form, we present an example that is a bit suggestive of the connection between these quantities.

Example 2.4.1

Geometric Multiplicities of a Matrix in Jordan Canonical Form

Compute the geometric multiplicities of each of the eigenvalues of

$$A = \left[\begin{array}{ccc|cc} 2 & 1 & 0 & 0 & 0 \\ 0 & 2 & 1 & 0 & 0 \\ 0 & 0 & 2 & 0 & 0 \\ \hline 0 & 0 & 0 & 2 & 1 \\ 0 & 0 & 0 & 0 & 2 \end{array}\right].$$

Solution:

Since A is triangular (in fact, it is in Jordan canonical form), we see immediately that its only eigenvalue is 2 with algebraic multiplicity 5. We then compute

We partition A as a block matrix in this way just to better highlight what happens to its Jordan blocks when computing the powers $(A - 2I)^k$.

$$A - 2I = \left[\begin{array}{ccc|cc} 0 & 1 & 0 & 0 & 0 \\ 0 & 0 & 1 & 0 & 0 \\ 0 & 0 & 0 & 0 & 0 \\ \hline 0 & 0 & 0 & 0 & 1 \\ 0 & 0 & 0 & 0 & 0 \end{array}\right], \quad (A - 2I)^2 = \left[\begin{array}{ccc|cc} 0 & 0 & 1 & 0 & 0 \\ 0 & 0 & 0 & 0 & 0 \\ 0 & 0 & 0 & 0 & 0 \\ \hline 0 & 0 & 0 & 0 & 0 \\ 0 & 0 & 0 & 0 & 0 \end{array}\right],$$

and $(A - 2I)^k = O$ whenever $k \geq 3$. The geometric multiplicities of the eigenvalue 2 of A are the nullities of these matrices, which are $\gamma_1 = 2$, $\gamma_2 = 4$, and $\gamma_k = 5$ whenever $k \geq 3$.

In the above example, each Jordan block contributed one dimension to $\text{null}(A - 2I)$, each Jordan block of order at least 2 contributed one extra dimension to $\text{null}((A - 2I)^2)$, and the Jordan block of order 3 contributed one extra dimension to $\text{null}((A - 2I)^3)$. The following theorem says that this relationship between geometric multiplicities and Jordan blocks holds in general, and we can thus use geometric multiplicities to determine how many Jordan blocks of each size a matrix's Jordan canonical form has.

Theorem 2.4.2

Jordan Canonical Form from Geometric Multiplicities

Suppose $A \in \mathcal{M}_n(\mathbb{C})$ has eigenvalue λ with order-k geometric multiplicity γ_k. Then for each $k \geq 1$, every Jordan canonical form of A has

a) $\gamma_k - \gamma_{k-1}$ Jordan blocks $J_j(\lambda)$ of order $j \geq k$, and

b) $2\gamma_k - \gamma_{k+1} - \gamma_{k-1}$ Jordan blocks $J_k(\lambda)$ of order exactly k.

Before proving this theorem, we note that properties (a) and (b) are actually equivalent to each other—each one can be derived from the other via straightforward algebraic manipulations. We just present them both because property (a) is a bit simpler to work with, but property (b) is what we actually want.

Proof of Theorem 2.4.2. Suppose A has Jordan decomposition $A = PJP^{-1}$. Since $\gamma_k = \text{nullity}((PJP^{-1} - \lambda I)^k) = \text{nullity}((J - \lambda I)^k)$ for all $k \geq 0$, we assume without loss of generality that $A = J$.

Since J is block diagonal, so is $(J - \lambda I)^k$ for each $k \geq 0$. Furthermore, since the nullity of a block diagonal matrix is just the sum of the nullities of its diagonal blocks, it suffices to prove the follow two claims:

- nullity$\big((J_j(\mu) - \lambda I)^k\big) = 0$ whenever $\lambda \neq \mu$. To see why this is the case, simply notice that $J_j(\mu) - \lambda I$ has diagonal entries (and thus eigenvalues, since it is upper triangular) $\lambda - \mu \neq 0$. It follows that $J_j(\mu) - \lambda I$ is invertible whenever $\lambda \neq \mu$, so it and its powers all have full rank (and thus nullity 0).

- nullity$\big((J_j(\lambda) - \lambda I)^k\big) = k$ whenever $0 \leq k \leq j$. To see why this is the case, notice that

That is,
$J_j(\lambda) - \lambda I = J_j(0)$. In Exercise 2.4.17, we call this matrix N_1.

$$J_j(\lambda) - \lambda I = \begin{bmatrix} 0 & 1 & 0 & \cdots & 0 & 0 \\ 0 & 0 & 1 & \cdots & 0 & 0 \\ 0 & 0 & 0 & \cdots & 0 & 0 \\ \vdots & \vdots & \vdots & \ddots & \vdots & \vdots \\ 0 & 0 & 0 & \cdots & 0 & 1 \\ 0 & 0 & 0 & \cdots & 0 & 0 \end{bmatrix},$$

which is a simple enough matrix that we can compute the nullities of its powers fairly directly (this computation is left to Exercise 2.4.17).

Indeed, if we let m_k $(1 \leq k \leq n)$ denote the number of occurrences of the Jordan block $J_k(\lambda)$ along the diagonal of J, the above two claims tell us that $\gamma_1 = m_1 + m_2 + m_3 + \cdots + m_n$, $\gamma_2 = m_1 + 2m_2 + 2m_3 + \cdots + 2m_n$, $\gamma_3 = m_1 + 2m_2 + 3m_3 + \cdots + 3m_n$, and in general

$$\gamma_k = \sum_{j=1}^{k} jm_j + \sum_{j=k+1}^{n} km_j.$$

Subtracting these formulas from each other gives $\gamma_k - \gamma_{k-1} = \sum_{j=k}^{n} m_j$, which is exactly statement (a) of the theorem. Statement (b) of the theorem then follows from subtracting the formula in statement (a) from a shifted version of itself:

$$2\gamma_k - \gamma_{k+1} - \gamma_{k-1} = (\gamma_k - \gamma_{k-1}) - (\gamma_{k+1} - \gamma_k)$$

$$= \sum_{j=k}^{n} m_j - \sum_{j=k+1}^{n} m_j = m_k. \qquad \blacksquare$$

The above theorem has the following immediate corollaries that can sometimes be used to reduce the amount of work needed to construct a matrix's Jordan canonical form:

- The geometric multiplicity γ_1 of the eigenvalue λ counts the number of Jordan blocks corresponding to λ.

- If $\gamma_k = \gamma_{k+1}$ for a particular value of k then $\gamma_k = \gamma_{k+1} = \gamma_{k+2} = \cdots$.

Furthermore, if we make use of the fact that the sum of the orders of the Jordan blocks corresponding to a particular eigenvalue λ must equal its algebraic multiplicity (i.e., the number of times that λ appears along the diagonal of the Jordan canonical form), we get a bound on how many geometric multiplicities we have to compute in order to construct a Jordan canonical form:

Furthermore,
$A \in \mathcal{M}_n(\mathbb{C})$ is diagonalizable if and only if, for each of its eigenvalues, the multiplicities satisfy
$\gamma_1 = \gamma_2 = \cdots = \mu$.

(!) If λ is an eigenvalue of a matrix with algebraic multiplicity μ and geometric multiplicities $\gamma_1, \gamma_2, \gamma_3, \ldots$, then $\gamma_k \leq \mu$ for each $k \geq 1$. Furthermore, $\gamma_k = \mu$ whenever $k \geq \mu$.

As one final corollary of the above theorem, we are now able to show that Jordan canonical forms are indeed unique (up to re-ordering their Jordan blocks), assuming that they exist in the first place:

Proof of uniqueness in Theorem 2.4.1. Theorem 2.4.2 tells us exactly what the Jordan blocks in any Jordan canonical form of a matrix $A \in \mathcal{M}_n(\mathbb{C})$ must be. ∎

Example 2.4.2

Our First Jordan Canonical Form

Compute the Jordan canonical form of $A = \begin{bmatrix} -6 & 0 & -2 & 1 \\ 0 & -3 & -2 & 1 \\ 3 & 0 & 1 & 1 \\ -3 & 0 & 2 & 2 \end{bmatrix}$.

Solution:

The eigenvalues of this matrix are (listed according to algebraic multiplicity) 3, -3, -3, and -3. Since $\lambda = 3$ has algebraic multiplicity 1, we know that $J_1(3) = [3]$ is one of the Jordan blocks in the Jordan canonical form J of A. Similarly, since $\lambda = -3$ has algebraic multiplicity 3, we know that the orders of the Jordan blocks for $\lambda = -3$ must sum to 3. That is, the Jordan canonical form of A must be one of

$$\left[\begin{array}{c|ccc} 3 & 0 & 0 & 0 \\ \hline 0 & -3 & 1 & 0 \\ 0 & 0 & -3 & 1 \\ 0 & 0 & 0 & -3 \end{array}\right], \quad \left[\begin{array}{c|c|cc} 3 & 0 & 0 & 0 \\ \hline 0 & -3 & 0 & 0 \\ \hline 0 & 0 & -3 & 1 \\ 0 & 0 & 0 & -3 \end{array}\right], \quad \left[\begin{array}{c|c|c|c} 3 & 0 & 0 & 0 \\ \hline 0 & -3 & 0 & 0 \\ \hline 0 & 0 & -3 & 0 \\ \hline 0 & 0 & 0 & -3 \end{array}\right].$$

We could also compute $\gamma_2 = 2$ and $\gamma_k = 3$ whenever $k \geq 3$, so that A has $2\gamma_1 - \gamma_2 - \gamma_0 = 0$ copies of $J_1(-3)$, $2\gamma_2 - \gamma_3 - \gamma_1 = 0$ copies of $J_2(-3)$, and $2\gamma_3 - \gamma_4 - \gamma_2 = 1$ copy of $J_3(-3)$ as Jordan blocks.

To determine which of these canonical forms is correct, we simply note that $\lambda = -3$ has geometric multiplicity $\gamma_1 = 1$ (its corresponding eigenvectors are all scalar multiples of $(0, 1, 0, 0)$), so it must have just one corresponding Jordan block. That is, the Jordan canonical form J of A is

$$J = \left[\begin{array}{c|ccc} 3 & 0 & 0 & 0 \\ \hline 0 & -3 & 1 & 0 \\ 0 & 0 & -3 & 1 \\ 0 & 0 & 0 & -3 \end{array}\right].$$

Example 2.4.3

A Hideous Jordan Canonical Form

Compute the Jordan canonical form of

$$A = \begin{bmatrix} 1 & -1 & -1 & 0 & 0 & 1 & -1 & 0 \\ 1 & 2 & 1 & 0 & 0 & -1 & 1 & -1 \\ 2 & 1 & 2 & 0 & 1 & -1 & 0 & 0 \\ 1 & 1 & 1 & 1 & 1 & -1 & 0 & -1 \\ -2 & 0 & 0 & 0 & 0 & 0 & 1 & 1 \\ 1 & 1 & 1 & 0 & 0 & 0 & 1 & 0 \\ -2 & -1 & -1 & 0 & -1 & 1 & 1 & 1 \\ 0 & -1 & -1 & 0 & 0 & 1 & -1 & 1 \end{bmatrix}.$$

The important take-away from this example is how to construct the Jordan canonical form from the geometric multiplicities. Don't worry about the details of computing those multiplicities.

Solution:

The only eigenvalue of A is $\lambda = 1$, which has algebraic multiplicity $\mu = 8$. The geometric multiplicities of this eigenvalue can be computed via standard techniques, but we omit the details here due to the size of this matrix. In particular, the geometric multiplicity of $\lambda = 1$ is $\gamma_1 = 4$, and we furthermore have

$$\gamma_2 = \text{nullity}\big((A-I)^2\big)$$

$$= \text{nullity}\left(\begin{bmatrix} 0 & 0 & 0 & 0 & 0 & 0 & 0 & 0 \\ 0 & 0 & 0 & 0 & 0 & 0 & 0 & 0 \\ 0 & -1 & -1 & 0 & 0 & 1 & -1 & 0 \\ 0 & 1 & 1 & 0 & 0 & -1 & 1 & 0 \\ 0 & 0 & 0 & 0 & 0 & 0 & 0 & 0 \\ 0 & -1 & -1 & 0 & 0 & 1 & -1 & 0 \\ 0 & 0 & 0 & 0 & 0 & 0 & 0 & 0 \\ 0 & 0 & 0 & 0 & 0 & 0 & 0 & 0 \end{bmatrix}\right) = 7,$$

and $(A-I)^3 = O$ so $\gamma_k = 8$ whenever $k \geq 3$. It follows that the Jordan canonical form J of A has

- $2\gamma_1 - \gamma_2 - \gamma_0 = 8 - 7 - 0 = 1$ Jordan block $J_1(1)$ of order 1,
- $2\gamma_2 - \gamma_3 - \gamma_1 = 14 - 8 - 4 = 2$ Jordan blocks $J_2(1)$ of order 2,
- $2\gamma_3 - \gamma_4 - \gamma_2 = 16 - 8 - 7 = 1$ Jordan block $J_3(1)$ of order 3, and
- $2\gamma_k - \gamma_{k+1} - \gamma_{k-1} = 16 - 8 - 8 = 0$ Jordan blocks $J_k(1)$ of order k when $k \geq 4$.

In other words, the Jordan canonical form J of A is

We use dots ($\cdot$) to denote some of the zero entries, for ease of visualization.

$$J = \left[\begin{array}{ccc|cc|cc|c} 1 & 1 & 0 & \cdot & \cdot & \cdot & \cdot & \cdot \\ 0 & 1 & 1 & \cdot & \cdot & \cdot & \cdot & \cdot \\ 0 & 0 & 1 & \cdot & \cdot & \cdot & \cdot & \cdot \\ \hline \cdot & \cdot & \cdot & 1 & 1 & \cdot & \cdot & \cdot \\ \cdot & \cdot & \cdot & 0 & 1 & \cdot & \cdot & \cdot \\ \hline \cdot & \cdot & \cdot & \cdot & \cdot & 1 & 1 & \cdot \\ \cdot & \cdot & \cdot & \cdot & \cdot & 0 & 1 & \cdot \\ \hline \cdot & \cdot & \cdot & \cdot & \cdot & \cdot & \cdot & 1 \end{array}\right]$$

A Return to Similarity

Recall that two matrices $A, B \in \mathcal{M}_n(\mathbb{C})$ are called **similar** if there exists an invertible matrix $P \in \mathcal{M}_n(\mathbb{C})$ such that $A = PBP^{-1}$. Two matrices are similar if and only if there is a common linear transformation T between n-dimensional vector spaces such that A and B are both standard matrices of T (with respect to different bases). Many tools from introductory linear algebra can help determine whether or not two matrices are similar in certain special cases (e.g., if two matrices are similar then they have the same characteristic polynomial, and thus the same eigenvalues, trace, and determinant), but a complete characterization of similarity relies on the Jordan canonical form:

Theorem 2.4.3

Similarity via the Jordan Decomposition

Suppose $A, B \in \mathcal{M}_n(\mathbb{C})$. The following are equivalent:

a) A and B are similar,

b) the Jordan canonical forms of A and B have the same Jordan blocks, and

c) A and B have the same eigenvalues, and those eigenvalues have the same order-k geometric multiplicities for each k.

Proof. The equivalence of conditions (b) and (c) follows immediately from Theorem 2.4.2, which tells us that we can determine the orders of a matrix's Jordan blocks from its geometric multiplicities, and vice-versa. We thus just focus on demonstrating that conditions (a) and (b) are equivalent.

To see that (a) implies (b), suppose A and B are similar so that we can write $A = QBQ^{-1}$ for some invertible Q. If $B = PJP^{-1}$ is a Jordan decomposition of B (i.e., the Jordan canonical form of B is J) then $A = Q(PJP^{-1})Q^{-1} = (QP)J(QP)^{-1}$ is a Jordan decomposition of A with the same Jordan canonical form J.

If $A = PJP^{-1}$ is a Jordan decomposition, we can permute the columns of P to put the diagonal blocks of J in any order we like.

For the reverse implication, suppose that A and B have identical Jordan canonical forms. That is, we can find invertible P and Q such that the Jordan decompositions of A and B are $A = PJP^{-1}$ and $B = QJQ^{-1}$, where J is their (shared) Jordan canonical form. Rearranging these equations gives $J = P^{-1}AP$ and $J = Q^{-1}BQ$, so $P^{-1}AP = Q^{-1}BQ$. Rearranging one more time then gives $A = P(Q^{-1}BQ)P^{-1} = (PQ^{-1})B(PQ^{-1})^{-1}$, so A and B are similar. ∎

Example 2.4.4

Checking Similarity

Determine whether or not the following matrices are similar:

$$A = \begin{bmatrix} 1 & 1 & 0 & 0 \\ 0 & 1 & 0 & 0 \\ 0 & 0 & 1 & 1 \\ 0 & 0 & 0 & 1 \end{bmatrix} \quad \text{and} \quad B = \begin{bmatrix} 1 & 0 & 0 & 0 \\ 0 & 1 & 1 & 0 \\ 0 & 0 & 1 & 1 \\ 0 & 0 & 0 & 1 \end{bmatrix}.$$

This example is tricky if we do not use Theorem 2.4.3, since A and B have the same rank and characteristic polynomial.

Solution:

These matrices are already in Jordan canonical form. In particular, the Jordan blocks of A are

$$\begin{bmatrix} 1 & 1 \\ 0 & 1 \end{bmatrix} \quad \text{and} \quad \begin{bmatrix} 1 & 1 \\ 0 & 1 \end{bmatrix},$$

whereas the Jordan blocks of B are

$$\begin{bmatrix} 1 \end{bmatrix} \quad \text{and} \quad \begin{bmatrix} 1 & 1 & 0 \\ 0 & 1 & 1 \\ 0 & 0 & 1 \end{bmatrix}.$$

Since they have different Jordan blocks, A and B are not similar.

Example 2.4.5

Checking Similarity
(Again)

Determine whether or not the following matrices are similar:

$$A = \begin{bmatrix} -6 & 0 & -2 & 1 \\ 0 & -3 & -2 & 1 \\ 3 & 0 & 1 & 1 \\ -3 & 0 & 2 & 2 \end{bmatrix} \quad \text{and} \quad B = \begin{bmatrix} -3 & 0 & 3 & 0 \\ -2 & 2 & -5 & 2 \\ 2 & 1 & -4 & -2 \\ -2 & 2 & 1 & -1 \end{bmatrix}.$$

Solution:

We already saw in Example 2.4.2 that A has Jordan canonical form

$$J = \left[\begin{array}{c|ccc} 3 & 0 & 0 & 0 \\ \hline 0 & -3 & 1 & 0 \\ 0 & 0 & -3 & 1 \\ 0 & 0 & 0 & -3 \end{array} \right].$$

It follows that, to see whether or not A and B are similar, we need to check whether or not B has the same Jordan canonical form J.

To this end, we note that B also has eigenvalues (listed according to algebraic multiplicity) $3, -3, -3,$ and -3. Since $\lambda = 3$ has algebraic multiplicity 1, we know that $J_1(3) = [3]$ is one of the Jordan blocks in the Jordan canonical form B. Similarly, since $\lambda = -3$ has algebraic multiplicity 3, we know that the orders of the Jordan blocks for $\lambda = -3$ sum to 3.

It is straightforward to show that the eigenvalue $\lambda = -3$ has geometric multiplicity $\gamma_1 = 1$ (its corresponding eigenvectors are all scalar multiples of $(1,0,0,1)$), so it must have just one corresponding Jordan block. That is, the Jordan canonical form of B is indeed the matrix J above, so A and B are similar.

Example 2.4.6

Checking Similarity
(Yet Again)

Determine whether or not the following matrices are similar:

$$A = \begin{bmatrix} 1 & 2 & 3 & 4 \\ 2 & 1 & 0 & 3 \\ 3 & 0 & 1 & 2 \\ 4 & 3 & 2 & 1 \end{bmatrix} \quad \text{and} \quad B = \begin{bmatrix} 1 & 0 & 0 & 4 \\ 0 & 2 & 3 & 0 \\ 0 & 3 & 2 & 0 \\ 4 & 0 & 0 & 1 \end{bmatrix}.$$

Solution:

These matrices do note even have the same trace ($\operatorname{tr}(A) = 4$, but $\operatorname{tr}(B) = 6$), so they cannot possibly be similar. We could have also computed their Jordan canonical forms to show that they are not similar, but it saves a lot of work to do the easier checks based on the trace, determinant, eigenvalues, and rank first!

2.4.2 Existence and Computation

The goal of this subsection is to develop a method of computing a matrix's Jordan decomposition $A = PJP^{-1}$, rather than just its Jordan canonical form J. One useful by-product of this method will be that it works for every complex matrix and thus (finally) proves that every matrix does indeed have a Jordan decomposition—a fact that we have been taking for granted up until now.

In order to get an idea of how we might compute a matrix's Jordan decomposition (or even just convince ourselves that it exists), suppose for a moment that the Jordan canonical form of a matrix $A \in \mathcal{M}_k(\mathbb{C})$ has just a single Jordan block. That is, suppose we could write $A = PJ_k(\lambda)P^{-1}$ for some invertible $P \in \mathcal{M}_k(\mathbb{C})$ and some scalar $\lambda \in \mathbb{C}$ (necessarily equal to the one and only eigenvalue of A). Our usual block matrix multiplication techniques show that if $\mathbf{v}^{(j)}$ is the j-th column of P then

> Recall that $P\mathbf{e}_j = \mathbf{v}^{(j)}$, and multiplying both sides on the left by P^{-1} shows that $P^{-1}\mathbf{v}^{(j)} = \mathbf{e}_j$.

$$A\mathbf{v}^{(1)} = PJ_k(\lambda)P^{-1}\mathbf{v}^{(1)} = P\begin{bmatrix} \lambda & 1 & \cdots & 0 \\ 0 & \lambda & \cdots & 0 \\ \vdots & \vdots & \ddots & \vdots \\ 0 & 0 & \cdots & \lambda \end{bmatrix}\mathbf{e}_1 = \lambda P\mathbf{e}_1 = \lambda\mathbf{v}^{(1)},$$

and

> We use superscripts for $\mathbf{v}^{(j)}$ here since we will use subscripts for something else shortly.

$$A\mathbf{v}^{(j)} = PJ_k(\lambda)P^{-1}\mathbf{v}^{(j)} = P\begin{bmatrix} \lambda & 1 & \cdots & 0 \\ 0 & \lambda & \cdots & 0 \\ \vdots & \vdots & \ddots & \vdots \\ 0 & 0 & \cdots & \lambda \end{bmatrix}\mathbf{e}_j = P(\lambda\mathbf{e}_j + \mathbf{e}_{j-1})$$
$$= \lambda\mathbf{v}^{(j)} + \mathbf{v}^{(j-1)}$$

for all $2 \le j \le k$.

Phrased slightly differently, $\mathbf{v}^{(1)}$ is an eigenvector of A with eigenvalue λ, (i.e., $(A - \lambda I)\mathbf{v}^{(1)} = \mathbf{0}$) and $\mathbf{v}^{(2)}, \mathbf{v}^{(3)}, \ldots, \mathbf{v}^{(k)}$ are vectors that form a "chain" with the property that

$$(A - \lambda I)\mathbf{v}^{(k)} = \mathbf{v}^{(k-1)}, \quad (A - \lambda I)\mathbf{v}^{(k-1)} = \mathbf{v}^{(k-2)}, \quad \cdots \quad (A - \lambda I)\mathbf{v}^{(2)} = \mathbf{v}^{(1)}.$$

It is perhaps useful to picture this relationship between the vectors slightly more diagrammatically:

$$\mathbf{v}^{(k)} \xrightarrow{A - \lambda I} \mathbf{v}^{(k-1)} \xrightarrow{A - \lambda I} \cdots \xrightarrow{A - \lambda I} \mathbf{v}^{(2)} \xrightarrow{A - \lambda I} \mathbf{v}^{(1)} \xrightarrow{A - \lambda I} \mathbf{0}.$$

We thus guess that vectors that form a "chain" leading down to an eigenvector are the key to constructing a matrix's Jordan decomposition, which leads naturally to the following definition:

Definition 2.4.3
Jordan Chains

Suppose $A \in \mathcal{M}_n(\mathbb{C})$ is a matrix with eigenvalue λ and corresponding eigenvector $\mathbf{v}$. A sequence of non-zero vectors $\mathbf{v}^{(1)}, \mathbf{v}^{(2)}, \ldots, \mathbf{v}^{(k)}$ is called a **Jordan chain** of **order k** corresponding to $\mathbf{v}$ if $\mathbf{v}^{(1)} = \mathbf{v}$ and

$$(A - \lambda I)\mathbf{v}^{(j)} = \mathbf{v}^{(j-1)} \quad \text{for all} \quad 2 \le j \le k.$$

By mimicking the block matrix multiplication techniques that we performed above, we arrive immediately at the following theorem:

Theorem 2.4.4

Multiplication by a Single Jordan Chain

Suppose $\mathbf{v}^{(1)}, \mathbf{v}^{(2)}, \ldots, \mathbf{v}^{(k)}$ is a Jordan chain corresponding to an eigenvalue λ of $A \in \mathcal{M}_n(\mathbb{C})$. If we define

$$P = \left[\, \mathbf{v}^{(1)} \mid \mathbf{v}^{(2)} \mid \cdots \mid \mathbf{v}^{(k)} \,\right] \in \mathcal{M}_{n,k}(\mathbb{C})$$

then $AP = PJ_k(\lambda)$.

Proof. Just use block matrix multiplication to compute

$$AP = \left[\, A\mathbf{v}^{(1)} \mid A\mathbf{v}^{(2)} \mid \cdots \mid A\mathbf{v}^{(k)} \,\right]$$
$$= \left[\, \lambda\mathbf{v}^{(1)} \mid \mathbf{v}^{(1)} + \lambda\mathbf{v}^{(2)} \mid \cdots \mid \mathbf{v}^{(k-1)} + \lambda\mathbf{v}^{(k)} \,\right]$$

and

> If $n = k$ and the members of the Jordan chain form a linearly independent set then P is invertible and this theorem tells us that $A = PJ_k(\lambda)P^{-1}$, which is a Jordan decomposition.

$$PJ_k(\lambda) = \left[\, \mathbf{v}^{(1)} \mid \mathbf{v}^{(2)} \mid \cdots \mid \mathbf{v}^{(k)} \,\right] \begin{bmatrix} \lambda & 1 & \cdots & 0 \\ 0 & \lambda & \cdots & 0 \\ \vdots & \vdots & \ddots & \vdots \\ 0 & 0 & \cdots & \lambda \end{bmatrix}$$

$$= \left[\, \lambda\mathbf{v}^{(1)} \mid \mathbf{v}^{(1)} + \lambda\mathbf{v}^{(2)} \mid \cdots \mid \mathbf{v}^{(k-1)} + \lambda\mathbf{v}^{(k)} \,\right],$$

so $AP = PJ_k(\lambda)$. ∎

With the above theorem in mind, our goal is now to demonstrate how to construct Jordan chains, which we will then place as columns into the matrix P in a Jordan decomposition $A = PJP^{-1}$.

One Jordan Block per Eigenvalue

> In other words, we are first considering the case when each eigenvalue has geometric multiplicity $\gamma_1 = 1$.

Since the details of Jordan chains and the Jordan decomposition simplify considerably in the case when there is just one Jordan block (and thus just one Jordan chain) corresponding to each of the matrix's eigenvalues, that is the situation that we explore first. In this case, we can find a matrix's Jordan chains just by finding an eigenvector corresponding to each of its eigenvalues, and then solving some additional linear systems to find the other members of the Jordan chains.

Example 2.4.7

Computing Jordan Chains and the Jordan Decomposition

Find the Jordan chains of the matrix

$$A = \begin{bmatrix} 5 & 1 & -1 \\ 1 & 3 & -1 \\ 2 & 0 & 2 \end{bmatrix},$$

and use them to construct its Jordan decomposition.

Solution:

> At this point, we know that the Jordan canonical form of A must be
> $$\left[\begin{array}{c|cc} 2 & 0 & 0 \\ \hline 0 & 4 & 1 \\ 0 & 0 & 4 \end{array}\right].$$

The eigenvalues of this matrix are (listed according to their algebraic multiplicity) 2, 4, and 4. An eigenvector corresponding to $\lambda = 2$ is $\mathbf{v}_1 = (0, 1, 1)$, and an eigenvector corresponding to $\lambda = 4$ is $\mathbf{v}_2 = (1, 0, 1)$. However, the eigenspace corresponding to the eigenvalue $\lambda = 4$ is just 1-dimensional (i.e., its geometric multiplicity is $\gamma_1 = 1$), so it is not possible to find a linearly independent set of two eigenvectors corresponding to $\lambda = 4$.

Instead, we must find a Jordan chain $\{\mathbf{v}_2^{(1)}, \mathbf{v}_2^{(2)}\}$ with the property that

$$(A - 4I)\mathbf{v}_2^{(2)} = \mathbf{v}_2^{(1)} \quad \text{and} \quad (A - 4I)\mathbf{v}_2^{(1)} = \mathbf{0}.$$

We choose $\mathbf{v}_2^{(1)} = \mathbf{v}_2 = (1, 0, 1)$ to be the eigenvector corresponding to $\lambda = 4$ that we already found, and we can then find $\mathbf{v}_2^{(2)}$ by solving the linear system $(A - 4I)\mathbf{v}_2^{(2)} = \mathbf{v}_2^{(1)}$ for $\mathbf{v}_2^{(2)}$:

$$\left[\begin{array}{ccc|c} 1 & 1 & -1 & 1 \\ 1 & -1 & -1 & 0 \\ 2 & 0 & -2 & 1 \end{array}\right] \xrightarrow{\text{row reduce}} \left[\begin{array}{ccc|c} 1 & 0 & -1 & 1/2 \\ 0 & 1 & 0 & 1/2 \\ 0 & 0 & 0 & 0 \end{array}\right].$$

It follows that the third entry of $\mathbf{v}_2^{(2)}$ is free, while its other two entries are not. If we choose the third entry to be 0 then we get $\mathbf{v}_2^{(2)} = (1/2, 1/2, 0)$ as one possible solution.

Now that we have a set of 3 vectors $\{\mathbf{v}_1, \mathbf{v}_2^{(1)}, \mathbf{v}_2^{(2)}\}$, we can place them as columns in a 3×3 matrix, and that matrix *should* bring A into its Jordan canonical form:

$$P = \left[\, \mathbf{v}_1 \mid \mathbf{v}_2^{(1)} \mid \mathbf{v}_2^{(2)} \,\right] = \begin{bmatrix} 0 & 1 & 1/2 \\ 1 & 0 & 1/2 \\ 1 & 1 & 0 \end{bmatrix}.$$

Straightforward calculation then reveals that

Keep in mind that $P^{-1}AP = J$ is equivalent to $A = PJP^{-1}$, which is why the order of P and P^{-1} is reversed from that of Theorem 2.4.1.

$$P^{-1}AP = \begin{bmatrix} -1/2 & 1/2 & 1/2 \\ 1/2 & -1/2 & 1/2 \\ 1 & 1 & -1 \end{bmatrix} \begin{bmatrix} 5 & 1 & -1 \\ 1 & 3 & -1 \\ 2 & 0 & 2 \end{bmatrix} \begin{bmatrix} 0 & 1 & 1/2 \\ 1 & 0 & 1/2 \\ 1 & 1 & 0 \end{bmatrix}$$

$$= \left[\begin{array}{c|cc} 2 & 0 & 0 \\ \hline 0 & 4 & 1 \\ 0 & 0 & 4 \end{array}\right],$$

which is indeed in Jordan canonical form.

The procedure carried out in the above example works as long as each eigenspace is 1-dimensional—to find the Jordan decomposition of a matrix A, we just find an eigenvector for each eigenvalue and then extend it to a Jordan chain so as to fill up the columns of a square matrix P. Doing so results in P being invertible (a fact which is not obvious, but we will prove shortly) and $P^{-1}AP$ being the Jordan canonical form of A. The following example illustrates this procedure again with a longer Jordan chain.

Example 2.4.8

Finding Longer Jordan Chains

Find the Jordan chains of the matrix

$$A = \begin{bmatrix} -6 & 0 & -2 & 1 \\ 0 & -3 & -2 & 1 \\ 3 & 0 & 1 & 1 \\ -3 & 0 & 2 & 2 \end{bmatrix},$$

and use them to construct its Jordan decomposition.

Solution:

We noted in Example 2.4.2 that this matrix has eigenvalues 3 (with algebraic multiplicity 1) and -3 (with algebraic multiplicity 3 and geometric multiplicity 1). An eigenvector corresponding to $\lambda = 3$ is $\mathbf{v}_1 = (0,0,1,2)$, and an eigenvector corresponding to $\lambda = -3$ is $\mathbf{v}_2 = (0,1,0,0)$.

To construct a Jordan decomposition of A, we do what we did in the previous example: we find a Jordan chain $\{\mathbf{v}_2^{(1)}, \mathbf{v}_2^{(2)}, \mathbf{v}_2^{(3)}\}$ corresponding to $\lambda = -3$. We choose $\mathbf{v}_2^{(1)} = \mathbf{v}_2 = (0,1,0,0)$ to be the eigenvector that we already computed, and then we find $\mathbf{v}_2^{(2)}$ by solving the linear system $(A + 3I)\mathbf{v}_2^{(2)} = \mathbf{v}_2^{(1)}$ for $\mathbf{v}_2^{(2)}$:

$$\left[\begin{array}{cccc|c} -3 & 0 & -2 & 1 & 0 \\ 0 & 0 & -2 & 1 & 1 \\ 3 & 0 & 4 & 1 & 0 \\ -3 & 0 & 2 & 5 & 0 \end{array}\right] \xrightarrow{\text{row reduce}} \left[\begin{array}{cccc|c} 1 & 0 & 0 & 0 & 1/3 \\ 0 & 0 & 1 & 0 & -1/3 \\ 0 & 0 & 0 & 1 & 1/3 \\ 0 & 0 & 0 & 0 & 0 \end{array}\right].$$

It follows that the second entry of $\mathbf{v}_2^{(2)}$ is free, while its other entries are not. If we choose the second entry to be 0 then we get $\mathbf{v}_2^{(2)} = (1/3, 0, -1/3, 1/3)$ as one possible solution.

The same sequence of row operations can be used to find $\mathbf{v}_2^{(3)}$ as was used to find $\mathbf{v}_2^{(2)}$. This happens in general, since the only difference between these linear systems is the right-hand side.

We still need one more vector to complete this Jordan chain, so we just repeat this procedure: we solve the linear system $(A + 3I)\mathbf{v}_2^{(3)} = \mathbf{v}_2^{(2)}$ for $\mathbf{v}_2^{(3)}$:

$$\left[\begin{array}{cccc|c} -3 & 0 & -2 & 1 & 1/3 \\ 0 & 0 & -2 & 1 & 0 \\ 3 & 0 & 4 & 1 & -1/3 \\ -3 & 0 & 2 & 5 & 1/3 \end{array}\right] \xrightarrow{\text{row reduce}} \left[\begin{array}{cccc|c} 1 & 0 & 0 & 0 & -1/9 \\ 0 & 0 & 1 & 0 & 0 \\ 0 & 0 & 0 & 1 & 0 \\ 0 & 0 & 0 & 0 & 0 \end{array}\right].$$

Similar to before, the second entry of $\mathbf{v}_2^{(3)}$ is free, while its other entries are not. If we choose the second entry to be 0, then we get $\mathbf{v}_2^{(3)} = (-1/9, 0, 0, 0)$ as one possible solution.

Be careful—we might be tempted to normalize $\mathbf{v}_2^{(3)}$ to $\mathbf{v}_2^{(3)} = (1,0,0,0)$, but we cannot do this! We can choose the free variable (the second entry of $\mathbf{v}_2^{(3)}$ in this case), but not the overall scaling in Jordan chains.

Now that we have a set of 4 vectors $\{\mathbf{v}_1, \mathbf{v}_2^{(1)}, \mathbf{v}_2^{(2)}, \mathbf{v}_2^{(3)}\}$, we can place them as columns in a 4×4 matrix $P = [\mathbf{v}_1 \mid \mathbf{v}_2^{(1)} \mid \mathbf{v}_2^{(2)} \mid \mathbf{v}_2^{(3)}]$, and that matrix *should* bring A into its Jordan canonical form:

$$P = \begin{bmatrix} 0 & 0 & 1/3 & -1/9 \\ 0 & 1 & 0 & 0 \\ 1 & 0 & -1/3 & 0 \\ 2 & 0 & 1/3 & 0 \end{bmatrix} \quad \text{and} \quad P^{-1} = \begin{bmatrix} 0 & 0 & 1/3 & 1/3 \\ 0 & 1 & 0 & 0 \\ 0 & 0 & -2 & 1 \\ -9 & 0 & -6 & 3 \end{bmatrix}.$$

Straightforward calculation then shows that

$$P^{-1}AP = \left[\begin{array}{c|ccc} 3 & 0 & 0 & 0 \\ \hline 0 & -3 & 1 & 0 \\ 0 & 0 & -3 & 1 \\ 0 & 0 & 0 & -3 \end{array}\right],$$

which is indeed the Jordan canonical form of A.

Multiple Jordan Blocks per Eigenvalue

Unfortunately, the method described above for constructing the Jordan decomposition of a matrix only works if all of its eigenvalues have geometric multiplicity equal to 1. To illustrate how this method can fail in more complicated situations, we try to use this method to construct a Jordan decomposition of the matrix

$$A = \begin{bmatrix} -2 & -1 & 1 \\ 3 & 2 & -1 \\ -6 & -2 & 3 \end{bmatrix}.$$

In particular, $\lambda = 1$ has geometric multiplicity $\gamma_1 = 2$.

The only eigenvalue of this matrix is $\lambda = 1$, and one basis of the corresponding eigenspace is $\{\mathbf{v}_1, \mathbf{v}_2\}$ where $\mathbf{v}_1 = (1, -3, 0)$ and $\mathbf{v}_2 = (1, 0, 3)$. However, if we try to extend either of these eigenvectors $\mathbf{v}_1$ or $\mathbf{v}_2$ to a Jordan chain of order larger than 1 then we quickly get stuck: the linear systems $(A - I)\mathbf{w} = \mathbf{v}_1$ and $(A - I)\mathbf{w} = \mathbf{v}_2$ each have no solutions.

There are indeed Jordan chains of degree 2 corresponding to $\lambda = 1$, but the eigenvectors that they correspond to are non-trivial linear combinations of $\mathbf{v}_1$ and $\mathbf{v}_2$, not $\mathbf{v}_1$ or $\mathbf{v}_2$ themselves. Furthermore, it is not immediately obvious how we could find which eigenvectors (i.e., linear combinations) can be extended. For this reason, we now introduce another method of constructing Jordan chains that avoids problems like this one.

Example 2.4.9

Jordan Chains with a Repeated Eigenvalue

Compute a Jordan decomposition of the matrix $A = \begin{bmatrix} -2 & -1 & 1 \\ 3 & 2 & -1 \\ -6 & -2 & 3 \end{bmatrix}$.

Solution:

As we noted earlier, the only eigenvalue of this matrix is $\lambda = 1$, with geometric multiplicity 2 and $\{\mathbf{v}_1, \mathbf{v}_2\}$ (where $\mathbf{v}_1 = (1, -3, 0)$ and $\mathbf{v}_2 = (1, 0, 3)$) forming a basis of the corresponding eigenspace.

In order to extend some eigenvector $\mathbf{w}_1^{(1)}$ of A to a degree-2 Jordan chain, we would like to find a vector $\mathbf{w}_1^{(2)}$ with the property that $(A - I)\mathbf{w}_1^{(2)}$ is an eigenvector of A corresponding to $\lambda = 1$. That is, we want

$$(A - I)\mathbf{w}_1^{(2)} \neq \mathbf{0} \quad \text{but} \quad (A - I)\big((A - I)\mathbf{w}_1^{(2)}\big) = (A - I)^2\mathbf{w}_1^{(2)} = \mathbf{0}.$$

Direct computation shows that $(A - I)^2 = O$, so $\text{null}\big((A - I)^2\big) = \mathbb{C}^3$, so we can just pick $\mathbf{w}_1^{(2)}$ to be *any* vector that is not in $\text{null}(A - I)$: we arbitrarily choose $\mathbf{w}_1^{(2)} = (1, 0, 0)$ to give us the Jordan chain $\mathbf{w}_1^{(1)}, \mathbf{w}_1^{(2)}$, where

Notice that $\mathbf{w}_1^{(1)} = -\mathbf{v}_1 - 2\mathbf{v}_2$ really is an eigenvector of A corresponding to $\lambda = 1$.

$$\mathbf{w}_1^{(2)} = (1, 0, 0) \quad \text{and} \quad \mathbf{w}_1^{(1)} = (A - I)\mathbf{w}_1^{(2)} = (-3, 3, -6).$$

To complete the Jordan decomposition of A, we now just pick any other eigenvector $\mathbf{w}_2$ with the property that $\{\mathbf{w}_1^{(1)}, \mathbf{w}_1^{(2)}, \mathbf{w}_2\}$ is linearly independent—we can choose $\mathbf{w}_2 = \mathbf{v}_2 = (1, 0, 3)$, for example. Then A has Jordan decomposition $A = PJP^{-1}$, where

We could have chosen $\mathbf{w}_2 = \mathbf{v}_1 = (1, -3, 0)$ too, or any number of other choices.

$$J = \begin{bmatrix} 1 & 1 & 0 \\ 0 & 1 & 0 \\ 0 & 0 & 1 \end{bmatrix} \quad \text{and} \quad P = \begin{bmatrix} \mathbf{w}_1^{(1)} & | & \mathbf{w}_1^{(2)} & | & \mathbf{w}_2 \end{bmatrix} = \begin{bmatrix} -3 & 1 & 1 \\ 3 & 0 & 0 \\ -6 & 0 & 3 \end{bmatrix}.$$

The above example suggests that instead of constructing a matrix's Jordan decomposition from the "bottom up" (i.e., starting with the eigenvectors, which are at the "bottom" of a Jordan chain, and working our way "up" the chain by solving linear systems, as we did earlier), we should instead compute it from the "top down" (i.e., starting with vectors at the "top" of a Jordan chain and then multiplying them by $(A - \lambda I)$ to work our way "down" the chain).

Indeed, this "top down" approach works in general—a fact that we now start proving. To this end, note that if $\mathbf{v}^{(1)}, \mathbf{v}^{(2)}, \ldots, \mathbf{v}^{(k)}$ is a Jordan chain of A corresponding to the eigenvalue λ, then multiplying their defining equation (i.e., $(A - \lambda I)\mathbf{v}^{(j)} = \mathbf{v}^{(j-1)}$ for all $2 \leq j \leq k$ and $(A - \lambda I)\mathbf{v}^{(1)} = \mathbf{0}$) on the left by powers of $A - \lambda I$ shows that

For any sets X and Y, $X \backslash Y$ is the set of all members of X that are not in Y.

$$\mathbf{v}^{(j)} \in \text{null}\big((A - \lambda_j I)^j\big) \setminus \text{null}\big((A - \lambda_j I)^{j-1}\big) \quad \text{for all} \quad 1 \leq j \leq k.$$

Indeed, these subspaces $\text{null}\big((A - \lambda_j I)^j\big)$ (for $1 \leq j \leq k$) play a key role in the construction of the Jordan decomposition—after all, their dimensions are the geometric multiplicities from Definition 2.4.2, which we showed determine the sizes of A's Jordan blocks in Theorem 2.4.2. These spaces are called the **generalized eigenspaces** of A corresponding to λ (and their members are called the **generalized eigenvectors** of A corresponding to λ). Notice that if $j = 1$ then they are just standard eigenspace and eigenvectors, respectively.

Our first main result that helps us toward a proof of the Jordan decomposition is the fact that bases of the generalized eigenspaces of a matrix can always be "stitched together" to form a basis of the entire space $\mathbb{C}^n$.

Theorem 2.4.5

Generalized Eigenbases

Suppose $A \in \mathcal{M}_n(\mathbb{C})$ has distinct eigenvalues $\lambda_1, \lambda_2, \ldots, \lambda_m$ with algebraic multiplicities $\mu_1, \mu_2, \ldots, \mu_m$, respectively, and

$$B_j \quad \text{is a basis of} \quad \text{null}\big((A - \lambda_j I)^{\mu_j}\big) \quad \text{for each} \quad 1 \leq j \leq m.$$

Then $B_1 \cup B_2 \cup \cdots \cup B_m$ is a basis of $\mathbb{C}^n$.

Proof. We can use Schur triangularization to write $A = UTU^*$, where $U \in \mathcal{M}_n(\mathbb{C})$ is unitary and $T \in \mathcal{M}_n(\mathbb{C})$ is upper triangular, and we can choose T so that its diagonal entries (i.e., the eigenvalues of A) are arranged so that any repeated entries are next to each other. That is, T can be chosen to have the form

Asterisks ($*$) denote blocks whose entries are potentially non-zero, but irrelevant (i.e., not important enough to give names to).

$$T = \begin{bmatrix} T_1 & * & \cdots & * \\ O & T_2 & \cdots & * \\ \vdots & \vdots & \ddots & \vdots \\ O & O & \cdots & T_m \end{bmatrix},$$

where each T_j ($1 \leq j \leq m$) is upper triangular and has every diagonal entry equal to λ_j.

Our next goal is to show that if we are allowed to make use of arbitrary similarity transformations (rather than just unitary similarity transformations, as in Schur triangularization) then T can be chosen to be even simpler still. To this end, we make the following claim:

Claim: If two matrices $A \in \mathcal{M}_{d_1}(\mathbb{C})$ and $C \in \mathcal{M}_{d_2}(\mathbb{C})$ do not have any eigenvalues in common then, for all $B \in \mathcal{M}_{d_1, d_2}(\mathbb{C})$, the following block matrices

This claim fails if A and C share eigenvalues (if it didn't, it would imply that all matrices are diagonalizable).

are similar:

$$\begin{bmatrix} A & B \\ O & C \end{bmatrix} \quad \text{and} \quad \begin{bmatrix} A & O \\ O & C \end{bmatrix}.$$

We prove this claim as Theorem B.2.1 in Appendix B.2, as it is slightly technical and the proof of this theorem is already long enough without it.

By making use of this claim repeatedly, we see that each of the following matrices are similar to each other:

$$\begin{bmatrix} T_1 & * & * & \cdots & * \\ O & T_2 & * & \cdots & * \\ O & O & T_3 & \cdots & * \\ \vdots & \vdots & \vdots & \ddots & \vdots \\ O & O & O & \cdots & T_m \end{bmatrix}, \begin{bmatrix} T_1 & O & O & \cdots & O \\ O & T_2 & * & \cdots & * \\ O & O & T_3 & \cdots & * \\ \vdots & \vdots & \vdots & \ddots & \vdots \\ O & O & O & \cdots & T_m \end{bmatrix},$$

$$\begin{bmatrix} T_1 & O & O & \cdots & O \\ O & T_2 & O & \cdots & O \\ O & O & T_3 & \cdots & * \\ \vdots & \vdots & \vdots & \ddots & \vdots \\ O & O & O & \cdots & T_m \end{bmatrix}, \dots, \begin{bmatrix} T_1 & O & O & \cdots & O \\ O & T_2 & O & \cdots & O \\ O & O & T_3 & \cdots & O \\ \vdots & \vdots & \vdots & \ddots & \vdots \\ O & O & O & \cdots & T_m \end{bmatrix}.$$

It follows that T, and thus A, is similar to the block-diagonal matrix at the bottom-right. That is, there exists an invertible matrix $P \in \mathcal{M}_n(\mathbb{C})$ such that

$$A = P \begin{bmatrix} T_1 & O & \cdots & O \\ O & T_2 & \cdots & O \\ \vdots & \vdots & \ddots & \vdots \\ O & O & \cdots & T_m \end{bmatrix} P^{-1}.$$

With this decomposition of A in hand, it is now straightforward to verify that

$$(A - \lambda_1 I)^{\mu_1} = P \begin{bmatrix} (T_1 - \lambda_1 I)^{\mu_1} & O & \cdots & O \\ O & (T_2 - \lambda_1 I)^{\mu_1} & \cdots & O \\ \vdots & \vdots & \ddots & \vdots \\ O & O & \cdots & (T_m - \lambda_1 I)^{\mu_1} \end{bmatrix} P^{-1}$$

$$= P \begin{bmatrix} O & O & \cdots & O \\ O & (T_2 - \lambda_1 I)^{\mu_1} & \cdots & O \\ \vdots & \vdots & \ddots & \vdots \\ O & O & \cdots & (T_m - \lambda_1 I)^{\mu_1} \end{bmatrix} P^{-1},$$

In this theorem and proof, unions like $C_1 \cup \cdots \cup C_m$ are meant as *multisets*, so that if there were any vector in multiple C_j's then their union would necessarily be linearly dependent.

with the second equality following from the fact that $T_1 - \lambda_1 I$ is an upper triangular $\mu_1 \times \mu_1$ matrix with all diagonal entries equal to 0, so $(T_1 - \lambda_1 I)^{\mu_1} = O$ by Exercise 2.4.16. Similarly, for each $2 \le j \le m$ the matrix $(T_j - \lambda_1 I)^{\mu_1}$ has non-zero diagonal entries and is thus invertible, so we see that the first μ_1 columns of P form a basis of $\text{null}\big((A - \lambda_1 I)^{\mu_1}\big)$. A similar argument shows that the next μ_2 columns of P form a basis of $\text{null}\big((A - \lambda_2 I)^{\mu_2}\big)$, and so on.

We have thus proved that, for each $1 \le j \le m$, there *exists* a basis C_j of $\text{null}\big((A - \lambda_j I)^{\mu_j}\big)$ such that $C_1 \cup C_2 \cup \cdots \cup C_m$ is a basis of $\mathbb{C}^n$ (in particular, we can choose C_1 to consist of the first μ_1 columns of P, C_2 to consist of its next μ_2 columns, and so on). To see that the same result holds *no matter which* basis B_j of $\text{null}\big((A - \lambda_j I)^{\mu_j}\big)$ is chosen, simply note that since B_j and C_j are bases of the same space, we must have $\text{span}(B_j) = \text{span}(C_j)$. It follows that

$$\text{span}(B_1 \cup B_2 \cup \cdots \cup B_m) = \text{span}(C_1 \cup C_2 \cup \cdots \cup C_m) = \mathbb{C}^n.$$

Since we also have the $|B_j| = |C_j|$ for all $1 \leq j \leq m$, it must be the case that $|B_1 \cup B_2 \cup \cdots \cup B_m| = |C_1 \cup C_2 \cup \cdots \cup C_m| = n$, which implies that $B_1 \cup B_2 \cup \cdots \cup B_m$ is indeed a basis of $\mathbb{C}^n$. ∎

Remark 2.4.1 **The Jordan Decomposition and Direct Sums**	Another way of phrasing Theorem 2.4.5 is as saying that, for every matrix $A \in \mathcal{M}_n(\mathbb{C})$, we can write $\mathbb{C}^n$ as a direct sum (see Section 1.B) of its generalized eigenspaces: $$\mathbb{C}^n = \bigoplus_{j=1}^{m} \text{null}\big((A - \lambda_j I)^{\mu_j}\big).$$ Furthermore, A is diagonalizable if and only if we can replace each generalized eigenspace by its non-generalized counterpart: $$\mathbb{C}^n = \bigoplus_{j=1}^{m} \text{null}(A - \lambda_j I).$$

We are now in a position to rigorously prove that every matrix has a Jordan decomposition. We emphasize that it really is worth reading through this proof (even if we did not do so for the proof of Theorem 2.4.5), since it describes an explicit procedure for actually constructing the Jordan decomposition in general.

Proof of existence in Theorem 2.4.1. Suppose λ is an eigenvalue of A with algebraic multiplicity μ. We start by showing that we can construct a basis B of $\text{null}\big((A - \lambda I)^{\mu}\big)$ that is made up of Jordan chains. We do this via the following iterative procedure that works by starting at the "tops" of the longest Jordan chains and working its way "down":

Step 1. Set $k = \mu$ and $B_{\mu+1} = \{\}$ (the empty set).

Step 2. Set $C_k = B_{k+1}$ and then add any vector from

$$\text{null}\big((A - \lambda I)^k\big) \setminus \text{span}\big(C_k \cup \text{null}\big((A - \lambda I)^{k-1}\big)\big)$$

In words, B_{k+1} is the piece of the basis B that contains Jordan chains of order at least $k + 1$, and the vectors that we add to C_k are the "tops" of the Jordan chains of order exactly k.

to C_k. Continue adding vectors to C_k in this way until no longer possible, so that

$$\text{null}\big((A - \lambda I)^k\big) \subseteq \text{span}\big(C_k \cup \text{null}\big((A - \lambda I)^{k-1}\big)\big). \tag{2.4.1}$$

Notice that C_k is necessarily linearly independent since B_{k+1} is linearly independent, and each vector that we added to C_k was specifically chosen to not be in the span of its other members.

Step 3. Construct all Jordan chains that have "tops" at the members of $C_k \setminus B_{k+1}$. That is, if we denote the members of $C_k \setminus B_{k+1}$ by $\mathbf{v}_1^{(k)}, \mathbf{v}_2^{(k)}, \ldots$, then for each $1 \leq i \leq |C_k \setminus B_{k+1}|$ we set

$$\mathbf{v}_i^{(j-1)} = (A - \lambda I)\mathbf{v}_i^{(j)} \quad \text{for} \quad j = k, k-1, \ldots, 2.$$

Then define $B_k = B_{k+1} \cup \left\{ \mathbf{v}_i^{(j)} : 1 \leq i \leq |C_k \backslash B_{k+1}|,\ 1 \leq j \leq k \right\}$. We claim that B_k is linearly independent, but pinning down this claim is quite technical, so we leave the details to Theorem B.2.2 in Appendix B.2.

Step 4. If $k = 1$ then stop. Otherwise, decrease k by 1 and then return to Step 2.

After performing the above procedure, we set $B = B_1$ (which also equals C_1, since Step 3 does nothing in the $k = 1$ case). We note that B consists of Jordan chains of A corresponding to the eigenvalue λ by construction, and Step 3 tells us that B is linearly independent. Furthermore, the inclusion (2.4.1) from Step 2 tells us that $\text{span}(C_1) \supseteq \text{null}(A - \lambda I)$. Repeating this argument for C_2, and using the fact that $C_2 \subseteq C_1$, shows that

$$\text{span}(C_1) = \text{span}(C_2 \cup C_1) \supseteq \text{span}\big(C_2 \cup \text{null}(A - \lambda I)\big) \supseteq \text{null}\big((A - \lambda I)^2\big).$$

Carrying on in this way for $C_3, C_4, \ldots, C_\mu$ shows that $\text{span}(B) = \text{span}(C_1) \supseteq \text{null}\big((A - \lambda I)^\mu\big)$. Since B is contained within $\text{null}\big((A - \lambda I)^\mu\big)$, we conclude that B must be a basis of it, as desired.

Since we now know how to construct a basis of each generalized eigenspace $\text{null}\big((A - \lambda I)^\mu\big)$ consisting of Jordan chains of A, Theorem 2.4.5 tells us that we can construct a basis of all of $\mathbb{C}^n$ consisting of Jordan chains of A. If there are m Jordan chains in such a basis and the j-th Jordan chain has order k_j and corresponding eigenvalue λ_j for all $1 \leq j \leq m$, then placing the members of that j-th Jordan chain as columns into a matrix $P_j \in \mathcal{M}_{n,k_j}(\mathbb{C})$ in the manner suggested by Theorem 2.4.4 tells us that $AP_j = P_j J_{k_j}(\lambda_j)$. We can then construct a (necessarily invertible) matrix

$$P = \big[\, P_1 \mid P_2 \mid \cdots \mid P_m \,\big] \in \mathcal{M}_n(\mathbb{C}).$$

Our usual block matrix multiplication techniques show that

$$
\begin{aligned}
AP &= \big[\, AP_1 \mid AP_2 \mid \cdots \mid AP_m \,\big] && \text{(block matrix mult.)} \\
&= \big[\, P_1 J_{k_1}(\lambda_1) \mid P_2 J_{k_2}(\lambda_2) \mid \cdots \mid P_m J_{k_m}(\lambda_m) \,\big] && \text{(Theorem 2.4.4)} \\
&= P \begin{bmatrix} J_{k_1}(\lambda_1) & \cdots & O \\ \vdots & \ddots & \vdots \\ O & \cdots & J_{k_m}(\lambda_m) \end{bmatrix}. && \text{(block matrix mult.)}
\end{aligned}
$$

Multiplying on the right by P^{-1} gives a Jordan decomposition of A.   ∎

It is worth noting that the vectors that we add to the set C_k in Step 2 of the above proof are the "tops" of the Jordan chains of order exactly k. Since each Jordan chain corresponds to one Jordan block in the Jordan canonical form, Theorem 2.4.2 tells us that we must add $|C_k \backslash B_{k+1}| = 2\gamma_k - \gamma_{k+1} - \gamma_{k-1}$ such vectors for all $k \geq 1$, where γ_k is the corresponding geometric multiplicity of order k.

While this procedure for constructing the Jordan decomposition likely seems quite involved, we emphasize that things "usually" are simple enough that we can just use our previous worked examples as a guide. For now though, we work through a rather large example so as to illustrate the full algorithm in a bit more generality.

Example 2.4.10

A Hideous Jordan Decomposition

Construct a Jordan decomposition of the 8×8 matrix from Example 2.4.3.

Solution:

As we noted in Example 2.4.3, the only eigenvalue of A is $\lambda = 1$, which has algebraic multiplicity $\mu = 8$ and geometric multiplicities $\gamma_1 = 4$, $\gamma_2 = 7$, and $\gamma_k = 8$ whenever $k \geq 3$. In the language of the proof of Theorem 2.4.1, Step 1 thus tells us to set $k = 8$ and $B_9 = \{\}$.

In Step 2, nothing happens when $k = 8$ since we add $|C_8 \backslash B_9| = 2\gamma_8 - \gamma_9 - \gamma_7 = 16 - 8 - 8 = 0$ vectors. It follows that nothing happens in Step 3 as well, so $B_8 = C_8 = B_9 = \{\}$. A similar argument shows that $B_k = B_{k+1} = \{\}$ for all $k \geq 4$, so we proceed directly to the $k = 3$ case of Step 2.

$k = 3$: In Step 2, we start by setting $C_3 = B_4 = \{\}$, and then we need to add $2\gamma_3 - \gamma_4 - \gamma_2 = 16 - 8 - 7 = 1$ vector from

$$\text{null}\big((A - \lambda I)^3\big) \setminus \text{null}\big((A - \lambda I)^2\big) = \mathbb{C}^8 \setminus \text{null}\big((A - \lambda I)^2\big)$$

to C_3. One vector that works is $(0, 0, -1, 0, 0, -1, 1, 0)$, so we choose C_3 to be the set containing this single vector, which we call $\mathbf{v}_1^{(3)}$.

In Step 3, we just extend $\mathbf{v}_1^{(3)}$ to a Jordan chain by computing

$$\mathbf{v}_1^{(2)} = (A - I)\mathbf{v}_1^{(3)} = (-1, 1, 0, 0, 1, 1, 0, -1) \quad \text{and}$$
$$\mathbf{v}_1^{(1)} = (A - I)\mathbf{v}_1^{(2)} = (0, 0, -1, 1, 0, -1, 0, 0),$$

and then we let B_3 be the set containing these three vectors.

$k = 2$: In Step 2, we start by setting $C_2 = B_3$, and then we need to add any $2\gamma_2 - \gamma_3 - \gamma_1 = 14 - 8 - 4 = 2$ vectors from $\text{null}\big((A - \lambda I)^2\big) \setminus \text{null}(A - \lambda I)$ to C_2 while preserving linear independence. Two vectors that work are

$$\mathbf{v}_2^{(2)} = (0, 0, 0, -1, 1, 0, 0, 1) \quad \text{and}$$
$$\mathbf{v}_3^{(2)} = (-1, 0, 1, 1, 0, 0, -1, 0),$$

so we add these vectors to C_2, which now contains 5 vectors total.

In Step 3, we just extend $\mathbf{v}_2^{(2)}$ and $\mathbf{v}_3^{(2)}$ to Jordan chains by multiplying by $A - I$:

$$\mathbf{v}_2^{(1)} = (A - I)\mathbf{v}_2^{(2)} = (0, -1, 1, 0, 0, 0, 0, 0) \quad \text{and}$$
$$\mathbf{v}_3^{(1)} = (A - I)\mathbf{v}_3^{(2)} = (0, -1, -1, 0, 1, -1, 1, 0),$$

and let $B_2 = C_2 \cup \{\mathbf{v}_2^{(1)}, \mathbf{v}_3^{(1)}\}$ be the set containing all 7 vectors discussed so far.

Just try to follow along with the steps of the algorithm presented here, without worrying about the nasty calculations needed to compute any particular vectors or matrices we present.

In practice, large computations like this one are done by computers.

Recall that $\gamma_0 = 0$ always.

$k = 1$: In Step 2, we start by setting $C_1 = B_2$, and then we need to add any $2\gamma_1 - \gamma_2 - \gamma_0 = 8 - 7 - 0 = 1$ vector from $\text{null}(A - I)$ to C_1 while preserving linear independence. One vector that works is

$$\mathbf{v}_4^{(1)} = (0, 1, 0, 0, -1, 0, -1, 0),$$

so we add this vector to C_1, which now contains 8 vectors total (so we are done and can set $B_1 = C_1$).

To complete the Jordan decomposition of A, we simply place all of these Jordan chains as columns into a matrix P:

$$P = \left[\mathbf{v}_1^{(1)} \mid \mathbf{v}_1^{(2)} \mid \mathbf{v}_1^{(3)} \mid \mathbf{v}_2^{(1)} \mid \mathbf{v}_2^{(2)} \mid \mathbf{v}_3^{(1)} \mid \mathbf{v}_3^{(2)} \mid \mathbf{v}_4^{(1)} \right]$$

$$= \begin{bmatrix} 0 & -1 & 0 & 0 & 0 & 0 & -1 & 0 \\ 0 & 1 & 0 & -1 & 0 & -1 & 0 & 1 \\ -1 & 0 & -1 & 1 & 0 & -1 & 1 & 0 \\ 1 & 0 & 0 & 0 & -1 & 0 & 1 & 0 \\ 0 & 1 & 0 & 0 & 1 & 1 & 0 & -1 \\ -1 & 1 & -1 & 0 & 0 & -1 & 0 & 0 \\ 0 & 0 & 1 & 0 & 0 & 1 & -1 & -1 \\ 0 & -1 & 0 & 0 & 1 & 0 & 0 & 0 \end{bmatrix}.$$

It is straightforward (but arduous) to verify that $A = PJP^{-1}$, where J is the Jordan canonical form of A that we computed in Example 2.4.3.

2.4.3 Matrix Functions

One of the most useful applications of the Jordan decomposition is that it provides us with a method of applying functions to arbitrary matrices. We of course could apply functions to matrices entrywise, but doing so is a bit silly and does not agree with how we compute matrix powers or polynomials of matrices.

In order to apply functions to matrices "properly", we exploit the fact that many functions equal their Taylor series (functions with this property are called **analytic**). For example, recall that

See Appendix A.2.2 for a refresher on Taylor series.

$$e^x = 1 + x + \frac{x^2}{2!} + \frac{x^3}{3!} + \frac{x^4}{4!} + \cdots = \sum_{j=0}^{\infty} \frac{x^j}{j!} \quad \text{for all} \quad x \in \mathbb{C}.$$

Since this definition of e^x only depends on addition, scalar multiplication, and powers of x, and all of these operations are well-behaved when applied to matrices, it seems reasonable to analogously define e^A, the exponential of a *matrix*, via

When we talk about a sum of matrices converging, we just mean that every entry in that sum converges.

$$e^A = \sum_{j=0}^{\infty} \frac{1}{j!} A^j \quad \text{for all} \quad A \in \mathcal{M}_n(\mathbb{C}),$$

as long as this sum converges.

More generally, we define the "matrix version" of an analytic function as follows:

Definition 2.4.4

Matrix Functions

Suppose $A \in \mathcal{M}_n(\mathbb{C})$ and $f : \mathbb{C} \to \mathbb{C}$ is analytic on an open disc centered at some scalar $a \in \mathbb{C}$. Then

$$f(A) \overset{\text{def}}{=} \sum_{j=0}^{\infty} \frac{f^{(j)}(a)}{j!} (A - aI)^j,$$

as long as the sum on the right converges.

An "open" disc is a filled-in circle that does not include its boundary. We need it to be open so that things like derivatives make sense everywhere in it.

The above definition implicitly makes the (very not obvious) claim that the matrix that the sum on the right converges to does not depend on which value $a \in \mathbb{C}$ is chosen. That is, the value of $a \in \mathbb{C}$ might affect whether or not the sum on the right converges, but not what it converges *to*. If $k = 1$ (so $A \in \mathcal{M}_1(\mathbb{C})$ is a scalar, which we relabel as $A = x \in \mathbb{C}$) then this fact follows from the sum in Definition 2.4.4 being the usual Taylor series of f centered at a, which we know converges to $f(x)$ (if it converges at all) since f is analytic. However, if $k \geq 2$ then it is not so obvious that this sum is so well-behaved, nor is it obvious how we could possibly compute what it converges to.

The next two theorems solve these problems—they tell us when the sum in Definition 2.4.4 converges, that the scalar $a \in \mathbb{C}$ has no effect on what it converges to (though it may affect whether or not it converges in the first place), and that we can use the Jordan decomposition of A to compute $f(A)$. We start by showing how to compute $f(A)$ in the special case when A is a Jordan block.

Theorem 2.4.6

Functions of Jordan Blocks

Suppose $J_k(\lambda) \in \mathcal{M}_k(\mathbb{C})$ is a Jordan block and $f : \mathbb{C} \to \mathbb{C}$ is analytic on some open set containing λ. Then

The notation $f^{(j)}(\lambda)$ means the j-th derivative of f evaluated at λ.

$$f\big(J_k(\lambda)\big) = \begin{bmatrix} f(\lambda) & \frac{f'(\lambda)}{1!} & \frac{f''(\lambda)}{2!} & \cdots & \frac{f^{(k-2)}(\lambda)}{(k-2)!} & \frac{f^{(k-1)}(\lambda)}{(k-1)!} \\ 0 & f(\lambda) & \frac{f'(\lambda)}{1!} & \cdots & \frac{f^{(k-3)}(\lambda)}{(k-3)!} & \frac{f^{(k-2)}(\lambda)}{(k-2)!} \\ 0 & 0 & f(\lambda) & \cdots & \frac{f^{(k-4)}(\lambda)}{(k-4)!} & \frac{f^{(k-3)}(\lambda)}{(k-3)!} \\ \vdots & \vdots & \vdots & \ddots & \vdots & \vdots \\ 0 & 0 & 0 & \cdots & f(\lambda) & \frac{f'(\lambda)}{1!} \\ 0 & 0 & 0 & \cdots & 0 & f(\lambda) \end{bmatrix}.$$

Proof. For each $1 \leq n < k$, let $N_n \in \mathcal{M}_k(\mathbb{C})$ denote the matrix with ones on its n-th superdiagonal and zeros elsewhere (i.e., $[N_n]_{i,j} = 1$ if $j - i = n$ and $[N_n]_{i,j} = 0$ otherwise). Then $J_k(\lambda) = \lambda I + N_1$, and we show in Exercise 2.4.17 that these matrices satisfy $N_1^n = N_n$ for all $1 \leq n < k$, and $N_1^n = O$ when $n \geq k$.

Recall that $[N_n]_{i,j}$ refers to the (i, j)-entry of N_n.

We now prove the statement of this theorem when we choose to center the Taylor series from Definition 2.4.4 at $a = \lambda$. That is, we write f as a Taylor series centered at λ, so that

$$f(x) = \sum_{n=0}^{\infty} \frac{f^{(n)}(\lambda)}{n!}(x - \lambda)^n, \quad \text{so} \quad f\big(J_k(\lambda)\big) = \sum_{n=0}^{\infty} \frac{f^{(n)}(\lambda)}{n!}(J_k(\lambda) - \lambda I)^n.$$

By making use of the fact that $J_k(\lambda) - \lambda I = N_1$, together with our earlier observation about powers of N_1, we see that

This sum becomes
finite because
$N_1^n = O$ when $n \geq k$.

$$f(J_k(\lambda)) = \sum_{n=0}^{\infty} \frac{f^{(n)}(\lambda)}{n!} N_1^n = \sum_{n=0}^{k-1} \frac{f^{(n)}(\lambda)}{n!} N_n,$$

which is exactly the formula given in the statement of the theorem.

To complete this proof, we must show that Definition 2.4.4 is actually well-defined (i.e., *no matter which* value a in the open set on which f is analytic we center the Taylor series of f at, the formula provided by this theorem still holds). This extra argument is very similar to the one we just went through, but with some extra ugly details, so we defer it to Appendix B.2 (see Theorem B.2.3 in particular). ∎

Theorem 2.4.7

Matrix Functions via the Jordan Decomposition

Suppose $A \in \mathcal{M}_n(\mathbb{C})$ has Jordan decomposition as in Theorem 2.4.1, and $f : \mathbb{C} \to \mathbb{C}$ is analytic on some open disc containing all of the eigenvalues of A. Then

$$f(A) = P \begin{bmatrix} f(J_{k_1}(\lambda_1)) & O & \cdots & O \\ O & f(J_{k_2}(\lambda_2)) & \cdots & O \\ \vdots & \vdots & \ddots & \vdots \\ O & O & \cdots & f(J_{k_m}(\lambda_m)) \end{bmatrix} P^{-1}.$$

Proof. We just exploit the fact that matrix powers, and thus Taylor series, behave very well with block diagonal matrices and the Jordan decomposition. For any a in the open disc on which f is analytic, we have

$$f(A) = \sum_{j=0}^{\infty} \frac{f^{(j)}(a)}{j!} (A - aI)^j$$

In the second
equality, we use the
fact that
$(PJP^{-1})^j = PJ^jP^{-1}$ for
all j.

$$= P \left(\sum_{j=0}^{\infty} \frac{f^{(j)}(a)}{j!} \begin{bmatrix} (J_{k_1}(\lambda_1) - aI)^j & \cdots & O \\ \vdots & \ddots & \vdots \\ O & \cdots & (J_{k_m}(\lambda_m) - aI)^j \end{bmatrix} \right) P^{-1}$$

$$= P \begin{bmatrix} f(J_{k_1}(\lambda_1)) & \cdots & O \\ \vdots & \ddots & \vdots \\ O & \cdots & f(J_{k_m}(\lambda_m)) \end{bmatrix} P^{-1},$$

as claimed. ∎

By combining the previous two theorems, we can explicitly apply any analytic function f to any matrix by first constructing that matrix's Jordan decomposition, applying f to each of the Jordan blocks in its Jordan canonical form via the formula of Theorem 2.4.6, and then stitching those computations together via Theorem 2.4.7.

Example 2.4.11

A Matrix Exponential via the Jordan Decomposition

Compute e^A if $A = \begin{bmatrix} 5 & 1 & -1 \\ 1 & 3 & -1 \\ 2 & 0 & 2 \end{bmatrix}$.

Solution:

We computed the following Jordan decomposition $A = PJP^{-1}$ of this matrix in Example 2.4.7:

$$P = \frac{1}{2}\begin{bmatrix} 0 & 2 & 1 \\ 2 & 0 & 1 \\ 2 & 2 & 0 \end{bmatrix}, \; J = \begin{bmatrix} 2 & 0 & 0 \\ 0 & 4 & 1 \\ 0 & 0 & 4 \end{bmatrix}, \; P^{-1} = \frac{1}{2}\begin{bmatrix} -1 & 1 & 1 \\ 1 & -1 & 1 \\ 2 & 2 & -2 \end{bmatrix}.$$

Applying the function $f(x) = e^x$ to the two Jordan blocks in J via the formula of Theorem 2.4.6 gives

Here we use the fact that $f'(x) = e^x$ too.

$$f(2) = e^2 \quad \text{and} \quad f\left(\begin{bmatrix} 4 & 1 \\ 0 & 4 \end{bmatrix}\right) = \begin{bmatrix} f(4) & f'(4) \\ 0 & f(4) \end{bmatrix} = \begin{bmatrix} e^4 & e^4 \\ 0 & e^4 \end{bmatrix}.$$

Stitching everything together via Theorem 2.4.7 then tells us that

$$f(A) = e^A = \frac{1}{4}\begin{bmatrix} 0 & 2 & 1 \\ 2 & 0 & 1 \\ 2 & 2 & 0 \end{bmatrix}\begin{bmatrix} e^2 & 0 & 0 \\ 0 & e^4 & e^4 \\ 0 & 0 & e^4 \end{bmatrix}\begin{bmatrix} -1 & 1 & 1 \\ 1 & -1 & 1 \\ 2 & 2 & -2 \end{bmatrix}$$

$$= \frac{1}{2}\begin{bmatrix} 4e^4 & 2e^4 & -2e^4 \\ e^4 - e^2 & e^4 + e^2 & e^2 - e^4 \\ 3e^4 - e^2 & e^2 + e^4 & e^2 - e^4 \end{bmatrix}.$$

When we define matrix functions in this way, they interact with matrix multiplication how we would expect them to. For example, the principal square root function $f(x) = \sqrt{x}$ is analytic everywhere except on the set of non-positive real numbers, so Theorems 2.4.6 and 2.4.7 tell us how to compute the principal square root $\sqrt{A}$ of any *matrix* $A \in \mathcal{M}_n(\mathbb{C})$ whose eigenvalues can be placed in an open disc avoiding that strip in the complex plane. As we would hope, this matrix satisfies $(\sqrt{A})^2 = A$, and if A is positive definite then this method produces exactly the positive definite principal square root $\sqrt{A}$ described by Theorem 2.2.11.

Example 2.4.12

The Principal Square Root of a Non-Positive Semidefinite Matrix

Compute $\sqrt{A}$ if $A = \begin{bmatrix} 3 & 2 & 3 \\ 1 & 2 & -3 \\ -1 & -1 & 4 \end{bmatrix}$.

Solution:

We can use the techniques of this section to see that A has Jordan decomposition $A = PJP^{-1}$, where

Try computing a Jordan decomposition of A on your own.

$$P = \begin{bmatrix} -2 & -1 & 2 \\ 2 & 1 & -1 \\ 0 & -1 & 1 \end{bmatrix}, \; J = \begin{bmatrix} 1 & 0 & 0 \\ 0 & 4 & 1 \\ 0 & 0 & 4 \end{bmatrix}, \; P^{-1} = \frac{1}{2}\begin{bmatrix} 0 & 1 & 1 \\ 2 & 2 & -2 \\ 2 & 2 & 0 \end{bmatrix}.$$

To apply the function $f(x) = \sqrt{x}$ to the two Jordan blocks in J via the formula of Theorem 2.4.6, we note that $f'(x) = \frac{1}{2\sqrt{x}}$, so that

$$f(1) = 1 \quad \text{and} \quad f\left(\begin{bmatrix} 4 & 1 \\ 0 & 4 \end{bmatrix}\right) = \begin{bmatrix} f(4) & f'(4) \\ 0 & f(4) \end{bmatrix} = \begin{bmatrix} 2 & 1/4 \\ 0 & 2 \end{bmatrix}.$$

Double-check on
your own that
$(\sqrt{A})^2 = A$.

Stitching everything together via Theorem 2.4.7 then tells us that

$$f(A) = \sqrt{A} = \frac{1}{2}\begin{bmatrix} -2 & -1 & 2 \\ 2 & 1 & -1 \\ 0 & -1 & 1 \end{bmatrix}\begin{bmatrix} 1 & 0 & 0 \\ 0 & 2 & 1/4 \\ 0 & 0 & 2 \end{bmatrix}\begin{bmatrix} 0 & 1 & 1 \\ 2 & 2 & -2 \\ 2 & 2 & 0 \end{bmatrix}$$

$$= \frac{1}{4}\begin{bmatrix} 7 & 3 & 4 \\ 1 & 5 & -4 \\ -1 & -1 & 8 \end{bmatrix}.$$

Remark 2.4.2

Is Analyticity Needed?

If a function f is not analytic on any disc containing a certain scalar $\lambda \in \mathbb{C}$ then the results of this section say nothing about how to compute $f(A)$ for matrices with λ as an eigenvalue. Indeed, it may be the case that there is no sensible way to even *define* $f(A)$ for these matrices, as the sum in Definition 2.4.4 may not converge. For example, if $f(x) = \sqrt{x}$ is the principal square root function, then f is not analytic at $x = 0$ (despite being defined there), so it is not obvious how to compute (or even define) $\sqrt{A}$ if A has 0 as an eigenvalue.

We can sometimes get around this problem by just applying the formulas of Theorems 2.4.6 and 2.4.7 anyway, as long as the quantities used in those formulas all exist. For example, we could say that if

$$A = \begin{bmatrix} 2i & 0 \\ 0 & 0 \end{bmatrix} \quad \text{then} \quad \sqrt{A} = \begin{bmatrix} \sqrt{2i} & 0 \\ 0 & \sqrt{0} \end{bmatrix} = \begin{bmatrix} 1+i & 0 \\ 0 & 0 \end{bmatrix},$$

even though A has 0 as an eigenvalue. On the other hand, this method says that

We are again using
the fact that if
$f(x) = \sqrt{x}$ then
$f'(x) = 1/(2\sqrt{x})$ here.

$$B = \begin{bmatrix} 0 & 1 \\ 0 & 0 \end{bmatrix} \quad \text{then} \quad \sqrt{B} = \begin{bmatrix} \sqrt{0} & 1/(2\sqrt{0}) \\ 0 & \sqrt{0} \end{bmatrix},$$

which makes no sense (notice the division by 0 in the top-right corner of $\sqrt{B}$, corresponding to the fact that the square root function is not differentiable at 0). Indeed, this matrix B does not have a square root at all, nor does any matrix with a Jordan block of size 2×2 or larger corresponding to the eigenvalue 0.

Similarly, Definition 2.4.4 does not tell us what $\sqrt{C}$ means if

$$C = \begin{bmatrix} i & 0 \\ 0 & -i \end{bmatrix},$$

even though the principal square root function $f = \sqrt{x}$ is analytic on open discs containing the eigenvalues i and $-i$. The problem here is that there is no *common* disc D containing i and $-i$ on which f is analytic, since any such disc must contain 0 as well. In practice, we just ignore this problem and apply the formulas of Theorems 2.4.6 and 2.4.7 anyway to get

$$\sqrt{C} = \begin{bmatrix} \sqrt{i} & 0 \\ 0 & \sqrt{-i} \end{bmatrix} = \frac{1}{\sqrt{2}}\begin{bmatrix} 1+i & 0 \\ 0 & 1-i \end{bmatrix}.$$

Example 2.4.13

Geometric Series for Matrices

Compute $I + A + A^2 + A^3 + \cdots$ if

$$A = \frac{1}{2} \begin{bmatrix} 3 & 4 & 0 \\ 0 & 1 & -2 \\ 1 & 2 & -1 \end{bmatrix}.$$

Solution:

We can use the techniques of this section to see that A has Jordan decomposition $A = PJP^{-1}$, where

$$P = \begin{bmatrix} 2 & 2 & 0 \\ -1 & 0 & 1 \\ 0 & 1 & 0 \end{bmatrix}, \quad J = \begin{bmatrix} 1/2 & 1 & 0 \\ 0 & 1/2 & 1 \\ 0 & 0 & 1/2 \end{bmatrix}, \quad P^{-1} = \frac{1}{2} \begin{bmatrix} 1 & 0 & -2 \\ 0 & 0 & 2 \\ 1 & 2 & -2 \end{bmatrix}.$$

We then recall that

$$1 + x + x^2 + x^3 + \cdots = \frac{1}{1-x} \quad \text{whenever} \quad |x| < 1,$$

Before doing this computation, it was not even clear that the sum $I + A + A^2 + A^3 + \cdots$ converges.

so we can compute $I + A + A^2 + A^3 + \cdots$ by applying Theorems 2.4.6 and 2.4.7 to the function $f(x) = 1/(1-x)$, which is analytic on the open disc $D = \{x \in \mathbb{C} : |x| < 1\}$ containing the eigenvalue $1/2$ of A. In particular, $f'(x) = 1/(1-x)^2$ and $f''(x) = 2/(1-x)^3$, so we have

$$f(A) = I + A + A^2 + A^3 + \cdots$$

$$= P \begin{bmatrix} f(1/2) & f'(1/2) & \frac{1}{2}f''(1/2) \\ 0 & f(1/2) & f'(1/2) \\ 0 & 0 & f(1/2) \end{bmatrix} P^{-1}$$

$$= \frac{1}{2} \begin{bmatrix} 2 & 2 & 0 \\ -1 & 0 & 1 \\ 0 & 1 & 0 \end{bmatrix} \begin{bmatrix} 2 & 4 & 8 \\ 0 & 2 & 4 \\ 0 & 0 & 2 \end{bmatrix} \begin{bmatrix} 1 & 0 & -2 \\ 0 & 0 & 2 \\ 1 & 2 & -2 \end{bmatrix} = \begin{bmatrix} 14 & 24 & -16 \\ -4 & -6 & 4 \\ 2 & 4 & -2 \end{bmatrix}.$$

We close this section by noting that if A is diagonalizable then each of its Jordan blocks is 1×1, so Theorems 2.4.6 and 2.4.7 tell us that we can compute matrix functions in the manner that we already knew from introductory linear algebra:

Corollary 2.4.8

Functions of Diagonalizable Matrices

If $A \in \mathcal{M}_n(\mathbb{C})$ is diagonalizable via $A = PDP^{-1}$, and $f : \mathbb{C} \to \mathbb{C}$ is analytic on some open disc containing the eigenvalues of D, then $f(A) = Pf(D)P^{-1}$, where $f(D)$ is obtained by applying f to each diagonal entry of D.

In particular, since the spectral decomposition (Theorem 2.1.4) is a special case of diagonalization, we see that if $A \in \mathcal{M}_n(\mathbb{C})$ is normal with spectral decomposition $A = UDU^*$, then $f(A) = Uf(D)U^*$.

Exercises

solutions to starred exercises on page 472

2.4.1 Compute the Jordan canonical form of each of the following matrices.

* (a) $\begin{bmatrix} 3 & -2 \\ 2 & -1 \end{bmatrix}$
(b) $\begin{bmatrix} 4 & 2 \\ -3 & -1 \end{bmatrix}$

* (c) $\begin{bmatrix} 0 & -1 \\ 4 & 4 \end{bmatrix}$
(d) $\begin{bmatrix} -1 & 1 \\ -2 & 1 \end{bmatrix}$

* (e) $\begin{bmatrix} 1 & -1 & 2 \\ -1 & -1 & 4 \\ -1 & -2 & 5 \end{bmatrix}$
(f) $\begin{bmatrix} 3 & 0 & 0 \\ 1 & 3 & -1 \\ 1 & 0 & 2 \end{bmatrix}$

* (g) $\begin{bmatrix} 5 & 2 & -2 \\ -1 & 2 & 1 \\ 1 & 1 & 2 \end{bmatrix}$
(h) $\begin{bmatrix} 1 & 2 & 0 \\ -1 & 5 & 2 \\ 1 & -1 & 3 \end{bmatrix}$

2.4.2 Compute a Jordan decomposition of each of the matrices from Exercise 2.4.1.

2.4.3 Determine whether or not the given matrices are similar.

* (a) $\begin{bmatrix} 2 & 1 \\ -4 & 6 \end{bmatrix}$ and $\begin{bmatrix} 3 & -1 \\ -5 & 7 \end{bmatrix}$

(b) $\begin{bmatrix} 1 & -2 \\ 2 & -3 \end{bmatrix}$ and $\begin{bmatrix} 0 & 1 \\ -1 & -2 \end{bmatrix}$

* (c) $\begin{bmatrix} 1 & 2 & 0 \\ 0 & 1 & -1 \\ 0 & -1 & 1 \end{bmatrix}$ and $\begin{bmatrix} 2 & 1 & 1 \\ 0 & 1 & 0 \\ 2 & 0 & 0 \end{bmatrix}$

(d) $\begin{bmatrix} 2 & 1 & -1 \\ -2 & -1 & 2 \\ -1 & -1 & 2 \end{bmatrix}$ and $\begin{bmatrix} 2 & 1 & 3 \\ -1 & 0 & -4 \\ 0 & 0 & 1 \end{bmatrix}$

* (e) $\begin{bmatrix} 2 & -1 & 1 \\ 0 & -1 & 0 \\ -1 & -5 & 4 \end{bmatrix}$ and $\begin{bmatrix} 1 & -2 & 2 \\ 1 & 6 & 1 \\ -1 & -7 & -2 \end{bmatrix}$

(f) $\begin{bmatrix} 1 & 1 & 0 & 0 \\ 0 & 1 & 0 & 0 \\ 0 & 0 & 1 & 1 \\ 0 & 0 & 0 & 1 \end{bmatrix}$ and $\begin{bmatrix} 1 & 1 & 0 & 0 \\ 0 & 1 & 0 & 0 \\ 0 & 0 & 1 & 0 \\ 0 & 0 & 0 & 1 \end{bmatrix}$

2.4.4 Compute the indicated matrix.

* (a) $\sqrt{A}$, where $A = \begin{bmatrix} 2 & 0 \\ 1 & 2 \end{bmatrix}$.

(b) e^A, where $A = \begin{bmatrix} 1 & -1 \\ 1 & -1 \end{bmatrix}$.

(c) $\sin(A)$, where $A = \frac{1}{3}\begin{bmatrix} 2 & -1 \\ 4 & -2 \end{bmatrix}$.

(d) $\cos(A)$, where $A = \frac{1}{3}\begin{bmatrix} 2 & -1 \\ 4 & -2 \end{bmatrix}$.

* (e) e^A, where $A = \begin{bmatrix} 1 & -2 & -1 \\ 0 & 3 & 1 \\ 0 & -4 & -1 \end{bmatrix}$.

(f) $\sqrt{A}$, where $A = \begin{bmatrix} 1 & -2 & -1 \\ 0 & 3 & 1 \\ 0 & -4 & -1 \end{bmatrix}$.

2.4.5 Determine which of the following statements are true and which are false.

* (a) Every matrix $A \in \mathcal{M}_n(\mathbb{C})$ can be diagonalized.
(b) Every matrix $A \in \mathcal{M}_n(\mathbb{C})$ has a Jordan decomposition.
* (c) If $A = P_1 J_1 P_1^{-1}$ and $A = P_2 J_2 P_2^{-1}$ are two Jordan decompositions of the same matrix A then $J_1 = J_2$.
(d) Two matrices $A, B \in \mathcal{M}_n(\mathbb{C})$ are similar if and only if there is a Jordan canonical form J such that $A = PJP^{-1}$ and $B = QJQ^{-1}$.
* (e) Two diagonalizable matrices $A, B \in \mathcal{M}_n(\mathbb{C})$ are similar if and only if they have the same eigenvalues (counting with algebraic multiplicity).
(f) There exist matrices $A, B \in \mathcal{M}_n(\mathbb{C})$ such that A is diagonalizable, B is not diagonalizable, and A and B are similar.
* (g) The series $e^A = \sum_{j=0}^{\infty} A^j / j!$ converges for all $A \in \mathcal{M}_n(\mathbb{C})$.

2.4.6 Suppose $A \in \mathcal{M}_4(\mathbb{C})$ has all four of its eigenvalues equal to 2. What are its possible Jordan canonical forms (do not list Jordan canonical forms that have the same Jordan blocks as each other in a different order)?

2.4.7 Compute the eigenvalues of $\sin(A)$, where

$$A = \begin{bmatrix} 3 & -1 & -1 \\ 0 & 4 & 0 \\ -1 & -1 & 3 \end{bmatrix}.$$

*2.4.8** Find bases C and D of $\mathbb{R}^2$, and a linear transformation $T : \mathbb{R}^2 \to \mathbb{R}^2$, such that $A = [T]_C$ and $B = [T]_D$, where

$$A = \begin{bmatrix} 1 & 2 \\ 5 & 4 \end{bmatrix} \quad \text{and} \quad B = \begin{bmatrix} 13 & 7 \\ -14 & -8 \end{bmatrix}.$$

2.4.9 Let $A \in \mathcal{M}_n(\mathbb{C})$.

(a) Show that if A is Hermitian then e^A is positive definite.
(b) Show that if A is skew-Hermitian then e^A is unitary.

2.4.10 Show that $\det(e^A) = e^{\text{tr}(A)}$ for all $A \in \mathcal{M}_n(\mathbb{C})$.

*2.4.11** Show that e^A is invertible for all $A \in \mathcal{M}_n(\mathbb{C})$.

2.4.12 Suppose $A, B \in \mathcal{M}_n(\mathbb{C})$.

(a) Show that if $AB = BA$ then $e^{A+B} = e^A e^B$.
[Hint: It is probably easier to use the definition of e^{A+B} rather than the Jordan decomposition. You may use the fact that you can rearrange infinite sums arising from analytic functions just like finite sums.]
(b) Provide an example to show that if e^{A+B} may not equal $e^A e^B$ if $AB \neq BA$.

2.4.13 Show that a matrix $A \in \mathcal{M}_n(\mathbb{C})$ is invertible if and only if it can be written in the form $A = Ue^X$, where $U \in \mathcal{M}_n(\mathbb{C})$ is unitary and $X \in \mathcal{M}_n(\mathbb{C})$ is Hermitian.

∗∗2.4.14 Show that $\sin^2(A) + \cos^2(A) = I$ for all $A \in \mathcal{M}_n(\mathbb{C})$.

2.4.15 Show that every matrix $A \in \mathcal{M}_n(\mathbb{C})$ can be written in the form $A = D + N$, where $D \in \mathcal{M}_n(\mathbb{C})$ is diagonalizable, $N \in \mathcal{M}_n(\mathbb{C})$ is **nilpotent** (i.e., $N^k = O$ for some integer $k \geq 1$), and $DN = ND$.

[Side note: This is called the **Jordan–Chevalley decomposition** of A.]

∗∗2.4.16 Suppose $A \in \mathcal{M}_n$ is strictly upper triangular (i.e., it is upper triangular with diagonal entries equal to 0).

    (a) Show that, for each $1 \leq k \leq n$, the first k superdiagonals of A^k consist entirely of zeros. That is, show that $\left[A^k\right]_{i,j} = 0$ whenever $j - i < k$.

    (b) Show that $A^n = O$.

∗∗2.4.17 For each $1 \leq j < k$, let $N_n \in \mathcal{M}_k$ denote the matrix with ones on its n-th superdiagonal and zeros elsewhere (i.e., $[N_n]_{i,j} = 1$ if $j - i = n$ and $[N_n]_{i,j} = 0$ otherwise).

    (a) Show that $N_1^n = N_n$ for all $1 \leq n < k$, and $N_1^n = O$ when $n \geq k$.

    (b) Show that nullity$(N_n) = \min\{k, n\}$.

2.5   Summary and Review

In this chapter, we learned about several new matrix decompositions, and how they fit in with and generalize the matrix decompositions that we already knew about, like diagonalization. For example, we learned about a generalization of diagonalization that applies to all matrices, called the Jordan decomposition (Theorem 2.4.1), and we learned about a special case of diagonalization called the spectral decomposition (Theorems 2.1.4 and 2.1.6) that applies to normal or symmetric matrices (depending on whether the field is $\mathbb{C}$ or $\mathbb{R}$, respectively). See Figure 2.10 for a reminder of which decompositions from this chapter are special cases of each other.

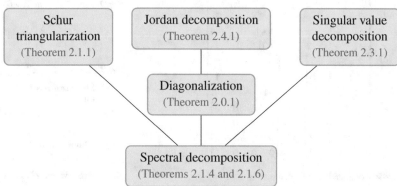

Figure 2.10: Some matrix decompositions from this chapter. The decompositions in the top row apply to any square complex matrix (and the singular value decomposition even applies to rectangular matrices). Black lines between two decompositions indicate that the lower decomposition is a special case of the one above it that applies to a smaller set of matrices.

One common way of thinking about some matrix decompositions is as providing us with a **canonical form** for matrices that is (a) unique, and (b) captures all of the "important" information about that matrix (where the exact meaning of "important" depends on what we want to do with the matrix or what it represents). For example,

We call a property of a matrix A "basis independent" if a change of basis does not affect it (i.e., A and PAP^{-1} share that property for all invertible P).

- If we are thinking of $A \in \mathcal{M}_{m,n}$ as representing a linear system, the relevant canonical form is its reduced row echelon form (RREF), which is unique and contains all information about the solutions of that linear system.
- If we are thinking of $A \in \mathcal{M}_n$ as representing a linear transformation and are only interested in basis-independent properties of it (e.g., its

unique and contains all basis-independent information about that linear transformation.

These canonical forms can also be thought of as answering the question of how simple a matrix can be made upon multiplying it on the left and/or right by invertible matrices, and they are summarized in Table 2.2 for easy reference.

All three of the forms reached by these decompositions are canonical. For example, every matrix can be put into one, and only one, RREF by multiplying it on the left by an invertible matrix.

Type	Decomposition	Name and notes
One-sided	$A = PR$ • P is invertible • R is the RREF of A	See Appendix A.1.3.
Two-sided	$A = PDQ$ • P, Q are invertible • $D = \begin{bmatrix} I_{\text{rank}(A)} & O \\ O & O \end{bmatrix}$	From Exercise 2.3.6. Either P or Q (but not both) can be chosen to be unitary.
Similarity	$A = PJP^{-1}$ • P is invertible • J is Jordan form	Jordan decomposition—see Section 2.4. J is block diagonal and each block has constant diagonal and superdiagonal equal to 1.

Table 2.2: A summary of the matrix decompositions that answer the question of how simple a matrix can be made upon multiplying it by an invertible matrix on the left and/or right. These decompositions are all canonical and apply to every matrix (with the understanding that the matrix must be square for similarity to make sense).

We can similarly ask how simple we can make a matrix upon multiplying it on the left and/or right by unitary matrices, but it turns out that the answers to this question are not so straightforward. For example, Schur triangularization (Theorem 2.1.1) tells us that by applying a *unitary* similarity to a matrix $A \in \mathcal{M}_n(\mathbb{C})$, we can make it upper triangular (i.e., we can find a unitary matrix $U \in \mathcal{M}_n(\mathbb{C})$ such that $A = UTU^*$ for some upper triangular matrix $T \in \mathcal{M}_n(\mathbb{C})$). However, this upper triangular form is *not* canonical, since most matrices have numerous different Schur triangularizations that look nothing like one another.

Since Schur triangularization is not canonical, we cannot use it to answer the question of, given two matrices $A, B \in \mathcal{M}_n(\mathbb{C})$, whether or not there exists a unitary matrix $U \in \mathcal{M}_n(\mathbb{C})$ such that $A = UBU^*$.

A related phenomenon happens when we consider one-sided multiplication by a unitary matrix. There are two decompositions that give completely different answers for what type of matrix can be reached in this way—the polar decomposition (Theorem 2.2.12) says that every matrix can be made positive semidefinite, whereas the QR decomposition (Theorem 1.C.1) says that every matrix can be made upper triangular. Furthermore, both of these forms are "not quite canonical"—they are unique as long as the original matrix is invertible, but they are not unique otherwise.

Of the decompositions that consider multiplication by unitary matrices, only the singular value decomposition (Theorem 2.3.1) is canonical. We summarize these observations in Table 2.3 for easy reference.

We also learned about the special role that the set of normal matrices, as well as its subsets of unitary, Hermitian, skew-Hermitian, and positive semidefinite matrices play in the realm of matrix decompositions. For example, the spectral

Type	Decomposition	Name and notes
One-sided	$A = UT$ • U is unitary • T is triangular	QR decomposition—see Section 1.C and note that T can be chosen to have non-negative diagonal entries.
One-sided	$A = UP$ • U is unitary • P is PSD	Polar decomposition—see Theorem 2.2.12. P is positive *definite* if and only if A is invertible.
Two-sided	$A = U\Sigma V$ • U, V are unitary • Σ is diagonal	Singular value decomposition (SVD)—see Section 2.3. The diagonal entries of Σ can be chosen to be non-negative and in non-increasing order.
Unitary similarity	$A = UTU^*$ • U is unitary • T is triangular	Schur triangularization—see Section 2.1.1. The spectral decomposition (Theorem 2.1.4) is a special case.

Table 2.3: A summary of the matrix decompositions that answer the question of how simple a matrix can be made upon multiplying it by unitary matrices on the left and/or right. These decompositions all apply to every matrix A (with the understanding that A must be square for unitary similarity to make sense).

decomposition tells us that normal matrices are exactly the matrices that can not only be diagonalized, but can be diagonalized via a unitary matrix (rather than just an invertible matrix).

While we have already given characterization theorems for unitary matrices (Theorem 1.4.9) and positive semidefinite matrices (Theorem 2.2.1) that provide numerous different equivalent conditions that could be used to define them, we have not yet done so for normal matrices. For ease of reference, we provide such a characterization here, though we note that most of these properties were already proved earlier in various exercises.

Theorem 2.5.1

Characterization of Normal Matrices

Suppose $A \in \mathcal{M}_n(\mathbb{C})$ has eigenvalues $\lambda_1, \ldots, \lambda_n$. The following are equivalent:

    a) A is normal,
    b) $A = UDU^*$ for some unitary $U \in \mathcal{M}_n(\mathbb{C})$ and diagonal $D \in \mathcal{M}_n(\mathbb{C})$,
    c) there is an orthonormal basis of $\mathbb{C}^n$ consisting of eigenvectors of A,
    d) $\|A\|_F^2 = |\lambda_1|^2 + \cdots + |\lambda_n|^2$,
    e) the singular values of A are $|\lambda_1|, \ldots, |\lambda_n|$,
    f) $(A\mathbf{v}) \cdot (A\mathbf{w}) = (A^*\mathbf{v}) \cdot (A^*\mathbf{w})$ for all $\mathbf{v}, \mathbf{w} \in \mathbb{C}^n$, and
    g) $\|A\mathbf{v}\| = \|A^*\mathbf{v}\|$ for all $\mathbf{v} \in \mathbb{C}^n$.

Compare
statements (g) and
(h) of this theorem
to the
characterizations of
unitary matrices
given in
Theorem 1.4.9.

Proof. We have already discussed the equivalence of (a), (b), and (c) extensively, as (b) and (c) are just different statements of the spectral decomposition. The equivalence of (a) and (d) was proved in Exercise 2.1.12, the fact that (a) implies (e) was proved in Theorem 2.3.4 and its converse in Exercise 2.3.20. Finally, the fact that (a), (f), and (g) are equivalent follows from taking $B = A^*$ in Exercise 1.4.19. ∎

It is worth noting, however, that there are even more equivalent characterizations of normality, though they are somewhat less important than those discussed above. See Exercises 2.1.14, 2.2.14, and 2.5.4, for example.

Exercises

solutions to starred exercises on page 473

2.5.1 For each of the following matrices, say which of the following matrix decompositions can be applied to it: (i) diagonalization (i.e., $A = PDP^{-1}$ with P invertible and D diagonal), (ii) Schur triangularization, (iii) spectral decomposition, (iv) singular value decomposition, and/or (v) Jordan decomposition.

* (a) $\begin{bmatrix} 1 & 1 \\ -1 & 1 \end{bmatrix}$
  (b) $\begin{bmatrix} 0 & -i \\ i & 0 \end{bmatrix}$

* (c) $\begin{bmatrix} 1 & 0 \\ 1 & 1 \end{bmatrix}$
  (d) $\begin{bmatrix} 1 & 2 \\ 3 & 4 \end{bmatrix}$

* (e) $\begin{bmatrix} 1 & 0 \\ 1 & 1.0001 \end{bmatrix}$
  (f) $\begin{bmatrix} 1 & 2 & 3 \\ 4 & 5 & 6 \end{bmatrix}$

* (g) $\begin{bmatrix} 0 & 0 \\ 0 & 0 \end{bmatrix}$
  (h) The 75×75 matrix with every entry equal to 1.

2.5.2 Determine which of the following statements are true and which are false.

* (a) Any two eigenvectors that come from different eigenspaces of a normal matrix must be orthogonal.
  (b) Any two distinct eigenvectors of a Hermitian matrix must be orthogonal.
* (c) If $A \in \mathcal{M}_n$ has polar decomposition $A = UP$ then P is the principal square root of A^*A.

2.5.3 Suppose $A \in \mathcal{M}_n(\mathbb{C})$ is normal and $B \in \mathcal{M}_n(\mathbb{C})$ commutes with A (i.e., $AB = BA$).

  (a) Show that A^* and B commute as well. [Hint: Compute $\|A^*B - BA^*\|_F^2$.] [Side note: This result is called **Fuglede's theorem**.]
  (b) Show that if B is also normal then so is AB.

∗∗2.5.4 Suppose $A \in \mathcal{M}_n(\mathbb{C})$ has eigenvalues $\lambda_1, \ldots, \lambda_k$ (listed according to *geometric* multiplicity) with corresponding eigenvectors $\mathbf{v}_1, \ldots, \mathbf{v}_k$, respectively, that form a linearly independent set.

Show that A is normal if and only if A^* has eigenvalues $\overline{\lambda_1}, \ldots, \overline{\lambda_k}$ with corresponding eigenvectors $\mathbf{v}_1, \ldots, \mathbf{v}_k$, respectively.

[Hint: One direction is much harder than the other. For the difficult direction, be careful not to assume that $k = n$ without actually proving it—Schur triangularization might help.]

2.5.5 Two Hermitian matrices $A, B \in \mathcal{M}_n^H$ are called **∗-congruent** if there exists an invertible matrix $S \in \mathcal{M}_n(\mathbb{C})$ such that $A = SBS^*$.

  (a) Show that for every Hermitian matrix $A \in \mathcal{M}_n^H$ there exist non-negative integers p and n such that A is ∗-congruent to

  $$\left[\begin{array}{c|c|c} I_p & O & O \\ \hline O & -I_n & O \\ \hline O & O & O \end{array} \right].$$

  [Hint: This follows quickly from a decomposition that we learned about in this chapter.]
  (b) Show that every Hermitian matrix is ∗-congruent to *exactly one* matrix of the form described by part (a) (i.e., p and n are determined by A).
  [Hint: Exercise 2.2.17 might help here.]
  [Side note: This shows that the form described by part (a) is a canonical form for ∗-congruence, a fact that is called **Sylvester's law of inertia**.]

2.5.6 Suppose $A, B \in \mathcal{M}_n(\mathbb{C})$ are diagonalizable and A has distinct eigenvalues.

  (a) Show that A and B commute if and only if $B \in \text{span}\{I, A, A^2, A^3, \ldots\}$. [Hint: Use Exercise 2.1.29 and think about interpolating polynomials.]
  (b) Provide an example to show that if A does *not* have distinct eigenvalues then it may be the case that A and B commute even though $B \notin \text{span}\{I, A, A^2, A^3, \ldots\}$.

∗2.5.7 Suppose that $A \in \mathcal{M}_n$ is a positive definite matrix for which each entry is either 0 or 1. In this exercise, we show that the only such matrix is $A = I$.

  (a) Show that $\text{tr}(A) \leq n$.
  (b) Show that $\det(A) \geq 1$.
  [Hint: First show that $\det(A)$ is an integer.]
  (c) Show that $\det(A) \leq 1$.
  [Hint: Use the AM–GM inequality (Theorem A.5.3) with the eigenvalues of A.]
  (d) Use parts (a), (b), and (c) to show that every eigenvalue of A equals 1 and thus $A = I$.

2.A Extra Topic: Quadratic Forms and Conic Sections

Refer back to Section 1.3.2 if you need a refresher on linear or bilinear forms.

One of the most useful applications of the real spectral decomposition is a characterization of quadratic forms, which can be thought of as a generalization of quadratic functions of one input variable (e.g., $q(x) = 3x^2 + 2x - 7$) to multiple variables, in the same way that linear forms generalize linear functions from one input variable to multiple variables.

Definition 2.A.1

Quadratic Forms

Suppose $\mathcal{V}$ is a vector space over $\mathbb{R}$. Then a function $q : \mathcal{V} \to \mathbb{R}$ is called a **quadratic form** if there exists a bilinear form $f : \mathcal{V} \times \mathcal{V} \to \mathbb{R}$ such that

$$q(\mathbf{v}) = f(\mathbf{v}, \mathbf{v}) \quad \text{for all} \quad \mathbf{v} \in \mathcal{V}.$$

For example, the function $q : \mathbb{R}^2 \to \mathbb{R}$ defined by $q(x,y) = 3x^2 + 2xy + 5y^2$ is a quadratic form, since if we let $\mathbf{v} = (x,y)$ and

> The **degree** of a term is the sum of the exponents of variables being multiplied together (e.g., x^2 and xy each have degree 2).

$$A = \begin{bmatrix} 3 & 1 \\ 1 & 5 \end{bmatrix} \quad \text{then} \quad q(\mathbf{v}) = \mathbf{v}^T A \mathbf{v}.$$

That is, q looks like the "piece" of the bilinear form $f(\mathbf{v}, \mathbf{w}) = \mathbf{v}^T A \mathbf{w}$ that we get if we plug $\mathbf{v}$ into both inputs. More generally, every polynomial $q : \mathbb{R}^n \to \mathbb{R}$ in which every term has degree exactly equal to 2 is a quadratic form, since this same procedure can be carried out by simply placing the coefficients of the squared terms along the diagonal of the matrix A and half of the coefficients of the cross terms in the corresponding off-diagonal entries.

Example 2.A.1

Writing a Degree-2 Polynomial as a Matrix

Suppose $q : \mathbb{R}^3 \to \mathbb{R}$ is the function defined by

$$q(x,y,z) = 2x^2 - 2xy + 3y^2 + 2xz + 3z^2.$$

Find a symmetric matrix $A \in \mathcal{M}_3(\mathbb{R})$ such that if $\mathbf{v} = (x,y,z)$ then q has the form $q(\mathbf{v}) = \mathbf{v}^T A \mathbf{v}$, and thus show that q is a quadratic form.

Solution:

Direct computation shows that if $A \in \mathcal{M}_3(\mathbb{R})$ is symmetric then

$$\mathbf{v}^T A \mathbf{v} = a_{1,1} x^2 + 2a_{1,2} xy + 2a_{1,3} xz + a_{2,2} y^2 + 2a_{2,3} yz + a_{3,3} z^2.$$

Simply matching up this form with the coefficients of q shows that we can choose

> The fact that the $(2,3)$-entry of A equals 0 is a result of the fact that q has no "yz" term.

$$A = \begin{bmatrix} 2 & -1 & 1 \\ -1 & 3 & 0 \\ 1 & 0 & 3 \end{bmatrix} \quad \text{so that} \quad q(\mathbf{v}) = \mathbf{v}^T A \mathbf{v}.$$

It follows that q is a quadratic form, since $q(\mathbf{v}) = f(\mathbf{v}, \mathbf{v})$, where $f(\mathbf{v}, \mathbf{w}) = \mathbf{v}^T A \mathbf{w}$ is a bilinear form.

In fact, the converse holds as well—not only are polynomials with degree-2 terms quadratic forms, but every quadratic form on $\mathbb{R}^n$ can be written as a polynomial with degree-2 terms. We now state and prove this observation.

Theorem 2.A.1

Characterization of Quadratic Forms

Suppose $q : \mathbb{R}^n \to \mathbb{R}$ is a function. The following are equivalent:

a) q is a quadratic form,
b) q is an n-variable polynomial in which every term has degree 2,
c) there is a matrix $A \in \mathcal{M}_n(\mathbb{R})$ such that $q(\mathbf{v}) = \mathbf{v}^T A \mathbf{v}$, and
d) there is a *symmetric* matrix $A \in \mathcal{M}_n^S(\mathbb{R})$ such that $q(\mathbf{v}) = \mathbf{v}^T A \mathbf{v}$.

Furthermore, the vector space of quadratic forms is isomorphic to the vector space $\mathcal{M}_n^S$ of symmetric matrices.

Proof. We have already discussed why conditions (b) and (d) are equivalent, and the equivalence of (c) and (a) follows immediately from applying

Theorem 1.3.5 to the bilinear form f associated with q (i.e., the bilinear form with the property that $q(\mathbf{v}) = f(\mathbf{v}, \mathbf{v})$ for all $\mathbf{v} \in \mathbb{R}^n$). Furthermore, the fact that (d) implies (c) is trivial, so the only remaining implication to prove is that (c) implies (d).

To this end, simply notice that if $q(\mathbf{v}) = \mathbf{v}^T A \mathbf{v}$ for all $\mathbf{v} \in \mathbb{R}^n$ then

> $\mathbf{v}^T A \mathbf{v}$ is a scalar and thus equals its own transpose.

$$\mathbf{v}^T(A + A^T)\mathbf{v} = \mathbf{v}^T A \mathbf{v} + \mathbf{v}^T A^T \mathbf{v} = q(\mathbf{v}) + (\mathbf{v}^T A \mathbf{v})^T = q(\mathbf{v}) + \mathbf{v}^T A \mathbf{v} = 2q(\mathbf{v}).$$

We can thus replace A by the symmetric matrix $(A + A^T)/2$ without changing the quadratic form q.

For the "furthermore" claim, suppose $A \in \mathcal{M}_n^S$ is symmetric and define a linear transformation T by $T(A) = q$, where $q(\mathbf{v}) = \mathbf{v}^T A \mathbf{v}$. The equivalence of conditions (a) and (d) shows that q is a quadratic form, and every quadratic form is in the range of T. To see that T is invertible (and thus an isomorphism), we just need to show that $T(A) = 0$ implies $A = O$. In other words, we need to show that if $q(\mathbf{v}) = \mathbf{v}^T A \mathbf{v} = 0$ for all $\mathbf{v} \in \mathbb{R}^n$ then $A = O$. This fact follows immediately from Exercise 1.4.28(b), so we are done. ∎

> Recall from Example 1.B.3 that the matrix $(A + A^T)/2$ from the above proof is the symmetric part of A.

Since the above theorem tells us that every quadratic form can be written in terms of a symmetric matrix, we can use any tools that we know of for manipulating symmetric matrices to help us better understand quadratic forms. For example, the quadratic form $q(x, y) = (3/2)x^2 + xy + (3/2)y^2$ can be written in the form $q(\mathbf{v}) = \mathbf{v}^T A \mathbf{v}$, where

$$\mathbf{v} = \begin{bmatrix} x \\ y \end{bmatrix} \quad \text{and} \quad A = \frac{1}{2}\begin{bmatrix} 3 & 1 \\ 1 & 3 \end{bmatrix}.$$

> We investigate higher-degree generalizations of linear and quadratic forms in Section 3.B.

Since A has real spectral decomposition $A = UDU^T$ with

$$U = \frac{1}{\sqrt{2}}\begin{bmatrix} 1 & 1 \\ 1 & -1 \end{bmatrix} \quad \text{and} \quad D = \begin{bmatrix} 2 & 0 \\ 0 & 1 \end{bmatrix},$$

we can multiply out $q(\mathbf{v}) = \mathbf{v}^T A \mathbf{v} = (U^T \mathbf{v})D(U^T \mathbf{v})$ to see that

$$q(\mathbf{v}) = \frac{1}{2}[x+y, \ x-y]\begin{bmatrix} 2 & 0 \\ 0 & 1 \end{bmatrix}\begin{bmatrix} x+y \\ x-y \end{bmatrix} = (x+y)^2 + \frac{1}{2}(x-y)^2. \quad (2.A.1)$$

In other words, the real spectral decomposition tells us how to write q as a sum (or, if some of the eigenvalues of A are negative, difference) of squares. We now state this observation explicitly.

Corollary 2.A.2

Diagonalization of Quadratic Forms

Suppose $q : \mathbb{R}^n \to \mathbb{R}$ is a quadratic form. Then there exist scalars $\lambda_1, \ldots, \lambda_n \in \mathbb{R}$ and an orthonormal basis $\{\mathbf{u}_1, \mathbf{u}_2, \ldots, \mathbf{u}_n\}$ of $\mathbb{R}^n$ such that

$$q(\mathbf{v}) = \lambda_1(\mathbf{u}_1 \cdot \mathbf{v})^2 + \lambda_2(\mathbf{u}_2 \cdot \mathbf{v})^2 + \cdots + \lambda_n(\mathbf{u}_n \cdot \mathbf{v})^2 \quad \text{for all} \quad \mathbf{v} \in \mathbb{R}^n.$$

Proof. We know from Theorem 2.A.1 that there exists a symmetric matrix $A \in \mathcal{M}_n(\mathbb{R})$ such that $q(\mathbf{v}) = \mathbf{v}^T A \mathbf{v}$ for some all $\mathbf{v} \in \mathbb{R}^n$. Applying the real spectral decomposition (Theorem 2.1.6) to A gives us a unitary matrix $U \in \mathcal{M}_n(\mathbb{R})$ and a diagonal matrix $D \in \mathcal{M}_n(\mathbb{R})$ such that $A = UDU^T$, so $q(\mathbf{v}) = (U^T \mathbf{v})D(U^T \mathbf{v})$, as before.

If we write $U = \begin{bmatrix} \mathbf{u}_1 & | & \mathbf{u}_2 & | & \cdots & | & \mathbf{u}_n \end{bmatrix}$ and let $\lambda_1, \lambda_2, \ldots, \lambda_n$ denote the eigenvalues of A, listed in the order in which they appear on the diagonal of D, then

$$q(\mathbf{v}) = (U^T\mathbf{v})D(U^T\mathbf{v})$$

Here we use the fact that $\mathbf{u}_j^T\mathbf{v} = \mathbf{u}_j \cdot \mathbf{v}$ for all $1 \leq j \leq n$.

$$= \begin{bmatrix} \mathbf{u}_1 \cdot \mathbf{v}, & \mathbf{u}_2 \cdot \mathbf{v}, & \cdots, & \mathbf{u}_n \cdot \mathbf{v} \end{bmatrix} \begin{bmatrix} \lambda_1 & 0 & \cdots & 0 \\ 0 & \lambda_2 & \cdots & 0 \\ \vdots & \vdots & \ddots & \vdots \\ 0 & 0 & \cdots & \lambda_n \end{bmatrix} \begin{bmatrix} \mathbf{u}_1 \cdot \mathbf{v} \\ \mathbf{u}_2 \cdot \mathbf{v} \\ \vdots \\ \mathbf{u}_n \cdot \mathbf{v} \end{bmatrix}$$

$$= \lambda_1(\mathbf{u}_1 \cdot \mathbf{v})^2 + \lambda_2(\mathbf{u}_2 \cdot \mathbf{v})^2 + \cdots + \lambda_n(\mathbf{u}_n \cdot \mathbf{v})^2$$

for all $\mathbf{v} \in \mathbb{R}^n$, as claimed. ■

The real magic of the above theorem thus comes from the fact that, not only does it let us write any quadratic form as a sum or difference of squares, but it lets us do so through an orthonormal change of variables. To see what we mean by this, recall from Theorem 1.4.5 that if $B = \{\mathbf{u}_1, \mathbf{u}_2, \ldots, \mathbf{u}_n\}$ is an orthonormal basis then $[\mathbf{v}]_B = (\mathbf{u}_1 \cdot \mathbf{v}, \mathbf{u}_2 \cdot \mathbf{v}, \ldots, \mathbf{u}_n \cdot \mathbf{v})$. In other words, the above theorem says that every quadratic form really just looks like a function of the form

Just like diagonalization of a matrix gets rid of its off-diagonal entries, diagonalization of a quadratic form gets rid of its cross terms.

$$f(x_1, x_2, \ldots, x_n) = \lambda_1 x_1^2 + \lambda_2 x_2^2 + \cdots + \lambda_n x_n^2, \tag{2.A.2}$$

but rotated and/or reflected. We now investigate what effect different types of eigenvalues $\lambda_1, \lambda_2, \ldots, \lambda_n$, have on the graph of these functions.

2.A.1 Definiteness, Ellipsoids, and Paraboloids

In the case when a all eigenvalues of a symmetric matrix have the same sign (i.e., either all positive or all negative), the associated quadratic form becomes much easier to analyze.

Positive Eigenvalues

If the symmetric matrix $A \in \mathcal{M}_n^S(\mathbb{R})$ corresponding to a quadratic form $q(\mathbf{v}) = \mathbf{v}^T A\mathbf{v}$ is positive semidefinite then, by definition, we must have $q(\mathbf{v}) \geq 0$ for all $\mathbf{v}$. For this reason, we say in this case that q itself is **positive semidefinite (PSD)**, and we say that it is furthermore **positive definite (PD)** if $q(\mathbf{v}) > 0$ whenever $\mathbf{v} \neq \mathbf{0}$.

All of our results about positive semidefinite matrices carry over straightforwardly to the setting of PSD quadratic forms. In particular, the following fact follows immediately from Theorem 2.2.1:

Recall that the scalars in Corollary 2.A.2 are the eigenvalues of A.

(!) A quadratic form $q : \mathbb{R}^n \to \mathbb{R}$ is positive semidefinite if and only if each of the scalars in Corollary 2.A.2 are non-negative. It is positive *definite* if and only if those scalars are all strictly positive.

A **level set** of q is the set of solutions to $q(\mathbf{v}) = c$, where c is a given scalar. They are horizontal slices of q's graph.

If q is positive definite then has it ellipses (or ellipsoids, or hyperellipsoids, depending on the dimension) as its level sets. The principal radii of the level set $q(\mathbf{v}) = 1$ are equal to $1/\sqrt{\lambda_1}, 1/\sqrt{\lambda_2}, \ldots, 1/\sqrt{\lambda_n}$, where $\lambda_1, \lambda_2, \ldots, \lambda_n$ are the eigenvalues of its associated matrix A, and the orthonormal eigenvectors $\{\mathbf{u}_1, \mathbf{u}_2, \ldots, \mathbf{u}_n\}$ described by Corollary 2.A.2 specify the directions of their corresponding principal axes. Furthermore, in this case when the level sets of q are ellipses, its graph must be a paraboloid (which looks like an infinitely-deep bowl).

For example, the level sets of $q(x, y) = x^2 + y^2$ are circles, the level sets of $q(x, y) = x^2 + 2y^2$ are ellipses that are squished by a factor of $1/\sqrt{2}$ in the y-direction, and the level sets of $h(x, y, z) = x^2 + 2y^2 + 3z^2$ are ellipsoids in $\mathbb{R}^3$ that are squished by factors of $1/\sqrt{2}$ and $1/\sqrt{3}$ in the directions of the y- and z-axes, respectively. We now work through an example that is rotated and thus cannot be "eyeballed" so easily.

Example 2.A.2 **A Positive Definite** **Quadratic Form**	Plot the level sets of the quadratic form $q(x, y) = (3/2)x^2 + xy + (3/2)y^2$ and then graph it.

Solution:

We already diagonalized this quadratic form back in Equation (2.A.1). In particular, if we let $\mathbf{v} = (x, y)$ then

$$q(\mathbf{v}) = 2(\mathbf{u}_1 \cdot \mathbf{v})^2 + (\mathbf{u}_2 \cdot \mathbf{v})^2, \quad \text{where} \quad \mathbf{u}_1 = (1, 1)/\sqrt{2}, \ \mathbf{u}_2 = (1, -1)/\sqrt{2}.$$

It follows that the level sets of q are ellipses rotated so that their principal axes point in the directions of $\mathbf{u}_1 = (1, 1)/\sqrt{2}$ and $\mathbf{u}_2 = (1, -1)/\sqrt{2}$, and the level set $q(\mathbf{v}) = 1$ has corresponding principal radii equal to $1/\sqrt{2}$ and 1, respectively. These level sets, as well as the resulting graph of q, are displayed below:

The level sets are shown on the left. The graph is shown on the right.

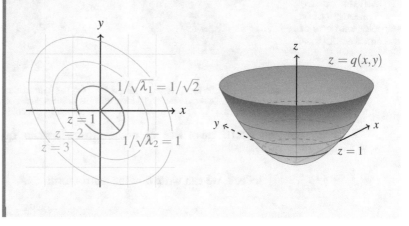

If a quadratic form acts on $\mathbb{R}^3$ instead of $\mathbb{R}^2$ then its level sets are ellipsoids (rather than ellipses), and its graph is a "hyperparaboloid" living in $\mathbb{R}^4$ that is a bit difficult to visualize. For example, applying the spectral decomposition to the quadratic form

$$q(x, y, z) = 3x^2 + 4y^2 - 2xz + 3z^2$$

reveals that if we write $\mathbf{v} = (x, y, z)$ then

$$q(\mathbf{v}) = 4(\mathbf{u}_1 \cdot \mathbf{v})^2 + 4(\mathbf{u}_2 \cdot \mathbf{v})^2 + 2(\mathbf{u}_3 \cdot \mathbf{v})^2, \quad \text{where}$$
$$\mathbf{u}_1 = (1, 0, -1)/\sqrt{2}, \ \mathbf{u}_2 = (0, 1, 0), \ \text{and} \ \mathbf{u}_3 = (1, 0, 1)/\sqrt{2}.$$

It follows that the level sets of q are ellipsoids with principal axes pointing in the directions of $\mathbf{u}_1$, $\mathbf{u}_2$, and $\mathbf{u}_3$. Furthermore, the corresponding principal radii for the level set $q(\mathbf{v}) = 1$ are $1/2$, $1/2$, and $1/\sqrt{2}$, respectively, as displayed in Figure 2.11(a).

Negative Eigenvalues

A symmetric matrix with all-negative eigenvalues is called **negative definite**.

If all of the eigenvalues of a symmetric matrix are strictly negative then the level sets of the associated quadratic form are still ellipses (or ellipsoids, or hyperellipsoids), but its graph is instead a (hyper)paraboloid that opens down (not up). These observations follow simply from noticing that if q has negative eigenvalues then $-q$ (which has the same level sets as q) has positive eigenvalues and is thus positive definite.

Some Zero Eigenvalues

The shapes that arise in this section are the conic sections and their higher-dimensional counterparts.

If a quadratic form is just positive *semi*definite (i.e., one or more of the eigenvalues of the associated symmetric matrix equal zero) then its level sets are degenerate—they look like lower-dimensional ellipsoids that are stretched into higher-dimensional space. For example, the quadratic form

$$q(x,y,z) = 3x^2 + 2xy + 3y^2$$

does not actually depend on z at all, so its level sets extend arbitrarily far in the z direction, as in Figure 2.11(b).

These elliptical cylinders can be thought of as ellipsoids with one of their radii equal to ∞.

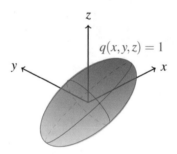
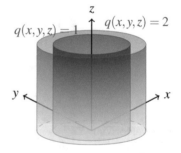

(a) The quadratic form $q(x,y,z) = 3x^2 + 4y^2 - 2xz + 3z^2$ is positive definite, so its level sets are ellipsoids.

(b) The quadratic form $q(x,y,z) = 3x^2 + 2xy + 3y^2$ is positive *semi*definite, so its level sets are degenerate ellipsoids (elliptical cylinders).

Figure 2.11: Level sets of positive (semi)definite quadratic forms are (hyper)ellipsoids.

Indeed, we can write q in the matrix form

$$q(\mathbf{v}) = \mathbf{v}^T A \mathbf{v}, \quad \text{where} \quad A = \begin{bmatrix} 3 & 1 & 0 \\ 1 & 3 & 0 \\ 0 & 0 & 0 \end{bmatrix},$$

which has eigenvalues 4, 2, and 0. Everything on the z-axis is an eigenvector corresponding to the 0 eigenvalue, which explains why nothing interesting happens in that direction.

Example 2.A.3

A Positive *Semi*definite Quadratic Form

Plot the level sets of the quadratic form $q(x,y) = x^2 - 2xy + y^2$ and then graph it.

Solution:

This quadratic form is simple enough that we can simply eyeball a diagonalization of it:

$$q(x,y) = x^2 - 2xy + y^2 = (x-y)^2.$$

The fact that we just need one term in a sum-of-squares decomposition of q tells us right away that the associated symmetric matrix A has at most one non-zero eigenvalue. We can verify this explicitly by noting that if $\mathbf{v} = (x,y)$ then

$$q(\mathbf{v}) = \mathbf{v}^T A \mathbf{v}, \quad \text{where} \quad A = \begin{bmatrix} 1 & -1 \\ -1 & 1 \end{bmatrix},$$

which has eigenvalues 2 and 0. The level sets of q are thus degenerate ellipses (which are just pairs of lines, since "ellipses" in $\mathbb{R}^1$ are just pairs of points). Furthermore, the graph of this quadratic form is degenerate paraboloid (i.e., a parabolic sheet), as shown below:

We can think of these level sets (i.e., pairs of parallel lines) as ellipses with one of their principal radii equal to ∞.

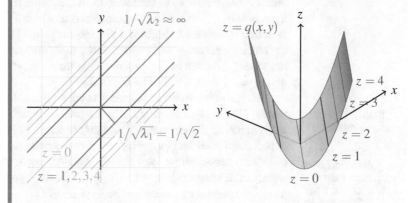

2.A.2   Indefiniteness and Hyperboloids

If a quadratic form or symmetric matrix is neither positive semidefinite nor negative semidefinite, we say that it is **indefinite**.

If the symmetric matrix associated with a quadratic form has both positive and negative eigenvalues then its graph looks like a "saddle"—there are directions on in along which it opens up (i.e., the directions of its eigenvectors corresponding to positive eigenvalues), and other directions along which it opens down (i.e., the directions of its eigenvectors corresponding to negative eigenvalues). The level sets of such a shape are hyperbolas (or hyperboloids, depending on the dimension).

Example 2.A.4

An Indefinite Quadratic Form

Plot the level sets of the quadratic form $q(x,y) = xy$ and then graph it.

Solution:
   Applying the spectral decomposition to the symmetric matrix associated with q reveals that it has eigenvalues $\lambda_1 = 1/2$ and $\lambda_2 = -1/2$, with corresponding eigenvectors $(1,1)$ and $(1,-1)$, respectively. It follows that q can be diagonalized as

$$q(x,y) = \frac{1}{4}(x+y)^2 - \frac{1}{4}(x-y)^2.$$

Since there are terms being both added and subtracted, we conclude that the level sets of q are hyperbolas and its graph is a saddle, as shown below:

The $z = 0$ level set is exactly the x- and y-axes, which separate the other two families of hyperbolic level sets.

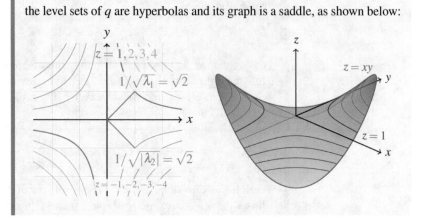

As suggested by the image in the above example, if λ_1 is a positive eigenvalue of the symmetric matrix associated with a quadratic form $q : \mathbb{R}^2 \to \mathbb{R}$ then $1/\sqrt{\lambda_1}$ measures the distance from the origin to the closest point on the hyperbola $q(x,y) = 1$, and the eigenvectors to which λ_1 corresponds specify the direction between the origin and that closest point. The eigenvectors to which the other (necessarily negative) eigenvalue λ_2 corresponds similarly specifies the "open" direction of this hyperbola. Furthermore, if we look at the hyperbola $q(x,y) = -1$ then the roles of λ_1 and λ_2 swap.

In higher dimensions, the level sets are similarly hyperboloids (i.e., shapes whose 2D cross-sections are ellipses and/or hyperbolas), and a similar interpretation of the eigenvalues and eigenvectors holds. In particular, for the hyperboloid level set $q(\mathbf{v}) = 1$, the positive eigenvalues specify the radii of its elliptical cross-sections, and the eigenvectors corresponding to the negative eigenvalues specify directions along which the hyperboloid is "open" (and these interpretations switch for the level set $q(\mathbf{v}) = -1$).

There is also one other possibility: if one eigenvalue is positive, one is negative, and one equals zero, then the level sets look like "hyperbolic sheets"—2D hyperbolas stretched along a third dimension.

For example, if $q : \mathbb{R}^3 \to \mathbb{R}$ is the quadratic form

$$q(x,y,z) = x^2 + y^2 - z^2$$

then the level set $q(x,y,z) = 1$ is a hyperboloid that looks like the circle $x^2 + y^2 = 1$ in the xy-plane, the hyperbola $x^2 - z^2 = 1$ in the xz-plane, and the hyperbola $y^2 - z^2 = 1$ in the yz-plane. It is thus open along the z-axis and has radius 1 along the x- and y-axes (see Figure 2.12(a)). A similar analysis of the level set $q(x,y,z) = -1$ reveals that it is open along the x- and y-axes and thus is not even connected (the xy-plane separates its two halves), as shown in Figure 2.12(c). For this reason, we call this set a "hyperboloid of two sheets" (whereas we call the level set $q(x,y,z) = 1$ a "hyperboloid of one sheet").

Exercises

solutions to starred exercises on page 474

* (g) $q(x,y,z) = 3x^2 - xz + z^2$

2.A.1 Classify each of the following quadratic forms as positive (semi)definite, negative (semi)definite, or indefinite.

* (a) $q(x,y) = x^2 + 3y^2$
  (b) $q(x,y) = 3y^2 - 2x^2$
* (c) $q(x,y) = x^2 + 4xy + 3y^2$
  (d) $q(x,y) = x^2 + 4xy + 4y^2$
* (e) $q(x,y) = x^2 + 4xy + 5y^2$
  (f) $q(x,y) = 2xy - 2x^2 - y^2$

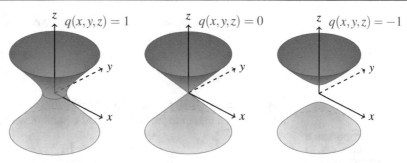

(a) The hyperboloid of one sheet $x^2 + y^2 - z^2 = 1$.

(b) The double cone $x^2 + y^2 - z^2 = 0$.

(c) The hyperboloid of two sheets $x^2 + y^2 - z^2 = -1$.

Figure 2.12: The quadratic form $q(x,y,z) = x^2 + y^2 - z^2$ is indefinite, so its level sets are hyperboloids. The one exception is the level set $q(x,y,z) = 0$, which is a double cone that serves as a boundary between the two types of hyperboloids that exist (one-sheeted and two-sheeted).

(h) $q(x,y,z) = 3x^2 + 3y^2 + 3z^2 - 2xy - 2xz - 2yz$
∗ (i) $q(x,y,z) = x^2 + 2y^2 + 2z^2 - 2xy + 2xz - 4yz$
(j) $q(w,x,y,z) = w^2 + 2x^2 - y^2 - 2xy + xz - 2wx + yz$

∗ (e) $x^2 + y^2 + 2z^2 - 2xz - 2yz = 3$
(f) $x^2 + y^2 + 4z^2 - 2xy - 4xz = 2$
∗ (g) $3x^2 + y^2 + 2z^2 - 2xy - 2xz - yz = 1$
(h) $2x^2 + y^2 + 4z^2 - 2xy - 4xz = 0$

2.A.2 Determine what type of object the graph of the given equation in $\mathbb{R}^2$ is (e.g., ellipse, hyperbola, two lines, or maybe even nothing at all).

∗ (a) $x^2 + 2y^2 = 1$
(b) $x^2 - 2y^2 = -4$
∗ (c) $x^2 + 2xy + 2y^2 = 1$
(d) $x^2 + 2xy + 2y^2 = -2$
∗ (e) $2x^2 + 4xy + y^2 = 3$
(f) $2x^2 + 4xy + y^2 = 0$
∗ (g) $2x^2 + 4xy + y^2 = -1$
(h) $2x^2 + xy + 3y^2 = 0$

2.A.3 Determine what type of object the graph of the given equation in $\mathbb{R}^3$ is (e.g., ellipsoid, hyperboloid of one sheet, hyperboloid of two sheets, two touching cones, or maybe even nothing at all).

∗ (a) $x^2 + 3y^2 + 2z^2 = 2$
(b) $x^2 + 3y^2 + 2z^2 = -3$
∗ (c) $2x^2 + 2xy - 2xz + 2yz = 1$
(d) $2x^2 + 2xy - 2xz + 2yz = -2$

2.A.4 Determine which of the following statements are true and which are false.

∗ (a) Quadratic forms are bilinear forms.
(b) The sum of two quadratic forms is a quadratic form.
∗ (c) A quadratic form $q(\mathbf{v}) = \mathbf{v}^T A \mathbf{v}$ is positive semidefinite if and only if A is positive semidefinite.
(d) The graph of a non-zero quadratic form $q : \mathbb{R}^2 \to \mathbb{R}$ is either an ellipse, a hyperbola, or two lines.
∗ (e) The function $f : \mathbb{R} \to \mathbb{R}$ defined by $f(x) = x^2$ is a quadratic form.

2.A.5 Determine which values of $a \in \mathbb{R}$ make the following quadratic form positive definite:
$$q(x,y,z) = x^2 + y^2 + z^2 - a(xy + xz + yz).$$

2.B Extra Topic: Schur Complements and Cholesky

One common technique for making large matrices easier to work with is to break them up into 2×2 block matrices and then try to use properties of the smaller blocks to determine corresponding properties of the large matrix. For example, solving a large linear system $Q\mathbf{x} = \mathbf{b}$ directly might be quite time-consuming, but we can make it smaller and easier to solve by writing Q as a 2×2 block matrix, and similarly writing $\mathbf{x}$ and $\mathbf{b}$ as "block vectors", as follows:

$$Q = \begin{bmatrix} A & B \\ C & D \end{bmatrix}, \quad \mathbf{x} = \begin{bmatrix} \mathbf{x}_1 \\ \mathbf{x}_2 \end{bmatrix}, \quad \text{and} \quad \mathbf{b} = \begin{bmatrix} \mathbf{b}_1 \\ \mathbf{b}_2 \end{bmatrix}.$$

Then $Q\mathbf{x} = \mathbf{b}$ can be written in either of the equivalent forms

$$\begin{bmatrix} A & B \\ C & D \end{bmatrix} \begin{bmatrix} \mathbf{x}_1 \\ \mathbf{x}_2 \end{bmatrix} = \begin{bmatrix} \mathbf{b}_1 \\ \mathbf{b}_2 \end{bmatrix} \qquad \Longleftrightarrow \qquad \begin{matrix} A\mathbf{x}_1 + B\mathbf{x}_2 = \mathbf{b}_1 \\ C\mathbf{x}_1 + D\mathbf{x}_2 = \mathbf{b}_2. \end{matrix}$$

If A is invertible then one way to solve this linear system is to subtract CA^{-1} times the first equation from the second equation, which puts it into the form

This procedure is sometimes called "block Gaussian elimination", and it really can be thought of as a block matrix version of the Gaussian elimination algorithm that we already know.

$$A\mathbf{x}_1 + B\mathbf{x}_2 = \mathbf{b}_1$$
$$(D - CA^{-1}B)\mathbf{x}_2 = \mathbf{b}_2 - CA^{-1}\mathbf{b}_1.$$

This version of the linear system perhaps looks uglier at first glance, but it has the useful property of being block upper triangular: we can solve it by first solving for $\mathbf{x}_2$ in the smaller linear system $(D - CA^{-1}B)\mathbf{x}_2 = \mathbf{b}_2 - CA^{-1}\mathbf{b}_1$ and then solving for $\mathbf{x}_1$ in the (also smaller) linear system $A\mathbf{x}_1 + B\mathbf{x}_2 = \mathbf{b}_1$. By using this technique, we can solve a $2n \times 2n$ linear system $Q\mathbf{x} = \mathbf{b}$ entirely via matrices that are $n \times n$.

This type of reasoning can also be applied directly to the block matrix Q (rather than the corresponding linear system), and doing so leads fairly quickly to the following very useful block matrix decomposition:

Theorem 2.B.1

2 × 2 Block Matrix LDU Decomposition

Suppose $A \in \mathcal{M}_n$ is invertible. Then

$$\begin{bmatrix} A & B \\ C & D \end{bmatrix} = \begin{bmatrix} I & O \\ CA^{-1} & I \end{bmatrix} \begin{bmatrix} A & O \\ O & D - CA^{-1}B \end{bmatrix} \begin{bmatrix} I & A^{-1}B \\ O & I \end{bmatrix}.$$

We do not require A and D to have the same size in this theorem (and thus B and C need not be square).

Proof. We simply multiply together the block matrices on the right:

$$\begin{bmatrix} I & O \\ CA^{-1} & I \end{bmatrix} \begin{bmatrix} A & O \\ O & D - CA^{-1}B \end{bmatrix} \begin{bmatrix} I & A^{-1}B \\ O & I \end{bmatrix}$$

$$= \begin{bmatrix} I & O \\ CA^{-1} & I \end{bmatrix} \begin{bmatrix} A & A(A^{-1}B) \\ O & D - CA^{-1}B \end{bmatrix}$$

$$= \begin{bmatrix} A & B \\ (CA^{-1})A & (CA^{-1}B) + (D - CA^{-1}B) \end{bmatrix} = \begin{bmatrix} A & B \\ C & D \end{bmatrix},$$

as claimed. ∎

We will use the above decomposition throughout this section to come up with formulas for things like $\det(Q)$ and Q^{-1}, or find a way to determine positive semidefiniteness of Q, based only on corresponding properties of its blocks A, B, C, and D.

2.B.1 The Schur Complement

The beauty of Theorem 2.B.1 is that it lets us write almost any 2×2 block matrix as a product of triangular matrices with ones on their diagonals and a block-diagonal matrix. We can thus determine many properties of Q just from the corresponding properties of those diagonal blocks. Since it will be appearing repeatedly, we now give a name to the ugly diagonal block $D - CA^{-1}B$.

Definition 2.B.1

Schur Complement

If $A \in \mathcal{M}_n$ is invertible then the **Schur complement** of A in the 2×2 block matrix

$$\begin{bmatrix} A & B \\ C & D \end{bmatrix}$$

is $D - CA^{-1}B$.

Example 2.B.1

Computing a Schur Complement

Partition the following matrix Q as a 2×2 block matrix of 2×2 matrices and compute the Schur complement of the top-left block in Q:

$$Q = \begin{bmatrix} 2 & 1 & 1 & 0 \\ 1 & 2 & 1 & 1 \\ 1 & 1 & 2 & 1 \\ 0 & 1 & 1 & 2 \end{bmatrix}.$$

To help remember the formula for the Schur complement, keep in mind that $CA^{-1}B$ is always defined as long as the block matrix Q makes sense, whereas the incorrect $BA^{-1}C$ will not be if B and C are not square.

Solution:

The 2×2 blocks of this matrix are

$$A = \begin{bmatrix} 2 & 1 \\ 1 & 2 \end{bmatrix}, \quad B = \begin{bmatrix} 1 & 0 \\ 1 & 1 \end{bmatrix}, \quad C = \begin{bmatrix} 1 & 1 \\ 0 & 1 \end{bmatrix}, \quad \text{and} \quad D = \begin{bmatrix} 2 & 1 \\ 1 & 2 \end{bmatrix}.$$

Since A is invertible (its determinant is 3), the Schur complement of A in Q is defined and it and equals

$$D - CA^{-1}B = \begin{bmatrix} 2 & 1 \\ 1 & 2 \end{bmatrix} - \frac{1}{3}\begin{bmatrix} 1 & 1 \\ 0 & 1 \end{bmatrix}\begin{bmatrix} 2 & -1 \\ -1 & 2 \end{bmatrix}\begin{bmatrix} 1 & 0 \\ 1 & 1 \end{bmatrix} = \frac{1}{3}\begin{bmatrix} 4 & 2 \\ 2 & 4 \end{bmatrix}.$$

To illustrate what we can do with the Schur complement and the block matrix decomposition of Theorem 2.B.1, we now present formulas for the determinant and inverse of a 2×2 block matrix.

Theorem 2.B.2

Determinant of a 2×2 Block Matrix

Suppose $A \in \mathcal{M}_n$ is invertible and $S = D - CA^{-1}B$ is the Schur complement of A in the 2×2 block matrix

$$Q = \begin{bmatrix} A & B \\ C & D \end{bmatrix}.$$

Then $\det(Q) = \det(A)\det(S)$.

Notice that if the blocks are all 1×1 then this theorem says $\det(Q) = a(d - c(1/a)b) = ad - bc$, which is the formula we already know for 2×2 matrices.

Proof. We just take the determinant of both sides of the block matrix decomposition of Theorem 2.B.1:

Recall that
$\det(XY) = \det(X)\det(Y)$
and that the
determinant of a
triangular matrix is
the product of its
diagonal entries.

$$\det\left(\begin{bmatrix} A & B \\ C & D \end{bmatrix}\right) = \det\left(\begin{bmatrix} I & O \\ CA^{-1} & I \end{bmatrix}\begin{bmatrix} A & O \\ O & S \end{bmatrix}\begin{bmatrix} I & A^{-1}B \\ O & I \end{bmatrix}\right)$$

$$= \det\left(\begin{bmatrix} I & O \\ CA^{-1} & I \end{bmatrix}\right)\det\left(\begin{bmatrix} A & O \\ O & S \end{bmatrix}\right)\det\left(\begin{bmatrix} I & A^{-1}B \\ O & I \end{bmatrix}\right)$$

$$= 1\cdot\det\left(\begin{bmatrix} A & O \\ O & S \end{bmatrix}\right)\cdot 1$$

$$= \det(A)\det(S).$$

Note that in the final step we used the fact that the determinant of a block diagonal matrix equals the product of the determinants of its blocks (this fact is often covered in introductory linear algebra texts—see the end of Appendix A.1.5). ∎

If A is not invertible
then these block
matrix formulas do
not work. For one
way to sometimes
get around this
problem, see
Exercise 2.B.8.

In particular, notice that $\det(Q) = 0$ if and only if $\det(S) = 0$ (since we are assuming that A is invertible, so we know that $\det(A) \neq 0$). By recalling that a matrix has determinant 0 if and only if it is not invertible, this tells us that Q is invertible if and only if its Schur complement is invertible. This fact is re-stated in the following theorem, along with an explicit formula for its inverse.

Theorem 2.B.3

Inverse of a 2 × 2 Block Matrix

Suppose $A \in \mathcal{M}_n$ is invertible and $S = D - CA^{-1}B$ is the Schur complement of A in the 2×2 block matrix

$$Q = \begin{bmatrix} A & B \\ C & D \end{bmatrix}.$$

Then Q is invertible if and only if S is invertible, and its inverse is

$$Q^{-1} = \begin{bmatrix} I & -A^{-1}B \\ O & I \end{bmatrix}\begin{bmatrix} A^{-1} & O \\ O & S^{-1} \end{bmatrix}\begin{bmatrix} I & O \\ -CA^{-1} & I \end{bmatrix}.$$

Proof. We just take the inverse of both sides of the block matrix decomposition of Theorem 2.B.1:

Recall that
$(XYZ)^{-1} = Z^{-1}Y^{-1}X^{-1}$.

$$\begin{bmatrix} A & B \\ C & D \end{bmatrix}^{-1} = \left(\begin{bmatrix} I & O \\ CA^{-1} & I \end{bmatrix}\begin{bmatrix} A & O \\ O & S \end{bmatrix}\begin{bmatrix} I & A^{-1}B \\ O & I \end{bmatrix}\right)^{-1}$$

$$= \begin{bmatrix} I & A^{-1}B \\ O & I \end{bmatrix}^{-1}\begin{bmatrix} A & O \\ O & S \end{bmatrix}^{-1}\begin{bmatrix} I & O \\ CA^{-1} & I \end{bmatrix}^{-1}$$

If the blocks are all
1×1 then the
formula provided by
this theorem
simplifies to the
familiar formula

$$= \begin{bmatrix} I & -A^{-1}B \\ O & I \end{bmatrix}\begin{bmatrix} A^{-1} & O \\ O & S^{-1} \end{bmatrix}\begin{bmatrix} I & O \\ -CA^{-1} & I \end{bmatrix}.$$

Note that in the final step we used the fact that the inverse of a block diagonal matrix is just the matrix with inverted diagonal blocks and the fact that for any matrix X we have

$$Q^{-1} = \frac{1}{\det(Q)}\begin{bmatrix} d & -b \\ -c & a \end{bmatrix}$$

that we already
know for 2×2
matrices.

$$\begin{bmatrix} I & X \\ O & I \end{bmatrix}^{-1} = \begin{bmatrix} I & -X \\ O & I \end{bmatrix}. \qquad \blacksquare$$

If we wanted to, we could explicitly multiply out the formula provided by Theorem 2.B.3 to see that

$$\begin{bmatrix} A & B \\ C & D \end{bmatrix}^{-1} = \begin{bmatrix} A^{-1} + A^{-1}BS^{-1}CA^{-1} & -A^{-1}BS^{-1} \\ -S^{-1}CA^{-1} & S^{-1} \end{bmatrix}.$$

However, this formula seems rather ugly and cumbersome, so we typically prefer the factored form provided by the theorem.

The previous theorems are useful because they let us compute properties of large matrices just by computing the corresponding properties of matrices that are half as large. The following example highlights the power of this technique—we will be able to compute the determinant of a 4×4 matrix just by doing some computations with some 2×2 matrices (a much easier task).

Example 2.B.2

Using the Schur Complement

Use the Schur complement to compute the determinant of the following matrix:

$$Q = \begin{bmatrix} 2 & 1 & 1 & 0 \\ 1 & 2 & 1 & 1 \\ 1 & 1 & 2 & 1 \\ 0 & 1 & 1 & 2 \end{bmatrix}.$$

Solution:

We recall the top-left A block and the Schur complement S of A in Q from Example 2.B.1:

$$A = \begin{bmatrix} 2 & 1 \\ 1 & 2 \end{bmatrix} \quad \text{and} \quad S = \frac{1}{3} \begin{bmatrix} 4 & 2 \\ 2 & 4 \end{bmatrix}.$$

Since $\det(A) = 3$ and $\det(S) = 4/3$, it follows that

$$\det(Q) = \det(A)\det(S) = 3 \cdot (4/3) = 4.$$

Finally, as an application of the Schur complement that is a bit more relevant to our immediate interests in this chapter, we now show that its positive (semi)definiteness completely determines positive (semi)definiteness of the full block matrix as well.

Theorem 2.B.4

Positive (Semi)definiteness of 2×2 Block Matrices

Suppose $A \in \mathcal{M}_n$ is invertible and $S = C - B^*A^{-1}B$ is the Schur complement of A in the self-adjoint 2×2 block matrix

$$Q = \begin{bmatrix} A & B \\ B^* & C \end{bmatrix}.$$

Then Q is positive (semi)definite if and only if A and S are positive (semi)definite.

Proof. We notice that in this case, the decomposition of Theorem 2.B.1 simplifies slightly to $Q = P^*DP$, where

$$P = \begin{bmatrix} I & A^{-1}B \\ O & I \end{bmatrix} \quad \text{and} \quad D = \begin{bmatrix} A & O \\ O & S \end{bmatrix}.$$

It follows from Exercise 2.2.13 that D is positive (semi)definite if and only if A and S both are. It also follows from Theorem 2.2.3(d) that if D is positive (semi)definite then so is Q.

On the other hand, we know that P is invertible since all of its eigenvalues equal 1 (and in particular are thus all non-zero), so we can rearrange the above decomposition of Q into the form

$$(P^{-1})^* Q (P^{-1}) = D.$$

By the same logic as above, if Q is positive (semi)definite then so is D, which is equivalent to A and S being positive (semi)definite. ∎

For example, if we return to the matrix Q from Example 2.B.2, it is straightforward to check that both A and S are positive definite (via Theorem 2.2.7, for example), so Q is positive definite too.

2.B.2   The Cholesky Decomposition

Recall that one of the central questions that we asked in Section 2.2.3 was how "simple" we can make the matrix $B \in \mathcal{M}_{m,n}$ in a positive semidefinite decomposition of a matrix $A = B^* B \in \mathcal{M}_n$. One possible answer to this question was provided by the principal square root (Theorem 2.2.11), which says that we can always choose B to be positive semidefinite (as long as $m = n$ so that positive semidefiniteness is a concept that makes sense). We now make use of Schur complements to show that, alternatively, we can always make B upper triangular:

Theorem 2.B.5 **Cholesky** **Decomposition**	Suppose $\mathbb{F} = \mathbb{R}$ or $\mathbb{F} = \mathbb{C}$, and $A \in \mathcal{M}_n(\mathbb{F})$ is positive semidefinite with $m = \text{rank}(A)$. There exists a unique matrix $T \in \mathcal{M}_{m,n}$ in row echelon form with real strictly positive leading entries such that $$A = T^* T.$$

A "leading entry" is the first non-zero entry in a row.

Before proving this theorem, we recall that T being in row echelon form implies that it is upper triangular, but is actually a slightly stronger requirement than just upper triangularity. For example, the matrices

$$\begin{bmatrix} 0 & 1 \\ 0 & 0 \end{bmatrix} \quad \text{and} \quad \begin{bmatrix} 0 & 1 \\ 0 & 1 \end{bmatrix}$$

are both upper triangular, but only the one on the left is in row echelon form. We also note that the choice of $m = \text{rank}(A)$ in this theorem is optimal in some sense:

If A is positive *definite* then T is square, upper triangular, and its diagonal entries are strictly positive.

- If $m < \text{rank}(A)$ then no such decomposition of A is possible (even if we ignore the upper triangular requirement) since if $T \in \mathcal{M}_{m,n}$ then $\text{rank}(A) = \text{rank}(T^* T) = \text{rank}(T) \leq m$.
- If $m > \text{rank}(A)$ then decompositions of this type exist (for example, we can just pad the matrix T from the $m = \text{rank}(A)$ case with extra rows of zeros at the bottom), but they are no longer unique—see Remark 2.B.1.

Proof of Theorem 2.B.5. We prove the result by induction on n (the size of A). For the base case, the result is clearly true if $n = 1$ since we can choose $T = [\sqrt{a}]$, which is an upper triangular 1×1 matrix with a non-negative diagonal entry. For the inductive step, suppose that every $(n-1) \times (n-1)$ positive semidefinite matrix has a Cholesky decomposition—we want to show that if $A \in \mathcal{M}_n$ is positive semidefinite then it has one too. We split into two cases:

Case 1 can only happen if A is positive semidefinite but not positive definite.

Case 1: $a_{1,1} = 0$. We know from Exercise 2.2.11 that the entire first row and column of A must equal 0, so we can write A as the block matrix

$$A = \begin{bmatrix} 0 & \mathbf{0}^T \\ \mathbf{0} & A_{2,2} \end{bmatrix},$$

where $A_{2,2} \in \mathcal{M}_{n-1}$ is positive semidefinite and has $\text{rank}(A_{2,2}) = \text{rank}(A)$. By the inductive hypothesis, $A_{2,2}$ has a Cholesky decomposition $A_{2,2} = T^*T$, so

$$A = \begin{bmatrix} 0 & \mathbf{0}^T \\ \mathbf{0} & A_{2,2} \end{bmatrix} = \begin{bmatrix} \mathbf{0} \mid T \end{bmatrix}^* \begin{bmatrix} \mathbf{0} \mid T \end{bmatrix}$$

is a Cholesky decomposition of A.

Case 2: $a_{1,1} \neq 0$. We can write A as the block matrix

$$A = \begin{bmatrix} a_{1,1} & \mathbf{a}_{2,1}^* \\ \mathbf{a}_{2,1} & A_{2,2} \end{bmatrix},$$

where $A_{2,2} \in \mathcal{M}_{n-1}$ and $\mathbf{a}_{2,1} \in \mathbb{F}^{n-1}$ is a column vector. By applying Theorem 2.B.1, we see that if $S = A_{2,2} - \mathbf{a}_{2,1}\mathbf{a}_{2,1}^*/a_{1,1}$ is the Schur complement of $a_{1,1}$ in A then we can decompose A in the form

The Schur complement exists because $a_{1,1} \neq 0$ in this case, so $a_{1,1}$ is invertible.

$$A = \begin{bmatrix} 1 & \mathbf{0}^T \\ \mathbf{a}_{2,1}/a_{1,1} & I \end{bmatrix} \begin{bmatrix} a_{1,1} & \mathbf{0}^T \\ \mathbf{0} & S \end{bmatrix} \begin{bmatrix} 1 & \mathbf{a}_{2,1}^*/a_{1,1} \\ \mathbf{0} & I \end{bmatrix}.$$

Since A is positive semidefinite, it follows that S is positive semidefinite too, so it has a Cholesky decomposition $S = T^*T$ by the inductive hypothesis. Furthermore, since $a_{1,1} \neq 0$ we conclude that $\text{rank}(A) = \text{rank}(S) + 1$, and we see that

$$A = \begin{bmatrix} 1 & \mathbf{0}^T \\ \mathbf{a}_{2,1}/a_{1,1} & I \end{bmatrix} \begin{bmatrix} \sqrt{a_{1,1}}\sqrt{a_{1,1}} & \mathbf{0}^T \\ \mathbf{0} & T^*T \end{bmatrix} \begin{bmatrix} 1 & \mathbf{a}_{2,1}^*/a_{1,1} \\ \mathbf{0} & I \end{bmatrix}$$

$$= \begin{bmatrix} \sqrt{a_{1,1}} & \mathbf{a}_{2,1}^*/\sqrt{a_{1,1}} \\ \mathbf{0} & T \end{bmatrix}^* \begin{bmatrix} \sqrt{a_{1,1}} & \mathbf{a}_{2,1}^*/\sqrt{a_{1,1}} \\ \mathbf{0} & T \end{bmatrix},$$

is a Cholesky decomposition of A. This completes the inductive step and the proof of the fact that every matrix has a Cholesky decomposition. We leave the proof of uniqueness to Exercise 2.B.12. ∎

This argument can be reversed (to derive the QR decomposition from the Cholesky decomposition) via Theorem 2.2.10.

It is worth noting that the Cholesky decomposition is essentially equivalent to the QR decomposition of Section 1.C, which said that every matrix $B \in \mathcal{M}_{m,n}$ can be written in the form $B = UT$, where $U \in \mathcal{M}_m$ is unitary and $T \in \mathcal{M}_{m,n}$ is upper triangular with non-negative real entries on its diagonal. Indeed, if A is positive semidefinite then we can use the QR decomposition to write

$$A = B^*B = (UT)^*(UT) = T^*U^*UT = T^*T,$$

which is basically a Cholesky decomposition of A.

Example 2.B.3

Finding a Cholesky Decomposition

Find the Cholesky decomposition of the matrix $A = \begin{bmatrix} 4 & -2 & 2 \\ -2 & 2 & -2 \\ 2 & -2 & 3 \end{bmatrix}$.

Solution:

To construct A's Cholesky decomposition, we mimic the proof of Theorem 2.B.5. We start by writing A as a block matrix with 1×1 top-left block:

$$A = \begin{bmatrix} a_{1,1} & \mathbf{a}_{2,1}^* \\ \mathbf{a}_{2,1} & A_{2,2} \end{bmatrix}, \quad \text{where } a_{1,1} = 4, \ \mathbf{a}_{2,1} = \begin{bmatrix} -2 \\ 2 \end{bmatrix}, \ A_{2,2} = \begin{bmatrix} 2 & -2 \\ -2 & 3 \end{bmatrix}.$$

The Schur complement of $a_{1,1}$ in A is then

$$S = A_{2,2} - \mathbf{a}_{2,1}\mathbf{a}_{2,1}^*/a_{1,1}$$

$$= \begin{bmatrix} 2 & -2 \\ -2 & 3 \end{bmatrix} - \frac{1}{4}\begin{bmatrix} -2 \\ 2 \end{bmatrix}\begin{bmatrix} -2 & 2 \end{bmatrix} = \begin{bmatrix} 1 & -1 \\ -1 & 2 \end{bmatrix}.$$

At this point, we have done one step of the induction in the proof that A has a Cholesky decomposition. The next step is to apply the same procedure to S, and so on.

It follows from the proof of Theorem 2.B.5 that the Cholesky decomposition of A is

$$A = \begin{bmatrix} \sqrt{a_{1,1}} & \mathbf{a}_{2,1}^*/\sqrt{a_{1,1}} \\ \mathbf{0} & T \end{bmatrix}^* \begin{bmatrix} \sqrt{a_{1,1}} & \mathbf{a}_{2,1}^*/\sqrt{a_{1,1}} \\ \mathbf{0} & T \end{bmatrix}$$

$$= \begin{bmatrix} 2 & -1 & 1 \\ 0 & & \\ 0 & & T \end{bmatrix}^* \begin{bmatrix} 2 & -1 & 1 \\ 0 & & \\ 0 & & T \end{bmatrix}, \quad (2.B.1)$$

where $S = T^*T$ is a Cholesky decomposition of S.

This is a step in the right direction—we now know the top row of the triangular matrix in the Cholesky decomposition of A. To proceed from here, we perform the same procedure on the Schur complement S—we write S as a block matrix with 1×1 top-left block:

If A was even larger, we would just repeat this procedure. Each iteration finds us one more row in its Cholesky decomposition.

$$S = \begin{bmatrix} s_{1,1} & \mathbf{s}_{2,1}^* \\ \mathbf{s}_{2,1} & S_{2,2} \end{bmatrix}, \quad \text{where} \quad s_{1,1} = 1, \quad \mathbf{s}_{2,1} = \begin{bmatrix} -1 \end{bmatrix}, \quad S_{2,2} = \begin{bmatrix} 2 \end{bmatrix}.$$

The Schur complement of $s_{1,1}$ in S is then

$$S_{2,2} - \mathbf{s}_{2,1}\mathbf{s}_{2,1}^*/s_{1,1} = 2 - (-1)^2 = 1.$$

It follows from the proof of Theorem 2.B.5 that a Cholesky decomposition of S is

The term $\sqrt{S_{2,2}}$ here comes from the fact that $S_{2,2} = \sqrt{S_{2,2}}\sqrt{S_{2,2}}$ is a Cholesky decomposition of the 1×1 matrix $S_{2,2}$.

$$S = \begin{bmatrix} \sqrt{s_{1,1}} & \mathbf{s}_{2,1}^*/\sqrt{s_{1,1}} \\ 0 & \sqrt{S_{2,2}} \end{bmatrix}^* \begin{bmatrix} \sqrt{s_{1,1}} & \mathbf{s}_{2,1}^*/\sqrt{s_{1,1}} \\ 0 & \sqrt{S_{2,2}} \end{bmatrix} = \begin{bmatrix} 1 & -1 \\ 0 & 1 \end{bmatrix}^* \begin{bmatrix} 1 & -1 \\ 0 & 1 \end{bmatrix}.$$

If we plug this decomposition of S into Equation (2.B.1), we get the following Cholesky decomposition of A:

$$A = \begin{bmatrix} 2 & -1 & 1 \\ 0 & 1 & -1 \\ 0 & 0 & 1 \end{bmatrix}^* \begin{bmatrix} 2 & -1 & 1 \\ 0 & 1 & -1 \\ 0 & 0 & 1 \end{bmatrix}.$$

Remark 2.B.1

(Non-)Uniqueness of the Cholesky Decomposition

It is worth emphasizing that decompositions of the form $A = T^*T$ are no longer necessarily unique if we only require that T be upper triangular (rather than in row echelon form) or if $m > \text{rank}(A)$. One reason for this is that, in Case 1 of the proof of Theorem 2.B.5, the matrix $\begin{bmatrix} \mathbf{0} \mid T \end{bmatrix}$ may be upper triangular even if T is not. Furthermore, instead of writing

$$A = \begin{bmatrix} 0 & \mathbf{0}^T \\ \mathbf{0} & A_{2,2} \end{bmatrix} = \begin{bmatrix} \mathbf{0} \mid T_1 \end{bmatrix}^* \begin{bmatrix} \mathbf{0} \mid T_1 \end{bmatrix}$$

where $A_{2,2} = T_1^* T_1$ is a Cholesky decomposition of $A_{2,2}$, we could instead write

$$\begin{bmatrix} 0 & \mathbf{0}^T \\ \mathbf{0} & A_{2,2} \end{bmatrix} = \begin{bmatrix} 0 & \mathbf{x}^* \\ \mathbf{0} & T_2 \end{bmatrix}^* \begin{bmatrix} 0 & \mathbf{x}^* \\ \mathbf{0} & T_2 \end{bmatrix},$$

where $A_{2,2} = \mathbf{x}\mathbf{x}^* + T_2^* T_2$. We could thus just choose $\mathbf{x} \in \mathbb{F}^{n-1}$ small enough so that $A_{2,2} - \mathbf{x}\mathbf{x}^*$ is positive semidefinite and thus has a Cholesky decomposition $A_{2,2} - \mathbf{x}\mathbf{x}^* = T_2^* T_2$.

The Cholesky decomposition is a special case of the LU decomposition $A = LU$. If A is positive semidefinite, we can choose $L = U^*$.

For example, it is straightforward to verify that if

$$A = \begin{bmatrix} 0 & 0 & 0 \\ 0 & 2 & 3 \\ 0 & 3 & 5 \end{bmatrix}$$

then $A = T_1^* T_1 = T_2^* T_2 = T_3^* T_3$, where

$$T_1 = \begin{bmatrix} 0 & \sqrt{2} & 3/\sqrt{2} \\ 0 & 0 & 1/\sqrt{2} \end{bmatrix}, \quad T_2 = \begin{bmatrix} 0 & 1 & 1 \\ 0 & 1 & 2 \end{bmatrix}, \quad T_3 = \begin{bmatrix} 0 & \sqrt{2} & 3/\sqrt{2} \\ 0 & 0 & 1/\sqrt{2} \\ 0 & 0 & 0 \end{bmatrix}.$$

However, only the decomposition involving T_1 is a valid Cholesky decomposition, since T_2 is not in row echelon form (despite being upper triangular) and T_3 has $3 = m > \text{rank}(A) = 2$ rows.

Exercises

solutions to starred exercises on page 474

2.B.1 Use the Schur complement to help you solve the following linear system by only ever doing computations with 2×2 matrices:

$$\begin{aligned} 2w + x + y + z &= -3 \\ w - x + 2y - z &= 3 \\ -w + x + 2y + z &= 1 \\ 3w \quad + y + 2z &= 2 \end{aligned}$$

2.B.2 Compute the Schur complement of the top-left 2×2 block in each of the following matrices, and use it to compute the determinant of the given matrix and determine whether or not it is positive (semi)definite.

* (a) $\begin{bmatrix} 2 & 1 & 1 & 0 \\ 1 & 3 & 1 & 1 \\ 1 & 1 & 2 & 0 \\ 0 & 1 & 0 & 1 \end{bmatrix}$ (b) $\begin{bmatrix} 3 & 2 & 0 & -1 \\ 2 & 2 & 1 & 0 \\ 0 & 1 & 2 & 0 \\ -1 & 0 & 0 & 3 \end{bmatrix}$

2.B.3 Determine which of the following statements are true and which are false.

* (a) The Schur complement of the top-left 2×2 block of a 5×5 matrix is a 3×3 matrix.
 (b) If a matrix is positive definite then so are the Schur complements of any of its top-left blocks.
* (c) Every matrix has a Cholesky decomposition.

*2.B.4 Find infinitely many different decompositions of the matrix
$$A = \begin{bmatrix} 0 & 0 & 0 \\ 0 & 1 & 1 \\ 0 & 1 & 2 \end{bmatrix}$$
of the form $A = T^*T$, where $T \in \mathcal{M}_3$ is upper triangular.

2.B.5 Suppose $X \in \mathcal{M}_{m,n}$ and $c \in \mathbb{R}$ is a scalar. Use the Schur complement to show that the block matrix
$$\begin{bmatrix} cI_m & X \\ X^* & cI_n \end{bmatrix}$$
is positive semidefinite if and only if $\|X\| \leq c$.

[Side note: You were asked to prove this directly in Exercise 2.3.15.]

2.B.6 This exercise shows that it is *not* possible to determine the eigenvalues of a 2×2 block matrix from its A block and its Schur complement. Let
$$Q = \begin{bmatrix} 1 & 2 \\ 2 & 1 \end{bmatrix} \quad \text{and} \quad R = \begin{bmatrix} 1 & 3 \\ 3 & 6 \end{bmatrix}.$$

 (a) Compute the Schur complement of the top-left 1×1 block in each of Q and R.
  [Side note: They are the same.]
 (b) Compute the eigenvalues of each of Q and R.
  [Side note: They are different.]

*2.B.7 Suppose $A \in \mathcal{M}_n$ is invertible and $S = D - CA^{-1}B$ is the Schur complement of A in the 2×2 block matrix
$$Q = \begin{bmatrix} A & B \\ C & D \end{bmatrix}.$$
Show that $\text{rank}(Q) = \text{rank}(A) + \text{rank}(S)$.

2.B.8 Consider the 2×2 block matrix
$$Q = \begin{bmatrix} A & B \\ C & D \end{bmatrix}.$$
In this exercise, we show how to come up with block matrix formulas for Q if the $D \in \mathcal{M}_n$ block is invertible (rather than the A block).

Suppose that D is invertible and let $S = A - BD^{-1}C$ (S is called the **Schur complement** of D in Q).

 (a) Show how to write
$$Q = U \begin{bmatrix} S & O \\ O & D \end{bmatrix} L,$$
  where U is an upper triangular block matrix with ones on its diagonal and L is a lower triangular block matrix with ones on its diagonal.
 (b) Show that $\det(Q) = \det(D)\det(S)$.
 (c) Show that Q is invertible if and only if S is invertible, and find a formula for its inverse.
 (d) Show that Q is positive (semi)definite if and only if S is positive (semi)definite.

2.B.9 Suppose $A \in \mathcal{M}_{m,n}$ and $B \in \mathcal{M}_{n,m}$. Show that
$$\det(I_m + AB) = \det(I_n + BA).$$

[Side note: This is called **Sylvester's determinant identity**.]

[Hint: Compute the determinant of the block matrix
$$Q = \begin{bmatrix} I_m & -A \\ B & I_n \end{bmatrix}$$
using both Schur complements (see Exercise 2.B.8).]

2.B.10 Suppose $A \in \mathcal{M}_{m,n}$ and $B \in \mathcal{M}_{n,m}$.
 (a) Use the result of Exercise 2.B.9 to show that $I_m + AB$ is invertible if and only if $I_n + BA$ is invertible.
 (b) Find a formula for $(I_m + AB)^{-1}$ in terms of A, B, and $(I_n + BA)^{-1}$. [Hint: Try using Schur complements just like in Exercise 2.B.9.]

2.B.11 Suppose $A \in \mathcal{M}_{m,n}, B \in \mathcal{M}_{n,m}$, and $m \geq n$. Use the result of Exercise 2.B.9 to show that the characteristic polynomials of AB and BA satisfy
$$p_{AB}(\lambda) = (-\lambda)^{m-n} p_{BA}(\lambda).$$

[Side note: In other words, AB and BA have the same eigenvalues, counting algebraic multiplicity, but with AB having $m - n$ extra zero eigenvalues.]

2.B.12 Show that the Cholesky decomposition described by Theorem 2.B.5 is unique.

[Hint: Follow along with the given proof of that theorem and show inductively that if Cholesky decompositions in $\mathcal{M}_{n-1}$ are unique then they are also unique in $\mathcal{M}_n$.]

2.C   Extra Topic: Applications of the SVD

In this section, we explore two particularly useful and interesting applications of the singular value decomposition (Theorem 2.3.1).

2.C.1   The Pseudoinverse and Least Squares

Recall that if a matrix $A \in \mathcal{M}_n$ is invertible then the linear system $A\mathbf{x} = \mathbf{b}$ has unique solution $\mathbf{x} = A^{-1}\mathbf{b}$. However, that linear system might have a solution even if A is *not* invertible (or even square). For example, the linear system

$$\begin{bmatrix} 1 & 2 & 3 \\ -1 & 0 & 1 \\ 3 & 2 & 1 \end{bmatrix} \begin{bmatrix} x \\ y \\ z \end{bmatrix} = \begin{bmatrix} 6 \\ 0 \\ 6 \end{bmatrix} \qquad (2.\text{C}.1)$$

has infinitely many solutions, like $(x, y, z) = (1, 1, 1)$ and $(x, y, z) = (2, -1, 2)$, even though its coefficient matrix has rank 2 (which we showed in Example 2.3.2) and is thus not invertible.

With this example in mind, it seems natural to ask whether or not there exists a matrix $A^\dagger$ with the property that we can find a solution to the linear system $A\mathbf{x} = \mathbf{b}$ (when it exists, but even if A is not invertible) by setting $\mathbf{x} = A^\dagger \mathbf{b}$.

Definition 2.C.1 **Pseudoinverse of a Matrix**	Suppose $\mathbb{F} = \mathbb{R}$ or $\mathbb{F} = \mathbb{C}$, and $A \in \mathcal{M}_{m,n}(\mathbb{F})$ has orthogonal rank-one sum decomposition $$A = \sum_{j=1}^{r} \sigma_j \mathbf{u}_j \mathbf{v}_j^*.$$ Then the **pseudoinverse** of A, denoted by $A^\dagger \in \mathcal{M}_{n,m}$, is the matrix $$A^\dagger \overset{\text{def}}{=} \sum_{j=1}^{r} \frac{1}{\sigma_j} \mathbf{v}_j \mathbf{u}_j^*.$$

The orthogonal rank-one sum decomposition was introduced in Theorem 2.3.3.

Equivalently, if A has singular value decomposition $A = U\Sigma V^*$ then its pseudoinverse is the matrix $A^\dagger = V\Sigma^\dagger U^*$, where $\Sigma^\dagger \in \mathcal{M}_{n,m}$ is the diagonal matrix whose non-zero entries are the reciprocals of the non-zero entries of Σ (and its zero entries are unchanged). It is straightforward to see that $(A^\dagger)^\dagger = A$ for all $A \in \mathcal{M}_{m,n}$. Furthermore, if A is square and all of its singular values are non-zero (i.e., it is invertible) then $\Sigma^\dagger = \Sigma^{-1}$, so

Be careful: some other books (particularly physics books) use $A^\dagger$ to mean the conjugate transpose of A instead of its pseudoinverse.

$$A^\dagger A = V\Sigma^\dagger U^* U \Sigma V^* = V\Sigma^\dagger \Sigma V^* = VV^* = I.$$

That is, we have proved that the pseudoinverse really does generalize the inverse:

> (!) If $A \in \mathcal{M}_n$ is invertible then $A^\dagger = A^{-1}$.

Recall that singular values are unique, but singular vectors are not.

The advantage of the pseudoinverse over the regular inverse is that every matrix has one. Before we can properly explore the pseudoinverse and see what we can do with it though, we have to prove that it is well-defined. That is, we have to show that no matter which orthogonal rank-one sum decomposition (i.e., no matter which singular value decomposition) of $A \in \mathcal{M}_{m,n}$ we use, the formula provided by Definition 2.C.1 results in the same matrix $A^\dagger$. The following theorem provides us with a first step in this direction.

Theorem 2.C.1

The Pseudoinverse and Fundamental Subspaces

Suppose $A \in \mathcal{M}_{m,n}$ has pseudoinverse $A^\dagger$. Then

a) $AA^\dagger$ is the orthogonal projection onto range(A),

b) $I - AA^\dagger$ is the orthogonal projection onto null$(A^*) = $ null$(A^\dagger)$,

c) $A^\dagger A$ is the orthogonal projection onto range$(A^*) = $ range$(A^\dagger)$, and

d) $I - A^\dagger A$ is the orthogonal projection onto null(A).

Proof. We start by writing A in its orthogonal rank-one sum decomposition

$$A = \sum_{j=1}^{r} \sigma_j \mathbf{u}_j \mathbf{v}_j^*, \quad \text{so} \quad A^\dagger = \sum_{j=1}^{r} \frac{1}{\sigma_j} \mathbf{v}_j \mathbf{u}_j^*.$$

To see why part (a) of the theorem is true, we multiply A by $A^\dagger$ to get

The third equality here is the tricky one—the double sum collapses into a single sum because all of the terms with $i \neq j$ equal 0.

$$AA^\dagger = \left(\sum_{i=1}^{r} \sigma_i \mathbf{u}_i \mathbf{v}_i^* \right) \left(\sum_{j=1}^{r} \frac{1}{\sigma_j} \mathbf{v}_j \mathbf{u}_j^* \right)$$

$$= \sum_{i,j=1}^{r} \frac{\sigma_i}{\sigma_j} \mathbf{u}_i (\mathbf{v}_i^* \mathbf{v}_j) \mathbf{u}_j^* \qquad \text{(product of two sums is a double sum)}$$

$$= \sum_{j=1}^{r} \frac{\sigma_j}{\sigma_j} \mathbf{u}_j \mathbf{u}_j^* \qquad (\mathbf{v}_i^* \mathbf{v}_j = 1 \text{ if } i = j, \ \mathbf{v}_i^* \mathbf{v}_j = 0 \text{ otherwise})$$

$$= \sum_{j=1}^{r} \mathbf{u}_j \mathbf{u}_j^*. \qquad (\sigma_j / \sigma_j = 1)$$

The fact that this is the orthogonal projection onto range(A) follows from the fact that $\{\mathbf{u}_1, \mathbf{u}_2, \ldots, \mathbf{u}_r\}$ forms an orthonormal basis of range(A) (Theorem 2.3.2(a)), together with Theorem 1.4.10.

The fact that $A^\dagger A$ is the orthogonal projection onto range(A^*) follows from computing $A^\dagger A = \sum_{j=1}^{r} \mathbf{v}_j \mathbf{v}_j^*$ in a manner similar to above, and then recalling from Theorem 2.3.2(c) that $\{\mathbf{v}_1, \mathbf{v}_2, \ldots, \mathbf{v}_r\}$ forms an orthonormal basis of range(A^*). On the other hand, the fact that $A^\dagger A$ is the orthogonal projection onto range$(A^\dagger)$ (and thus range$(A^*) = $ range$(A^\dagger)$) follows from swapping the roles of A and $A^\dagger$ (and using the fact that $(A^\dagger)^\dagger = A$) in part (a).

The proof of parts (b) and (d) of the theorem are all almost identical, so we leave them to Exercise 2.C.7. ∎

We now show the converse of the above theorem—the pseudoinverse $A^\dagger$ is the *only* matrix with the property that $A^\dagger A$ and $AA^\dagger$ are the claimed orthogonal projections. In particular, this shows that the pseudoinverse is well-defined and does not depend on which orthogonal rank-one sum decomposition of A was used to construct it—each one of them results in a matrix with the properties of Theorem 2.C.1, and there is only one such matrix.

Theorem 2.C.2

Well-Definedness of the Pseudoinverse

If $A \in \mathcal{M}_{m,n}$ and $B \in \mathcal{M}_{n,m}$ are such that AB is the orthogonal projection onto range(A) and BA is the orthogonal projection onto range$(B) = $ range(A^*) then $B = A^\dagger$.

Proof. We know from Theorem 2.C.1 that $AA^\dagger$ is the orthogonal projection onto range(A), we know from Theorem 1.4.10 that orthogonal projections are uniquely determined by their range, so $AB = AA^\dagger$. A similar argument (making use of the fact that range$(A^\dagger) = $ range$(A^*) = $ range(B)) shows that $BA = A^\dagger A$.

Since projections leave everything in their range unchanged, we conclude that $(BA)B\mathbf{x} = B\mathbf{x}$ for all $\mathbf{x} \in \mathbb{F}^n$, so $BAB = B$, and a similar argument shows that $A^\dagger AA^\dagger = A^\dagger$. Putting these facts together shows that

$$B = BAB = (BA)B = (A^\dagger A)B = A^\dagger(AB) = A^\dagger(AA^\dagger) = A^\dagger AA^\dagger = A^\dagger. \quad \blacksquare$$

Example 2.C.1 **Computing a Pseudoinverse**	Compute the pseudoinverse of the matrix $A = \begin{bmatrix} 1 & 2 & 3 \\ -1 & 0 & 1 \\ 3 & 2 & 1 \end{bmatrix}$.

Solution:

We already saw in Example 2.3.2 that the singular value decomposition of this matrix is $A = U\Sigma V^*$, where

$$U = \frac{1}{\sqrt{6}} \begin{bmatrix} \sqrt{3} & \sqrt{2} & 1 \\ 0 & \sqrt{2} & -2 \\ \sqrt{3} & -\sqrt{2} & -1 \end{bmatrix}, \quad V = \frac{1}{\sqrt{6}} \begin{bmatrix} \sqrt{2} & -\sqrt{3} & -1 \\ \sqrt{2} & 0 & 2 \\ \sqrt{2} & \sqrt{3} & -1 \end{bmatrix},$$

$$\Sigma = \begin{bmatrix} 2\sqrt{6} & 0 & 0 \\ 0 & \sqrt{6} & 0 \\ 0 & 0 & 0 \end{bmatrix}.$$

*The pseudoinverse is sometimes called the **Moore–Penrose pseudoinverse** of A.*

It follows that

$$A^\dagger = V\Sigma^\dagger U^*$$

$$= \frac{1}{6} \begin{bmatrix} \sqrt{2} & -\sqrt{3} & -1 \\ \sqrt{2} & 0 & 2 \\ \sqrt{2} & \sqrt{3} & -1 \end{bmatrix} \begin{bmatrix} 1/(2\sqrt{6}) & 0 & 0 \\ 0 & 1/\sqrt{6} & 0 \\ 0 & 0 & 0 \end{bmatrix} \begin{bmatrix} \sqrt{3} & 0 & \sqrt{3} \\ \sqrt{2} & \sqrt{2} & -\sqrt{2} \\ 1 & -2 & -1 \end{bmatrix}$$

$$= \frac{1}{12} \begin{bmatrix} -1 & -2 & 3 \\ 1 & 0 & 1 \\ 3 & 2 & -1 \end{bmatrix}.$$

Example 2.C.2 **Computing a Rectangular Pseudoinverse**	Compute the pseudoinverse of the matrix $A = \begin{bmatrix} 1 & 1 & 1 & -1 \\ 0 & 1 & 1 & 0 \\ -1 & 1 & 1 & 1 \end{bmatrix}$.

Solution:

This is the same matrix from Example 2.3.3, which has singular value decomposition $A = U\Sigma V^*$, where

$$U = \frac{1}{\sqrt{6}} \begin{bmatrix} \sqrt{2} & -\sqrt{3} & -1 \\ \sqrt{2} & 0 & 2 \\ \sqrt{2} & \sqrt{3} & -1 \end{bmatrix}, \quad \Sigma = \begin{bmatrix} \sqrt{6} & 0 & 0 & 0 \\ 0 & 2 & 0 & 0 \\ 0 & 0 & 0 & 0 \end{bmatrix} \quad \text{and}$$

$$V = \frac{1}{\sqrt{2}} \begin{bmatrix} 0 & -1 & 0 & 1 \\ 1 & 0 & 1 & 0 \\ 1 & 0 & -1 & 0 \\ 0 & 1 & 0 & 1 \end{bmatrix}.$$

It follows that

Verify this matrix multiplication on your own. It builds character.

$$A^\dagger = V\Sigma^\dagger U^*$$

$$= \frac{1}{\sqrt{12}} \begin{bmatrix} 0 & -1 & 0 & 1 \\ 1 & 0 & 1 & 0 \\ 1 & 0 & -1 & 0 \\ 0 & 1 & 0 & 1 \end{bmatrix} \begin{bmatrix} 1/\sqrt{6} & 0 & 0 \\ 0 & 1/2 & 0 \\ 0 & 0 & 0 \\ 0 & 0 & 0 \end{bmatrix} \begin{bmatrix} \sqrt{2} & \sqrt{2} & \sqrt{2} \\ -\sqrt{3} & 0 & \sqrt{3} \\ -1 & 2 & -1 \end{bmatrix}$$

$$= \frac{1}{12} \begin{bmatrix} 3 & 0 & -3 \\ 2 & 2 & 2 \\ 2 & 2 & 2 \\ -3 & 0 & 3 \end{bmatrix}.$$

Solving Linear Systems

Now that we know how to construct the pseudoinverse of a matrix, we return to the linear system $A\mathbf{x} = \mathbf{b}$ from Equation (2.C.1). If we (very naïvely for now) try to use the pseudoinverse to solve this linear system by setting $\mathbf{x} = A^\dagger \mathbf{b}$, then we get

We computed this pseudoinverse $A^\dagger$ in Example 2.C.1.

$$\mathbf{x} = A^\dagger \begin{bmatrix} 6 \\ 0 \\ 6 \end{bmatrix} = \frac{1}{12} \begin{bmatrix} -1 & -2 & 3 \\ 1 & 0 & 1 \\ 3 & 2 & -1 \end{bmatrix} \begin{bmatrix} 6 \\ 0 \\ 6 \end{bmatrix} = \begin{bmatrix} 1 \\ 1 \\ 1 \end{bmatrix}.$$

This is indeed a solution of the original linear system, as we might hope. The following theorem shows that this is always the case—if a linear system has a solution, then the pseudoinverse finds one. Furthermore, if there are multiple solutions to the linear system, it finds the "best" one:

Theorem 2.C.3

Pseudoinverses Solve Linear Systems

Suppose $\mathbb{F} = \mathbb{R}$ or $\mathbb{F} = \mathbb{C}$, and $A \in \mathcal{M}_{m,n}(\mathbb{F})$ and $\mathbf{b} \in \mathbb{F}^m$ are such that the linear system $A\mathbf{x} = \mathbf{b}$ has at least one solution. Then $\mathbf{x} = A^\dagger \mathbf{b}$ is a solution, and furthermore if $\mathbf{y} \in \mathbb{F}^n$ is any other solution then $\|A^\dagger \mathbf{b}\| < \|\mathbf{y}\|$.

Proof. The linear system $A\mathbf{x} = \mathbf{b}$ has a solution if and only if $\mathbf{b} \in \text{range}(A)$. To see that $\mathbf{x} = A^\dagger \mathbf{b}$ is a solution in this case, we simply notice that

$$A\mathbf{x} = A(A^\dagger \mathbf{b}) = (AA^\dagger)\mathbf{b} = \mathbf{b},$$

since $AA^\dagger$ is the orthogonal projection onto range(A), by Theorem 2.C.1(a).

To see that $\|A^\dagger \mathbf{b}\| \leq \|\mathbf{y}\|$ for all solutions $\mathbf{y}$ of the linear system (i.e., all $\mathbf{y}$ such that $A\mathbf{y} = \mathbf{b}$), we note that

$$A^\dagger \mathbf{b} = A^\dagger(A\mathbf{y}) = (A^\dagger A)\mathbf{y}.$$

Since $A^\dagger A$ is an orthogonal projection, it follows from the fact that orthogonal projections cannot increase the norm of a vector (Theorem 1.4.11) that $\|A^\dagger \mathbf{b}\| = \|(A^\dagger A)\mathbf{y}\| \leq \|\mathbf{y}\|$, and furthermore equality holds if and only if $(A^\dagger A)\mathbf{y} = \mathbf{y}$ (i.e., if and only if $\mathbf{y} = A^\dagger \mathbf{b}$).   ∎

To get a rough idea for why it's desirable to find the solution with smallest norm, as in the previous theorem, we again return to the linear system

$$\begin{bmatrix} 1 & 2 & 3 \\ -1 & 0 & 1 \\ 3 & 2 & 1 \end{bmatrix} \begin{bmatrix} x \\ y \\ z \end{bmatrix} = \begin{bmatrix} 6 \\ 0 \\ 6 \end{bmatrix}$$

that we originally introduced in Equation (2.C.1). The solution set of this linear system consists of the vectors of the form $(0,3,0) + z(1,-2,1)$, where z is a free variable. This set contains some vectors that are hideous (e.g., choosing $z = 341$ gives the solution $(x,y,z) = (341,-679,341)$), but it also contains some vectors that are not so hideous (e.g., choosing $z = 1$ gives the solution $(x,y,z) = (1,1,1)$, which is the solution found by the pseudoinverse). The guarantee that the pseudoinverse finds the smallest-norm solution means that we do not have to worry about it returning "large and ugly" solutions like $(341,-679,341)$. Geometrically, it means that it finds the solution closest to the origin (see Figure 2.13).

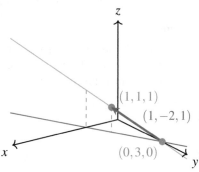

Figure 2.13: Every point on the line $(0,3,0) + z(1,-2,1)$ is a solution of the linear system from Equation (2.C.1). The pseudoinverse finds the solution $(x,y,z) = (1,1,1)$, which is the point on that line closest to the origin.

Not only does the pseudoinverse find the "best" solution when a solution exists, it even find the "best" non-solution when no solution exists. To make sense of this statement, we again think in terms of norms and distances—if no solution to a linear system $A\mathbf{x} = \mathbf{b}$ exists, then it seems reasonable that the "next best thing" to a solution would be the vector that makes $A\mathbf{x}$ as close to $\mathbf{b}$ as possible. In other words, we want to find the vector $\mathbf{x}$ that minimizes $\|A\mathbf{x} - \mathbf{b}\|$. The following theorem shows that choosing $\mathbf{x} = A^\dagger \mathbf{b}$ also solves this problem:

> Think of $A\mathbf{x} - \mathbf{b}$ as the error in our solution $\mathbf{x}$: it is the difference between what $A\mathbf{x}$ actually is and what we want it to be ($\mathbf{b}$).

Theorem 2.C.4

Linear Least Squares

Suppose $\mathbb{F} = \mathbb{R}$ or $\mathbb{F} = \mathbb{C}$, and $A \in \mathcal{M}_{m,n}(\mathbb{F})$ and $\mathbf{b} \in \mathbb{F}^m$. If $\mathbf{x} = A^\dagger \mathbf{b}$ then

$$\|A\mathbf{x} - \mathbf{b}\| \leq \|A\mathbf{y} - \mathbf{b}\| \quad \text{for all} \quad \mathbf{y} \in \mathbb{F}^n.$$

> Here, $A\mathbf{y}$ is just an arbitrary element of range(A).

Proof. We know from Theorem 1.4.11 that the closest point to $\mathbf{b}$ in range(A) is $P\mathbf{b}$, where P is the orthogonal projection onto range(A). Well,

Theorem 2.C.1(a) tells us that $P = AA^\dagger$, so

$$\|A\mathbf{x} - \mathbf{b}\| = \|(AA^\dagger)\mathbf{b} - \mathbf{b}\| \leq \|A\mathbf{y} - \mathbf{b}\| \quad \text{for all} \quad \mathbf{y} \in \mathbb{F}^n,$$

as desired.   ∎

This method of finding the closest thing to a solution of a linear system is called **linear least squares**, and it is particularly useful when trying to fit data to a model, such as finding a line (or plane, or curve) of best fit for a set of data. For example, suppose we have many data points $(x_1, y_1), (x_2, y_2), \ldots, (x_n, y_n)$, and we want to find the line that best describes the relationship between x and y as in Figure 2.14.

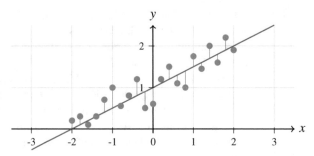

Figure 2.14: The line of best fit for a set of points is the line that minimizes the sum of squares of vertical displacements of the points from the line (highlighted here in orange).

To find this line, we consider the "ideal" scenario—we try (and typically fail) to find a line that passes exactly through all n data points by setting up the corresponding linear system:

$$y_1 = mx_1 + b$$
$$y_2 = mx_2 + b$$
$$\vdots$$
$$y_n = mx_n + b.$$

Keep in mind that $x_1, \ldots, x_n$ and $y_1, \ldots, y_n$ are given to us—the variables that we are trying to solve for in this linear system are m and b.

Since this linear system has n equations, but only 2 variables (m and b), we do not expect to find an exact solution, but we can find the closest thing to a solution by using the pseudoinverse, as in the following example.

Example 2.C.3

Finding a Line of Best Fit

Find the line of best fit for the points $(-2,0), (-1,1), (0,0), (1,2)$, and $(2,2)$.

Solution:

To find the line of best fit, we set up the system of linear equations that we would *like* to solve. Ideally, we would like to find a line $y = mx + b$ that goes through all 5 data points. That is, we want to find m and b such that

$$0 = -2m + b, \qquad 1 = -1m + b, \qquad 0 = 0m + b,$$
$$2 = 1m + b, \qquad 2 = 2m + b.$$

It's not difficult to see that this linear system has no solution, but we can find the closest thing to a solution (i.e., the line of best fit) by using the pseudoinverse. Specifically, we write the linear system in matrix form as

$$Ax = b \quad \text{where} \quad A = \begin{bmatrix} -2 & 1 \\ -1 & 1 \\ 0 & 1 \\ 1 & 1 \\ 2 & 1 \end{bmatrix}, \quad x = \begin{bmatrix} m \\ b \end{bmatrix}, \quad \text{and} \quad b = \begin{bmatrix} 0 \\ 1 \\ 0 \\ 2 \\ 2 \end{bmatrix}.$$

Try computing this pseudoinverse yourself: the singular values of A are $\sqrt{10}$ and $\sqrt{5}$.

Well, the pseudoinverse of A is

$$A^\dagger = \frac{1}{10} \begin{bmatrix} -2 & -1 & 0 & 1 & 2 \\ 2 & 2 & 2 & 2 & 2 \end{bmatrix},$$

so the coefficients of the line of best fit are given by

$$x = \begin{bmatrix} m \\ b \end{bmatrix} = A^\dagger b = \frac{1}{10} \begin{bmatrix} -2 & -1 & 0 & 1 & 2 \\ 2 & 2 & 2 & 2 & 2 \end{bmatrix} \begin{bmatrix} 0 \\ 1 \\ 0 \\ 2 \\ 2 \end{bmatrix} = \begin{bmatrix} 1/2 \\ 1 \end{bmatrix}.$$

In other words, the line of best fit is $y = x/2 + 1$, as shown below:

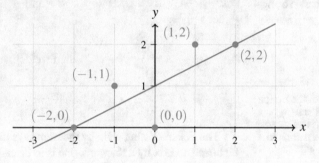

This exact same method also works for finding the "plane of best fit" for data points $(x_1, y_1, z_1), (x_2, y_2, z_2), \ldots, (x_n, y_n, z_n)$, and so on for higher-dimensional data as well (see Exercise 2.C.5). We can even do things like find quadratics of best fit, exponentials of best fit, or other weird functions of best fit (see Exercise 2.C.6).

By putting together all of the results of this section, we see that the pseudoinverse gives the "best solution" to a system of linear equations $Ax = b$ in all cases:

- If the system has a unique solution, it is $x = A^\dagger b$.
- If the system has infinitely many solutions, then $x = A^\dagger b$ is the smallest solution—it minimizes $\|x\|$.
- If the system has no solutions, then $x = A^\dagger b$ is the closest thing to a solution—it minimizes $\|Ax - b\|$.

2.C.2   Low-Rank Approximation

As one final application of the singular value decomposition, we consider the problem of approximating a matrix by another matrix with small rank. One of the primary reasons why we would do this is that it allows us to compress the data that is represented by a matrix, since a full (real) $n \times n$ matrix requires us to store n^2 real numbers, but a rank-k matrix only requires us to store $2kn + k$ real numbers. To see this, note that we can store a low-rank matrix via its orthogonal rank-one sum decomposition

$$A = \sum_{j=1}^{k} \sigma_j \mathbf{u}_j \mathbf{v}_j^*,$$

In fact, it suffices to store just $2kn - k$ real numbers and k signs (which each require only a single bit), since each $\mathbf{u}_j$ and $\mathbf{v}_j$ has norm 1.

which consists of $2k$ vectors with n entries each, as well as k singular values. Since $2kn + k$ is much smaller than n^2 when k is small, it is much less resource-intensive to store low-rank matrices than general matrices. This observation is useful in practice—instead of storing the exact matrix A that contains our data of interest, we can sometimes find a nearby matrix with small rank and store that instead.

To actually find a nearby low-rank matrix, we use the following theorem, which says that the singular value decomposition tells us what to do. Specifically, the closest rank-k matrix to a given matrix A is the one that is obtained by replacing all except for the k largest singular values of A by 0.

Theorem 2.C.5

Eckart–Young–Mirsky

Suppose $\mathbb{F} = \mathbb{R}$ or $\mathbb{F} = \mathbb{C}$, and $A \in \mathcal{M}_{m,n}(\mathbb{F})$ has singular values $\sigma_1 \geq \sigma_2 \geq \cdots \geq \sigma_r > 0$. If $1 \leq k \leq r$ and A has orthogonal rank-one sum decomposition

$$A = \sum_{j=1}^{r} \sigma_j \mathbf{u}_j \mathbf{v}_j^*, \quad \text{then we define} \quad A_k = \sum_{j=1}^{k} \sigma_j \mathbf{u}_j \mathbf{v}_j^*.$$

In particular, notice that rank$(A_k) = k$, so A_k is the closest rank-k matrix to A.

Then $\|A - A_k\| \leq \|A - B\|$ for all $B \in \mathcal{M}_{m,n}(\mathbb{F})$ with rank$(B) = k$.

The Eckart–Young–Mirsky theorem says that the best way to approximate a high-rank matrix by a low-rank one is to discard the pieces of its singular value decomposition corresponding to its smallest singular values. In other words, the singular value decomposition organizes a matrix into its "most important" and "least important" pieces—the largest singular values (and their corresponding singular vectors) describe the broad strokes of the matrix, while the smallest singular values (and their corresponding singular vectors) just fill in its fine details.

Proof of Theorem 2.C.5. Pick any matrix $B \in \mathcal{M}_{m,n}$ with rank$(B) = k$, which necessarily has $(n - k)$-dimensional null space by the rank-nullity theorem (Theorem A.1.2(e)). Also, consider the vector space $\mathcal{V}_{k+1} = \text{span}\{\mathbf{v}_1, \mathbf{v}_2, \ldots, \mathbf{v}_{k+1}\}$, which is $(k + 1)$-dimensional. Since $(n - k) + (k + 1) = n + 1$, we know from Exercise 1.5.2(b) that null$(B) \cap \mathcal{V}_{k+1}$ is at least 1-dimensional, so there exists a unit vector $\mathbf{w} \in \text{null}(B) \cap \mathcal{V}_{k+1}$. Then we have

$$\begin{aligned}
\|A - B\| &\geq \|(A - B)\mathbf{w}\| &&\text{(since } \|\mathbf{w}\| = 1) \\
&= \|A\mathbf{w}\| &&\text{(since } \mathbf{w} \in \text{null}(B)) \\
&= \left\| \left(\sum_{j=1}^{r} \sigma_j \mathbf{u}_j \mathbf{v}_j^* \right) \mathbf{w} \right\| &&\text{(orthogonal rank-one sum decomp. of } A) \\
&= \left\| \sum_{j=1}^{k+1} \sigma_j (\mathbf{v}_j^* \mathbf{w}) \mathbf{u}_j \right\|. &&(\mathbf{w} \in \mathcal{V}_{k+1}, \text{ so } \mathbf{v}_j^* \mathbf{w} = 0 \text{ when } j > k+1)
\end{aligned}$$

At this point, we note that Theorem 1.4.5 tells us that $(\mathbf{v}_1^*\mathbf{w}, \ldots, \mathbf{v}_{k+1}^*\mathbf{w})$ is the coefficient vector of $\mathbf{w}$ in the basis $\{\mathbf{v}_1, \mathbf{v}_2, \ldots, \mathbf{v}_{k+1}\}$ of $\mathcal{V}_{k+1}$. This then implies, via Corollary 1.4.4, that $\|\mathbf{w}\|^2 = \sum_{j=1}^{k+1} |\mathbf{v}_j^*\mathbf{w}|^2$, and similarly that

$$\left\| \sum_{j=1}^{k+1} \sigma_j(\mathbf{v}_j^*\mathbf{w})\mathbf{u}_j \right\|^2 = \sum_{j=1}^{k+1} \sigma_j^2 |\mathbf{v}_j^*\mathbf{w}|^2.$$

With this observation in hand, we now continue the chain of inequalities from above:

$$\|A - B\| \geq \sqrt{\sum_{j=1}^{k+1} \sigma_j^2 |\mathbf{v}_j^*\mathbf{w}|^2} \qquad \text{(since } \{\mathbf{u}_1, \ldots, \mathbf{u}_{k+1}\} \text{ is an orthonormal set)}$$

$$\geq \sigma_{k+1} \sqrt{\sum_{j=1}^{k+1} |\mathbf{v}_j^*\mathbf{w}|^2} \quad \text{(since } \sigma_j \geq \sigma_{k+1})$$

$$= \sigma_{k+1} \qquad\qquad \text{(since } \sum_{j=1}^{k+1} |\mathbf{v}_j^*\mathbf{w}|^2 = \|\mathbf{w}\|^2 = 1)$$

$$= \|A - A_k\|, \qquad \text{(since } A - A_k = \sum_{j=k+1}^{r} \sigma_j \mathbf{u}_j \mathbf{v}_j^*)$$

as desired. ∎

Example 2.C.4

Closest Rank-1 Approximation

Find the closest rank-1 approximation to

$$A = \begin{bmatrix} 1 & 2 & 3 \\ -1 & 0 & 1 \\ 3 & 2 & 1 \end{bmatrix}.$$

That is, find the matrix B with $\text{rank}(B) = 1$ that minimizes $\|A - B\|$.

Solution:

Recall from Example 2.3.2 that the singular value decomposition of this matrix is $A = U\Sigma V^*$, where

$$U = \frac{1}{\sqrt{6}} \begin{bmatrix} \sqrt{3} & \sqrt{2} & 1 \\ 0 & \sqrt{2} & -2 \\ \sqrt{3} & -\sqrt{2} & -1 \end{bmatrix}, \quad V = \frac{1}{\sqrt{6}} \begin{bmatrix} \sqrt{2} & -\sqrt{3} & -1 \\ \sqrt{2} & 0 & 2 \\ \sqrt{2} & \sqrt{3} & -1 \end{bmatrix},$$

$$\Sigma = \begin{bmatrix} 2\sqrt{6} & 0 & 0 \\ 0 & \sqrt{6} & 0 \\ 0 & 0 & 0 \end{bmatrix}.$$

To get the closest rank-1 approximation of A, we simply replace all except for the largest singular value by 0, and then multiply the singular value decomposition back together. This procedure gives us the following closest rank-1 matrix B:

$$B = U \begin{bmatrix} 2\sqrt{6} & 0 & 0 \\ 0 & 0 & 0 \\ 0 & 0 & 0 \end{bmatrix} V^*$$

As always,
multiplying matrices
like these together
is super fun.

$$= \frac{1}{6} \begin{bmatrix} \sqrt{3} & \sqrt{2} & 1 \\ 0 & \sqrt{2} & -2 \\ \sqrt{3} & -\sqrt{2} & -1 \end{bmatrix} \begin{bmatrix} 2\sqrt{6} & 0 & 0 \\ 0 & 0 & 0 \\ 0 & 0 & 0 \end{bmatrix} \begin{bmatrix} \sqrt{2} & \sqrt{2} & \sqrt{2} \\ -\sqrt{3} & 0 & \sqrt{3} \\ -1 & 2 & -1 \end{bmatrix}$$

$$= \begin{bmatrix} 2 & 2 & 2 \\ 0 & 0 & 0 \\ 2 & 2 & 2 \end{bmatrix}.$$

We can use this method to compress pretty much any information that we can represent with a matrix, but it works best when there is some correlation between the entries in the rows and columns of the matrix (e.g., this method does not help much if we just place inherently 1-dimensional data like a text file into a matrix of some arbitrary shape). For example, we can use it to compress black-and-white images by representing the brightness of each pixel in the image by a number, arranging those numbers in a matrix of the same size and shape as the image, and then applying the Eckart–Young–Mirsky theorem to that matrix.

Similarly, we can compress color images by using the fact that every color can be obtained from mixing red, green, and blue, so we can use three matrices: one for each of those primary colors. Figure 2.15 shows the result of applying a rank-k approximation of this type to an image for $k = 1, 5, 20,$ and 100.

Remark 2.C.1

Low-Rank Approximation in other Matrix Norms

It seems natural to ask how low-rank matrix approximation changes if we use a matrix norm other than the operator norm. It turns out that, for a wide variety of matrix norms (including the Frobenius norm), nothing changes at all. For example, one rank-k matrix B that minimizes $\|A - B\|_F$ is exactly the same as the one that minimizes $\|A - B\|$. That is, the closest rank-k approximation does not change at all, even if we measure "closeness" in this very different way.

This fact about
unitarily-invariant
norms is beyond the
scope of this
book—see (Mir60)
for a proof.

More generally, low-rank approximation works the same way for every matrix norm that is unitarily-invariant (i.e., if Theorem 2.3.6 holds for a particular matrix norm, then so does the Eckart–Young–Mirsky theorem). For example, Exercise 2.3.19 shows that something called the "trace norm" is unitarily-invariant, so the Eckart–Young–Mirsky theorem still works if we replace the operator norm with it.

Exercises

solutions to starred exercises on page 475

2.C.1 For each of the following linear systems $Ax = b$, determine whether or not it has a solution. If it does, find the smallest one (i.e., find the solution x that minimizes $\|x\|$). If it doesn't, find the closest thing to a solution (i.e., find the vector x that minimizes $\|Ax - b\|$).

$*$ (a) $A = \begin{bmatrix} 1 & 2 \\ 3 & 6 \end{bmatrix}$, $b = \begin{bmatrix} 2 \\ 1 \end{bmatrix}$

(b) $A = \begin{bmatrix} 1 & 2 \\ 3 & 6 \end{bmatrix}$, $b = \begin{bmatrix} 1 \\ 3 \end{bmatrix}$

$*$ (c) $A = \begin{bmatrix} 1 & 2 & 3 \\ 2 & 3 & 4 \end{bmatrix}$, $b = \begin{bmatrix} 2 \\ -1 \end{bmatrix}$

(a) The author's cats.

The rank-1 approximation is interesting because we can actually *see* that it has rank 1: every row and column is a multiple of every other row and column, which is what creates the banding effect in the image.

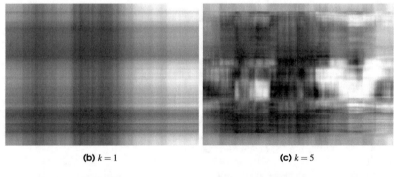

(b) $k = 1$ **(c)** $k = 5$

The rank-100 approximation is almost indistinguishable from the original image.

(d) $k = 20$ **(e)** $k = 100$

Figure 2.15: A picture of the author's cats that has been compressed via the Eckart–Young–Mirsky theorem. The images are (a) uncompressed, (b) use a rank-1 approximation, as well as a (c) rank-5, (d) rank-20, and a (e) rank-100 approximation. The original image is 500×700 with full rank 500.

(d) $A = \begin{bmatrix} 1 & 1 & 1 \\ 1 & 0 & 1 \\ 0 & 1 & 0 \end{bmatrix}$, $\mathbf{b} = \begin{bmatrix} 1 \\ 1 \\ 1 \end{bmatrix}$

2.C.2 Find the line of best fit (in the sense of Example 2.C.3) for the following collections of data points.

[Side note: Exercise 2.C.10 provides a way of solving this problem that avoids computing the pseudoinverse.]

* (a) $(1,1), (2,4)$.
  (b) $(-1,4), (0,1), (1,-1)$.
* (c) $(1,-1), (3,2), (4,7)$.
  (d) $(-1,3), (0,1), (1,-1), (2,2)$.

2.C.3 Find the best rank-k approximations of each of the following matrices for the given value of k.

* (a) $A = \begin{bmatrix} 3 & 1 \\ 1 & 3 \end{bmatrix}$, $k = 1$.

  (b) $A = \begin{bmatrix} 1 & -2 & 3 \\ 3 & 2 & 1 \end{bmatrix}$, $k = 1$.

* (c) $A = \begin{bmatrix} 1 & 0 & 2 \\ 0 & 1 & 1 \\ -2 & -1 & 1 \end{bmatrix}$, $k = 1$.

  (d) $A = \begin{bmatrix} 1 & 0 & 2 \\ 0 & 1 & 1 \\ -2 & -1 & 1 \end{bmatrix}$, $k = 2$.

2.C.4 Determine which of the following statements are true and which are false.

* (a) $I^\dagger = I$.
  (b) The function $T : \mathcal{M}_3(\mathbb{R}) \to \mathcal{M}_3(\mathbb{R})$ defined by $T(A) = A^\dagger$ (the pseudoinverse of A) is a linear transformation.
* (c) For all $A \in \mathcal{M}_4(\mathbb{C})$, it is the case that $\text{range}(A^\dagger) = \text{range}(A)$.
  (d) If $A \in \mathcal{M}_{m,n}(\mathbb{R})$ and $\mathbf{b} \in \mathbb{R}^m$ are such that the linear system $A\mathbf{x} = \mathbf{b}$ has a solution then there is a unique solution vector $\mathbf{x} \in \mathbb{R}^n$ that minimizes $\|\mathbf{x}\|$.
* (e) For every $A \in \mathcal{M}_{m,n}(\mathbb{R})$ and $\mathbf{b} \in \mathbb{R}^m$ there is a unique vector $\mathbf{x} \in \mathbb{R}^n$ that minimizes $\|A\mathbf{x} - \mathbf{b}\|$.

∗∗2.C.5 Find the plane $z = ax + by + c$ of best fit for the following 4 data points (x, y, z):

$$(0,-1,-1), (0,0,0), (0,1,3), (2,0,3).$$

∗∗ 2.C.6 Find the curve of the form $y = c_1 \sin(x) + c_2 \cos(x)$ that best fits the following 3 data points (x, y):

$$(0,-1), (\pi/2, 1), (\pi, 0).$$

∗∗2.C.7 Prove parts (b) and (d) of Theorem 2.C.1.

2.C.8 Suppose $\mathbb{F} = \mathbb{R}$ or $\mathbb{F} = \mathbb{C}$, and $A \in \mathcal{M}_{m,n}(\mathbb{F})$, $B \in \mathcal{M}_{n,p}(\mathbb{F})$, and $C \in \mathcal{M}_{p,r}(\mathbb{F})$. Explain how to compute ABC if we know AB, B, and BC, but not necessarily A or C themselves.

[Hint: This would be trivial if B were square and invertible.]

∗2.C.9 In this exercise, we derive explicit formulas for the pseudoinverse in some special cases.

  (a) Show that if $A \in \mathcal{M}_{m,n}$ has linearly independent columns then $A^\dagger = (A^*A)^{-1}A^*$.
  (b) Show that if $A \in \mathcal{M}_{m,n}$ has linearly independent rows then $A^\dagger = A^*(AA^*)^{-1}$.

[Side note: A^*A and AA^* are indeed invertible in these cases. See Exercise 1.4.30, for example.]

2.C.10 Show that if $x_1, x_2, \ldots, x_n$ are not all the same then the line of best fit $y = mx + b$ for the points $(x_1, y_1), (x_2, y_2), \ldots, (x_n, y_n)$ is the unique solution of the linear system

$$A^*A\mathbf{x} = A^*\mathbf{b},$$

where

$$A = \begin{bmatrix} x_1 & 1 \\ x_2 & 1 \\ \vdots & \vdots \\ x_n & 1 \end{bmatrix}, \quad \mathbf{b} = \begin{bmatrix} y_1 \\ y_2 \\ \vdots \\ y_n \end{bmatrix}, \quad \text{and} \quad \mathbf{x} = \begin{bmatrix} m \\ b \end{bmatrix}.$$

[Hint: Use Exercise 2.C.9.]

2.D　Extra Topic: Continuity and Matrix Analysis

The pseudoinverse of Section 2.C.1 and the Jordan decomposition of Section 2.4 can also help us cope with the fact that some matrices are not invertible or diagonalizable.

Much of introductory linear algebra is devoted to defining and investigating classes of matrices that are easier to work with than general matrices. For example, invertible matrices are easier to work with than typical matrices since we can do algebra with them like we are used to (e.g., if A is invertible then we can solve the matrix equation $AX = B$ via $X = A^{-1}B$). Similarly, diagonalizable matrices are easier to work with than typical matrices since we can easily compute powers and analytic functions of them.

This section explores some techniques that can be used to treat all matrices as if they were invertible and/or diagonalizable, even though they aren't. The rough idea is to exploit the fact that every matrix is extremely close to one that is invertible and diagonalizable, and many linear algebraic properties do not change much if we change the entries of a matrix by just a tiny bit.

2.D.1   Dense Sets of Matrices

One particularly useful technique from mathematical analysis is to take advantage of the fact that every continuous function $f : \mathbb{R} \to \mathbb{R}$ is completely determined by how it acts on the rational numbers (in fact, we made use of this technique when we proved Theorem 1.D.8). For example, if we know that $f(x) = \sin(x)$ whenever $x \in \mathbb{Q}$ and we furthermore know that f is continuous, then it must be the case that $f(x) = \sin(x)$ for *all* $x \in \mathbb{R}$ (see Figure 2.16).

> Recall that $\mathbb{Q}$ denotes the set of rational numbers (i.e., numbers that can be written as a ratio p/q of two integers).

Intuitively, this follows because f being continuous means that $f(\tilde{x})$ must be close to $f(x)$ whenever $\tilde{x} \in \mathbb{R}$ is close to $x \in \mathbb{Q}$, and we can always find an $x \in \mathbb{Q}$ that is as close to $\tilde{x}$ as we like. Roughly speaking, $\mathbb{Q}$ has no "holes" of width greater than 0 along the real number line, so defining how f behaves on $\mathbb{Q}$ leaves no "wiggle room" for what its values outside of $\mathbb{Q}$ can be.

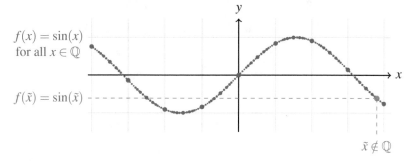

Figure 2.16: If $f(x) = \sin(x)$ for all $x \in \mathbb{Q}$ and f is continuous, then it must be the case that $f(x) = \sin(x)$ for *all* $x \in \mathbb{R}$.

Remark 2.D.1

Defining Exponential Functions

> Recall that "irrational" just means "not rational". Some well-known irrational numbers include $\sqrt{2}$, π, and e.

This idea of extending a function from $\mathbb{Q}$ to all of $\mathbb{R}$ via continuity is actually how some common functions are *defined*. For example, what does the expression 2^x even mean if x is irrational? We build up to the answer to this question one step at a time:

- If x is a positive integer then $2^x = 2 \cdot 2 \cdots 2$ (x times). If x is a negative integer then $2^x = 1/2^{-x}$. If $x = 0$ then $2^x = 1$.
- If x is rational, so we can write $x = p/q$ for some integers p, q ($q \neq 0$), then $2^x = 2^{p/q} = \sqrt[q]{2^p}$.
- If x is irrational, we define 2^x by requiring that the function $f(x) = 2^x$ be continuous. That is, we set

$$2^x = \lim_{\substack{r \to x \\ r \text{ is rational}}} 2^r.$$

For example, a number like 2^π is defined as the limit of the sequence of numbers $2^3, 2^{3.1}, 2^{3.14}, 2^{3.141}, \ldots$, where we take better and better decimal approximations of π (which are rational) in the exponent.

The property of $\mathbb{Q}$ that is required to make this sort of argument work is the fact that it is **dense** in $\mathbb{R}$: every $x \in \mathbb{R}$ can be written as a limit of rational numbers. We now define the analogous property for sets of matrices.

Definition 2.D.1	Suppose $\mathbb{F} = \mathbb{R}$ or $\mathbb{F} = \mathbb{C}$. A set of matrices $B \subseteq \mathcal{M}_{m,n}(\mathbb{F})$ is called **dense** in $\mathcal{M}_{m,n}(\mathbb{F})$ if, for every matrix $A \in \mathcal{M}_{m,n}(\mathbb{F})$, there exist matrices $A_1, A_2, \ldots \in B$ such that $$\lim_{k \to \infty} A_k = A.$$
Dense Set of Matrices	

Recall that $[A_k]_{i,j}$ is the (i,j)-entry of A_k.

Before proceeding, we clarify that limits of matrices are simply meant entrywise: $\lim_{k \to \infty} A_k = A$ means that $\lim_{k \to \infty} [A_k]_{i,j} = [A]_{i,j}$ for all i and j. We illustrate with an example.

Example 2.D.1
Limits of Matrices

Let $A_k = \dfrac{1}{k} \begin{bmatrix} k & 2k-1 \\ 2k-1 & 4k+1 \end{bmatrix}$ and $B = \begin{bmatrix} 2 & 1 \\ 4 & 2 \end{bmatrix}$. Compute $\lim_{k \to \infty} \left(A_k^{-1} B \right)$.

Solution:

We start by computing

Recall that the inverse of a 2×2 matrix is
$$\begin{bmatrix} a & b \\ c & d \end{bmatrix}^{-1} =$$
$$\frac{1}{ad - bc} \begin{bmatrix} d & -b \\ -c & a \end{bmatrix}.$$

$$A_k^{-1} = \frac{k}{5k-1} \begin{bmatrix} 4k+1 & 1-2k \\ 1-2k & k \end{bmatrix}.$$

In particular, it is worth noting that $\lim_{k \to \infty} A_k^{-1}$ does not exist (the entries of A_k^{-1} get larger and larger as k increases), which should not be surprising since

$$\lim_{k \to \infty} A_k = \lim_{k \to \infty} \left(\frac{1}{k} \begin{bmatrix} k & 2k-1 \\ 2k-1 & 4k+1 \end{bmatrix} \right) = \begin{bmatrix} 1 & 2 \\ 2 & 4 \end{bmatrix}$$

is not invertible. However, if we multiply on the right by B then

$$A_k^{-1} B = \frac{k}{5k-1} \begin{bmatrix} 4k+1 & 1-2k \\ 1-2k & k \end{bmatrix} \begin{bmatrix} 2 & 1 \\ 4 & 2 \end{bmatrix} = \frac{k}{5k-1} \begin{bmatrix} 6 & 3 \\ 2 & 1 \end{bmatrix},$$

so

$$\lim_{k \to \infty} \left(A_k^{-1} B \right) = \lim_{k \to \infty} \left(\frac{k}{5k-1} \begin{bmatrix} 6 & 3 \\ 2 & 1 \end{bmatrix} \right) = \frac{1}{5} \begin{bmatrix} 6 & 3 \\ 2 & 1 \end{bmatrix}.$$

Intuitively, a set $B \subseteq \mathcal{M}_{m,n}(\mathbb{F})$ is dense if every matrix is arbitrarily close to a member of B. As suggested earlier, we are already very familiar with two very useful dense sets of matrices: the sets of invertible and diagonalizable matrices. In both cases, the rough reason why these sets are dense is that changing the entries of a matrix just slightly changes its eigenvalues just slightly, and we can make this change so as to have its eigenvalues avoid 0 (and thus be invertible) and avoid repetitions (and thus be diagonalizable). We now make these ideas precise.

Theorem 2.D.1
Invertible Matrices are Dense

Suppose $\mathbb{F} = \mathbb{R}$ or $\mathbb{F} = \mathbb{C}$. The set of invertible matrices is dense in $\mathcal{M}_n(\mathbb{F})$.

Proof. Suppose $A \in \mathcal{M}_n(\mathbb{F})$ and then define, for each integer $k \geq 1$, the matrix $A_k = A + \frac{1}{k} I$. It is clear that

$$\lim_{k \to \infty} A_k = A,$$

and we claim that A_k is invertible whenever k is sufficiently large.

The idea here is that we can wiggle the zero eigenvalues of A away from zero without wiggling any of the others into zero.

To see why this claim holds, recall that if A has eigenvalues $\lambda_1, \ldots, \lambda_n$ then A_k has eigenvalues $\lambda_1 + 1/k, \ldots, \lambda_n + 1/k$. Well, if $k > 1/|\lambda_j|$ for each non-zero eigenvalue λ_j of A, then we can show that all of the eigenvalues of A_k are non-zero by considering two cases:

- If $\lambda_j = 0$ then $\lambda_j + 1/k = 1/k \neq 0$, and
- if $\lambda_j \neq 0$ then $|\lambda_j + 1/k| \geq |\lambda_j| - 1/k > 0$, so $\lambda_j + 1/k \neq 0$.

It follows that each eigenvalue of A_k is non-zero, and thus A_k is invertible, whenever k is sufficiently large. ∎

For example, the matrix

$$A = \begin{bmatrix} 1 & 1 & 2 \\ 2 & 3 & 5 \\ -1 & 0 & -1 \end{bmatrix}$$

is not invertible, since it has 0 as an eigenvalue (as well as two other eigenvalues equal to $(3 \pm \sqrt{13})/2$, which are approximately 3.30278 and -0.30278). However, the matrix

In fact, this A_k is invertible when $k = 1, 2, 3$ too, since none of the eigenvalues of A equal -1, $-1/2$, or $-1/3$. However, we just need invertibility when k is large.

$$A_k = A + \frac{1}{k}I = \begin{bmatrix} 1 + 1/k & 1 & 2 \\ 2 & 3 + 1/k & 5 \\ -1 & 0 & -1 + 1/k \end{bmatrix}$$

is invertible whenever $k \geq 4$ since adding a number in the interval $(0, 1/4] = (0, 0.25]$ to each of the eigenvalues of A ensures that none of them equal 0.

Theorem 2.D.2

Diagonalizable Matrices are Dense

Suppose $\mathbb{F} = \mathbb{R}$ or $\mathbb{F} = \mathbb{C}$. The set of diagonalizable matrices is dense in $\mathcal{M}_n(\mathbb{F})$.

Proof. Suppose $A \in \mathcal{M}_n(\mathbb{F})$, which we Schur triangularize as $A = UTU^*$. We then let D be the diagonal matrix with diagonal entries $1, 2, \ldots, n$ (in that order) and define, for each integer $k \geq 1$, the matrix $A_k = A + \frac{1}{k}UDU^*$. It is clear that

$$\lim_{k \to \infty} A_k = A,$$

and we claim that A_k has distinct eigenvalues, and is thus diagonalizable, whenever k is sufficiently large.

To see why this claim holds, recall that the eigenvalues $\lambda_1, \lambda_2, \ldots, \lambda_n$ of A are the diagonal entries of T (which we assume are arranged in that order). Similarly, the eigenvalues of

$$A_k = UTU^* + \frac{1}{k}UDU^* = U\left(T + \frac{1}{k}D\right)U^*$$

are the diagonal entries of $T + \frac{1}{k}D$, which are $\lambda_1 + 1/k, \lambda_2 + 2/k, \ldots, \lambda_n + n/k$. Well, if $k > (n-1)/|\lambda_i - \lambda_j|$ for each distinct pair of eigenvalues $\lambda_i \neq \lambda_j$, then we can show that the eigenvalues of A_k are distinct as follows:

- If $\lambda_i = \lambda_j$ (but $i \neq j$) then $\lambda_i + i/k \neq \lambda_j + j/k$, and

- if $\lambda_i \neq \lambda_j$ then

$$\begin{aligned}
\left|(\lambda_i + i/k) - (\lambda_j + j/k)\right| &= \left|(\lambda_i - \lambda_j) + (i/k - j/k)\right| \\
&\geq |\lambda_i - \lambda_j| - |i - j|/k \\
&\geq |\lambda_i - \lambda_j| - (n-1)/k > 0.
\end{aligned}$$

It follows that the eigenvalues of A_k are distinct, and it is thus diagonalizable, whenever k is sufficiently large. ∎

Remark 2.D.2	Limits can actually be defined in any normed vector space—we have just restricted attention to $\mathcal{M}_{m,n}$ since that is the space where these concepts are particularly useful for us, and the details simplify in this case since matrices have entries that we can latch onto.
Limits in Normed Vector Spaces	

In general, as long as a vector space $\mathcal{V}$ is finite-dimensional, we can define limits in $\mathcal{V}$ by first fixing a basis B of $\mathcal{V}$ and then saying that

$$\lim_{k\to\infty} \mathbf{v}_k = \mathbf{v} \quad \text{whenever} \quad \lim_{k\to\infty} [\mathbf{v}_k]_B = [\mathbf{v}]_B,$$

where the limit on the right is just meant entrywise. It turns out that this notion of limit does not depend on which basis B of $\mathcal{V}$ we choose (i.e., for any two bases B and C of $\mathcal{V}$, it is the case that $\lim_{k\to\infty} [\mathbf{v}_k]_B = [\mathbf{v}]_B$ if and only if $\lim_{k\to\infty} [\mathbf{v}_k]_C = [\mathbf{v}]_C$).

If $\mathcal{V}$ is infinite-dimensional then this approach does not work, since we may not be able to construct a basis of $\mathcal{V}$ in the first place, so we may not have any notion of "entrywise" limits to work with. We can get around this problem by picking some norm on $\mathcal{V}$ (see Section 1.D) and saying that

$$\lim_{k\to\infty} \mathbf{v}_k = \mathbf{v} \quad \text{whenever} \quad \lim_{k\to\infty} \|\mathbf{v}_k - \mathbf{v}\| = 0.$$

In the finite-dimensional case, this definition of limits based on norms turns out to be equivalent to the earlier one based on coordinate vectors, and furthermore it does not matter which norm we use on $\mathcal{V}$ (a fact that follows from all norms on finite-dimensional vector spaces being equivalent to each other—see Theorem 1.D.1).

On the other hand, in infinite-dimensional vector spaces, it is no longer necessarily the case that all norms are equivalent (see Exercise 1.D.26), so it is also no longer the case that limits are independent of the norm being used. For example, consider the 1-norm and ∞-norm on $\mathcal{P}[0,1]$ (the space of real-valued polynomials on the interval $[0,1]$):

We introduced these norms back in Section 1.D.1.

$$\|f\|_1 = \int_0^1 |f(x)|\, dx \quad \text{and} \quad \|f\|_\infty = \max_{0\le x\le 1} |f(x)|.$$

We claim that the sequence of polynomials $\{f_k\}_{k=1}^\infty \subset \mathcal{P}[0,1]$ defined by $f_k(x) = x^k$ for all $k \geq 1$ converges to the zero function in the 1-norm, but does not in the ∞-norm. To see why this is the case, we note that

$$\|f_k - \mathbf{0}\|_1 = \int_0^1 |x^k| \, dx = \left.\frac{x^{k+1}}{k+1}\right|_{x=0}^1 = \frac{1}{k+1} \quad \text{and}$$

$$\|f_k - \mathbf{0}\|_\infty = \max_{x \in [0,1]} |x^k| = 1^k = 1 \quad \text{for all} \quad k \geq 1.$$

If we take limits of these quantities, we see that

$$\lim_{k \to \infty} \|f_k - \mathbf{0}\|_1 = \lim_{k \to \infty} \frac{1}{k+1} = 0, \quad \text{so} \quad \lim_{k \to \infty} f_k = \mathbf{0}, \quad \text{but}$$

$$\lim_{k \to \infty} \|f_k - \mathbf{0}\|_\infty = \lim_{k \to \infty} 1 = 1 \neq 0, \quad \text{so} \quad \lim_{k \to \infty} f_k \neq \mathbf{0}.$$

> A sequence of functions $\{f_k\}_{k=1}^\infty$ converging in the ∞-norm means that the maximum difference between $f_k(x)$ and the limit function $f(x)$ goes to 0. That is, these functions converge **pointwise**, which is probably the most intuitive notion of convergence for functions.

Geometrically, all this is saying is that the area under the graphs of the polynomials $f_k(x) = x^k$ on the interval $[0, 1]$ decreases toward zero, yet the maximum values of these polynomials on that interval does not similarly converge toward zero (it is fixed at 1):

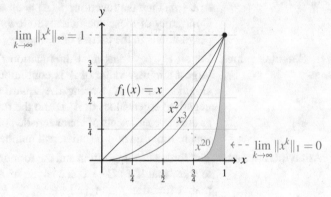

2.D.2 Continuity of Matrix Functions

Just as with functions on $\mathbb{R}$, the idea behind continuity of functions acting on $\mathcal{M}_{m,n}$ is that we should be able to make two outputs of the function as close together as we like simply by making their inputs sufficiently close together. We can make this idea precise via limits:

Definition 2.D.2

Continuous Functions

Suppose $\mathbb{F} = \mathbb{R}$ or $\mathbb{F} = \mathbb{C}$. We say that a function $f : \mathcal{M}_{m,n}(\mathbb{F}) \to \mathcal{M}_{r,s}(\mathbb{F})$ is **continuous** if it is the case that

$$\lim_{k \to \infty} f(A_k) = f(A) \quad \text{whenever} \quad \lim_{k \to \infty} A_k = A.$$

Phrased slightly differently, a function $f : \mathcal{M}_{m,n}(\mathbb{F}) \to \mathcal{M}_{r,s}(\mathbb{F})$ is continuous if we can pull limits in or out of it:

$$f\left(\lim_{k \to \infty} A_k\right) = \lim_{k \to \infty} f(A_k)$$

whenever the limit on the left exists.

Many of the functions of matrices that we have learned about are continuous.

Some important examples (which we do not rigorously prove, since we are desperately trying to avoid epsilons and deltas) include

Linear transformations: Roughly speaking, continuity of linear transformations follows from combining two facts: (i) if f is a linear transformation then the entries of $f(A)$ are linear combinations of the entries of A, and (ii) if each entry of A_k is close to the corresponding entry of A, then any linear combination of the entries of A_k is close to the corresponding linear combination of the entries of A.

Matrix multiplication: The fact that the function $f(A) = BAC$ (where B and C are fixed matrices) is continuous follows from it being a linear transformation.

The trace: Again, the trace is a linear transformation and is thus continuous.

Polynomials: Matrix polynomials like $f(A) = A^2 - 3A + 2I$, and also polynomials of a matrix's entries like $g(A) = a_{1,1}^3 a_{1,2} - a_{2,1}^2 a_{2,2}^2$, are all continuous.

The determinant: Continuity of the determinant follows from the fact that Theorem A.1.4 shows that it is a polynomial in the matrix's entries.

Coefficients of char. poly.: If A has characteristic polynomial $p_A(\lambda) = (-1)^n \lambda^n + a_{n-1}\lambda^{n-1} + \cdots + a_1\lambda + a_0$ then the functions $f_k(A) = a_k$ are continuous for all $0 \le k < n$. Continuity of these coefficients follows from the fact that they are all polynomials in the entries of A, since $p_A(\lambda) = \det(A - \lambda I)$.

Singular values: For all $1 \le k \le \min\{m, n\}$ the function $f_k(A) = \sigma_k$ that outputs the k-th largest singular value of A is continuous. This fact follows from the singular values of A being the square roots of the (necessarily non-negative) eigenvalues of A^*A, and the function $A \mapsto A^*A$ is continuous, as are the coefficients of characteristic polynomials and the non-negative square root of a non-negative real number.

Analytic functions: The Jordan decomposition and the formula of Theorem 2.4.6 can be used to show that if $f : \mathbb{C} \to \mathbb{C}$ is analytic on all of $\mathbb{C}$ then the corresponding matrix function $f : \mathcal{M}_n(\mathbb{C}) \to \mathcal{M}_n(\mathbb{C})$ is continuous. For example, the functions e^A, $\sin(A)$, and $\cos(A)$ are continuous.

We also note that the sum, difference, and composition of any continuous functions are again continuous. With all of these examples taken care of, we now state the theorem that is the main reason that we have introduced dense sets of matrices and continuity of matrix functions in the first place.

Theorem 2.D.3	Suppose $\mathbb{F} = \mathbb{R}$ or $\mathbb{F} = \mathbb{C}$, $B \subseteq \mathcal{M}_{m,n}(\mathbb{F})$ is dense in $\mathcal{M}_{m,n}(\mathbb{F})$, and $f, g :$
Continuity Plus Density	$\mathcal{M}_{m,n}(\mathbb{F}) \to \mathcal{M}_{r,s}(\mathbb{F})$ are continuous. If $f(A) = g(A)$ whenever $A \in B$ then $f(A) = g(A)$ for all $A \in \mathcal{M}_{m,n}(\mathbb{F})$.

Proof. Since B is dense in $\mathcal{M}_{m,n}(\mathbb{F})$, for any $A \in \mathcal{M}_{m,n}(\mathbb{F})$ we can find matrices $A_1, A_2, \ldots \in B$ such that $\lim_{k \to \infty} A_k = A$. Then

$$f(A) = \lim_{k \to \infty} f(A_k) = \lim_{k \to \infty} g(A_k) = g(A),$$

with the middle equality following from the fact that $A_k \in B$ so $f(A_k) = g(A_k)$. ∎

For example, the above theorem tells us that if $f : \mathcal{M}_n(\mathbb{C}) \to \mathbb{C}$ is a continuous function for which $f(A) = \det(A)$ whenever A is invertible, then f must in fact be the determinant function, and that if $g : \mathcal{M}_n(\mathbb{C}) \to \mathcal{M}_n(\mathbb{C})$ is a continuous function for which $g(A) = A^2 - 2A + 3I$ whenever A is diagonalizable, then $g(A) = A^2 - 2A + 3I$ for *all* $A \in \mathcal{M}_n(\mathbb{C})$. The remainder of this section is devoted to exploring somewhat more exotic applications of this theorem.

2.D.3   Working with Non-Invertible Matrices

We now start illustrating the utility of the concepts introduced in the previous subsections by showing how they can help us extend results from the set of invertible matrices to the set of all matrices. These methods are useful because it is often much easier to prove that some nice property holds for invertible matrices, and then use continuity and density arguments to extend the result to all matrices, than it is to prove that it holds for all matrices directly.

Similarity-Invariance and the Trace

We stated back in Corollary 1.A.2 that the trace is the unique (up to scaling) linear form on matrices that is similarity invariant. We now use continuity to finally prove this result, which we re-state here for ease of reference.

Theorem 2.D.4

Similarity-Invariance Defines the Trace

Suppose $\mathbb{F} = \mathbb{R}$ or $\mathbb{F} = \mathbb{C}$, and $f : \mathcal{M}_n(\mathbb{F}) \to \mathbb{F}$ is a linear form with the following properties:

   a) $f(A) = f(PAP^{-1})$ for all $A, P \in \mathcal{M}_n(\mathbb{F})$ with P invertible, and

   b) $f(I) = n$.

Then $f(A) = \mathrm{tr}(A)$ for all $A \in \mathcal{M}_n(\mathbb{F})$.

We originally proved that AB is similar to BA (when A or B is invertible) back in Exercise 1.A.5.

Proof. We start by noticing that if A is invertible then AB and BA are similar, since

$$AB = A(BA)A^{-1}.$$

In particular, this tells us that $f(AB) = f\big(A(BA)A^{-1}\big) = f(BA)$ whenever A is invertible. If we could show that $f(AB) = f(BA)$ for *all* A and B then we would be done, since Theorem 1.A.1 would then tell us that $f(A) = \mathrm{tr}(A)$ for all $A \in \mathcal{M}_n$.

However, this final claim follows immediately from continuity of f and density of the set of invertible matrices. In particular, if we fix any matrix $B \in \mathcal{M}_n$ and define $f_B(A) = f(AB)$ and $g_B(A) = f(BA)$ then we just showed that $f_B(A) = g_B(A)$ for all invertible $A \in \mathcal{M}_n$ and all (not necessarily invertible) $B \in \mathcal{M}_n$. Since f_B and g_B are continuous, it follows from Theorem 2.D.3 that $f_B(A) = g_B(A)$ (so $f(AB) = f(BA)$) for *all* $A, B \in \mathcal{M}_n$, which completes the proof. ∎

If the reader is uncomfortable with the introduction of the function f_B and g_B at the end of the above proof, it can instead be finished a bit more directly by making use of some of the ideas from the proof of Theorem 2.D.1. In particular, for any (potentially non-invertible) matrix $A \in \mathcal{M}_n$ and integer $k > 1$, we define $A_k = A + \frac{1}{k}I$ and note that $\lim_{k \to \infty} A_k = A$ and A_k is invertible when k is large.

We then compute

$$f(AB) = f\left(\lim_{k\to\infty} A_k B\right) \qquad \text{(since } \lim_{k\to\infty} A_k = A\text{)}$$

$$= \lim_{k\to\infty} f(A_k B) \qquad \text{(since } f \text{ is linear, thus continuous)}$$

$$= \lim_{k\to\infty} f(B A_k) \qquad \text{(since } A_k \text{ is invertible when } k \text{ is large)}$$

$$= f\left(\lim_{k\to\infty} B A_k\right) \qquad \text{(since } f \text{ is linear, thus continuous)}$$

$$= f(BA), \qquad \text{(since } \lim_{k\to\infty} A_k = A\text{)}$$

as desired.

Finishing the QR Decomposition

We now use continuity and density arguments to complete the proof that every matrix $A \in \mathcal{M}_{m,n}$ has a QR decomposition, which we originally stated as Theorem 1.C.1 and only proved in the special case when its leftmost $\min\{m,n\}$ columns form a linearly independent set. We re-state this theorem here for ease of reference.

Theorem 2.D.5 **QR Decomposition** **(Again)**	Suppose $\mathbb{F} = \mathbb{R}$ or $\mathbb{F} = \mathbb{C}$, and $A \in \mathcal{M}_{m,n}(\mathbb{F})$. There exists a unitary matrix $U \in \mathcal{M}_m(\mathbb{F})$ and an upper triangular matrix $T \in \mathcal{M}_{m,n}(\mathbb{F})$ with non-negative real entries on its diagonal such that $$A = UT.$$

In addition to the techniques that we have already presented in this section, the proof of this theorem also relies on a standard result from analysis called the Bolzano–Weierstrass theorem, which says that every sequence in a closed and bounded subset of a finite-dimensional vector space over $\mathbb{R}$ or $\mathbb{C}$ has a convergent subsequence. For our purposes, we note that the set of unitary matrices is a closed and bounded subset of $\mathcal{M}_n$, so any sequence of unitary matrices has a convergent subsequence.

Proof of Theorem 2.D.5. We assume that $n \geq m$ and simply note that a completely analogous argument works if $n < m$. We write $A = [\, B \mid C \,]$ where $B \in \mathcal{M}_m$ and define $A_k = [\, B + \frac{1}{k}I \mid C \,]$ for each integer $k \geq 1$. Since $B + \frac{1}{k}I$ is invertible (i.e., its columns are linearly independent) whenever k is sufficiently large, the proof of Theorem 1.C.1 tells us that A_k has a QR decomposition $A_k = U_k T_k$ whenever k is sufficiently large. We now use limit arguments to show that A itself also has such a decomposition.

Since the set of unitary matrices is closed and bounded, the Bolzano–Weierstrass theorem tells us that there is a sequence $k_1 < k_2 < k_3 < \cdots$ with the property that $U = \lim_{j\to\infty} U_{k_j}$ exists and is unitary. Similarly, if $\lim_{j\to\infty} T_{k_j}$ exists then it must be upper triangular since each T_{k_j} is upper triangular. To see that this limit *does* exist, we compute

> Recall that limits are taken entrywise, and the limit of the 0 entries in the lower triangular portion of each T_{k_j} is just 0.

$$\lim_{j\to\infty} T_{k_j} = \lim_{j\to\infty} U_{k_j}^* A_{k_j} = \left(\lim_{j\to\infty} U_{k_j}\right)^* \left(\lim_{j\to\infty} A_{k_j}\right) = U^* \left(\lim_{k\to\infty} A_k\right) = U^* A.$$

It follows that $A = UT$, where $T = \lim_{j\to\infty} T_{k_j}$ is upper triangular, which completes the proof. ∎

If the reader does not like having to make use of the Bolzano–Weierstrass theorem as we did above, another proof of the QR decomposition is outlined in Exercise 2.D.7.

From the Inverse to the Adjugate

One final method of extending a property of matrices from those that are invertible to those that perhaps are not is to make use of something called the "adjugate" of a matrix:

Definition 2.D.3

The Adjugate

Suppose that $\mathbb{F} = \mathbb{R}$ or $\mathbb{F} = \mathbb{C}$ and $A \in \mathcal{M}_n(\mathbb{F})$ has characteristic polynomial

$$p_A(\lambda) = \det(A - \lambda I) = (-1)^n \lambda^n + a_{n-1} \lambda^{n-1} + \cdots + a_1 \lambda + a_0.$$

Then the **adjugate** of A, denoted by $\mathrm{adj}(A)$, is the matrix

$$\mathrm{adj}(A) = -\big((-1)^n A^{n-1} + a_{n-1} A^{n-2} + \cdots + a_2 A + a_1 I\big).$$

While the above definition likely seems completely bizarre at first glance, it is motivated by the following two properties:

- The function $\mathrm{adj} : \mathcal{M}_n \to \mathcal{M}_n$ is continuous, since matrix multiplication, addition, and the coefficients of characteristic polynomials are all continuous, and

The adjugate is sometimes defined in terms of the cofactors of A instead (see (Joh20, Section 3.A), for example). These two definitions are equivalent.

- The adjugate satisfies $A^{-1} = \mathrm{adj}(A)/\det(A)$ whenever $A \in \mathcal{M}_n$ is invertible. To verify this claim, first recall that the constant term of the characteristic polynomial is $a_0 = p_A(0) = \det(A - 0I) = \det(A)$. Using the Cayley–Hamilton theorem (Theorem 2.1.3) then tells us that

$$p_A(A) = (-1)^n A^n + a_{n-1} A^{n-1} + \cdots + a_1 A + \det(A) I = O,$$

and multiplying this equation by A^{-1} (if it exists) shows that

$$(-1)^n A^{n-1} + a_{n-1} A^{n-2} + \cdots + a_1 I + \det(A) A^{-1} = O.$$

Rearranging slightly shows exactly that $\det(A) A^{-1} = \mathrm{adj}(A)$, as claimed.

It follows that if we have a formula or property of matrices that involves A^{-1}, we can often extend it to an analogous formula that holds for *all* square matrices by simply making the substitution $A^{-1} = \mathrm{adj}(A)/\det(A)$ and invoking continuity. We illustrate this method with an example.

Theorem 2.D.6

Jacobi's Formula

Suppose that $\mathbb{F} = \mathbb{R}$ or $\mathbb{F} = \mathbb{C}$ and $A(t) \in \mathcal{M}_n(\mathbb{F})$ is a matrix whose entries depend in a continuously differentiable way on a parameter $t \in \mathbb{F}$. If we let $\mathrm{d}A/\mathrm{d}t$ denote the matrix that is obtained by taking the derivative of each entry of A with respect to t, then

$$\frac{\mathrm{d}}{\mathrm{d}t} \det\big(A(t)\big) = \mathrm{tr}\left(\mathrm{adj}\big(A(t)\big) \frac{\mathrm{d}A}{\mathrm{d}t} \right).$$

Proof. We proved in Exercise 1.A.8 that if $A(t)$ is invertible then

$$\frac{\mathrm{d}}{\mathrm{d}t} \det\big(A(t)\big) = \det\big(A(t)\big) \mathrm{tr}\left(A(t)^{-1} \frac{\mathrm{d}A}{\mathrm{d}t} \right).$$

If we make the substitution $A(t)^{-1} = \operatorname{adj}(A(t))/\det(A(t))$ in the above formula then we see that

$$\frac{d}{dt}\det(A(t)) = \operatorname{tr}\left(\operatorname{adj}(A(t))\frac{dA}{dt}\right)$$

whenever $A(t)$ is invertible. Since the set of invertible matrices is dense in $\mathcal{M}_n(\mathbb{F})$ and this function is continuous, we conclude that it must in fact hold for all $A(t)$ (invertible or not).   ∎

2.D.4   Working with Non-Diagonalizable Matrices

We can also apply continuity and density arguments to the set of diagonalizable matrices in much the same way that we applied them to the set of invertible matrices in the previous subsection. We can often use Schur triangularization (Theorem 2.1.1) or the Jordan decomposition (Theorem 2.4.1) to achieve the same end result as the methods of this section, but it is nonetheless desirable to avoid those decompositions when possible (after all, diagonal matrices are much simpler to work with than triangular ones).

We start by illustrating how these techniques can simplify arguments involving matrix functions.

Theorem 2.D.7

Trigonometric Identities for Matrices

For all $A \in \mathcal{M}_n(\mathbb{C})$ it is the case that $\sin^2(A) + \cos^2(A) = I$.

Proof. Recall that $\sin^2(x) + \cos^2(x) = 1$ for all $x \in \mathbb{C}$. It follows that if A is diagonalizable via $A = PDP^{-1}$, where D has diagonal entries $\lambda_1, \lambda_2, \ldots, \lambda_n$, then

> We proved this theorem via the Jordan decomposition in Exercise 2.4.14.

$$\sin^2(A) + \cos^2(A) = P\begin{bmatrix} \sin^2(\lambda_1) & 0 & \cdots & 0 \\ 0 & \sin^2(\lambda_2) & \cdots & 0 \\ \vdots & \vdots & \ddots & \vdots \\ 0 & 0 & \cdots & \sin^2(\lambda_n) \end{bmatrix}P^{-1}$$

$$+ P\begin{bmatrix} \cos^2(\lambda_1) & 0 & \cdots & 0 \\ 0 & \cos^2(\lambda_2) & \cdots & 0 \\ \vdots & \vdots & \ddots & \vdots \\ 0 & 0 & \cdots & \cos^2(\lambda_n) \end{bmatrix}P^{-1}$$

$$= PP^{-1} = I.$$

Since $f(x) = \sin^2(x) + \cos^2(x)$ is analytic on $\mathbb{C}$ and thus continuous on $\mathcal{M}_n(\mathbb{C})$, it follows from Theorem 2.D.3 that $\sin^2(A) + \cos^2(A) = I$ for all (not necessarily diagonalizable) $A \in \mathcal{M}_n(\mathbb{C})$.   ∎

As another example to illustrate the utility of this technique, we now provide an alternate proof of the Cayley–Hamilton theorem (Theorem 2.1.3) that avoids the technical argument that we originally used to prove it via Schur triangularization (which also has a messy, technical proof).

Theorem 2.D.8
Cayley–Hamilton
(Again)

Suppose $A \in \mathcal{M}_n(\mathbb{C})$ has characteristic polynomial $p_A(\lambda) = \det(A - \lambda I)$. Then $p_A(A) = O$.

Proof. Since A is complex, the fundamental theorem of algebra (Theorem A.3.1) tells us that the characteristic polynomial of A has a root and can thus be factored as

$$p_A(\lambda) = (\lambda_1 - \lambda)(\lambda_2 - \lambda) \cdots (\lambda_n - \lambda),$$

where $\lambda_1, \lambda_2, \ldots, \lambda_n$ are the eigenvalues of A, listed according to algebraic multiplicity. Our goal is thus to show that $(\lambda_1 I - A)(\lambda_2 I - A) \cdots (\lambda_n I - A) = O$.

To see why this is the case, notice that if A is diagonalizable via $A = PDP^{-1}$ then

$$(\lambda_1 I - PDP^{-1}) \cdots (\lambda_n I - PDP^{-1}) = P(\lambda_1 I - D) \cdots (\lambda_n I - D)P^{-1},$$

and it is clear that $(\lambda_1 I - D)(\lambda_2 I - D) \cdots (\lambda_n I - D) = O$, since D is diagonal with diagonal entries $\lambda_1, \lambda_2, \ldots, \lambda_n$. It follows that $p_A(A) = O$ whenever A is diagonalizable.

To see that this same conclusion must hold even when A is not diagonalizable, simply notice that the function $f(A) = p_A(A)$ is continuous (since the coefficients in the characteristic polynomial are continuous) and then apply Theorem 2.D.3 to it. ∎

Exercises

solutions to starred exercises on page 475

2.D.1 Compute $\lim\limits_{k \to \infty} A^k$ if $A = \frac{1}{10} \begin{bmatrix} 9 & 2 \\ 1 & 8 \end{bmatrix}$.

2.D.2 Determine which of the following statements are true and which are false.

* (a) The Frobenius norm of a matrix is continuous.
  (b) The function $T : \mathcal{M}_{m,n} \to \mathcal{M}_{n,m}$ defined by $T(A) = A^{\dagger}$ (the pseudoinverse of A—see Section 2.C.1) is continuous.

∗∗2.D.3 Show that if $A_1, A_2, \ldots \in \mathcal{M}_{m,n}$ are matrices with rank$(A_k) \leq r$ for all k then

$$\text{rank}\left(\lim\limits_{k \to \infty} A_k\right) \leq r$$

too, as long as this limit exists.

[Side note: For this reason, the rank of a matrix is said to be a **lower semicontinuous** function.]

2.D.4 Show that adj$(AB) = $ adj(B)adj(A) for all $A, B \in \mathcal{M}_n$.

[Hint: This is easy to prove if A and B are invertible.]

∗ 2.D.5 Show that every positive semidefinite matrix $A \in \mathcal{M}_n$ can be written as a limit of positive *definite* matrices.

[Side note: In other words, the set of positive definite matrices is dense in the set of positive semidefinite matrices.]

2.D.6 Suppose that $A, B \in \mathcal{M}_n$ are positive semidefinite. Show that all eigenvalues of AB are real and non-negative.

[Hint: We proved this fact for positive *definite* matrices back in Exercise 2.2.26.]

∗∗2.D.7 Provide an alternate proof of the QR decomposition (Theorem 2.D.5) by combining the Cholesky decomposition (Theorem 2.B.5) and Theorem 2.2.10.

3. Tensors and Multilinearity

> Do not think about what tensors are, but rather what the whole construction of a tensor product space can do for you.
>
> Keith Conrad

Up until now, all of our explorations in linear algebra have focused on vectors, matrices, and linear transformations—all of our matrix decompositions, change of basis techniques, algorithms for solving linear systems or constructing orthonormal bases, and so on have had the purpose of deepening our understanding of these objects. We now introduce a common generalization of all of these objects, called "tensors", and investigate which of our tools and techniques do and do not still work in this new setting.

For example, just like we can think of vectors (that is, the type of vectors that live in $\mathbb{F}^n$) as 1-dimensional lists of numbers and matrices as 2-dimensional arrays of numbers, we can think of tensors as any-dimensional *arrays* of numbers. Perhaps more usefully though, recall that we can think of vectors geometrically as arrows in space, and matrices as linear transformations that move those arrows around. We can similarly think of tensors as functions that move vectors, matrices, or even other more general tensors themselves around in a linear way. In fact, they can even move *multiple* vectors, matrices, or tensors around (much like bilinear forms and multilinear forms did in Section 1.3.3).

In a sense, this chapter is where we really push linear algebra to its ultimate limit, and see just how far our techniques can extend. Tensors provide a common generalization of almost every single linear algebraic object that we have seen—not only are vectors, matrices, linear transformations, linear forms, and bilinear forms examples of tensors, but so are more exotic operations like the cross product, the determinant, and even matrix multiplication itself.

3.1 The Kronecker Product

Before diving into the full power of tensors, we start by considering a new operation on $\mathbb{F}^n$ and $\mathcal{M}_{m,n}(\mathbb{F})$ that contains much of their essence. After all, tensors themselves are quite abstract and will take some time to get our heads around, so it will be useful to have this very concrete motivating operation in our minds when we are introduced to them.

© Springer Nature Switzerland AG 2021
N. Johnston, *Advanced Linear and Matrix Algebra*,
https://doi.org/10.1007/978-3-030-52815-7_3

3.1.1 Definition and Basic Properties

The Kronecker product can be thought of simply as a way of multiplying matrices together (regardless of their sizes) so as to get much larger matrices.

Definition 3.1.1

Kronecker Product

Recall that $a_{i,j}$ is the (i,j)-entry of A.

The **Kronecker product** of matrices $A \in \mathcal{M}_{m,n}$ and $B \in \mathcal{M}_{p,q}$ is the block matrix

$$A \otimes B \overset{\text{def}}{=} \begin{bmatrix} a_{1,1}B & a_{1,2}B & \cdots & a_{1,n}B \\ a_{2,1}B & a_{2,2}B & \cdots & a_{2,n}B \\ \vdots & \vdots & \ddots & \vdots \\ a_{m,1}B & a_{m,2}B & \cdots & a_{m,n}B \end{bmatrix} \in \mathcal{M}_{mp,nq}.$$

In particular, notice that $A \otimes B$ is defined no matter what the sizes of A and B are (i.e., we do not need to make sure that the sizes of A and B are compatible with each other like we do with standard matrix multiplication). As a result of this, we can also apply the Kronecker product to vectors simply by thinking of them as $1 \times n$ or $m \times 1$ matrices. For this reason, we similarly say that if $\mathbf{v} \in \mathbb{F}^m$ and $\mathbf{w} \in \mathbb{F}^n$ then the Kronecker product of $\mathbf{v}$ and $\mathbf{w}$ is

$$\mathbf{v} \otimes \mathbf{w} \overset{\text{def}}{=} (v_1\mathbf{w}, v_2\mathbf{w}, \ldots, v_m\mathbf{w})$$
$$= (v_1w_1, \ldots, v_1w_n,\ v_2w_1, \ldots, v_2w_n,\ \ldots,\ v_mw_1, \ldots, v_mw_n).$$

We now compute a few Kronecker products to make ourselves more comfortable with these ideas.

Example 3.1.1

Numerical Examples of the Kronecker Product

Suppose $A = \begin{bmatrix} 1 & 2 \\ 3 & 4 \end{bmatrix}$, $B = \begin{bmatrix} 2 & 1 \\ 0 & 1 \end{bmatrix}$, $\mathbf{v} = \begin{bmatrix} 1 \\ 2 \end{bmatrix}$, and $\mathbf{w} = \begin{bmatrix} 3 \\ 4 \end{bmatrix}$. Compute:

a) $A \otimes B$,
b) $B \otimes A$, and
c) $\mathbf{v} \otimes \mathbf{w}$.

As always, the bars that we use to partition these block matrices are just provided for ease of visualization—they have no mathematical meaning.

Solutions:

a) $A \otimes B = \begin{bmatrix} B & 2B \\ 3B & 4B \end{bmatrix} = \left[\begin{array}{cc|cc} 2 & 1 & 4 & 2 \\ 0 & 1 & 0 & 2 \\ \hline 6 & 3 & 8 & 4 \\ 0 & 3 & 0 & 4 \end{array} \right].$

b) $B \otimes A = \begin{bmatrix} 2A & A \\ O & A \end{bmatrix} = \left[\begin{array}{cc|cc} 2 & 4 & 1 & 2 \\ 6 & 8 & 3 & 4 \\ \hline 0 & 0 & 1 & 2 \\ 0 & 0 & 3 & 4 \end{array} \right].$

c) $\mathbf{v} \otimes \mathbf{w} = \begin{bmatrix} \mathbf{w} \\ 2\mathbf{w} \end{bmatrix} = \begin{bmatrix} 3 \\ 4 \\ 6 \\ 8 \end{bmatrix}.$

We will see in Theorem 3.1.8 that, even though $A \otimes B$ and $B \otimes A$ are typically not equal, they share many of the same properties.

The above example shows that the Kronecker product is not commutative in general: $A \otimes B$ might not equal $B \otimes A$. However, it does have most of the other "standard" properties that we would expect a matrix product to have, and in particular it interacts with matrix addition and scalar multiplication exactly how we would hope that it does:

Suppose A, B, and C are matrices with sizes such that the operations below make sense, and let $c \in \mathbb{F}$ be a scalar. Then

a) $(A \otimes B) \otimes C = A \otimes (B \otimes C)$ (associativity)
b) $A \otimes (B + C) = A \otimes B + A \otimes C$ (left distributivity)
c) $(A + B) \otimes C = A \otimes C + B \otimes C$ (right distributivity)
d) $(cA) \otimes B = A \otimes (cB) = c(A \otimes B)$

Proof. The proofs of all of these statements are quite similar to each other, so we only explicitly prove part (b)—the remaining parts of the theorem are left to Exercise 3.1.20.

To this end, we just fiddle around with block matrices a bit:

$$A \otimes (B + C) = \begin{bmatrix} a_{1,1}(B+C) & a_{1,2}(B+C) & \cdots & a_{1,n}(B+C) \\ a_{2,1}(B+C) & a_{2,2}(B+C) & \cdots & a_{2,n}(B+C) \\ \vdots & \vdots & \ddots & \vdots \\ a_{m,1}(B+C) & a_{m,2}(B+C) & \cdots & a_{m,n}(B+C) \end{bmatrix}$$

$$= \begin{bmatrix} a_{1,1}B & a_{1,2}B & \cdots & a_{1,n}B \\ a_{2,1}B & a_{2,2}B & \cdots & a_{2,n}B \\ \vdots & \vdots & \ddots & \vdots \\ a_{m,1}B & a_{m,2}B & \cdots & a_{m,n}B \end{bmatrix} + \begin{bmatrix} a_{1,1}C & a_{1,2}C & \cdots & a_{1,n}C \\ a_{2,1}C & a_{2,2}C & \cdots & a_{2,n}C \\ \vdots & \vdots & \ddots & \vdots \\ a_{m,1}C & a_{m,2}C & \cdots & a_{m,n}C \end{bmatrix}$$

$$= A \otimes B + A \otimes C,$$

as desired. ∎

In particular, associativity of the Kronecker product (i.e., property (a) of the above theorem) tells us that we can unambiguously define **Kronecker powers** of matrices by taking the Kronecker product of a matrix with itself repeatedly, without having to worry about the exact order in which we perform those products. That is, for any integer $p \geq 1$ we can define

> Associativity of the Kronecker product also tells us that expressions like $A \otimes B \otimes C$ make sense.

$$A^{\otimes p} \stackrel{\text{def}}{=} \underbrace{A \otimes A \otimes \cdots \otimes A}_{p \text{ copies}}.$$

Kronecker powers increase in size extremely quickly, since increasing the power by 1 multiplies the number of rows and columns in the result by the number of rows and columns in A, respectively.

Example 3.1.2

Numerical Examples of Kronecker Powers

Suppose $H = \begin{bmatrix} 1 & 1 \\ 1 & -1 \end{bmatrix}$ and $I = \begin{bmatrix} 1 & 0 \\ 0 & 1 \end{bmatrix}$. Compute:

a) $I^{\otimes 2}$,
b) $H^{\otimes 2}$, and
c) $H^{\otimes 3}$.

Solutions:

a) $I^{\otimes 2} = I \otimes I = \begin{bmatrix} I & O \\ O & I \end{bmatrix} = \begin{bmatrix} 1 & 0 & 0 & 0 \\ 0 & 1 & 0 & 0 \\ 0 & 0 & 1 & 0 \\ 0 & 0 & 0 & 1 \end{bmatrix}$.

b) $H^{\otimes 2} = H \otimes H = \begin{bmatrix} H & H \\ H & -H \end{bmatrix} = \begin{bmatrix} 1 & 1 & 1 & 1 \\ 1 & -1 & 1 & -1 \\ 1 & 1 & -1 & -1 \\ 1 & -1 & -1 & 1 \end{bmatrix}$.

We could also compute $H^{\otimes 3} = (H^{\otimes 2}) \otimes H$. We would get the same answer.

c) $H^{\otimes 3} = H \otimes (H^{\otimes 2}) = \begin{bmatrix} H^{\otimes 2} & H^{\otimes 2} \\ H^{\otimes 2} & -H^{\otimes 2} \end{bmatrix}$

$$= \begin{bmatrix} 1 & 1 & 1 & 1 & 1 & 1 & 1 & 1 \\ 1 & -1 & 1 & -1 & 1 & -1 & 1 & -1 \\ 1 & 1 & -1 & -1 & 1 & 1 & -1 & -1 \\ 1 & -1 & -1 & 1 & 1 & -1 & -1 & 1 \\ 1 & 1 & 1 & 1 & -1 & -1 & -1 & -1 \\ 1 & -1 & 1 & -1 & -1 & 1 & -1 & 1 \\ 1 & 1 & -1 & -1 & -1 & -1 & 1 & 1 \\ 1 & -1 & -1 & 1 & -1 & 1 & 1 & -1 \end{bmatrix}.$$

Remark 3.1.1

Hadamard Matrices

Notice that the matrices H, $H^{\otimes 2}$, and $H^{\otimes 3}$ from Example 3.1.2 all have the following two properties: their entries are all ± 1, and their columns are mutually orthogonal. Matrices with these properties are called **Hadamard matrices**, and the Kronecker product gives one method of constructing them: $H^{\otimes k}$ is a Hadamard matrix for all $k \geq 1$.

One of the longest-standing unsolved questions in linear algebra asks which values of n are such that there exists an $n \times n$ Hadamard matrix. The above argument shows that they exist whenever $n = 2^k$ for some $k \geq 1$ (since $H^{\otimes k}$ is a $2^k \times 2^k$ matrix), but it is expected that they exist whenever n is a multiple of 4. For example, here is a 12×12 Hadamard matrix that cannot be constructed via the Kronecker product:

Entire books have been written about Hadamard matrices and the various ways of constructing them (Aga85, Hor06).

$$\begin{bmatrix} 1 & -1 & -1 & -1 & -1 & -1 & -1 & -1 & -1 & -1 & -1 & -1 \\ 1 & 1 & -1 & 1 & -1 & -1 & -1 & 1 & 1 & 1 & -1 & 1 \\ 1 & 1 & 1 & -1 & 1 & -1 & -1 & -1 & 1 & 1 & 1 & -1 \\ 1 & -1 & 1 & 1 & -1 & 1 & -1 & -1 & -1 & 1 & 1 & 1 \\ 1 & 1 & -1 & 1 & 1 & -1 & 1 & -1 & -1 & -1 & 1 & 1 \\ 1 & 1 & 1 & -1 & 1 & 1 & -1 & 1 & -1 & -1 & -1 & 1 \\ 1 & 1 & 1 & 1 & -1 & 1 & 1 & -1 & 1 & -1 & -1 & -1 \\ 1 & -1 & 1 & 1 & 1 & -1 & 1 & 1 & -1 & 1 & -1 & -1 \\ 1 & -1 & -1 & 1 & 1 & 1 & -1 & 1 & 1 & -1 & 1 & -1 \\ 1 & -1 & -1 & -1 & 1 & 1 & 1 & -1 & 1 & 1 & -1 & 1 \\ 1 & 1 & -1 & -1 & -1 & 1 & 1 & 1 & -1 & 1 & 1 & -1 \\ 1 & -1 & 1 & -1 & -1 & -1 & 1 & 1 & 1 & -1 & 1 & 1 \end{bmatrix}.$$

Numerous different methods of constructing Hadamard matrices are now known, and currently the smallest n for which it is not known if there exists a Hadamard matrix is $n = 668$.

Notice that, in part (a) of the above example, the Kronecker product of two identity matrices was simply a larger identity matrix—this happens in general, regardless of the sizes of the identity matrices in the product. Similarly, the Kronecker product of two diagonal matrices is always diagonal.

We look at some other sets of matrices that are preserved by the Kronecker product shortly, in Theorem 3.1.4.

The Kronecker product also plays well with usual matrix multiplication and other operations like the transpose and inverse. We summarize these additional properties here:

Theorem 3.1.2

Algebraic Properties of the Kronecker Product

Suppose A, B, C, and D are matrices with sizes such that the operations below make sense. Then

a) $(A \otimes B)(C \otimes D) = (AC) \otimes (BD)$,
b) $(A \otimes B)^{-1} = A^{-1} \otimes B^{-1}$, if either side of this expression exists
c) $(A \otimes B)^T = A^T \otimes B^T$, and
d) $(A \otimes B)^* = A^* \otimes B^*$ if the matrices are complex.

Proof. The proofs of all of these statements are quite similar to each other and follow directly from the definitions of the relevant operations, so we only explicitly prove part (a)—the remaining parts of the theorem are left to Exercise 3.1.21.

To see why part (a) of the theorem holds, we compute $(A \otimes B)(C \otimes D)$ via block matrix multiplication:

So that these matrix multiplications actually make sense, we are assuming that $A \in \mathcal{M}_{m,n}$ and $C \in \mathcal{M}_{n,p}$.

$$(A \otimes B)(C \otimes D) = \begin{bmatrix} a_{1,1}B & a_{1,2}B & \cdots & a_{1,n}B \\ a_{2,1}B & a_{2,2}B & \cdots & a_{2,n}B \\ \vdots & \vdots & \ddots & \vdots \\ a_{m,1}B & a_{m,2}B & \cdots & a_{m,n}B \end{bmatrix} \begin{bmatrix} c_{1,1}D & c_{1,2}D & \cdots & c_{1,p}D \\ c_{2,1}D & c_{2,2}D & \cdots & c_{2,p}D \\ \vdots & \vdots & \ddots & \vdots \\ c_{n,1}D & c_{n,2}D & \cdots & c_{n,p}D \end{bmatrix}$$

This calculation looks ugly, but really it's just applying the definition of matrix multiplication multiple times.

$$= \begin{bmatrix} \sum_{j=1}^n a_{1,j}c_{j,1}BD & \sum_{j=1}^n a_{1,j}c_{j,2}BD & \cdots & \sum_{j=1}^n a_{1,j}c_{j,p}BD \\ \sum_{j=1}^n a_{2,j}c_{j,1}BD & \sum_{j=1}^n a_{2,j}c_{j,2}BD & \cdots & \sum_{j=1}^n a_{2,j}c_{j,p}BD \\ \vdots & \vdots & \ddots & \vdots \\ \sum_{j=1}^n a_{m,j}c_{j,1}BD & \sum_{j=1}^n a_{m,j}c_{j,2}BD & \cdots & \sum_{j=1}^n a_{m,j}c_{j,p}BD \end{bmatrix}$$

$$= \begin{bmatrix} \sum_{j=1}^n a_{1,j}c_{j,1} & \sum_{j=1}^n a_{1,j}c_{j,2} & \cdots & \sum_{j=1}^n a_{1,j}c_{j,p} \\ \sum_{j=1}^n a_{2,j}c_{j,1} & \sum_{j=1}^n a_{2,j}c_{j,2} & \cdots & \sum_{j=1}^n a_{2,j}c_{j,p} \\ \vdots & \vdots & \ddots & \vdots \\ \sum_{j=1}^n a_{m,j}c_{j,1} & \sum_{j=1}^n a_{m,j}c_{j,2} & \cdots & \sum_{j=1}^n a_{m,j}c_{j,p} \end{bmatrix} \otimes (BD)$$

$$= (AC) \otimes (BD),$$

as desired. ■

It is worth noting that Theorems 3.1.1 and 3.1.2 still work if we replace all of the matrices by vectors, since we can think of those vectors as $1 \times n$ or $m \times 1$ matrices. Doing this in parts (a) and (d) of the above theorem shows us that if $\mathbf{v}, \mathbf{w} \in \mathbb{F}^n$ and $\mathbf{x}, \mathbf{y} \in \mathbb{F}^m$ are (column) vectors, then

$$(\mathbf{v} \otimes \mathbf{x}) \cdot (\mathbf{w} \otimes \mathbf{y}) = (\mathbf{v} \otimes \mathbf{x})^*(\mathbf{w} \otimes \mathbf{y}) = (\mathbf{v}^*\mathbf{w})(\mathbf{x}^*\mathbf{y}) = (\mathbf{v} \cdot \mathbf{w})(\mathbf{x} \cdot \mathbf{y}).$$

In other words, the dot product of two Kronecker products is just the product of the individual dot products. In particular, this means that $\mathbf{v} \otimes \mathbf{x}$ is orthogonal to $\mathbf{w} \otimes \mathbf{y}$ if and only if $\mathbf{v}$ is orthogonal to $\mathbf{w}$ or $\mathbf{x}$ is orthogonal to $\mathbf{y}$ (or both).

Properties Preserved by the Kronecker Product

Because the Kronecker product and Kronecker powers can create such large matrices so quickly, it is important to understand how properties of $A \otimes B$ relate to the corresponding properties of A and B themselves. For example, the following theorem shows that we can compute the eigenvalues, determinant, and trace of $A \otimes B$ directly from A and B themselves.

Theorem 3.1.3 **Eigenvalues, Trace, and Determinant of the Kronecker Product**	Suppose $A \in \mathcal{M}_m$ and $B \in \mathcal{M}_n$. a) If λ and μ are eigenvalues of A and B, respectively, with corresponding eigenvectors $\mathbf{v}$ and $\mathbf{w}$, then $\lambda\mu$ is an eigenvalue of $A \otimes B$ with corresponding eigenvector $\mathbf{v} \otimes \mathbf{w}$, b) $\operatorname{tr}(A \otimes B) = \operatorname{tr}(A)\operatorname{tr}(B)$, and c) $\det(A \otimes B) = \big(\det(A)\big)^n \big(\det(B)\big)^m$.

Proof. Part (a) of the theorem follows almost immediately from the fact that the Kronecker product plays well with matrix multiplication and scalar multiplication (i.e., Theorems 3.1.1 and 3.1.2):

$$(A \otimes B)(\mathbf{v} \otimes \mathbf{w}) = (A\mathbf{v}) \otimes (B\mathbf{w}) = (\lambda\mathbf{v}) \otimes (\mu\mathbf{w}) = \lambda\mu(\mathbf{v} \otimes \mathbf{w}),$$

so $\mathbf{v} \otimes \mathbf{w}$ is an eigenvector of $A \otimes B$ corresponding to eigenvalue $\lambda\mu$, as claimed.

Note in particular that parts (b) and (c) of this theorem imply that $\operatorname{tr}(A \otimes B) = \operatorname{tr}(B \otimes A)$ and $\det(A \otimes B) = \det(B \otimes A)$.

Part (b) follows directly from the definition of the Kronecker product:

$$\operatorname{tr}(A \otimes B) = \operatorname{tr}\left(\begin{bmatrix} a_{1,1}B & a_{1,2}B & \cdots & a_{1,n}B \\ a_{2,1}B & a_{2,2}B & \cdots & a_{2,n}B \\ \vdots & \vdots & \ddots & \vdots \\ a_{m,1}B & a_{m,2}B & \cdots & a_{m,n}B \end{bmatrix}\right)$$

$$= a_{1,1}\operatorname{tr}(B) + a_{2,2}\operatorname{tr}(B) + \cdots + a_{n,n}\operatorname{tr}(B) = \operatorname{tr}(A)\operatorname{tr}(B).$$

Finally, for part (c) we note that

$$\det(A \otimes B) = \det\big((A \otimes I_n)(I_m \otimes B)\big) = \det(A \otimes I_n)\det(I_m \otimes B).$$

We will see another way to prove this determinant equality (as long as $\mathbb{F} = \mathbb{R}$ or $\mathbb{F} = \mathbb{C}$) in Exercise 3.1.9.

Since $I_m \otimes B$ is block diagonal, its determinant just equals the product of the determinants of its diagonal blocks:

$$\det(I_m \otimes B) = \det\left(\begin{bmatrix} B & O & \cdots & O \\ O & B & \cdots & O \\ \vdots & \vdots & \ddots & \vdots \\ O & O & \cdots & B \end{bmatrix}\right) = \big(\det(B)\big)^m.$$

A similar argument shows that $\det(A \otimes I_n) = \det(A)^n$, so we get $\det(A \otimes B) = \det(A)^n \det(B)^m$, as claimed. $\blacksquare$

The Kronecker product also preserves several useful families of matrices. For example, it follows straightforwardly from the definition of the Kronecker product that if $A \in \mathcal{M}_m$ and $B \in \mathcal{M}_n$ are both upper triangular, then so is $A \otimes B$. We summarize some observations of this type in the following theorem:

Theorem 3.1.4

Matrix Properties Preserved by the Kronecker Product

Suppose $A \in \mathcal{M}_m$ and $B \in \mathcal{M}_n$.

a) If A and B are upper (lower) triangular, so is $A \otimes B$,
b) If A and B are diagonal, so is $A \otimes B$,
c) If A and B are normal, so is $A \otimes B$,
d) If A and B are unitary, so is $A \otimes B$,
e) If A and B are symmetric or Hermitian, so is $A \otimes B$, and
f) If A and B are positive (semi)definite, so is $A \otimes B$.

We leave the proof of the above theorem to Exercise 3.1.22, as none of it is very difficult or enlightening—the claims can each be proved in a line or two simply by invoking properties of the Kronecker product that we saw earlier.

In particular, because the matrix properties described by Theorem 3.1.4 are exactly the ones used in most of the matrix decompositions we have seen, it follows that the Kronecker product interacts with most matrix decompositions just the way we would hope for it to. For example, if $A \in \mathcal{M}_m(\mathbb{C})$ and $B \in \mathcal{M}_n(\mathbb{C})$ have Schur triangularizations

$$A = U_1 T_1 U_1^* \quad \text{and} \quad B = U_2 T_2 U_2^*$$

(where U_1 and U_2 are unitary, and T_1 and T_2 are upper triangular), then to find a Schur triangularization of $A \otimes B$ we can simply compute the Kronecker products $U_1 \otimes U_2$ and $T_1 \otimes T_2$, since

$$A \otimes B = (U_1 T_1 U_1^*) \otimes (U_2 T_2 U_2^*) = (U_1 \otimes U_2)(T_1 \otimes T_2)(U_1 \otimes U_2)^*.$$

An analogous argument shows that the Kronecker product also preserves diagonalizations of matrices (in the sense of Theorem 2.0.1), as well as spectral decompositions, QR decompositions, singular value decompositions, and polar decompositions.

Because these decompositions behave so well under the Kronecker product, we can use them to get our hands on any matrix properties that can be inferred from these decompositions. For example, by looking at how the Kronecker product interacts with the singular value decomposition, we arrive at the following theorem:

Theorem 3.1.5

Kronecker Product of Singular Value Decompositions

Suppose $A \in \mathcal{M}_{m,n}$ and $B \in \mathcal{M}_{p,q}$.

a) If σ and τ are singular values of A and B, respectively, then $\sigma\tau$ is a singular value of $A \otimes B$,
b) $\text{rank}(A \otimes B) = \text{rank}(A)\text{rank}(B)$,
c) $\text{range}(A \otimes B) = \text{span}\{\mathbf{v} \otimes \mathbf{w} : \mathbf{v} \in \text{range}(A), \mathbf{w} \in \text{range}(B)\}$,
d) $\text{null}(A \otimes B) = \text{span}\{\mathbf{v} \otimes \mathbf{w} : \mathbf{v} \in \text{null}(A), \mathbf{w} \in \text{null}(B)\}$, and
e) $\|A \otimes B\| = \|A\|\|B\|$ and $\|A \otimes B\|_F = \|A\|_F\|B\|_F$.

Again, all of these properties follow fairly quickly from the relevant definitions and the fact that if A and B have singular value decompositions $A = U_1 \Sigma_1 V_1^*$ and $B = U_2 \Sigma_2 V_2^*$, then

$$A \otimes B = (U_1 \Sigma_1 V_1^*) \otimes (U_2 \Sigma_2 V_2^*) = (U_1 \otimes U_2)(\Sigma_1 \otimes \Sigma_2)(V_1 \otimes V_2)^*$$

is a singular value decomposition of $A \otimes B$. We thus leave the proof of the above theorem to Exercise 3.1.23.

The one notable matrix decomposition that does not behave quite so cleanly under the Kronecker product is the Jordan decomposition. In particular, if $J_1 \in \mathcal{M}_m(\mathbb{C})$ and $J_2 \in \mathcal{M}_n(\mathbb{C})$ are two matrices in Jordan canonical form then $J_1 \otimes J_2$ may not be in Jordan canonical form. For example, if

$$J_1 = J_2 = \begin{bmatrix} 1 & 1 \\ 0 & 1 \end{bmatrix} \quad \text{then} \quad J_1 \otimes J_2 = \left[\begin{array}{cc|cc} 1 & 1 & 1 & 1 \\ 0 & 1 & 0 & 1 \\ \hline 0 & 0 & 1 & 1 \\ 0 & 0 & 0 & 1 \end{array} \right],$$

which is not in Jordan canonical form.

Bases and Linear Combinations of Kronecker Products

The fact that some vectors cannot be written in the form $\mathbf{v} \otimes \mathbf{w}$ is exactly why we needed the "span" in conditions (c) and (d) of Theorem 3.1.5.

The Kronecker product of vectors in $\mathbb{F}^m$ and $\mathbb{F}^n$ can only be used to construct a tiny portion of all vectors in $\mathbb{F}^{mn}$ (and similarly, most matrices in $\mathcal{M}_{mp,nq}$ cannot be written in the form $A \otimes B$). For example, there do not exist vectors $\mathbf{v}, \mathbf{w} \in \mathbb{C}^2$ such that $\mathbf{v} \otimes \mathbf{w} = (1, 0, 0, 1)$, since that would imply $v_1 \mathbf{w} = (1, 0)$ and $v_2 \mathbf{w} = (0, 1)$, but $(1, 0)$ and $(0, 1)$ are not scalar multiples of each other.

However, it *is* the case that every vector in $\mathbb{F}^{mn}$ can be written as a *linear combination* of vectors of the form $\mathbf{v} \otimes \mathbf{w}$ with $\mathbf{v} \in \mathbb{F}^m$ and $\mathbf{w} \in \mathbb{F}^n$. In fact, if B and C are bases of $\mathbb{F}^m$ and $\mathbb{F}^n$, respectively, then the set

$$B \otimes C \stackrel{\text{def}}{=} \{ \mathbf{v} \otimes \mathbf{w} : \mathbf{v} \in B, \ \mathbf{w} \in C \}$$

is a basis of $\mathbb{F}^{mn}$ (see Exercise 3.1.17, and note that a similar statement holds for matrices in $\mathcal{M}_{mp,nq}$ being written as a linear combination of matrices of the form $A \otimes B$). In the special case when B and C are the *standard* bases of $\mathbb{F}^m$ and $\mathbb{F}^n$, respectively, $B \otimes C$ is also the standard basis of $\mathbb{F}^{mn}$. Furthermore, ordering the basis vectors $\mathbf{e}_i \otimes \mathbf{e}_j$ by placing their subscripts in lexicographical order produces exactly the "usual" ordering of the standard basis vectors of $\mathbb{F}^{mn}$. For example, if $m = 2$ and $n = 3$ then

In other words, if we "count" in the subscripts of $\mathbf{e}_i \otimes \mathbf{e}_j$, but with each digit starting at 1 instead of 0, we get the usual ordering of these basis vectors.

$$\begin{aligned}
\mathbf{e}_1 \otimes \mathbf{e}_1 &= (1, 0) \otimes (1, 0, 0) = (1, 0, 0, 0, 0, 0), \\
\mathbf{e}_1 \otimes \mathbf{e}_2 &= (1, 0) \otimes (0, 1, 0) = (0, 1, 0, 0, 0, 0), \\
\mathbf{e}_1 \otimes \mathbf{e}_3 &= (1, 0) \otimes (0, 0, 1) = (0, 0, 1, 0, 0, 0), \\
\mathbf{e}_2 \otimes \mathbf{e}_1 &= (0, 1) \otimes (1, 0, 0) = (0, 0, 0, 1, 0, 0), \\
\mathbf{e}_2 \otimes \mathbf{e}_2 &= (0, 1) \otimes (0, 1, 0) = (0, 0, 0, 0, 1, 0), \\
\mathbf{e}_2 \otimes \mathbf{e}_3 &= (0, 1) \otimes (0, 0, 1) = (0, 0, 0, 0, 0, 1).
\end{aligned}$$

In fact, this same observation works when taking the Kronecker product of 3 or more standard basis vectors as well.

When working with vectors that are (linear combinations of) Kronecker products of other vectors, we typically want to know what the dimensions of the different factors of the Kronecker product are. For example, if we say that $\mathbf{v} \otimes \mathbf{w} \in \mathbb{F}^6$, we might wonder whether $\mathbf{v}$ and $\mathbf{w}$ live in 2- and 3-dimensional spaces, respectively, or 3- and 2-dimensional spaces. To alleviate this issue, we use the notation $\mathbb{F}^m \otimes \mathbb{F}^n$ to mean $\mathbb{F}^{mn}$, but built out of Kronecker products of vectors from $\mathbb{F}^m$ and $\mathbb{F}^n$, in that order (and we similarly use the notation $\mathcal{M}_{m,n} \otimes \mathcal{M}_{p,q}$ for $\mathcal{M}_{mp,nq}$). When working with the Kronecker product of many vectors, we often use the shorthand notation

$$(\mathbb{F}^n)^{\otimes p} \stackrel{\text{def}}{=} \underbrace{\mathbb{F}^n \otimes \mathbb{F}^n \otimes \cdots \otimes \mathbb{F}^n}_{p \text{ copies}}.$$

We will clarify what the word "tensor" means in the coming sections.

Furthermore, we say that any vector $\mathbf{v} = \mathbf{v}_1 \otimes \mathbf{v}_2 \otimes \cdots \otimes \mathbf{v}_p \in (\mathbb{F}^n)^{\otimes p}$ that can be written as a Kronecker product (rather than as a *linear combination* of Kronecker products) is an **elementary tensor**.

3.1.2 Vectorization and the Swap Matrix

We now note that the space $\mathbb{F}^m \otimes \mathbb{F}^n$ is "essentially the same" as $\mathcal{M}_{m,n}(\mathbb{F})$. To clarify what we mean by this, we recall that we can represent matrices in $\mathcal{M}_{m,n}$ via their coordinate vectors with respect to a given basis. In particular, if we use the standard basis $E = \{E_{1,1}, E_{1,2}, \ldots, E_{m,n}\}$, then this coordinate vector has a very special form—it just contains the entries of the matrix, read row-by-row. This leads to the following definition:

Recall that $E_{i,j}$ has a 1 in its (i,j)-entry and zeros elsewhere.

Definition 3.1.2

Vectorization

Suppose $A \in \mathcal{M}_{m,n}$. The **vectorization** of A, denoted by $\text{vec}(A)$, is the vector in $\mathbb{F}^{mn}$ that is obtained by reading the entries of A row-by-row.

We typically think of vectorization as the "most natural" isomorphism from $\mathcal{M}_{m,n}(\mathbb{F})$ to $\mathbb{F}^{mn}$—we do not do anything "fancy" to transform the matrix into a vector, but rather we just read it as we see it. In the other direction, we define the **matricization** of a vector $\mathbf{v} \in \mathbb{F}^{mn}$, denoted by $\text{mat}(\mathbf{v})$, to be the matrix whose vectorization if $\mathbf{v}$ (i.e., the linear transformation $\text{mat} : \mathbb{F}^{mn} \to \mathcal{M}_{m,n}$ is defined by $\text{mat} = \text{vec}^{-1}$). In other words, matricization is the operation that places the entries of a vector row-by-row into a matrix. For example, in the $m = n = 2$ case we have

Some other books define vectorization as reading the entries of the matrix column-by-column instead of row-by-row, but that convention makes some of the upcoming formulas a bit uglier.

$$\text{vec}\left(\begin{bmatrix} a & b \\ c & d \end{bmatrix}\right) = \begin{bmatrix} a \\ b \\ c \\ d \end{bmatrix} \quad \text{and} \quad \text{mat}\left(\begin{bmatrix} a \\ b \\ c \\ d \end{bmatrix}\right) = \begin{bmatrix} a & b \\ c & d \end{bmatrix}.$$

While vectorization is easy to work with when given an explicit matrix, in this "read a matrix row-by-row" form it is a bit difficult to prove things about it and work with it abstractly. The following theorem provides an alternate characterization of vectorization that is often more convenient.

Theorem 3.1.6

Vectorization and the Kronecker Product

If $\mathbf{v} \in \mathbb{F}^m$ and $\mathbf{w} \in \mathbb{F}^n$ are column vectors then $\text{vec}(\mathbf{v}\mathbf{w}^T) = \mathbf{v} \otimes \mathbf{w}$.

Proof. One way to see why this holds is to compute $\mathbf{v}\mathbf{w}^T$ via block matrix multiplication as follows:

Yes, we really want $\mathbf{w}^T$ (not $\mathbf{w}^*$) in this theorem, even if $\mathbb{F} = \mathbb{C}$. The reason is that vectorization is linear (like transposition), not conjugate linear.

$$\mathbf{v}\mathbf{w}^T = \begin{bmatrix} v_1 \mathbf{w}^T \\ v_2 \mathbf{w}^T \\ \vdots \\ v_m \mathbf{w}^T \end{bmatrix}.$$

Since $\text{vec}(\mathbf{v}\mathbf{w}^T)$ places the entries of $\mathbf{v}\mathbf{w}^T$ into a vector one row at a time, we see that its first n entries are the same as those of $v_1\mathbf{w}^T$ (which are the same as those of $v_1\mathbf{w}$), its next n entries are the same as those of $v_2\mathbf{w}$, and so on. But this is exactly the definition of $\mathbf{v} \otimes \mathbf{w}$, so we are done. ∎

The above theorem could be flipped around to (equivalently) say that $\text{mat}(\mathbf{v} \otimes \mathbf{w}) = \mathbf{v}\mathbf{w}^T$. From this characterization of the Kronecker product, it is perhaps clearer why not every vector in $\mathbb{F}^m \otimes \mathbb{F}^n$ is an elementary tensor $\mathbf{v} \otimes \mathbf{w}$: those vectors correspond via matricization to the rank-1 matrices $\mathbf{v}\mathbf{w}^T \in \mathcal{M}_{m,n}(\mathbf{F})$, and not every matrix has rank 1.

Vectorization is useful because it lets us think about matrices as vectors, and linear transformations acting on matrices as (larger) matrices themselves. This is nothing new—we can think of a linear transformation on *any* finite-dimensional vector space in terms of its standard matrix. However, in this particular case the exact details of the computation are carried out by vectorization and the Kronecker product, which makes them much simpler to actually work with in practice.

> **Look back at Section 1.2.1 if you need a refresher on how to represent linear transformations as matrices.**

Theorem 3.1.7 **Vectorization of a Product**	Suppose $A \in \mathcal{M}_{p,m}$ and $B \in \mathcal{M}_{r,n}$. Then $$\text{vec}(AXB^T) = (A \otimes B)\text{vec}(X) \quad \text{for all} \quad X \in \mathcal{M}_{m,n}.$$

In other words, this theorem tells us about the standard matrix of the linear transformation $T_{A,B} : \mathcal{M}_{m,n} \to \mathcal{M}_{p,r}$ defined by $T_{A,B}(X) = AXB^T$, where $A \in \mathcal{M}_{p,m}$ and $B \in \mathcal{M}_{r,n}$ are fixed. In particular, it says that the standard matrix of $T_{A,B}$ (with respect to the standard basis $E = \{E_{1,1}, E_{1,2}, \ldots, E_{m,n}\}$) is simply $[T_{A,B}] = A \otimes B$.

> **Recall that when working with standard matrices with respect to the standard basis E, we use the shorthand $[T] = [T]_E$.**

Proof of Theorem 3.1.7 We start by showing that if $X = E_{i,j}$ for some i, j then $\text{vec}(AE_{i,j}B^T) = (A \otimes B)\text{vec}(E_{i,j})$. To see this, note that $E_{i,j} = \mathbf{e}_i\mathbf{e}_j^T$, so using Theorem 3.1.6 twice tells us that

$$\text{vec}(AE_{i,j}B^T) = \text{vec}(A\mathbf{e}_i\mathbf{e}_j^T B^T) = \text{vec}((A\mathbf{e}_i)(B\mathbf{e}_j)^T) = (A\mathbf{e}_i) \otimes (B\mathbf{e}_j)$$

$$= (A \otimes B)(\mathbf{e}_i \otimes \mathbf{e}_j) = (A \otimes B)\text{vec}(\mathbf{e}_i\mathbf{e}_j^T) = (A \otimes B)\text{vec}(E_{i,j}).$$

If we then use the fact that we can write as $X = \sum_{i,j} x_{i,j}E_{i,j}$, the result follows from the fact that vectorization is linear:

> **This proof technique is quite common when we want to show that a linear transformation acts in a certain way: we show that it acts that way on a basis, and then use linearity to show that it must do the same thing on the entire vector space.**

$$\text{vec}(AXB^T) = \text{vec}\left(A\left(\sum_{i,j} x_{i,j}E_{i,j}\right)B^T\right) = \sum_{i,j} x_{i,j}\text{vec}(AE_{i,j}B^T)$$

$$= \sum_{i,j} x_{i,j}(A \otimes B)\text{vec}(E_{i,j}) = (A \otimes B)\text{vec}\left(\sum_{i,j} x_{i,j}E_{i,j}\right)$$

$$= (A \otimes B)\text{vec}(X),$$

as desired. ∎

The above theorem is nothing revolutionary, but it is useful because it provides an explicit and concrete way of solving many problems that we have (at least implicitly) encountered before. For example, suppose we had a fixed matrix $A \in \mathcal{M}_n$ and we wanted to find all matrices $X \in \mathcal{M}_n$ that commute with it. One way to do this would be to multiply out AX and XA, set entries of those matrices equal to each other, and solve the resulting linear system. To make the details of this linear system more explicit, however, we can notice that $AX = XA$ if and only if $AX - XA = O$, which (by taking the vectorization of both sides of the equation and applying Theorem 3.1.7) is equivalent to

$$(A \otimes I - I \otimes A^T)\text{vec}(X) = \mathbf{0}.$$

This is a linear system that we can solve "directly", as we now illustrate with an example.

Example 3.1.3

Finding Matrices that Commute

Find all matrices $X \in \mathcal{M}_2$ that commute with $A = \begin{bmatrix} 1 & 1 \\ 0 & 0 \end{bmatrix}$.

Solution:

As noted above, one way to tackle this problem is to solve the linear system $(A \otimes I - I \otimes A^T)\text{vec}(X) = \mathbf{0}$. The coefficient matrix of this linear system is

$$A \otimes I - I \otimes A^T = \begin{bmatrix} 1 & 0 & 1 & 0 \\ 0 & 1 & 0 & 1 \\ 0 & 0 & 0 & 0 \\ 0 & 0 & 0 & 0 \end{bmatrix} - \begin{bmatrix} 1 & 0 & 0 & 0 \\ 1 & 0 & 0 & 0 \\ 0 & 0 & 1 & 0 \\ 0 & 0 & 1 & 0 \end{bmatrix} = \begin{bmatrix} 0 & 0 & 1 & 0 \\ -1 & 1 & 0 & 1 \\ 0 & 0 & -1 & 0 \\ 0 & 0 & -1 & 0 \end{bmatrix}.$$

To solve the corresponding linear system, we apply Gaussian elimination to it:

$$\begin{bmatrix} 0 & 0 & 1 & 0 & | & 0 \\ -1 & 1 & 0 & 1 & | & 0 \\ 0 & 0 & -1 & 0 & | & 0 \\ 0 & 0 & -1 & 0 & | & 0 \end{bmatrix} \xrightarrow{\text{row reduce}} \begin{bmatrix} 1 & -1 & 0 & -1 & | & 0 \\ 0 & 0 & 1 & 0 & | & 0 \\ 0 & 0 & 0 & 0 & | & 0 \\ 0 & 0 & 0 & 0 & | & 0 \end{bmatrix}.$$

From here we can see that, if we label the entries of $\text{vec}(X)$ as $\text{vec}(X) = (x_1, x_2, x_3, x_4)$, then $x_3 = 0$ and $-x_1 + x_2 + x_4 = 0$, so $x_1 = x_2 + x_4$ (x_2 and x_4 are free). It follows that $\text{vec}(X) = (x_2 + x_4, x_2, 0, x_4)$, so the matrices X that commute with A are exactly the ones of the form

$$X = \text{mat}\big((x_2 + x_4, x_2, 0, x_4)\big) = \begin{bmatrix} x_2 + x_4 & x_2 \\ 0 & x_4 \end{bmatrix},$$

where $x_2, x_4 \in \mathbb{F}$ are arbitrary.

As another application of Theorem 3.1.7, we now start building towards an explanation of exactly what the relationship between $A \otimes B$ and $B \otimes A$ is. In particular, we will show that there is a specific matrix W with the property that $A \otimes B = W(B \otimes A)W^T$:

Definition 3.1.3

Swap Matrix

Given positive integers m and n, the **swap matrix** $W_{m,n} \in \mathcal{M}_{mn}$ is the matrix defined in any of the three following (equivalent) ways:

a) $W_{m,n} = [T]$, the standard matrix of the transposition map $T : \mathcal{M}_{m,n} \to \mathcal{M}_{n,m}$ with respect to the standard basis E,

b) $W_{m,n}(\mathbf{e}_i \otimes \mathbf{e}_j) = \mathbf{e}_j \otimes \mathbf{e}_i$ for all $1 \leq i \leq m, 1 \leq j \leq n$, and

c) $W_{m,n} = \begin{bmatrix} E_{1,1} & E_{2,1} & \cdots & E_{m,1} \\ E_{1,2} & E_{2,2} & \cdots & E_{m,2} \\ \vdots & \vdots & \ddots & \vdots \\ E_{1,n} & E_{2,n} & \cdots & E_{m,n} \end{bmatrix}$.

The entries of $W_{m,n}$ are all just 0 or 1, and this result works over any field $\mathbb{F}$. If $\mathbb{F} = \mathbb{R}$ or $\mathbb{F} = \mathbb{C}$ then $W_{m,n}$ is unitary.

If the dimensions m and n are clear from context or irrelevant, we denote this matrix simply by W.

For example, we already showed in Example 1.2.10 that the standard matrix of the transpose map $T : \mathcal{M}_2 \to \mathcal{M}_2$ is

$$[T] = \begin{bmatrix} 1 & 0 & 0 & 0 \\ 0 & 0 & 1 & 0 \\ 0 & 1 & 0 & 0 \\ 0 & 0 & 0 & 1 \end{bmatrix},$$

We use W (instead of S) to denote the swap matrix because the letter "S" is going to become overloaded later in this section. We can think of "W" as standing for "s**W**ap".

so this is the swap matrix $W_{2,2}$. We can check that it satisfies definition (b) by directly computing each of $W_{2,2}(\mathbf{e}_1 \otimes \mathbf{e}_1)$, $W_{2,2}(\mathbf{e}_1 \otimes \mathbf{e}_2)$, $W_{2,2}(\mathbf{e}_2 \otimes \mathbf{e}_1)$, and $W_{2,2}(\mathbf{e}_2 \otimes \mathbf{e}_2)$. Similarly, it also satisfies definition (c) since we can write it as the block matrix

$$W_{2,2} = \begin{bmatrix} E_{1,1} & E_{2,1} \\ E_{1,2} & E_{2,2} \end{bmatrix} = \left[\begin{array}{cc|cc} 1 & 0 & 0 & 0 \\ 0 & 0 & 1 & 0 \\ \hline 0 & 1 & 0 & 0 \\ 0 & 0 & 0 & 1 \end{array} \right].$$

To see that these three definitions agree in general (i.e., when m and n do not necessarily both equal 2), we first compute

$$\begin{aligned} [T] &= \big[\, [E_{1,1}^T]_E \mid [E_{1,2}^T]_E \mid \cdots \mid [E_{m,n}^T]_E \,\big] & \text{(definition of } [T]) \\ &= \big[\, \text{vec}(E_{1,1}) \mid \text{vec}(E_{2,1}) \mid \cdots \mid \text{vec}(E_{n,m}) \,\big] & \text{(definition of vec)} \\ &= \big[\, \text{vec}(\mathbf{e}_1 \mathbf{e}_1^T) \mid \text{vec}(\mathbf{e}_2 \mathbf{e}_1^T) \mid \cdots \mid \text{vec}(\mathbf{e}_n \mathbf{e}_m^T) \,\big] & (E_{i,j} = \mathbf{e}_i \mathbf{e}_j^T \text{ for all } i, j) \\ &= \big[\, \mathbf{e}_1 \otimes \mathbf{e}_1 \mid \mathbf{e}_2 \otimes \mathbf{e}_1 \mid \cdots \mid \mathbf{e}_n \otimes \mathbf{e}_m \,\big]. & \text{(Theorem 3.1.6)} \end{aligned}$$

Some books call the swap matrix the **commutation matrix** and denote it by $K_{m,n}$.

The equivalence of definitions (a) and (b) of the swap matrix follows fairly quickly now, since multiplying a matrix by $\mathbf{e}_i \otimes \mathbf{e}_j$ just results in one of the columns of that matrix, and in this case we will have $[T](\mathbf{e}_i \otimes \mathbf{e}_j) = \mathbf{e}_j \otimes \mathbf{e}_i$. Similarly, the equivalence of definitions (a) and (c) follows just by explicitly writing out what the columns of the block matrix (c) are: they are exactly $\mathbf{e}_1 \otimes \mathbf{e}_1, \mathbf{e}_2 \otimes \mathbf{e}_1, \ldots, \mathbf{e}_n \otimes \mathbf{e}_m$, which we just showed are also the columns of $[T]$.

Example 3.1.4

Constructing Swap Matrices

Construct the swap matrix $W_{m,n} \in \mathcal{M}_{mn}$ when

a) $m = 2$ and $n = 3$, and when
b) $m = n = 3$.

Solutions:

a) We could use any of the three defining characterizations of the swap matrix to construct it, but the easiest one to use is likely characterization (c):

Note that swap matrices are *always* square, even if $m \neq n$.

$$W_{2,3} = \begin{bmatrix} E_{1,1} & E_{2,1} \\ E_{1,2} & E_{2,2} \\ E_{1,3} & E_{2,3} \end{bmatrix} = \left[\begin{array}{ccc|ccc} 1 & \cdot & \cdot & \cdot & \cdot & \cdot \\ \cdot & \cdot & \cdot & 1 & \cdot & \cdot \\ \cdot & 1 & \cdot & \cdot & \cdot & \cdot \\ \cdot & \cdot & \cdot & \cdot & 1 & \cdot \\ \cdot & \cdot & 1 & \cdot & \cdot & \cdot \\ \cdot & \cdot & \cdot & \cdot & \cdot & 1 \end{array} \right].$$

Here we use dots (·) instead of zeros for ease of visualization.

b) Again, we use characterization (c) to construct this swap matrix:

$$W_{3,3} = \begin{bmatrix} E_{1,1} & E_{2,1} & E_{3,1} \\ E_{1,2} & E_{2,2} & E_{3,2} \\ E_{1,3} & E_{2,3} & E_{3,3} \end{bmatrix} = \left[\begin{array}{ccc|ccc|ccc} 1 & \cdot & \cdot & \cdot & \cdot & \cdot & \cdot & \cdot & \cdot \\ \cdot & \cdot & \cdot & 1 & \cdot & \cdot & \cdot & \cdot & \cdot \\ \cdot & \cdot & \cdot & \cdot & \cdot & \cdot & 1 & \cdot & \cdot \\ \cdot & 1 & \cdot & \cdot & \cdot & \cdot & \cdot & \cdot & \cdot \\ \cdot & \cdot & \cdot & \cdot & 1 & \cdot & \cdot & \cdot & \cdot \\ \cdot & \cdot & \cdot & \cdot & \cdot & \cdot & \cdot & 1 & \cdot \\ \cdot & \cdot & 1 & \cdot & \cdot & \cdot & \cdot & \cdot & \cdot \\ \cdot & \cdot & \cdot & \cdot & \cdot & 1 & \cdot & \cdot & \cdot \\ \cdot & \cdot & \cdot & \cdot & \cdot & \cdot & \cdot & \cdot & 1 \end{array} \right]$$

Matrices (like the swap matrix) with a single 1 in each row and column, and zeros elsewhere, are called **permutation matrices**. Compare this with signed or complex permutation matrices from Theorem 1.D.10.

The swap matrix $W_{m,n}$ has some very nice properties, which we prove in Exercise 3.1.18. In particular, every row and column has a single non-zero entry (equal to 1), if $\mathbb{F} = \mathbb{R}$ or $\mathbb{F} = \mathbb{C}$ then it is unitary, and if $m = n$ then it is symmetric. If the dimensions m and n are clear from context or irrelevant, we just denote this matrix by W for simplicity.

The name "swap matrix" comes from the fact that it swaps the two factors in any Kronecker product: $W(\mathbf{v} \otimes \mathbf{w}) = \mathbf{w} \otimes \mathbf{v}$ for all $\mathbf{v} \in \mathbb{F}^m$ and $\mathbf{w} \in \mathbb{F}^n$. To see this, just write each of $\mathbf{v}$ and $\mathbf{w}$ as linear combinations of the standard basis vectors ($\mathbf{v} = \sum_i v_i \mathbf{e}_i$ and $\mathbf{w} = \sum_j w_j \mathbf{e}_j$) and then use characterization (b) of swap matrices:

$$W(\mathbf{v} \otimes \mathbf{w}) = W\left(\left(\sum_i v_i \mathbf{e}_i \right) \otimes \left(\sum_j w_j \mathbf{e}_j \right) \right) = \sum_{i,j} v_i w_j W(\mathbf{e}_i \otimes \mathbf{e}_j)$$

$$= \sum_{i,j} v_i w_j \mathbf{e}_j \otimes \mathbf{e}_i = \left(\sum_j w_j \mathbf{e}_j \right) \otimes \left(\sum_i v_i \mathbf{e}_i \right) = \mathbf{w} \otimes \mathbf{v}.$$

More generally, the following theorem shows that swap matrices also solve exactly the problem that we introduced them to solve—they can be used to transform $A \otimes B$ into $B \otimes A$:

Theorem 3.1.8

Almost-Commutativity of the Kronecker Product

Suppose $A \in \mathcal{M}_{m,n}$ and $B \in \mathcal{M}_{p,q}$. Then $B \otimes A = W_{m,p}(A \otimes B)W_{n,q}^T$.

Proof. Notice that if we write A and B in terms of their columns as $A = \begin{bmatrix} \mathbf{a}_1 & \mathbf{a}_2 & \cdots & \mathbf{a}_n \end{bmatrix}$ and $B = \begin{bmatrix} \mathbf{b}_1 & \mathbf{b}_2 & \cdots & \mathbf{b}_q \end{bmatrix}$, respectively, then

Alternatively, we could use Theorem A.1.3 here to write A and B as a sum of rank-1 matrices.

$$A = \sum_{i=1}^{n} \mathbf{a}_i \mathbf{e}_i^T \quad \text{and} \quad B = \sum_{j=1}^{q} \mathbf{b}_j \mathbf{e}_j^T.$$

Then we can use the fact that $W(\mathbf{a} \otimes \mathbf{b}) = \mathbf{b} \otimes \mathbf{a}$ to see that

$$W_{m,p}(A \otimes B)W_{n,q}^T = W_{m,p}\left(\left(\sum_{i=1}^n \mathbf{a}_i \mathbf{e}_i^T \right) \otimes \left(\sum_{j=1}^q \mathbf{b}_j \mathbf{e}_j^T \right) \right) W_{n,q}^T$$

$$= \sum_{i=1}^n \sum_{j=1}^q W_{m,p}(\mathbf{a}_i \otimes \mathbf{b}_j)(\mathbf{e}_i \otimes \mathbf{e}_j)^T W_{n,q}^T$$

Throughout this theorem, $W_{m,p}$ and $W_{n,q}$ are swap matrices.

$$= \sum_{i=1}^n \sum_{j=1}^q (\mathbf{b}_j \otimes \mathbf{a}_i)(\mathbf{e}_j \otimes \mathbf{e}_i)^T$$

$$= \left(\left(\sum_{j=1}^q \mathbf{b}_j \mathbf{e}_j^T \right) \otimes \left(\sum_{i=1}^n \mathbf{a}_i \mathbf{e}_i^T \right) \right)$$

$$= B \otimes A,$$

which completes the proof. ∎

In the special case when $m = n$ and $p = q$ (i.e., A and B are each square), we have $W_{n,q} = W_{m,p}$, and since these matrices are real and unitary we furthermore have $W_{n,q}^T = W_{m,p}^{-1}$, which establishes the following corollary:

Corollary 3.1.9

Unitary Similarity of Kronecker Products

Suppose $\mathbb{F} = \mathbb{R}$ or $\mathbb{F} = \mathbb{C}$, and $A \in \mathcal{M}_m(\mathbb{F})$ and $B \in \mathcal{M}_n(\mathbb{F})$. Then $A \otimes B$ and $B \otimes A$ are unitarily similar.

In particular, this corollary tells us that if A and B are square, then $A \otimes B$ and $B \otimes A$ share all similarity-invariant properties, like their rank, trace, determinant, eigenvalues, and characteristic polynomial (though this claim is not true in general if A and B are not square, even if $A \otimes B$ is—see Exercise 3.1.4).

Example 3.1.5

Swapping a Kronecker Product

Suppose $A, B \in \mathcal{M}_2$ satisfy $A \otimes B = \begin{bmatrix} 2 & 1 & 4 & 2 \\ 0 & 1 & 0 & 2 \\ 6 & 3 & 8 & 4 \\ 0 & 3 & 0 & 4 \end{bmatrix}$. Compute $B \otimes A$.

Solution:

We know from Theorem 3.1.8 that $B \otimes A = W(A \otimes B)W^T$, where

$$W = \begin{bmatrix} 1 & 0 & 0 & 0 \\ 0 & 0 & 1 & 0 \\ 0 & 1 & 0 & 0 \\ 0 & 0 & 0 & 1 \end{bmatrix}$$

is the swap matrix. We thus just need to perform the indicated matrix multiplications:

$$B \otimes A = W(A \otimes B)W^T$$

$$= \begin{bmatrix} 1 & 0 & 0 & 0 \\ 0 & 0 & 1 & 0 \\ 0 & 1 & 0 & 0 \\ 0 & 0 & 0 & 1 \end{bmatrix} \begin{bmatrix} 2 & 1 & 4 & 2 \\ 0 & 1 & 0 & 2 \\ 6 & 3 & 8 & 4 \\ 0 & 3 & 0 & 4 \end{bmatrix} \begin{bmatrix} 1 & 0 & 0 & 0 \\ 0 & 0 & 1 & 0 \\ 0 & 1 & 0 & 0 \\ 0 & 0 & 0 & 1 \end{bmatrix}$$

$$= \begin{bmatrix} 2 & 4 & 1 & 2 \\ 6 & 8 & 3 & 4 \\ 0 & 0 & 1 & 2 \\ 0 & 0 & 3 & 4 \end{bmatrix}.$$

Note that this agrees with Example 3.1.1, where we computed each of $A \otimes B$ and $B \otimes A$ explicitly.

3.1.3 · The Symmetric and Antisymmetric Subspaces

In the previous section, we introduced the swap matrix $W \in \mathcal{M}_{n^2}$ as the (unique) matrix with the property that $W(\mathbf{v} \otimes \mathbf{w}) = \mathbf{w} \otimes \mathbf{v}$ for all $\mathbf{v}, \mathbf{w} \in \mathbb{F}^n$. However, since we can take the Kronecker product of three or more vectors as well, we can similarly discuss matrices that permute any combination of Kronecker product factors. For example, there is a unique matrix $W_{231} \in \mathcal{M}_{n^3}$ with the property that

Recall from Theorem 3.1.1(a) that $(\mathbf{v} \otimes \mathbf{w}) \otimes \mathbf{x} = \mathbf{v} \otimes (\mathbf{w} \otimes \mathbf{x})$, so it is okay to omit parentheses in expressions like $\mathbf{v} \otimes \mathbf{w} \otimes \mathbf{x}$.

$$W_{231}(\mathbf{v} \otimes \mathbf{w} \otimes \mathbf{x}) = \mathbf{w} \otimes \mathbf{x} \otimes \mathbf{v} \quad \text{for all} \quad \mathbf{v}, \mathbf{w}, \mathbf{x} \in \mathbb{F}^n, \tag{3.1.1}$$

and it can be constructed explicitly by requiring that $W_{231}(\mathbf{e}_i \otimes \mathbf{e}_j \otimes \mathbf{e}_k) = \mathbf{e}_j \otimes \mathbf{e}_k \otimes \mathbf{e}_i$ for all $1 \leq i, j, k \leq n$. For example, in the $n = 2$ case the swap matrix W_{231} has the form

As always, the gray bars in this block matrix has no mathematical meaning—they are just there to help us visualize the three different Kronecker factors that W_{231} acts on.

$$W_{231} = \left[\begin{array}{cccc|cccc} 1 & \cdot & \cdot & \cdot & \cdot & \cdot & \cdot & \cdot \\ \cdot & \cdot & \cdot & \cdot & 1 & \cdot & \cdot & \cdot \\ \cdot & 1 & \cdot & \cdot & \cdot & \cdot & \cdot & \cdot \\ \cdot & \cdot & \cdot & \cdot & \cdot & 1 & \cdot & \cdot \\ \hline \cdot & \cdot & 1 & \cdot & \cdot & \cdot & \cdot & \cdot \\ \cdot & \cdot & \cdot & \cdot & \cdot & \cdot & 1 & \cdot \\ \cdot & \cdot & \cdot & 1 & \cdot & \cdot & \cdot & \cdot \\ \cdot & \cdot & \cdot & \cdot & \cdot & \cdot & \cdot & 1 \end{array} \right].$$

More generally, we can consider swap matrices acting on the Kronecker product of any number p of vectors. If we fix a permutation $\sigma : \{1, 2, \ldots, p\} \to \{1, 2, \ldots, p\}$ (i.e., a function for which $\sigma(i) = \sigma(j)$ if and only if $i = j$), then we define $W_\sigma \in \mathcal{M}_{n^p}$ to be the (unique) matrix with the property that

These swap matrices can also be defined if $\mathbf{v}_1, \mathbf{v}_2, \ldots, \mathbf{v}_p$ have different dimensionalities, but for the sake of simplicity we only consider the case when they are all n-dimensional here.

$$W_\sigma(\mathbf{v}_1 \otimes \mathbf{v}_2 \otimes \cdots \otimes \mathbf{v}_p) = \mathbf{v}_{\sigma(1)} \otimes \mathbf{v}_{\sigma(2)} \otimes \cdots \otimes \mathbf{v}_{\sigma(p)}$$

for all $\mathbf{v}_1, \mathbf{v}_2, \ldots, \mathbf{v}_p \in \mathbb{F}^n$. From this definition is should be clear that if σ and τ are any two permutations then $W_\sigma W_\tau = W_{\sigma \circ \tau}$.

We typically denote permutations by their **one-line notation**, which means we list them simply by a string of digits such that, for each $1 \leq j \leq p$, the corresponds to the permutation $\sigma : \{1,2,3\} \to \{1,2,3\}$ for which $\sigma(1) = 2$, $\sigma(2) = 3$, and $\sigma(3) = 1$, and thus W_{231} is exactly the swap matrix described by Equation (3.1.1).

Recall that
$p! = p(p-1)\cdots 3\cdot 2\cdot 1.$

In general, there are $p!$ permutations acting on $\{1,2,\ldots,p\}$, and we denote the set consisting of all of these permutations by S_p (this set is typically called the **symmetric group**). There are thus exactly $p!$ swap matrices that permute the Kronecker factors of vectors like $\mathbf{v}_1 \otimes \mathbf{v}_2 \otimes \cdots \otimes \mathbf{v}_p$ as well. In the $p = 2$, in which case there are $p! = 2$ such swap matrices: the matrix that we called "the" swap matrix back in Section 3.1.2, which corresponds to the permutation $\sigma = 21$, and the identity matrix, which corresponds to the identity permutation $\sigma = 12$.

Some background
material on
permutations is
provided in
Appendix A.1.5.

In the $p > 2$ case, these more general swap matrices retain some of the nice properties of "the" swap matrix from the $p = 2$ case, but lose others. In particular, they are still permutation matrices and thus unitary, but they are no longer symmetric in general (after all, even W_{231} is not symmetric).

The Symmetric Subspace

We now introduce one particularly important subspace that arises somewhat naturally from the Kronecker product—the subspace of vectors that remain unchanged when their Kronecker factors are permuted.

Definition 3.1.4

The Symmetric Subspace

Suppose $n, p \geq 1$ are integers. The **symmetric subspace** $\mathcal{S}_n^p$ is the subspace of $(\mathbb{F}^n)^{\otimes p}$ consisting of vectors that are unchanged by swap matrices:

$$\mathcal{S}_n^p \overset{\text{def}}{=} \left\{ \mathbf{v} \in (\mathbb{F}^n)^{\otimes p} : W_\sigma \mathbf{v} = \mathbf{v} \text{ for all } \sigma \in S_p \right\}.$$

In the $p = 2$ case, the symmetric subspace is actually quite familiar. Since the swap matrix W is the standard matrix of the transpose map, we see that the equation $W\mathbf{v} = \mathbf{v}$ is equivalent to $\text{mat}(\mathbf{v})^T = \text{mat}(\mathbf{v})$. That is, the symmetric subspace $\mathcal{S}_n^2$ is isomorphic to the set of $n \times n$ symmetric matrices $\mathcal{M}_n^S$ via matricization.

We can thus think of the symmetric subspace $\mathcal{S}_n^p$ as a natural generalization of the set $\mathcal{M}_n^S$ of symmetric matrices. We remind ourselves of some of the properties of $\mathcal{M}_n^S$ here, as well as the corresponding properties of $\mathcal{S}_n^2$ that they imply:

There are many
other bases of $\mathcal{M}_n^S$
too, but the one
shown here is
particularly simple.

Properties of $\mathcal{M}_n^S$

basis: $\{E_{j,j} : 1 \leq j \leq n\} \cup \{E_{i,j} + E_{j,i} : 1 \leq i < j \leq n\}$

dimension: $\dbinom{n+1}{2} = \dfrac{n(n+1)}{2}$

Properties of $\mathcal{S}_n^2$

basis: $\{\mathbf{e}_j \otimes \mathbf{e}_j : 1 \leq j \leq n\} \cup \{\mathbf{e}_i \otimes \mathbf{e}_j + \mathbf{e}_j \otimes \mathbf{e}_i : 1 \leq i < j \leq n\}$

dimension: $\dbinom{n+1}{2} = \dfrac{n(n+1)}{2}$

For example, the members of $\mathcal{S}_2^2$ are the vectors of the form (a,b,b,c), which are isomorphic via matricization to the 2×2 (symmetric) matrices of the form

$$\begin{bmatrix} a & b \\ b & c \end{bmatrix}.$$

The following theorem generalizes these properties to higher values of n

The symmetric subspace $\mathcal{S}_n^p \subseteq (\mathbb{F}^n)^{\otimes p}$ has the following properties:

a) One projection onto $\mathcal{S}_n^p$ is given by $\dfrac{1}{p!} \displaystyle\sum_{\sigma \in S_p} W_\sigma$,

b) $\dim(\mathcal{S}_n^p) = \dbinom{n+p-1}{p}$, and

c) the following set is a basis of $\mathcal{S}_n^p$:

$$\left\{ \sum_{\sigma \in S_p} W_\sigma(\mathbf{e}_{j_1} \otimes \mathbf{e}_{j_2} \otimes \cdots \otimes \mathbf{e}_{j_p}) : 1 \leq j_1 \leq j_2 \leq \cdots \leq j_p \leq n \right\}.$$

> When we say "orthogonal" here, we mean with respect to the usual dot product on $(\mathbb{F}^n)^{\otimes p}$.

Furthermore, if $\mathbb{F} = \mathbb{R}$ or $\mathbb{F} = \mathbb{C}$ then the projection in part (a) and the basis in part (c) are each orthogonal.

Proof. We begin by proving property (a). Define $P = \sum_{\sigma \in S_p} W_\sigma / p!$ to be the proposed projection onto $\mathcal{S}_n^p$. It is straightforward to check that $P^2 = P = P^T$, so P is an (orthogonal, if $\mathbb{F} = \mathbb{R}$ or $\mathbb{F} = \mathbb{C}$) projection onto *some* subspace of $(\mathbb{F}^n)^{\otimes p}$—we leave the proof of those statements to Exercise 3.1.15.

It thus now suffices to show that $\mathrm{range}(P) = \mathcal{S}_n^p$. To this end, first notice that for all $\tau \in S_p$ we have

> The fact that composing all permutations by τ gives the set of all permutations follows from the fact that S_p is a group (i.e., every permutation is invertible). To write a particular permutation $\rho \in S_p$ as $\rho = \tau \circ \sigma$, just choose $\sigma = \tau^{-1} \circ \rho$.

$$W_\tau P = \frac{1}{p!} \sum_{\sigma \in S_p} W_\tau W_\sigma = \frac{1}{p!} \sum_{\sigma \in S_p} W_{\tau \circ \sigma} = P,$$

with the final equality following from the fact that every permutation in S_p can be written in the form $\tau \circ \sigma$ for some $\sigma \in S_p$. It follows that everything in $\mathrm{range}(P)$ is unchanged by W_τ (for all $\tau \in S_p$), so $\mathrm{range}(P) \subseteq \mathcal{S}_n^p$.

To prove the opposite inclusion, we just notice that if $\mathbf{v} \in \mathcal{S}_n^p$ then

$$P\mathbf{v} = \frac{1}{p!} \sum_{\sigma \in S_p} W_\sigma \mathbf{v} = \frac{1}{p!} \sum_{\sigma \in S_p} \mathbf{v} = \mathbf{v},$$

so $\mathbf{v} \in \mathrm{range}(P)$ and thus $\mathcal{S}_n^p \subseteq \mathrm{range}(P)$. Since we already proved the opposite inclusion, it follows that $\mathrm{range}(P) = \mathcal{S}_n^p$, so P is a projection onto $\mathcal{S}_n^p$ as claimed.

To prove property (c) (we will prove property (b) shortly), we first notice that the columns of the projection P from part (a) have the form

$$P(\mathbf{e}_{j_1} \otimes \mathbf{e}_{j_2} \otimes \cdots \otimes \mathbf{e}_{j_p}) = \frac{1}{p!} \sum_{\sigma \in S_p} W_\sigma(\mathbf{e}_{j_1} \otimes \mathbf{e}_{j_2} \otimes \cdots \otimes \mathbf{e}_{j_p}),$$

where $1 \leq j_1, j_2, \ldots, j_p \leq n$. To turn this set of vectors into a basis of $\mathrm{range}(P) = \mathcal{S}_n^p$, we omit the columns that are equal to each other by only considering the columns for which $1 \leq j_1 \leq j_2 \leq \cdots \leq j_p \leq n$. If $\mathbb{F} = \mathbb{R}$ or $\mathbb{F} = \mathbb{C}$ (so we have an inner product to work with) then these remaining vectors are mutually orthogonal and thus form an orthogonal basis of $\mathrm{range}(P)$, and otherwise they are linearly independent (and thus form a basis of $\mathrm{range}(P)$) since the coordinates of their non-zero entries in the standard basis form disjoint subsets of $\{1, 2, \ldots, n^p\}$. If we multiply these vectors each by $p!$ then they form the basis described in the statement of the theorem.

To demonstrate property (b), we simply notice that the basis from part (c) of
the theorem contains as many vectors as there are multisets $\{j_1, j_2, \ldots, j_p\} \subseteq$
$\{1, 2, \ldots, n\}$. A standard combinatorics result says that there are exactly

$$\binom{n+p-1}{p} = \frac{(n-p+1)!}{p!\,(n-1)!}$$

such multisets (see Remark 3.1.2), which completes the proof.   ∎

Remark 3.1.2

Counting Multisets

We now illustrate why there are exactly

$$\binom{n+p-1}{p} = \frac{(n-p+1)!}{p!\,(n-1)!}$$

p-element multisets with entries chosen from an n-element set (a fact that
we made use of at the end of the proof of Theorem 3.1.10). We represent
each multiset graphically via "stars and bars", where p stars represent the
members of a multiset and $n-1$ bars separate the values of those stars.

For example, in the $n = 5$, $p = 6$ case, the multisets $\{1, 2, 3, 3, 5, 5\}$ and
$\{1, 1, 1, 2, 4, 4\}$ would be represented by the stars and bars arrangements

$$\star\,|\,\star\,|\,\star\star\,|\,|\,\star\star \quad \text{and} \quad \star\star\star\,|\,\star\,|\,|\,\star\star\,|,$$

respectively.

Notice that there are a total of $n + p - 1$ positions in such an arrangement of stars and bars (p positions for the stars and $n - 1$ positions for the
bars), and each arrangement is completely determined by the positions
that we choose for the p stars. It follows that there are $\binom{n+p-1}{p}$ such con-
figurations of stars and bars, and thus exactly that many multisets of size
p chosen from a set of size n, as claimed.

The orthogonal basis vectors from part (c) of Theorem 3.1.10 do not form
an ortho*normal* basis because they are not properly normalized. For example,
in the $n = 2$, $p = 3$ case, we have $\dim(\mathcal{S}_2^3) = \binom{4}{3} = 4$ and the basis of $\mathcal{S}_2^3$
described by the theorem consists of the following 4 vectors:

Basis Vector	Tuple (j_1, j_2, j_3)
$6\,\mathbf{e}_1 \otimes \mathbf{e}_1 \otimes \mathbf{e}_1$	$(1, 1, 1)$
$2\big(\mathbf{e}_1 \otimes \mathbf{e}_1 \otimes \mathbf{e}_2 + \mathbf{e}_1 \otimes \mathbf{e}_2 \otimes \mathbf{e}_1 + \mathbf{e}_2 \otimes \mathbf{e}_1 \otimes \mathbf{e}_1\big)$	$(1, 1, 2)$
$2\big(\mathbf{e}_1 \otimes \mathbf{e}_2 \otimes \mathbf{e}_2 + \mathbf{e}_2 \otimes \mathbf{e}_1 \otimes \mathbf{e}_2 + \mathbf{e}_2 \otimes \mathbf{e}_2 \otimes \mathbf{e}_1\big)$	$(1, 2, 2)$
$6\,\mathbf{e}_2 \otimes \mathbf{e}_2 \otimes \mathbf{e}_2$	$(2, 2, 2)$

In order to turn this basis into an orthonormal one, we must divide each
vector in it by $\sqrt{p!\,m_1!\,m_2!\cdots m_n!}$, where m_j denotes the multiplicity of j
in the corresponding tuple $(j_1, j_2, \ldots, j_n)$. For example, for the basis vector
$2\big(\mathbf{e}_1 \otimes \mathbf{e}_1 \otimes \mathbf{e}_2 + \mathbf{e}_1 \otimes \mathbf{e}_2 \otimes \mathbf{e}_1 + \mathbf{e}_2 \otimes \mathbf{e}_1 \otimes \mathbf{e}_1\big)$ corresponding to the tuple $(1, 1, 2)$
above, we have $m_1 = 2$ and $m_2 = 1$, so we divide that vector by $\sqrt{p!\,m_1!\,m_2!} =$
$\sqrt{3!\,2!\,1!} = 2\sqrt{3}$ to normalize it:

$$\frac{1}{\sqrt{3}}\big(\mathbf{e}_1 \otimes \mathbf{e}_1 \otimes \mathbf{e}_2 + \mathbf{e}_1 \otimes \mathbf{e}_2 \otimes \mathbf{e}_1 + \mathbf{e}_2 \otimes \mathbf{e}_1 \otimes \mathbf{e}_1\big).$$

We close our discussion of the symmetric subspace by showing that it could
be defined in another (equivalent) way—as the span of Kronecker powers of
vectors in $\mathbb{F}^n$ (as long as $\mathbb{F} = \mathbb{R}$ or $\mathbb{F} = \mathbb{C}$).

Theorem 3.1.11

Tensor-Power Basis of the Symmetric Subspace

Suppose $\mathbb{F} = \mathbb{R}$ or $\mathbb{F} = \mathbb{C}$. The symmetric subspace $\mathcal{S}_n^p \subseteq (\mathbb{F}^n)^{\otimes p}$ is the span of Kronecker powers of vectors:

$$\mathcal{S}_n^p = \text{span}\{\mathbf{v}^{\otimes p} : \mathbf{v} \in \mathbb{F}^n\}.$$

However, a proof of this theorem requires some technicalities that we have not yet developed, so we defer it to Section 3.B.1. Furthermore, we need to be careful to keep in mind that this theorem does not hold over some other fields—see Exercise 3.1.13.

Recall that $\mathbf{v}^{\otimes p} = \mathbf{v} \otimes \cdots \otimes \mathbf{v}$, where there are p copies of $\mathbf{v}$ on the right-hand side.

For now, we just note that in the $p = 2$ case it says (via the usual isomorphism between $\mathcal{S}_n^2$ and $\mathcal{M}_n^S$) that every symmetric matrix can be written as linear combination of rank-1 symmetric matrices. This fact is not completely obvious, as the most natural basis of $\mathcal{M}_n^S$ contains rank-2 matrices. For example, in the $n = 3$ case the "typical" basis of $\mathcal{M}_3^S$ consists of the following 6 matrices:

$$\begin{bmatrix} 1 & 0 & 0 \\ 0 & 0 & 0 \\ 0 & 0 & 0 \end{bmatrix}, \begin{bmatrix} 0 & 0 & 0 \\ 0 & 1 & 0 \\ 0 & 0 & 0 \end{bmatrix}, \begin{bmatrix} 0 & 0 & 0 \\ 0 & 0 & 0 \\ 0 & 0 & 1 \end{bmatrix}, \underbrace{\begin{bmatrix} 0 & 1 & 0 \\ 1 & 0 & 0 \\ 0 & 0 & 0 \end{bmatrix}, \begin{bmatrix} 0 & 0 & 1 \\ 0 & 0 & 0 \\ 1 & 0 & 0 \end{bmatrix}, \begin{bmatrix} 0 & 0 & 0 \\ 0 & 0 & 1 \\ 0 & 1 & 0 \end{bmatrix}}_{\text{each has rank 2}}$$

In order to turn this basis into one consisting only of rank-1 symmetric matrices, we need to make it slightly uglier so that the non-zero entries of the basis matrices overlap somewhat:

Refer back to Exercise 1.2.8 to see a generalization of this basis to larger values of n.

$$\begin{bmatrix} 1 & 0 & 0 \\ 0 & 0 & 0 \\ 0 & 0 & 0 \end{bmatrix}, \begin{bmatrix} 0 & 0 & 0 \\ 0 & 1 & 0 \\ 0 & 0 & 0 \end{bmatrix}, \begin{bmatrix} 0 & 0 & 0 \\ 0 & 0 & 0 \\ 0 & 0 & 1 \end{bmatrix}, \begin{bmatrix} 1 & 1 & 0 \\ 1 & 1 & 0 \\ 0 & 0 & 0 \end{bmatrix}, \begin{bmatrix} 1 & 0 & 1 \\ 0 & 0 & 0 \\ 1 & 0 & 1 \end{bmatrix}, \begin{bmatrix} 0 & 0 & 0 \\ 0 & 1 & 1 \\ 0 & 1 & 1 \end{bmatrix}$$

Remark 3.1.3

The Spectral Decomposition in the Symmetric Subspace

Another way to see that symmetric matrices can be written as a linear combination of symmetric rank-1 matrices is to make use of the real spectral decomposition (Theorem 2.1.6, if $\mathbb{F} = \mathbb{R}$) or the Takagi factorization (Exercise 2.3.26, if $\mathbb{F} = \mathbb{C}$). In particular, if $\mathbb{F} = \mathbb{R}$ and $A \in \mathcal{M}_n^S$ has $\{\mathbf{u}_1, \mathbf{u}_2, \ldots, \mathbf{u}_n\}$ as an orthonormal basis of eigenvectors with corresponding eigenvalues $\lambda_1, \lambda_2, \ldots, \lambda_n$, then

$$A = \sum_{j=1}^n \lambda_j \mathbf{u}_j \mathbf{u}_j^T$$

is one way of writing A in the desired form. If we trace things back through the isomorphism between $\mathcal{S}_n^2$ and $\mathcal{M}_n^S$ then this shows in the $p = 2$ case that we can write every vector $\mathbf{v} \in \mathcal{S}_n^2$ in the form

When $\mathbb{F} = \mathbb{C}$, a symmetric (*not* Hermitian!) matrix might not be normal, so the complex spectral decomposition might not apply to it, which is why we must use the Takagi factorization.

$$\mathbf{v} = \sum_{j=1}^n \lambda_j \mathbf{u}_j \otimes \mathbf{u}_j,$$

and a similar argument works when $\mathbb{F} = \mathbb{C}$ if we use the Takagi factorization instead.

Notice that what we have shown here is stronger than the statement of Theorem 3.1.11 (in the $p = 2$ case), which does not require the set $\{\mathbf{u}_1, \mathbf{u}_2, \ldots, \mathbf{u}_n\}$ to be orthogonal. Indeed, when $p \geq 3$, this stronger claim is no longer true—not only do we lose orthogonality, but we even lose linear independence in general! For example, we will show in Example 3.3.4 that the vector $\mathbf{v} = (0,1,1,0,1,0,0,0) \in \mathcal{S}_2^3$ cannot be written in the form

$$\mathbf{v} = \lambda_1 \mathbf{v}_1 \otimes \mathbf{v}_1 \otimes \mathbf{v}_1 + \lambda_2 \mathbf{v}_2 \otimes \mathbf{v}_2 \otimes \mathbf{v}_2$$

for *any* choice of $\lambda_1, \lambda_2 \in \mathbb{R}$, and $\mathbf{v}_1, \mathbf{v}_2 \in \mathbb{R}^2$. Instead, at least 3 terms are needed in such a sum, and since each $\mathbf{v}_j$ is 2-dimensional, they cannot be chosen to be linearly independent.

The Antisymmetric Subspace

Just like the symmetric subspace can be thought of as a natural generalization of the set of symmetric matrices, there is also a natural generalization of the set of skew-symmetric matrices to higher Kronecker powers.

Definition 3.1.5

The Antisymmetric Subspace

Suppose $n, p \geq 1$ are integers. The **antisymmetric subspace** $\mathcal{A}_n^p$ is the following subspace of $(\mathbb{F}^n)^{\otimes p}$:

$$\mathcal{A}_n^p \stackrel{\text{def}}{=} \left\{ \mathbf{v} \in (\mathbb{F}^n)^{\otimes p} : W_\sigma \mathbf{v} = \text{sgn}(\sigma)\mathbf{v} \text{ for all } \sigma \in S_p \right\}.$$

Recall that the **sign** $\text{sgn}(\sigma)$ of a permutation σ is the number of transpositions needed to generate it (see Appendix A.1.5).

As suggested above, in the $p = 2$ case the antisymmetric subspace is isomorphic to the set $\mathcal{M}_n^{sS}$ of skew-symmetric matrices via matricization, since $W\mathbf{v} = -\mathbf{v}$ if and only if $\text{mat}(\mathbf{v})^T = -\text{mat}(\mathbf{v})$. With this in mind, we now remind ourselves of some of the properties of $\mathcal{M}_n^{sS}$ here, as well as the corresponding properties of $\mathcal{A}_n^2$ that they imply:

	Properties of $\mathcal{M}_n^{sS}$	**Properties of $\mathcal{A}_n^2$**
basis:	$\{E_{i,j} - E_{j,i} : 1 \leq i < j \leq n\}$	$\{\mathbf{e}_i \otimes \mathbf{e}_j - \mathbf{e}_j \otimes \mathbf{e}_i : 1 \leq i < j \leq n\}$
dim.:	$\binom{n}{2} = \dfrac{n(n-1)}{2}$	$\binom{n}{2} = \dfrac{n(n-1)}{2}$

For example, the members of $\mathcal{A}_2^2$ are the vectors of the form $(0, a, -a, 0)$, which are isomorphic via matricization to the 2×2 (skew-symmetric) matrices of the form

$$\begin{bmatrix} 0 & a \\ -a & 0 \end{bmatrix}.$$

The following theorem generalizes these properties to higher values of p.

Theorem 3.1.12

Properties of the Antisymmetric Subspace

The antisymmetric subspace $\mathcal{A}_n^p \subseteq (\mathbb{F}^n)^{\otimes p}$ has the following properties:

a) One projection onto $\mathcal{A}_n^p$ is given by $\dfrac{1}{p!}\displaystyle\sum_{\sigma \in S_p} \text{sgn}(\sigma) W_\sigma$,

b) $\dim(\mathcal{A}_n^p) = \dbinom{n}{p}$, and

c) the following set is a basis of $\mathcal{A}_n^p$:

$$\left\{ \sum_{\sigma \in S_p} \text{sgn}(\sigma) W_\sigma(\mathbf{e}_{j_1} \otimes \cdots \otimes \mathbf{e}_{j_p}) : 1 \leq j_1 < \cdots < j_p \leq n \right\}.$$

Furthermore, if $\mathbb{F} = \mathbb{R}$ or $\mathbb{F} = \mathbb{C}$ then the projection in part (a) and the basis in part (c) are each orthogonal.

We leave the proof of this theorem to Exercise 3.1.24, as it is almost identical to that of Theorem 3.1.10, except the details work out more cleanly here. For example, to normalize the basis vectors in part (c) of this theorem to make an ortho*normal* basis of $\mathcal{A}_n^p$, we just need to divide each of them by $\sqrt{p!}$ (whereas the normalization factor was somewhat more complicated for the orthogonal basis of the symmetric subspace).

To get a bit of a feel for what the antisymmetric subspace looks like when $p > 2$, we consider the $p = 3$, $n = 4$ case, where the basis described by the above theorem consists of the following $\binom{4}{3} = 4$ vectors:

Basis Vector	Tuple (j_1, j_2, j_3)
$\mathbf{e}_1 \otimes \mathbf{e}_2 \otimes \mathbf{e}_3 + \mathbf{e}_2 \otimes \mathbf{e}_3 \otimes \mathbf{e}_1 + \mathbf{e}_3 \otimes \mathbf{e}_1 \otimes \mathbf{e}_2$ $- \mathbf{e}_1 \otimes \mathbf{e}_3 \otimes \mathbf{e}_2 - \mathbf{e}_2 \otimes \mathbf{e}_1 \otimes \mathbf{e}_3 - \mathbf{e}_3 \otimes \mathbf{e}_2 \otimes \mathbf{e}_1$	$(1,2,3)$
$\mathbf{e}_1 \otimes \mathbf{e}_2 \otimes \mathbf{e}_4 + \mathbf{e}_2 \otimes \mathbf{e}_4 \otimes \mathbf{e}_1 + \mathbf{e}_4 \otimes \mathbf{e}_1 \otimes \mathbf{e}_2$ $- \mathbf{e}_1 \otimes \mathbf{e}_4 \otimes \mathbf{e}_2 - \mathbf{e}_2 \otimes \mathbf{e}_1 \otimes \mathbf{e}_4 - \mathbf{e}_4 \otimes \mathbf{e}_2 \otimes \mathbf{e}_1$	$(1,2,4)$
$\mathbf{e}_1 \otimes \mathbf{e}_3 \otimes \mathbf{e}_4 + \mathbf{e}_3 \otimes \mathbf{e}_4 \otimes \mathbf{e}_1 + \mathbf{e}_4 \otimes \mathbf{e}_1 \otimes \mathbf{e}_3$ $- \mathbf{e}_1 \otimes \mathbf{e}_4 \otimes \mathbf{e}_3 - \mathbf{e}_3 \otimes \mathbf{e}_1 \otimes \mathbf{e}_4 - \mathbf{e}_4 \otimes \mathbf{e}_3 \otimes \mathbf{e}_1$	$(1,3,4)$
$\mathbf{e}_2 \otimes \mathbf{e}_3 \otimes \mathbf{e}_4 + \mathbf{e}_3 \otimes \mathbf{e}_4 \otimes \mathbf{e}_2 + \mathbf{e}_4 \otimes \mathbf{e}_2 \otimes \mathbf{e}_3$ $- \mathbf{e}_2 \otimes \mathbf{e}_4 \otimes \mathbf{e}_3 - \mathbf{e}_3 \otimes \mathbf{e}_2 \otimes \mathbf{e}_4 - \mathbf{e}_4 \otimes \mathbf{e}_3 \otimes \mathbf{e}_2$	$(2,3,4)$

Direct sums and orthogonal complements were covered in Section 1.B.

The symmetric and antisymmetric subspaces are always orthogonal to each other (as long as $\mathbb{F} = \mathbb{R}$ or $\mathbb{F} = \mathbb{C}$) in the sense that if $\mathbf{v} \in \mathcal{S}_n^p$ and $\mathbf{w} \in \mathcal{A}_n^p$ then $\mathbf{v} \cdot \mathbf{w} = 0$ (see Exercise 3.1.14). Furthermore, if $p = 2$ then in fact they are orthogonal complements of each other:

$$(\mathcal{S}_n^2)^\perp = \mathcal{A}_n^2, \quad \text{or equivalently their direct sum is} \quad \mathcal{S}_n^2 \oplus \mathcal{A}_n^2 = \mathbb{F}^n \otimes \mathbb{F}^n,$$

which can be verified just by observing that their dimensions satisfy $\binom{n+1}{2} + \binom{n}{2} = n^2$. In fact, this direct sum decomposition of $\mathbb{F}^n \otimes \mathbb{F}^n$ is completely analogous (isomorphic?) to the fact that $\mathcal{M}_n = \mathcal{M}_n^S \oplus \mathcal{M}_n^{sS}$. However, this property fails when $p > 2$, in which case there are vectors in $(\mathbb{F}^n)^{\otimes p}$ that are orthogonal to everything in each of $\mathcal{S}_n^p$ and $\mathcal{A}_n^p$.

*Recall that the **zero vector space** is $\{0\}$.*

There are certain values of p and n that it is worth focusing a bit of attention on. If $p > n$ then $\mathcal{A}_n^p$ is the zero vector space, which can be verified simply by observing that $\binom{n}{p} = 0$ in this case. If $p = n$ then $\mathcal{A}_n^n$ is $\binom{n}{n} = 1$-dimensional, so up to scaling there is only one vector in the antisymmetric subspace, and it is

$$\sum_{\sigma \in S_n} \text{sgn}(\sigma) W_\sigma (\mathbf{e}_1 \otimes \mathbf{e}_2 \otimes \cdots \otimes \mathbf{e}_n) = \sum_{\sigma \in S_n} \text{sgn}(\sigma) \mathbf{e}_{\sigma(1)} \otimes \mathbf{e}_{\sigma(2)} \otimes \cdots \otimes \mathbf{e}_{\sigma(n)}.$$

For example, the unique (up to scaling) vectors in $\mathcal{A}_2^2$ and $\mathcal{A}_3^3$ are

$$\mathbf{e}_1 \otimes \mathbf{e}_2 - \mathbf{e}_2 \otimes \mathbf{e}_1 \quad \text{and} \quad \mathbf{e}_1 \otimes \mathbf{e}_2 \otimes \mathbf{e}_3 + \mathbf{e}_2 \otimes \mathbf{e}_3 \otimes \mathbf{e}_1 + \mathbf{e}_3 \otimes \mathbf{e}_1 \otimes \mathbf{e}_2$$
$$- \mathbf{e}_1 \otimes \mathbf{e}_3 \otimes \mathbf{e}_2 - \mathbf{e}_2 \otimes \mathbf{e}_1 \otimes \mathbf{e}_3 - \mathbf{e}_3 \otimes \mathbf{e}_2 \otimes \mathbf{e}_1,$$

respectively.

It is worth comparing these antisymmetric vectors to the formula for the determinant of a matrix, which for matrices $A \in \mathcal{M}_2$ and $B \in \mathcal{M}_3$ have the forms

$$\det(A) = a_{1,1}a_{2,2} - a_{1,2}a_{2,1}, \quad \text{and}$$
$$\det(B) = b_{1,1}b_{2,2}b_{3,3} + b_{1,2}b_{2,3}b_{3,1} + b_{1,3}b_{2,1}b_{3,2}$$
$$- b_{1,1}b_{2,3}b_{3,2} - b_{1,2}b_{2,1}b_{3,3} - b_{1,3}b_{2,2}b_{3,1},$$

This formula is stated as Theorem A.1.4 in Appendix A.1.5.

respectively, and which has the following form in general for matrices $C \in \mathcal{M}_n$:

$$\det(C) = \sum_{\sigma \in S_n} \text{sgn}(\sigma) c_{1,\sigma(1)} c_{2,\sigma(2)} \cdots c_{n,\sigma(n)}.$$

The fact that the antisymmetric vector in $\mathcal{A}_n^n$ looks so much like the formula for the determinant of an $n \times n$ matrix is no coincidence—we will see in Section 3.2 (Example 3.2.9 in particular) that there is a well-defined sense in which the determinant "is" this unique antisymmetric vector.

Exercises

solutions to starred exercises on page 476

3.1.1 Compute $A \otimes B$ for the following pairs of matrices:

*(a) $A = \begin{bmatrix} 1 & 2 \\ 3 & 0 \end{bmatrix}$, $B = \begin{bmatrix} -1 & 2 \\ 0 & -3 \end{bmatrix}$

(b) $A = \begin{bmatrix} 3 & 1 & 2 \\ 2 & 0 & 1 \end{bmatrix}$, $B = \begin{bmatrix} 2 & -1 \\ -1 & 3 \end{bmatrix}$

*(c) $A = \begin{bmatrix} 1 \\ 2 \\ 3 \end{bmatrix}$, $B = \begin{bmatrix} 2 & -3 & 1 \end{bmatrix}$

3.1.2 Use the method of Example 3.1.3 to find all matrices that commute with the given matrix.

*(a) $\begin{bmatrix} 1 & 0 \\ 0 & 0 \end{bmatrix}$ (b) $\begin{bmatrix} 0 & 1 \\ 0 & 0 \end{bmatrix}$

*(c) $\begin{bmatrix} 1 & 0 \\ 0 & 1 \end{bmatrix}$ (d) $\begin{bmatrix} 1 & 2 \\ -2 & 1 \end{bmatrix}$

3.1.3 Determine which of the following statements are true and which are false.

*(a) If A is 3×4 and B is 4×3 then $A \otimes B$ is 12×12.

(b) If $A, B \in \mathcal{M}_n$ are symmetric then so is $A \otimes B$.

*(c) If $A, B \in \mathcal{M}_n$ are skew-symmetric then so is $A \otimes B$.

(d) If $A \in \mathcal{M}_{m,1}$ and $B \in \mathcal{M}_{1,n}$ then $A \otimes B = B \otimes A$.

∗∗3.1.4 Construct an example to show that if $A \in \mathcal{M}_{2,3}$ and $B \in \mathcal{M}_{3,2}$ then it might be the case that $\text{tr}(A \otimes B) \neq \text{tr}(B \otimes A)$. Why does this not contradict Theorem 3.1.3 or Corollary 3.1.9?

3.1.5 Suppose $H \in \mathcal{M}_n$ is a matrix with every entry equal to 1 or -1.

(a) Show that $|\det(H)| \leq n^{n/2}$.
[Hint: Make use of Exercise 2.2.28.]

(b) Show that Hadamard matrices (see Remark 3.1.1) are exactly the ones for which equality is attained in part (a).

∗3.1.6 Suppose $A \in \mathcal{M}_{m,n}(\mathbb{C})$ and $B \in \mathcal{M}_{p,q}(\mathbb{C})$, and $A^\dagger$ denotes the pseudoinverse of A (introduced in Section 2.C.1). Show that $(A \otimes B)^\dagger = A^\dagger \otimes B^\dagger$.

3.1.7 Suppose that λ is an eigenvalue of $A \in \mathcal{M}_m(\mathbb{C})$ and μ is an eigenvalue of $B \in \mathcal{M}_n(\mathbb{C})$ with corresponding eigenvectors $\mathbf{v}$ and $\mathbf{w}$, respectively.

(a) Show that $\lambda + \mu$ is an eigenvalue of $A \otimes I_n + I_m \otimes B$ by finding a corresponding eigenvector.
[Side note: This matrix $A \otimes I_n + I_m \otimes B$ is sometimes called the **Kronecker sum** of A and B.]

(b) Show that *every* eigenvalue of $A \otimes I_n + I_m \otimes B$ is the sum of an eigenvalue of A and an eigenvalue of B.

∗∗3.1.8 A **Sylvester equation** is a matrix equation of the form

$$AX + XB = C,$$

where $A \in \mathcal{M}_m(\mathbb{C})$, $B \in \mathcal{M}_n(\mathbb{C})$, and $C \in \mathcal{M}_{m,n}(\mathbb{C})$ are given, and the goal is to solve for $X \in \mathcal{M}_{n,m}(\mathbb{C})$.

(a) Show that the equation $AX + XB = C$ is equivalent to $(A \otimes I + I \otimes B^T)\text{vec}(X) = \text{vec}(C)$.
(b) Show that a Sylvester equation has a unique solution if and only if A and $-B$ do not share a common eigenvalue. [Hint: Make use of part (a) and the result of Exercise 3.1.7.]

∗∗3.1.9 Let $A \in \mathcal{M}_m(\mathbb{C})$ and $B \in \mathcal{M}_n(\mathbb{C})$.

(a) Show that every eigenvalue of $A \otimes B$ is of the form $\lambda \mu$ for some eigenvalues λ of A and μ of B. [Side note: This exercise is sort of the converse of Theorem 3.1.3(a).]
(b) Use part (a) to show that $\det(A \otimes B) = \det(A)^n \det(B)^m$.

3.1.10 Suppose $\mathbb{F} = \mathbb{R}$ or $\mathbb{F} = \mathbb{C}$.

(a) Construct the orthogonal projection onto $\mathcal{S}_2^3$. That is, write this projection down as an 8×8 matrix.
(b) Construct the orthogonal projection onto $\mathcal{A}_3^2$.

∗∗3.1.11 Suppose $\mathbf{x} \in \mathbb{F}^m \otimes \mathbb{F}^n$.

(a) Show that there exist linearly independent sets $\{\mathbf{v}_j\} \subset \mathbb{F}^m$ and $\{\mathbf{w}_j\} \subset \mathbb{F}^n$ such that

$$\mathbf{x} = \sum_{j=1}^{\min\{m,n\}} \mathbf{v}_j \otimes \mathbf{w}_j.$$

(b) Show that if $\mathbb{F} = \mathbb{R}$ or $\mathbb{F} = \mathbb{C}$ then the sets $\{\mathbf{v}_j\}$ and $\{\mathbf{w}_j\}$ from part (a) can be chosen to be mutually orthogonal.
[Side note: This is sometimes called the **Schmidt decomposition** of $\mathbf{x}$.]

3.1.12 Compute a Schmidt decomposition (see Exercise 3.1.11) of $\mathbf{x} = (2, 1, 0, 0, 1, -2) \in \mathbb{R}^2 \otimes \mathbb{R}^3$.

∗∗3.1.13 Show that Theorem 3.1.11 does not hold when $\mathbb{F} = \mathbb{Z}_2$ is the field with 2 elements (see Appendix A.4) and $n = 2$, $p = 3$.

∗∗3.1.14 Suppose $\mathbb{F} = \mathbb{R}$ or $\mathbb{F} = \mathbb{C}$. Show that if $\mathbf{v} \in \mathcal{S}_n^p$ and $\mathbf{w} \in \mathcal{A}_n^p$ then $\mathbf{v} \cdot \mathbf{w} = 0$.

∗∗3.1.15 In this exercise, we complete the proof of Theorem 3.1.10. Let $P = \sum_{\sigma \in S_p} W_\sigma / p!$.

(a) Show that $P^T = P$.
(b) Show that $P^2 = P$.

∗∗3.1.16 Show that if $\{\mathbf{w}_1, \mathbf{w}_2, \ldots, \mathbf{w}_k\} \subseteq \mathbb{F}^n$ is linearly independent and $\{\mathbf{v}_1, \mathbf{v}_2, \ldots, \mathbf{v}_k\} \subseteq \mathbb{F}^m$ is any set then the equation

$$\sum_{j=1}^{k} \mathbf{v}_j \otimes \mathbf{w}_j = \mathbf{0}$$

implies $\mathbf{v}_1 = \mathbf{v}_2 = \cdots = \mathbf{v}_k = \mathbf{0}$.

∗∗3.1.17 Show that if B and C are bases of $\mathbb{F}^m$ and $\mathbb{F}^n$, respectively, then the set

$$B \otimes C = \{\mathbf{v} \otimes \mathbf{w} : \mathbf{v} \in B, \mathbf{w} \in C\}$$

is a basis of $\mathbb{F}^{mn}$.

[Hint: Use Exercises 1.2.27(a) and 3.1.16.]

∗∗3.1.18 Show that the swap matrix $W_{m,n}$ has the following properties:

(a) Each row and column of $W_{m,n}$ has exactly one non-zero entry, equal to 1.
(b) If $\mathbb{F} = \mathbb{R}$ or $\mathbb{F} = \mathbb{C}$ then $W_{m,n}$ is unitary.
(c) If $m = n$ then $W_{m,n}$ is symmetric.

3.1.19 Show that 1 and -1 are the only eigenvalues of the swap matrix $W_{n,n}$, and the corresponding eigenspaces are $\mathcal{S}_n^2$ and $\mathcal{A}_n^2$, respectively.

∗∗3.1.20 Recall Theorem 3.1.1, which established some of the basic properties of the Kronecker product.

(a) Prove part (a) of the theorem.
(b) Prove part (c) of the theorem.
(c) Prove part (d) of the theorem.

∗∗3.1.21 Recall Theorem 3.1.2, which established some of the ways that the Kronecker product interacts with other matrix operations.

(a) Prove part (b) of the theorem.
(b) Prove part (c) of the theorem.
(c) Prove part (d) of the theorem.

∗∗3.1.22 Prove Theorem 3.1.4.

∗∗3.1.23 Prove Theorem 3.1.5.

∗∗3.1.24 Prove Theorem 3.1.12.

3.1.25 Let $1 \leq p \leq \infty$ and let $\|\cdot\|_p$ denote the p-norm from Section 1.D.1. Show that $\|\mathbf{v} \otimes \mathbf{w}\|_p = \|\mathbf{v}\|_p \|\mathbf{w}\|_p$ for all $\mathbf{v} \in \mathbb{C}^m$, $\mathbf{w} \in \mathbb{C}^n$.

∗∗3.1.26 Let $1 \leq p, q \leq \infty$ be such that $1/p + 1/q = 1$. We now provide an alternate proof of **Hölder's inequality** (Theorem 1.D.5), which says that

$$|\mathbf{v} \cdot \mathbf{w}| \leq \|\mathbf{v}\|_p \|\mathbf{w}\|_q \quad \text{for all} \quad \mathbf{v}, \mathbf{w} \in \mathbb{C}^n.$$

(a) Explain why it suffices to prove this inequality in the case when $\|\mathbf{v}\|_p = \|\mathbf{w}\|_q = 1$. Make this assumption throughout the rest of this exercise.
(b) Show that, for each $1 \leq j \leq n$, either

$$|v_j w_j| \leq |v_j|^p \quad \text{or} \quad |v_j w_j| \leq |w_j|^q.$$

(c) Show that

$$|\mathbf{v} \cdot \mathbf{w}| \leq \|\mathbf{v}\|_p^p + \|\mathbf{w}\|_q^q = 2.$$

(d) This is not quite what we wanted (we wanted to show that $|\mathbf{v} \cdot \mathbf{w}| \leq 1$, not 2). To fix this problem, let $k \geq 1$ be an integer and replace $\mathbf{v}$ and $\mathbf{w}$ in part (c) by $\mathbf{v}^{\otimes k}$ and $\mathbf{w}^{\otimes k}$, respectively. What happens as k gets large?

3.2 Multilinear Transformations

When we first encountered matrix multiplication, it seemed like a strange and arbitrary operation, but it was defined specifically so as to capture how composition of linear transformations affects standard matrices. Similarly, we have now introduced the Kronecker product of matrices, and it perhaps seems like a strange operation to focus so much attention on. However, there is a very natural reason for introducing and exploring it—it lets us represent *multilinear* transformations, which are functions that act on *multiple* vectors, each in a linear way. We now explore this more general class of transformations and how they are related to the Kronecker product.

3.2.1 Definition and Basic Examples

Recall from Section 1.2.3 that a linear transformation is a function that sends vectors from one vector space to another in a linear way, and from Section 1.3.3 that a multilinear form is a function that sends a collection of vectors to a scalar in a manner that treats each input vector linearly. Multilinear transformations provide the natural generalization of both of these concepts—they can be thought of as the sweet spot in between linear transformations and multilinear forms where we have lots of input spaces *and* potentially have a non-trivial output space as well.

Definition 3.2.1

Multilinear Transformations

Suppose $\mathcal{V}_1, \mathcal{V}_2, \ldots, \mathcal{V}_p$ and $\mathcal{W}$ are vector spaces over the same field. A **multilinear transformation** is a function $T : \mathcal{V}_1 \times \mathcal{V}_2 \times \cdots \times \mathcal{V}_p \to \mathcal{W}$ with the property that, if we fix $1 \leq j \leq p$ and any $p-1$ vectors $\mathbf{v}_i \in \mathcal{V}_i$ $(1 \leq i \neq j \leq p)$, then the function $S : \mathcal{V}_j \to \mathcal{W}$ defined by

$$S(\mathbf{v}) = T(\mathbf{v}_1, \ldots, \mathbf{v}_{j-1}, \mathbf{v}, \mathbf{v}_{j+1}, \ldots, \mathbf{v}_p) \quad \text{for all} \quad \mathbf{v} \in \mathcal{V}_j$$

is a linear transformation.

The above definition is a bit of a mouthful, but the idea is simply that a multilinear transformation is a function that looks like a linear transformation on each of its inputs individually. When there are just $p = 2$ input spaces we refer to these functions as **bilinear transformations**, and we note that bilinear forms (refer back to Section 1.3.3) are the special case that arises when the output space is $\mathcal{W} = \mathbb{F}$. Similarly, we sometimes call a multilinear transformation with p input spaces a **p-linear transformation** (much like we sometimes called multilinear forms p-linear forms).

Example 3.2.1

The Cross Product is a Bilinear Transformation

Consider the function $C : \mathbb{R}^3 \times \mathbb{R}^3 \to \mathbb{R}^3$ defined by $C(\mathbf{v}, \mathbf{w}) = \mathbf{v} \times \mathbf{w}$ for all $\mathbf{v}, \mathbf{w} \in \mathbb{R}^3$, where $\mathbf{v} \times \mathbf{w}$ is the cross product of $\mathbf{v}$ and $\mathbf{w}$:

$$\mathbf{v} \times \mathbf{w} = (v_2 w_3 - v_3 w_2, \ v_3 w_1 - v_1 w_3, \ v_1 w_2 - v_2 w_1).$$

Show that C is a bilinear transformation.

Solution:

The following facts about the cross product are equivalent to C being bilinear:

- $(\mathbf{v}+\mathbf{w}) \times \mathbf{x} = \mathbf{v} \times \mathbf{x} + \mathbf{w} \times \mathbf{x}$,
- $\mathbf{v} \times (\mathbf{w}+\mathbf{x}) = \mathbf{v} \times \mathbf{w} + \mathbf{v} \times \mathbf{x}$, and
- $(c\mathbf{v}) \times \mathbf{w} = \mathbf{v} \times (c\mathbf{w}) = c(\mathbf{v} \times \mathbf{w})$ for all $\mathbf{v}, \mathbf{w}, \mathbf{x} \in \mathbb{R}^3$ and $c \in \mathbb{R}$.

These properties are all straightforward to prove directly from the definition of the cross product, so we just prove the first one here:

> We proved these properties in Section 1.A of (Joh20).

$$\begin{aligned}
(\mathbf{v}+\mathbf{w}) \times \mathbf{x} &= \big((v_2+w_2)x_3 - (v_3+w_3)x_2, \; (v_3+w_3)x_1 - (v_1+w_1)x_3, \\
&\qquad (v_1+w_1)x_2 - (v_2+w_2)x_1\big) \\
&= (v_2 x_3 - v_3 x_2, \; v_3 x_1 - v_1 x_3, \; v_1 x_2 - v_2 x_1) \\
&\quad + (w_2 x_3 - w_3 x_2, \; w_3 x_1 - w_1 x_3, \; w_1 x_2 - w_2 x_1) \\
&= \mathbf{v} \times \mathbf{x} + \mathbf{w} \times \mathbf{x},
\end{aligned}$$

as claimed.

In fact, the cross product being bilinear is exactly why we think of it as a "product" or "multiplication" in the first place—bilinearity (and more generally, multilinearity) corresponds to the "multiplication" distributing over vector addition. We already made note of this fact back in Section 1.3.3—the usual multiplication operation on $\mathbb{R}$ is bilinear, as is the dot product on $\mathbb{R}^n$. We now present some more examples that all have a similar interpretation.

- Matrix multiplication is a bilinear transformation. That is, if we define the function $T_\times : \mathcal{M}_{m,n} \times \mathcal{M}_{n,p} \to \mathcal{M}_{m,p}$ that multiplies two matrices together via

$$T_\times(A,B) = AB \quad \text{for all} \quad A \in \mathcal{M}_{m,n}, \; B \in \mathcal{M}_{n,p},$$

then $T_\times$ is bilinear. To verify this claim, we just have to recall that matrix multiplication is both left- and right-distributive, so for any matrices A, B, and C of appropriate size and any scalar c, we have

$$T_\times(A+cB,C) = (A+cB)C = AC + cBC = T_\times(A,C) + cT_\times(B,C) \quad \text{and}$$
$$T_\times(A,B+cC) = A(B+cC) = AB + cAC = T_\times(A,B) + cT_\times(A,C).$$

- The Kronecker product is bilinear. That is, the function $T_\otimes : \mathcal{M}_{m,n} \times \mathcal{M}_{p,q} \to \mathcal{M}_{mp,nq}$ that is defined by $T_\otimes(A,B) = A \otimes B$, is a bilinear transformation. This fact follows immediately from Theorem 3.1.1(b–d).

All of these "multiplications" also become multilinear transformations if we extend them to three or more inputs in the natural way. For example, the function $T_\times$ defined on *triples* of matrices via $T_\times(A,B,C) = ABC$ is a multilinear (trilinear?) transformation.

It is often useful to categorize multilinear transformations into groups based on how many input and output vector spaces they have. With this in mind, we say that a multilinear transformation $T : \mathcal{V}_1 \times \mathcal{V}_2 \times \cdots \times \mathcal{V}_p \to \mathcal{W}$ is of **type** $(p,0)$ if $\mathcal{W} = \mathbb{F}$ is the ground field, and we say that it is of type $(p,1)$ otherwise. That is, the first number in a multilinear transformation's type tells us how many input vector spaces it has, and its second number similarly tells us how many output vector spaces it has (with $\mathcal{W} = \mathbb{F}$ being interpreted as 0 output spaces, since the output space is trivial). The sum of these two numbers (i.e., either p or $p + 1$, depending on whether $\mathcal{W} = \mathbb{F}$ or not) is called its **order**.

For example, matrix multiplication (between two matrices) is a multilinear transformation of type $(2, 1)$ and order $2 + 1 = 3$, and the dot product has type $(2, 0)$ and order $2 + 0 = 2$. We will see shortly that the order of a multilinear transformation tells us the dimensionality of an array of numbers that should be used to represent that transformation (just like vectors can be represented via a 1D list of numbers and linear transformations can be represented via a 2D array of numbers/a matrix). For now though, we spend some time clarifying what types of multilinear transformations correspond to which sets of linear algebraic objects that we are already familiar with:

- Transformations of type $(1, 1)$ are functions $T : \mathcal{V} \to \mathcal{W}$ that act linearly on $\mathcal{V}$. In other words, they are exactly linear transformations, which we already know and love. Furthermore, the order of a linear transformation is $1 + 1 = 2$, which corresponds to the fact that we can represent them via matrices, which are 2D arrays of numbers.

- Transformations of type $(1, 0)$ are linear transformations $f : \mathcal{V} \to \mathbb{F}$, which are linear forms. The order of a linear form is $1 + 0 = 1$, which corresponds to the fact that we can represent them via vectors (via Theorem 1.3.3), which are 1D arrays of numbers. In particular, recall that we think of these as row vectors.

As a slightly more trivial special case, scalars can be thought of as multilinear transformations of type $(0,0)$.

- Transformations of type $(2, 0)$ are bilinear forms $T : \mathcal{V}_1 \times \mathcal{V}_2 \to \mathbb{F}$, which have order $2 + 0 = 2$. The order once again corresponds to the fact (Theorem 1.3.5) that they can be represented naturally by matrices (2D arrays of numbers).

- Slightly more generally, transformations of type $(p, 0)$ are multilinear forms $T : \mathcal{V}_1 \times \cdots \times \mathcal{V}_p \to \mathbb{F}$. The fact that they have order $p + 0 = p$ corresponds to the fact that we can represent them via Theorem 1.3.6 as p-dimensional arrays of scalars.

- Transformations of type $(0, 1)$ are linear transformations $T : \mathbb{F} \to \mathcal{W}$, which are determined completely by the value of $T(1)$. In particular, for every such linear transformation T, there exists a vector $\mathbf{w} \in \mathcal{W}$ such that $T(c) = c\mathbf{w}$ for all $c \in \mathbb{F}$. The fact that the order of these transformations is $0 + 1 = 1$ corresponds to the fact that we can represent them via the vector $\mathbf{w}$ (i.e., a 1D array of numbers) in this way, though this time we think of it as a *column* vector.

We summarize the above special cases, as well the earlier examples of bilinear transformations like matrix multiplication, in Figure 3.1 for easy reference.

Remark 3.2.1

Tensors

Multilinear transformations can be generalized even further to something called **tensors**, but doing so requires some technicalities that we enjoy avoiding. While multilinear transformations allow multilinearity (rather than just linearity) on their input, sometimes it is useful to allow their output to behave in a multilinear (rather than just linear) way as well. The simplest way to make this happen is to make use of dual spaces.

Recall from Definition 1.3.3 that $\mathcal{W}_j^*$ is the dual space of $\mathcal{W}_j$, which consists of all linear forms acting on $\mathcal{W}_j$. Dual spaces were explored back in Section 1.3.2.

Specifically, if $\mathcal{V}_1, \mathcal{V}_2, \ldots, \mathcal{V}_p$ and $\mathcal{W}_1, \mathcal{W}_2, \ldots, \mathcal{W}_q$ are vector spaces over a field $\mathbb{F}$, then a **tensor** of **type** (p, q) is a multilinear form

$$f : \mathcal{V}_1 \times \mathcal{V}_2 \times \cdots \times \mathcal{V}_p \times \mathcal{W}_1^* \times \mathcal{W}_2^* \times \cdots \times \mathcal{W}_q^* \to \mathbb{F}.$$

Furthermore, its **order** is the quantity $p + q$. The idea is that attaching $\mathcal{W}_j^*$

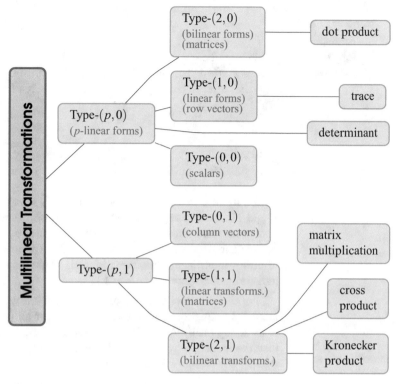

We do not give any explicit examples of type-$(0,0)$, $(0,1)$, or $(1,1)$ transformations here (i.e., scalars, column vectors, or linear transformations) since you are hopefully familiar enough with these objects by now that you can come up with some yourself.

Figure 3.1: A summary of the various multilinear transformations, and their types, that we have encountered so far.

We do not make use of any tensors of type (p,q) with $q \geq 2$, which is why we focus on multilinear transformations.

as an input vector space sort of mimics having $\mathcal{W}_j$ itself as an output vector space, for the exact same reason that the double-dual $\mathcal{W}_j^{**}$ is so naturally isomorphic to $\mathcal{W}_j$. In particular, if $q = 1$ then we can consider a tensor $f : \mathcal{V}_1 \times \mathcal{V}_2 \times \cdots \times \mathcal{V}_p \times \mathcal{W}^* \to \mathbb{F}$ as "the same thing" as a multilinear transformation

$$T : \mathcal{V}_1 \times \mathcal{V}_2 \times \cdots \times \mathcal{V}_p \to \mathcal{W}.$$

3.2.2 Arrays

We saw back in Theorem 1.3.6 that we can represent multilinear forms (i.e., type-$(p,0)$ multilinear transformations) on finite-dimensional vector spaces as multi-dimensional arrays of scalars, much like we represent linear transformations as matrices. We now show that we can similarly do this for multilinear transformations of type $(p,1)$. This fact should not be too surprising—the idea is simply that each multilinear transformation is determined completely by how it acts on basis vectors in each of the input arguments.

Theorem 3.2.1

Multilinear Transformations as Arrays

Suppose $\mathcal{V}_1, \ldots, \mathcal{V}_p$ and $\mathcal{W}$ are finite-dimensional vector spaces over the same field and $T : \mathcal{V}_1 \times \cdots \times \mathcal{V}_p \to \mathcal{W}$ is a multilinear transformation. Let $v_{1,j}, \ldots, v_{p,j}$ denote the j-th coordinate of $\mathbf{v}_1 \in \mathcal{V}_1, \ldots, \mathbf{v}_p \in \mathcal{V}_p$ with respect to some bases of $\mathcal{V}_1, \ldots, \mathcal{V}_p$, respectively, and let $\{\mathbf{w}_1, \mathbf{w}_2, \ldots\}$ be a basis of $\mathcal{W}$. Then there exists a unique family of scalars $\{a_{i;j_1,\ldots,j_p}\}$ such that

$$T(\mathbf{v}_1, \ldots, \mathbf{v}_p) = \sum_{i, j_1, \ldots, j_p} a_{i;j_1,\ldots,j_p} v_{1,j_1} \cdots v_{p,j_p} \mathbf{w}_i$$

for all $\mathbf{v}_1 \in \mathcal{V}_1, \ldots, \mathbf{v}_p \in \mathcal{V}_p$.

Proof. In the $p = 1$ case, this theorem simply says that, once we fix bases of $\mathcal{V}$ and $\mathcal{W}$, we can represent every linear transformation $T : \mathcal{V} \to \mathcal{W}$ via its standard matrix, whose (i, j)-entry we denote here by $a_{i;j}$. This is a fact that we already know to be true from Theorem 1.2.6. We now prove this more general result for multilinear transformations via induction on p, and note that linear transformations provide the $p = 1$ base case.

For the inductive step, we proceed exactly as we did in the proof of Theorem 1.3.6. Suppose that the result is true for all $(p-1)$-linear transformations acting on $\mathcal{V}_2 \times \cdots \times \mathcal{V}_p$. If we let $B = \{\mathbf{x}_1, \mathbf{x}_2, \ldots, \mathbf{x}_m\}$ be a basis of $\mathcal{V}_1$ then the inductive hypothesis tells us that the $(p-1)$-linear transformations $S_{j_1} : \mathcal{V}_2 \times \cdots \times \mathcal{V}_p \to \mathcal{W}$ defined by

> We use j_1 here instead of just j since it will be convenient later on.

$$S_{j_1}(\mathbf{v}_2, \ldots, \mathbf{v}_p) = T(\mathbf{x}_{j_1}, \mathbf{v}_2, \ldots, \mathbf{v}_p)$$

can be written as

> The scalar $a_{i;j_1,j_2,\ldots,j_p}$ here depends on j_1 (not just $i, j_2, \ldots, j_p$) since each choice of $\mathbf{x}_{j_1}$ gives a different $(p-1)$-linear transformation S_{j_1}.

$$T(\mathbf{x}_{j_1}, \mathbf{v}_2, \ldots, \mathbf{v}_p) = S_{j_1}(\mathbf{v}_2, \ldots, \mathbf{v}_p) = \sum_{i, j_2, \ldots, j_p} a_{i;j_1,j_2,\ldots,j_p} v_{2,j_2} \cdots v_{p,j_p} \mathbf{w}_i$$

for some fixed family of scalars $\{a_{i;j_1,j_2,\ldots,j_p}\}$. If we write an arbitrary vector $\mathbf{v}_1 \in \mathcal{V}_1$ as a linear combination of the basis vectors $\mathbf{x}_1, \mathbf{x}_2, \ldots, \mathbf{x}_m$ (i.e., $\mathbf{v}_1 = v_{1,1}\mathbf{x}_1 + v_{1,2}\mathbf{x}_2 + \cdots + v_{1,m}\mathbf{x}_m$), it then follows via linearity that

$$
\begin{aligned}
&T(\mathbf{v}_1, \mathbf{v}_2, \ldots, \mathbf{v}_p) \\
&= T\left(\sum_{j_1=1}^{m} v_{1,j_1} \mathbf{x}_{j_1}, \mathbf{v}_2, \ldots, \mathbf{v}_p \right) && \left(\mathbf{v}_1 = \sum_{j_1=1}^{m} v_{1,j_1} \mathbf{x}_{j_1} \right) \\
&= \sum_{j_1=1}^{m} v_{1,j_1} T(\mathbf{x}_{j_1}, \mathbf{v}_2, \ldots, \mathbf{v}_p) && \text{(multilinearity of } T) \\
&= \sum_{j_1=1}^{m} v_{1,j_1} \sum_{i, j_2, \ldots, j_p} a_{i;j_1,j_2,\ldots,j_p} v_{2,j_2} \cdots v_{p,j_p} \mathbf{w}_i && \text{(inductive hypothesis)} \\
&= \sum_{i, j_1, \ldots, j_p} a_{j_1,\ldots,j_p} v_{1,j_1} \cdots v_{p,j_p} \mathbf{w}_i && \text{(group sums together)}
\end{aligned}
$$

for all $\mathbf{v}_1 \in \mathcal{V}_1, \mathbf{v}_2 \in \mathcal{V}_2, \ldots, \mathbf{v}_p \in \mathcal{V}_p$, which completes the inductive step and shows that the family of scalars $\{a_{i;j_1,\ldots,j_p}\}$ exists.

To see that the scalars $\{a_{i;j_1,\ldots,j_p}\}$ are unique, just note that if we choose $\mathbf{v}_1$ to be the j_1-th member of the basis of $\mathcal{V}_1$, $\mathbf{v}_2$ to be the j_2-th member of the basis of $\mathcal{V}_2$, and so on, then we get

$$T(\mathbf{v}_1, \ldots, \mathbf{v}_p) = \sum_{i} a_{i;j_1,\ldots,j_p} \mathbf{w}_i.$$

Since the coefficients of $T(\mathbf{v}_1, \ldots, \mathbf{v}_p)$ in the basis $\{\mathbf{w}_1, \mathbf{w}_2, \ldots\}$ are unique, it follows that the scalars $\{a_{i;j_1,\ldots,j_p}\}_i$ are as well. ∎

The above theorem generalizes numerous theorems for representing (multi)-linear objects that we have seen in the past. In the case of type-$(1,1)$ transformations, it gives us exactly the fact (Theorem 1.2.6) that every linear transformation can be represented by a matrix (with the (i, j)-entry denoted by $a_{i;j}$ instead of $a_{i,j}$ this time). It also generalizes Theorem 1.3.6, which told us that every multilinear form (i.e., type-$(p,0)$ transformation) can be represented by an array $\{a_{j_1,j_2,\ldots,j_p}\}$.

> This result also generalizes Theorem 1.3.5 for bilinear forms, since that was a special case of Theorem 1.3.6 for multilinear forms.

Just as was the case when representing linear transformations via standard matrices, the "point" of the above theorem is that, once we know the values of the scalars $\{a_{i;j_1,\ldots,j_p}\}$, we can compute the value of the multilinear transformation T on any tuple of input vectors just by writing each of those vectors in terms of the given basis and then using multilinearity. It is perhaps most natural to arrange those scalars in a multi-dimensional array (with one dimension for each subscript), which we call the **standard array** of the transformation, just as we are used to arranging the scalars $\{a_{i;j}\}$ into a 2D matrix for linear transformations. However, the details are somewhat uglier here since it is not so easy to write these multi-dimensional arrays on two-dimensional paper or computer screens.

Since the order of a multilinear transformation tells us how many subscripts are needed to specify the scalars $a_{i;j_1,\ldots,j_p}$, it also tells us the dimensionality of the standard array of the multilinear transformation. For example, this is why both linear transformations (which have type $(1,1)$ and order $1+1=2$) and bilinear forms (which have type $(2,0)$ and order $2+0=2$) can be represented by 2-dimensional matrices. The **size** of an array is an indication of how many rows it has in each of its dimensions. For example, an array $\{a_{i;j_1,j_2}\}$ for which $1 \le i \le 2$, $1 \le j_1 \le 3$, and $1 \le j_2 \le 4$ has size $2 \times 3 \times 4$.

We already saw one example of an order 3 multilinear transformation (in particular, the determinant of a 3×3 matrix, which has type $(3,0)$) represented as a 3D array back in Example 1.3.14. We now present two more such examples, but this time for bilinear transformations (i.e., order-3 multilinear transformations of type $(2,1)$).

Example 3.2.2

The Cross Product as a 3D Array

Let $C : \mathbb{R}^3 \times \mathbb{R}^3 \to \mathbb{R}^3$ be the cross product (defined in Example 3.2.1). Construct the standard array of C with respect to the standard basis of $\mathbb{R}^3$.

Solution:

Since C is an order-3 tensor, its standard array is 3-dimensional and specifically of size $3 \times 3 \times 3$. To compute the scalars $\{a_{i;j_1,j_2}\}$, we do exactly the same thing that we do to compute the standard matrix of a linear transformation: plug the basis vectors of the input spaces into C and write the results in terms of the basis vectors of the output space. For example, direct calculation shows that

$$C(\mathbf{e}_1, \mathbf{e}_2) = \mathbf{e}_1 \times \mathbf{e}_2 = (1,0,0) \times (0,1,0) = (0,0,1) = \mathbf{e}_3,$$

and the other 8 cross products can be similarly computed to be

Some of these computations can be sped up by recalling that $\mathbf{v} \times \mathbf{v} = \mathbf{0}$ and $\mathbf{v} \times \mathbf{w} = -(\mathbf{w} \times \mathbf{v})$ for all $\mathbf{v}, \mathbf{w} \in \mathbb{R}^3$.

$$C(\mathbf{e}_1, \mathbf{e}_1) = \mathbf{0} \qquad C(\mathbf{e}_1, \mathbf{e}_2) = \mathbf{e}_3 \qquad C(\mathbf{e}_1, \mathbf{e}_3) = -\mathbf{e}_2$$
$$C(\mathbf{e}_2, \mathbf{e}_1) = -\mathbf{e}_3 \qquad C(\mathbf{e}_2, \mathbf{e}_2) = \mathbf{0} \qquad C(\mathbf{e}_2, \mathbf{e}_3) = \mathbf{e}_1$$
$$C(\mathbf{e}_3, \mathbf{e}_1) = \mathbf{e}_2 \qquad C(\mathbf{e}_3, \mathbf{e}_2) = -\mathbf{e}_1 \qquad C(\mathbf{e}_3, \mathbf{e}_3) = \mathbf{0}.$$

It follows that the scalars $\{a_{i;j_1,j_2}\}$ are given by

$$a_{3;1,2} = 1 \qquad\qquad a_{2;1,3} = -1$$
$$a_{3;2,1} = -1 \qquad\qquad a_{1;2,3} = 1$$
$$a_{2;3,1} = 1 \qquad a_{1;3,2} = -1,$$

Recall that the first subscript in $a_{i;j_1,j_2}$ corresponds to the output of the transformation and the next two subscripts correspond to the inputs.

and $a_{i;j_1,j_2} = 0$ otherwise. We can arrange these scalars into a 3D array, which we display below with i indexing the rows, j_1 indexing the columns, and j_2 indexing the "layers" as indicated:

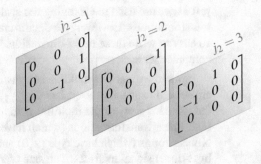

If you ever used a mnemonic like
$$\det\left(\begin{bmatrix} \mathbf{e}_1 & \mathbf{e}_2 & \mathbf{e}_3 \\ v_1 & v_2 & v_3 \\ w_1 & w_2 & w_3 \end{bmatrix}\right)$$
to compute $\mathbf{v} \times \mathbf{w}$, this is why it works.

It is worth noting that the standard array of the cross product that we computed in the above example is *exactly* the same as the standard array of the determinant (of a 3×3 matrix) that we constructed back in Example 1.3.14. In a sense, the determinant and cross product are the exact same thing, just written in a different way. They are both order-3 multilinear transformations, but the determinant on $\mathcal{M}_3$ is of type $(3,0)$ whereas the cross product is of type $(2,1)$, which explains why they have such similar properties (e.g., they can both be used to find the area of parallelograms in $\mathbb{R}^2$ and volume of parallelepipeds in $\mathbb{R}^3$).

Example 3.2.3

Matrix Multiplication as a 3D Array

Construct the standard array of the matrix multiplication map $T_\times : \mathcal{M}_2 \times \mathcal{M}_2 \to \mathcal{M}_2$ with respect to the standard basis of $\mathcal{M}_2$.

Solution:

Since $T_\times$ is an order-3 tensor, its corresponding array is 3-dimensional and specifically of size $4 \times 4 \times 4$. To compute the scalars $\{a_{i;j_1,j_2}\}$, we again plug the basis vectors (matrices) of the input spaces into $T_\times$ and write the results in terms of the basis vectors of the output space. Rather than compute all 16 of the required matrix products explicitly, we simply recall that

$$T_\times(E_{k,a}, E_{b,\ell}) = E_{k,a}E_{b,\ell} = \begin{cases} E_{k,\ell}, & \text{if } a = b \text{ or} \\ O, & \text{otherwise.} \end{cases}$$

For example, $a_{1;2,3} = 1$ because $E_{1,2}$ (the 2nd basis vector) times $E_{2,1}$ (the 3rd basis vector) equals $E_{1,1}$ (the 1st basis vector).

We thus have

$$a_{1;1,1} = 1 \qquad a_{1;2,3} = 1 \qquad a_{2;1,2} = 1 \qquad a_{2;2,4} = 1$$
$$a_{3;3,1} = 1 \qquad a_{3;4,3} = 1 \qquad a_{4;3,2} = 1 \qquad a_{4;4,4} = 1,$$

and $a_{i;j_1,j_2} = 0$ otherwise. We can arrange these scalars into a 3D array, which we display below with i indexing the rows, j_1 indexing the columns, and j_2 indexing the "layers" as indicated:

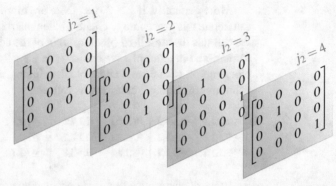

Even though there is no standard order which which dimensions correspond to the subscripts $j_1, \ldots, j_k$, we always use i (which corresponded to the single output vector space) to index the rows of the standard array.

It is worth noting that there is no "standard" order for which of the three subscripts should represent the rows, columns, and layers of a 3D array, and swapping the roles of the subscripts can cause the array to look quite different. For example, if we use i to index the rows, j_2 to index the columns, and j_1 to index the layers, then this array instead looks like

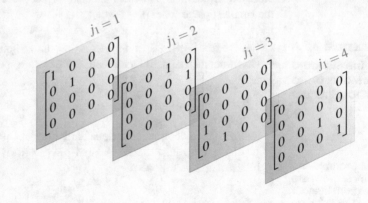

One of the author's greatest failings is his inability to draw 4D objects on 2D paper.

Arrays as Block Matrices

When the order of a multilinear transformation is 4 or greater, it becomes extremely difficult to effectively visualize its (4- or higher-dimensional) standard array. To get around this problem, we can instead write an array $\{a_{i;j_1,\ldots,j_p}\}$ as a block matrix. To illustrate how this procedure works, suppose for now that $p = 2$, so the array has the form $\{a_{i;j_1,j_2}\}$. We can represent this array as a block matrix by letting i index the rows, j_1 index the block columns, and j_2 index the columns within the blocks. For example, we could write a $3 \times 2 \times 4$ array $\{a_{i;j_1,j_2}\}$ as a matrix with 3 rows, 2 block columns, and 4 columns in each block:

Reading the subscripts in this matrix from left-to-right, top-to-bottom, is like counting: the final digit increases to its maximum value (4 in this case) and then "rolls over" to increase the next-to-last digit by 1, and so on.

$$A = \left[\begin{array}{cccc|cccc} a_{1;1,1} & a_{1;1,2} & a_{1;1,3} & a_{1;1,4} & a_{1;2,1} & a_{1;2,2} & a_{1;2,3} & a_{1;2,4} \\ a_{2;1,1} & a_{2;1,2} & a_{2;1,3} & a_{2;1,4} & a_{2;2,1} & a_{2;2,2} & a_{2;2,3} & a_{2;2,4} \\ a_{3;1,1} & a_{3;1,2} & a_{3;1,3} & a_{3;1,4} & a_{3;2,1} & a_{3;2,2} & a_{3;2,3} & a_{3;2,4} \end{array}\right].$$

As a more concrete example, we can write the array corresponding to the matrix multiplication tensor $T_\times : \mathcal{M}_2 \times \mathcal{M}_2 \to \mathcal{M}_2$ from Example 3.2.3 as the following 4×1 block matrix of 4×4 matrices:

$$
\left[
\begin{array}{cccc|cccc|cccc|cccc}
1 & 0 & 0 & 0 & 0 & 0 & 1 & 0 & 0 & 0 & 0 & 0 & 0 & 0 & 0 & 0 \\
0 & 1 & 0 & 0 & 0 & 0 & 0 & 1 & 0 & 0 & 0 & 0 & 0 & 0 & 0 & 0 \\
0 & 0 & 0 & 0 & 0 & 0 & 0 & 0 & 1 & 0 & 0 & 0 & 0 & 0 & 1 & 0 \\
0 & 0 & 0 & 0 & 0 & 0 & 0 & 0 & 0 & 1 & 0 & 0 & 0 & 0 & 0 & 1
\end{array}
\right].
$$

More generally, if $k > 2$ (i.e., there are more than two "j" subscripts), we can iterate this procedure to create a block matrix that has blocks within blocks. We call this the **standard block matrix** of the corresponding multilinear transformation. For example, we could write a $3 \times 2 \times 2 \times 2$ array $\{a_{i;j_1,j_2,j_3}\}$ as a block matrix with 3 rows and 2 block columns, each of which contain 2 block columns containing 2 columns, as follows:

$$
\left[
\begin{array}{cc|cc|cc|cc}
a_{1;1,1,1} & a_{1;1,1,2} & a_{1;1,2,1} & a_{1;1,2,2} & a_{1;2,1,1} & a_{1;2,1,2} & a_{1;2,2,1} & a_{1;2,2,2} \\
a_{2;1,1,1} & a_{2;1,1,2} & a_{2;1,2,1} & a_{2;1,2,2} & a_{2;2,1,1} & a_{2;2,1,2} & a_{2;2,2,1} & a_{2;2,2,2} \\
a_{3;1,1,1} & a_{3;1,1,2} & a_{3;1,2,1} & a_{3;1,2,2} & a_{3;2,1,1} & a_{3;2,1,2} & a_{3;2,2,1} & a_{3;2,2,2}
\end{array}
\right]
$$

In particular, notice that if we arrange an array into a block matrix in this way, the index j_1 indicates the "most significant" block column, j_2 indicates the next most significant block column, and so on. This is completely analogous to how the digits of a number are arranged left-to-right from most significant to least significant (e.g., in the number 524, the digit "5" indicates the largest piece (hundreds), "2" indicates the next largest piece (tens), and "4" indicates the smallest piece (ones)).

Example 3.2.4

The Standard Block Matrix of the Cross Product

Let $C : \mathbb{R}^3 \times \mathbb{R}^3 \to \mathbb{R}^3$ be the cross product (defined in Example 3.2.1). Construct the standard block matrix of C with respect to the standard basis of $\mathbb{R}^3$.

Solution:

We already computed the standard array of C in Example 3.2.2, so we just have to arrange the 27 scalars from that array into its 3×9 standard block matrix as follows:

For example, the leftmost "-1" here comes from the fact that, in the standard array A of C, $a_{2;1,3} = -1$ (the subscripts say row 2, block column 1, column 3.

$$
A = \left[
\begin{array}{ccc|ccc|ccc}
0 & 0 & 0 & 0 & 0 & 1 & 0 & -1 & 0 \\
0 & 0 & -1 & 0 & 0 & 0 & 1 & 0 & 0 \\
0 & 1 & 0 & -1 & 0 & 0 & 0 & 0 & 0
\end{array}
\right].
$$

As one particular case that requires some special attention, consider what the standard block matrix of a bilinear form $f : \mathcal{V}_1 \times \mathcal{V}_2 \to \mathbb{F}$ looks like. Since bilinear forms are multilinear forms of order 2, we could arrange the entries of their standard arrays into a matrix of size $\dim(\mathcal{V}_1) \times \dim(\mathcal{V}_2)$, just as we did back in Theorem 1.3.5. However, if we follow the method described above for turning a standard array into a standard block matrix, we notice that we instead get a $1 \times \dim(\mathcal{V}_1)$ block matrix (i.e., row vector) whose entries are themselves $1 \times \dim(\mathcal{V}_2)$ row vectors. For example, the standard block matrix $\{a_{1;j_1,j_2}\}$ of a bilinear form acting on two 3-dimensional vector spaces would have the form

$$
\left[
\begin{array}{ccc|ccc|ccc}
a_{1;1,1} & a_{1;1,2} & a_{1;1,3} & a_{1;2,1} & a_{1;2,2} & a_{1;2,3} & a_{1;3,1} & a_{1;3,2} & a_{1;3,3}
\end{array}
\right].
$$

Example 3.2.5

The Standard Block Matrix of the Dot Product

Construct the standard block matrix of the dot product on $\mathbb{R}^4$ with respect to the standard basis.

Solution:

Recall from Example 1.3.9 that the dot product is a bilinear form that thus a multilinear transformation of type $(2,0)$. To compute the entries of its standard block matrix, we compute its value when all 16 possible combinations of standard vectors are plugged into it:

$$\mathbf{e}_{j_1} \cdot \mathbf{e}_{j_2} = \begin{cases} 1, & \text{if } j_1 = j_2 \text{ or} \\ 0, & \text{otherwise.} \end{cases}$$

It follows that

$$a_{1;1,1} = 1 \qquad a_{1;2,2} = 1 \qquad a_{1;3,3} = 1 \qquad a_{1;4,4} = 1,$$

and $a_{1;j_1,j_2} = 0$ otherwise. If we let j_1 index which block column an entry is placed in and j_2 index which column within that block it is placed in, we arrive at the following standard block matrix:

$$A = \begin{bmatrix} 1 & 0 & 0 & 0 \mid 0 & 1 & 0 & 0 \mid 0 & 0 & 1 & 0 \mid 0 & 0 & 0 & 1 \end{bmatrix}.$$

Notice that the standard block matrix of the dot product that we constructed in the previous example is simply the identity matrix, read row-by-row. This is not a coincidence—it follows from the fact that the matrix representation of the dot product (in the sense of Theorem 1.3.5) is the identity matrix. We now state this observation slightly more generally and more prominently:

(!) If $f : \mathcal{V}_1 \times \mathcal{V}_2 \to \mathbb{F}$ is a bilinear form acting on finite-dimensional vector spaces then its standard block matrix is the vectorization of its matrix representation from Theorem 1.3.5, as a row vector.

Remark 3.2.2

Same Order but Different Type

One of the advantages of the standard block matrix over a standard array is that its shape tells us its type (rather than just its order). For example, we saw in Examples 3.2.2 and 1.3.14 that the cross product and the determinant have the same standard arrays. However, their standard block matrices are different—they are matrices of size 3×9 and 1×27, respectively.

This distinction can be thought of as telling us that the cross product and determinant of a 3×3 matrix are the same object (e.g., anything that can be done with the determinant of a 3×3 matrix can be done with the cross product, and vice-versa), but the way that they act on *other* objects is different (the cross product takes in 2 vectors and outputs 1 vector, whereas the determinant takes in 3 vectors and outputs a scalar).

Similarly, while matrices represent both bilinear forms and linear transformations, the way that they act in those two settings is different (since their types are different but their orders are not.

We can similarly construct standard block matrices of tensors (see Remark 3.2.1), and the only additional wrinkle is that they can have block rows too (not just block columns). For example, the standard arrays of bilinear forms $f : \mathcal{V}_1 \times \mathcal{V}_2 \to \mathbb{F}$ of type $(2,0)$, linear transformations $T : \mathcal{V} \to \mathcal{W}$ and type-$(0,2)$ tensors $f : \mathcal{W}_1^* \times \mathcal{W}_2^* \to \mathbb{F}$ have the following forms, respectively, if all vectors spaces involved are 3-dimensional:

$$\left[\begin{array}{ccc|ccc|ccc} a_{1;1,1} & a_{1;1,2} & a_{1;1,3} & a_{1;2,1} & a_{1;2,2} & a_{1;2,3} & a_{1;3,1} & a_{1;3,2} & a_{1;3,3} \end{array}\right],$$

$$\begin{bmatrix} a_{1;1} & a_{1;2} & a_{1;3} \\ a_{2;1} & a_{2;2} & a_{2;3} \\ a_{3;1} & a_{3;2} & a_{3;3} \end{bmatrix}, \quad \text{and} \quad \begin{bmatrix} a_{1,1;1} \\ a_{1,2;1} \\ a_{1,3;1} \\ \hline a_{2,1;1} \\ a_{2,2;1} \\ a_{2,3;1} \\ \hline a_{3,1;1} \\ a_{3,2;1} \\ a_{3,3;1} \end{bmatrix}.$$

In general, just like there are two possible shapes for tensors of order 1 (row vectors and column vectors), there are three possible shapes for tensors of order 2 (corresponding to types $(2,0)$, $(1,1)$, and $(0,2)$) and $p+1$ possible shapes for tensors of order p. Each of these different types corresponds to a different shape of the standard block matrix and a different way in which the tensor acts on vectors.

Refer back to Theorem 1.2.6 for a precise statement about how a linear transformations relate to matrix multiplication.

Just like the action of a linear transformation on a vector can be represented via multiplication by that transformation's standard matrix, so too can the action of a multilinear transformation on multiple vectors. However, we need one additional piece of machinery to make this statement actually work (after all, what does it even mean to multiply a matrix by *multiple* vectors?), and that machinery is the Kronecker product.

Theorem 3.2.2

Standard Block Matrix of a Multilinear Transformation

Suppose $V_1, \ldots, V_p$ and W are finite-dimensional vector spaces with bases $B_1, \ldots, B_p$ and D, respectively, and $T : V_1 \times \cdots \times V_p \to W$ is a multilinear transformation. If A is the standard block matrix of T with respect to those bases, then

$$\big[T(\mathbf{v}_1, \ldots, \mathbf{v}_p)\big]_D = A\big([\mathbf{v}_1]_{B_1} \otimes \cdots \otimes [\mathbf{v}_p]_{B_p}\big) \quad \text{for all} \quad \mathbf{v}_1 \in V_1, \ldots, \mathbf{v}_p \in V_p.$$

We could use notation like $[T]_{D \leftarrow B_1, \ldots, B_p}$ to denote the standard block matrix of T with respect to the bases $B_1, \ldots, B_p$ and D. That seems like a bit much, though.

Before proving this result, we look at some examples to clarify exactly what it is saying. Just as with linear transformations, the main idea here is that we can now represent arbitrary *multi*linear transformations (on finite-dimensional vector spaces) via explicit matrix calculations. Even more amazingly though, this theorem tells us that the Kronecker product turns multilinear things (the transformation T) into linear things (multiplication by the matrix A). In a sense, the Kronecker product can be used to absorb all of the multilinearity of multilinear transformations, turning them into linear transformations (which are much easier to work with).

Example 3.2.6

The Cross Product via the Kronecker Product

Verify Theorem 3.2.2 for the cross product $C : \mathbb{R}^3 \times \mathbb{R}^3 \to \mathbb{R}^3$ with respect to the standard basis of $\mathbb{R}^3$.

Solution:

Our goal is simply to show that if A is the standard block matrix of the cross product then $C(\mathbf{v}, \mathbf{w}) = A(\mathbf{v} \otimes \mathbf{w})$ for all $\mathbf{v}, \mathbf{w} \in \mathbb{R}^3$. If we recall this standard block matrix from Example 3.2.4 then we can simply compute

In this setting, we interpret $\mathbf{v} \otimes \mathbf{w}$ as a column vector, just like we do for $\mathbf{v}$ in a matrix equation like $A\mathbf{v} = \mathbf{b}$.

$$A(\mathbf{v} \otimes \mathbf{w}) = \begin{bmatrix} 0 & 0 & 0 & | & 0 & 0 & 1 & | & 0 & -1 & 0 \\ 0 & 0 & -1 & | & 0 & 0 & 0 & | & 1 & 0 & 0 \\ 0 & 1 & 0 & | & -1 & 0 & 0 & | & 0 & 0 & 0 \end{bmatrix} \begin{bmatrix} v_1 w_1 \\ v_1 w_2 \\ v_1 w_3 \\ \hline v_2 w_1 \\ v_2 w_2 \\ v_2 w_3 \\ \hline v_3 w_1 \\ v_3 w_2 \\ v_3 w_3 \end{bmatrix}$$

$$= \begin{bmatrix} v_2 w_3 - v_3 w_2 \\ v_3 w_1 - v_1 w_3 \\ v_1 w_2 - v_2 w_1 \end{bmatrix},$$

which is indeed exactly the cross product $C(\mathbf{v}, \mathbf{w})$, as expected.

Example 3.2.7

The Dot Product via the Kronecker Product

Verify Theorem 3.2.2 for the dot product on $\mathbb{R}^4$ with respect to the standard basis.

Solution:

Our goal is to show that if A is the standard block matrix of the dot product (which we computed in Example 3.2.5) then $\mathbf{v} \cdot \mathbf{w} = A(\mathbf{v} \otimes \mathbf{w})$ for all $\mathbf{v}, \mathbf{w} \in \mathbb{R}^4$. If we recall that this standard block matrix is

For space reasons, we do not explicitly list the product $A(\mathbf{v} \otimes \mathbf{w})$—it is a 1×16 matrix times a 16×1 matrix. The four terms in the dot product correspond to the four "1"s in A.

$$A = \begin{bmatrix} 1 & 0 & 0 & 0 & | & 0 & 1 & 0 & 0 & | & 0 & 0 & 1 & 0 & | & 0 & 0 & 0 & 1 \end{bmatrix},$$

then we see that

$$A(\mathbf{v} \otimes \mathbf{w}) = v_1 w_1 + v_2 w_2 + v_3 w_3 + v_4 w_4,$$

which is indeed the dot product $\mathbf{v} \cdot \mathbf{w}$.

Example 3.2.8

Matrix Multiplication via the Kronecker Product

Verify Theorem 3.2.2 for matrix multiplication on $\mathcal{M}_2$ with respect to the standard basis.

Solution:

Once again, our goal is to show that if $E = \{E_{1,1}, E_{1,2}, E_{2,1}, E_{2,2}\}$ is the standard basis of $\mathcal{M}_2$ and

$$A = \begin{bmatrix} 1 & 0 & 0 & 0 & | & 0 & 0 & 1 & 0 & | & 0 & 0 & 0 & 0 & | & 0 & 0 & 0 & 0 \\ 0 & 1 & 0 & 0 & | & 0 & 0 & 0 & 1 & | & 0 & 0 & 0 & 0 & | & 0 & 0 & 0 & 0 \\ 0 & 0 & 0 & 0 & | & 0 & 0 & 0 & 0 & | & 1 & 0 & 0 & 0 & | & 0 & 0 & 1 & 0 \\ 0 & 0 & 0 & 0 & | & 0 & 0 & 0 & 0 & | & 0 & 1 & 0 & 0 & | & 0 & 0 & 0 & 1 \end{bmatrix}$$

is the standard block matrix of matrix multiplication on $\mathcal{M}_2$, then $[BC]_E = A([B]_E \otimes [C]_E)$ for all $B, C \in \mathcal{M}_2$. To this end, we first recall that if $b_{i,j}$ and $c_{i,j}$ denote the (i, j)-entries of B and C (as usual), respectively, then

$$[B]_E = \begin{bmatrix} b_{1,1} \\ b_{1,2} \\ b_{2,1} \\ b_{2,2} \end{bmatrix} \quad \text{and} \quad [C]_E = \begin{bmatrix} c_{1,1} \\ c_{1,2} \\ c_{2,1} \\ c_{2,2} \end{bmatrix}.$$

Directly computing the Kronecker product and matrix multiplication then shows that

$$A([B]_E \otimes [C]_E) = \begin{bmatrix} b_{1,1}c_{1,1} + b_{1,2}c_{2,1} \\ b_{1,1}c_{1,2} + b_{1,2}c_{2,2} \\ b_{2,1}c_{1,1} + b_{2,2}c_{2,1} \\ b_{2,1}c_{1,2} + b_{2,2}c_{2,2} \end{bmatrix},$$

On the other hand, it is similarly straightforward to compute

$$BC = \begin{bmatrix} b_{1,1}c_{1,1} + b_{1,2}c_{2,1} & b_{1,1}c_{1,2} + b_{1,2}c_{2,2} \\ b_{2,1}c_{1,1} + b_{2,2}c_{2,1} & b_{2,1}c_{1,2} + b_{2,2}c_{2,2} \end{bmatrix},$$

so $[BC]_E = A([B]_E \otimes [C]_E)$, as expected.

Example 3.2.9

The Standard Block Matrix of the Determinant is the Antisymmetric Kronecker Product

Construct the standard block matrix of the determinant (as a multilinear form from $\mathbb{F}^n \times \cdots \times \mathbb{F}^n$ to $\mathbb{F}$) with respect to the standard basis of $\mathbb{F}^n$.

Solution:

Recall that if $A, B \in \mathcal{M}_n$ and B is obtained from A by interchanging two of its columns, then $\det(B) = -\det(A)$. As a multilinear form acting on the *columns* of a matrix, this means that the determinant is **antisymmetric**:

$$\det(\mathbf{v}_1, \ldots, \mathbf{v}_{j_1}, \ldots, \mathbf{v}_{j_2}, \ldots, \mathbf{v}_n) = -\det(\mathbf{v}_1, \ldots, \mathbf{v}_{j_2}, \ldots, \mathbf{v}_{j_1}, \ldots, \mathbf{v}_n)$$

for all $\mathbf{v}_1, \ldots, \mathbf{v}_n$ and $1 \le j_1 \ne j_2 \le n$. By applying Theorem 3.2.2 to both sides of this equation, we see that if A is the standard block matrix of the determinant (which is $1 \times n^n$ and can thus be thought of as a row vector) then

$$A(\mathbf{v}_1 \otimes \cdots \otimes \mathbf{v}_{j_1} \otimes \cdots \otimes \mathbf{v}_{j_2} \otimes \cdots \otimes \mathbf{v}_n)$$
$$= -A(\mathbf{v}_1 \otimes \cdots \otimes \mathbf{v}_{j_2} \otimes \cdots \otimes \mathbf{v}_{j_1} \otimes \cdots \otimes \mathbf{v}_n).$$

Now we notice that, since interchanging two Kronecker factors multiplies this result by -1, it is in fact the case that permuting the Kronecker factors according to a permutation $\sigma \in S_n$ multiplies the result by $\mathrm{sgn}(\sigma)$. Furthermore, we use the fact that the elementary tensors $\mathbf{v}_1 \otimes \cdots \otimes \mathbf{v}_n$ span all of $(\mathbb{F}^n)^{\otimes n}$, so we have

Recall that S_n is the symmetric group, which consists of all permutations acting on $\{1, 2, \ldots, n\}$.

$$A\mathbf{v} = \mathrm{sgn}(\sigma)(A(W_\sigma \mathbf{v})) = (\mathrm{sgn}(\sigma) W_\sigma^T A^T)^T \mathbf{v}$$

for all $\sigma \in S_n$ and $\mathbf{v} \in (\mathbb{F}^n)^{\otimes n}$. By using the fact that $W_\sigma^T = W_{\sigma^{-1}}$, we then see that $A^T = \mathrm{sgn}(\sigma) W_{\sigma^{-1}} A^T$ for all $\sigma \in S_n$, which exactly means that A lives in the antisymmetric subspace of $(\mathbb{F}^n)^{\otimes n}$: $A \in \mathcal{A}^{n,n}$.

We now recall from Section 3.1.3 that $\mathcal{A}^{n,n}$ is 1-dimensional, and in particular A must have the form

$$A = c \sum_{\sigma \in S_n} \mathrm{sgn}(\sigma) \mathbf{e}_{\sigma(1)} \otimes \mathbf{e}_{\sigma(2)} \otimes \cdots \otimes \mathbf{e}_{\sigma(n)}$$

This example also shows that the determinant is the only function with the following three properties: multilinearity, antisymmetry, and $\det(I) = 1$.

for some $c \in \mathbb{F}$. Furthermore, since $\det(I) = 1$, we see that $A^T(\mathbf{e}_1 \otimes \cdots \otimes \mathbf{e}_n) = 1$, which tells us that $c = 1$. We have thus shown that the standard block matrix of the determinant is

$$A = \sum_{\sigma \in S_n} \text{sgn}(\sigma) \mathbf{e}_{\sigma(1)} \otimes \mathbf{e}_{\sigma(2)} \otimes \cdots \otimes \mathbf{e}_{\sigma(n)},$$

as a row vector.

Proof of Theorem 3.2.2. There actually is not much that needs to be done to prove this theorem—all of the necessary bits and pieces just fit together via naïve (but messy) direct computation. For ease of notation, we let $v_{1,j}$, ..., $v_{p,j}$ denote the j-th entry of $[\mathbf{v}_1]_{B_1}$, ..., $[\mathbf{v}_p]_{B_p}$, respectively, just like in Theorem 3.2.1.

On the one hand, using the definition of the Kronecker product tells us that, for each integer i, the i-th entry of $A([\mathbf{v}_1]_{B_1} \otimes \cdots \otimes [\mathbf{v}_p]_{B_p})$ is

The notation here is a bit unfortunate. $[\cdot]_B$ refers to a coordinate vector and $[\cdot]_i$ refers to the i-th entry of a vector.

$$\left[A\left([\mathbf{v}_1]_{B_1} \otimes \cdots \otimes [\mathbf{v}_p]_{B_p}\right)\right]_i = \underbrace{\sum_{j_1, j_2, \ldots, j_p}}_{\text{sum across } i\text{-th row of } A} a_{i; j_1, j_2, \ldots, j_p} \underbrace{v_{1,j_1} v_{2,j_2} \cdots v_{p,j_p}}_{\substack{\text{entries of} \\ [\mathbf{v}_1]_{B_1} \otimes \cdots \otimes [\mathbf{v}_p]_{B_p}}}.$$

On the other hand, this is exactly what Theorem 3.2.1 tells us that the i-th entry of $\left[T(\mathbf{v}_1, \ldots, \mathbf{v}_p)\right]_D$ is (in the notation of that theorem, this was the coefficient of $\mathbf{w}_i$, the i-th basis vector in D), which is all we needed to observe to complete the proof. ∎

3.2.3 Properties of Multilinear Transformations

Most of the usual properties of linear transformations that we are familiar with can be extended to multilinear transformations, but the details almost always become significantly uglier and more difficult to work with. We now discuss how to do this for some of the most basic properties like the operator norm, null space, range, and rank.

Operator Norm

The operator norm of a linear transformation $T : \mathcal{V} \to \mathcal{W}$ between finite-dimensional inner product spaces is

A similar definition works if $\mathcal{V}$ and $\mathcal{W}$ are infinite-dimensional, but we must replace "max" with "sup", and even then the value of the supremum might be ∞.

$$\|T\| = \max_{\mathbf{v} \in \mathcal{V}} \left\{ \|T(\mathbf{v})\| : \|\mathbf{v}\| = 1 \right\},$$

where $\|\mathbf{v}\|$ refers to the norm of $\mathbf{v}$ induced by the inner product on $\mathcal{V}$, and $\|T(\mathbf{v})\|$ refers to the norm of $T(\mathbf{v})$ induced by the inner product on $\mathcal{W}$. In the special case when $\mathcal{V} = \mathcal{W} = \mathbb{F}^n$, this is just the operator norm of a matrix from Section 2.3.3.

To extend this idea to *multi*linear transformations, we just optimize over unit vectors in each input argument. That is, if $T : \mathcal{V}_1 \times \mathcal{V}_2 \times \cdots \times \mathcal{V}_p \to \mathcal{W}$ is a multilinear transformation between finite-dimensional inner product spaces, then we define its **operator norm** as follows:

$$\|T\| \stackrel{\text{def}}{=} \max_{\substack{\mathbf{v}_j \in \mathcal{V}_j \\ 1 \le j \le p}} \left\{ \|T(\mathbf{v}_1, \mathbf{v}_2, \ldots, \mathbf{v}_p)\| : \|\mathbf{v}_1\| = \|\mathbf{v}_2\| = \cdots = \|\mathbf{v}_p\| = 1 \right\}.$$

While the operator norm of a linear transformation can be computed quickly via the singular value decomposition (refer back to Theorem 2.3.7), no quick or easy method of computing the operator norm of a *multi*linear transformation is known. However, we can carry out the computation for certain special multilinear transformations if we are willing to work hard enough.

Example 3.2.10

The Operator Norm of the Cross Product

Let $C : \mathbb{R}^3 \times \mathbb{R}^3 \to \mathbb{R}^3$ be the cross product (defined in Example 3.2.1). Compute $\|C\|$.

Solution:

One of the basic properties of the cross product is that the norm (induced by the dot product) of its output satisfies

This formula is proved in (Joh20, Section 1.A), for example.

$$\|C(\mathbf{v}, \mathbf{w})\| = \|\mathbf{v}\| \|\mathbf{w}\| \sin(\theta) \quad \text{for all} \quad \mathbf{v}, \mathbf{w} \in \mathbb{R}^3,$$

where θ is the angle between $\mathbf{v}$ and $\mathbf{w}$. Since $|\sin(\theta)| \leq 1$, this implies $\|C(\mathbf{v}, \mathbf{w})\| \leq \|\mathbf{v}\| \|\mathbf{w}\|$. If we normalize $\mathbf{v}$ and $\mathbf{w}$ to be unit vectors then we get $\|C(\mathbf{v}, \mathbf{w})\| \leq 1$, with equality if and only if $\mathbf{v}$ and $\mathbf{w}$ are orthogonal. It follows that $\|C\| = 1$.

Example 3.2.11

The Operator Norm of the Determinant

Suppose $\mathbb{F} = \mathbb{R}$ or $\mathbb{F} = \mathbb{C}$, and $\det : \mathbb{F}^n \times \cdots \times \mathbb{F}^n \to \mathbb{F}$ is the determinant. Compute $\|\det\|$.

Solution:

Recall from Exercise 2.2.28 (Hadamard's inequality) that

$$\left| \det(\mathbf{v}_1, \mathbf{v}_2, \ldots, \mathbf{v}_n) \right| \leq \|\mathbf{v}_1\| \|\mathbf{v}_2\| \cdots \|\mathbf{v}_n\| \quad \text{for all} \quad \mathbf{v}_1, \mathbf{v}_2, \ldots, \mathbf{v}_n \in \mathbb{F}^n.$$

If we normalize each $\mathbf{v}_1, \mathbf{v}_2, \ldots, \mathbf{v}_n$ to have norm 1 then this inequality becomes $|\det(\mathbf{v}_1, \mathbf{v}_2, \ldots, \mathbf{v}_n)| \leq 1$. Furthermore, equality is sometimes attained (e.g., when $\{\mathbf{v}_1, \mathbf{v}_2, \ldots, \mathbf{v}_n\}$ is an orthonormal basis of $\mathbb{F}^n$), so we conclude that $\|\det\| = 1$.

It is perhaps worth noting that if a multilinear transformation acts on finite-dimensional inner product spaces, then Theorem 3.2.2 tells us that we can express its operator norm in terms of its standard block matrix A (with respect to orthonormal bases of all of the vector spaces). In particular, if we let $d_j = \dim(\mathcal{V}_j)$ for all $1 \leq j \leq k$, then

$$\|T\| = \max_{\substack{\mathbf{v}_j \in \mathbb{F}^{d_j} \\ 1 \leq j \leq p}} \left\{ \|A(\mathbf{v}_1 \otimes \mathbf{v}_2 \otimes \cdots \otimes \mathbf{v}_p)\| : \|\mathbf{v}_1\| = \|\mathbf{v}_2\| = \cdots = \|\mathbf{v}_p\| = 1 \right\}.$$

While this quantity is difficult to compute in general, there is an easy-to-compute upper bound of it: if we just optimize $\|A\mathbf{v}\|$ over *all* unit vectors in $\mathbb{F}^{d_1} \otimes \cdots \otimes \mathbb{F}^{d_p}$ (instead of just over elementary tensors, as above), we get exactly the "usual" operator norm of A. In other words, we have shown the following:

Be careful: $\|T\|$ is the *multilinear* operator norm (which is hard to compute), while $\|A\|$ is the operator norm of a matrix (which is easy to compute).

(!) If $T : \mathcal{V}_1 \times \cdots \times \mathcal{V}_p \to \mathcal{W}$ is a multilinear transformation with standard block matrix A then $\|T\| \leq \|A\|$.

For example, it is straightforward to verify that the standard block matrix A of the cross product C (see Example 3.2.4) has all three of its non-zero singular values equal to $\sqrt{2}$, so $\|C\| \leq \|A\| = \sqrt{2}$. Similarly, the standard block matrix

B of the determinant $\det : \mathbb{F}^n \times \cdots \times \mathbb{F}^n \to \mathbb{F}$ (see Example 3.2.9) is just a row vector with $n!$ non-zero entries all equal to ± 1, so $\|\det\| \leq \|B\| = \sqrt{n!}$. Of course, both of these bounds agree with (but are weaker than) the facts from Examples 3.2.10 and 3.2.11 that $\|C\| = 1$ and $\|\det\| = 1$.

Range and Kernel

In Section 1.2.4, we introduced the range and null space of a linear transformation $T : \mathcal{V} \to \mathcal{W}$ to be the following subspaces of $\mathcal{W}$ and $\mathcal{V}$, respectively:

$$\text{range}(T) = \big\{ T(\mathbf{x}) : \mathbf{x} \in \mathcal{V} \big\} \quad \text{and} \quad \text{null}(T) = \big\{ \mathbf{x} \in \mathcal{V} : T(\mathbf{x}) = \mathbf{0} \big\}.$$

The natural generalizations of these sets to a *multi*linear transformation $T : \mathcal{V}_1 \times \mathcal{V}_2 \times \cdots \times \mathcal{V}_p \to \mathcal{W}$ are called its **range** and **kernel**, respectively, and they are defined by

$$\text{range}(T) \stackrel{\text{def}}{=} \big\{ T(\mathbf{v}_1, \mathbf{v}_2, \ldots, \mathbf{v}_p) : \mathbf{v}_1 \in \mathcal{V}_1, \mathbf{v}_2 \in \mathcal{V}_2, \ldots, \mathbf{v}_p \in \mathcal{V}_p \big\} \quad \text{and}$$

$$\ker(T) \stackrel{\text{def}}{=} \big\{ (\mathbf{v}_1, \mathbf{v}_2, \ldots, \mathbf{v}_p) \in \mathcal{V}_1 \times \mathcal{V}_2 \times \cdots \times \mathcal{V}_p : T(\mathbf{v}_1, \mathbf{v}_2, \ldots, \mathbf{v}_p) = \mathbf{0} \big\}.$$

In other words, the range of a multilinear transformation contains all of the possible outputs of that transformation, just as was the case for linear transformations, and its kernel consists of all inputs that result in an output of $\mathbf{0}$, just as was the case for the null space of linear transformations. The reason for changing terminology from "null space" to "kernel" in this multilinear setting is that these sets are no longer subspaces if $p \geq 2$. We illustrate this fact with an example.

Example 3.2.12 **The Range of the Kronecker Product**	Describe the range of the bilinear transformation $T_\otimes : \mathbb{R}^2 \times \mathbb{R}^2 \to \mathbb{R}^4$ that computes the Kronecker product of two vectors, and show that it is not a subspace of $\mathbb{R}^4$. **Solution:** The range of $T_\otimes$ is the set of all vectors of the form $T_\otimes(\mathbf{v}, \mathbf{w}) = \mathbf{v} \otimes \mathbf{w}$, which are the elementary tensors in $\mathbb{R}^2 \otimes \mathbb{R}^2 \cong \mathbb{R}^4$. We noted near the end of Section 3.1.1 that $\mathbb{R}^2 \otimes \mathbb{R}^2$ is spanned by elementary tensors, but not every vector in this space *is* an elementary tensor.

Recall from Theorem 3.1.6 that if $\mathbf{v}$ is an elementary tensor then $\text{rank}(\text{mat}(\mathbf{v})) = 1$.

    For example, $(1,0) \otimes (1,0) = (1,0,0,0)$ and $(0,1) \otimes (0,1) = (0,0,0,1)$ are both in range$(T_\otimes)$, but $(1,0,0,0) + (0,0,0,1) = (1,0,0,1)$ is not (after all, its matricization is the identity matrix I_2, which does not have rank 1), so range$(T_\otimes)$ is not a subspace of $\mathbb{R}^4$.

To even talk about whether or not $\ker(T)$ is a subspace, we need to clarify what it could be a subspace *of*. The members of $\ker(T)$ come from the set $\mathcal{V}_1 \times \mathcal{V}_2 \times \cdots \times \mathcal{V}_p$, but this set is not a vector space (so it does not make sense for it to have subspaces). We can turn it into a vector space by defining a vector addition and scalar multiplication on it—one way of doing this gives us the external direct sum from Section 1.B. However, $\ker(T)$ fails to be a subspace under pretty much any reasonable choice of vector operations.

Example 3.2.13 **Kernel of the Determinant**	Describe the kernel of the determinant $\det : \mathbb{F}^n \times \cdots \times \mathbb{F}^n \to \mathbb{F}$. **Solution:** Recall that the determinant of a matrix equals 0 if and only if that matrix is not invertible. Also recall that a matrix is not invertible if and

only if its columns form a linearly dependent set. By combining these two facts, we see that $\det(\mathbf{v}_1, \mathbf{v}_2, \ldots, \mathbf{v}_n) = 0$ if and only if $\{\mathbf{v}_1, \mathbf{v}_2, \ldots, \mathbf{v}_n\}$ is linearly dependent. In other words,

$$\ker(\det) = \big\{(\mathbf{v}_1, \mathbf{v}_2, \ldots, \mathbf{v}_n) \in \mathbb{F}^n \times \cdots \times \mathbb{F}^n :$$
$$\{\mathbf{v}_1, \mathbf{v}_2, \ldots, \mathbf{v}_n\} \text{ is linearly dependent}\big\}.$$

To get our hands on $\text{range}(T)$ and $\ker(T)$ in general, we can make use of the standard block matrix A of T, just like we did for the operator norm. Specifically, making use of Theorem 3.2.2 immediately gives us the following result:

Theorem 3.2.3 **Range and Kernel in Terms of Standard Block Matrices**	Suppose $\mathcal{V}_1, \ldots, \mathcal{V}_p$ and $\mathcal{W}$ are finite-dimensional vector spaces with bases $B_1, \ldots, B_p$ and D, respectively, and $T : \mathcal{V}_1 \times \cdots \times \mathcal{V}_p \to \mathcal{W}$ is a multilinear transformation. If A is the standard block matrix of T with respect to these bases, then $$\text{range}(T) = \big\{\mathbf{w} \in \mathcal{W} : [\mathbf{w}]_D = A([\mathbf{v}_1]_{B_1} \otimes \cdots \otimes [\mathbf{v}_p]_{B_p})$$ $$\text{for some } \mathbf{v}_1 \in \mathcal{V}_1, \ldots, \mathbf{v}_p \in \mathcal{V}_p\big\}, \quad \text{and}$$ $$\ker(T) = \big\{(\mathbf{v}_1, \ldots, \mathbf{v}_p) \in \mathcal{V}_1 \times \cdots \times \mathcal{V}_p : A([\mathbf{v}_1]_{B_1} \otimes \cdots \otimes [\mathbf{v}_p]_{B_p}) = \mathbf{0}\big\}.$$

That is, the range and kernel of a multilinear transformation consist of the elementary tensors in the range and null space of its standard block matrix, respectively (up to fixing bases of the vector spaces so that this statement actually makes sense). This fact is quite analogous to our earlier observation that the operator norm of a multilinear transformation T is obtained by maximizing $\|A\mathbf{v}\|$ over elementary tensors $\mathbf{v} = \mathbf{v}_1 \otimes \cdots \otimes \mathbf{v}_p$.

Example 3.2.14 **Kernel of the Cross Product**	Compute the kernel of the cross product $C : \mathbb{R}^3 \times \mathbb{R}^3 \to \mathbb{R}^3$ directly, and also via Theorem 3.2.3. **Solution:** To compute the kernel directly, recall that $\|C(\mathbf{v}, \mathbf{w})\| = \|\mathbf{v}\| \|\mathbf{w}\| |\sin(\theta)|$, where θ is the angle between $\mathbf{v}$ and $\mathbf{w}$ (we already used this formula in Example 3.2.10). It follows that $C(\mathbf{v}, \mathbf{w}) = \mathbf{0}$ if and only if $\mathbf{v} = \mathbf{0}$, $\mathbf{w} = \mathbf{0}$, or $\sin(\theta) = 0$. That is, $\ker(C)$ consists of all pairs $(\mathbf{v}, \mathbf{w})$ for which $\mathbf{v}$ and $\mathbf{w}$ lie on a common line.

In other words, $\ker(C)$ consists of sets $\{\mathbf{v}, \mathbf{w}\}$ that are linearly dependent (compare with Example 3.2.13).

To instead arrive at this result via Theorem 3.2.3, we first recall that C has standard block matrix

$$A = \begin{bmatrix} 0 & 0 & 0 & 0 & 0 & 1 & 0 & -1 & 0 \\ 0 & 0 & -1 & 0 & 0 & 0 & 1 & 0 & 0 \\ 0 & 1 & 0 & -1 & 0 & 0 & 0 & 0 & 0 \end{bmatrix}.$$

Standard techniques from introductory linear algebra show that the null space of A is the symmetric subspace: $\text{null}(A) = \mathcal{S}_3^2$. The kernel of C thus consists of the elementary tensors in $\mathcal{S}_3^2$ (once appropriately re-interpreted just as pairs of vectors, rather than as elementary tensors).

We know from Theorem 3.1.11 that these elementary tensors are exactly the ones of the form $c(\mathbf{v} \otimes \mathbf{v})$ for some $\mathbf{v} \in \mathbb{R}^3$. Since $c(\mathbf{v} \otimes \mathbf{v}) = (c\mathbf{v}) \otimes \mathbf{v} = \mathbf{v} \otimes (c\mathbf{v})$, it follows that $\ker(C)$ consists of all pairs $(\mathbf{v}, \mathbf{w})$ for which $\mathbf{v}$ is a multiple of $\mathbf{w}$ or $\mathbf{w}$ is a multiple of $\mathbf{v}$ (i.e., $\mathbf{v}$ and $\mathbf{w}$ lie on a common line).

Rank

Recall that the rank of a linear transformation is the dimension of its range. Since the range of a multilinear transformation is not necessarily a subspace, we cannot use this same definition in this more general setting. Instead, we treat the rank-one sum decomposition (Theorem A.1.3) as the definition of the rank of a matrix, and we then generalize it to linear transformations and then multilinear transformations.

Once we do this, we see that one way of expressing the rank of a linear transformation $T : V \to W$ is as the minimal integer r such that there exist linear forms $f_1, f_2, \ldots, f_r : V \to \mathbb{F}$ and vectors $\mathbf{w}_1, \ldots, \mathbf{w}_r \in W$ with the property that

> Notice that Equation (3.2.1) implies that the range of T is contained within the span of $\mathbf{w}_1, \ldots, \mathbf{w}_r$, so $\mathrm{rank}(T) \leq r$.

$$T(\mathbf{v}) = \sum_{j=1}^{r} f_j(\mathbf{v})\mathbf{w}_j \quad \text{for all} \quad \mathbf{v} \in V. \tag{3.2.1}$$

Indeed, this way of defining the rank of T makes sense if we recall that, in the case when $V = \mathbb{F}^n$ and $W = \mathbb{F}^m$, each f_j looks like a row vector: there exist vectors $\mathbf{x}_1, \ldots, \mathbf{x}_r \in \mathbb{F}^n$ such that $f_j(\mathbf{v}) = \mathbf{x}_j^T \mathbf{v}$ for all $1 \leq j \leq r$. In this case, Equation (3.2.1) says that

$$T(\mathbf{v}) = \sum_{j=1}^{r} (\mathbf{x}_j^T \mathbf{v})\mathbf{w}_j = \sum_{j=1}^{r} \left(\mathbf{w}_j \mathbf{x}_j^T\right)\mathbf{v} \quad \text{for all} \quad \mathbf{v} \in \mathbb{F}^n,$$

which simply means that the standard matrix of T is $\sum_{j=1}^{r} \mathbf{w}_j \mathbf{x}_j^T$. In other words, Equation (3.2.1) just extends the idea of a rank-one sum decomposition from matrices to linear transformations.

To generalize Equation (3.2.1) to *multi*linear transformations, we just introduce additional linear forms—one for each of the input vector spaces.

Definition 3.2.2 **The Rank of a Multilinear Transformation**	Suppose $V_1, \ldots, V_p$ and W are vector spaces and $T : V_1 \times \cdots \times V_p \to W$ is a multilinear transformation. The **rank** of T, denoted by $\mathrm{rank}(T)$, is the minimal integer r such that there exist linear forms $f_1^{(i)}, f_2^{(i)}, \ldots, f_r^{(i)} : V_i \to \mathbb{F}$ $(1 \leq i \leq p)$ and vectors $\mathbf{w}_1, \ldots, \mathbf{w}_r \in W$ with the property that $$T(\mathbf{v}_1, \ldots, \mathbf{v}_p) = \sum_{j=1}^{r} f_j^{(1)}(\mathbf{v}_1) \cdots f_j^{(p)}(\mathbf{v}_p)\mathbf{w}_j \quad \text{for all} \quad \mathbf{v}_1 \in V_1, \ldots, \mathbf{v}_p \in V_p.$$

Just like the previous properties of multilinear transformations that we looked at (the operator norm, range, and kernel), the rank of a multilinear transformation is extremely difficult to compute in general. In fact, the rank is so difficult to compute that we do not even know the rank of many of the most basic multilinear transformations that we have been working with, like matrix multiplication and the determinant.

Example 3.2.15 **The Rank of 2×2 Matrix Multiplication**	Show that if $T_\times : M_2 \times M_2 \to M_2$ is the bilinear transformation that multiplies two 2×2 matrices, then $\mathrm{rank}(T_\times) \leq 8$. **Solution:** Our goal is to construct a sum of the form described by Definition 3.2.2 for $T_\times$ that consists of no more than 8 terms. Fortunately, one such sum

is very easy to construct—we just write down the definition of matrix multiplication.

More explicitly, recall that the top-left entry in the product $T_\times(A,B) = AB$ is $a_{1,1}b_{1,1} + a_{1,2}b_{2,1}$, so 2 of the 8 terms in the sum that defines AB are $a_{1,1}b_{1,1}E_{1,1}$ and $a_{1,2}b_{2,1}E_{1,1}$. The other entries of AB similarly contribute 2 terms each to the sum, for a total of 8 terms in the sum. Even more explicitly, if we define

$$f_1^{(1)}(A) = a_{1,1} \quad f_2^{(1)}(A) = a_{1,2} \quad f_3^{(1)}(A) = a_{1,1} \quad f_4^{(1)}(A) = a_{1,2}$$
$$f_5^{(1)}(A) = a_{2,1} \quad f_6^{(1)}(A) = a_{2,2} \quad f_7^{(1)}(A) = a_{2,1} \quad f_8^{(1)}(A) = a_{2,2}$$

$$f_1^{(2)}(B) = b_{1,1} \quad f_2^{(2)}(B) = b_{2,1} \quad f_3^{(2)}(B) = b_{1,2} \quad f_4^{(2)}(B) = b_{2,2}$$
$$f_5^{(2)}(B) = b_{1,1} \quad f_6^{(2)}(B) = b_{2,1} \quad f_7^{(2)}(B) = b_{1,2} \quad f_8^{(2)}(B) = b_{2,2}$$

$$\mathbf{w}_1 = E_{1,1} \qquad \mathbf{w}_2 = E_{1,1} \qquad \mathbf{w}_3 = E_{1,2} \qquad \mathbf{w}_4 = E_{1,2}$$
$$\mathbf{w}_5 = E_{2,1} \qquad \mathbf{w}_6 = E_{2,1} \qquad \mathbf{w}_7 = E_{2,2} \qquad \mathbf{w}_8 = E_{2,2},$$

A similar argument shows that matrix multiplication $T_\times :$ $\mathcal{M}_{m,n} \times \mathcal{M}_{n,p} \to \mathcal{M}_{m,p}$ has $\text{rank}(T_\times) \le mnp$.

then it is straightforward (but tedious) to check that

$$T_\times(A,B) = AB = \sum_{j=1}^{8} f_j^{(1)}(A) f_j^{(2)}(B)\mathbf{w}_j \quad \text{for all} \quad A, B \in \mathcal{M}_2.$$

After all, this sum can be written more explicitly as
$$AB = (a_{1,1}b_{1,1} + a_{1,2}b_{2,1})E_{1,1} + (a_{1,1}b_{1,2} + a_{1,2}b_{2,2})E_{1,2}$$
$$+ (a_{2,1}b_{1,1} + a_{2,2}b_{2,1})E_{2,1} + (a_{2,1}b_{1,2} + a_{2,2}b_{2,2})E_{2,2}.$$

One of the reasons that the rank of a multilinear transformation is so difficult to compute is that we do not have a method of reliably detecting whether or not a sum of the type described by Definition 3.2.2 is optimal.

For example, it seems reasonable to expect that the matrix multiplication transformation $T_\times : \mathcal{M}_2 \times \mathcal{M}_2 \to \mathcal{M}_2$ has $\text{rank}(T_\times) = 8$ (i.e., we might expect that the rank-one sum decomposition from Example 3.2.15 is optimal). Rather surprisingly, however, this is not the case—the following decomposition of $T_\times$ as $T_\times(A,B) = AB = \sum_{j=1}^{7} f_j^{(1)}(A) f_j^{(2)}(B)\mathbf{w}_j$ shows that $\text{rank}(T_\times) \le 7$:

It turns out that, for 2×2 matrix multiplication, $\text{rank}(T_\times) = 7$, but this is not obvious. That is, it is hard to show that there is no sum of the type described by Definition 3.2.2 consisting of only 6 terms.

$$f_1^{(1)}(A) = a_{1,1} \qquad f_2^{(1)}(A) = a_{2,2} \qquad f_3^{(1)}(A) = a_{1,2} - a_{2,2}$$
$$f_4^{(1)}(A) = a_{2,1} + a_{2,2} \qquad f_5^{(1)}(A) = a_{1,1} + a_{1,2} \qquad f_6^{(1)}(A) = a_{2,1} - a_{1,1}$$
$$f_7^{(1)}(A) = a_{1,1} + a_{2,2}$$

$$f_1^{(2)}(B) = b_{1,2} - b_{2,2} \qquad f_2^{(2)}(B) = b_{2,1} - b_{1,1} \qquad f_3^{(2)}(B) = b_{2,1} + b_{2,2}$$
$$f_4^{(2)}(B) = b_{1,1} \qquad f_5^{(2)}(B) = b_{2,2} \qquad f_6^{(2)}(B) = b_{1,1} + b_{1,2} \qquad (\dagger)$$
$$f_7^{(2)}(B) = b_{1,1} + b_{2,2}$$

$$\mathbf{w}_1 = E_{1,2} + E_{2,2} \qquad \mathbf{w}_2 = E_{1,1} + E_{2,1} \qquad \mathbf{w}_3 = E_{1,1}$$
$$\mathbf{w}_4 = E_{2,1} - E_{2,2} \qquad \mathbf{w}_5 = E_{1,2} - E_{1,1} \qquad \mathbf{w}_6 = E_{2,2}$$
$$\mathbf{w}_7 = E_{1,1} + E_{2,2}.$$

The process of verifying that this decomposition works is just straightforward (but tedious) algebra. However, actually discovering this decomposition in the first place is much more difficult, and beyond the scope of this book.

Remark 3.2.3

Algorithms for Faster Matrix Multiplication

When multiplying two 2×2 matrices A and B together, we typically perform 8 scalar multiplications:

$$AB = \begin{bmatrix} a_{1,1} & a_{1,2} \\ a_{2,1} & a_{2,2} \end{bmatrix} \begin{bmatrix} b_{1,1} & b_{1,2} \\ b_{2,1} & b_{2,2} \end{bmatrix}$$

$$= \begin{bmatrix} a_{1,1}b_{1,1} + a_{1,2}b_{2,1} & a_{1,1}b_{1,2} + a_{1,2}b_{2,2} \\ a_{2,1}b_{1,1} + a_{2,2}b_{2,1} & a_{2,1}b_{1,2} + a_{2,2}b_{2,2} \end{bmatrix}.$$

However, it is possible to compute this same product a bit more efficiently—only 7 scalar multiplications are required if we first compute the 7 matrices

This method of computing AB requires more additions, but that is OK—the number of additions grows much more slowly than the number of multiplications when we apply it to larger matrices.

$$M_1 = a_{1,1}(b_{1,2} - b_{2,2}) \qquad M_2 = a_{2,2}(b_{2,1} - b_{1,1})$$
$$M_3 = (a_{1,2} - a_{2,2})(b_{2,1} + b_{2,2}) \qquad M_4 = (a_{2,1} + a_{2,2})b_{1,1}$$
$$M_5 = (a_{1,1} + a_{1,2})b_{2,2} \qquad M_6 = (a_{2,1} - a_{1,1})(b_{1,1} + b_{1,2})$$
$$M_7 = (a_{1,1} + a_{2,2})(b_{1,1} + b_{2,2}).$$

It is straightforward to check that if $\mathbf{w}_1, \ldots, \mathbf{w}_7$ are as in the decomposition (†), then

$$AB = \sum_{j=1}^{7} M_j \otimes \mathbf{w}_j = \begin{bmatrix} M_2 + M_3 - M_5 + M_7 & M_1 + M_5 \\ M_2 + M_4 & M_1 - M_4 + M_6 + M_7 \end{bmatrix}.$$

By iterating this procedure, we can compute the product of large matrices much more quickly than we can via the definition of matrix multiplication. For example, computing the product of two 4×4 matrices directly from the definition of matrix multiplication requires 64 scalar multiplications. However, if we partition those matrices as 2×2 block matrices and multiply them via the clever method above, we just need to perform 7 multiplications of 2×2 matrices, each of which can be implemented via 7 scalar multiplications in the same way, for a total of $7^2 = 49$ scalar multiplications.

To apply Strassen's algorithm to matrices of size that is not a power of 2, just pad it with rows and columns of zeros as necessary.

This faster method of matrix multiplication is called **Strassen's algorithm**, and it requires only $O(n^{\log_2(7)}) \approx O(n^{2.8074})$ scalar operations, versus the standard matrix multiplication algorithm's $O(n^3)$ scalar operations. The fact that we can multiply two 2×2 matrices together via just 7 scalar multiplications (i.e., the fact that Strassen's algorithm exists) follows immediately from the 7-term rank sum decomposition (†) for the 2×2 matrix multiplication transformation. After all, the matrices M_j ($1 \leq j \leq 7$) are simply the products $f_{1,j}(A)f_{2,j}(B)$.

The expressions like "$O(n^3)$" here are examples of big-O notation. For example, "$O(n^3)$ operations" means "no more than Cn^3 operations, for some scalar C".

Similarly, finding clever rank sum decompositions of larger matrix multiplication transformations leads to even faster algorithms for matrix multiplication, and these techniques have been used to construct an algorithm that multiplies two $n \times n$ matrices in $O(n^{2.3729})$ scalar operations. One of the most important open questions in all of linear algebra

Even the rank of the 3×3 matrix multiplication transformation is currently unknown—all we know is that it is between 19 and 23, inclusive. See (Sto10) and references therein for details.

asks whether or not, for every $\varepsilon > 0$, there exists such an algorithm that requires only $O(n^{2+\varepsilon})$ scalar operations.

The rank of a multilinear transformation can also be expressed in terms of its standard block matrix. By identifying the linear forms $\{f_{i,j}\}$ with row vectors in the usual way and making use of Theorem 3.2.2, we get the following theorem:

Theorem 3.2.4

Rank via Standard Block Matrices

Suppose $\mathcal{V}_1, \ldots, \mathcal{V}_p$ and $\mathcal{W}$ are d_1-, ..., d_p-, and $d_{\mathcal{W}}$-dimensional vector spaces, respectively, $T : \mathcal{V}_1 \times \cdots \times \mathcal{V}_p \to \mathcal{W}$ is a multilinear transformation, and A is the standard block matrix of T. Then the smallest integer r for which there exist sets of vectors $\{\mathbf{w}_j\}_{j=1}^r \subset \mathbb{F}^{d_{\mathcal{W}}}$ and $\{\mathbf{v}_j^{(1)}\}_{j=1}^r \subset \mathbb{F}^{d_1}$, ..., $\{\mathbf{v}_j^{(p)}\}_{j=1}^r \subset \mathbb{F}^{d_p}$ with

$$A = \sum_{j=1}^r \mathbf{w}_j \big(\mathbf{v}_j^{(1)} \otimes \mathbf{v}_j^{(2)} \otimes \cdots \otimes \mathbf{v}_j^{(p)} \big)^T$$

is exactly $r = \text{rank}(T)$.

The sets of vectors in this theorem *cannot* be chosen to be linearly independent in general, in contrast with Theorem A.1.3 (where they could).

If we recall from Theorem A.1.3 that $\text{rank}(A)$ (where now we are just thinking of A as a matrix—not as anything to do with multilinearity) equals the smallest integer r for which we can write

$$A = \sum_{j=1}^r \mathbf{w}_j \mathbf{v}_j^T,$$

we see learn immediately the important fact that the rank of a multilinear transformation is bounded from below by the rank of its standard block matrix (after all, every rank-one sum decomposition of the form described by Theorem 3.2.4 is also of the form described by Theorem A.1.3, but not vice-versa):

> (!) If $T : \mathcal{V}_1 \times \cdots \times \mathcal{V}_p \to \mathcal{W}$ is a multilinear transformation with standard block matrix A then $\text{rank}(T) \geq \text{rank}(A)$.

For example, it is straightforward to check that if $C : \mathbb{R}^3 \times \mathbb{R}^3 \to \mathbb{R}^3$ is the cross product, with standard block matrix $A \in \mathcal{M}_{3,9}$ as in Example 3.2.4, then $\text{rank}(C) \geq \text{rank}(A) = 3$ (in fact, $\text{rank}(C) = 5$, but this is difficult to show—see Exercise 3.2.13). Similarly, if $T_\times : \mathcal{M}_2 \times \mathcal{M}_2 \to \mathcal{M}_2$ is the matrix multiplication transformation with standard matrix $A \in \mathcal{M}_{4,16}$ from Example 3.2.8, then $\text{rank}(T_\times) \geq \text{rank}(A) = 4$ (recall that $\text{rank}(T_\times)$ actually equals 7).

Exercises

solutions to starred exercises on page 478

3.2.1 Determine which of the following functions are and are not multilinear transformations.

*(a) The function $T : \mathbb{R}^2 \times \mathbb{R}^2 \to \mathbb{R}^2$ defined by $T(\mathbf{v}, \mathbf{w}) = (v_1 + v_2, w_1 - w_2)$.

(b) The function $T : \mathbb{R}^2 \times \mathbb{R}^2 \to \mathbb{R}^2$ defined by $T(\mathbf{v}, \mathbf{w}) = (v_1 w_2 + v_2 w_1, v_1 w_1 - 2v_2 w_2)$.

*(c) The function $T : \mathbb{R}^n \times \mathbb{R}^n \to \mathbb{R}^{n^2}$ defined by $T(\mathbf{v}, \mathbf{w}) = \mathbf{v} \otimes \mathbf{w}$.

(d) The function $T : \mathcal{M}_n \times \mathcal{M}_n \to \mathcal{M}_n$ defined by $T(A, B) = A + B$.

*(e) Given a fixed matrix $X \in \mathcal{M}_n$, the function $T_X : \mathcal{M}_n \to \mathcal{M}_n$ defined by $T_X(A, B) = AXB^T$.

(f) The function per : $\mathbb{F}^n \times \cdots \times \mathbb{F}^n \to \mathbb{F}$ (with n copies of $\mathbb{F}^n$) defined by

$$\text{per}\big(\mathbf{v}^{(1)}, \mathbf{v}^{(2)}, \ldots, \mathbf{v}^{(n)}\big) = \sum_{\sigma \in S_n} v_{\sigma(1)}^{(1)} v_{\sigma(2)}^{(2)} \cdots v_{\sigma(n)}^{(n)}.$$

[Side note: This function is called the **permanent**, and its formula is similar to that of the determinant, but with the signs of permutations ignored.]

3.2.2 Compute the standard block matrix (with respect to the standard bases of the given spaces) of each of the following multilinear transformations:

(a) The function $T : \mathbb{R}^2 \times \mathbb{R}^2 \to \mathbb{R}^2$ defined by $T(\mathbf{v}, \mathbf{w}) = (2v_1 w_1 - v_2 w_2, v_1 w_1 + 2v_2 w_2 + 3v_2 w_2)$.

*(b) The function $T : \mathbb{R}^3 \times \mathbb{R}^2 \to \mathbb{R}^3$ defined by $T(\mathbf{v}, \mathbf{w}) = (v_2 w_1 + 2v_1 w_2 + 3v_3 w_1, v_1 w_1 + v_2 w_2 + v_3 w_1, v_3 w_2 - v_1 w_1)$.

*(c) The Kronecker product $T_\otimes : \mathbb{R}^2 \times \mathbb{R}^2 \to \mathbb{R}^4$ (i.e., the bilinear transformation defined by $T_\otimes(\mathbf{v}, \mathbf{w}) = \mathbf{v} \otimes \mathbf{w}$).

(d) Given a fixed matrix $X \in \mathcal{M}_2$, the function $T_X : \mathcal{M}_2 \times \mathbb{R}^2 \to \mathbb{R}^2$ defined by $T_X(A, \mathbf{v}) = AX\mathbf{v}$.

3.2.3 Determine which of the following statements are true and which are false.

*(a) The dot product $D : \mathbb{R}^n \times \mathbb{R}^n \to \mathbb{R}$ is a multilinear transformation of type $(n, n, 1)$.

(b) The multilinear transformation from Exercise 3.2.2(b) has type $(2, 1)$.

*(c) The standard block matrix of the matrix multiplication transformation $T_\times : \mathcal{M}_{m,n} \times \mathcal{M}_{n,p} \to \mathcal{M}_{m,p}$ has size $mn^2 p \times mp$.

(d) The standard block matrix of the Kronecker product $T_\otimes : \mathbb{R}^m \times \mathbb{R}^n \to \mathbb{R}^{mn}$ (i.e., the bilinear transformation defined by $T_\otimes(\mathbf{v}, \mathbf{w}) = \mathbf{v} \otimes \mathbf{w}$) is the $mn \times mn$ identity matrix.

3.2.4 Suppose $T : \mathcal{V}_1 \times \cdots \times \mathcal{V}_p \to \mathcal{W}$ is a multilinear transformation. Show that if there exists an index j for which $\mathbf{v}_j = \mathbf{0}$ then $T(\mathbf{v}_1, \ldots, \mathbf{v}_p) = \mathbf{0}$.

3.2.5 Suppose $T_\times : \mathcal{M}_{m,n} \times \mathcal{M}_{n,p} \to \mathcal{M}_{m,p}$ is the bilinear transformation that multiplies two matrices. Compute $\|T_\times\|$.

[Hint: Keep in mind that the norm induced by the inner product on $\mathcal{M}_{m,n}$ is the Frobenius norm. What inequalities do we know involving that norm?]

3.2.6 Show that the operator norm of a multilinear transformation is in fact a norm (i.e., satisfies the three properties of Definition 1.D.1).

3.2.7 Show that the range of a multilinear form $f : \mathcal{V}_1 \times \cdots \times \mathcal{V}_p \to \mathbb{F}$ is always a subspace of $\mathbb{F}$ (i.e., $\text{range}(f) = \{\mathbf{0}\}$ or $\text{range}(f) = \mathbb{F}$).

[Side note: Recall that this is *not* true of multilinear *transformations*.]

*3.2.8 Describe the range of the cross product (i.e., the bilinear transformation $C : \mathbb{R}^3 \times \mathbb{R}^3 \to \mathbb{R}^3$ from Example 3.2.1).

*3.2.9 Suppose $T : \mathcal{V}_1 \times \cdots \times \mathcal{V}_p \to \mathcal{W}$ is a multilinear transformation. Show that if its standard array has k nonzero entries then $\text{rank}(T) \leq k$.

*3.2.10 Suppose $T_\times : \mathcal{M}_{m,n} \times \mathcal{M}_{n,p} \to \mathcal{M}_{m,p}$ is the bilinear transformation that multiplies two matrices. Show that $mp \leq \text{rank}(T_\times) \leq mnp$.

*3.2.11 We motivated many properties of multilinear transformations so as to generalize properties of linear transformations. In this exercise, we show that they also generalize properties of bilinear forms.

Suppose $\mathcal{V}$ and $\mathcal{W}$ are finite-dimensional vector spaces over a field $\mathbb{F}$, $f : \mathcal{V} \times \mathcal{W} \to \mathbb{F}$ is a bilinear form, and A is the matrix associated with f via Theorem 1.3.5.

(a) Show that $\text{rank}(f) = \text{rank}(A)$.

(b) Show that if $\mathbb{F} = \mathbb{R}$ or $\mathbb{F} = \mathbb{C}$, and we choose the bases B and C in Theorem 1.3.5 to be orthonormal, then $\|f\| = \|A\|$.

3.2.12 Let $D : \mathbb{R}^n \times \mathbb{R}^n \to \mathbb{R}$ be the dot product.

(a) Show that $\|D\| = 1$.

(b) Show that $\text{rank}(D) = n$.

3.2.13 Let $C : \mathbb{R}^3 \times \mathbb{R}^3 \to \mathbb{R}^3$ be the cross product.

(a) Construct the "naïve" rank sum decomposition that shows that $\text{rank}(C) \leq 6$.
[Hint: Mimic Example 3.2.15.]

(b) Show that $C(\mathbf{v}, \mathbf{w}) = \sum_{j=1}^5 f_{1,j}(\mathbf{v}) f_{2,j}(\mathbf{w}) \mathbf{x}_j$ for all $\mathbf{v}, \mathbf{w} \in \mathbb{R}^3$, and thus $\text{rank}(C) \leq 5$, where

$$\begin{aligned} f_{1,1}(\mathbf{v}) &= v_1 & f_{1,2}(\mathbf{v}) &= v_1 + v_3 \\ f_{1,3}(\mathbf{v}) &= -v_2 & f_{1,4}(\mathbf{v}) &= v_2 + v_3 \\ f_{1,5}(\mathbf{v}) &= v_2 - v_1 \end{aligned}$$

$$\begin{aligned} f_{2,1}(\mathbf{w}) &= w_2 + w_3 & f_{2,2}(\mathbf{w}) &= -w_2 \\ f_{2,3}(\mathbf{w}) &= w_1 + w_3 & f_{2,4}(\mathbf{w}) &= w_1 \\ f_{2,5}(\mathbf{w}) &= w_3 \end{aligned}$$

$$\begin{aligned} \mathbf{x}_1 &= \mathbf{e}_1 + \mathbf{e}_3 & \mathbf{x}_2 &= \mathbf{e}_1 \\ \mathbf{x}_3 &= \mathbf{e}_2 + \mathbf{e}_3 & \mathbf{x}_4 &= \mathbf{e}_2 \\ \mathbf{x}_5 &= \mathbf{e}_1 + \mathbf{e}_2 + \mathbf{e}_3. \end{aligned}$$

[Side note: It is actually the case that $\text{rank}(C) = 5$, but this is quite difficult to show.]

3.2.14 Use the rank sum decomposition from Example 3.2.13(b) to come up with a formula for the cross product $C : \mathbb{R}^3 \times \mathbb{R}^3 \to \mathbb{R}^3$ that involves only 5 real number multiplications, rather than 6.

[Hint: Mimic Remark 3.2.3.]

*3.2.15 Let $\det : \mathbb{R}^n \times \cdots \times \mathbb{R}^n \to \mathbb{R}$ be the determinant.

(a) Suppose $n = 2$. Compute $\text{rank}(\det)$.

(b) Suppose $n = 3$. What is $\text{rank}(\det)$? You do not need to rigorously justify your answer—it is given away in one of the other exercises in this section.

[Side note: $\text{rank}(\det)$ is not known in general. For example, if $n = 5$ then the best bounds we know are $17 \leq \text{rank}(\det) \leq 20$.]

3.2.16 Provide an example to show that in the rank sum decomposition of Theorem 3.2.4, if $r = \text{rank}(T)$ then the sets of vectors $\{\mathbf{w}_j\}_{j=1}^r \subset \mathbb{F}^{d_{\mathcal{W}}}$ and $\{\mathbf{v}_{1,j}\}_{j=1}^r \subset \mathbb{F}^{d_1}, \ldots,$ $\{\mathbf{v}_{k,j}\}_{j=1}^r \subset \mathbb{F}^{d_k}$ cannot necessarily be chosen to be linearly independent (in contrast with Theorem A.1.3, where they could).

[Hint: Consider one of the multilinear transformations whose rank we mentioned in this section.]

3.3   The Tensor Product

We now introduce an operation, called the tensor product, that combines two vector spaces into a new vector space. This operation can be thought of as a generalization of the Kronecker product that not only lets us multiply together vectors (from $\mathbb{F}^n$) and matrices of different sizes, but also allows us to multiply together vectors from *any* vector spaces. For example, we can take the tensor product of a polynomial from $\mathcal{P}^3$ with a matrix from $\mathcal{M}_{2,7}$ while retaining a vector space structure. Perhaps more usefully though, it provides us with a systematic way of treating multilinear transformations as linear transformations, without having to explicitly construct a standard block matrix representation as in Theorem 3.2.2.

3.3.1   Motivation and Definition

Recall that we defined vector spaces in Section 1.1 so as to generalize $\mathbb{F}^n$ to other settings in which we can apply our linear algebraic techniques. We similarly would like to define the tensor product of two vectors spaces to behave "like" the Kronecker product does on $\mathbb{F}^n$ or $\mathcal{M}_{m,n}$. More specifically, if $\mathcal{V}$ and $\mathcal{W}$ are vector spaces over the same field $\mathbb{F}$, then we would like to define their tensor product $\mathcal{V} \otimes \mathcal{W}$ to consist of vectors that we denote by $\mathbf{v} \otimes \mathbf{w}$ and their linear combinations, so that the following properties hold for all $\mathbf{v}, \mathbf{x} \in \mathcal{V}$, $\mathbf{w}, \mathbf{y} \in \mathcal{W}$, and $c \in \mathbb{F}$:

- $\mathbf{v} \otimes (\mathbf{w} + \mathbf{y}) = (\mathbf{v} \otimes \mathbf{w}) + (\mathbf{v} \otimes \mathbf{y})$
- $(\mathbf{v} + \mathbf{x}) \otimes \mathbf{w} = (\mathbf{v} \otimes \mathbf{w}) + (\mathbf{x} \otimes \mathbf{w})$,   and   (3.3.1)
- $(c\mathbf{v}) \otimes \mathbf{w} = \mathbf{v} \otimes (c\mathbf{w}) = c(\mathbf{v} \otimes \mathbf{w})$.

Theorem 3.1.1 also had the associativity property $(\mathbf{v} \otimes \mathbf{w}) \otimes \mathbf{x} = \mathbf{v} \otimes (\mathbf{w} \otimes \mathbf{x})$. We will see shortly that associativity comes "for free", so we do not worry about it right now.

In other words, we would like the tensor product to be bilinear. Indeed, if we return to the Kronecker product on $\mathbb{F}^n$ or $\mathcal{M}_{m,n}$, then these are exactly the properties from Theorem 3.1.1 that describe how it interacts with vector addition and scalar multiplication—it does so in a bilinear way.

However, this definition of $\mathcal{V} \otimes \mathcal{W}$ is insufficient, as there are typically *many* ways of constructing a vector space in a bilinear way out of two vector spaces $\mathcal{V}$ and $\mathcal{W}$, and they may look quite different from each other. For example, if $\mathcal{V} = \mathcal{W} = \mathbb{R}^2$ then each of the following vector spaces $\mathcal{X}$ and operations $\otimes : \mathbb{R}^2 \times \mathbb{R}^2 \to \mathcal{X}$ satisfy the three bilinearity properties (3.3.1):

1) $\mathcal{X} = \mathbb{R}^4$, with $\mathbf{v} \otimes \mathbf{w} = (v_1 w_1, v_1 w_2, v_2 w_1, v_2 w_2)$. This is just the usual Kronecker product.

2) $\mathcal{X} = \mathbb{R}^3$, with $\mathbf{v} \otimes \mathbf{w} = (v_1 w_1, v_1 w_2, v_2 w_2)$. This can be thought of as the orthogonal projection of the usual Kronecker product down onto $\mathbb{R}^3$ by omitting its third coordinate.

In cases (2)–(5) here, "⊗" does not refer to the Kronecker product, but rather to new operations that we are defining so as to mimic its properties.

3) $\mathcal{X} = \mathbb{R}^2$, with $\mathbf{v} \otimes \mathbf{w} = (v_1 w_1 + v_2 w_2, v_1 w_2 - v_2 w_1)$. This is a projected and rotated version of the Kronecker product—it is obtained by multiplying the Kronecker product by the matrix

$$\begin{bmatrix} 1 & 0 & 0 & 1 \\ 0 & 1 & -1 & 0 \end{bmatrix}.$$

4) $\mathcal{X} = \mathbb{R}$, with $\mathbf{v} \otimes \mathbf{w} = \mathbf{v} \cdot \mathbf{w}$ for all $\mathbf{v}, \mathbf{w} \in \mathbb{R}^2$.

5) The zero subspace $\mathcal{X} = \{\mathbf{0}\}$ with $\mathbf{v} \otimes \mathbf{w} = \mathbf{0}$ for all $\mathbf{v}, \mathbf{w} \in \mathbb{R}^2$.

Notice that $\mathcal{X} = \mathbb{R}^4$ with the Kronecker product is the largest of the above vector spaces, and all of the others can be thought of as projections of the Kronecker product down onto a smaller space. In fact, it really is the largest example possible—it is not possible to construct a 5-dimensional vector space via two copies of $\mathbb{R}^2$ in a bilinear way, since every vector must be a linear combination of $\mathbf{e}_1 \otimes \mathbf{e}_1$, $\mathbf{e}_1 \otimes \mathbf{e}_2$, $\mathbf{e}_2 \otimes \mathbf{e}_1$, and $\mathbf{e}_2 \otimes \mathbf{e}_2$, no matter how we define the bilinear "⊗" operation.

The above examples suggest that the extra property that we need to add in order to make tensor products unique is the fact that they are as large as possible while still satisfying the bilinearity properties (3.3.1). We call this feature the **universal property** of the tensor product, and the following definition pins all of these ideas down. In particular, property (d) is this universal property.

Definition 3.3.1

Tensor Product

Suppose $\mathcal{V}$ and $\mathcal{W}$ are vector spaces over a field $\mathbb{F}$. Their **tensor product** is the (unique up to isomorphism) vector space $\mathcal{V} \otimes \mathcal{W}$, also over the field $\mathbb{F}$, with vectors and operations satisfying the following properties:

a) For every pair of vectors $\mathbf{v} \in \mathcal{V}$ and $\mathbf{w} \in \mathcal{W}$, there is an associated vector (called an **elementary tensor**) $\mathbf{v} \otimes \mathbf{w} \in \mathcal{V} \otimes \mathcal{W}$, and every vector in $\mathcal{V} \otimes \mathcal{W}$ can be written as a linear combination of these elementary tensors.

b) Vector addition satisfies

$$\mathbf{v} \otimes (\mathbf{w} + \mathbf{y}) = (\mathbf{v} \otimes \mathbf{w}) + (\mathbf{v} \otimes \mathbf{y}) \quad \text{and}$$
$$(\mathbf{v} + \mathbf{x}) \otimes \mathbf{w} = (\mathbf{v} \otimes \mathbf{w}) + (\mathbf{x} \otimes \mathbf{w}) \quad \text{for all} \quad \mathbf{v}, \mathbf{x} \in \mathcal{V}, \ \mathbf{w}, \mathbf{y} \in \mathcal{W}.$$

c) Scalar multiplication satisfies

$$c(\mathbf{v} \otimes \mathbf{w}) = (c\mathbf{v}) \otimes \mathbf{w} = \mathbf{v} \otimes (c\mathbf{w}) \quad \text{for all} \quad c \in \mathbb{F}, \ \mathbf{v} \in \mathcal{V}, \ \mathbf{w} \in \mathcal{W}.$$

d) For every vector space $\mathcal{X}$ over $\mathbb{F}$ and every bilinear transformation $T : \mathcal{V} \times \mathcal{W} \to \mathcal{X}$, there exists a linear transformation $S : \mathcal{V} \otimes \mathcal{W} \to \mathcal{X}$ such that

$$T(\mathbf{v}, \mathbf{w}) = S(\mathbf{v} \otimes \mathbf{w}) \quad \text{for all} \quad \mathbf{v} \in \mathcal{V} \text{ and } \mathbf{w} \in \mathcal{W}.$$

Unique "up to isomorphism" means that all vector spaces satisfying these properties are isomorphic to each other.

Property (d) is the universal property that forces $\mathcal{V} \otimes \mathcal{W}$ to be as large as possible.

The way to think about property (d) above (the universal property) is that it forces $\mathcal{V} \otimes \mathcal{W}$ to be so large that we can squash it down (via the linear transformation S) onto any other vector space $\mathcal{X}$ that is similarly constructed from $\mathcal{V}$ and $\mathcal{W}$ in a bilinear way. In this sense, the tensor product can be thought of as "containing" (up to isomorphism) every bilinear combination of vectors from $\mathcal{V}$ and $\mathcal{W}$.

To make the tensor product seem more concrete, it is good to keep the Kronecker product in the back of our minds as the motivating example. For

Throughout this entire section, every time we encounter a new theorem, ask yourself what it means for the Kronecker product of vectors or matrices.

example, $\mathbb{F}^m \otimes \mathbb{F}^n = \mathbb{F}^{mn}$ and $\mathcal{M}_{m,n} \otimes \mathcal{M}_{p,q} = \mathcal{M}_{mp,nq}$ contain all vectors of the form $\mathbf{v} \otimes \mathbf{w}$ and all matrices of the form $A \otimes B$, respectively, as well as their linear combinations (this is property (a) of the above definition). Properties (b) and (c) of the above definition are then just chosen to be analogous to the properties that we proved for the Kronecker product in Theorem 3.1.1, and property (d) gets us all linear combinations of Kronecker products (rather than just some subspace of them). In fact, Theorem 3.2.2 is essentially equivalent to property (d) in this case: the standard matrix of S equals the standard block matrix of T.

We now work through another example to try to start building up some intuition for how the tensor product works more generally.

Example 3.3.1

The Tensor Product of Polynomials

Show that $\mathcal{P} \otimes \mathcal{P} = \mathcal{P}_2$, the space of polynomials in two variables, if we define the elementary tensors $f \otimes g$ by $(f \otimes g)(x,y) = f(x)g(y)$.

Solution:

We just need to show that $\mathcal{P}_2$ satisfies all four properties of Definition 3.3.1 if we define $f \otimes g$ in the indicated way:

a) If $h \in \mathcal{P}_2$ is any 2-variable polynomial then we can write it in the form

$$h(x,y) = \sum_{j,k=1}^{\infty} a_{j,k} x^j y^k$$

For example, if $f(x) = 2x^2 - 4$ and $g(x) = x^3 + x - 1$ then $(f \otimes g)(x,y) = f(x)g(y) = (2x^2 - 4)(y^3 + y - 1)$.

where there are only finitely many non-zero terms in the sum. Since each term $x^j y^k$ is an elementary tensor (it is $x^j \otimes y^k$), it follows that every $h \in \mathcal{P}_2$ is a linear combination of elementary tensors, as desired.

b), c) These properties follow almost immediately from bilinearity of usual multiplication: if $f, g, h \in \mathcal{P}$ and $c \in \mathbb{R}$ then

$$\big(f \otimes (g+ch)\big)(x,y) = f(x)\big(g(y)+ch(y)\big)$$
$$= f(x)g(y) + cf(x)h(y)$$
$$= (f \otimes g)(x,y) + c(f \otimes h)(x,y)$$

for all $x,y \in \mathbb{R}$, so $f \otimes (g+ch) = (f \otimes g) + c(f \otimes h)$. The fact that $(f+cg) \otimes h = (f \otimes h) + c(g \otimes h)$ can be proved similarly.

This example relies on the fact that every function in $\mathcal{P}_2$ can be written as a *finite* sum of polynomials of the form $f(x)g(y)$. Analogous statements in other function spaces fail (see Exercise 3.3.6).

d) Given any vector space $\mathcal{X}$ and bilinear form $T : \mathcal{P} \times \mathcal{P} \to \mathcal{X}$, we just define $S(x^j y^k) = T(x^j, y^k)$ for all integers $j, k \geq 0$ and then extend via linearity (i.e., we are defining how S acts on a basis of $\mathcal{P}_2$ and then using linearity to determine how it acts on other members of $\mathcal{P}_2$).

Once we move away from the spaces $\mathbb{F}^n$, $\mathcal{M}_{m,n}$, and $\mathcal{P}$, there are quite a few details that need to be explained in order for Definition 3.3.1 to make sense. In fact, that definition appears at first glance to be almost completely content-free. It does not give us an explicit definition of either the vector addition or scalar multiplication operations in $\mathcal{V} \otimes \mathcal{W}$, nor has it even told us what its vectors "look like"—we denote some of them by $\mathbf{v} \otimes \mathbf{w}$, but so what? It is also not clear that it exists *or* that it is unique up to isomorphism.

3.3.2 Existence and Uniqueness

We now start clearing up the aforementioned issues with the tensor product—we show that it exists, that it is unique up to isomorphism, and we discuss how to actually construct it.

Uniqueness

We first show that the tensor product (if it exists) is unique up to isomorphism. Fortunately, this property follows almost immediately from the universal property of the tensor product.

Suppose there are two vector spaces $\mathcal{Y}$ and $\mathcal{Z}$ satisfying the four defining properties from Definition 3.3.1, so we are thinking of each of them as the tensor product $\mathcal{V} \otimes \mathcal{W}$. In order to distinguish these two tensor products, we denote them by $\mathcal{Y} = \mathcal{V} \otimes_\mathcal{Y} \mathcal{W}$ and $\mathcal{Z} = \mathcal{V} \otimes_\mathcal{Z} \mathcal{W}$, respectively, and similarly for the vectors in these spaces (for example, $\mathbf{v} \otimes_\mathcal{Y} \mathbf{w}$ refers to the elementary tensor in $\mathcal{Y}$ obtained from $\mathbf{v}$ and $\mathbf{w}$).

> Our goal here is to show that $\mathcal{Y}$ and $\mathcal{Z}$ are isomorphic (i.e., there is an invertible linear transformation from one to the other).

Since $\otimes_\mathcal{Z} : \mathcal{V} \times \mathcal{W} \to \mathcal{Z}$ is a bilinear transformation, the universal property (property (d) of Definition 3.3.1, with $\mathcal{X} = \mathcal{Z}$ and $T = \otimes_\mathcal{Z}$) for $\mathcal{Y} = \mathcal{V} \otimes_\mathcal{Y} \mathcal{W}$ tells us that there exists a linear transformation $S_{\mathcal{Y} \to \mathcal{Z}} : \mathcal{V} \otimes_\mathcal{Y} \mathcal{W} \to \mathcal{Z}$ such that

$$S_{\mathcal{Y} \to \mathcal{Z}}(\mathbf{v} \otimes_\mathcal{Y} \mathbf{w}) = \mathbf{v} \otimes_\mathcal{Z} \mathbf{w} \quad \text{for all} \quad \mathbf{v} \in \mathcal{V}, \ \mathbf{w} \in \mathcal{W}.$$

> In fact, the linear transformation S in Definition 3.3.1(d) is necessarily unique (see Exercise 3.3.4).

A similar argument via the universal property of $\mathcal{Z} = \mathcal{V} \otimes_\mathcal{Z} \mathcal{W}$ shows that there exists a linear transformation $S_{\mathcal{Z} \to \mathcal{Y}} : \mathcal{V} \otimes_\mathcal{Z} \mathcal{W} \to \mathcal{Y}$ such that

$$S_{\mathcal{Z} \to \mathcal{Y}}(\mathbf{v} \otimes_\mathcal{Z} \mathbf{w}) = \mathbf{v} \otimes_\mathcal{Y} \mathbf{w} \quad \text{for all} \quad \mathbf{v} \in \mathcal{V}, \ \mathbf{w} \in \mathcal{W}.$$

Since every vector in $\mathcal{Y}$ or $\mathcal{Z}$ is a linear combination of elementary tensors $\mathbf{v} \otimes_\mathcal{Y} \mathbf{w}$ or $\mathbf{v} \otimes_\mathcal{Z} \mathbf{w}$, respectively, we conclude that $S_{\mathcal{Y} \to \mathcal{Z}} = S_{\mathcal{Z} \to \mathcal{Y}}^{-1}$ is an invertible linear transformation from $\mathcal{Y}$ to $\mathcal{Z}$. It follows that $\mathcal{Y}$ and $\mathcal{Z}$ are isomorphic, as desired.

Existence and Bases

Showing that the tensor product space $\mathcal{V} \otimes \mathcal{W}$ actually exists in general (i.e., there is a vector space whose vector addition and scalar multiplication operations satisfy the required properties) requires some abstract algebra machinery that we have not developed here. For this reason, we just show how to construct $\mathcal{V} \otimes \mathcal{W}$ in the special case when $\mathcal{V}$ and $\mathcal{W}$ each have a basis, which is true at least in the case when $\mathcal{V}$ and $\mathcal{W}$ are each finite-dimensional. In particular, the following theorem tells us that we can construct a basis of $\mathcal{V} \otimes \mathcal{W}$ in exactly the same way we did when working with the Kronecker product—simply tensor together bases of $\mathcal{V}$ and $\mathcal{W}$.

> Look back at Remark 1.2.1 for a discussion of which vector spaces have bases.

Theorem 3.3.1	Suppose $\mathcal{V}$ and $\mathcal{W}$ are vector spaces over the same field with bases B
Bases of Tensor Products	and C, respectively. Then their tensor product $\mathcal{V} \otimes \mathcal{W}$ exists and has the following set as a basis:

$$B \otimes C \stackrel{\text{def}}{=} \{\mathbf{e} \otimes \mathbf{f} \mid \mathbf{e} \in B, \ \mathbf{f} \in C\}.$$

Proof. We start with the trickiest part of this proof to get our heads around—we simply *define* $B \otimes C$ to be a linearly independent set and $\mathcal{V} \otimes \mathcal{W}$ to be its span. That is, we are not defining $\mathcal{V} \otimes \mathcal{W}$ via Definition 3.3.1, but rather we are defining it as the span of a collection of vectors, and then we will show that it has all of the properties of Definition 3.3.1 (thus establishing that a vector space with those properties really does exist). Before proceeding, we make some points of clarification:

If you are uncomfortable with defining $\mathcal{V} \otimes \mathcal{W}$ to be a vector space made up of abstract symbols, look ahead to Remark 3.3.1, which might clear things up a bit.

- Throughout this proof, we are just thinking of "$\mathbf{e} \otimes \mathbf{f}$" as a symbol that means nothing more than "the ordered pair of vectors $\mathbf{e}$ and $\mathbf{f}$".
- Similarly, we do not yet know that $\mathcal{V} \otimes \mathcal{W}$ really is the tensor product of $\mathcal{V}$ and $\mathcal{W}$, but rather we are just thinking of it as the vector space that has $B \otimes C$ as a basis:

$$\mathcal{V} \otimes \mathcal{W} = \left\{ \sum_{i,j} c_i d_j (\mathbf{e}_i \otimes \mathbf{f}_j) : b_i, c_j \in \mathbb{F}, \ \mathbf{e}_i \in B, \ \mathbf{f}_j \in C \text{ for all } i,j \right\},$$

where the sum is finite.

For each $\mathbf{v} = \sum_i c_i \mathbf{e}_i \in \mathcal{V}$ and $\mathbf{w} = \sum_j d_j \mathbf{f}_j \in \mathcal{W}$ (where $c_i, d_j \in \mathbb{F}$, $\mathbf{e}_i \in B$, and $\mathbf{f}_j \in C$ for all i and j) we then similarly define

This definition of $\mathbf{v} \otimes \mathbf{w}$ implicitly relies on uniqueness of linear combinations when representing $\mathbf{v}$ and $\mathbf{w}$ with respect to the bases B and C, respectively (Theorem 1.1.4).

$$\mathbf{v} \otimes \mathbf{w} = \sum_{i,j} c_i d_j (\mathbf{e}_i \otimes \mathbf{f}_j). \tag{3.3.2}$$

Our goal is now to show that when we define $\mathcal{V} \otimes \mathcal{W}$ and $\mathbf{v} \otimes \mathbf{w}$ in this way, each of the properties (a)–(d) of Definition 3.3.1 holds.

Property (a) holds trivially since we have defined $\mathbf{v} \otimes \mathbf{w} \in \mathcal{V} \otimes \mathcal{W}$ in general, and we constructed $\mathcal{V} \otimes \mathcal{W}$ so that it is spanned by vectors of the form $\mathbf{e} \otimes \mathbf{f}$, where $\mathbf{e} \in B$ and $\mathbf{f} \in C$, so it is certainly spanned by vectors of the form $\mathbf{v} \otimes \mathbf{w}$, where $\mathbf{v} \in \mathcal{V}$ and $\mathbf{w} \in \mathcal{W}$.

To see why properties (b) and (c) hold, suppose $\mathbf{v} = \sum_i c_i \mathbf{e}_i \in \mathcal{V}$, $\mathbf{w} = \sum_j d_j \mathbf{f}_j \in \mathcal{W}$, and $\mathbf{y} = \sum_j b_j \mathbf{f}_j \in \mathcal{W}$. Then

$$\mathbf{v} \otimes (\mathbf{w} + \mathbf{y}) = \mathbf{v} \otimes \left(\sum_j d_j \mathbf{f}_j + \sum_j b_j \mathbf{f}_j \right) \quad \text{(expand } \mathbf{w} \text{ and } \mathbf{y})$$

$$= \mathbf{v} \otimes \left(\sum_j (d_j + b_j) \mathbf{f}_j \right) \quad \text{(distributivity in } \mathcal{W})$$

$$= \sum_{i,j} c_i (d_j + b_j)(\mathbf{e}_i \otimes \mathbf{f}_j) \quad \text{(use Equation (3.3.2))}$$

$$= \sum_{i,j} c_i d_j (\mathbf{e}_i \otimes \mathbf{f}_j) + \sum_{i,j} c_i b_j (\mathbf{e}_i \otimes \mathbf{f}_j) \quad \text{(distributivity in } \mathcal{V} \otimes \mathcal{W})$$

$$= (\mathbf{v} \otimes \mathbf{w}) + (\mathbf{v} \otimes \mathbf{y}), \quad \text{(Equation (3.3.2) again)}$$

and a similar argument shows that

$$\mathbf{v} \otimes (c\mathbf{w}) = \mathbf{v} \otimes \left(c \sum_j d_j \mathbf{f}_j \right) = \mathbf{v} \otimes \left(\sum_j (cd_j) \mathbf{f}_j \right)$$

$$= \sum_{i,j} c_i (cd_j)(\mathbf{e}_i \otimes \mathbf{f}_j) = c \sum_{i,j} c_i d_j (\mathbf{e}_i \otimes \mathbf{f}_j) = c(\mathbf{v} \otimes \mathbf{w}).$$

The proofs that $(\mathbf{v} + \mathbf{x}) \otimes \mathbf{w} = (\mathbf{v} \otimes \mathbf{w}) + (\mathbf{x} \otimes \mathbf{w})$ and $(c\mathbf{w}) \otimes \mathbf{w} = c(\mathbf{v} \otimes \mathbf{w})$ are almost identical to the ones presented above, and are thus omitted.

Finally, for property (d) we simply define $S : \mathcal{V} \otimes \mathcal{W} \to \mathcal{X}$ by setting $S(\mathbf{e} \otimes \mathbf{f}) = T(\mathbf{e}, \mathbf{f})$ for each $\mathbf{e} \in B$ and $\mathbf{f} \in C$, and extending via linearity (recall that every linear transformation is determined completely by how it acts on a basis of the input space). ∎

Since the dimension of a vector space is defined to equal the number of vectors in any of its bases, we immediately get the following corollary that tells us that the dimension of the tensor product of two finite-dimensional vector spaces is just the product of their individual dimensions.

Corollary 3.3.2

Dimensionality of Tensor Products

Suppose V and W are finite-dimensional vector spaces. Then

$$\dim(V \otimes W) = \dim(V)\dim(W).$$

Remark 3.3.1

The Direct Sum and Tensor Product

The tensor product is actually quite analogous to the external direct sum from Section 1.B.3. Just like the direct sum $V \oplus W$ can be thought of as a way of constructing a new vector space by "adding" two vector spaces V and W, the tensor product $V \otimes W$ can be thought of as a way of constructing a new vector space by "multiplying" V and W.

More specifically, both of these vector spaces $V \oplus W$ and $V \otimes W$ are constructed out of ordered pairs of vectors from V and W, along with a specification of how to add those ordered pairs and multiply them by scalars. In $V \oplus W$ we denoted those ordered pairs by $(\mathbf{v}, \mathbf{w})$, whereas in $V \otimes W$ we denote those ordered pairs by $\mathbf{v} \otimes \mathbf{w}$. The biggest difference between these two constructions is that these ordered pairs are the *only* members of $V \oplus W$, whereas some members of $V \otimes W$ are just *linear combinations* of these ordered pairs.

This analogy between the direct sum and tensor product is even more explicit if we look at what they do to bases B and C of V and W, respectively: a basis of $V \oplus W$ is the disjoint union $B \cup C$, whereas a basis of $V \otimes W$ is the Cartesian product $B \times C$ (up to relabeling the members of these sets appropriately). For example, if $V = \mathbb{R}^2$ and $W = \mathbb{R}^3$ have bases $B = \{\mathbf{v}_1, \mathbf{v}_2\}$ and $C = \{\mathbf{v}_3, \mathbf{v}_4, \mathbf{v}_5\}$, respectively, then $\mathbb{R}^2 \oplus \mathbb{R}^3 \cong \mathbb{R}^5$ has basis

> In this basis of $\mathbb{R}^2 \oplus \mathbb{R}^3$, each subscript from B and C appears exactly once. In the basis of $\mathbb{R}^2 \otimes \mathbb{R}^3$, each *ordered pair* of subscripts from B and C appears exactly once.

$$\{(\mathbf{v}_1, \mathbf{0}), (\mathbf{v}_2, \mathbf{0}), (\mathbf{0}, \mathbf{v}_3), (\mathbf{0}, \mathbf{v}_4), (\mathbf{0}, \mathbf{v}_5)\},$$

which is just the disjoint union of B and C, but with each vector turned into an ordered pair so that this statement makes sense. Similarly, $\mathbb{R}^2 \otimes \mathbb{R}^3 \cong \mathbb{R}^6$ has basis

$$\{\mathbf{v}_1 \otimes \mathbf{v}_3, \mathbf{v}_1 \otimes \mathbf{v}_4, \mathbf{v}_1 \otimes \mathbf{v}_5, \mathbf{v}_2 \otimes \mathbf{v}_3, \mathbf{v}_2 \otimes \mathbf{v}_4, \mathbf{v}_2 \otimes \mathbf{v}_5\},$$

which is just the Cartesian product of B and C, but with each ordered pair written as an elementary tensor so that this statement makes sense.

Even after working through the proof of Theorem 3.3.1, tensor products *still* likely feel very abstract—how do we actually *construct* them? The reason that they feel this way is that they are only defined up to isomorphism, so it's somewhat impossible to say what they look like. We could say that $\mathbb{F}^m \otimes \mathbb{F}^n = \mathbb{F}^{mn}$ and that $\mathbf{v} \otimes \mathbf{w}$ is just the Kronecker product of $\mathbf{v} \in \mathbb{F}^m$ and $\mathbf{w} \in \mathbb{F}^n$, but we could just as well say that $\mathbb{F}^m \otimes \mathbb{F}^n = \mathcal{M}_{m,n}(\mathbb{F})$ and that $\mathbf{v} \otimes \mathbf{w} = \mathbf{v}\mathbf{w}^T$ is the outer product of $\mathbf{v}$ and $\mathbf{w}$. After all, these spaces are isomorphic via vectorization/matricization, so they look the same when we just describe what linear algebraic properties they satisfy.

For this reason, when we construct "the" tensor product of two vector spaces, we have some freedom in how we represent it. In the following example, we pick one particularly natural representation of the tensor product, but we emphasize that it is not the only one possible.

Example 3.3.2

The Tensor Product of Matrices and Polynomials

Describe the vector space $\mathcal{M}_2(\mathbb{R}) \otimes \mathcal{P}^2$ and its elementary tensors.

Solution:

If we take inspiration from the Kronecker product, which places a copy of one vector space on each basis vector (i.e., entry) of the other vector space, it seems natural to guess that we can represent $\mathcal{M}_2 \otimes \mathcal{P}^2$ as the vector space of 2×2 matrices whose entries are polynomials of degree at most 2. That is, we guess that

Recall that $\mathcal{P}^2$ is the space of polynomials of degree at most 2.

$$\mathcal{M}_2 \otimes \mathcal{P}^2 = \left\{ \begin{bmatrix} f_{1,1}(x) & f_{1,2}(x) \\ f_{2,1}(x) & f_{2,2}(x) \end{bmatrix} \middle| f_{i,j} \in \mathcal{P}^2 \text{ for } 1 \le i, j \le 2 \right\}.$$

The elementary tensors in $\mathcal{M}_2 \otimes \mathcal{P}^2$ are those that can be written in the form $Af(x)$ for some $A \in \mathcal{M}_2$ and $f \in \mathcal{P}^2$ (i.e., the elementary tensors are the ones for which the polynomials $f_{i,j}$ are all multiples of each other).

Alternatively, by grouping the coefficients of each $f_{i,j}$, we could think of the members of $\mathcal{M}_2 \otimes \mathcal{P}^2$ as degree-2 polynomials from $\mathbb{R}$ to $\mathcal{M}_2$:

$$\mathcal{M}_2 \otimes \mathcal{P}^2 =$$
$$\left\{ f : \mathbb{R} \to \mathcal{M}_2 \mid f(x) = Ax^2 + Bx + C \text{ for some } A, B, C \in \mathcal{M}_2 \right\}.$$

When viewed in this way, the elementary tensors in $\mathcal{M}_2 \otimes \mathcal{P}^2$ are again those of the form $Af(x)$ for some $A \in \mathcal{M}_2$ and $f \in \mathcal{P}^2$ (i.e., the ones for which A, B, and C are all multiples of each other).

To verify that this set really is the tensor product $\mathcal{M}_2 \otimes \mathcal{P}^2$, we could proceed as we did in Example 3.3.1 and check the four defining properties of Definition 3.3.1. All four of these properties are straightforward and somewhat unenlightening to check, so we leave them to Exercise 3.3.5.

Higher-Order Tensor Products

Just as was the case with the Kronecker product, one of the main purposes of the tensor product is that it lets us turn bilinear transformations into linear ones. This is made explicit by the universal property (property (d) of Definition 3.3.1)—if we have some bilinear transformation $T : V \times W \to X$ that we wish to know more about, we can instead construct the tensor product space $V \otimes W$ and investigate the *linear* transformation $S : V \otimes W \to X$ defined by $S(\mathbf{v} \otimes \mathbf{w}) = T(\mathbf{v}, \mathbf{w})$.

Slightly more generally, we can also consider the tensor product of three or more vector spaces, and doing so lets us represent *multi*linear transformations as linear transformations acting on the tensor product space. We now pin down the details needed to show that this is indeed the case.

Theorem 3.3.3

Associativity of the Tensor Product

If V, W, and X are vector spaces over the same field then

$$(V \otimes W) \otimes X \cong V \otimes (W \otimes X).$$

Recall that "$\cong$" means "is isomorphic to"

Proof. For each $\mathbf{x} \in X$, define a bilinear map $T_{\mathbf{x}} : V \times W \to V \otimes (W \otimes X)$ by $T_{\mathbf{x}}(\mathbf{v}, \mathbf{w}) = \mathbf{v} \otimes (\mathbf{w} \otimes \mathbf{x})$. Since $T_{\mathbf{x}}$ is bilinear, the universal property of the tensor product (i.e., property (d) of Definition 3.3.1) says that there exists a linear transformation $S_{\mathbf{x}} : \mathcal{V} \otimes \mathcal{W} \to \mathcal{V} \otimes (\mathcal{W} \otimes \mathcal{X})$ that acts via $S_{\mathbf{x}}(\mathbf{v} \otimes \mathbf{w}) = T_{\mathbf{x}}(\mathbf{v}, \mathbf{w}) = \mathbf{v} \otimes (\mathbf{w} \otimes \mathbf{x})$.

Next, define a bilinear map $\widetilde{T} : (\mathcal{V} \otimes \mathcal{W}) \times \mathcal{X} \to \mathcal{V} \otimes (\mathcal{W} \otimes \mathcal{X})$ via $\widetilde{T}(\mathbf{u}, \mathbf{x}) = S_{\mathbf{x}}(\mathbf{u})$ for all $\mathbf{u} \in \mathcal{V} \otimes \mathcal{W}$ and $\mathbf{x} \in \mathcal{X}$. By using the universal property again, we see that there exists a linear transformation $\widetilde{S} : (\mathcal{V} \otimes \mathcal{W}) \otimes \mathcal{X} \to \mathcal{V} \otimes (\mathcal{W} \otimes \mathcal{X})$ that acts via $\widetilde{S}(\mathbf{u} \otimes \mathbf{x}) = \widetilde{T}(\mathbf{u}, \mathbf{x}) = S_{\mathbf{x}}(\mathbf{u})$. If $\mathbf{u} = \mathbf{v} \otimes \mathbf{w}$ then this says that

$$\widetilde{S}\big((\mathbf{v} \otimes \mathbf{w}) \otimes \mathbf{x}\big) = S_{\mathbf{x}}(\mathbf{v} \otimes \mathbf{w}) = \mathbf{v} \otimes (\mathbf{w} \otimes \mathbf{x}).$$

This argument can also be reversed to find a linear transformation that sends $\mathbf{v} \otimes (\mathbf{w} \otimes \mathbf{x})$ to $(\mathbf{v} \otimes \mathbf{w}) \otimes \mathbf{x}$, so $\widetilde{S}$ is invertible and thus an isomorphism. ∎

Now that we know that the tensor product is associative, we can unambiguously refer to the tensor product of 3 vector spaces $\mathcal{V}$, $\mathcal{W}$, and $\mathcal{X}$ as $\mathcal{V} \otimes \mathcal{W} \otimes \mathcal{X}$, since it does not matter whether we take this to mean $(\mathcal{V} \otimes \mathcal{W}) \otimes \mathcal{X}$ or $\mathcal{V} \otimes (\mathcal{W} \otimes \mathcal{X})$. The same is true when we tensor together 4 or more vector spaces, and in this setting we say that an **elementary tensor** in $\mathcal{V}_1 \otimes \mathcal{V}_2 \otimes \cdots \otimes \mathcal{V}_p$ is a vector of the form $\mathbf{v}_1 \otimes \mathbf{v}_2 \otimes \cdots \otimes \mathbf{v}_p$, where $\mathbf{v}_j \in \mathcal{V}_j$ for each $1 \le j \le p$.

It is similarly the case that the tensor product is commutative in the sense that $\mathcal{V} \otimes \mathcal{W} \cong \mathcal{W} \otimes \mathcal{V}$ (see Exercise 3.3.10). For example, if $\mathcal{V} = \mathbb{F}^m$ and $\mathcal{W} = \mathbb{F}^n$ then the swap operator $W_{m,n}$ of Definition 3.1.3 is an isomorphism from $\mathcal{V} \otimes \mathcal{W} = \mathbb{F}^m \otimes \mathbb{F}^n$ to $\mathcal{W} \otimes \mathcal{V} = \mathbb{F}^n \otimes \mathbb{F}^m$. More generally, higher-order tensor product spaces $\mathcal{V}_1 \otimes \mathcal{V}_2 \otimes \cdots \otimes \mathcal{V}_p$ are also isomorphic to the tensor product of the same spaces in any other order. That is, if $\sigma : \{1, 2, \ldots, p\} \to \{1, 2, \ldots, p\}$ is a permutation then

> Keep in mind that $\mathcal{V} \otimes \mathcal{W} \cong \mathcal{W} \otimes \mathcal{V}$ does *not* mean that $\mathbf{v} \otimes \mathbf{w} = \mathbf{w} \otimes \mathbf{v}$ for all $\mathbf{v} \in \mathcal{V}$, $\mathbf{w} \in \mathcal{W}$. Rather, it just means that there is an isomorphism that sends each $\mathbf{v} \otimes \mathbf{w}$ to $\mathbf{w} \otimes \mathbf{v}$.

$$\mathcal{V}_1 \otimes \mathcal{V}_2 \otimes \cdots \otimes \mathcal{V}_p \cong \mathcal{V}_{\sigma(1)} \otimes \mathcal{V}_{\sigma(2)} \otimes \cdots \otimes \mathcal{V}_{\sigma(p)}.$$

Again, if each of these vector spaces is $\mathbb{F}^n$ (and thus the tensor product is the Kronecker product) then the swap matrix W_σ introduced in Section 3.1.3 is the standard isomorphism between these spaces.

We motivated the tensor product as a generalization of the Kronecker product that applies to any vector spaces (as opposed to just $\mathbb{F}^n$ and/or $\mathcal{M}_{m,n}$). We now note that if we fix bases of the vector spaces that we are working with then the tensor product really does look like the Kronecker product of coordinate vectors, as we would hope:

> **Theorem 3.3.4**
>
> **Kronecker Product of Coordinate Vectors**

Suppose $\mathcal{V}_1, \mathcal{V}_2, \ldots, \mathcal{V}_p$ are finite-dimensional vector spaces over the same field with bases $B_1, B_2, \ldots, B_p$, respectively. Then

$$B_1 \otimes B_2 \otimes \cdots \otimes B_p \stackrel{\text{def}}{=} \big\{ \mathbf{b}^{(1)} \otimes \mathbf{b}^{(2)} \otimes \cdots \otimes \mathbf{b}^{(p)} \mid \mathbf{b}^{(j)} \in B_j \text{ for all } 1 \le j \le p \big\}$$

is a basis of $\mathcal{V}_1 \otimes \mathcal{V}_2 \otimes \cdots \otimes \mathcal{V}_p$, and if we order it lexicographically then

$$\big[\mathbf{v}_1 \otimes \mathbf{v}_2 \otimes \cdots \otimes \mathbf{v}_p \big]_{B_1 \otimes B_2 \otimes \cdots \otimes B_p} = [\mathbf{v}_1]_{B_1} \otimes [\mathbf{v}_2]_{B_2} \otimes \cdots \otimes [\mathbf{v}_p]_{B_p}$$

for all $\mathbf{v}_1 \in \mathcal{V}_1$, $\mathbf{v}_2 \in \mathcal{V}_2$, ..., $\mathbf{v}_p \in \mathcal{V}_p$.

> In the concluding line of this theorem, the "$\otimes$" on the left is the tensor product and the "$\otimes$" on the right is the Kronecker product.

In the above theorem, when we say that we are ordering the basis $B_1 \otimes B_2 \otimes \cdots \otimes B_p$ lexicographically, we mean that we order it so as to "count" its basis vectors in the most natural way, much like we did for bases of Kronecker product spaces in Section 3.1 and for standard block matrices of multilinear transformations in Section 3.2.2. For example, if $B_1 = \{\mathbf{v}_1, \mathbf{v}_2\}$ and $B_2 = \{\mathbf{w}_1, \mathbf{w}_2, \mathbf{w}_3\}$ then we order $B_1 \otimes B_2$ as

$$B_1 \otimes B_2 = \{\mathbf{v}_1 \otimes \mathbf{w}_1, \mathbf{v}_1 \otimes \mathbf{w}_2, \mathbf{v}_1 \otimes \mathbf{w}_3, \mathbf{v}_2 \otimes \mathbf{w}_1, \mathbf{v}_2 \otimes \mathbf{w}_2, \mathbf{v}_2 \otimes \mathbf{w}_3\}.$$

Proof of Theorem 3.3.4. To see that $B_1 \otimes B_2 \otimes \cdots \otimes B_p$ is a basis of $\mathcal{V}_1 \otimes \mathcal{V}_2 \otimes \cdots \otimes \mathcal{V}_p$, we note that repeated application of Corollary 3.3.2 tells us that

$$\dim(\mathcal{V}_1 \otimes \mathcal{V}_2 \otimes \cdots \otimes \mathcal{V}_p) = \dim(\mathcal{V}_1)\dim(\mathcal{V}_2)\cdots\dim(\mathcal{V}_p),$$

which is exactly the number of vectors in $B_1 \otimes B_2 \otimes \cdots \otimes B_p$. It is straightforward to see that every elementary tensor is in $\mathrm{span}(B_1 \otimes B_2 \otimes \cdots \otimes B_p)$, so $\mathcal{V}_1 \otimes \mathcal{V}_2 \otimes \cdots \otimes \mathcal{V}_p = \mathrm{span}(B_1 \otimes B_2 \otimes \cdots \otimes B_p)$, and thus Exercise 1.2.27(b) tells us that $B_1 \otimes B_2 \otimes \cdots \otimes B_p$ is indeed a basis of $\mathcal{V}_1 \otimes \mathcal{V}_2 \otimes \cdots \otimes \mathcal{V}_p$.

The fact that $\left[\mathbf{v}_1 \otimes \mathbf{v}_2 \otimes \cdots \otimes \mathbf{v}_p\right]_{B_1 \otimes B_2 \otimes \cdots \otimes B_p} = [\mathbf{v}_1]_{B_1} \otimes [\mathbf{v}_2]_{B_2} \otimes \cdots \otimes [\mathbf{v}_p]_{B_p}$ follows just by matching up all of the relevant definitions: if we write $B_j = \{\mathbf{b}_1^{(j)}, \mathbf{b}_2^{(j)}, \ldots, \mathbf{b}_{k_j}^{(j)}\}$ and $\mathbf{v}_j = \sum_{i=1}^{k_j} c_i^{(j)} \mathbf{b}_i^{(j)}$ (so that $[\mathbf{v}_j]_{B_j} = (c_1^{(j)}, c_2^{(j)}, \ldots, c_{k_j}^{(j)})$) for each $1 \le j \le p$, then

> This proof is the epitome of "straightforward but hideous". Nothing that we are doing here is clever—it is all just definition chasing, but it is ugly because there are so many objects to keep track of.

$$\mathbf{v}_1 \otimes \mathbf{v}_2 \otimes \cdots \otimes \mathbf{v}_p = \left(\sum_{i=1}^{k_1} c_i^{(1)} \mathbf{b}_i^{(1)}\right) \otimes \left(\sum_{i=1}^{k_2} c_i^{(2)} \mathbf{b}_i^{(2)}\right) \otimes \cdots \otimes \left(\sum_{i=1}^{k_p} c_i^{(p)} \mathbf{b}_i^{(p)}\right)$$

$$= \sum_{i_1, i_2, \ldots, i_p} c_{i_1}^{(1)} c_{i_2}^{(2)} \cdots c_{i_p}^{(p)} \left(\mathbf{b}_{i_1}^{(1)} \otimes \mathbf{b}_{i_2}^{(2)} \otimes \cdots \otimes \mathbf{b}_{i_p}^{(p)}\right).$$

Since the vectors $\left\{\mathbf{b}_{i_1}^{(1)} \otimes \mathbf{b}_{i_2}^{(2)} \otimes \cdots \otimes \mathbf{b}_{i_p}^{(p)}\right\}$ on the right are the members of $B_1 \otimes B_2 \otimes \cdots \otimes B_p$ and the scalars $\left\{c_{i_1}^{(1)} c_{i_2}^{(2)} \cdots c_{i_p}^{(p)}\right\}$ are the entries of the Kronecker product $[\mathbf{v}_1]_{B_1} \otimes [\mathbf{v}_2]_{B_2} \otimes \cdots \otimes [\mathbf{v}_p]_{B_p}$, in the same order, the result follows. ∎

It might sometimes be convenient to instead arrange the entries of the coordinate vector $\left[\mathbf{v}_1 \otimes \mathbf{v}_2 \otimes \cdots \otimes \mathbf{v}_p\right]_{B_1 \otimes B_2 \otimes \cdots \otimes B_p}$ into a p-dimensional $\dim(\mathcal{V}_1) \times \dim(\mathcal{V}_2) \times \cdots \times \dim(\mathcal{V}_p)$ array, rather than a long $\dim(\mathcal{V}_1)\dim(\mathcal{V}_2)\cdots\dim(\mathcal{V}_p)$-entry vector as we did here. What the most convenient representation of $\mathbf{v}_1 \otimes \mathbf{v}_2 \otimes \cdots \otimes \mathbf{v}_p$ is depends heavily on context—what the tensor product represents and what we are trying to do with it.

3.3.3 Tensor Rank

Although not every vector in a tensor product space $\mathcal{V}_1 \otimes \mathcal{V}_2 \otimes \cdots \otimes \mathcal{V}_p$ is an elementary tensor, every vector can (by definition) be written as a *linear combination* of elementary tensors. It seems natural to ask how many terms are required in such a linear combination, and we give this minimal number of terms a name:

Definition 3.3.2
Tensor Rank

Suppose $V_1, V_2, \ldots, V_p$ are finite-dimensional vector spaces over the same field and $\mathbf{v} \in V_1 \otimes V_2 \otimes \cdots \otimes V_p$. The **tensor rank** (or simply the **rank**) of $\mathbf{v}$, denoted by $\mathrm{rank}(\mathbf{v})$, is the minimal integer r such that $\mathbf{v}$ can be written as a sum of r elementary tensors:

$$\mathbf{v} = \sum_{i=1}^{r} \mathbf{v}_i^{(1)} \otimes \mathbf{v}_i^{(2)} \otimes \cdots \otimes \mathbf{v}_i^{(p)},$$

where $\mathbf{v}_i^{(j)} \in V_j$ for each $1 \le i \le r$ and $1 \le j \le p$.

The tensor rank generalizes the rank of a matrix in the following sense: if $p = 2$ and $V_1 = \mathbb{F}^m$ and $V_2 = \mathbb{F}^n$, then $\mathbb{F}^m \otimes \mathbb{F}^n \cong M_{m,n}(\mathbb{F})$, and the tensor rank in this space really is just the usual matrix rank. After all, when we represent $\mathbb{F}^m \otimes \mathbb{F}^n$ in this way, its elementary tensors are the matrices $\mathbf{v} \otimes \mathbf{w} = \mathbf{v}\mathbf{w}^T$, and we know from Theorem A.1.3 that the rank of a matrix is the fewest number of these (rank-1) matrices that are needed to sum to it.

In fact, a similar argument shows that when all of the spaces are finite-dimensional, the tensor rank is equivalent to the rank of a multilinear transformation. After all, we showed in Theorem 3.2.4 that the rank of a multilinear transformation T is the least integer r such that its standard block matrix A can be written in the form

$$A = \sum_{j=1}^{r} \mathbf{w}_j \big(\mathbf{v}_j^{(1)} \otimes \mathbf{v}_j^{(2)} \otimes \cdots \otimes \mathbf{v}_j^{(p)} \big)^T.$$

Well, the vectors in the above sum are exactly the elementary tensors if we represent $\mathbb{F}_{d_1} \otimes \mathbb{F}_{d_2} \otimes \cdots \otimes \mathbb{F}_{d_p} \otimes \mathbb{F}_{d_W}$ as $M_{d_W, d_1 d_2 \cdots d_p}(\mathbb{F})$ in the natural way.

Flattenings and Bounds

Since the rank of a multilinear transformation is difficult to compute, so is tensor rank. However, there are a few bounds that we can use to help narrow it down somewhat. The simplest of these bounds comes from just forgetting about part of the tensor product structure of $V_1 \otimes V_2 \otimes \cdots \otimes V_p$. That is, if we let $\{S_1, S_2, \ldots, S_k\}$ be any partition of $\{1, 2, \ldots, p\}$ (i.e., $S_1, S_2, \ldots, S_k$ are sets such that $S_1 \cup S_2 \cup \cdots \cup S_k = \{1, 2, \ldots, p\}$ and $S_i \cap S_j = \{\}$ whenever $i \ne j$) then

The notation

$$\bigotimes_{i \in S_j} V_i$$

means the tensor product of each V_i where $i \in S_j$. It is analogous to big-Σ notation for sums.

$$V_1 \otimes V_2 \otimes \cdots \otimes V_p \cong \left(\bigotimes_{i \in S_1} V_i \right) \otimes \left(\bigotimes_{i \in S_2} V_i \right) \otimes \cdots \otimes \left(\bigotimes_{i \in S_k} V_i \right).$$

To be clear, we are thinking of the vector space on the right as a tensor product of just k vector spaces—its elementary tensors are the ones of the form $\mathbf{v}_1 \otimes \mathbf{v}_2 \otimes \cdots \otimes \mathbf{v}_k$, where $\mathbf{v}_j \in \bigotimes_{i \in S_j} V_i$ for each $1 \le j \le k$.

Perhaps the most natural isomorphism between these two spaces comes from recalling that we can write each $\mathbf{v} \in V_1 \otimes V_2 \otimes \cdots \otimes V_p$ as a sum of elementary tensors

$$\mathbf{v} = \sum_{j=1}^{r} \mathbf{v}_j^{(1)} \otimes \mathbf{v}_j^{(2)} \otimes \cdots \otimes \mathbf{v}_j^{(p)},$$

and if we simply regroup the those products we get the following vector

$$\tilde{\mathbf{v}} \in \left(\bigotimes_{i \in S_1} \mathcal{V}_i\right) \otimes \left(\bigotimes_{i \in S_2} \mathcal{V}_i\right) \otimes \cdots \otimes \left(\bigotimes_{i \in S_k} \mathcal{V}_i\right):$$

$$\tilde{\mathbf{v}} = \sum_{j=1}^{r} \left(\bigotimes_{i \in S_1} \mathbf{v}_j^{(i)}\right) \otimes \left(\bigotimes_{i \in S_2} \mathbf{v}_j^{(i)}\right) \otimes \cdots \otimes \left(\bigotimes_{i \in S_k} \mathbf{v}_j^{(i)}\right).$$

We call $\tilde{\mathbf{v}}$ a **flattening** of $\mathbf{v}$, and we note that the rank of $\tilde{\mathbf{v}}$ never exceeds the rank of $\mathbf{v}$, simply because the procedure that we used to construct $\tilde{\mathbf{v}}$ from $\mathbf{v}$ turns a sum of r elementary tensors into another sum of r elementary tensors. We state this observation as the following theorem:

Theorem 3.3.5 **Tensor Rank of Flattenings**	Suppose $\mathcal{V}_1, \mathcal{V}_2, \ldots, \mathcal{V}_p$ are finite-dimensional vector spaces over the same field and $\mathbf{v} \in \mathcal{V}_1 \otimes \mathcal{V}_2 \otimes \cdots \otimes \mathcal{V}_p$. If $\tilde{\mathbf{v}}$ is a flattening of $\mathbf{v}$ then $$\text{rank}(\mathbf{v}) \geq \text{rank}(\tilde{\mathbf{v}}).$$

Flattenings are most useful when $k = 2$, as we can then easily compute their rank by thinking of them as matrices.

We emphasize that the opposite inequality does not holds in general, since the coarser tensor product structure of some of the elementary tensors in $\left(\bigotimes_{i \in S_1} \mathcal{V}_i\right) \otimes \left(\bigotimes_{i \in S_2} \mathcal{V}_i\right) \otimes \cdots \otimes \left(\bigotimes_{i \in S_k} \mathcal{V}_i\right)$ do not correspond to elementary tensors in $\mathcal{V}_1 \otimes \mathcal{V}_2 \otimes \cdots \otimes \mathcal{V}_p$, since the former space is a "coarser" tensor product of just $k \leq p$ spaces.

For example, if $\mathbf{v} \in \mathbb{F}^{d_1} \otimes \mathbb{F}^{d_2} \otimes \mathbb{F}^{d_3} \otimes \mathbb{F}^{d_4}$ then we obtain flattenings of $\mathbf{v}$ by just grouping some of these tensor product factors together. There are many ways to do this, so these flattenings of $\mathbf{v}$ may live in many different spaces, some of which are listed below along with the partition $\{S_1, S_2, \ldots, S_k\}$ of $\{1, 2, 3, 4\}$ that they correspond to:

Partition $\{S_1, S_2, \ldots, S_k\}$	Flattened Space
$S_1 = \{1, 2\}, S_2 = \{3\}, S_3 = \{4\}$	$\mathbb{F}^{d_1 d_2} \otimes \mathbb{F}^{d_3} \otimes \mathbb{F}^{d_4}$
$S_1 = \{1, 3\}, S_2 = \{2\}, S_3 = \{4\}$	$\mathbb{F}^{d_1 d_3} \otimes \mathbb{F}^{d_2} \otimes \mathbb{F}^{d_4}$
$S_1 = \{1, 2\}, S_2 = \{3, 4\}$	$\mathbb{F}^{d_1 d_2} \otimes \mathbb{F}^{d_3 d_4}$
$S_1 = \{1, 2, 3\}, S_2 = \{4\}$	$\mathbb{F}^{d_1 d_2 d_3} \otimes \mathbb{F}^{d_4}$

In fact, if we flatten the space $\mathbb{F}^{d_1} \otimes \mathbb{F}^{d_2} \otimes \cdots \otimes \mathbb{F}^{d_p} \otimes \mathbb{F}^{d_W}$ by choosing $S_1 = \{p + 1\}$ and $S_2 = \{1, 2, \ldots, p\}$ then we see that the tensor rank of $\mathbf{v} \in \mathbb{F}^{d_1} \otimes \mathbb{F}^{d_2} \otimes \cdots \otimes \mathbb{F}^{d_p} \otimes \mathbb{F}^{d_W}$ is lower-bounded by the tensor rank of its flattening $\tilde{\mathbf{v}} \in \mathbb{F}^{d_W} \otimes \mathbb{F}^{d_1 d_2 \cdots d_p}$, which equals the usual (matrix) rank of $\text{mat}(\tilde{\mathbf{v}}) \in \mathcal{M}_{d_W, d_1 d_2 \cdots d_p}(\mathbb{F})$. We have thus recovered exactly our observation from the end of Section 3.2.3 that if T is a multilinear transformation with standard block matrix A then $\text{rank}(T) \geq \text{rank}(A)$; we can think of standard block matrices as flattenings of arrays.

However, flattenings are somewhat more general than standard block matrices, as we are free to partition the tensor factors in any way we like, and some choices of partition may give better lower bounds than others. As long as we form the partition via $k = 2$ subsets though, we can interpret the resulting flattening just as a matrix and compute its rank using standard linear algebra machinery, thus obtaining an easily-computed lower bound on tensor rank.

Example 3.3.3 **Computing Tensor Rank Bounds via Flattenings**	Show that the following vector $\mathbf{v} \in (\mathbb{C}^2)^{\otimes 4}$ has tensor rank 4: $$\mathbf{v} = \mathbf{e}_1 \otimes \mathbf{e}_1 \otimes \mathbf{e}_1 \otimes \mathbf{e}_1 + \mathbf{e}_1 \otimes \mathbf{e}_2 \otimes \mathbf{e}_1 \otimes \mathbf{e}_2 +$$ $$\mathbf{e}_2 \otimes \mathbf{e}_1 \otimes \mathbf{e}_2 \otimes \mathbf{e}_1 + \mathbf{e}_2 \otimes \mathbf{e}_2 \otimes \mathbf{e}_2 \otimes \mathbf{e}_2.$$

Solution:

It is clear that $\text{rank}(\mathbf{v}) \leq 4$, since $\mathbf{v}$ was provided to us as a sum of 4 elementary tensors. To compute lower bounds of $\text{rank}(\mathbf{v})$, we could use any of its many flattenings. For example, if we choose the partition $(\mathbb{C}^2)^{\otimes 4} \cong \mathbb{C}^2 \otimes \mathbb{C}^8$ then we get the following flattening $\tilde{\mathbf{v}}$ of $\mathbf{v}$:

In the final line here, the first factor of terms like $\mathbf{e}_1 \otimes \mathbf{e}_6$ lives in $\mathbb{C}^2$ while the second lives in $\mathbb{C}^8$.

$$\tilde{\mathbf{v}} = \mathbf{e}_1 \otimes (\mathbf{e}_1 \otimes \mathbf{e}_1 \otimes \mathbf{e}_1) + \mathbf{e}_1 \otimes (\mathbf{e}_2 \otimes \mathbf{e}_1 \otimes \mathbf{e}_2) +$$
$$\mathbf{e}_2 \otimes (\mathbf{e}_1 \otimes \mathbf{e}_2 \otimes \mathbf{e}_1) + \mathbf{e}_2 \otimes (\mathbf{e}_2 \otimes \mathbf{e}_2 \otimes \mathbf{e}_2)$$
$$= \mathbf{e}_1 \otimes \mathbf{e}_1 + \mathbf{e}_1 \otimes \mathbf{e}_6 + \mathbf{e}_2 \otimes \mathbf{e}_3 + \mathbf{e}_2 \otimes \mathbf{e}_8,$$

To compute $\text{rank}(\tilde{\mathbf{v}})$, we note that its matricization is

As another way to see that $\text{rank}(\tilde{\mathbf{v}}) = 2$, we can rearrange $\tilde{\mathbf{v}}$ as $\tilde{\mathbf{v}} = \mathbf{e}_1 \otimes (\mathbf{e}_1 + \mathbf{e}_6) + \mathbf{e}_2 \otimes (\mathbf{e}_3 + \mathbf{e}_8)$.

$$\text{mat}(\tilde{\mathbf{v}}) = \mathbf{e}_1 \mathbf{e}_1^T + \mathbf{e}_1 \mathbf{e}_6^T + \mathbf{e}_2 \mathbf{e}_3^T + \mathbf{e}_2 \mathbf{e}_8^T = \begin{bmatrix} 1 & 0 & 0 & 0 & 0 & 1 & 0 & 0 \\ 0 & 0 & 1 & 0 & 0 & 0 & 0 & 1 \end{bmatrix},$$

so $\text{rank}(\tilde{\mathbf{v}}) = \text{rank}(\text{mat}(\tilde{\mathbf{v}})) = 2$ and thus $\text{rank}(\mathbf{v}) \geq 2$.

To get a better lower bound, we just try another flattening of $\mathbf{v}$: we instead choose the partition $(\mathbb{C}^2)^{\otimes 4} \cong \mathbb{C}^4 \otimes \mathbb{C}^4$ so that we get the following flattening $\mathbf{v}'$ of $\mathbf{v}$:

$$\mathbf{v}' = (\mathbf{e}_1 \otimes \mathbf{e}_1) \otimes (\mathbf{e}_1 \otimes \mathbf{e}_1) + (\mathbf{e}_1 \otimes \mathbf{e}_2) \otimes (\mathbf{e}_1 \otimes \mathbf{e}_2) +$$
$$(\mathbf{e}_2 \otimes \mathbf{e}_1) \otimes (\mathbf{e}_2 \otimes \mathbf{e}_1) + (\mathbf{e}_2 \otimes \mathbf{e}_2) \otimes (\mathbf{e}_2 \otimes \mathbf{e}_2)$$
$$= \mathbf{e}_1 \otimes \mathbf{e}_1 + \mathbf{e}_2 \otimes \mathbf{e}_2 + \mathbf{e}_3 \otimes \mathbf{e}_3 + \mathbf{e}_4 \otimes \mathbf{e}_4.$$

To compute $\text{rank}(\mathbf{v}')$, we note that its matricization is

$$\text{mat}(\mathbf{v}') = \mathbf{e}_1 \mathbf{e}_1^T + \mathbf{e}_2 \mathbf{e}_2^T + \mathbf{e}_3 \mathbf{e}_3^T + \mathbf{e}_4 \mathbf{e}_4^T = \begin{bmatrix} 1 & 0 & 0 & 0 \\ 0 & 1 & 0 & 0 \\ 0 & 0 & 1 & 0 \\ 0 & 0 & 0 & 1 \end{bmatrix},$$

so $\text{rank}(\mathbf{v}') = \text{rank}(\text{mat}(\mathbf{v}')) = 4$ and thus $\text{rank}(\mathbf{v}) \geq 4$, as desired.

There are many situations where none of a vector's flattenings have rank equal to the vector itself, so it may be the case that none of the lower bounds obtained in this way are tight. We thus introduce one more way of bounding tensor rank that is sometimes a bit stronger than these bounds based on flattenings. The rough idea behind this bound is that if $f \in \mathcal{V}_1^*$ is a linear form then the linear transformation $f \otimes I \otimes \cdots \otimes I : \mathcal{V}_1 \otimes \mathcal{V}_2 \otimes \cdots \otimes \mathcal{V}_p \to \mathcal{V}_2 \otimes \cdots \otimes \mathcal{V}_p$ defined by

We only specified how $f \otimes I \otimes \cdots \otimes I$ acts on elementary tensors here. How it acts on the rest of space is determined via linearity.

$$(f \otimes I \otimes \cdots \otimes I)(\mathbf{v}_1 \otimes \mathbf{v}_2 \otimes \cdots \otimes \mathbf{v}_p) = f(\mathbf{v}_1)(\mathbf{v}_2 \otimes \cdots \otimes \mathbf{v}_p)$$

sends elementary tensors to elementary tensors, and can thus be used to help us investigate tensor rank. We note that the universal property of the tensor product (Definition 3.3.1(d)) tells us that this function $f \otimes I \otimes \cdots \otimes I$ actually exists and is well-defined: existence of $f \otimes I \otimes \cdots \otimes I$ follows from first constructing a multilinear transformation $g : \mathcal{V}_1 \times \mathcal{V}_2 \times \cdots \times \mathcal{V}_p \to \mathcal{V}_2 \otimes \cdots \otimes \mathcal{V}_p$ that acts via $g(\mathbf{v}_1, \mathbf{v}_2, \ldots, \mathbf{v}_p) = f(\mathbf{v}_1)(\mathbf{v}_2 \otimes \cdots \otimes \mathbf{v}_p)$ and then using the universal property to see that there exists a function $f \otimes I \otimes \cdots \otimes I$ satisfying $(f \otimes I \otimes \cdots \otimes I)(\mathbf{v}_1 \otimes \mathbf{v}_2 \otimes \cdots \otimes \mathbf{v}_p) = g(\mathbf{v}_1, \mathbf{v}_2, \ldots, \mathbf{v}_p))$.

Theorem 3.3.6

**Another Lower Bound
on Tensor Rank**

Suppose $\mathcal{V}_1, \mathcal{V}_2, \ldots, \mathcal{V}_p$ are finite-dimensional vector spaces over the same field and $\mathbf{v} \in \mathcal{V}_1 \otimes \mathcal{V}_2 \otimes \cdots \otimes \mathcal{V}_p$. Define

$$\mathcal{S}_{\mathbf{v}} = \{(f \otimes I \otimes \cdots \otimes I)(\mathbf{v}) \mid f \in \mathcal{V}_1^*\},$$

which is a subspace of $\mathcal{V}_2 \otimes \cdots \otimes \mathcal{V}_p$. If there does not exist a basis of $\mathcal{S}_{\mathbf{v}}$ consisting entirely of elementary tensors then $\mathrm{rank}(\mathbf{v}) > \dim(\mathcal{S}_{\mathbf{v}})$.

Proof. We prove the contrapositive of the statement of the theorem. Let $r = \mathrm{rank}(\mathbf{v})$ so that we can write $\mathbf{v}$ as a sum of r elementary tensors:

This theorem also works if we instead apply a linear form $f \in \mathcal{V}_k^*$ to the k-th tensor factor for any $1 \le k \le p$ (we just state it in the $k = 1$ case here for simplicity).

$$\mathbf{v} = \sum_{j=1}^{r} \mathbf{v}_j^{(1)} \otimes \mathbf{v}_j^{(2)} \otimes \cdots \otimes \mathbf{v}_j^{(p)}.$$

Since

$$(f \otimes I \otimes \cdots \otimes I)(\mathbf{v}) = \sum_{j=1}^{r} f(\mathbf{v}_j^{(1)})(\mathbf{v}_j^{(2)} \otimes \cdots \otimes \mathbf{v}_j^{(p)}),$$

for all $f \in \mathcal{V}_1^*$, it follows that the set $B = \{\mathbf{v}_j^{(2)} \otimes \cdots \otimes \mathbf{v}_j^{(p)} : 1 \le j \le r\}$ satisfies $\mathrm{span}(B) \supseteq \mathcal{S}_{\mathbf{v}}$. If $r \le \dim(\mathcal{S}_{\mathbf{v}})$ then we must actually have $r = \dim(\mathcal{S}_{\mathbf{v}})$ and $\mathrm{span}(B) = \mathcal{S}_{\mathbf{v}}$, as that is the only way for an r-dimensional vector space to be contained in a vector space of dimension at most r. It then follows from Exercise 1.2.27(b) that B is a basis of $\mathcal{S}_{\mathbf{v}}$ consisting entirely of elementary tensors. ∎

In the $p = 2$ case, the above theorem says nothing at all, since the elementary tensors in $\mathcal{S}_{\mathbf{v}} \subseteq \mathcal{V}_2$ are simply all of the members of $\mathcal{S}_{\mathbf{v}}$, and there is of course a basis of $\mathcal{S}_{\mathbf{v}}$ within $\mathcal{S}_{\mathbf{v}}$. However, when $p \ge 3$, the above theorem can sometimes be used to prove bounds on tensor rank that are better than any bound possible via flattenings.

For example, all non-trivial flattenings of a vector $\mathbf{v} \in (\mathbb{C}^2)^{\otimes 3}$ live in either $\mathbb{C}^2 \otimes \mathbb{C}^4$, so flattenings can never provide a better lower bound than $\mathrm{rank}(\mathbf{v}) \ge 2$ in this case. However, the following example shows that some vectors in $(\mathbb{C}^2)^{\otimes 3}$ have tensor rank 3.

Example 3.3.4

**Tensor Rank Can
Exceed Local
Dimension**

Show that the following vector $\mathbf{v} \in (\mathbb{C}^2)^{\otimes 3}$ has tensor rank 3:

$$\mathbf{v} = \mathbf{e}_1 \otimes \mathbf{e}_1 \otimes \mathbf{e}_2 + \mathbf{e}_1 \otimes \mathbf{e}_2 \otimes \mathbf{e}_1 + \mathbf{e}_2 \otimes \mathbf{e}_1 \otimes \mathbf{e}_1.$$

Solution:
    It is clear that $\mathrm{rank}(\mathbf{v}) \le 3$, since $\mathbf{v}$ was provided to us as a sum of 3 elementary tensors. On the other hand, to see that $\mathrm{rank}(\mathbf{v}) \ge 3$, we construct the subspace $\mathcal{S}_{\mathbf{v}} \subseteq \mathbb{C}^2 \otimes \mathbb{C}^2$ described by Theorem 3.3.6:

Here we have associated the linear form f with a row vector $\mathbf{w}^T$ (via Theorem 1.3.3) and defined $a = \mathbf{w}^T \mathbf{e}_2$ and $b = \mathbf{w}^T \mathbf{e}_1$ for simplicity of notation.

$$\mathcal{S}_{\mathbf{v}} = \{(\mathbf{w}^T \mathbf{e}_1)(\mathbf{e}_1 \otimes \mathbf{e}_2 + \mathbf{e}_2 \otimes \mathbf{e}_1) + (\mathbf{w}^T \mathbf{e}_2)(\mathbf{e}_1 \otimes \mathbf{e}_1) \mid \mathbf{w} \in \mathbb{C}^2\}$$
$$= \{(a, b, b, 0) \mid a, b \in \mathbb{C}\}.$$

It is clear that $\dim(\mathcal{S}_{\mathbf{v}}) = 2$. Furthermore, we can see that the only elementary tensors in $\mathcal{S}_{\mathbf{v}}$ are those of the form $(a, 0, 0, 0)$, since the matricization of $(a, b, b, 0)$ is

$$\begin{bmatrix} a & b \\ b & 0 \end{bmatrix},$$

which has rank 2 whenever $b \neq 0$. It follows that $\mathcal{S}_{\mathbf{v}}$ does not have a basis consisting of elementary tensors, so Theorem 3.3.6 tells us that $\operatorname{rank}(\mathbf{v}) > \dim(\mathcal{S}_{\mathbf{v}}) = 2$, so $\operatorname{rank}(\mathbf{v}) = 3$.

Tensor Rank is a Nightmare

We close this section by demonstrating two ways in which tensor rank is less well-behaved than usual matrix rank (besides just being more difficult to compute), even when we just restrict to the space $\mathbb{F}^{d_1} \otimes \cdots \otimes \mathbb{F}^{d_p}$ when $\mathbb{F} = \mathbb{R}$ or $\mathbb{F} = \mathbb{C}$. The first of these unfortunate aspects of tensor rank is that some vectors can be approximated arbitrarily well by vectors with strictly smaller rank.

For example, if we define

$$
\begin{aligned}
\mathbf{v}_k = {} & k(\mathbf{e}_1 + \mathbf{e}_2/k) \otimes (\mathbf{e}_1 + \mathbf{e}_2/k) \otimes (\mathbf{e}_1 + \mathbf{e}_2/k) - k\mathbf{e}_1 \otimes \mathbf{e}_1 \otimes \mathbf{e}_1 \\
= {} & \mathbf{e}_1 \otimes \mathbf{e}_1 \otimes \mathbf{e}_2 + \mathbf{e}_1 \otimes \mathbf{e}_2 \otimes \mathbf{e}_1 + \mathbf{e}_2 \otimes \mathbf{e}_1 \otimes \mathbf{e}_1 \\
& + \frac{1}{k}\big(\mathbf{e}_2 \otimes \mathbf{e}_2 \otimes \mathbf{e}_1 + \mathbf{e}_2 \otimes \mathbf{e}_1 \otimes \mathbf{e}_2 + \mathbf{e}_1 \otimes \mathbf{e}_2 \otimes \mathbf{e}_2\big) \\
& + \frac{1}{k^2}\big(\mathbf{e}_2 \otimes \mathbf{e}_2 \otimes \mathbf{e}_2\big)
\end{aligned}
$$

then it is straightforward to see that $\operatorname{rank}(\mathbf{v}_k) = 2$ for all k. However,

$$
\lim_{k \to \infty} \mathbf{v}_k = \mathbf{e}_1 \otimes \mathbf{e}_1 \otimes \mathbf{e}_2 + \mathbf{e}_1 \otimes \mathbf{e}_2 \otimes \mathbf{e}_1 + \mathbf{e}_2 \otimes \mathbf{e}_1 \otimes \mathbf{e}_1
$$

is the vector with tensor rank 3 from Example 3.3.4. To deal with issues like this, we define the **border rank** of a vector $\mathbf{v} \in \mathbb{F}^{d_1} \otimes \cdots \otimes \mathbb{F}^{d_p}$ to be the smallest integer r such that $\mathbf{v}$ can be written as a limit of vectors with tensor rank no larger than r. We just showed that the vector from Example 3.3.4 has border rank ≤ 2, despite having tensor rank 3.

In fact, its border rank is exactly 2. See Exercise 3.3.12.

Nothing like this happens when $p = 2$ (i.e., when we can think of tensor rank in $\mathbb{F}^{d_1} \otimes \mathbb{F}^{d_2}$ as the usual rank of a matrix in $\mathcal{M}_{d_1,d_2}(\mathbb{F})$): if $A_1, A_2, \ldots \in \mathcal{M}_{d_1,d_2}(\mathbb{F})$ each have $\operatorname{rank}(A_k) \leq r$ then

In other words, the rank of a matrix can "jump down" in a limit, but it cannot "jump up". When $p \geq 3$, tensor rank can jump either up or down in limits.

$$
\operatorname{rank}\left(\lim_{k \to \infty} A_k\right) \leq r
$$

too (as long as this limit exists). This fact can be verified using the techniques of Section 2.D—recall that the singular values of a matrix are continuous in its entries, so if each A_k has at most r non-zero singular values then the same must be true of their limit (in fact, this was exactly Exercise 2.D.3).

The other unfortunate aspect to tensor rank is that it is field-dependent. For example, if we let $\mathbf{e}_+ = \mathbf{e}_1 + \mathbf{e}_2 \in \mathbb{R}^2$ and $\mathbf{e}_- = \mathbf{e}_1 - \mathbf{e}_2 \in \mathbb{R}^2$ then the vector

$$
\mathbf{v} = \mathbf{e}_+ \otimes \mathbf{e}_+ \otimes \mathbf{e}_1 - \mathbf{e}_1 \otimes \mathbf{e}_1 \otimes \mathbf{e}_- - \mathbf{e}_2 \otimes \mathbf{e}_2 \otimes \mathbf{e}_+ \in (\mathbb{R}^2)^{\otimes 3} \tag{3.3.3}
$$

The tensor rank over $\mathbb{C}$ is always a lower bound of the tensor rank over $\mathbb{R}$, since every real tensor sum decomposition is also a complex one (but not vice-versa).

has tensor rank 3 (see Exercise 3.3.2). However, direct calculation shows that if we instead interpret $\mathbf{v}$ as a member of $(\mathbb{C}^2)^{\otimes 3}$ then it has tensor rank 2 thanks to the decomposition

$$
\mathbf{v} = \frac{1}{2}\big(\mathbf{w} \otimes \mathbf{w} \otimes \mathbf{w} + \overline{\mathbf{w}} \otimes \overline{\mathbf{w}} \otimes \overline{\mathbf{w}}\big), \quad \text{where} \quad \mathbf{w} = (i, -1).
$$

Situations like this do not arise for the rank of a matrix (and thus the tensor rank when $p = 2$), since the singular value decomposition tells us that the singular values (and thus the rank) of a real matrix do not depend on whether we consider it as a member of $\mathcal{M}_{m,n}(\mathbb{R})$ or $\mathcal{M}_{m,n}(\mathbb{C})$. Another way to see that matrix rank does not suffer from this problem is to just notice that allowing complex arithmetic in Gaussian elimination does not increase the number of zero rows that we can obtain in a row echelon form of a real matrix.

Remark 3.3.2	Tensors are one of the most ubiquitous and actively-researched areas in all of science right now. We have provided a brief introduction to the topic and its basic motivation, but there are entire textbooks devoted to exploring properties of tensors and what can be done with them, and we cannot possibly do the subject justice here. See [KB09] and [Lan12], for example, for a more thorough treatment.
We Are Out of Our Depth	

Exercises

solutions to starred exercises on page 479

3.3.1 Compute the tensor rank of each of the following vectors.

*(a) $\mathbf{e}_1 \otimes \mathbf{e}_1 + \mathbf{e}_2 \otimes \mathbf{e}_2 \in \mathbb{C}^2 \otimes \mathbb{C}^2$
 (b) $\mathbf{e}_1 \otimes \mathbf{e}_1 \otimes \mathbf{e}_1 + \mathbf{e}_2 \otimes \mathbf{e}_2 \otimes \mathbf{e}_2 \in (\mathbb{C}^2)^{\otimes 3}$
*(c) $\mathbf{e}_1 \otimes \mathbf{e}_1 \otimes \mathbf{e}_1 + \mathbf{e}_2 \otimes \mathbf{e}_2 \otimes \mathbf{e}_2 + \mathbf{e}_3 \otimes \mathbf{e}_3 \otimes \mathbf{e}_3 \in (\mathbb{C}^3)^{\otimes 3}$

**3.3.2 Show that the vector $\mathbf{v} \in (\mathbb{R}^2)^{\otimes 3}$ from Equation (3.3.3) has tensor rank 3.

[Hint: Mimic Example 3.3.4. Where does this argument break down if we use complex numbers instead of real numbers?]

3.3.3 Determine which of the following statements are true and which are false.

*(a) $\mathbb{C} \otimes \mathbb{C} \cong \mathbb{C}$.
 (b) If we think of $\mathbb{C}$ as a 2-dimensional vector space over $\mathbb{R}$ (e.g., with basis $\{1,i\}$) then $\mathbb{C} \otimes \mathbb{C} \cong \mathbb{C}$.
*(c) If $\mathbf{v} \in \mathbb{F}^m \otimes \mathbb{F}^n$ then $\operatorname{rank}(\mathbf{v}) \leq \min\{m,n\}$.
 (d) If $\mathbf{v} \in \mathbb{F}^m \otimes \mathbb{F}^n \otimes \mathbb{F}^p$ then $\operatorname{rank}(\mathbf{v}) \leq \min\{m,n,p\}$.
*(e) The tensor rank of $\mathbf{v} \in \mathbb{R}^{d_1} \otimes \mathbb{R}^{d_2} \otimes \cdots \otimes \mathbb{R}^{d_p}$ is at least as large as its tensor rank as a member of $\mathbb{C}^{d_1} \otimes \mathbb{C}^{d_2} \otimes \cdots \otimes \mathbb{C}^{d_p}$.

**3.3.4 Show that the linear transformation $S : \mathcal{V} \otimes \mathcal{W} \to \mathcal{X}$ from Definition 3.3.1(d) is necessarily unique.

**3.3.5 Verify the claim of Example 3.3.2 that we can represent $\mathcal{M}_2 \otimes \mathcal{P}^2$ as the set of functions $f : \mathbb{F} \to \mathcal{M}_2$ of the form $f(x) = Ax^2 + Bx + C$ for some $A,B,C \in \mathcal{M}_2$, with elementary tensors defined by $(A \otimes f)(x) = Af(x)$. That is, verify that the four defining properties of Definition 3.3.1 hold for this particular representation of $\mathcal{M}_2 \otimes \mathcal{P}^2$.

**3.3.6 Show that if $\mathcal{C}$ is the vector space of 1-variable continuous functions then $\mathcal{C} \otimes \mathcal{C}$ is *not* the space of 2-variable continuous functions with elementary tensors of the form $f(x)g(y)$ (in contrast with Example 3.3.1).

[Hint: Consider the function $f(x,y) = e^{xy}$ and use Exercise 1.1.22.]

3.3.7 Suppose that $\mathcal{V}_1, \ldots, \mathcal{V}_p$, and $\mathcal{W}_1, \ldots, \mathcal{W}_p$ are vector spaces over the same field, and $T_j : \mathcal{V}_j \to \mathcal{W}_j$ is an isomorphism for each $1 \leq j \leq p$. Let $T_1 \otimes \cdots \otimes T_p : \mathcal{V}_1 \otimes \cdots \otimes \mathcal{V}_p \to \mathcal{W}_1 \otimes \cdots \otimes \mathcal{W}_p$ be the linear transformation defined on elementary tensors by $(T_1 \otimes \cdots \otimes T_p)(\mathbf{v}_1 \otimes \cdots \otimes \mathbf{v}_p) = T_1(\mathbf{v}_1) \otimes \cdots \otimes T_p(\mathbf{v}_p)$. Show that

$$\operatorname{rank}\big((T_1 \otimes \cdots \otimes T_p)(\mathbf{v})\big) = \operatorname{rank}(\mathbf{v})$$

for all $\mathbf{v} \in \mathcal{V}_1 \otimes \cdots \otimes \mathcal{V}_p$.

3.3.8 In this exercise, we generalize some of the observations that we made about the vector from Example 3.3.4. Suppose $\{\mathbf{x}_1,\mathbf{x}_2\}$, $\{\mathbf{y}_1,\mathbf{y}_2\}$, and $\{\mathbf{z}_1,\mathbf{z}_2\}$ are bases of $\mathbb{C}^2$ and let

$$\mathbf{v} = \mathbf{x}_1 \otimes \mathbf{y}_1 \otimes \mathbf{z}_2 + \mathbf{x}_1 \otimes \mathbf{y}_2 \otimes \mathbf{z}_1 + \mathbf{x}_2 \otimes \mathbf{y}_1 \otimes \mathbf{z}_1 \in (\mathbb{C}^2)^{\otimes 3}.$$

 (a) Show that $\mathbf{v}$ has tensor rank 3.
 (b) Show that $\mathbf{v}$ has border rank 2.

3.3.9 Show that the 3-variable polynomial $f(x,y,z) = x + y + z$ cannot be written in the form $f(x,y,z) = p_1(x)q_1(y)r_1(z) + p_2(x)q_2(y)r_2(z)$ for any single-variable polynomials $p_1, p_2, q_1, q_2, r_1, r_2$.

[Hint: You can prove this directly, but it might be easier to leech off of some vector that we showed has tensor rank 3.]

**3.3.10 Show that if $\mathcal{V}$ and $\mathcal{W}$ are vector spaces over the same field then $\mathcal{V} \otimes \mathcal{W} \cong \mathcal{W} \otimes \mathcal{V}$.

3.3.11 Show that if $\mathcal{V}$, $\mathcal{W}$, and $\mathcal{X}$ are vector spaces over the same field then the tensor product distributes over the external direct sum (see Section 1.B.3) in the sense that

$$(\mathcal{V} \oplus \mathcal{W}) \otimes \mathcal{X} \cong (\mathcal{V} \otimes \mathcal{X}) \oplus (\mathcal{V} \otimes \mathcal{W})$$

**3.3.12 Suppose $\mathbb{F} = \mathbb{R}$ or $\mathbb{F} = \mathbb{C}$. Show that if a nonzero vector $\mathbf{v} \in \mathbb{F}^{d_1} \otimes \cdots \otimes \mathbb{F}^{d_p}$ has border rank 1 then it also has tensor rank 1.

[Hint: The Bolzano–Weierstrass theorem from analysis might help.]

3.4   Summary and Review

In this chapter, we introduced multilinear transformations, which are functions that act on *tuples* of vectors in such a way that, if we keep all except for one of the vectors in that tuple constant, then it looks like a *linear* transformation. Multilinear transformations are an extremely wide class of functions that generalize and contain as special cases many objects that we have seen earlier in linear algebra:

- Linear transformations are multilinear transformations with just one input space;

- Multilinear forms (and thus bilinear forms) are multilinear transformation for which the output vector space is just $\mathbb{F}$; and

- Most "multiplications", including the dot product, cross product, matrix multiplication, and the Kronecker product, are bilinear transformations (i.e., multilinear transformations with 2 input spaces).

We introduced multilinear forms back in Section 1.3.2.

We also presented a new way of multiplying two vectors (in $\mathbb{F}^n$) or matrices called the Kronecker product. While this product has many useful properties, the "point" of it is that it lets us represent multilinear transformations via matrices in much the same way that we represent linear transformations via matrices. In particular, Theorem 3.2.2 tells us that if $T : \mathcal{V}_1 \times \cdots \times \mathcal{V}_p \to \mathcal{W}$ is a multilinear transformation acting on finite-dimensional vector spaces with bases $B_1, \ldots, B_p$, and D, respectively, then there exists a matrix A (called the standard block matrix of T) such that

That is, $B_1, \ldots, B_p$ are bases of $\mathcal{V}_1, \ldots, \mathcal{V}_p$, respectively, and D is a basis of $\mathcal{W}$.

$$\left[T(\mathbf{v}_1, \ldots, \mathbf{v}_p)\right]_D = A\left([\mathbf{v}_1]_{B_1} \otimes \cdots \otimes [\mathbf{v}_p]_{B_p}\right) \quad \text{for all} \quad \mathbf{v}_1 \in \mathcal{V}_1, \ldots, \mathbf{v}_p \in \mathcal{V}_p.$$

In particular, if $p = 1$ (i.e., T is just a *linear* transformation) then this theorem reduces to the usual statement that linear transformations have standard matrices.

Finally, we finished this chapter by introducing the tensor product, which generalizes the Kronecker product to arbitrary vector spaces. It can roughly be thought of as a way of "multiplying" two vector spaces together, much like we think of the external direct sum from Section 1.B.3 as a way of "adding" two vector spaces together. Just like the Kronecker product lets us represent multilinear transformations via matrices (once we have chosen bases of the given vector spaces), the tensor product lets us represent multilinear transformations via linear transformations on the tensor product space. In particular, if $T : \mathcal{V}_1 \times \cdots \times \mathcal{V}_p \to \mathcal{W}$ is a multilinear transformation then there exists a linear transformation $S : \mathcal{V}_1 \otimes \cdots \otimes \mathcal{V}_p \to \mathcal{W}$ such that

This is the "universal property" (d) from Definition 3.3.1.

$$T(\mathbf{v}_1, \ldots, \mathbf{v}_p) = S(\mathbf{v}_1 \otimes \cdots \otimes \mathbf{v}_p) \quad \text{for all} \quad \mathbf{v}_1 \in \mathcal{V}_1, \ldots, \mathbf{v}_p \in \mathcal{V}_p.$$

The advantage of the tensor product in this regard is that we do not have to choose bases of the vector spaces. In particular, this means that we can even apply this technique to infinite-dimensional vector spaces.

Exercises

solutions to starred exercises on page 481

3.4.1 Determine which of the following statements are true and which are false.

*(a) If $A, B \in \mathcal{M}_n(\mathbb{C})$ are normal then so is $A \otimes B$.

(b) If $A \in \mathcal{M}_{m,n}, B \in \mathcal{M}_{n,m}$ then $\text{tr}(A \otimes B) = \text{tr}(B \otimes A)$.

*(c) If $A, B \in \mathcal{M}_n$ then $\det(A \otimes B) = \det(B \otimes A)$.

(d) The cross product is a multilinear transformation with type $(2, 1)$.

*(e) If $\mathbf{v}, \mathbf{w} \in \mathbb{R}^7$ then $\mathbf{v} \otimes \mathbf{w} = \mathbf{w} \otimes \mathbf{v}$.

(f) If $\mathcal{V}$ and $\mathcal{W}$ are vector spaces over the same field then $\mathcal{V} \otimes \mathcal{W} \cong \mathcal{W} \otimes \mathcal{V}$.

*(g) The tensor rank of $\mathbf{v} \in \mathbb{R}^{d_1} \otimes \mathbb{R}^{d_2} \otimes \cdots \otimes \mathbb{R}^{d_p}$ equals its tensor rank as a member of $\mathbb{C}^{d_1} \otimes \mathbb{C}^{d_2} \otimes \cdots \otimes \mathbb{C}^{d_p}$.

*3.4.2 Let $W_{n,n} \in \mathcal{M}_{n^2}$ be the swap matrix. Find a formula (depending on n) for $\det(W_{n,n})$.

3.4.3 Let $\mathcal{V}$ and $\mathcal{W}$ be finite-dimensional inner product spaces.

(a) Show that there is an inner product $\langle \cdot, \cdot \rangle_{\mathcal{V} \otimes \mathcal{W}}$ on $\mathcal{V} \otimes \mathcal{W}$ that satisfies

$$\langle \mathbf{v}_1 \otimes \mathbf{w}_1, \mathbf{v}_2 \otimes \mathbf{w}_2 \rangle_{\mathcal{V} \otimes \mathcal{W}} = \langle \mathbf{v}_1, \mathbf{v}_2 \rangle_{\mathcal{V}} \langle \mathbf{w}_1, \mathbf{w}_2 \rangle_{\mathcal{W}}$$

for all $\mathbf{v}_1, \mathbf{v}_2 \in \mathcal{V}$ and $\mathbf{w}_1, \mathbf{w}_2 \in \mathcal{W}$.

(b) Construct an example to show that not all inner products on $\mathcal{V} \otimes \mathcal{W}$ arise from inner products on $\mathcal{V}$ and $\mathcal{W}$ in the manner described by part (a), even in the simple case when $\mathcal{V} = \mathcal{W} = \mathbb{R}^2$.

3.A   Extra Topic: Matrix-Valued Linear Maps

We now provide a more thorough treatment of linear transformations that act on the vector space of matrices. In a sense, there is nothing special about these linear transformations—we know from way back in Section 1.2 that we can represent them by their standard matrix and do all of our usual linear algebra trickery on them. However, many of their interesting properties are most easily unearthed by investigating how they interact with the Kronecker product, so it makes sense to revisit them now.

3.A.1   Representations

We start by precisely defining the types linear transformations that we are now going to focus our attention on.

Definition 3.A.1

Matrix-Valued Linear Map

A **matrix-valued linear map** is a linear transformation

$$\Phi : \mathcal{M}_{m,n} \to \mathcal{M}_{p,q}.$$

Since matrix-valued linear maps are linear transformations, we can represent them by their standard matrices. However, there are also several other ways of representing them that are often much more useful, so it is worthwhile to make the details explicit and see how these different representations relate to each other.

The Standard Matrix

Matrix-valued linear maps are sometimes called **superoperators**, since they can be thought of as operators (functions) acting on operators (matrices).

We start by considering the most basic representation that a matrix-valued linear map $\Phi : \mathcal{M}_{m,n} \to \mathcal{M}_{p,q}$ (or any linear transformation) can have—its standard matrix. In particular, we focus on its standard matrix with respect to the standard basis E of $\mathcal{M}_{m,n}$ and $\mathcal{M}_{p,q}$, which we recall is denoted simply by $[\Phi]$.

If we recall from Theorem 1.2.6 that the standard matrix $[\Phi]$ is constructed so that its columns are $[\Phi(E_{i,j})]_E$ $(1 \leq i \leq m, 1 \leq j \leq n)$ and that $[\Phi(E_{i,j})]_E = \text{vec}(\Phi(E_{i,j}))$, we immediately arrive at the following formula for the standard matrix of Φ:

$$[\Phi] = \left[\text{vec}(\Phi(E_{1,1})) \mid \text{vec}(\Phi(E_{1,2})) \mid \cdots \mid \text{vec}(\Phi(E_{m,n})) \right].$$

If we are interested in standard linear-algebraic properties of Φ like its eigenvalues, rank, invertibility, or finding a matrix square root of it, this standard matrix is likely the simplest tool to make use of, since it satisfies $\text{vec}(\Phi(X)) = [\Phi]\text{vec}(X)$ for all $X \in \mathcal{M}_{m,n}$. For example, we showed back in Example 1.2.10 that the standard matrix of the transposition map $T : \mathcal{M}_2 \to \mathcal{M}_2$ is

More generally, the standard matrix of the transposition map $T : \mathcal{M}_{m,n} \to \mathcal{M}_{n,m}$ is the swap matrix (this is what property (a) of Definition 3.1.3 says).

$$[T] = \begin{bmatrix} 1 & 0 & 0 & 0 \\ 0 & 0 & 1 & 0 \\ 0 & 1 & 0 & 0 \\ 0 & 0 & 0 & 1 \end{bmatrix}.$$

We then used this standard matrix in Example 1.2.17 to find the eigenvalues and eigenvectors of T, and we used it in Example 1.2.18 to find a square root of it.

Example 3.A.1

The Reduction Map

Construct the standard matrix of the linear map $\Phi_R : \mathcal{M}_n \to \mathcal{M}_n$ (called the **reduction map**) defined by $\Phi_R(X) = \text{tr}(X)I - X$.

Solution:
We first compute

$$\Phi_R(E_{i,j}) = \text{tr}(E_{i,j})I - E_{i,j} = \begin{cases} I - E_{j,j} & \text{if } i = j, \\ -E_{i,j} & \text{otherwise.} \end{cases}$$

If we define $\mathbf{e}_+ \overset{\text{def}}{=} \text{vec}(I) = \sum_{i=1}^{n} \mathbf{e}_i \otimes \mathbf{e}_i$ then

$$\text{vec}(\Phi_R(E_{i,j})) = \begin{cases} \mathbf{e}_+ - \mathbf{e}_j \otimes \mathbf{e}_j & \text{if } i = j, \\ -\mathbf{e}_i \otimes \mathbf{e}_j & \text{otherwise.} \end{cases}$$

It follows that the standard matrix of Φ_R is

$$\begin{aligned} [\Phi_R] &= \big[\text{vec}(\Phi_R(E_{1,1})) \mid \text{vec}(\Phi_R(E_{1,2})) \mid \cdots \mid \text{vec}(\Phi_R(E_{n,n})) \big] \\ &= \big[\mathbf{e}_+ - \mathbf{e}_1 \otimes \mathbf{e}_1 \mid -\mathbf{e}_1 \otimes \mathbf{e}_2 \mid \cdots \mid \mathbf{e}_+ - \mathbf{e}_n \otimes \mathbf{e}_n \big] \\ &= \mathbf{e}_+ \mathbf{e}_+^T - I. \end{aligned}$$

For example, in the $n = 3$ case this standard matrix has the form

As usual, we use dots ($\cdot$) to denote entries equal to 0.

$$[\Phi_R] = \begin{bmatrix} \cdot & \cdot & \cdot & \cdot & 1 & \cdot & \cdot & \cdot & 1 \\ \cdot & -1 & \cdot & \cdot & \cdot & \cdot & \cdot & \cdot & \cdot \\ \cdot & \cdot & -1 & \cdot & \cdot & \cdot & \cdot & \cdot & \cdot \\ \cdot & \cdot & \cdot & -1 & \cdot & \cdot & \cdot & \cdot & \cdot \\ 1 & \cdot & \cdot & \cdot & \cdot & \cdot & \cdot & \cdot & 1 \\ \cdot & \cdot & \cdot & \cdot & \cdot & -1 & \cdot & \cdot & \cdot \\ \cdot & \cdot & \cdot & \cdot & \cdot & \cdot & -1 & \cdot & \cdot \\ \cdot & \cdot & \cdot & \cdot & \cdot & \cdot & \cdot & -1 & \cdot \\ 1 & \cdot & \cdot & \cdot & 1 & \cdot & \cdot & \cdot & \cdot \end{bmatrix}.$$

For example, the (unique up to scalar multiplication) eigenvector of Φ_R with eigenvalue $n-1$ is I: $\Phi_R(I) = nI - I = (n-1)I$.

From the standard matrix of a matrix-valued linear map Φ, it is straightforward to deduce many of its properties. For example, since $\mathbf{e}_+ \mathbf{e}_+^T$ has rank 1 and

$$\|\mathbf{e}_+\| = \left\| \sum_{i=1}^{n} \mathbf{e}_i \otimes \mathbf{e}_i \right\| = \sqrt{n},$$

we conclude that if Φ_R is the reduction map from Example 3.A.1, then $[\Phi_R] = e_+ e_+^T - I$ (and thus Φ_R itself) has one eigenvalue equal to $n-1$ and the other $n-1$ eigenvalues (counted according to multiplicity) equal to -1.

The Choi Matrix

There is another matrix representation of a matrix-valued linear map Φ that has some properties that often make it easier to work with than the standard matrix. The idea here is that $\Phi : \mathcal{M}_{m,n} \to \mathcal{M}_{p,q}$ (like every linear transformation) is completely determined by how it acts on a basis of the input space, so it is completely determined by the mn matrices $\Phi(E_{1,1})$, $\Phi(E_{1,2})$, ..., $\Phi(E_{m,n})$, and it is often convenient to arrange these matrices into a single large block matrix.

Definition 3.A.2	The **Choi matrix** of a matrix-valued linear map $\Phi : \mathcal{M}_{m,n} \to \mathcal{M}_{p,q}$ is the $mp \times nq$ matrix
Choi Matrix	$$C_\Phi \overset{\text{def}}{=} \sum_{i=1}^{m} \sum_{j=1}^{n} \Phi(E_{i,j}) \otimes E_{i,j}.$$

We can think of the Choi matrix as a $p \times q$ block matrix whose blocks are $m \times n$. When partitioned in this way, the (i,j)-entry of each block is determined by the corresponding entry of $\Phi(E_{i,j})$. For example, if $\Phi : \mathcal{M}_3 \to \mathcal{M}_3$ is such that

We use asterisks $(*)$ here to denote entries whose values we do not care about right now.

$$\Phi(E_{1,1}) = \begin{bmatrix} a & b & c \\ d & e & f \\ g & h & i \end{bmatrix}, \quad \text{then } C_\Phi = \left[\begin{array}{ccc|ccc|ccc} a & * & * & b & * & * & c & * & * \\ * & * & * & * & * & * & * & * & * \\ * & * & * & * & * & * & * & * & * \\ \hline d & * & * & e & * & * & f & * & * \\ * & * & * & * & * & * & * & * & * \\ * & * & * & * & * & * & * & * & * \\ \hline g & * & * & h & * & * & i & * & * \\ * & * & * & * & * & * & * & * & * \\ * & * & * & * & * & * & * & * & * \end{array}\right].$$

Equivalently, Theorem 3.1.8 tells us that C_Φ is just the swapped version of the block matrix whose (i,j)-block equals $\Phi(E_{i,j})$:

Notice that $E_{i,j} \otimes A$ is a block matrix with A in its (i,j)-block, whereas $A \otimes E_{i,j}$ is a block matrix with the entries of A in the (i,j)-entries of each block.

$$C_\Phi = W_{m,p} \left(\sum_{i=1}^{m} \sum_{j=1}^{n} E_{i,j} \otimes \Phi(E_{i,j}) \right) W_{n,q}^T$$

$$= W_{m,p} \begin{bmatrix} \Phi(E_{1,1}) & \Phi(E_{1,2}) & \cdots & \Phi(E_{1,n}) \\ \Phi(E_{2,1}) & \Phi(E_{2,2}) & \cdots & \Phi(E_{2,n}) \\ \vdots & \vdots & \ddots & \vdots \\ \Phi(E_{m,1}) & \Phi(E_{m,2}) & \cdots & \Phi(E_{m,n}) \end{bmatrix} W_{n,q}^T.$$

Example 3.A.2	Construct the Choi matrix of the linear map $\Phi_R : \mathcal{M}_n \to \mathcal{M}_n$ defined by $\Phi_R(X) = \text{tr}(X)I - X$ (i.e., the reduction map from Example 3.A.1).
The Choi Matrix of the Reduction Map	

Solution:

As we noted in Example 3.A.1,

$$\Phi_R(E_{i,j}) = \text{tr}(E_{i,j})I - E_{i,j} = \begin{cases} I - E_{j,j} & \text{if } i = j, \\ -E_{i,j} & \text{otherwise.} \end{cases}$$

> The fact that $C_{\Phi_R} = -[\Phi_R]$ in this case is just a coincidence, but the fact that C_{Φ_R} and $[\Phi_R]$ have the same entries as each other in different positions is not (see Theorem 3.A.3).

It follows that the Choi matrix of Φ_R is

$$C_{\Phi_R} = W_{3,3} \begin{bmatrix} \Phi(E_{1,1}) & \Phi(E_{1,2}) & \cdots & \Phi(E_{1,n}) \\ \Phi(E_{2,1}) & \Phi(E_{2,2}) & \cdots & \Phi(E_{2,n}) \\ \vdots & \vdots & \ddots & \vdots \\ \Phi(E_{m,1}) & \Phi(E_{m,2}) & \cdots & \Phi(E_{m,n}) \end{bmatrix} W_{3,3}^T$$

$$= W_{3,3} \begin{bmatrix} I - E_{1,1} & -E_{1,2} & \cdots & -E_{1,n} \\ -E_{2,1} & I - E_{2,2} & \cdots & -E_{2,n} \\ \vdots & \vdots & \ddots & \vdots \\ -E_{m,1} & -E_{m,2} & \cdots & I - E_{m,n} \end{bmatrix} W_{3,3}^T$$

$$= W_{3,3} \left(I - \mathbf{e}_+ \mathbf{e}_+^T\right) W_{3,3}^T = I - \mathbf{e}_+ \mathbf{e}_+^T,$$

where $\mathbf{e}_+ = \sum_{i=1}^{n} \mathbf{e}_i \otimes \mathbf{e}_i$ as before.

> Definition 3.1.3(c) tells us that the Choi matrix of the transpose is the swap matrix (just like its standard matrix).

We emphasize that the Choi matrix C_Φ of a linear map Φ does *not* act as a linear transformation in the same way that Φ does. That is, it is *not* the case that $\text{vec}(\Phi(X)) = C_\Phi \text{vec}(X)$; the Choi matrix C_Φ behaves *as a linear transformation* very differently than Φ does. However, C_Φ does make it easier to identify some important properties of Φ. For example, we say that $\Phi : \mathcal{M}_n \to \mathcal{M}_m$ is **transpose-preserving** if $\Phi(X^T) = \Phi(X)^T$ for all $X \in \mathcal{M}_n$—a property that is encoded very naturally in C_Φ:

Theorem 3.A.1

Transpose-Preserving Linear Maps

A linear map $\Phi : \mathcal{M}_n \to \mathcal{M}_m$ is transpose-preserving if and only if C_Φ is symmetric.

Proof. Since Φ is linear, its behavior is determined entirely by how it acts on the standard basis matrices. In particular, it is transpose-preserving if and only if $\Phi(E_{i,j}^T) = \Phi(E_{i,j})^T$ for all $1 \leq i, j \leq n$. On the other hand, C_Φ is symmetric if and only if

> In the final equality here, we just swap the names of i and j.

$$\sum_{i,j=1}^{n} \Phi(E_{i,j}) \otimes E_{i,j} = \left(\sum_{i,j=1}^{n} \Phi(E_{i,j}) \otimes E_{i,j} \right)^T$$

$$= \sum_{i,j=1}^{n} \Phi(E_{i,j})^T \otimes E_{j,i} = \sum_{i,j=1}^{n} \Phi(E_{j,i})^T \otimes E_{i,j}.$$

Since $E_{i,j}^T = E_{j,i}$, these two conditions are equivalent to each other. ∎

For example, the reduction map from Example 3.A.2 is transpose-preserving—a fact that can be verified in a straightforward manner from its definition or by noticing that its Choi matrix is symmetric.

In the case when the ground field is $\mathbb{F} = \mathbb{C}$ and we focus on the *conjugate* transpose (i.e., adjoint), things work out even more cleanly. We say that $\Phi : \mathcal{M}_n \to \mathcal{M}_m$ is **adjoint preserving** if $\Phi(X^*) = \Phi(X)^*$ for all $X \in \mathcal{M}_n$, and we say that it is **Hermiticity-preserving** if $\Phi(X)$ is Hermitian whenever $X \in \mathcal{M}_n(\mathbb{C})$ is Hermitian. The following result says that these two families of maps coincide with each other when $\mathbb{F} = \mathbb{C}$ and can each be easily identified from their Choi matrices.

Theorem 3.A.2
 	
Hermiticity-Preserving
Linear Maps
	Suppose $\Phi : \mathcal{M}_n(\mathbb{C}) \to \mathcal{M}_m(\mathbb{C})$ is a linear map. The following are equivalent:
	
	  a) Φ is Hermiticity-preserving,
	  b) Φ is adjoint-preserving, and
	  c) C_Φ is Hermitian.

Proof. The equivalence of properties (b) and (c) follows from the same argument as in the proof of Theorem 3.A.1, just with a complex conjugation thrown on top of the transposes. We thus focus on the equivalence of properties (a) and (b).

To see that (b) implies (a) we notice that if Φ is adjoint-preserving and X is Hermitian then $\Phi(X)^* = \Phi(X^*) = \Phi(X)$, so $\Phi(X)$ is Hermitian too. For the converse, notice that we can write every matrix $X \in \mathcal{M}_n(\mathbb{C})$ as a linear combination of Hermitian matrices:

It is worth comparing this with the Cartesian decomposition of Remark 1.B.1.

$$X = \frac{1}{2}\underbrace{(X + X^*)}_{\text{Hermitian}} + \frac{1}{2i}\underbrace{(iX - iX^*)}_{\text{Hermitian}}.$$

It follows that if Φ is Hermiticity-preserving then $\Phi(X + X^*)$ and $\Phi(iX - iX^*)$ are each Hermitian, so

$$\Phi(X)^* = \Phi\left(\frac{1}{2}(X + X^*) + \frac{1}{2i}(iX - iX^*)\right)^*$$

$$= \left(\frac{1}{2}\Phi(X + X^*) + \frac{1}{2i}\Phi(iX - iX^*)\right)^*$$

$$= \frac{1}{2}\Phi(X + X^*) - \frac{1}{2i}\Phi(iX - iX^*)$$

$$= \Phi\left(\frac{1}{2}(X + X^*) - \frac{1}{2i}(iX - iX^*)\right)$$

$$= \Phi(X^*)$$

for all $X \in \mathcal{M}_n(\mathbb{C})$, so Φ is adjoint-preserving too. $\blacksquare$

Just like standard matrices, Choi matrices are isomorphic to matrix-valued linear maps (i.e., the linear transformation that sends Φ to C_Φ is invertible).

It is worth emphasizing that the equivalence of conditions (a) and (b) in the above result really is specific to the field $\mathbb{C}$. While it is possible to define a matrix-valued linear map on $\mathcal{M}_n(\mathbb{F})$ (regardless of $\mathbb{F}$) to be **symmetry-preserving** if $\Phi(X)$ is symmetric whenever X is symmetric, such a map may not be transpose-preserving (see Exercise 3.A.12). The reason for this difference in behavior is that the set of symmetric matrices $\mathcal{M}_n^S$ is a subspace of $\mathcal{M}_n$, so specifying how a matrix-valued linear map acts on symmetric matrices does not specify how it acts on all of $\mathcal{M}_n$. In contrast, the set of Hermitian matrices $\mathcal{M}_n^H$ spans all of $\mathcal{M}_n(\mathbb{C})$, so Hermiticity-preservation of Φ restricts how it acts on all of $\mathcal{M}_n(\mathbb{C})$.

It turns out that symmetry-preserving maps do not actually come up in practice much, nor do they have many nice mathematical properties, so we instead focus on yet *another* closely-related family of maps, called **bisymmetric** maps. These are the matrix-valued linear maps Φ with the property that $\Phi = T \circ \Phi = \Phi \circ T$ (whereas transpose-preserving maps just require $T \circ \Phi = \Phi \circ T$). The output of such a map is always symmetric, and it only depends on the symmetric part of the input, so we can think of a bisymmetric map Φ as sending $\mathcal{M}_n^S$ to $\mathcal{M}_m^S$ (much like we can think of Hermiticity-preserving maps as sending $\mathcal{M}_n^H$ to $\mathcal{M}_m^H$).

> Here, T denotes the transpose map.

These different families of maps that can be thought of as preserving the transpose and/or symmetry in slightly different ways, and are summarized in Table 3.1 for ease of reference.

> Γ refers to the **partial transpose**, which we introduce in Section 3.A.2.

	Transpose-Preserving	Bisymmetric
Characterizations:	$\Phi \circ T = T \circ \Phi$ $C_\Phi = C_\Phi^T$	$\Phi = \Phi \circ T = T \circ \Phi$ $C_\Phi = C_\Phi^T = \Gamma(C_\Phi)$

Table 3.1: A comparison of transpose-preserving maps and bisymmetric maps. The characterizations in terms of Choi matrices come from Theorem 3.A.1 and Exercise 3.A.14.

Operator-Sum Representations

The third and final representation of matrix-valued linear maps that we will use is one that is a bit more "direct"—it does not aim to represent $\Phi : \mathcal{M}_{m,n} \to \mathcal{M}_{p,q}$ as a matrix, but rather provides a formula to clarify how Φ acts *on* matrices.

> **Definition 3.A.3**
>
> **Operator-Sum Representation**

An **operator-sum representation** of a linear map $\Phi : \mathcal{M}_{m,n} \to \mathcal{M}_{p,q}$ is a formula of the form

$$\Phi(X) = \sum_i A_i X B_i^T \quad \text{for all} \quad X \in \mathcal{M}_{m,n},$$

where $\{A_i\} \subseteq \mathcal{M}_{p,m}$ and $\{B_i\} \subseteq \mathcal{M}_{q,n}$ are fixed families of matrices.

> If the input and output spaces are square (i.e., $m = n$ and $p = q$) then the A_i and B_i matrices have the same size.

It is perhaps not immediately obvious that every matrix-valued linear map even *has* an operator-sum representation, so before delving into any examples, we present a theorem that tells us how to convert a standard matrix or Choi matrix into an operator-sum representation (and vice-versa).

> **Theorem 3.A.3**
>
> **Converting Between Representations**

Suppose $\Phi : \mathcal{M}_{m,n} \to \mathcal{M}_{p,q}$ is a matrix-valued linear map. The following are equivalent:

a) $[\Phi] = \sum_i A_i \otimes B_i$.

b) $C_\Phi = \sum_i \text{vec}(A_i)\text{vec}(B_i)^T$.

c) $\Phi(X) = \sum_i A_i X B_i^T$ for all $X \in \mathcal{M}_{m,n}$.

Proof. The equivalence of the representations (a) and (c) follows immediately from Theorem 3.1.7, which tells us that

$$\text{vec}\left(\sum_i A_i X B_i^T\right) = \left(\sum_i A_i \otimes B_i\right)\text{vec}(X)$$

Here we use the fact
that standard
matrices are unique.

for all $X \in \mathcal{M}_{m,n}$. Since it is similarly the case that $\text{vec}(\Phi(X)) = [\Phi]\text{vec}(X)$ for all $X \in \mathcal{M}_{m,n}$, we conclude that $\Phi(X) = \sum_i A_i X B_i^T$ for all $X \in \mathcal{M}_{m,n}$ if and only if $[\Phi] = \sum_i A_i \otimes B_i$.

To see that an operator-sum representation (c) corresponds to a Choi matrix of the form (b), we first write each A_i and B_i in terms of their columns: $A_i = \begin{bmatrix} \mathbf{a}_{i,1} & | & \mathbf{a}_{i,2} & | & \cdots & | & \mathbf{a}_{i,m} \end{bmatrix}$ and $B_i = \begin{bmatrix} \mathbf{b}_{i,1} & | & \mathbf{b}_{i,2} & | & \cdots & | & \mathbf{b}_{i,n} \end{bmatrix}$, so that $\text{vec}(A_i) = \sum_{j=1}^{m} \mathbf{a}_{i,j} \otimes \mathbf{e}_j$ and $\text{vec}(B_i) = \sum_{j=1}^{n} \mathbf{b}_{i,j} \otimes \mathbf{e}_j$. Then

$$
\begin{aligned}
C_\Phi &= \sum_{j=1}^{m} \sum_{k=1}^{n} \Phi(E_{j,k}) \otimes E_{j,k} \\
&= \sum_i \sum_{j=1}^{m} \sum_{k=1}^{n} \left(A_i E_{j,k} B_i^T \right) \otimes E_{j,k} \\
&= \sum_i \sum_{j=1}^{m} \sum_{k=1}^{n} \mathbf{a}_{i,j} \mathbf{b}_{i,k}^T \otimes \mathbf{e}_j \mathbf{e}_k^T \\
&= \sum_i \left(\sum_{j=1}^{m} \mathbf{a}_{i,j} \otimes \mathbf{e}_j \right) \left(\sum_{k=1}^{n} \mathbf{b}_{i,k} \otimes \mathbf{e}_k \right)^T \\
&= \sum_i \text{vec}(A_i) \text{vec}(B_i)^T,
\end{aligned}
$$

as claimed. Furthermore, each of these steps can be reversed to see that (b) implies (c) as well. ∎

In particular, part (b) of this theorem tells us that we can convert any rank-one sum decomposition of the Choi matrix of Φ into an operator-sum representation. Since every matrix has a rank-one sum decomposition, every matrix-valued linear map has an operator-sum representation as well. In fact, by leeching directly off of the many things that we know about rank-one sum decompositions, we immediately arrive at the following corollary:

Corollary 3.A.4

The Size of Operator-Sum Decompositions

Every matrix-valued linear map $\Phi: \mathcal{M}_{m,n}(\mathbb{F}) \to \mathcal{M}_{p,q}(\mathbb{F})$ has an operator-sum representation of the form

$$
\Phi(X) = \sum_{i=1}^{\text{rank}(C_\Phi)} A_i X B_i^T \quad \text{for all} \quad X \in \mathcal{M}_{m,n}
$$

In particular,
$\text{rank}(C_\Phi) \leq$
$\min\{mp, nq\}$ is the
minimum number of
terms in any
operator-sum
representation of Φ.
This quantity is
sometimes called
the **Choi rank** of Φ.

with both sets $\{A_i\}$ and $\{B_i\}$ linearly independent. Furthermore, if $\mathbb{F} = \mathbb{R}$ or $\mathbb{F} = \mathbb{C}$ then the sets $\{A_i\}$ and $\{B_i\}$ can be chosen to be mutually orthogonal with respect to the Frobenius inner product.

Proof. This result for an arbitrary field $\mathbb{F}$ follows immediately from the fact (see Theorem A.1.3) that C_Φ has a rank-one sum decomposition of the form

$$
C_\Phi = \sum_{i=1}^{\text{rank}(C_\Phi)} \mathbf{v}_i \mathbf{w}_i^T,
$$

where $\{\mathbf{v}_i\}_{i=1}^{r} \subset \mathbb{F}^{mp}$ and $\{\mathbf{w}_i\}_{i=1}^{r} \subset \mathbb{F}^{nq}$ are linearly independent sets of column vectors. Theorem 3.A.3 then gives us the desired operator-sum representation of $\Phi(X)$ by choosing $A_i = \text{mat}(\mathbf{v}_i)$ and $B_i = \text{mat}(\mathbf{w}_i)$ for all i. To see that the orthogonality claim holds when $\mathbb{F} = \mathbb{R}$ or $\mathbb{F} = \mathbb{C}$, we instead use the *orthogonal* rank-one sum decomposition provided by the singular value decomposition (Theorem 2.3.3). ∎

Example 3.A.3

Operator-Sum Representation of the Transpose Map

Construct an operator-sum representation of the transposition map $T : \mathcal{M}_{m,n} \to \mathcal{M}_{n,m}$.

Solution:

Recall that the Choi matrix of T is the swap matrix, which satisfies $C_T(\mathbf{e}_i \otimes \mathbf{e}_j) = \mathbf{e}_j \otimes \mathbf{e}_i$ for all $1 \leq i \leq m$, $1 \leq j \leq n$. It follows that C_T has the rank-one sum decomposition

This rank-one sum decomposition is just another way of saying that the column of C_T corresponding to the basis vector $\mathbf{e}_i \otimes \mathbf{e}_j$ is $\mathbf{e}_j \otimes \mathbf{e}_i$.

$$C_T = \sum_{i=1}^{m} \sum_{j=1}^{n} (\mathbf{e}_j \otimes \mathbf{e}_i)(\mathbf{e}_i \otimes \mathbf{e}_j)^T.$$

Since $\mathbf{e}_i \otimes \mathbf{e}_j = \text{vec}(E_{i,j})$, it follows from Theorem 3.A.3 that one operator-sum representation of T is

$$X^T = T(X) = \sum_{i=1}^{m} \sum_{j=1}^{n} E_{j,i} X E_{i,j}^T = \sum_{i=1}^{m} \sum_{j=1}^{n} E_{j,i} X E_{j,i} \quad \text{for all} \quad X \in \mathcal{M}_{m,n}.$$

Adjoints were introduced in Section 1.4.2.

We close this subsection by briefly noting how the various representations of the adjoint of a matrix-valued linear map $\Phi : \mathcal{M}_{m,n} \to \mathcal{M}_{p,q}$ (i.e., the linear map $\Phi^* : \mathcal{M}_{p,q} \to \mathcal{M}_{m,n}$ defined by $\langle \Phi(A), B \rangle = \langle A, \Phi^*(B) \rangle$ for all $A \in \mathcal{M}_{m,n}$ and $B \in \mathcal{M}_{p,q}$, where the inner product is the standard Frobenius inner product) are related to the corresponding representations of the original map.

Corollary 3.A.5

The Adjoint of a Matrix-Valued Linear Map

Suppose $\mathbb{F} = \mathbb{R}$ or $\mathbb{F} = \mathbb{C}$. The adjoint of a linear map $\Phi : \mathcal{M}_{m,n}(\mathbb{F}) \to \mathcal{M}_{p,q}(\mathbb{F})$ has the following representations:

a) $[\Phi^*] = [\Phi]^*$.

b) $C_{\Phi^*} = W_{p,m} \overline{C_\Phi} W_{q,n}^T$.

c) If $\Phi(X) = \sum_i A_i X B_i^T$ then $\Phi^*(Y) = \sum_i A_i^* Y \overline{B_i}$.

The overline on $\overline{C_\Phi}$ in this theorem just means complex conjugation (which has no effect if $\mathbb{F} = \mathbb{R}$).

Proof. Property (a) is just the special case of Theorem 1.4.8 that arises when working with the standard bases of $\mathcal{M}_{m,n}(\mathbb{F})$ and $\mathcal{M}_{p,q}(\mathbb{F})$, which are orthonormal. Property (c) then follows almost immediately from Theorem 3.A.3: if $\Phi(X) = \sum_i A_i X B_i^T$ then

$$[\Phi^*] = [\Phi]^* = \left(\sum_i A_i \otimes B_i \right)^* = \sum_i A_i^* \otimes B_i^*,$$

so $\Phi^*(Y) = \sum_i A_i^* Y (B_i^*)^T = \sum_i A_i^* Y \overline{B_i}$, as claimed.

Property (b) follows similarly by noticing that if $C_\Phi = \sum_i \text{vec}(A_i) \text{vec}(B_i)^T$ then

$$C_{\Phi^*} = \sum_i \text{vec}(A_i^*) \text{vec}(B_i^*)^T = \sum_i W_{p,m} \overline{\text{vec}(A_i) \text{vec}(B_i)}^T W_{q,n}^T = W_{p,m} \overline{C_\Phi} W_{q,n}^T,$$

which completes the proof. ∎

In particular, if Φ maps between square matrix spaces (i.e., $m = n$ and $p = q$) then part (b) of the above corollary tells us that $\overline{C_\Phi}$ and C_{Φ^*} are unitarily similar, so equalities involving similarity-invariant quantities like $\det(C_{\Phi^*}) = \overline{\det(C_\Phi)}$ follow almost immediately.

3.A.2   The Kronecker Product of Matrix-Valued Maps

So far, we have defined the Kronecker product on vectors (i.e., members of $\mathbb{F}^n$) and linear transformations on $\mathbb{F}^n$ (i.e., members of $\mathcal{M}_{m,n}(\mathbb{F})$). There is a natural way to ramp this definition up to matrix-valued linear maps acting on $\mathcal{M}_{m,n}(\mathbb{F})$ as well:

Definition 3.A.4

Kronecker Product of Matrix-Valued Linear Maps

> Suppose Φ and Ψ are matrix-valued linear maps acting on $\mathcal{M}_{m,n}$ and $\mathcal{M}_{p,q}$, respectively. Then $\Phi \otimes \Psi$ is the matrix-valued linear map defined by
>
> $$(\Phi \otimes \Psi)(A \otimes B) = \Phi(A) \otimes \Psi(B) \quad \text{for all} \quad A \in \mathcal{M}_{m,n}, \ B \in \mathcal{M}_{p,q}.$$

To apply this map $\Phi \otimes \Psi$ to a matrix $C \in \mathcal{M}_{mp,nq}$ that is not of the form $A \otimes B$, we just use linearity and the fact that every such matrix can be written in the form

To see that C can be written in the form, we can either use Theorem 3.A.3 or recall that $\{E_{i,j} \otimes E_{k,\ell}\}$ is a basis of $\mathcal{M}_{mp,nq}$.

$$C = \sum_i A_i \otimes B_i, \quad \text{so} \quad (\Phi \otimes \Psi)(C) = \sum_i \Phi(A_i) \otimes \Psi(B_i).$$

It is then perhaps not completely clear why this linear map $\Phi \otimes \Psi$ is well-defined—we can write C as a sum of Kronecker products in numerous ways, and why should different decompositions give the same value of $(\Phi \otimes \Psi)(C)$? It turns out that this is not actually a problem, which can be seen either by using the universal property of the tensor product (Definition 3.3.1(d)) or by considering how $\Phi \otimes \Psi$ acts on standard basis matrices (see Exercise 3.A.16). In fact, this is completely analogous to the fact that the matrix $A \otimes B$ is the unique one for which $(A \otimes B)(\mathbf{v} \otimes \mathbf{w}) = (A\mathbf{v}) \otimes (B\mathbf{w})$ for all (column) vectors $\mathbf{v}$ and $\mathbf{w}$.

In the special case when $\Phi = I_{m,n}$ is the identity map on $\mathcal{M}_{m,n}$ (which we denote by I_m if $m = n$ or simply I if the dimensions are clear from context or unimportant), the map $I \otimes \Psi$ just acts independently on each block of a $p \times q$ block matrix:

$$(I \otimes \Psi)\left(\begin{bmatrix} A_{1,1} & A_{1,2} & \cdots & A_{1,s} \\ A_{2,1} & A_{2,2} & \cdots & A_{2,s} \\ \vdots & \vdots & \ddots & \vdots \\ A_{p,1} & A_{p,2} & \cdots & A_{p,q} \end{bmatrix}\right) = \begin{bmatrix} \Psi(A_{1,1}) & \Psi(A_{1,2}) & \cdots & \Psi(A_{1,s}) \\ \Psi(A_{2,1}) & \Psi(A_{2,2}) & \cdots & \Psi(A_{2,s}) \\ \vdots & \vdots & \ddots & \vdots \\ \Psi(A_{p,1}) & \Psi(A_{p,2}) & \cdots & \Psi(A_{p,q}) \end{bmatrix}.$$

The map $\Phi \otimes I$ is perhaps a bit more difficult to think of in terms of block matrices, but it works by having Φ act independently on the matrix made up the $(1,1)$-entries of each block, the matrix made up of the $(1,2)$-entries of each block, and so on.

Two particularly important maps of this type are the **partial transpose** maps $T \otimes I$ and $I \otimes T$, which act on block matrices by either transposing the block structure or transposing the blocks themselves, respectively (where as the usual "full" transpose T transposes both):

$$(T \otimes I)\left(\begin{bmatrix} A_{1,1} & A_{1,2} & \cdots & A_{1,q} \\ A_{2,1} & A_{2,2} & \cdots & A_{2,q} \\ \vdots & \vdots & \ddots & \vdots \\ A_{p,1} & A_{p,2} & \cdots & A_{p,q} \end{bmatrix}\right) = \begin{bmatrix} A_{1,1} & A_{2,1} & \cdots & A_{p,1} \\ A_{1,2} & A_{2,2} & \cdots & A_{p,2} \\ \vdots & \vdots & \ddots & \vdots \\ A_{1,q} & A_{2,q} & \cdots & A_{p,q} \end{bmatrix} \quad \text{and}$$

$$(I \otimes T)\left(\begin{bmatrix} A_{1,1} & A_{1,2} & \cdots & A_{1,q} \\ A_{2,1} & A_{2,2} & \cdots & A_{2,q} \\ \vdots & \vdots & \ddots & \vdots \\ A_{p,1} & A_{p,2} & \cdots & A_{p,q} \end{bmatrix}\right) = \begin{bmatrix} A_{1,1}^T & A_{1,2}^T & \cdots & A_{1,q}^T \\ A_{2,1}^T & A_{2,2}^T & \cdots & A_{2,q}^T \\ \vdots & \vdots & \ddots & \vdots \\ A_{p,1}^T & A_{p,2}^T & \cdots & A_{p,q}^T \end{bmatrix}.$$

The notations $\intercal$ and Γ were chosen to each look like half of a T.

For brevity, we often denote these maps by $\intercal \stackrel{\text{def}}{=} T \otimes I$ and $\Gamma \stackrel{\text{def}}{=} I \otimes T$, respectively, and we think of each of them as one "half" of the transpose of a block matrix—after all, $\intercal \circ \Gamma = T$.

We can similarly define the **partial trace** maps as follows:

$$\text{tr}_1 \stackrel{\text{def}}{=} \text{tr} \otimes I_n : \mathcal{M}_{mn} \to \mathcal{M}_n \quad \text{and} \quad \text{tr}_2 \stackrel{\text{def}}{=} I_m \otimes \text{tr} : \mathcal{M}_{mn} \to \mathcal{M}_m.$$

Explicitly, these maps act on block matrices via

Alternatively, we could denote the partial transposes by $T_1 = T \otimes I$ and $T_2 = I \otimes T$, just like we use tr_1 and tr_2 to denote the partial traces. However, we find the $\intercal$ and Γ notation too cute to pass up.

$$\text{tr}_1\left(\begin{bmatrix} A_{1,1} & A_{1,2} & \cdots & A_{1,m} \\ A_{2,1} & A_{2,2} & \cdots & A_{2,m} \\ \vdots & \vdots & \ddots & \vdots \\ A_{m,1} & A_{m,2} & \cdots & A_{m,m} \end{bmatrix}\right) = A_{1,1} + A_{2,2} + \cdots + A_{m,m}, \quad \text{and}$$

$$\text{tr}_2\left(\begin{bmatrix} A_{1,1} & A_{1,2} & \cdots & A_{1,m} \\ A_{2,1} & A_{2,2} & \cdots & A_{2,m} \\ \vdots & \vdots & \ddots & \vdots \\ A_{m,1} & A_{m,2} & \cdots & A_{m,m} \end{bmatrix}\right) = \begin{bmatrix} \text{tr}(A_{1,1}) & \text{tr}(A_{1,2}) & \cdots & \text{tr}(A_{1,m}) \\ \text{tr}(A_{2,1}) & \text{tr}(A_{2,2}) & \cdots & \text{tr}(A_{2,m}) \\ \vdots & \vdots & \ddots & \vdots \\ \text{tr}(A_{m,1}) & \text{tr}(A_{m,2}) & \cdots & \text{tr}(A_{m,m}) \end{bmatrix}.$$

Example 3.A.4

Partial Transposes, Traces, and Maps

Compute each of the following matrices if $A = \left[\begin{array}{cc|cc} 1 & 2 & 0 & -1 \\ 2 & -1 & 3 & 2 \\ \hline -3 & 2 & 2 & 1 \\ 1 & 0 & 2 & 3 \end{array}\right]$.

a) $\Gamma(A)$,
b) $\text{tr}_1(A)$, and
c) $(I \otimes \Phi)(A)$, where $\Phi(X) = \text{tr}(X)I - X$ is the map from Example 3.A.1.

Solutions:

a) To compute $\Gamma(A) = (I \otimes T)(A)$, we just transpose each block of A:

$$\Gamma(A) = \left[\begin{array}{cc|cc} 1 & 2 & 0 & 3 \\ 2 & -1 & -1 & 2 \\ \hline -3 & 1 & 2 & 2 \\ 2 & 0 & 1 & 3 \end{array}\right].$$

b) $\mathrm{tr}_1(A)$ is just the sum of the diagonal blocks of A:

$$\mathrm{tr}_1(A) = \begin{bmatrix} 1 & 2 \\ 2 & -1 \end{bmatrix} + \begin{bmatrix} 2 & 1 \\ 2 & 3 \end{bmatrix} = \begin{bmatrix} 3 & 3 \\ 4 & 2 \end{bmatrix}.$$

c) We just apply the map Φ to each block of A independently:

$$(I \otimes \Phi)(A) = \left[\begin{array}{c|c} \Phi\left(\begin{bmatrix} 1 & 2 \\ 2 & -1 \end{bmatrix}\right) & \Phi\left(\begin{bmatrix} 0 & -1 \\ 3 & 2 \end{bmatrix}\right) \\ \hline \Phi\left(\begin{bmatrix} -3 & 2 \\ 1 & 0 \end{bmatrix}\right) & \Phi\left(\begin{bmatrix} 2 & 1 \\ 2 & 3 \end{bmatrix}\right) \end{array} \right]$$

$$= \left[\begin{array}{cc|cc} -1 & -2 & 2 & 1 \\ -2 & 1 & -3 & 0 \\ \hline 0 & -2 & 3 & -1 \\ -1 & -3 & -2 & 2 \end{array} \right].$$

This special vector $\mathbf{e}_+$ was originally introduced in Example 3.A.1.

Finally, it is worthwhile to notice that if $\mathbf{e}_+ \overset{\text{def}}{=} \mathrm{vec}(I) = \sum_i \mathbf{e}_i \otimes \mathbf{e}_i$, then

$$\mathbf{e}_+ \mathbf{e}_+^T = \left(\sum_i \mathbf{e}_i \otimes \mathbf{e}_i \right) \left(\sum_j \mathbf{e}_j \otimes \mathbf{e}_j \right)^T = \sum_{i,j} (\mathbf{e}_i \mathbf{e}_j^T) \otimes (\mathbf{e}_i \mathbf{e}_j^T) = \sum_{i,j} E_{i,j} \otimes E_{i,j}.$$

In particular, this means that the Choi matrix of a matrix-valued linear map Φ can be constructed by applying Φ to one half of this special rank-1 matrix $\mathbf{e}_+ \mathbf{e}_+^T$:

$$(\Phi \otimes I)(\mathbf{e}_+ \mathbf{e}_+^T) = \sum_{i,j} \Phi(E_{i,j}) \otimes E_{i,j} = C_\Phi.$$

3.A.3   Positive and Completely Positive Maps

One of the most interesting things that we can do with a matrix-valued linear map Φ that we cannot do with an arbitrary linear transformation is ask what sorts of matrix properties it preserves. For example, we saw in Theorem 3.A.2 that $\Phi : M_n(\mathbb{C}) \to M_m(\mathbb{C})$ is Hermiticity-preserving (i.e., $\Phi(X)$ is Hermitian whenever X is Hermitian) if and only if C_Φ is Hermitian.

Isometries (see Section 1.D.3) are linear preservers that preserve a norm. See Exercise 3.A.5 for an example.

The problem of characterizing which matrix-valued linear maps preserve a given matrix property is called a **linear preserver problem**, and the transpose map $T : M_{m,n} \to M_{n,m}$ plays a particularly important role in this setting. After all, it is straightforward to verify that A^T and A have the same rank and determinant as each other, as well as the same eigenvalues and the same singular values. Furthermore, A^T is normal, unitary, Hermitian, and/or positive semidefinite if and only if A itself has the same property, so we say that the transpose map is rank-preserving, eigenvalue-preserving, normality-preserving, and so on.

On the other hand, determining which maps *other* than the transpose (if any) preserve these matrix properties is often quite difficult.

Remark 3.A.1

Linear Preserver Problems

Many standard linear preserver problems have answers that are very similar to each other. For example, the linear maps $\Phi : M_n \to M_n$ that preserve the determinant (i.e., the maps Φ that satisfy $\det(\Phi(X)) = \det(X)$ for all

It is straightforward to
see that every map
of this form preserves
the determinant; the
hard part is showing
that there are no
other determinant-
preserving linear
maps.

$X \in \mathcal{M}_n$) are exactly those of the form

$$\Phi(X) = AXB \quad \text{for all} \quad X \in \mathcal{M}_n, \quad \text{or}$$
$$\Phi(X) = AX^T B \quad \text{for all} \quad X \in \mathcal{M}_n, \tag{‡}$$

where $A, B \in \mathcal{M}_n$ have $\det(AB) = 1$. Various other types of linear preservers have the same general form, but with different restrictions placed
on A and B:

- Rank-preserving maps have the form (‡) with A and B invertible.
- Singular value preservers have the form (‡) with A and B unitary.
- Eigenvalue preservers have the form (‡) with $B = A^{-1}$.

Proving these results is outside of the scope of this book (see survey papers
like [LT92, LP01] instead), but it is good to be familiar with them so that
we see just how special the transpose map is in this setting.

    The linear preserver problem that we are most interested in is preservation
of positive semidefiniteness.

Definition 3.A.5

Positive Maps

Suppose $\mathbb{F} = \mathbb{R}$ or $\mathbb{F} = \mathbb{C}$. An adjoint-preserving linear map $\Phi : \mathcal{M}_n(\mathbb{F}) \to \mathcal{M}_m(\mathbb{F})$ is called **positive** if $\Phi(X)$ is positive semidefinite whenever X is
positive semidefinite.

This is analogous to
the definition of PSD
matrices
(Definition 2.2.1). We
don't need A to be
self-adjoint for the
property $\mathbf{v}^* A \mathbf{v} \geq 0$ to
make sense—we
add it in so that PSD
matrices have nice
properties.

We note that if $\mathbb{F} = \mathbb{C}$ then we can omit adjoint-preservation from the above
definition, since any map that sends positive semidefinite matrices to positive
semidefinite matrices is necessarily adjoint-reserving (see Exercise 3.A.13).
However, if $\mathbb{F} = \mathbb{R}$ then the adjoint- (i.e., transpose)-preservation property is
required to make this family of maps have "nice" properties (e.g., we want the
Choi matrix of every positive map to be self-adjoint).

    Since the transposition map preserves eigenvalues, it necessarily preserves
positive semidefiniteness as well, so it is positive (and it is possibly the most
important example of such a map). The following example presents another
positive linear map.

Example 3.A.5

**The Reduction Map
is Positive**

Show that the linear map $\Phi_R : \mathcal{M}_n \to \mathcal{M}_n$ defined by $\Phi_R(X) = \text{tr}(X)I - X$
(i.e., the reduction map from Example 3.A.1) is positive.

Solution:
    Just notice that if the eigenvalues of X are $\lambda_1, \ldots, \lambda_n$ then the eigenval-
ues of $\Phi_R(X) = \text{tr}(X)I - X$ are $\text{tr}(X) - \lambda_n, \ldots, \text{tr}(X) - \lambda_1$. Since $\text{tr}(X) = \lambda_1 + \cdots + \lambda_n$, it follows that the eigenvalues of $\Phi_R(X)$ are non-negative
(since they are each the sum of $n - 1$ of the eigenvalues of X), so it is
positive semidefinite whenever X is positive semidefinite. It follows that
Φ_R is positive.

    We of course would like to be able to characterize positive maps in some
way (e.g., via some easily-testable condition on their Choi matrix or operator-
sum representations), but doing so is extremely difficult. Since we are ill-
equipped to make any substantial progress on this problem at this point, we first
introduce another notation of positivity of a matrix-valued linear map that has
somewhat simpler mathematical properties, which will help us in turn better
understand positive maps.

Definition 3.A.6

**k-Positive and
Completely Positive
Maps**

A matrix-valued linear map $\Phi : \mathcal{M}_n \to \mathcal{M}_m$ is called

a) **k-positive** if $\Phi \otimes I_k$ is positive, and

b) **completely positive (CP)** if Φ is k-positive for all $k \geq 1$.

It is worth noting that $\Phi \otimes I_k$ is positive if and only if $I_k \otimes \Phi$ is positive.

In particular, 1-positive linear maps are exactly the positive maps themselves, and if a map is $(k+1)$-positive then it is necessarily k-positive as well (see Exercise 3.A.18). We now look at some examples of linear maps that are and are not k-positive.

Example 3.A.6

The Transposition Map is Not 2-Positive

Show that the transposition map $T : \mathcal{M}_n \to \mathcal{M}_n$ $(n \geq 2)$ is not 2-positive.

Solution:

To see that the map $T \otimes I_2$ is not positive, lets consider how it acts in the $n = 2$ case on $\mathbf{e}_+ \mathbf{e}_+^T$, which is positive semidefinite:

$$(T \otimes I_2)(\mathbf{e}_+ \mathbf{e}_+^T) = (T \otimes I_2)\left(\begin{bmatrix} 1 & 0 & 0 & 1 \\ 0 & 0 & 0 & 0 \\ 0 & 0 & 0 & 0 \\ 1 & 0 & 0 & 1 \end{bmatrix}\right) = \begin{bmatrix} 1 & 0 & 0 & 0 \\ 0 & 0 & 1 & 0 \\ 0 & 1 & 0 & 0 \\ 0 & 0 & 0 & 1 \end{bmatrix}.$$

Recall that the transposition map *is* 1-positive.

The matrix on the right has eigenvalues 1 (with multiplicity 3) and -1 (with multiplicity 1), so it is not positive semidefinite. It follows that $T \otimes I_2$ is not positive, so $T : \mathcal{M}_2 \to \mathcal{M}_2$ is not 2-positive.

To see that T is not 2-positive when $n > 2$, we can just embed this same example into higher dimensions by padding the input matrix $\mathbf{e}_+ \mathbf{e}_+^T$ with extra rows and columns of zeros.

One useful technique that sometimes makes working with positive maps simpler is the observation that a linear map $\Phi : \mathcal{M}_n \to \mathcal{M}_m$ is positive if and only if $\Phi(\mathbf{vv}^*)$ is positive semidefinite for all $\mathbf{v} \in \mathbb{F}^n$. The reason for this is the spectral decomposition, which says that every positive semidefinite matrix $X \in \mathcal{M}_n$ can be written in the form

The spectral decomposition is either of Theorems 2.1.4 (if $\mathbb{F} = \mathbb{C}$) or 2.1.6 (if $\mathbb{F} = \mathbb{R}$).

$$X = \sum_{j=1}^n \mathbf{v}_j \mathbf{v}_j^* \quad \text{for some} \quad \{\mathbf{v}_j\} \subset \mathbb{F}^n.$$

In particular, if $\Phi(\mathbf{vv}^*)$ is PSD for all $\mathbf{v} \in \mathbb{F}^n$ then $\Phi(X) = \sum_j \Phi(\mathbf{v}_j \mathbf{v}_j^*)$ is PSD whenever X is PSD, so Φ is positive.

The following example makes use of this technique to show that the set of k-positive linear maps is not just contained within the set of $(k+1)$-positive maps, but in fact this inclusion is strict as long as $1 \leq k < n$.

Example 3.A.7

A k-Positive Linear Map

When $k = 1$, this is exactly the reduction map that we already showed is positive in Example 3.A.5. Part (b) here shows that it is not 2-positive.

Suppose $k \geq 1$ is an integer and consider the linear map $\Phi : \mathcal{M}_n \to \mathcal{M}_n$ defined by $\Phi(X) = k\mathrm{tr}(X)I - X$.

a) Show that Φ is k-positive, and

b) show that if $k < n$ then Φ is not $(k+1)$-positive.

Solutions:

a) As noted earlier, it suffices to just show that $(\Phi \otimes I_k)(\mathbf{vv}^*)$ is positive semidefinite for all $\mathbf{v} \in \mathbb{F}^n \otimes \mathbb{F}^k$. We can use the Schmidt decomposition (Exercise 3.1.11) to write

$$\mathbf{v} = \sum_{i=1}^{k} \mathbf{w}_i \otimes \mathbf{x}_i,$$

where $\{\mathbf{w}_i\} \subset \mathbb{F}^n$ and $\{\mathbf{x}_i\} \subset \mathbb{F}^k$ are mutually orthogonal sets (and by absorbing a scalar from $\mathbf{w}_i$ to $\mathbf{x}_i$, we can furthermore assume that $\|\mathbf{w}_i\| = 1$ for all i). Then we just observe that we can write $(\Phi \otimes I_k)(\mathbf{v}\mathbf{v}^*)$ as a sum of positive semidefinite matrices in the following way:

> Verifying the second equality here is ugly, but routine—we do not need to be clever.

$$(\Phi \otimes I_k)(\mathbf{v}\mathbf{v}^*) = \sum_{i,j=1}^{k} \left(k(\mathbf{w}_j^*\mathbf{w}_i)I - \mathbf{w}_i\mathbf{w}_j^* \right) \otimes \mathbf{x}_i\mathbf{x}_j^*$$

$$= \sum_{i=1}^{k} \sum_{j=i+1}^{k} \left(\mathbf{w}_i \otimes \mathbf{x}_i - \mathbf{w}_j \otimes \mathbf{x}_j \right)\left(\mathbf{w}_i \otimes \mathbf{x}_i - \mathbf{w}_j \otimes \mathbf{x}_j \right)^*$$

> Since $\|\mathbf{w}_i\| = 1$, the term $I - \mathbf{w}_i\mathbf{w}_i^*$ is positive semidefinite.

$$+ k \sum_{i=1}^{k} \left(I - \mathbf{w}_i\mathbf{w}_i^* \right) \otimes \mathbf{x}_i\mathbf{x}_i^*.$$

b) To see that Φ is *not* $(k+1)$-positive (when $k < n$), we consider how $\Phi \otimes I_{k+1}$ acts on the positive semidefinite matrix $\mathbf{e}_+\mathbf{e}_+^*$, where $\mathbf{e}_+ = \sum_{i=1}^{k+1} \mathbf{e}_i \otimes \mathbf{e}_i \in \mathbb{F}^n \otimes \mathbb{F}^{k+1}$. It is straightforward to check that

$$(\Phi \otimes I_{k+1})(\mathbf{e}_+\mathbf{e}_+^*) = k(I_n \otimes I_{k+1}) - \mathbf{e}_+\mathbf{e}_+^*,$$

which has -1 as an eigenvalue (with corresponding eigenvector $\mathbf{e}_+$) and is thus not positive semidefinite. It follows that $\Phi \otimes I_{k+1}$ is not positive, so Φ is not $(k+1)$-positive.

The above example raises a natural question—if the sets of k-positive linear maps are strict subsets of each other when $1 \leq k \leq n$, what about when $k > n$? For example, is there a matrix-valued linear map that is n-positive but not $(n+1)$-positive? The following theorem shows that no, this is not possible—any map that is n-positive is actually necessarily *completely* positive (i.e., k-positive for *all* $k \geq 1$). Furthermore, there is a simple characterization of these maps in terms of either their Choi matrices and operator-sum representations.

Theorem 3.A.6

Characterization of Completely Positive Maps

Suppose $\Phi : \mathcal{M}_n(\mathbb{F}) \to \mathcal{M}_m(\mathbb{F})$ is a linear map. The following are equivalent:

a) Φ is completely positive.
b) Φ is $\min\{m, n\}$-positive.
c) C_Φ is positive semidefinite.
d) There exist matrices $\{A_i\} \subset \mathcal{M}_{m,n}(\mathbb{F})$ such that $\Phi(X) = \sum_i A_i X A_i^*$.

> This result is sometimes called **Choi's theorem**.

> Recall that
> $$\mathbf{e}_+ = \sum_{i=1}^{n} \mathbf{e}_i \otimes \mathbf{e}_i.$$

Proof. We prove this theorem via the cycle of implications (a) $\implies$ (b) $\implies$ (c) $\implies$ (d) $\implies$ (a). The fact that (a) $\implies$ (b) is immediate from the relevant definitions, so we start by showing that (b) $\implies$ (c).

To this end, first notice that if $n \leq m$ then Φ being n-positive tells us that $(\Phi \otimes I_n)(\mathbf{e}_+\mathbf{e}_+^*)$ is positive semidefinite. However, we noted earlier that $(\Phi \otimes I_n)(\mathbf{e}_+\mathbf{e}_+^*) = C_\Phi$, so C_Φ is positive semidefinite. On the other hand, if

$n > m$ then Φ being m-positive tells us that $(\Phi \otimes I_m)(X)$ is positive semidefinite whenever $X \in M_n \otimes M_m$ is positive semidefinite. It follows that

Here we are using
the fact that
$\mathbf{v}^*A\mathbf{v} = \text{tr}(A\mathbf{v}\mathbf{v}^*)$.

$$\mathbf{e}_+^*(\Phi \otimes I_m)(X)\mathbf{e}_+ \geq 0, \quad \text{so} \quad \text{tr}\big(X(\Phi^* \otimes I_m)(\mathbf{e}_+\mathbf{e}_+^*)\big) \geq 0$$

for all positive semidefinite X. By Exercise 2.2.20, it follows that $(\Phi^* \otimes I_m)(\mathbf{e}_+\mathbf{e}_+^*) = C_{\Phi^*}$ is positive semidefinite, so Corollary 3.A.5 tells us that $C_\Phi = W_{m,n}\overline{C_{\Phi^*}}W_{m,n}^*$ is positive semidefinite too.

The fact that (c) $\implies$ (d) follows from using the spectral decomposition to write $C_\Phi = \sum_i \mathbf{v}_i\mathbf{v}_i^*$. Theorem 3.A.3 tells us that if we define $A_i = \text{mat}(\mathbf{v}_i)$ then Φ has the operator-sum representation $\Phi(X) = \sum_i A_i X A_i^*$, as desired.

To see that (d) $\implies$ (a) and complete the proof, we notice that if $k \geq 1$ and $X \in M_n \otimes M_k$ is positive semidefinite, then so is

$$\sum_i (A_i \otimes I_k)X(A_i \otimes I_k)^* = (\Phi \otimes I_k)(X).$$

It follows that Φ is k-positive for all k, so it is completely positive. ∎

The equivalence of properties (a) and (c) in the above theorem is rather extraordinary. It is trivially the case that (a) implies (c), since if Φ is completely positive then $\Phi \otimes I_n$ is positive, so $(\Phi \otimes I_n)(\mathbf{e}_+\mathbf{e}_+^*) = C_\Phi$ is positive semidefinite. However, the converse says that in order to conclude that $(\Phi \otimes I_k)(X)$ is positive semidefinite for *all* $k \geq 1$ and *all* positive semidefinite $X \in M_n \otimes M_k$, it suffices to just check what happens when $k = n$ and $X = \mathbf{e}_+\mathbf{e}_+^*$. In this sense, the matrix $\mathbf{e}_+\mathbf{e}_+^*$ is extremely special and can roughly be thought of as the "least positive semidefinite" PSD matrix—if a map of the form $\Phi \otimes I_k$ ever creates a non-PSD output from a PSD input, it does so at $\mathbf{e}_+\mathbf{e}_+^*$.

With completely positive maps taken care of, we now return to the problem of trying to characterize (not-necessarily-completely) positive maps. It is straightforward to show that the composition of positive maps is again positive, as is their sum. It follows that if a linear map $\Phi : M_n \to M_m$ can be written in the form

Again, T is the
transposition map
here.

$$\Phi = \Psi_1 + T \circ \Psi_2 \quad \text{for some CP maps} \quad \Psi_1, \Psi_2 : M_n \to M_m, \qquad (3.A.1)$$

then it is positive. We say that any linear map Φ of this form is **decomposable**. It is straightforward to see that every CP map is decomposable, and the above argument shows that every decomposable map is positive. The following example shows that the reduction map Φ_R, which we showed is positive in Example 3.A.5, is also decomposable.

Example 3.A.8
 	
The Reduction Map is Decomposable
 	
This map Ψ is sometimes called the **Werner–Holevo** map.
	Show that the linear map $\Psi : M_n \to M_n$ defined by $\Psi(X) = \text{tr}(X)I - X^T$ is completely positive.
	
	Solution:
	    We just compute the Choi matrix of Ψ. Since the Choi matrix of the map $X \mapsto \text{tr}(X)I$ is $I \otimes I$ and the Choi matrix of the transpose map T is the swap matrix $W_{n,n}$, we conclude that $C_\Psi = I \otimes I - W_{n,n}$. Since $W_{n,n}$ is Hermitian and unitary, its eigenvalues all equal 1 and -1, so C_Ψ is positive semidefinite. Theorem 3.A.6 then implies that Ψ is completely positive.
	
	    In particular, notice that if $\Phi_R(X) = \text{tr}(X)I - X$ is the reduction map from Example 3.A.5 then $\Phi_R = T \circ \Psi$ has the form (3.A.1) and is thus decomposable.

In fact, every positive map $\Phi : \mathcal{M}_n \to \mathcal{M}_m$ that we have seen so far is decomposable, so it seems natural to ask whether or not there are positive maps that are *not*. Remarkably, it turns out that the answer to this question depends on the dimensions m and n. We start by showing that the answer is "no" when $m, n \geq 3$—there are ugly positive linear maps out there that cannot be written in the form (3.A.1).

Theorem 3.A.7 **A Non-Decomposable** **Positive Linear Map**	Show that the linear map $\Phi_C : \mathcal{M}_3 \to \mathcal{M}_3$ defined by $$\Phi_C(X) = \begin{bmatrix} x_{1,1} + x_{3,3} & -x_{1,2} & -x_{1,3} \\ -x_{2,1} & x_{2,2} + x_{1,1} & -x_{2,3} \\ -x_{3,1} & -x_{3,2} & x_{3,3} + x_{2,2} \end{bmatrix}$$ is positive but not decomposable.

The linear map Φ_C is typically called the **Choi map**.

It is worth noting that the map described by this theorem is quite similar to the reduction map Φ_R of Example 3.A.1—the diagonal entries of its output are just shuffled up a bit (e.g., the top-left entry is $x_{1,1} + x_{3,3}$ instead of $x_{2,2} + x_{3,3}$).

Proof of Theorem 3.A.7. We have to prove two distinct claims: we have to show that Φ_C is positive and we have to show that it cannot be written in the form (3.A.1).

To see that Φ_C is positive, recall that it suffices to show that

$$\Phi_C(\mathbf{v}\mathbf{v}^*) = \begin{bmatrix} |v_1|^2 + |v_3|^2 & -v_1\overline{v_2} & -v_1\overline{v_3} \\ -\overline{v_1}v_2 & |v_2|^2 + |v_1|^2 & -v_2\overline{v_3} \\ -\overline{v_1}v_3 & -\overline{v_2}v_3 & |v_3|^2 + |v_2|^2 \end{bmatrix}$$

is positive semidefinite for all $\mathbf{v} \in \mathbb{F}^3$. To this end, recall that we can prove that this matrix is positive semidefinite by checking that all of its principal minors are nonnegative (see Remark 2.2.1).

A **principal minor** of a square matrix A is the determinant of a square submatrix that is obtained by deleting some rows and the same columns of A.

There are three 1×1 principal minors of $\Phi_C(\mathbf{v}\mathbf{v}^*)$:

$$|v_1|^2 + |v_3|^2, \quad |v_2|^2 + |v_1|^2, \quad \text{and} \quad |v_3|^2 + |v_2|^2,$$

which are clearly all nonnegative. Similarly, one of the three 2×2 principal minors of $\Phi_C(\mathbf{v}\mathbf{v}^*)$ is

$$\det\left(\begin{bmatrix} |v_1|^2 + |v_3|^2 & -\overline{v_1}v_2 \\ -v_1\overline{v_2} & |v_2|^2 + |v_1|^2 \end{bmatrix}\right) = (|v_1|^2 + |v_3|^2)(|v_2|^2 + |v_1|^2) - |v_1|^2|v_2|^2$$
$$= |v_1|^4 + |v_1|^2|v_3|^2 + |v_2|^2|v_3|^2$$
$$\geq 0.$$

The calculation for the other two 2×2 principal minors is almost identical, so we move right on to the one and only 3×3 principal minor of $\Phi_C(\mathbf{v}\mathbf{v}^*)$:

$$\det(\Phi_C(\mathbf{v}\mathbf{v}^*)) = (|v_1|^2 + |v_3|^2)(|v_2|^2 + |v_1|^2)(|v_3|^2 + |v_2|^2) - 2|v_1|^2|v_2|^2|v_3|^2$$
$$- (|v_1|^2 + |v_3|^2)|v_2|^2|v_3|^2 - (|v_2|^2 + |v_1|^2)|v_1|^2|v_3|^2$$
$$- (|v_3|^2 + |v_2|^2)|v_1|^2|v_2|^2$$
$$= |v_1|^2|v_3|^4 + |v_2|^2|v_1|^4 + |v_3|^2|v_2|^4 - 3|v_1|^2|v_2|^2|v_3|^2.$$

In order to show that this quantity is nonnegative, we recall that the AM–GM inequality (Theorem A.5.3) tells us that if $x, y, z \geq 0$ then $(x + y + z)/3 \geq \sqrt[3]{xyz}$. Choosing $x = |v_1|^2 |v_3|^4$, $y = |v_2|^2 |v_1|^4$ and $z = |v_3|^2 |v_2|^4$ gives

> The AM–GM inequality is introduced in Appendix A.5.2.

$$\frac{|v_1|^2 |v_3|^4 + |v_2|^2 |v_1|^4 + |v_3|^2 |v_2|^4}{3} \geq \sqrt[3]{|v_1|^6 |v_2|^6 |v_3|^6} = |v_1|^2 |v_2|^2 |v_3|^2,$$

which is exactly what we wanted. It follows that Φ_C is positive, as claimed.

To see that Φ_C cannot be written in the form $\Phi_C = \Psi_1 + T \circ \Psi_2$ with Ψ_1 and Ψ_2 completely positive, we first note that constructing the Choi matrix of both sides of this equation shows thats it is equivalent to show that we cannot write $C_{\Phi_C} = X + Y^{T_1}$ with both X and Y positive semidefinite. Suppose for the sake of establishing a contradiction that C_{Φ_C} can be written in this form. A straightforward computation shows that the Choi matrix of Φ_C is

> Recall that Y^{T_1} refers to $(T \otimes I)(Y)$; the first partial transpose of Y.

> Here, X and Y are the Choi matrices of Ψ_1 and Ψ_2, respectively.

$$C_{\Phi_C} = \left[\begin{array}{ccc|ccc|ccc} 1 & \cdot & \cdot & \cdot & -1 & \cdot & \cdot & \cdot & -1 \\ \cdot & \cdot & \cdot & \cdot & \cdot & \cdot & \cdot & \cdot & \cdot \\ \cdot & \cdot & 1 & \cdot & \cdot & \cdot & \cdot & \cdot & \cdot \\ \hline \cdot & \cdot & \cdot & 1 & \cdot & \cdot & \cdot & \cdot & \cdot \\ -1 & \cdot & \cdot & \cdot & 1 & \cdot & \cdot & \cdot & -1 \\ \cdot & \cdot & \cdot & \cdot & \cdot & \cdot & \cdot & \cdot & \cdot \\ \hline \cdot & \cdot & \cdot & \cdot & \cdot & \cdot & \cdot & \cdot & \cdot \\ \cdot & \cdot & \cdot & \cdot & \cdot & \cdot & \cdot & 1 & \cdot \\ -1 & \cdot & \cdot & \cdot & -1 & \cdot & \cdot & \cdot & 1 \end{array} \right].$$

Since $[C_{\Phi_C}]_{2,2} = 0$ and X and Y are positive semidefinite, it is necessarily the case that $y_{2,2} = 0$. The fact that $[C_{\Phi_C}]_{6,6} = [C_{\Phi_C}]_{7,7} = 0$ similarly implies $y_{6,6} = y_{7,7} = 0$. If a diagonal entry of a PSD matrix equals 0 then its entire row and column must equal 0 by Exercise 2.2.11, so we conclude in particular that

$$y_{2,4} = y_{4,2} = y_{3,7} = y_{7,3} = y_{6,8} = y_{8,6} = 0.$$

> It is also worth noting that this linear map Φ_C is not 2-positive—see Exercise 3.A.4.

The reason that we focused on these entries of Y is that, under partial transposition, they are the entries that are moved to the locations of the "-1" entries of C_{Φ_C} above. That is,

$$-1 = [C_{\Phi_C}]_{1,5} = x_{1,5} + [Y^{T_1}]_{1,5} = x_{1,5} + y_{4,2} = x_{1,5},$$

and a similar argument shows that $x_{1,9} = x_{5,1} = x_{5,9} = x_{9,1} = x_{9,5} = -1$ as well.

Positive semidefiniteness of Y also tells us that $y_{1,1}, y_{5,5}, y_{9,9} \geq 0$, which (since $[C_{\Phi_C}]_{1,1} = [C_{\Phi_C}]_{5,5} = [C_{\Phi_C}]_{9,9} = 1$) implies $x_{1,1}, x_{5,5}, x_{9,9} \leq 1$. We have now learned enough about X to see that it is not positive semidefinite: it is straightforward to check that

> Recall that $e_+ = \sum_{i=1}^{3} e_i \otimes e_i = (1, 0, 0, 0, 1, 0, 0, 0, 1)$.

$$e_+^T X e_+ = x_{1,1} + x_{5,5} + x_{9,9} - 6 \leq 3 - 6 = -3,$$

so X is not positive semidefinite. This is a contradiction that completes the proof. ∎

On the other hand, if the dimensions m and n are both sufficiently small, then a somewhat deep result (which we now state without proof) says that every positive map is indeed decomposable.

Theorem 3.A.8	Suppose $\Phi : \mathcal{M}_n \to \mathcal{M}_m$ is a linear map and that $mn \leq 6$. Then Φ is
Positive Maps in Small Dimensions	positive if and only if it is decomposable.

With all of the results of this section taken care of, it is worthwhile at this point to remind ourselves of the relationships between the various types of positivity of linear maps that we have now seen. A visual summary of these relationships is provided in Figure 3.2.

positive linear maps $\Phi : \mathcal{M}_n \to \mathcal{M}_n$

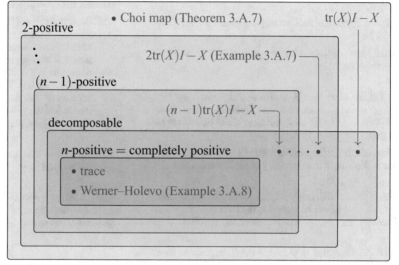

In certain dimensions, the relationship of the decomposable set with the others "collapses". For example, Theorem 3.A.8 says that if $n = 2$ then the decomposable set equals the entire positive set. Also, if $n = 3$ then every 2-positive map is decomposable (YYT16).

Figure 3.2: A schematic that depicts the relationships between the sets of positive, k-positive, completely positive, and decomposable linear maps, as well the locations of the various positive linear maps that we have seen so far within these sets.

Remark 3.A.2	The proof of Theorem 3.A.8 is quite technical—it was originally proved in
The Set of Positive Linear Maps	the $m = n = 2$ case in [Stø63] (though there are now some slightly simpler proofs of this case available [MO15]) and in the $\{m,n\} = \{2,3\}$ case in [Wor76]. If $n = 2$ and $m \geq 4$ then there are positive linear maps that are not decomposable (see Exercise 3.A.22), just like we showed in the $m,n \geq 3$ case in Theorem 3.A.7.

In higher dimensions, the structure of the set of positive linear maps is not understood very well, and constructing "strange" positive linear maps like the one from Theorem 3.A.7 is an active research area. The fact that this set is so much simpler when $mn \leq 6$ can be thought of roughly as a statement that there just is not enough room in those small dimensions for the "true nature" of the set of positive maps to become apparent.

Exercises

solutions to starred exercises on page 481

3.A.1 Determine which of the following statements are true and which are false.

∗**(a)** If $A \in \mathcal{M}_{m,n}(\mathbb{C})$ then the linear map $\Phi : \mathcal{M}_n(\mathbb{C}) \to \mathcal{M}_m(\mathbb{C})$ defined by $\Phi(X) = AXA^*$ is positive.

(b) If $\Phi : \mathcal{M}_n(\mathbb{C}) \to \mathcal{M}_m(\mathbb{C})$ is Hermiticity-preserving then $\Phi^* = \Phi$.

∗**(c)** If $\Phi, \Psi : \mathcal{M}_n \to \mathcal{M}_m$ are positive linear maps then so is $\Phi + \Psi$.

∗∗3.A.2 Let $\Psi : \mathcal{M}_3 \to \mathcal{M}_3$ be the (completely positive) Werner–Holevo map $\Psi(X) = \text{tr}(X)I - X^T$ from Example 3.A.8. Find matrices $\{A_i\} \subset \mathcal{M}_3$ such that $\Psi(X) = \sum_i A_i X A_i^*$ for all $X \in \mathcal{M}_3$.

3.A.3 Suppose $p \in \mathbb{R}$. Show that the linear map $\Psi_p : \mathcal{M}_n \to \mathcal{M}_n$ defined by $\Psi_p(X) = \text{tr}(X)I + pX^T$ is completely positive if and only if $-1 \leq p \leq 1$.

[Side note: If $p = -1$ then this is the Werner–Holevo map from Example 3.A.8.]

∗∗3.A.4 Show that the Choi map Φ_C from Theorem 3.A.7 is not 2-positive.

[Hint: Let $\mathbf{x} = (\mathbf{e}_1 + \mathbf{e}_2) \otimes \mathbf{e}_1 + \mathbf{e}_3 \otimes \mathbf{e}_2 \in \mathbb{F}^3 \otimes \mathbb{F}^2$. What is $(\Phi_C \otimes I_2)(\mathbf{x}\mathbf{x}^*)$?]

∗∗ 3.A.5 Show that a matrix-valued linear map $\Phi : \mathcal{M}_{m,n} \to \mathcal{M}_{m,n}$ preserves the Frobenius norm (i.e., $\|\Phi(X)\|_F = \|X\|_F$ for all $X \in \mathcal{M}_{m,n}$) if and only if $[\Phi]$ is unitary.

[Side note: In the language of Section 1.D.3, we would say that such a map Φ is an **isometry** of the Frobenius norm.]

3.A.6 Show that every adjoint-preserving linear map $\Phi : \mathcal{M}_n \to \mathcal{M}_m$ can be written in the form $\Phi = \Psi_1 - \Psi_2$, where $\Psi_1, \Psi_2 : \mathcal{M}_n \to \mathcal{M}_m$ are completely positive.

∗3.A.7 Show that if $\Phi : \mathcal{M}_n \to \mathcal{M}_m$ is positive then so is Φ^*.

∗3.A.8 A matrix-valued linear map $\Phi : \mathcal{M}_n \to \mathcal{M}_m$ is called **unital** if $\Phi(I_n) = I_m$. Show that the following are equivalent:

i) Φ is unital,
ii) $\text{tr}_2(C_\Phi) = I_m$, and
iii) if $\Phi(X) = \sum_j A_j X B_j^T$ then $\sum_j A_j B_j^T = I_m$.

3.A.9 A matrix-valued linear map $\Phi : \mathcal{M}_n \to \mathcal{M}_m$ is called **trace-preserving** if $\text{tr}(\Phi(X)) = \text{tr}(X)$ for all $X \in \mathcal{M}_n$. Show that the following are equivalent:

i) Φ is trace-preserving,
ii) $\text{tr}_1(C_\Phi) = I_n$, and
iii) if $\Phi(X) = \sum_j A_j X B_j^T$ then $\sum_j B_j^T A_j = I_n$.

[Hint: This can be proved directly, or it can be proved by noting that Φ is trace-preserving if and only if Φ^* is unital and then invoking Exercise 3.A.8.]

3.A.10 Suppose $\Phi, \Psi : \mathcal{M}_n \to \mathcal{M}_n$ are linear maps.

(a) Show that $C_{\Psi \circ \Phi} = (\Psi \otimes I)(C_\Phi)$.
(b) Show that $C_{\Phi \circ \Psi^*} = (I \otimes \overline{\Psi})(C_\Phi)$, where $\overline{\Psi}$ is the linear map defined by $\overline{\Psi}(X) = \overline{\Psi(\overline{X})}$ for all $X \in \mathcal{M}_n$.

3.A.11 Suppose $\Phi : \mathcal{M}_n \to \mathcal{M}_m$ is a linear map. Show that Φ is completely positive if and only if $T \circ \Phi \circ T$ is completely positive.

∗∗ 3.A.12 A linear map $\Phi : \mathcal{M}_n \to \mathcal{M}_m$ is called **symmetry-preserving** if $\Phi(X)$ is symmetric whenever X is symmetric.

(a) Show that every transpose-preserving linear map is symmetry-preserving.
(b) Provide an example to show that a map can be symmetry-preserving without being transpose-preserving.

∗∗3.A.13 Recall from Definition 3.A.5 that we required positive linear maps to be adjoint-preserving.

(a) Show that if $\mathbb{F} = \mathbb{C}$ then adjoint-preservation comes for free. That is, show that if $\Phi(X)$ is positive semidefinite whenever X is positive semidefinite, then Φ is adjoint-preserving.
(b) Show that if $\mathbb{F} = \mathbb{R}$ then adjoint-preservation does *not* come for free. That is, find a matrix-valued linear map Φ with the property that $\Phi(X)$ is positive semidefinite whenever X is positive semidefinite, but Φ is not adjoint-preserving.

∗∗3.A.14 Show that a matrix-valued linear map Φ is bisymmetric if and only if $C_\Phi = C_\Phi^T = \Gamma(C_\Phi)$.

∗∗3.A.15 Show that $\Phi : \mathcal{M}_n(\mathbb{R}) \to \mathcal{M}_m(\mathbb{R})$ is both bisymmetric and decomposable if and only if there exists a completely positive map $\Psi : \mathcal{M}_n(\mathbb{R}) \to \mathcal{M}_m(\mathbb{R})$ such that $\Phi = \Psi + T \circ \Psi$.

∗∗3.A.16 Show that the linear map $\Phi \otimes \Psi$ from Definition 3.A.4 is well-defined. That is, show that it is uniquely determined by Φ and Ψ, and that the value of $(\Phi \otimes \Psi)(C)$ does not depend on the particular way in which we write

$$C = \sum_i A_i \otimes B_i.$$

3.A.17 Suppose $\mathbb{F} = \mathbb{R}$ or $\mathbb{F} = \mathbb{C}$, and let $f : \mathcal{M}_n(\mathbb{F}) \to \mathbb{F}$ be a linear form, which we can think of as a matrix-valued linear map by noting that $\mathbb{F} \cong \mathcal{M}_1(\mathbb{F})$ in the obvious way. Show that the following are equivalent:

i) f is positive,
ii) f is completely positive, and
iii) there exists a positive semidefinite matrix $A \in \mathcal{M}_n(\mathbb{F})$ such that $f(X) = \text{tr}(AX)$ for all $X \in \mathcal{M}_n(\mathbb{F})$.

∗∗3.A.18 Show that if a linear map $\Phi : \mathcal{M}_n \to \mathcal{M}_m$ is $(k+1)$-positive for some integer $k \geq 1$ then it is also k-positive.

∗∗3.A.19 Suppose a linear map $\Phi : \mathcal{M}_n \to \mathcal{M}_m$ is such that $\Phi(X)$ is positive definite whenever X is positive definite. Use the techniques from Section 2.D to show that Φ is positive.

3.A.20 Suppose $U \in \mathcal{M}_n(\mathbb{C})$ is a skew-symmetric unitary matrix (i.e., $U^T = -U$ and $U^*U = I$). Show that the map $\Phi_U : \mathcal{M}_n(\mathbb{C}) \to \mathcal{M}_n(\mathbb{C})$ defined by $\Phi_U(X) = \operatorname{tr}(X)I - X - UX^TU^*$ is positive.

[Hint: Show that $\mathbf{x} \cdot (U\overline{\mathbf{x}}) = 0$ for all $\mathbf{x} \in \mathbb{C}^n$. Exercise 2.1.22 might help, but is not necessary.]

[Side note: Φ_U is called a **Breuer–Hall map**, and it is worth comparing it to the (also positive) reduction map of Example 3.A.1. While the reduction map is decomposable, Breuer–Hall maps are not.]

3.A.21 Consider the linear map $\Phi : \mathcal{M}_2 \to \mathcal{M}_4$ defined by

$$\Phi(X) = \operatorname{tr}(X)I_4 - \begin{bmatrix} 0 & 0 & x_{1,2} & \operatorname{tr}(X)/2 \\ 0 & 0 & x_{2,1} & x_{1,2} \\ x_{2,1} & x_{1,2} & 0 & 0 \\ \operatorname{tr}(X)/2 & x_{2,1} & 0 & 0 \end{bmatrix}.$$

(a) Show that Φ is positive.
[Hint: Show that $\Phi(X)$ is diagonally dominant.]
[Side note: In fact, Φ is decomposable—see Exercise 3.C.15.]

(b) Construct C_Φ.

(c) Show that Φ is not completely positive.

**3.A.22 Consider the linear map $\Phi : \mathcal{M}_2 \to \mathcal{M}_4$ defined by

$$\Phi\left(\begin{bmatrix} a & b \\ c & d \end{bmatrix}\right) = \begin{bmatrix} 4a - 2b - 2c + 3d & 2b - 2a & 0 & 0 \\ 2c - 2a & 2a & b & 0 \\ 0 & c & 2d & -2b - d \\ 0 & 0 & -2c - d & 4a + 2d \end{bmatrix}.$$

(a) Show that if X is positive definite then the determinant of the top-left 1×1, 2×2, and 3×3 blocks of $\Phi(X)$ are strictly positive.

(b) Show that if X is positive definite then $\det(\Phi(X)) > 0$ as well.
[Hint: This is *hard*. Try factoring this determinant as $\det(X)p(X)$, where p is a polynomial in the entries of X. Computer software might help.]

(c) Use Sylvester's Criterion (Theorem 2.2.6) and Exercise 3.A.19 to show that Φ is positive.
[Side note: However, Φ is *not* decomposable—see Exercise 3.C.17.]

3.B   Extra Topic: Homogeneous Polynomials

It might be a good idea to refer to Appendix A.2.1 for some basic facts about multivariable polynomials before reading this section.

Recall from Section 1.3.2 that if $\mathbb{F} = \mathbb{R}$ or $\mathbb{F} = \mathbb{C}$ then every linear form $f : \mathbb{F}^n \to \mathbb{F}$ can be represented in the form $f(\mathbf{x}) = \mathbf{v} \cdot \mathbf{x}$, where $\mathbf{v} \in \mathbb{F}^n$ is a fixed vector. If we just expand that expression out via the definition of the dot product, we see that it is equivalent to the statement that every linear form $f : \mathbb{F}^n \to \mathbb{F}$ can be represented as a multivariable polynomial in which each term has degree equal to exactly 1. That is, there exist scalars $a_1, a_2, \ldots, a_n \in \mathbb{F}$ such that

$$f(x_1, x_2, \ldots, x_n) = a_1 x_1 + a_2 x_2 + \cdots + a_n x_n = \sum_{j=1}^n a_j x_j.$$

The **degree** of a term (like $a_{i,j}x_ix_j$) is the sum of the exponents of its variables. For example, $7x^2yz^3$ has degree $2 + 1 + 3 = 6$.

Similarly, we saw in Section 2.A (Theorem 2.A.1 in particular) that every quadratic form $f : \mathbb{R}^n \to \mathbb{R}$ can be written as a multivariable polynomial in which each term has degree equal to exactly 2—there exist scalars $a_{i,j} \in \mathbb{R}$ for $1 \le i \le j \le n$ such that

$$f(x_1, x_2, \ldots, x_n) = \sum_{i=1}^n \sum_{j=i}^n a_{i,j} x_i x_j. \tag{3.B.1}$$

Every multivariable polynomial can be written as a sum of homogeneous polynomials (e.g., a degree-2 polynomial is a sum of a quadratic form, linear form, and a scalar).

We then used our various linear algebraic tools like the real spectral decomposition (Theorem 2.1.6) to learn more about the structure of these quadratic forms. Furthermore, analogous results hold for quadratic forms $f : \mathbb{C}^n \to \mathbb{C}$, and they can be proved simply by instead making use of the *complex* spectral decomposition (Theorem 2.1.4).

In general, a polynomial for which every term has the exact same degree is called a **homogeneous polynomial**. Linear and quadratic forms constitute the degree-1 and degree-2 homogeneous polynomials, respectively, and in this section we investigate their higher-degree brethren. In particular, we will see that homogeneous polynomials in general are intimately connected with tensors and the Kronecker product, so we can use the tools that we developed in this chapter to better understand these polynomials, and conversely we can use these polynomials to better understand the Kronecker product.

We denote the vector space of degree-p homogeneous polynomials in n variables by $\mathcal{HP}_n^p$ (or $\mathcal{HP}_n^p(\mathbb{F})$, if we wish to emphasize which field $\mathbb{F}$ these polynomials are acting on and have coefficients from). That is,

The "$\mathcal{H}$" in $\mathcal{HP}_n^p$ stands for "homogeneous".

$$\mathcal{HP}_n^p = \left\{ f : \mathbb{F}^n \to \mathbb{F} \,\middle|\, f(x_1, x_2, \ldots, x_n) = \sum_{k_1 + k_2 + \cdots + k_n = p} a_{k_1, k_2, \ldots, k_n} x_1^{k_1} x_2^{k_2} \cdots x_n^{k_n} \right.$$
$$\left. \text{for some scalars } \{a_{k_1, k_2, \ldots, k_n}\} \subset \mathbb{F} \right\}.$$

Degree-3 and degree-4 homogeneous polynomials are called **cubic forms** and **quartic forms**, respectively.

For example, $\mathcal{HP}_n^1$ is the space of n-variable linear forms, $\mathcal{HP}_n^2$ is the space of n-variable quadratic forms, and $\mathcal{HP}_2^3$ consists of all 2-variable polynomials in which every term has degree equal to exactly 3, and thus have the form

$$a_{3,0} x_1^3 + a_{2,1} x_1^2 x_2 + a_{1,2} x_1 x_2^2 + a_{0,3} x_1 x_2^3$$

for some $a_{3,0}, a_{2,1}, a_{1,2}, a_{0,3} \in \mathbb{F}$.

When working with homogeneous polynomials of degree higher than 2, it seems natural to ask which properties of low-degree polynomials carry over. We focus in particular on two problems that ask us how we can decompose these polynomials into powers of simpler ones.

3.B.1 Powers of Linear Forms

Recall from Corollary 2.A.2 that one way of interpreting the spectral decomposition of a real symmetric matrix is as saying that every quadratic form $f : \mathbb{R}^n \to \mathbb{R}$ can be written as a linear combination of squares of linear forms:

$$f(x_1, x_2, \ldots, x_n) = \sum_{i=1}^n \lambda_i (c_{i,1} x_1 + c_{i,2} x_2 + \cdots + c_{i,n} x_n)^2.$$

In particular, $\lambda_1, \lambda_2, \ldots, \lambda_n$ are the eigenvalues of the symmetric matrix A that represents f (in the sense of Equation (3.B.1)), and $(c_{i,1}, c_{i,2}, \ldots, c_{i,p})$ is a unit eigenvector corresponding to λ_i.

A natural question then is whether or not a similar technique can be applied to homogeneous polynomials of higher degree (and to polynomials with complex coefficients). That is, we would like to know whether or not it is possible to write every degree-p homogeneous polynomial as a linear combination of p-th powers of linear forms. The following theorem shows that the answer is yes, this is possible.

Theorem 3.B.1

Powers of Linear Forms

Suppose $\mathbb{F} = \mathbb{R}$ or $\mathbb{F} = \mathbb{C}$, and $f \in \mathcal{HP}_n^p(\mathbb{F})$. There exists an integer r and scalars $\lambda_i \in \mathbb{F}$ and $c_{i,j} \in \mathbb{F}$ for $1 \leq i \leq r$, $1 \leq j \leq n$ such that

$$f(x_1, x_2, \ldots, x_n) = \sum_{i=1}^r \lambda_i (c_{i,1} x_1 + c_{i,2} x_2 + \cdots + c_{i,n} x_n)^p.$$

By the spectral decomposition, if $p = 2$ then we can choose $r = n$ in this theorem. In general, we can choose $r = \dim(\mathcal{HP}_n^p) = \binom{n+p-1}{p}$ (we will see how this dimension is computed shortly).

Proof. In order to help us prove this result, we first define an inner product $\langle \cdot, \cdot \rangle : \mathcal{HP}_n^p \times \mathcal{HP}_n^p \to \mathbb{F}$ by setting $\langle f, g \rangle$ equal to a certain weighted dot product of the vectors of coefficients of $f, g \in \mathcal{HP}_n^p$. Specifically, if

$$f(x_1, x_2, \ldots, x_n) = \sum_{k_1+k_2+\cdots+k_n=p} a_{k_1,k_2,\ldots,k_n} x_1^{k_1} x_2^{k_2} \cdots x_n^{k_n} \quad \text{and}$$

$$g(x_1, x_2, \ldots, x_n) = \sum_{k_1+k_2+\cdots+k_n=p} b_{k_1,k_2,\ldots,k_n} x_1^{k_1} x_2^{k_2} \cdots x_n^{k_n},$$

then we define

$$\langle f, g \rangle = \sum_{k_1+\cdots+k_n=p} \frac{\overline{a_{k_1,k_2,\ldots,k_n}} b_{k_1,k_2,\ldots,k_n}}{\binom{p}{k_1,k_2,\ldots,k_n}}, \tag{3.B.2}$$

Multinomial coefficients and the multinomial theorem (Theorem A.2.2) are introduced in Appendix A.2.1.

where $\binom{p}{k_1,k_2,\ldots,k_n} = \frac{p!}{k_1! k_2! \cdots k_n!}$ is a multinomial coefficient. We show that this function is an inner product in Exercise 3.B.11.

Importantly, notice that if g is the p-th power of a linear form (i.e., there are scalars $c_1, c_2, \ldots, c_n \in \mathbb{F}$ such that $g(x_1, x_2, \ldots, x_n) = (c_1 x_1 + c_2 x_2 + \cdots + c_n x_n)^p$) then we can use the multinomial theorem (Theorem A.2.2) to see that

$$g(x_1, x_2, \ldots, x_n) = \sum_{k_1+k_2+\cdots+k_n=p} \left(\binom{p}{k_1, k_2, \ldots, k_n}\right) c_1^{k_1} c_2^{k_2} \cdots c_n^{k_n} x_1^{k_1} x_2^{k_2} \cdots x_n^{k_n},$$

so straightforward computation then shows us that

$$\langle f, g \rangle = \frac{\binom{p}{k_1,k_2,\ldots,k_n} \overline{a_{k_1,k_2,\ldots,k_n}} c_1^{k_1} c_2^{k_2} \cdots c_n^{k_n}}{\binom{p}{k_1,k_2,\ldots,k_n}} = \overline{a_{k_1,k_2,\ldots,k_n}} c_1^{k_1} c_2^{k_2} \cdots c_n^{k_n}$$

$$= \overline{f(\overline{c_1}, \overline{c_2}, \ldots, \overline{c_n})}. \tag{3.B.3}$$

With these details out of the way, we can now prove the theorem relatively straightforwardly. Another way of stating the theorem is as saying that

$$\mathcal{HP}_n^p = \text{span}\left\{g : \mathbb{F}^n \to \mathbb{F} \mid g(x_1, x_2, \ldots, x_n) = (c_1 x_1 + c_2 x_2 + \cdots + c_n x_n)^p \right.$$

$$\left. \text{for some } c_1, c_2, \ldots, c_n \in \mathbb{F}\right\}.$$

Our goal is thus to show that this span is not a *proper* subspace of $\mathcal{HP}_n^p$.

Suppose for the sake of establishing a contradiction that this span *were* a proper subspace of $\mathcal{HP}_n^p$. Then there would exist a non-zero homogeneous polynomial $f \in \mathcal{HP}_n^p$ orthogonal to each p-th power of a linear form:

For example, we could choose f to be any non-zero vector in the orthogonal complement (see Section 1.B.2) of the span of the g's.

$$\langle f, g \rangle = 0 \quad \text{whenever} \quad g(x_1, x_2, \ldots, x_n) = (c_1 x_1 + c_2 x_2 + \cdots + c_n x_n)^p.$$

Equation (3.B.3) then tells us that $\overline{f(\overline{c_1}, \overline{c_2}, \ldots, \overline{c_n})} = 0$ for all $c_1, c_2, \ldots, c_n \in \mathbb{F}$, which implies f is the zero polynomial, which is the desired contradiction. $\blacksquare$

In the case when p is odd, the linear combination of powers described by Theorem 3.B.1 really is just a sum of powers, since we can absorb any scalar (positive or negative) inside the linear forms. For example, every cubic form $f : \mathbb{F}^n \to \mathbb{F}$ can be written in the form

$$f(x_1, x_2, \ldots, x_n) = \sum_{i=1}^{r} (c_{i,1} x_1 + c_{i,2} x_2 + \cdots + c_{i,n} x_n)^3.$$

In fact, if $\mathbb{F} = \mathbb{C}$ then this can be done regardless of whether p is even or odd. However, if p is even and $\mathbb{F} = \mathbb{R}$ then the best we can do is absorb the absolute value of scalars λ_i into the powers of linear forms (i.e., we can assume without loss of generality that $\lambda_i = \pm 1$ for each $1 \le i \le m$).

Example 3.B.1

Decomposing a Cubic Form

Write the cubic form $f : \mathbb{R}^2 \to \mathbb{R}$ defined by

$$f(x,y) = 4x^3 + 12xy^2 - 6y^3$$

as a sum of cubes of linear forms.

Solution:

Unfortunately, the proof of Theorem 3.B.1 is non-constructive, so even though we know that such a sum-of-cubes way of writing f exists, we have not yet seen how to find one. The trick is to simply construct a basis of $\mathcal{HP}_2^3$ consisting of powers of linear forms, and then represent f in that basis.

We do not have a general method or formula for constructing such a basis, but one method that works in practice is to just choose powers of randomly-chosen linear forms until we have the right number of them to form a basis. For example, since $\mathcal{HP}_2^3$ is 4-dimensional, we know that any such basis must consist of 4 polynomials, and it is straightforward to verify that the set

> This basis is arbitrary. Many other similar sets like $\{x^3, (x-y)^3, (x+2y)^3, (x-3y)^3\}$ are also bases.

$$\{x^3, y^3, (x+y)^3, (x-y)^3\}$$

is one such basis.

Now our goal is to simply to represent f in this basis—we want to find c_1, c_2, c_3, c_4 such that

> The final line here just comes from expanding the terms $(x+y)^3$ and $(x-y)^3$ via the binomial theorem (Theorem A.2.1) and then grouping coefficients by which monomial (x^3, x^2y, xy^2, or y^3) they multiply.

$$\begin{aligned} f(x,y) &= 4x^3 + 12xy^2 - 6y^3 \\ &= c_1 x^3 + c_2 y^3 + c_3(x+y)^3 + c_4(x-y)^3 \\ &= (c_1 + c_3 + c_4)x^3 + (3c_3 - 3c_4)x^2y + \\ &\quad (3c_3 + 3c_4)xy^2 + (c_2 + c_3 - c_4)y^3. \end{aligned}$$

Matching up coefficients of these polynomials gives us the four linear equations

$$\begin{array}{lll} c_1 + c_3 + c_4 = 4, & & 3c_3 - 3c_4 = 0, \\ 3c_3 + 3c_4 = 12, & \text{and} & c_2 + c_3 - c_4 = -6. \end{array}$$

This linear system has $(c_1, c_2, c_3, c_4) = (1, -6, 2, 2)$ as its unique solution, which gives us the decomposition

$$f(x,y) = x^3 - 6y^3 + 2(x+y)^3 + 2(x-y)^3.$$

To turn this linear combination of cubes into a sum of cubes, we just absorb constants inside of the cubed terms:

$$f(x,y) = x^3 + \left(-\sqrt[3]{6}y\right)^3 + \left(\sqrt[3]{2}(x+y)\right)^3 + \left(\sqrt[3]{2}(x-y)\right)^3.$$

The key observation that connects homogeneous polynomials with the other topics of this chapter is that $\mathcal{HP}_n^p$ is isomorphic to the symmetric subspace $\mathcal{S}_n^p \subset (\mathbb{F}^n)^{\otimes p}$ that we explored in Section 3.1.3. To see why this is the case, recall that one basis of $\mathcal{S}_n^p$ consists of the vectors of the form

$$\left\{ \sum_{\sigma \in S_p} W_\sigma(e_{j_1} \otimes e_{j_2} \otimes \cdots \otimes e_{j_p}) : 1 \le j_1 \le j_2 \le \cdots \le j_p \le n \right\}.$$

In particular, each vector in this basis is determined by how many of the subscripts $j_1, j_2, \ldots, j_p$ are equal to each of $1, 2, \ldots, n$. If we let $k_1, k_2, \ldots, k_n$ denote the multiplicity of the subscripts $1, 2, \ldots, n$ in this manner then we can associate the vector

$$\sum_{\sigma \in S_p} W_\sigma(\mathbf{e}_{j_1} \otimes \mathbf{e}_{j_2} \otimes \cdots \otimes \mathbf{e}_{j_p})$$

with the tuple $(k_1, k_2, \ldots, k_n)$.

The fact that these spaces are isomorphic tells us that $\dim(\mathcal{HP}_n^p) = \dim(\mathcal{S}_n^p) = \binom{n+p-1}{p}$.

That is, the linear transformation $T : \mathcal{S}_n^p \to \mathcal{HP}_n^p$ defined by

$$T\left(\sum_{\sigma \in S_p} W_\sigma(\mathbf{e}_{j_1} \otimes \mathbf{e}_{j_2} \otimes \cdots \otimes \mathbf{e}_{j_p})\right) = x_1^{k_1} x_2^{k_2} \cdots x_n^{k_n},$$

where $k_1, k_2, \ldots, k_n$ count the number of occurrences of the subscripts $1, 2, \ldots, n$ (respectively) in the multiset $\{j_1, j_2, \ldots, j_p\}$, is an isomorphism. We have thus proved the following observation:

> ! $\mathcal{HP}_n^p$ is isomorphic to the symmetric subspace $\mathcal{S}_n^p \subset (\mathbb{F}^n)^{\otimes p}$.

The particular isomorphism that we constructed here perhaps seems ugly at first, but it really just associates the members of the "usual" basis of $\mathcal{S}_n^p$ with the monomials from $\mathcal{HP}_n^p$ in the simplest way possible. For example, if we return to the $n = 2$, $p = 3$ case, we get the following pairing of basis vectors for these two spaces:

Basis Vector in $\mathcal{S}_2^3$	Basis Vector in $\mathcal{HP}_2^3$
$6\mathbf{e}_1 \otimes \mathbf{e}_1 \otimes \mathbf{e}_1$	x_1^3
$2(\mathbf{e}_1 \otimes \mathbf{e}_1 \otimes \mathbf{e}_2 + \mathbf{e}_1 \otimes \mathbf{e}_2 \otimes \mathbf{e}_1 + \mathbf{e}_2 \otimes \mathbf{e}_1 \otimes \mathbf{e}_1)$	$x_1^2 x_2$
$2(\mathbf{e}_1 \otimes \mathbf{e}_2 \otimes \mathbf{e}_2 + \mathbf{e}_2 \otimes \mathbf{e}_1 \otimes \mathbf{e}_2 + \mathbf{e}_2 \otimes \mathbf{e}_2 \otimes \mathbf{e}_1)$	$x_1 x_2^2$
$6\mathbf{e}_2 \otimes \mathbf{e}_2 \otimes \mathbf{e}_2$	x_2^3

Compare this table with the one on page 314.

If we trace the statement of Theorem 3.B.1 through this isomorphism, we learn that every $\mathbf{w} \in \mathcal{S}_n^p$ can be written in the form

$$\mathbf{w} = \sum_{i=1}^{r} \lambda_i \mathbf{v}_i^{\otimes p}$$

for some $\lambda_1, \lambda_2, \ldots, \lambda_r \in \mathbb{F}$ and $\mathbf{v}_1, \mathbf{v}_2, \ldots, \mathbf{v}_r \in \mathbb{F}^n$. In other words, we have finally proved Theorem 3.1.11.

Example 3.B.2

Finding a Symmetric Kronecker Decomposition

Again, there is nothing special about this basis. Pick random tensor powers of 5 distinct vectors and there's a good chance we will get a basis.

Write the vector $\mathbf{w} = (2, 3, 3, 1, 3, 1, 1, 3, 3, 1, 1, 3, 1, 3, 3, 2) \in \mathcal{S}_2^4$ as a linear combination of vectors of the form $\mathbf{v}^{\otimes 4}$.

Solution:

We can solve this problem by mimicking what we did in Example 3.B.1. We start by constructing a basis of $\mathcal{S}_2^4$ consisting of vectors of the form $\mathbf{v}^{\otimes 4}$. We do not have an explicit method of carrying out this task, but in practice we can just pick random vectors of the form $\mathbf{v}^{\otimes 4}$ until we have enough of them to form a basis (so we need $\dim(\mathcal{S}_2^4) = 5$ of them). One such basis is

$$\left\{(1,0)^{\otimes 4}, (0,1)^{\otimes 4}, (1,1)^{\otimes 4}, (1,2)^{\otimes 4}, (2,1)^{\otimes 4}\right\}.$$

Now our goal is to simply to represent $\mathbf{w}$ in this basis—we want to find c_1, c_2, c_3, c_4, c_5 such that

$$\mathbf{w} = c_1(1,0)^{\otimes 4} + c_2(0,1)^{\otimes 4} + c_3(1,1)^{\otimes 4} + c_4(1,2)^{\otimes 4} + c_5(2,1)^{\otimes 4}.$$

If we explicitly write out what all of these vectors are (they have 16 entries each!), we get a linear system consisting of 16 equations and 5 variables. However, many of those equations are identical to each other, and after discarding those duplicates we arrive at the following linear system consisting of just 5 equations:

$$
\begin{aligned}
c_1 \quad\;\; + c_3 + \quad c_4 + 16c_5 &= 2 \\
c_2 + c_3 + 16c_4 + \quad c_5 &= 2 \\
c_3 + \quad 2c_4 + \quad 8c_5 &= 3 \\
c_3 + \quad 4c_4 + \quad 4c_5 &= 1 \\
c_3 + \quad 8c_4 + \quad 2c_5 &= 3
\end{aligned}
$$

This linear system has $(c_1, c_2, c_3, c_4, c_5) = (-8, -8, -7, 1, 1)$ as its unique solution, which gives us the decomposition

$$\mathbf{w} = (1,2)^{\otimes 4} + (2,1)^{\otimes 4} - 7(1,1)^{\otimes 4} - 8(1,0)^{\otimes 4} - 8(0,1)^{\otimes 4}.$$

Remark 3.B.1

Symmetric Tensor Rank

Recall from Section 3.3.3 that the tensor rank of $\mathbf{w} \in (\mathbb{F}^n)^{\otimes p}$ (denoted by $\mathrm{rank}(\mathbf{w})$) is the least integer r such that we can write

$$\mathbf{w} = \sum_{i=1}^{r} \mathbf{v}_i^{(1)} \otimes \mathbf{v}_i^{(2)} \otimes \cdots \otimes \mathbf{v}_i^{(p)},$$

where $\mathbf{v}_i^{(j)} \in \mathbb{F}^n$ for each $1 \le i \le r$ and $1 \le j \le p$.

Now that we know that Theorem 3.1.11 holds, we could similarly define (as long as $\mathbb{F} = \mathbb{R}$ or $\mathbb{F} = \mathbb{C}$) the **symmetric tensor rank** of any $\mathbf{w} \in \mathcal{S}_n^p$ (denoted by $\mathrm{rank}^S(\mathbf{w})$) to be the least integer r such that we can write

If $\mathbb{F} = \mathbb{C}$ then we can omit the λ_i scalars here since we can absorb $\sqrt[p]{\lambda_i}$ into $\mathbf{v}_i$. If $\mathbb{F} = \mathbb{R}$ then we can similarly assume that $\lambda_i = \pm 1$ for all i.

$$\mathbf{w} = \sum_{i=1}^{r} \lambda_i \mathbf{v}_i^{\otimes p},$$

where $\lambda_i \in \mathbb{F}$ and $\mathbf{v}_i \in \mathbb{F}^n$ for each $1 \le i \le r$.

Perhaps not surprisingly, only a few basic facts are known about the symmetric tensor rank:

- When $p = 2$, the symmetric tensor rank just equals the usual rank: $\mathrm{rank}(\mathbf{w}) = \mathrm{rank}^S(\mathbf{w})$ for all $\mathbf{w} \in \mathcal{S}_n^2$. This is yet another consequence of the spectral decomposition (see Exercise 3.B.6).

- In general, $\mathrm{rank}(\mathbf{w}) \le \mathrm{rank}^S(\mathbf{w})$ for all $\mathbf{w} \in \mathcal{S}_n^p$. This fact follows directly from the definitions of these two quantities—when computing the symmetric rank, we minimize over a strictly smaller set of sums than for the non-symmetric rank.

- There are cases when $\mathrm{rank}(\mathbf{w}) < \mathrm{rank}^S(\mathbf{w})$, even just when $p = 3$, but they are not easy to construct or explain [Shi18].

> - Since $\dim\left(\mathcal{S}_n^p\right) = \binom{n+p-1}{p}$, we have $\operatorname{rank}^S(\mathbf{w}) \le \binom{n+p-1}{p}$ for all $\mathbf{w} \in \mathcal{S}_n^p$.

3.B.2 Positive Semidefiniteness and Sums of Squares

Recall that if $\mathbb{F} = \mathbb{C}$, or $\mathbb{F} = \mathbb{R}$ and p is odd, then Theorem 3.B.1 says that every homogeneous polynomial $f \in \mathcal{HP}_n^p$ can be written as a *sum* (not just as a linear combination) of powers of linear forms, since we can absorb scalars inside those powers. However, in the case when $\mathbb{F} = \mathbb{R}$ and p is even, this is not always possible, since any such polynomial necessarily always produces non-negative output:

$$f(x_1, x_2, \ldots, x_n) = \sum_{i=1}^{r} (c_{i,1}x_1 + c_{i,2}x_2 + \cdots + c_{i,n}x_n)^p \ge 0 \text{ for all } x_1, x_2, \ldots, x_n.$$

We investigate polynomials of this type in this section, so we start by giving them a name:

Definition 3.B.1

Positive Semidefinite Polynomials

> We say that a real-valued polynomial f is **positive semidefinite (PSD)** if $f(x_1, x_2, \ldots, x_n) \ge 0$ for all $x_1, x_2, \ldots, x_n \in \mathbb{R}$.

In particular, we are interested in the converse to the observation that we made above—if a homogeneous polynomial is positive semidefinite, can we write it as a sum of even powers of linear forms? This question is certainly too restrictive, since we cannot even write the (clearly PSD) polynomial $f(x, y) = x^2 y^2$ in such a manner (see Exercise 3.B.5). We thus ask the following slightly weaker question about sums of squares of *polynomials*, rather than sums of even powers of *linear forms*:

> **(?)** For which values of n and p can the PSD polynomials in $\mathcal{HP}_n^p(\mathbb{R})$ be written as a sum of squares of polynomials?

We already know (again, as a corollary of the spectral decomposition) that this is possible when $p = 2$. The main result of this section shows that it is also possible when $n = 2$ (i.e., when there are only two variables).

The main technique that goes into proving this upcoming theorem is a trick called **dehomogenization**, which is a process for turning a homogeneous polynomial in n variables into a (not necessarily homogeneous) polynomial in $n - 1$ variables while preserving many of its important properties (like positive semidefiniteness). For a polynomial $f \in \mathcal{HP}_n^p$ acting on variables $x_1, x_2, \ldots, x_n$, this procedure works by dividing f through by x_n^p and then relabeling the ratios $x_1/x_n, x_2/x_n, \ldots, x_{n-1}/x_n$ as the variables. We illustrate with an example.

If you prefer, you can instead divide through by x_j^p for any $1 \le j \le n$ of your choosing.

Example 3.B.3

Dehomogenizing a Quartic

Dehomogenize the quartic polynomial $f \in \mathcal{HP}_3^4$ defined by

$$f(x_1, x_2, x_3) = x_1^4 + 2x_1^3 x_2 + 3x_1^2 x_3^2 + 4x_1 x_2^2 x_3 + 5x_3^4.$$

Solution:

If we divide f through by x_3^4, we get

$$\frac{f(x_1,x_2,x_3)}{x_3^4} = \frac{x_1^4}{x_3^4} + 2\frac{x_1^3 x_2}{x_3^4} + 3\frac{x_1^2}{x_3^2} + 4\frac{x_1 x_2^2}{x_3^3} + 5,$$

The terms in the dehomogenization do not necessarily all have the same degree. However, they do all have degree no larger than that of the original homogeneous polynomial.

which is not a polynomial in x_1, x_2, and x_3, but *is* a polynomial in the variables $x = x_1/x_3$ and $y = x_2/x_3$. That is, the dehomogenization of f is the two-variable (non-homogeneous) polynomial g defined by

$$g(x,y) = x^4 + 2x^3 y + 3x^2 + 4xy^2 + 5.$$

It should be reasonably clear that a polynomial is positive semidefinite if and only if its dehomogenization is positive semidefinite, since dividing by a positive term like x_n^p when p is even does not affect positive semidefiniteness. We now make use of this technique to answer our central question about PSD homogeneous polynomials in the 2-variable case.

Theorem 3.B.2

Two-Variable PSD Homogeneous Polynomials

Suppose $f \in \mathcal{HP}_2^p(\mathbb{R})$. Then f is positive semidefinite if and only if it can be written as a sum of squares of polynomials.

Proof. The "if" direction is trivial (even for polynomials of more than 2 variables), so we only show the "only if" direction. Also, thanks to dehomogenization, it suffices to just prove that every PSD (not necessarily homogeneous) polynomial in a *single* variable can be written as a sum of squares.

To this end, we induct on the degree p of the polynomial f. In the $p = 2$ base case, we can complete the square in order to write f in its vertex form

$$f(x) = a(x-h)^2 + k,$$

where $a > 0$ and the vertex of the graph of f is located at the point (h,k) (see Figure 3.3(a)). Since $f(x) \geq 0$ for all x we know that $f(h) = k \geq 0$, so this vertex form is really a sum of squares decomposition of f.

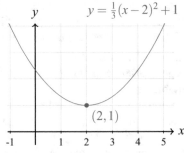

(a) Every quadratic can be written in vertex form. If it is PSD, its graph lies above the x-axis and that vertex form is a sum of squares.

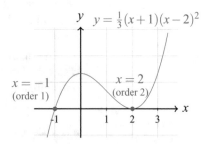

(b) The order of a root determines whether or not the polynomial crosses the x-axis (odd order) or just touches it (even order).

Figure 3.3: Illustrations of some basic facts about real-valued polynomials of a single variable.

For the inductive step, suppose f has even degree p (f can clearly not be positive semidefinite if p is odd). Let $k \geq 0$ be the minimal value of f (which exists, since p is even) and let h be a root of the polynomial $f(x) - k$. Since

$f(x) - k \geq 0$ for all x, the multiplicity of h as a root of $f(x) - k$ must be even (see Figure 3.3(b)). We can thus write

$$f(x) = (x - h)^{2q} g(x) + k$$

for some integer $q \geq 1$ and some polynomial g. Since $k \geq 0$ and $g(x) = (f(x) - k)/(x-h)^{2q} \geq 0$ is a PSD polynomial of degree $p - 2q < p$, it follows from the inductive hypothesis that g can be written as a sum of squares of polynomials, so f can as well. ∎

It is also known [Hil88] that every PSD polynomial in $\mathcal{HP}_3^4(\mathbb{R})$ can be written as a sum of squares of polynomials, though the proof is rather technical and outside of the scope of this book. Remarkably, this is the last case where such a property holds—for all other values of n and p, there exist PSD polynomials that cannot be written as a sum of squares of polynomials. The following example presents such a polynomial in $\mathcal{HP}_3^6(\mathbb{R})$.

Example 3.B.4

A Weird PSD Polynomial of Degree 6

Consider the real polynomial $f(x,y,z) = x^2 y^4 + y^2 z^4 + z^2 x^4 - 3x^2 y^2 z^2$.

a) Show that f is positive semidefinite.
b) Show that f cannot be written as a sum of squares of polynomials.

Solutions:

a) If we apply the AM–GM inequality (Theorem A.5.3) to the quantities $x^2 y^4$, $y^2 z^4$, and $z^2 x^4$, we learn that

$$\frac{x^2 y^4 + y^2 z^4 + z^2 x^4}{3} \geq \sqrt[3]{(x^2 y^4)(y^2 z^4)(z^2 x^4)} = x^2 y^2 z^2.$$

Multiplying this inequality through by 3 and rearranging gives us exactly the inequality $f(x,y,z) \geq 0$.

b) If f could be written as a sum of squares of polynomials, then it would have the form

The terms being squared and summed here are just general cubics. Recall that $\dim(\mathcal{HP}_3^3) = 10$, so there are 10 monomials in such a cubic.

$$f(x,y,z) = \sum_{i=1}^{r} \left(a_i xy^2 + b_i yz^2 + c_i zx^2 + d_i x^2 y + e_i y^2 z \right.$$
$$\left. + g_i z^2 x + h_i x^3 + j_i y^3 + k_i z^3 + \ell_i xyz \right)^2$$

for some scalars $a_i, b_i, \ldots, \ell_i \in \mathbb{R}$. By matching up terms of f with terms in this hypothetical sum-of-squares representation of f, we can learn about their coefficients.

For example, since the coefficient of x^6 in f is 0, whereas the coefficient of x^6 in this sum-of-squares representation is $\sum_{i=1}^{r} h_i^2$, we learn that $h_i = 0$ for all $1 \leq i \leq r$. A similar argument with the coefficients of y^6 and z^6 then shows that $j_i = k_i = 0$ for all $1 \leq i \leq r$ as well.

The coefficient of $x^4 y^2$ then similarly tells us that

$$0 = \sum_{i=1}^{r} d_i^2, \quad \text{so} \quad d_i = 0 \quad \text{for all} \quad 1 \leq i \leq r,$$

and the same argument with the coefficients of $y^4 z^2$ and $z^4 x^2$ tells us that $e_i = g_i = 0$ for all $1 \leq i \leq r$ as well.

At this point, our hypothetical sum-of-squares representation of f has been simplified down to the form

$$f(x,y,z) = \sum_{i=1}^{r} \left(a_i xy^2 + b_i yz^2 + c_i zx^2 + \ell_i xyz\right)^2.$$

Comparing the coefficient of $x^2 y^2 z^2$ in f to this decomposition of f then tells us that $-3 = 0$, which is (finally!) a contradiction that shows that no such sum-of-squares decomposition of f is possible in the first place.

Similar examples show that there are PSD polynomials that cannot be written as a sum of squares of polynomials in $\mathcal{HP}_n^p(\mathbb{R})$ whenever $n \geq 3$ and $p \geq 6$ is even. An example that demonstrates the existence of such polynomials in $\mathcal{HP}_4^4(\mathbb{R})$ (and thus $\mathcal{HP}_n^p(\mathbb{R})$ whenever $n \geq 4$ and $p \geq 4$ is even) is provided in Exercise 3.B.7. Finally, a summary of these various results about the connection between positive semidefiniteness and the ability to write a polynomial as a sum of squares of polynomials is provided by Table 3.2.

p (degree)	1	2	3	4	5
			n (number of variables)		
2			✓ (spectral decomp.)		
4			✓ [Hil88]	✗ (Exer. 3.B.7)	
6	✓ (trivial)	✓ (Thm. 3.B.2)	✗ (Exam. 3.B.4)		
8					
10					
12					

Table 3.2: A summary of which values of n and p lead to a real n-variable degree-p PSD homogeneous polynomial necessarily being expressible as a sum of squares of polynomials.

Remark 3.B.2

Sums of Squares of Rational Functions

Remarkably, even though some positive semidefinite polynomials cannot be written as a sum of squares of polynomials, they can all be written as a sum of squares of *rational functions* [Art27]. For example, we showed in Example 3.B.4 that the PSD homogeneous polynomial

$$f(x,y,z) = x^2 y^4 + y^2 z^4 + z^2 x^4 - 3x^2 y^2 z^2$$

cannot be written as a sum of squares of polynomials. However, if we define

$$g(x,y,z) = x^3 z - xzy^2 \quad \text{and} \quad h(x,y,z) = x^2(y^2 - z^2)$$

then we can write f in the following form that makes it "obvious" that it is positive semidefinite:

$$f(x,y,z) = \frac{g(x,y,z)^2 + g(y,z,x)^2 + g(z,x,y)^2}{x^2+y^2+z^2}$$
$$+ \frac{h(x,y,z)^2 + h(y,z,x)^2 + h(z,x,y)^2}{2(x^2+y^2+z^2)}.$$

This is not quite a sum of squares of rational functions, but it can be turned into one straightforwardly (see Exercise 3.B.8).

3.B.3 Biquadratic Forms

Recall from Section 2.A that one way of constructing a quadratic form is to plug the same vector into both input slots of a bilinear form. That is, if $f : \mathbb{R}^n \times \mathbb{R}^n \to \mathbb{R}$ is a bilinear form then the function $q : \mathbb{R}^n \to \mathbb{R}$ defined by $q(\mathbf{x}) = f(\mathbf{x},\mathbf{x})$ is a quadratic form (and the scalars that represent q are exactly the entries of the symmetric matrix A that represents f). The following facts follow via a similar argument:

- If $f : \mathbb{R}^n \times \mathbb{R}^n \times \mathbb{R}^n \to \mathbb{R}$ is a trilinear form then the function $q : \mathbb{R}^n \to \mathbb{R}$ defined by $q(\mathbf{x}) = f(\mathbf{x},\mathbf{x},\mathbf{x})$ is a cubic form.
- If $f : \mathbb{R}^n \times \mathbb{R}^n \times \mathbb{R}^n \times \mathbb{R}^n \to \mathbb{R}$ is a 4-linear (quadrilinear?) form then the function $q : \mathbb{R}^n \to \mathbb{R}$ defined by $q(\mathbf{x}) = f(\mathbf{x},\mathbf{x},\mathbf{x},\mathbf{x})$ is a quartic form.

$(\mathbb{R}^n)^{\times p}$ means the Cartesian product of p copies of $\mathbb{R}^n$ with itself.

- If $f : (\mathbb{R}^n)^{\times p} \to \mathbb{R}$ is a p-linear form then the function $q : \mathbb{R}^n \to \mathbb{R}$ defined by $q(\mathbf{x}) = f(\mathbf{x},\mathbf{x},\dots,\mathbf{x})$ is a degree-p homogeneous polynomial.

We now focus on what happens in the $p = 4$ case if we go halfway between quadrilinear and quartic forms: we consider functions $q : \mathbb{R}^m \times \mathbb{R}^n \to \mathbb{R}$ of the form $q(\mathbf{x},\mathbf{y}) = f(\mathbf{x},\mathbf{y},\mathbf{x},\mathbf{y})$, where $f : \mathbb{R}^m \times \mathbb{R}^n \times \mathbb{R}^m \times \mathbb{R}^n \to \mathbb{R}$ is a quadrilinear form. We call such functions **biquadratic forms**, and it is reasonably straightforward to show that they can all be written in the following form (see Exercise 3.B.9):

Definition 3.B.2

Biquadratic Forms

A **biquadratic form** is a degree-4 homogeneous polynomial $q : \mathbb{R}^m \times \mathbb{R}^n \to \mathbb{R}$ for which there exist scalars $\{a_{i,j;k,\ell}\}$ such that

$$q(\mathbf{x},\mathbf{y}) = \sum_{1=i\leq j}^{m} \sum_{1=k\leq\ell}^{n} a_{i,j;k,\ell} x_i x_j y_k y_\ell.$$

Analogously to how the set of bilinear forms is isomorphic to the set of matrices, quadratic forms are isomorphic to the set of *symmetric* matrices, and PSD quadratic forms are isomorphic to the set of *positive semidefinite* matrices, we can also represent quadrilinear and biquadratic forms in terms of matrix-valued linear maps in a natural way. In particular, notice that every linear map $\Phi : \mathcal{M}_n(\mathbb{R}) \to \mathcal{M}_m(\mathbb{R})$ can be thought of as a quadrilinear form $f : \mathbb{R}^m \times \mathbb{R}^n \times \mathbb{R}^m \times \mathbb{R}^n \to \mathbb{R}$ just by defining

$$f(\mathbf{v},\mathbf{w},\mathbf{x},\mathbf{y}) = \mathbf{v}^T \Phi(\mathbf{w}\mathbf{y}^T)\mathbf{x}.$$

Indeed, it is straightforward to show that Φ is linear if and only if f is quadrilinear, and that this relationship between f and Φ is actually an isomorphism. It follows that biquadratic forms similarly can be written in the form

$$q(\mathbf{x}, \mathbf{y}) = f(\mathbf{x}, \mathbf{y}, \mathbf{x}, \mathbf{y}) = \mathbf{x}^T \Phi(\mathbf{y}\mathbf{y}^T)\mathbf{x}.$$

The following theorem pins down what properties of Φ lead to various desirable properties of the associated biquadratic form q.

Theorem 3.B.3

Biquadratic Forms as Matrix-Valued Linear Maps

A function $q : \mathbb{R}^m \times \mathbb{R}^n \to \mathbb{R}$ is a biquadratic form if and only if there exists a bisymmetric linear map $\Phi : \mathcal{M}_n(\mathbb{R}) \to \mathcal{M}_m(\mathbb{R})$ (i.e., a map satisfying $\Phi = T \circ \Phi = \Phi \circ T$) such that $q(\mathbf{x}, \mathbf{y}) = \mathbf{x}^T \Phi(\mathbf{y}\mathbf{y}^T)\mathbf{x}$ for all $\mathbf{x} \in \mathbb{R}^m$ and $\mathbf{y} \in \mathbb{R}^n$. Properties of Φ correspond to properties of q as follows:

   a) Φ is positive if and only if q is positive semidefinite.

   b) Φ is decomposable if and only if q can be written as a sum of squares of bilinear forms.

Furthermore, this relationship between q and Φ is an isomorphism.

Bisymmetric maps were introduced in Section 3.A.1. Decomposable maps are ones that can be written in the form $\Phi = \Psi_1 + T \circ \Psi_2$ for some CP Ψ_1, Ψ_2.

Proof. We already argued why q is a biquadratic form if and only if we can write $q(\mathbf{x}, \mathbf{y}) = \mathbf{x}^T \Phi(\mathbf{y}\mathbf{y}^T)\mathbf{x}$ for some linear map Φ. To see why we can choose Φ to be bisymmetric, we just notice that if we define

$$\Psi = \frac{1}{4}\big(\Phi + (T \circ \Phi) + (\Phi \circ T) + (T \circ \Phi \circ T)\big)$$

then Ψ is bisymmetric and $q(\mathbf{x}, \mathbf{y}) = \mathbf{x}^T \Psi(\mathbf{y}\mathbf{y}^T)\mathbf{x}$ for all $\mathbf{x} \in \mathbb{R}^m$ and $\mathbf{y} \in \mathbb{R}^n$ as well.

To see that this association between biquadratic forms q and bisymmetric linear maps Φ is an isomorphism, we just notice that the dimensions of these two vector spaces match up. It is evident from Definition 3.B.2 that the space of biquadratic forms has dimension $\binom{m+1}{2}\binom{n+1}{2} = mn(m+1)(n+1)/4$, since each biquadratic form is determined by that many scalars $\{a_{i,j;k,\ell}\}$. On the other hand, recall from Exercise 3.A.14 that Φ is bisymmetric if and only if $C_\Phi = C_\Phi^T = \Gamma(C_\Phi)$. It follows that the set of bisymmetric maps also has dimension $\binom{m+1}{2}\binom{n+1}{2} = mn(m+1)(n+1)/4$, since every bisymmetric Φ depends only on the $\binom{m+1}{2} = m(m+1)/2$ entries in the upper-triangular portion of each of the $\binom{n+1}{2} = n(n+1)/2$ blocks in the upper-triangular portion of C_Φ.

To see why property (a) holds, we note that q is positive semidefinite if and only if

$$q(\mathbf{x}, \mathbf{y}) = \mathbf{x}^T \Phi(\mathbf{y}\mathbf{y}^T)\mathbf{x} \geq 0 \quad \text{for all} \quad \mathbf{x} \in \mathbb{R}^m, \mathbf{y} \in \mathbb{R}^n.$$

Positive semidefiniteness of q is thus equivalent to $\Phi(\mathbf{y}\mathbf{y}^T)$ being positive semidefinite for all $\mathbf{y} \in \mathbb{R}^n$, which in turn is equivalent to Φ being positive (since the real spectral decomposition tells us that every positive semidefinite $X \in \mathcal{M}_n(\mathbb{R})$ can be written as a sum of terms of the form $\mathbf{y}\mathbf{y}^T$).

To demonstrate property (b), we note that if Φ is decomposable then we can write it in the form $\Phi = \Psi_1 + T \circ \Psi_2$ for some completely positive maps Ψ_1 and Ψ_2 with operator-sum representations $\Psi_1(X) = \sum_i A_i X A_i^T$ and $\Psi_2(X) = \sum_i B_i X B_i^T$. Then

$$\begin{aligned}
q(\mathbf{x}, \mathbf{y}) &= \mathbf{x}^T \Psi_1(\mathbf{y}\mathbf{y}^T)\mathbf{x} + \mathbf{x}^T \big(\Psi_2(\mathbf{y}\mathbf{y}^T)\big)^T \mathbf{x} \\
&= \sum_i \mathbf{x}^T \big(A_i \mathbf{y}\mathbf{y}^T A_i^T\big)\mathbf{x} + \sum_i \mathbf{x}^T \big(B_i \mathbf{y}\mathbf{y}^T B_i^T\big)^T \mathbf{x} \\
&= \sum_i (\mathbf{x}^T A_i \mathbf{y})^2 + \sum_i (\mathbf{x}^T B_i \mathbf{y})^2,
\end{aligned}$$

which is a sum of squares of the bilinear forms $g_i(\mathbf{x}, \mathbf{y}) = \mathbf{x}^T A_i \mathbf{y}$ and $h_i(\mathbf{x}, \mathbf{y}) = \mathbf{x}^T B_i \mathbf{y}$.

Conversely, if q can be written as a sum of squares of bilinear forms $g_i : \mathbb{R}^m \times \mathbb{R}^n \to \mathbb{R}$ then we know from Theorem 1.3.5 that we can find matrices $\{A_i\}$ such that $g_i(\mathbf{x}, \mathbf{y}) = \mathbf{x}^T A_i \mathbf{y}$. It follows that

$$q(\mathbf{x}, \mathbf{y}) = \sum_i (\mathbf{x}^T A_i \mathbf{y})^2 = \sum_i \mathbf{x}^T \left(A_i \mathbf{y} \mathbf{y}^T A_i^T \right) \mathbf{x} = \mathbf{x}^T \Psi \left(\mathbf{y} \mathbf{y}^T \right) \mathbf{x},$$

where Ψ is the (completely positive) map with operator-sum representation $\Psi(X) = \sum_i A_i X A_i^T$. However, this map Ψ may not be bisymmetric. To see how we can represent q by a bisymmetric map (at the expense of weakening complete positivity to just decomposability), we introduce the decomposable map $\Phi = (\Psi + T \circ \Psi)/2$. It is straightforward to check that $q(\mathbf{x}, \mathbf{y}) = \mathbf{x}^T \Phi(\mathbf{y} \mathbf{y}^T) \mathbf{x}$ for all $\mathbf{x} \in \mathbb{R}^m$ and $\mathbf{y} \in \mathbb{R}^n$ as well, and the fact that it is bisymmetric follows from Exercise 3.A.15. ∎

This is analogous to the fact that a quadratic form $q(\mathbf{x}) = \mathbf{x}^T A \mathbf{x}$ being PSD does not imply A is PSD unless we choose A to be symmetric.

It is worth noting that many of the implications of the above theorem still work even if the linear map Φ is not bisymmetric. For example, if $\Phi : \mathcal{M}_n(\mathbb{R}) \to \mathcal{M}_m(\mathbb{R})$ is a positive linear map and we define $q(\mathbf{x}, \mathbf{y}) = \mathbf{x}^T \Phi(\mathbf{y} \mathbf{y}^T) \mathbf{x}$ then the proof given above shows that q must be positive semidefinite. However, the converse is not necessarily true unless Φ is bisymmetric—q might be positive semidefinite even if Φ is not positive (it might not even send symmetric matrices to symmetric matrices). Similarly, the proof of Theorem 3.B.3 shows that a biquadratic form can be written as a sum of squares if and only if it can be represented by a completely positive map, which is sometimes easier to work with than a decomposable bisymmetric map.

For ease of reference, we summarize the various ways that quadratic forms, biquadratic forms, and various other closely-related families of homogeneous polynomials are isomorphic to important sets of matrices and matrix-valued linear maps in Table 3.3.

The top half of this table has one fewer row than the bottom half since a quadratic form is PSD if and only if it is a sum of squares of linear forms.

Polynomial	Matrix(-Valued Map)
bilinear form	matrix
quadratic form	symmetric matrix
PSD quadratic form	PSD matrix
quadrilinear form	matrix-valued linear map
biquadratic form	bisymmetric matrix-valued map
PSD biquadratic form	positive bisymmetric linear map
sum of squares of bilin. forms	decomposable bisymmetric map

Table 3.3: A summary of various isomorphisms that relate certain families of homogeneous polynomials to certain families of matrices and matrix-valued linear maps.

Example 3.B.5

A Biquadratic Form as a Sum of Squares

Write the biquadratic form $q : \mathbb{R}^3 \times \mathbb{R}^3 \to \mathbb{R}$ defined by

$$q(\mathbf{x}, \mathbf{y}) = x_1^2(y_2^2 + y_3^2) + x_2^2(y_3^2 + y_1^2) + x_3^2(y_1^2 + y_2^2)$$
$$- 2x_1 x_2 y_1 y_2 - 2x_2 x_3 y_2 y_3 - 2x_3 x_1 y_3 y_1$$

as a sum of squares of bilinear forms.

We were lucky here that we could "eyeball" a CP map Ψ for which $q(\mathbf{x},\mathbf{y}) = \mathbf{x}^T \Psi(\mathbf{y}\mathbf{y}^T)\mathbf{x}$. In general, finding a suitable map Ψ to show that q is a sum of squares can be done via semidefinite programming (see Exercise 3.C.18).

Solution:

It is straightforward to show that $q(\mathbf{x},\mathbf{y}) = \mathbf{x}^T \Phi(\mathbf{y}\mathbf{y}^T)\mathbf{x}$, where $\Psi : \mathcal{M}_3 \to \mathcal{M}_3$ is the (completely positive) Werner–Holevo map defined by $\Psi(X) = \mathrm{tr}(X)I - X^T$ (see Example 3.A.8). It follows that if we can find an operator-sum representation $\Psi(X) = \sum_i A_i X A_i^T$ then we will get a sum-of-squares representation $q(\mathbf{x},\mathbf{y}) = \sum_i (\mathbf{x}^T A_i \mathbf{y})^2$ as an immediate corollary.

To construct such an operator-sum representation, we mimic the proof of Theorem 3.A.6: we construct C_Ψ and then let the A_i's be the matricizations of its scaled eigenvectors. We recall from Example 3.A.8 that $C_\Psi = I - W_{3,3}$, and applying the spectral decomposition to this matrix shows that $C_\Psi = \sum_{i=1}^3 \mathbf{v}_i \mathbf{v}_i^T$, where

$$\mathbf{v}_1 = (0,1,0,-1,0,0,0,0,0),$$
$$\mathbf{v}_2 = (0,0,1,0,0,0,-1,0,0), \quad \text{and}$$
$$\mathbf{v}_3 = (0,0,0,0,0,1,0,-1,0).$$

It follows that $\Psi(X) = \sum_{i=1}^3 A_i X A_i^T$, where

We already constructed this operator-sum representation back in Exercise 3.A.2.

$$A_1 = \begin{bmatrix} 0 & 1 & 0 \\ -1 & 0 & 0 \\ 0 & 0 & 0 \end{bmatrix}, \quad A_2 = \begin{bmatrix} 0 & 0 & 1 \\ 0 & 0 & 0 \\ -1 & 0 & 0 \end{bmatrix}, \quad \text{and } A_3 = \begin{bmatrix} 0 & 0 & 0 \\ 0 & 0 & 1 \\ 0 & -1 & 0 \end{bmatrix}.$$

This then gives us the sum-of-squares representation

$$q(\mathbf{x},\mathbf{y}) = \sum_{i=1}^3 \left(\mathbf{x}^T A_i \mathbf{y}\right)^2$$
$$= (x_1 y_2 - x_2 y_1)^2 + (x_1 y_3 - x_3 y_1)^2 + (x_2 y_3 - x_3 y_2)^2.$$

In light of examples like Example 3.B.4 and Exercise 3.B.7, it is perhaps not surprising that there are PSD biquadratic forms that cannot be written as a sum of squares of polynomials. What is interesting though is that this fact is so closely-related to the fact that there are positive linear maps that are not decomposable (i.e., that cannot be written in the form $\Phi = \Psi_1 + T \circ \Psi_2$ for some completely positive maps Ψ_1, Ψ_2)—a fact that we demonstrated back in Theorem 3.A.7.

Example 3.B.6

A PSD Biquadratic Form That is Not a Sum of Squares

Consider the biquadratic form $q : \mathbb{R}^3 \times \mathbb{R}^3 \to \mathbb{R}$ defined by

$$q(\mathbf{x},\mathbf{y}) = x_1^2(y_1^2 + y_2^2) + x_2^2(y_2^2 + y_3^2) + x_3^2(y_3^2 + y_1^2)$$
$$- 2x_1 x_2 y_1 y_2 - 2x_2 x_3 y_2 y_3 - 2x_3 x_1 y_3 y_1.$$

a) Show that q is positive semidefinite.
b) Show that q cannot be written as a sum of squares of bilinear forms.

Compare this biquadratic form with the one from Example 3.B.5.

Solutions:

a) We could prove that q is positive semidefinite directly in a manner analogous to that used in Example 3.B.4, but we instead make use of the relationship with matrix-valued linear maps that was described by Theorem 3.B.3. Specifically, we notice that if $\Phi_C : \mathcal{M}_3 \to \mathcal{M}_3$

This biquadratic form (as well as the Choi map from Theorem 3.A.7) was introduced by Choi (Cho75).

is the Choi map described by Theorem 3.A.7, then

$$y^T \Phi_C(xx^T)y = x_1^2(y_1^2 + y_2^2) + x_2^2(y_2^2 + y_3^2) + x_3^2(y_3^2 + y_1^2)$$
$$- 2x_1x_2y_1y_2 - 2x_2x_3y_2y_3 - 2x_3x_1y_3y_1 = q(\mathbf{x}, \mathbf{y}).$$

However, Φ_C is not bisymmetric, so we instead consider the bisymmetric map

$$\Psi = \frac{1}{2}(\Phi_C + T \circ \Phi_C),$$

which also has the property that $q(\mathbf{x}, \mathbf{y}) = y^T \Psi(xx^T)y$. Since Φ_C is positive, so is Ψ, so we know from Theorem 3.B.3 that q is positive semidefinite.

We show that q is PSD "directly" in Exercise 3.B.10.

b) On the other hand, even though we know that Φ_C is not decomposable, this does not directly imply that the map Ψ from part (a) is not decomposable, so we cannot directly use Theorem 3.B.3 to see that q cannot be written as a sum of squares of bilinear forms. We thus prove this property of q directly.

If q *could* be written as a sum of squares of bilinear forms, then it would have the form

$$q(\mathbf{x}, \mathbf{y}) = \sum_{i=1}^{r} \big(a_i x_1 y_1 + b_i x_1 y_2 + c_i x_1 y_3 + d_i x_2 y_1 + e_i x_2 y_2 + f_i x_2 y_3$$
$$+ h_i x_3 y_1 + j_i x_3 y_2 + k_i x_3 y_3\big)^2$$

for some families of scalars $\{a_i\}, \{b_i\}, \ldots, \{k_i\} \in \mathbb{R}$ and some integer r. By matching up terms of q with terms in this hypothetical sum-of-squares representation of q, we can learn about their coefficients.

For example, since the coefficient of $x_1^2 y_3^2$ in q is 0, whereas the coefficient of $x_1^2 y_3^2$ in this sum-of-squares representation is $\sum_{i=1}^{r} c_i^2$, we learn that $c_i = 0$ for all $1 \leq i \leq r$. A similar argument with the coefficients of $x_2^2 y_1^2$ and $x_3^2 y_2^2$ then shows that $d_i = j_i = 0$ for all $1 \leq i \leq r$ as well. It follows that our hypothetical sum-of-squares representation of q actually has the somewhat simpler form

$$q(\mathbf{x}, \mathbf{y}) = \sum_{i=1}^{r} \big(a_i x_1 y_1 + b_i x_1 y_2 + e_i x_2 y_2 + f_i x_2 y_3 + h_i x_3 y_1 + k_i x_3 y_3\big)^2.$$

Comparing the coefficients of $x_1^2 y_1^2$, $x_2^2 y_2^2$, and $x_1 x_2 y_1 y_2$ in q to this decomposition of q then tells us that

Specifically, we are applying the Cauchy–Schwarz inequality to the vectors $\mathbf{a} = (a_1, \ldots, a_r)$ and $\mathbf{e} = (e_1, \ldots, e_r)$ in $\mathbb{R}^r$. It says that $|\mathbf{a} \cdot \mathbf{e}| = \|\mathbf{a}\|\|\mathbf{e}\|$ if and only if $\mathbf{a}$ and $\mathbf{e}$ are multiples of each other.

$$\sum_{i=1}^{r} a_i^2 = 1, \quad \sum_{i=1}^{r} e_i^2 = 1, \quad \text{and} \quad \sum_{i=1}^{r} a_i e_i = -1.$$

The equality condition of the Cauchy–Schwarz inequality (Theorem 1.3.8) then tells us that $e_i = -a_i$ for all $1 \leq i \leq r$. A similar argument with the coefficients of $x_1^2 y_1^2$, $x_3^2 y_3^2$, and $x_1 x_3 y_1 y_3$ shows that $k_i = -a_i$ for all i, and the coefficients of $x_2^2 y_2^2$, $x_3^2 y_3^2$, and $x_2 x_3 y_2 y_3$ tell us that $k_i = -e_i$ for all i. In particular, we have shown that

> $k_i = -a_i$, but also that $k_i = -e_i = -(-a_i) = a_i$ for all $1 \le i \le r$. It follows that $a_i = 0$ for all $1 \le i \le r$, which contradicts the fact that $\sum_{i=1}^{r} a_i^2 = 1$ and shows that no such sum-of-squares decomposition of q is possible in the first place.

On the other hand, recall from Theorem 3.A.8 that all positive maps $\Phi : \mathcal{M}_n \to \mathcal{M}_m$ are decomposable when $(m,n) = (2,2)$, $(m,n) = (2,3)$, or $(m,n) = (3,2)$. It follows immediately that all biquadratic forms $q : \mathbb{R}^m \times \mathbb{R}^n \to \mathbb{R}$ can be written as a sum of squares of bilinear forms, under the same restrictions on m and n. We close this section by noting the following strengthening of this result: PSD biquadratic forms can be written as a sum of squares of bilinear forms as long as one of m or n equals 2.

Theorem 3.B.4 **PSD Biquadratic Forms of Few Variables**	Suppose $q : \mathbb{R}^m \times \mathbb{R}^n \to \mathbb{R}$ is a biquadratic form and $\min\{m,n\} = 2$. Then q can be written as a sum of squares of bilinear forms if and only if it is positive semidefinite.

This fact is not true when $(m,n) = (2,4)$ if Φ is not bisymmetric—see Exercise 3.C.17.

In particular, this result tells us via the isomorphism between biquadratic forms and bisymmetric maps that if $\Phi : \mathcal{M}_n(\mathbb{R}) \to \mathcal{M}_m(\mathbb{R})$ is bisymmetric with $\min\{m,n\} = 2$ then Φ being positive is equivalent to it being decomposable. We do not prove this theorem, however, as it is quite technical—the interested reader is directed to [Cal73].

Exercises

solutions to starred exercises on page 482

3.B.1 Write each of the following homogeneous polynomials as a linear combination of powers of linear forms, in the sense of Theorem 3.B.1.

*(a) $3x^2 + 3y^2 - 2xy$
(b) $2x^2 + 2y^2 - 3z^2 - 4xy + 6xz + 6yz$
*(c) $2x^3 - 9x^2y + 3xy^2 - y^3$
(d) $7x^3 + 3x^2y + 15xy^2$
*(e) $x^2y + y^2z + z^2x$
(f) $6xyz - x^3 - y^3 - z^3$
*(g) $2x^4 - 8x^3y - 12x^2y^2 - 32xy^3 - 10y^4$
(h) x^2y^2

3.B.2 Write each of the following vectors from the symmetric subspace $\mathcal{S}_n^p$ as a linear combination of vectors of the form $\mathbf{v}^{\otimes p}$.

*(a) $(2,3,3,5) \in \mathcal{S}_2^2$
(b) $(1,3,-1,3,-3,3,-1,3,1) \in \mathcal{S}_3^2$
*(c) $(1,-2,-2,0,-2,0,0,-1) \in \mathcal{S}_2^3$
(d) $(2,0,0,4,0,4,4,6,0,4,4,6,4,6,6,8) \in \mathcal{S}_2^4$

3.B.3 Determine which of the following statements are true and which are false.

*(a) If g is a dehomogenization of a homogeneous polynomial f then the degree of g equals that of f.
(b) If $f \in \mathcal{HP}_n^p(\mathbb{R})$ is non-zero and positive semidefinite then p must be even.

*(c) Every positive semidefinite polynomial in $\mathcal{HP}_2^6(\mathbb{R})$ can be written as a sum of squares of polynomials.
(d) Every positive semidefinite polynomial in $\mathcal{HP}_3^6(\mathbb{R})$ can be written as a sum of squares of polynomials.
*(e) Every positive semidefinite polynomial in $\mathcal{HP}_8^2(\mathbb{R})$ can be written as a sum of squares of polynomials.

3.B.4 Compute the dimension of $\mathcal{P}_n^p$, the vector space of (non-homogeneous) polynomials of degree at most p in n variables.

**3.B.5 Show that the polynomial $f(x,y) = x^2y^2$ cannot be written as a sum of 4-th powers of linear forms.

**3.B.6 We claimed in Remark 3.B.1 that the symmetric tensor rank of a vector $\mathbf{w} \in \mathcal{S}_n^2$ equals its usual tensor rank: $\text{rank}(\mathbf{w}) = \text{rank}^S(\mathbf{w})$.

a) Prove this claim if the ground field is $\mathbb{R}$.
   [Hint: Use the real spectral decomposition.]
b) Prove this claim if the ground field is $\mathbb{C}$.
   [Hint: Use the Takagi factorization of Exercise 2.3.26.]

∗∗3.B.7   In this exercise, we show that the polynomial $f \in \mathcal{HP}_4^4$ defined by

$$f(w,x,y,z) = w^4 + x^2y^2 + y^2z^2 + z^2x^2 - 4wxyz$$

is positive semidefinite, but cannot be written as a sum of squares of polynomials (much like we did for a polynomial in $\mathcal{HP}_3^6$ in Example 3.B.4).

   a) Show that f is positive semidefinite.
   b) Show that f cannot be written as a sum of squares of polynomials.

∗∗3.B.8   Write the polynomial from Remark 3.B.2 as a sum of squares of rational functions.

[Hint: Multiply and divide the "obvious" PSD decomposition of f by another copy of $x^2 + y^2 + z^2$.]

∗∗3.B.9   Show that a function $q : \mathbb{R}^m \times \mathbb{R}^n \to \mathbb{R}$ has the form described by Definition 3.B.2 if and only if there is a quadrilinear form $f : \mathbb{R}^m \times \mathbb{R}^n \times \mathbb{R}^m \times \mathbb{R}^n \to \mathbb{R}$ such that

$$q(\mathbf{x},\mathbf{y}) = f(\mathbf{x},\mathbf{y},\mathbf{x},\mathbf{y}) \quad \text{for all} \quad \mathbf{x} \in \mathbb{R}^m, \ \mathbf{y} \in \mathbb{R}^n.$$

∗∗3.B.10   Solve Example 3.B.6(a) "directly". That is, show that the biquadratic form $q : \mathbb{R}^3 \times \mathbb{R}^3 \to \mathbb{R}$ defined by

$$q(\mathbf{x},\mathbf{y}) = y_1^2(x_1^2 + x_3^2) + y_2^2(x_2^2 + x_1^2) + y_3^2(x_3^2 + x_2^2)$$
$$- 2x_1x_2y_1y_2 - 2x_2x_3y_2y_3 - 2x_3x_1y_3y_1$$

is positive semidefinite without appealing to Theorem 3.B.3 or the Choi map from Theorem 3.A.7.

[Hint: This is hard. One approach is to fix 5 of the 6 variables and use the quadratic formula on the other one.]

∗∗ 3.B.11   Show that the function defined in Equation (3.B.2) really is an inner product.

3.B.12   Let $\mathbf{v} \in (\mathbb{C}^2)^{\otimes 3}$ be the vector that we showed has tensor rank 3 in Example 3.3.4. Show that it also has *symmetric* tensor rank 3.

[Hint: The decompositions that we provided of the vector from Equation (3.3.3) might help.]

3.C   Extra Topic: Semidefinite Programming

One of the most useful tools in advanced linear algebra is an optimization method called semidefinite programming, which is a generalization of linear programming. Much like linear programming deals with maximizing or minimizing a linear function over a set of inequality constraints, semidefinite programming allows us to maximize or minimize a linear function over a set of positive semidefinite constraints.

Before getting into the details of what semidefinite programs look like or what we can do with them, we briefly introduce one piece of new notation that we use extensively throughout this section. Given Hermitian matrices $A, B \in \mathcal{M}_n^H$, we say that $A \succeq B$ (or equivalently, $B \preceq A$) if $A - B$ is positive semidefinite (in which case we could also write $A - B \succeq O$). For example, if

> We can similarly write $A \succ B$, $B \prec A$, or $A - B \succ O$ if $A - B$ is positive *definite*.

$$A = \begin{bmatrix} 1 & 1-i \\ 1+i & 3 \end{bmatrix} \quad \text{and} \quad B = \begin{bmatrix} 0 & 1 \\ 1 & 2 \end{bmatrix}$$

then $A \succeq B$, since

$$A - B = \begin{bmatrix} 1 & -i \\ i & 1 \end{bmatrix},$$

which is positive semidefinite (its eigenvalues are 2 and 0).

> We prove these basic properties of the Loewner partial order in Exercise 3.C.1.

This ordering of Hermitian matrices is called the **Loewner partial order**, and it shares many of the same properties as the usual ordering on $\mathbb{R}$ (in fact, for 1×1 Hermitian matrices it *is* the usual ordering on $\mathbb{R}$). For example, if $A, B, C \in \mathcal{M}_n^H$ then:

   Reflexive: it is the case that $A \succeq A$,

   Antisymmetric: if $A \succeq B$ and $B \succeq A$ then $A = B$, and

   Transitive: if $A \succeq B$ and $B \succeq C$ then $A \succeq C$.

In fact, these three properties are exactly the defining properties of a **partial order**—a function on a set (not necessarily a set of matrices) that behaves like we would expect something that we call an "ordering" to

Two matrices
$A, B \in \mathcal{M}_n^H$ for which
$A \not\succeq B$ and $B \not\succeq A$ are
called
non-comparable,
and their existence is
why this is called a
"partial" order (as
opposed to a "total"
order like the one on
$\mathbb{R}$).

However, there is one important property that the ordering on $\mathbb{R}$ has that is missing from the Loewner partial order: If $a, b \in \mathbb{R}$ then it is necessarily the case that either $a \geq b$ or $b \geq a$ (or both), but the analogous statement about the Loewner partial order on $\mathcal{M}_n^H$ does not hold. For example, if

$$A = \begin{bmatrix} 1 & 0 \\ 0 & 0 \end{bmatrix} \quad \text{and} \quad B = \begin{bmatrix} 0 & 0 \\ 0 & 1 \end{bmatrix}$$

then each of $A - B$ and $B - A$ have -1 as an eigenvalue, so $A \not\succeq B$ and $B \not\succeq A$.

3.C.1   The Form of a Semidefinite Program

The idea behind semidefinite programming is to construct a matrix–valued version of linear programming. We thus start by recalling that a linear program in an optimization program that can be put into the following form, where $A \in \mathcal{M}_{m,n}(\mathbb{R})$, $\mathbf{b} \in \mathbb{R}^m$, and $\mathbf{c} \in \mathbb{R}^n$ are fixed, $\mathbf{x} \in \mathbb{R}^n$ is a vector of variables that is being optimized over, and inequalities between vectors are meant entrywise:

Refer to any number
of other books, like
(Chv83) or (Joh20),
for a refresher on
linear programming.

$$
\begin{aligned}
\text{maximize:} \quad & \mathbf{c} \cdot \mathbf{x} \\
\text{subject to:} \quad & A\mathbf{x} \leq \mathbf{b} \\
& \mathbf{x} \geq \mathbf{0}
\end{aligned}
\tag{3.C.1}
$$

Recall (from Figure 2.6, for example) that we can think of Hermitian matrices and positive semidefinite matrices as the "matrix versions" of real numbers and non-negative real numbers, and the standard inner product on $\mathcal{M}_n^H$ is the Frobenius inner product $\langle C, X \rangle = \text{tr}(CX)$ (just like the standard inner product on $\mathbb{R}^n$ is the dot product $\langle \mathbf{c}, \mathbf{x} \rangle = \mathbf{c} \cdot \mathbf{x}$). Since we now similarly think of the Loewner partial order as the "matrix version" of the ordering of real numbers, the following definition hopefully seems like a somewhat natural generalization of linear programs—it just involves replacing every operation that is specific to $\mathbb{R}$ or $\mathbb{R}^n$ with its "natural" matrix-valued generalization.

Since we are
working with
Hermitian matrices
(i.e., $C, X \in \mathcal{M}_n^H$), we
do not need the
conjugate transpose
in the Frobenius inner
product $\text{tr}(C^*X)$.

Definition 3.C.1	Suppose $\Phi : \mathcal{M}_n^H \to \mathcal{M}_m^H$ is a linear transformation, and $B \in \mathcal{M}_m^H$ and $C \in \mathcal{M}_n^H$ are Hermitian matrices. The **semidefinite program** (SDP) associated with Φ, B, and C is the following optimization problem over the matrix variable $X \in \mathcal{M}_n^H$:

**Semidefinite Program
(Primal Standard Form)**

Recall from
Section 3.A that a
linear transformation
$\Phi : \mathcal{M}_n^H \to \mathcal{M}_m^H$ is
called **Hermiticity-
preserving**.

$$
\begin{aligned}
\text{maximize:} \quad & \text{tr}(CX) \\
\text{subject to:} \quad & \Phi(X) \preceq B \\
& X \succeq O
\end{aligned}
\tag{3.C.2}
$$

Furthermore, this is called the **primal standard form** of the semidefinite program.

Before jumping into specific examples of semidefinite programs, it is worth emphasizing that they really do generalize linear programs. In particular, we can write the linear program (3.C.1) in the form of the semidefinite program (3.C.2)

Semidefinite
programs over real
symmetric matrices
in $\mathcal{M}_n^S$ are fine
too—for example,
we could just add
the linear constraint
$X = X^T$ to a
complex-valued SDP
to make it
real-valued.

by defining $B = \mathrm{diag}(\mathbf{b})$, $C = \mathrm{diag}(\mathbf{c})$, and $\Phi : \mathcal{M}_n^H \to \mathcal{M}_m^H$ by

$$\Phi(X) = \begin{bmatrix} \sum_{j=1}^{n} a_{1,j}x_{j,j} & 0 & \cdots & 0 \\ 0 & \sum_{j=1}^{n} a_{2,j}x_{j,j} & \cdots & 0 \\ \vdots & \vdots & \ddots & \vdots \\ 0 & 0 & \cdots & \sum_{j=1}^{n} a_{m,j}x_{j,j} \end{bmatrix}.$$

A routine calculation then shows that the semidefinite program (3.C.2) is equivalent to the original linear program (3.C.1)—we have just stretched each vector in the original linear program out along the diagonal of a matrix and used the fact that a diagonal matrix is PSD if and only if its entries are non-negative.

Basic Manipulations into Primal Standard Form

Just as was the case with linear programs, the primal standard form of the semidefinite program (3.C.2) is not quite as restrictive as it appears at first glance. For example, we can allow for multiple constraints simply by making use of block matrices and matrix-valued linear maps that act on block matrices, as we now demonstrate.

Example 3.C.1

Turning Multiple Semidefinite Constraints Into One

Suppose $C \in \mathcal{M}_n^H$. Write the following optimization problem as a semidefinite program in primal standard form:

$$\begin{aligned} \text{maximize:} \quad & \mathrm{tr}(CX) \\ \text{subject to:} \quad & \mathrm{tr}(X) = 3 \\ & X \preceq I \\ & X \succeq O \end{aligned}$$

In fact, if $n \geq 3$ then
this semidefinite
program computes
the sum of the 3
largest eigenvalues
of C—see
Exercise 3.C.12.

Solution:
   We first split the constraint $\mathrm{tr}(X) = 3$ into the pair of "≤" constraints $\mathrm{tr}(X) \leq 3$ and $-\mathrm{tr}(X) \leq -3$, just like we would if we were trying to write a *linear* program in primal standard form. We can now rewrite the three inequality constraints $\mathrm{tr}(X) \leq 3$, $-\mathrm{tr}(X) \leq -3$, and $X \preceq I$ as the single matrix constraint $\Phi(X) \preceq B$ (and thus put the SDP into primal standard form) if we define

$$\Phi(X) = \begin{bmatrix} X & \mathbf{0} & \mathbf{0} \\ \mathbf{0}^T & \mathrm{tr}(X) & 0 \\ \mathbf{0}^T & 0 & -\mathrm{tr}(X) \end{bmatrix} \quad \text{and} \quad B = \begin{bmatrix} I & \mathbf{0} & \mathbf{0} \\ \mathbf{0}^T & 3 & 0 \\ \mathbf{0}^T & 0 & -3 \end{bmatrix}.$$

   More generally, multiple positive semidefinite constraints can be merged into a single positive semidefinite constraint simply by placing matrices along the diagonal blocks of a larger matrix—this works because a block diagonal matrix is positive semidefinite if and only if each of its diagonal blocks is positive semidefinite if (see Exercise 2.2.13).

   We make use of much of the same terminology when discussing semidefinite programs as we do for linear programs. The **objective function** of an SDP is the function being maximized or minimized (i.e., $\mathrm{tr}(CX)$ if it is written in standard form) and its **optimal value** is the maximal or minimal value that the objective function can attain subject to the constraints (i.e., it is the "solution" of the semidefinite program). A **feasible point** is a matrix $X \in \mathcal{M}_n^H$ that satisfies all of the constraints of the SDP (i.e., $\Phi(X) \preceq B$ and $X \succeq O$), and the **feasible region** is the set consisting of all feasible points.

One wrinkle that occurs for SDPs that did not occur for linear programs is that the maximum or minimum in a semidefinite program might not be attained (i.e., by "maximum" we really mean "supremum" and by "minimum" we really mean "infimum")—see Example 3.C.6.

In addition to turning multiple constraints and matrices into a single block-diagonal matrix and constraint, as in the previous example, we can also use techniques similar to those that are used for linear programs to transform a wide variety of optimization problems into the standard form of semidefinite programs. For example:

- We can turn a minimization problem into a maximization problem by multiplying the objective function by -1 and then multiplying the resulting optimal value by -1, as illustrated in Figure 3.4.

All of these modifications are completely analogous to how we can manipulate inequalities and equalities involving real numbers.

- We can turn an "$=$" constraint into a pair of "$\succeq$" and "$\preceq$" constraints (since the Loewner partial order is antisymmetric).

- We can turn a "$\succeq$" constraint into a "$\preceq$" constraint by multiplying it by -1 and flipping the direction of the inequality (see Exercise 3.C.1(e)).

- We can turn an unconstrained (i.e., not necessarily PSD) matrix variable X into a pair of PSD matrix variables by setting $X = X^+ - X^-$ where $X^+, X^- \succeq O$ (see Exercise 2.2.16).

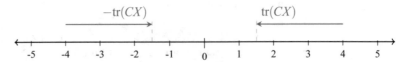

Figure 3.4: Minimizing $\mathrm{tr}(CX)$ is essentially the same as maximizing $-\mathrm{tr}(CX)$; the final answers just differ by a minus sign.

Example 3.C.2

Writing a Semidefinite Program in Primal Standard Form

Suppose $C, D \in \mathcal{M}_n^H$ and $\Phi, \Psi : \mathcal{M}_n^H \to \mathcal{M}_n^H$ are linear. Write the following semidefinite program (in the variables $X, Y \in \mathcal{M}_n^H$) in primal standard form:

$$
\begin{aligned}
\text{minimize:} \quad & \mathrm{tr}(CX) + \mathrm{tr}(DY) \\
\text{subject to:} \quad & X + \Psi(Y) = I \\
& \Phi(X) + \quad Y \succeq O \\
& X \qquad\qquad \succeq O
\end{aligned}
$$

Solution:

Since Y is unconstrained, we replace it by the pair of PSD variables Y^+ and Y^- via $Y = Y^+ - Y^-$ and $Y^+, Y^- \succeq O$. Making this change puts the SDP into the form

Here we used the fact that
$\Psi(Y^+ - Y^-) =$
$\Psi(Y^+) - \Psi(Y^-)$, since Ψ is linear.

$$
\begin{aligned}
\text{minimize:} \quad & \mathrm{tr}(CX) + \mathrm{tr}(DY^+) - \mathrm{tr}(DY^-) \\
\text{subject to:} \quad & X + \Psi(Y^+) - \Psi(Y^-) = I \\
& \Phi(X) + \quad Y^+ - \quad Y^- \succeq O \\
& X, \qquad\quad Y^+, \qquad Y^- \succeq O
\end{aligned}
$$

Next, we change the minimization into a maximization by multiplying the objective function by -1 and also placing a minus sign in front of the entire SDP. We also change the equality constraint $X + \Psi(Y^+) - \Psi(Y^-) = I$ into the pair of inequality constraints $X + \Psi(Y^+) - \Psi(Y^-) \succeq I$ and $X + \Psi(Y^+) - \Psi(Y^-) \preceq I$, and then convert both of the "$\succeq$" constraints into "$\preceq$" constraints by multiplying them through by -1. After making these changes, the SDP has the form

Converting an SDP into primal standard form perhaps makes it look uglier, but is useful for theoretical reasons.

$$
\begin{aligned}
-\,\text{maximize:} \quad & -\text{tr}(CX) - \text{tr}(DY^+) + \text{tr}(DY^-) \\
\text{subject to:} \quad & X + \Psi(Y^+) - \Psi(Y^-) \preceq I \\
& -\,X - \Psi(Y^+) + \Psi(Y^-) \preceq -I \\
& -\,\Phi(X) - Y^+ + Y^- \preceq O \\
& X, \quad Y^+, \quad Y^- \succeq O
\end{aligned}
$$

At this point, the SDP is essentially in primal standard form—all that remains is to collect the various pieces of it into block diagonal matrices. In particular, the version of this SDP in primal standard form optimizes over a positive semidefinite matrix variable $\widetilde{X} \in \mathcal{M}_{3n}^{\text{H}}$, which we partition as a block matrix as follows:

We use asterisks ($*$) to denote entries whose values we do not care about (they might be non-zero, but they are so unimportant that they do not deserve names).

$$
\widetilde{X} = \begin{bmatrix} X & * & * \\ * & Y^+ & * \\ * & * & Y^- \end{bmatrix}.
$$

We also define $\widetilde{B}, \widetilde{C} \in \mathcal{M}_{3n}^{\text{H}}$ and $\widetilde{\Phi} : \mathcal{M}_{3n}^{\text{H}} \to \mathcal{M}_{3n}^{\text{H}}$ as follows:

We use dots ($\cdot$) to denote entries equal to 0.

$$
\widetilde{B} = \begin{bmatrix} I & \cdot & \cdot \\ \cdot & -I & \cdot \\ \cdot & \cdot & O \end{bmatrix}, \quad \widetilde{C} = \begin{bmatrix} -C & \cdot & \cdot \\ \cdot & -D & \cdot \\ \cdot & \cdot & D \end{bmatrix}, \quad \text{and}
$$

$$
\widetilde{\Phi}(\widetilde{X}) = \begin{bmatrix} X + \Psi(Y^+ - Y^-) & \cdot & \cdot \\ \cdot & -X - \Psi(Y^+ - Y^-) & \cdot \\ \cdot & \cdot & -\Phi(X) - Y^+ + Y^- \end{bmatrix}
$$

After making this final substitution, the original linear program can be written in primal standard form as follows:

$$
\begin{aligned}
-\,\text{maximize:} \quad & \text{tr}\big(\widetilde{C}\widetilde{X}\big) \\
\text{subject to:} \quad & \widetilde{\Phi}(\widetilde{X}) \preceq \widetilde{B} \\
& \widetilde{X} \succeq O
\end{aligned}
$$

Some Less Obvious Conversions of SDPs

None of the transformations of optimization problems into SDPs in primal standard form that we have seen so far are particularly surprising—they pretty much just amount to techniques that are carried over from linear programming, together with some facts about block diagonal matrices and a lot of bookkeeping. However, one of the most remarkable things about semidefinite programming is that it can be used to compute quantities that at first glance do not even seem linear.

For example, recall the operator norm of a matrix $A \in \mathcal{M}_{m,n}(\mathbb{F})$ (where $\mathbb{F} = \mathbb{R}$ or $\mathbb{F} = \mathbb{C}$), which is defined by

We investigated the operator norm back in Section 2.3.3.

$$
\|A\| = \max_{\mathbf{v} \in \mathbb{F}^n} \big\{ \|A\mathbf{v}\| : \|\mathbf{v}\| = 1 \big\}.
$$

As written, this quantity does not appear to be amenable to semidefinite programming, since the quantity $\|A\mathbf{v}\|$ that is being maximized is not linear in $\mathbf{v}$. However, recall from Exercise 2.3.15 that if $x \in \mathbb{R}$ is a scalar then $\|A\| \leq x$ if and only if

$$\begin{bmatrix} xI_m & A \\ A^* & xI_n \end{bmatrix} \succeq O.$$

This leads immediately to the following semidefinite program for computing the operator norm of A:

$$\begin{aligned} \text{minimize:} \quad & x \\ \text{subject to:} \quad & \begin{bmatrix} xI_m & A \\ A^* & xI_n \end{bmatrix} \succeq O \\ & x \geq 0 \end{aligned} \qquad (3.C.3)$$

To truly convince ourselves that this is a semidefinite program, we could write it in the primal standard form of Definition 3.C.1 by defining $B \in \mathcal{M}_{m+n}^{\mathrm{H}}$, $C \in \mathcal{M}_1^{\mathrm{H}}$, and $\Phi : \mathcal{M}_1^{\mathrm{H}} \to \mathcal{M}_{m+n}^{\mathrm{H}}$ by

The set $\mathcal{M}_1^{\mathrm{H}}$ consists of the 1×1 Hermitian matrices. That is, it equals $\mathbb{R}$, the set of real numbers.

$$B = \begin{bmatrix} O & A \\ A^* & O \end{bmatrix}, \quad C = -1, \quad \text{and} \quad \Phi(x) = \begin{bmatrix} -xI_m & O \\ O & -xI_n \end{bmatrix}.$$

However, writing the SDP explicitly in standard form like this is perhaps not terribly useful—once an optimization problem has been written in a form involving a linear objective function and only linear entrywise and positive semidefinite constraints, the fact that it is an SDP (i.e., can be converted into primal standard form) is usually clear.

While computing the operator norm of a matrix via semidefinite programming is not really a wise choice (it is much quicker to just compute it via the fact that $\|A\|$ equals the largest singular value of A), this same technique lets us use the operator norm in the objective function or in the constraints of SDPs. For example, if $A \in \mathcal{M}_n$ then the optimization problem

In words, this SDP finds the closest (in the sense of the operator norm) PSD matrix to A that has the same diagonal entries as A.

$$\begin{aligned} \text{minimize:} \quad & \|Y - A\| \\ \text{subject to:} \quad & y_{j,j} = a_{j,j} \quad \text{for all} \quad 1 \leq j \leq n \\ & Y \succeq O \end{aligned} \qquad (3.C.4)$$

is a semidefinite program, since it can be written in the form

$$\begin{aligned} \text{minimize:} \quad & x \\ \text{subject to:} \quad & \begin{bmatrix} xI_m & Y - A \\ Y - A^* & xI_n \end{bmatrix} \succeq O \\ & y_{j,j} = a_{j,j} \quad \text{for all} \quad 1 \leq j \leq n \\ & x, Y \succeq 0, \end{aligned}$$

which in turn could be written in the primal standard form of Definition 3.C.1 if desired.

However, we have to be slightly more careful here than with our earlier manipulations of SDPs—the fact that the optimization problem (3.C.4) is a semidefinite program relies crucially on the fact that we are *minimizing* the norm in the objective function. The analogous maximization problem is *not* a semidefinite program, since it would involve a maximization over Y and a minimization over x—it would look something like

In the outer maximization here, we maximize over the variable Y, and in the inner minimization, we minimize over the variable x (for each particular choice of fixed Y).

$$\text{maximize:} \begin{cases} \text{minimize:} & x \\ \text{subject to:} & \begin{bmatrix} xI_m & Y-A \\ Y-A^* & xI_n \end{bmatrix} \succeq O \\ & x \geq 0 \end{cases}$$

$$\text{subject to:} \quad y_{j,j} = a_{j,j} \quad \text{for all} \quad 1 \leq j \leq n$$
$$Y \succeq O.$$

Optimization problems like this, which combine both maximizations and minimizations, typically cannot be represented as semidefinite programs. Instead, in an SDP we must maximize or minimize over *all* variables.

Remark 3.C.1

Semidefinite Programs and Convexity

One way to think about the fact that we can use semidefinite programming to *minimize* the operator norm, but not *maximize* it, is via the fact that it is **convex**:

$$\|(1-t)A+tB\| \leq (1-t)\|A\|+t\|B\| \quad \text{for all} \quad A,B \in \mathcal{M}_{m,n}, \ 0 \leq t \leq 1.$$

For an introduction to convex functions, see Appendix A.5.2. We can think of minimizing a convex function just like rolling a marble down the side of a bowl—the marble never does anything clever, but it finds the global minimum every time anyway.

Generally speaking, convex functions are easy to minimize since their graphs "open up". It follows that any local minimum that they have is necessarily a global minimum, so we can minimize them simply by following any path on the graph that leads down, as illustrated below:

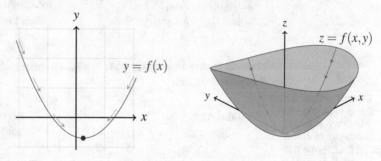

However, finding the *maximum* of a convex function might be hard, since there can be multiple different local maxima (for example, if we stand at the bottom of the parabola above on the left and can only see nearby, it is not clear if we should walk left or walk right to reach the highest point in the plotted domain). For a similar reason, **concave** functions (i.e., functions f for which $-f$ is convex) are easy to *maximize*, since their graphs "open down".

Some standard examples of concave functions include $f(x)=\sqrt{x}$ and $g(x)=\log(x)$.

Similarly, a constraint like $\|X\| \leq 2$ in a semidefinite program is OK, since the operator norm is convex and the upper bound of 2 just chops off the irrelevant top portion of its graph (i.e., it does not interfere with convexity, connectedness of the feasible region, or where its global minimum is located). However, constraints like $\|X\| \geq 2$ involving *lower* bounds on convex functions are not allowed in semidefinite programs, since the resulting feasible region might be extremely difficult to search over—it could have holes and many different local minima.

There are similar semidefinite programs for computing, maximizing, minimizing, and bounding many other linear algebraic quantities of interest, such as the Frobenius norm (well, its square anyway—see Exercise 3.C.9), the trace norm (see Exercise 2.3.17 and the upcoming Example 3.C.5), or the maximum or minimum eigenvalue (see Exercise 3.C.3). However, we must also be slightly careful about where we place these quantities in a semidefinite program.

Every norm is convex, but not every norm can be computed via semidefinite programming.

Norms are all convex, so they can be placed in a "minimize" objective function or on the smaller half of a "$\leq$" constraint, while the minimum eigenvalue is concave and thus can be placed in a "maximize" objective function or on the larger half of a "$\geq$" constraint. The possible placements of these quantities are summarized in Table 3.4.

We use $\lambda_{\max}(X)$ and $\lambda_{\min}(X)$ to refer to the maximal and minimal eigenvalues of X, respectively (in order to use these functions, X must be Hermitian so that its eigenvalues are real).

Function	Convexity	Objective func.	Constraint type
$\|X\|$	convex	min.	"$\leq$"
$\|X\|_{\mathrm{F}}^2$	convex	min.	"$\leq$"
$\|X\|_{\mathrm{tr}}$	convex	min.	"$\leq$"
$\lambda_{\max}(X)$	convex	min.	"$\leq$"
$\lambda_{\min}(X)$	concave	max.	"$\geq$"

Table 3.4: A summary of the convexity/concavity of some functions of a matrix variable X, as well as what type of objective function they can be placed in and what type of constraint they can be placed on the left-hand-side of.

For example, the two optimization problems (where $A \in \mathcal{M}_n^{\mathrm{H}}$ is fixed and $X \in \mathcal{M}_n^{\mathrm{H}}$ is the matrix variable)

$$
\begin{array}{ll}
\text{minimize:} & \lambda_{\max}(X) \\
\text{subject to:} & \|A - X\|_{\mathrm{F}} \leq 2 \\
& X \succeq O
\end{array}
\qquad
\begin{array}{ll}
\text{maximize:} & \lambda_{\min}(A - X) \\
\text{subject to:} & \lambda_{\max}(X) + \|X\|_{\mathrm{tr}} \leq 1 \\
& X \succeq O
\end{array}
$$

are both semidefinite programs, whereas neither of the following two optimization problems are:

The left problem is not an SDP due to the norm equality constraint and the right problem is not an SDP due to maximizing $\lambda_{\max}$.

$$
\begin{array}{ll}
\text{minimize:} & \mathrm{tr}(X) \\
\text{subject to:} & \|A - X\|_{\mathrm{tr}} = 1 \\
& X \succeq O
\end{array}
\qquad
\begin{array}{ll}
\text{maximize:} & \lambda_{\max}(A - X) \\
\text{subject to:} & X \preceq A \\
& X \succeq O.
\end{array}
$$

We also note that Table 3.4 is not remotely exhaustive. Many other linear algebraic quantities, particularly those involving eigenvalues, singular values, and/or norms, can be incorporated into semidefinite programs—see Exercises 3.C.10 and 3.C.13.

3.C.2 Geometric Interpretation and Solving

Semidefinite programs have a natural geometric interpretation that is completely analogous to that of linear programs. Recall that the feasible region of a linear program is a **convex polyhedron**—a convex shape with flat sides—and optimizing the objective function can be thought of as moving a line (or plane, or hyperplane, ...) in one direction as far as possible while still intersecting the feasible region. Furthermore, the optimal value of a linear program is always attained at a corner of the feasible region, as indicated in Figure 3.5.

For semidefinite programs, all that changes is that the feasible region might no longer be a polyhedron. Instead of having flat sides, the edges of the feasible region might "bow out" (though it must still be convex)—we call these shapes

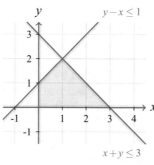

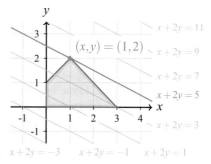

(a) The feasible region of a linear program is a polyhedron (convex with flat sides).

(b) Solving a linear program can be done by moving a line (or plane, ...) as far as possible while still intersecting the feasible region.

Figure 3.5: The feasible region of the linear program that asks us to maximize $x+2y$ subject to the constraints $x+y \leq 3$, $y-x \leq 1$, and $x, y \geq 0$. Its optimal value is 5, which is attained at $(x,y) = (1,2)$.

spectrahedra, and their boundaries are defined by polynomial equations (in much the same way that the sides of polyhedra are defined by linear equations).

Since the objective function $\operatorname{tr}(CX)$ of a semidefinite program is linear, we can still think of optimizing it as moving a line (or plane, or hyperplane, ...) in one direction as far as possible while still intersecting the feasible region. However, because of its bowed out sides, the optimal value might no longer be attained at a corner of the feasible region. For example, consider the following semidefinite program in the variables $x, y \in \mathbb{R}$:

In fact, the feasible region of an SDP might not even *have* any corners—it could be a filled circle or sphere, for example.

$$
\begin{aligned}
\text{maximize:} \quad & x \\
\text{subject to:} \quad & \begin{bmatrix} x+1 & y \\ y & 1 \end{bmatrix}, \begin{bmatrix} x & y-1 \\ y-1 & 3-x \end{bmatrix} \succeq O \\
& y \geq 0.
\end{aligned}
\tag{3.C.5}
$$

The feasible region of this SDP is displayed in Figure 3.6, and its optimal value is $x = 3$, which is attained on one of the curved boundaries of the feasible region (not at one of its corners).

Sylvester's criterion (Theorem 2.2.6) can turn any positive semidefinite constraint into a family of polynomial constraints in the matrix's entries. For example, Theorem 2.2.7 says that

$$
\begin{bmatrix} x+1 & y \\ y & 1 \end{bmatrix} \succeq O
$$

if and only if
$$ x+1 \geq y^2. $$

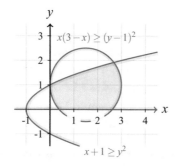

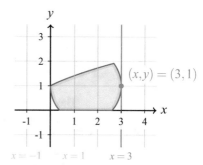

(a) The feasible region of a semidefinite program is a spectrahedron (convex with potentially curved sides).

(b) Solving a semidefinite program can be done by moving a line (or plane, ...) as far as possible while still intersecting the feasible region.

Figure 3.6: The feasible region of the semidefinite program (3.C.5). Its optimal value is 3, which is attained at $(x,y) = (3,1)$. Notably, this point is not a corner of the feasible region.

This geometric wrinkle that semidefinite programs have over linear programs has important implications when it comes to solving them. Recall that the standard method of solving a linear program is to use the simplex method, which works by jumping from corner to adjacent corner of the feasible region so as to increase the objective function by as much as possible at each step. This method relies crucially on the fact that the optimal value of the linear program occurs at a corner of the feasible region, so it does not generalize straightforwardly to semidefinite programs.

While the simplex method for linear programs always terminates in a finite number of steps and can produce an exact description of its optimal value, no such algorithm for semidefinite programs is known. There are efficient methods for numerically *approximating* the optimal value of an SDP to any desired accuracy, but these algorithms are not really practical to run by hand—they are instead implemented by various computer software packages, and we do not explore these methods here. There are entire books devoted to semidefinite programming (and convex optimization in general), so the interested reader is directed to [BV09] for a more thorough treatment. We are interested in semidefinite programming primarily for its duality theory, which often lets us find the optimal solution of an SDP analytically.

3.C.3   Duality

Just as was the case for linear programs, semidefinite programs have a robust duality theory. However, since we do not have simple methods for explicitly solving semidefinite programs by hand like we do for linear programs, we will find that duality plays an even more important role in this setting.

Suppose $B \in \mathcal{M}_m^H$, $C \in \mathcal{M}_n^H$, $\Phi : \mathcal{M}_n^H \to \mathcal{M}_m^H$ is linear, and $X \in \mathcal{M}_n^H$ and $Y \in \mathcal{M}_m^H$ are matrix variables. The **dual** of a semidefinite program

$$\begin{aligned} \text{maximize:} \quad & \text{tr}(CX) \\ \text{subject to:} \quad & \Phi(X) \preceq B \\ & X \succeq O \end{aligned}$$

is the semidefinite program

$$\begin{aligned} \text{minimize:} \quad & \text{tr}(BY) \\ \text{subject to:} \quad & \Phi^*(Y) \succeq C \\ & Y \succeq O \end{aligned}$$

The original semidefinite program in Definition 3.C.2 is called the **primal problem,** and the two of them together are called a **primal/dual pair.** Although constructing the dual of a semidefinite program is a rather routine and mechanical affair, keep in mind that the above definition only applies once the SDP is written in primal standard form (fortunately, we already know how to convert any SDP into that form). Also, even though we can construct the adjoint Φ^* of any linear map $\Phi : \mathcal{M}_n^H \to \mathcal{M}_m^H$ via Corollary 3.A.5, it is often quicker and easier to just "eyeball" a formula for the adjoint and then check that it is correct.

Example 3.C.3	Construct the dual of the semidefinite program (3.C.3) that computes the operator norm of a matrix $A \in \mathcal{M}_{m,n}$.
Constructing the Dual of an SDP	**Solution:**

Our first step is to write this SDP in primal standard form, which we recall can be done by defining

$$B = \begin{bmatrix} O & A \\ A^* & O \end{bmatrix}, \quad C = -1, \quad \text{and} \quad \Phi(x) = \begin{bmatrix} -xI_m & O \\ O & -xI_n \end{bmatrix}.$$

The adjoint of Φ is a linear map from $\mathcal{M}_{m+n}^H$ to $\mathcal{M}_1^H$, and it must satisfy

Again, we use asterisks (∗) to denote entries that are irrelevant for our purposes.

$$\mathrm{tr}\left(x\Phi^*\left(\begin{bmatrix} Y & * \\ * & Z \end{bmatrix} \right) \right) = \mathrm{tr}\left(\Phi(x) \begin{bmatrix} Y & * \\ * & Z \end{bmatrix} \right)$$

$$= \mathrm{tr}\left(\begin{bmatrix} -xI_m & O \\ O & -xI_n \end{bmatrix} \begin{bmatrix} Y & * \\ * & Z \end{bmatrix} \right) = -x\mathrm{tr}(Y) - x\mathrm{tr}(Z)$$

Recall from Theorem 1.4.8 that adjoints are unique.

for all $Y \in \mathcal{M}_m^H$ and $Z \in \mathcal{M}_n^H$. We can see from inspection that one map (and thus *the* map) Φ^* that works is given by the formula

$$\Phi^*\left(\begin{bmatrix} Y & * \\ * & Z \end{bmatrix} \right) = -\mathrm{tr}(Y) - \mathrm{tr}(Z).$$

The minus sign in front of this minimization comes from the fact that we had to convert the original SDP from a minimization to a maximization in order to put it into primal standard form (this is also why $C = -1$ instead of $C = 1$).

It follows that the dual of the original semidefinite program has the following form:

$$-\text{minimize:} \quad \mathrm{tr}\left(\begin{bmatrix} O & A \\ A^* & O \end{bmatrix} \begin{bmatrix} Y & X \\ X^* & Z \end{bmatrix} \right)$$

$$\text{subject to:} \quad \begin{bmatrix} Y & X \\ X^* & Z \end{bmatrix} \succeq O$$

$$-\mathrm{tr}(Y) - \mathrm{tr}(Z) \geq -1$$

After simplifying things somewhat, this dual problem can be written in the somewhat prettier (but equivalent) form

$$\text{maximize:} \quad \mathrm{Re}\left(\mathrm{tr}(AX^*)\right)$$

$$\text{subject to:} \quad \begin{bmatrix} Y & -X \\ -X^* & Z \end{bmatrix} \succeq O$$

$$\mathrm{tr}(Y) + \mathrm{tr}(Z) \leq 2$$

Just as is the case with linear programs, the dual of a semidefinite program is remarkable for the fact that it can provide us with upper bounds on the optimal value of the primal problem (and the primal problem similarly provides lower bounds to the optimal value of the dual problem):

Theorem 3.C.1	If $X \in \mathcal{M}_n^H$ is a feasible point of a semidefinite program in primal standard form, and $Y \in \mathcal{M}_m^H$ is a feasible point of its dual problem, then $\mathrm{tr}(CX) \leq$
Weak Duality	$\mathrm{tr}(BY)$.

Proof. Since Y and $B - \Phi(X)$ are both positive semidefinite, we know from Exercise 2.2.19 that

Recall that $\text{tr}(\Phi(X)Y) =$ $\text{tr}(X\Phi^*(Y))$ simply because we are working in the Frobenius inner product and Φ^* is (by definition) the adjoint of Φ in this inner product.

$$0 \leq \text{tr}\big((B - \Phi(X))Y\big) = \text{tr}(BY) - \text{tr}\big(\Phi(X)Y\big) = \text{tr}(BY) - \text{tr}\big(X\Phi^*(Y)\big).$$

It follows that $\text{tr}(BY) \geq \text{tr}\big(X\Phi^*(Y)\big)$. Then using the fact that X and $\Phi^*(Y) - C$ are both positive semidefinite, a similar argument shows that

$$0 \leq \text{tr}\big(X(\Phi^*(Y) - C)\big) = \text{tr}\big(X\Phi^*(Y)\big) - \text{tr}(XC),$$

so $\text{tr}\big(X\Phi^*(Y)\big) \geq \text{tr}(XC)$. Stringing these two inequalities together shows that

$$\text{tr}(BY) \geq \text{tr}\big(X\Phi^*(Y)\big) \geq \text{tr}(XC) = \text{tr}(CX),$$

as desired. ∎

Weak duality not only provides us with a way of establishing upper bounds on the optimal value of our semidefinite program, but it often also lets us easily determine when we have found its optimal value. In particular, if we can find feasible points of the primal and dual problems that give the same value when plugged into their respective objective functions, they must be optimal, since they cannot possibly be increased or decreased past each other (see Figure 3.7).

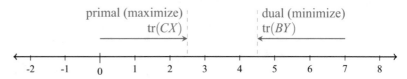

Figure 3.7: Weak duality says that the objective function of the primal (maximization) problem cannot be increased past the objective function of the dual (minimization) problem.

The following example illustrates how we can use this feature of weak duality to solve semidefinite programs, at least in the case when they are simple enough that we can spot what we think the solution should be. That is, weak duality can help us verify that a conjectured optimal solution really is optimal.

Example 3.C.4

Solving an SDP via Weak Duality

Use weak duality to solve the following SDP in the variable $X \in \mathcal{M}_3^H$:

$$\text{maximize:} \quad \text{tr}\left(\begin{bmatrix} 0 & 1 & 1 \\ 1 & 0 & 1 \\ 1 & 1 & 0 \end{bmatrix} X\right)$$

$$\text{subject to:} \quad O \preceq X \preceq I$$

Solution:
The objective function of this semidefinite program just adds up the off-diagonal entries of X, and the constraint $O \preceq X \preceq I$ says that every eigenvalue of X is between 0 and 1 (see Exercise 3.C.3). Roughly speaking, we thus want to find a PSD matrix X with small diagonal entries and large off-diagonal entries. One matrix that seems to work fairly well is

Actually, the objective function adds up the *real parts* of the off-diagonal entries of X.

$$X = \frac{1}{3}\begin{bmatrix} 1 & 1 & 1 \\ 1 & 1 & 1 \\ 1 & 1 & 1 \end{bmatrix},$$

which has eigenvalues 1, 0, and 0 (and is thus feasible), and produces a value of 2 in the objective function.

To show that this choice of X is optimal, we construct the dual of this SDP, which has the form

> The map Φ in this SDP is the identity, so Φ^* is the identity as well.

$$
\begin{aligned}
\text{minimize:} \quad & \operatorname{tr}(Y) \\
\text{subject to:} \quad & Y \succeq \begin{bmatrix} 0 & 1 & 1 \\ 1 & 0 & 1 \\ 1 & 1 & 0 \end{bmatrix} \\
& Y \succeq O
\end{aligned}
$$

Since the matrix on the right-hand-side of the constraint in this dual SDP has eigenvalues 2, -1, and -1, we can find a positive semidefinite matrix Y satisfying that constraint simply by increasing the negative eigenvalues to 0 (while keeping the corresponding eigenvectors the same). That is, if we have the spectral decomposition

> In other words, we choose Y to be the positive semidefinite part of
> $$\begin{bmatrix} 0 & 1 & 1 \\ 1 & 0 & 1 \\ 1 & 1 & 0 \end{bmatrix},$$
> in the sense of Exercise 2.2.16. Explicitly,
> $$Y = \frac{2}{3}\begin{bmatrix} 1 & 1 & 1 \\ 1 & 1 & 1 \\ 1 & 1 & 1 \end{bmatrix}.$$

$$
\begin{bmatrix} 0 & 1 & 1 \\ 1 & 0 & 1 \\ 1 & 1 & 0 \end{bmatrix} = U \begin{bmatrix} 2 & 0 & 0 \\ 0 & -1 & 0 \\ 0 & 0 & -1 \end{bmatrix} U^*
$$

then we choose

$$
Y = U \begin{bmatrix} 2 & 0 & 0 \\ 0 & 0 & 0 \\ 0 & 0 & 0 \end{bmatrix} U^*,
$$

which has $\operatorname{tr}(Y) = 2$.

Since we have now found feasible points of both the primal and dual problems that attain the same objective value of 2, we know that this must in fact be the optimal value of both problems.

As suggested by the previous example, weak duality is useful for the fact that it can often be used to solve a semidefinite program without actually performing any optimization at all, as long as the solution is simple enough that we can eyeball feasible points of each of the primal and dual problems that attain the same value. We now illustrate how this procedure can be used to give us new characterizations of linear algebraic objects.

Example 3.C.5

Using Weak Duality to Understand an SDP for the Trace Norm

Suppose $A \in \mathcal{M}_{m,n}$. Use weak duality to show that the following semidefinite program in the variables $X \in \mathcal{M}_m^H$ and $Y \in \mathcal{M}_n^H$ computes $\|A\|_{\operatorname{tr}}$ (the trace norm of A, which was introduced in Exercise 2.3.17):

$$
\begin{aligned}
\text{minimize:} \quad & (\operatorname{tr}(X) + \operatorname{tr}(Y))/2 \\
\text{subject to:} \quad & \begin{bmatrix} X & A \\ A^* & Y \end{bmatrix} \succeq O \\
& X, Y \succeq O
\end{aligned}
$$

Solution:

To start, we append rows or columns consisting entirely of zeros to A so as to make it square. Doing so does not affect $\|A\|_{\operatorname{tr}}$ or any substantial details of this SDP, but makes its analysis a bit cleaner.

To show that the optimal value of this SDP is bounded above by $\|A\|_{tr}$, we just need to find a feasible point that attains this quantity in the objective function. To this end, let $A = U\Sigma V^*$ be a singular value decomposition of A. We then define $X = U\Sigma U^*$ and $Y = V\Sigma V^*$. Then

$$\text{tr}(X) = \|U\Sigma U^*\|_{tr} = \|\Sigma\|_{tr} = \|A\|_{tr}, \quad \text{and}$$
$$\text{tr}(Y) = \|V\Sigma V^*\|_{tr} = \|\Sigma\|_{tr} = \|A\|_{tr},$$

so $(\text{tr}(X) + \text{tr}(Y))/2 = \|A\|_{tr}$ and thus X and Y produce the desired value in the objective function. Furthermore, X and Y are feasible since they are positive semidefinite and so is

If A (and thus Σ) is not square then either $U\sqrt{\Sigma}$ or $V\sqrt{\Sigma}^*$ needs to be padded with some zero columns for this decomposition to make sense.

$$\begin{bmatrix} X & A \\ A^* & Y \end{bmatrix} = \begin{bmatrix} U\Sigma U^* & U\Sigma V^* \\ V\Sigma^* U^* & V\Sigma V^* \end{bmatrix} = \begin{bmatrix} U\sqrt{\Sigma} \\ V\sqrt{\Sigma}^* \end{bmatrix} \begin{bmatrix} U\sqrt{\Sigma} \\ V\sqrt{\Sigma}^* \end{bmatrix}^*,$$

where $\sqrt{\Sigma}$ is the entrywise square root of Σ.

To prove the opposite inequality (i.e., to show that this SDP is bounded *below* by $\|A\|_{tr}$), we first need to construct its dual. To this end, we first explicitly list the components of its primal standard form (3.C.2):

Here, C is negative because the given SDP is a minimization problem, so we have to multiply the objective function by -1 to turn it into a maximization problem (i.e., to put it into primal standard form).

$$B = \begin{bmatrix} O & A \\ A^* & O \end{bmatrix}, \quad C = \frac{-1}{2}\begin{bmatrix} I_m & O \\ O & I_n \end{bmatrix}, \quad \text{and} \quad \Phi\left(\begin{bmatrix} X & * \\ * & Y \end{bmatrix}\right) = \begin{bmatrix} -X & O \\ O & -Y \end{bmatrix}.$$

It is straightforward to show that $\Phi^* = \Phi$, so the dual SDP (after simplifying a bit) has the form

maximize: $\text{Re}(\text{tr}(AZ))$
subject to: $\begin{bmatrix} X & Z \\ Z^* & Y \end{bmatrix} \succeq O$
              $X, Y \preceq I$

Next, we find a feasible point of this SDP that attains the desired value of $\|A\|_{tr}$ in the objective function. If A has the singular value decomposition $A = U\Sigma V^*$ then the matrix

Recall that the fact that $\Phi = \Phi^*$ means that it is called self-adjoint.

$$\begin{bmatrix} X & Z \\ Z^* & Y \end{bmatrix} = \begin{bmatrix} I & VU^* \\ UV^* & I \end{bmatrix}$$

is a feasible point of the dual SDP, since $X = I \preceq I$, $Y = I \preceq I$, and the fact that VU^* is unitary tells us that $\|VU^*\| = 1$ (by Exercise 2.3.5), so the given block matrix is positive semidefinite (by Exercise 2.3.15). Then

$$\text{Re}(\text{tr}(AZ)) = \text{Re}(\text{tr}(U\Sigma V^* VU^*)) = \text{Re}(\text{tr}(\Sigma)) = \text{tr}(\Sigma) = \|A\|_{tr},$$

which shows that $\|A\|_{tr}$ is a *lower* bound on the value of this SDP. Since we have now proved both bounds, we are done.

Remark 3.C.2

Numerics Combined with Duality Make a Powerful Combination

The previous examples of solving SDPs via duality might seem somewhat "cooked up" at first, especially since in Example 3.C.5 we were told what the optimal value is, and we just had to verify it. In practice, we are typically not told the optimal value of the SDP that we are working with ahead of time. However, this is actually not a huge restriction, since we can use computer software to numerically solve the SDP and then use that solution to help us eyeball the analytic (non-numerical) solution.

To illustrate what we mean by this, consider the following semidefinite program that maximizes over symmetric matrices $X \in \mathcal{M}_2^S$:

$$\begin{aligned}
\text{maximize:} \quad & x_{1,1} - x_{2,2} \\
\text{subject to:} \quad & x_{1,1} - x_{1,2} = 1/2 \\
& \operatorname{tr}(X) \leq 1 \\
& X \succeq O
\end{aligned}$$

It might be difficult to see what the optimal value of this SDP is, but we can get a helpful nudge by using computer software to solve it and find that the optimal value is approximately 0.7071, which is attained at a matrix

$$X \approx \begin{bmatrix} 0.8536 & 0.3536 \\ 0.3536 & 0.1464 \end{bmatrix}.$$

The optimal value looks like it is probably $1/\sqrt{2} = 0.7071\ldots$, and to prove it we just need to find feasible matrices that attain this value in the objective functions of the primal and dual problems.

Another way to guess the entries of X would be to plug the decimal approximations of its entries into a tool like the Inverse Symbolic Calculator (BBP95).

The matrix X that works in the primal problem must have $x_{1,1} + x_{2,2} \leq 1$ (and the numerics above suggest that equality holds) and $x_{1,1} - x_{2,2} = 1/\sqrt{2}$. Solving this 2×2 linear system gives $x_{1,1} = (2+\sqrt{2})/4$ and $x_{2,2} = (2-\sqrt{2})/4$, and the constraint $x_{1,1} - x_{1,2} = 1/2$ then tells us that $x_{1,2} = \sqrt{2}/4$. We thus guess that the maximum value is obtained at the matrix

$$X = \frac{1}{4} \begin{bmatrix} 2+\sqrt{2} & \sqrt{2} \\ \sqrt{2} & 2-\sqrt{2} \end{bmatrix},$$

which indeed satisfies all of the constraints and produces an objective value of $1/\sqrt{2}$, so this is a lower bound on the desired optimal value.

You are encouraged to work through the details of constructing this dual program yourself.

To see that we cannot do any better (i.e., to show that $1/\sqrt{2}$ is also an upper bound), we construct the dual problem, which has the form

$$\begin{aligned}
\text{minimize:} \quad & y_1 + y_2/2 \\
\text{subject to:} \quad & \begin{bmatrix} y_1 + y_2 & -y_2/2 \\ -y_2/2 & y_1 \end{bmatrix} \succeq \begin{bmatrix} 1 & 0 \\ 0 & -1 \end{bmatrix} \\
& y_1 \geq 0
\end{aligned}$$

Similar to before, we can solve this SDP numerically to find that its optimal value is also approximately 0.7071, and is attained when

$$y_1 \approx 0.2071 \quad \text{and} \quad y_2 \approx 1.000.$$

We can turn this into an exact answer by guessing $y_2 = 1$, which forces $y_1 = (\sqrt{2}-1)/2$ (since we want $y_1 + y_2/2 = 1/\sqrt{2}$). It is then a routine calculation to verify that this (y_1, y_2) pair satisfies all of the constraints in the dual problem and gives an objective value of $1/\sqrt{2}$, so this is indeed the optimal value of the SDP.

In the previous two examples, we saw that not only did the dual problem serve as an upper bound on the primal problem, but rather we were able to find particular feasible points of each problem that resulted in their objective functions taking on the same value, thus proving optimality. The following theorem shows that this phenomenon occurs with a great deal of generality—there are simple-to-check conditions that guarantee that there *must* be feasible points in each of the primal and dual problems that attain the optimal value in their objective functions.

Recall that $X \succ O$ means that X is positive *definite* (i.e., PSD and invertible) and $\Phi(X) \prec B$ means that $B - \Phi(X)$ is positive definite.

Before stating what these conditions are, we need one more piece of terminology: we say that an SDP in primal standard form (3.C.2) is **feasible** if there exists a matrix $X \in \mathcal{M}_n^H$ satisfying all of its constraints (i.e., $X \succeq O$ and $\Phi(X) \preceq B$), and we say that it is **strictly feasible** if X can be chosen to make both of those inequalities strict (i.e., $X \succ O$ and $\Phi(X) \prec B$). Feasibility and strict feasibility for problems written in the dual form of Definition 3.C.2 are defined analogously. Geometrically, strict feasibility of an SDP means that its feasible region has an interior—it is not a degenerate lower-dimensional shape that consists only of boundaries and edges.

Theorem 3.C.2

Strong Duality

Suppose that both problems in a primal/dual pair of SDPs are feasible, and at least one of them is strictly feasible. Then the optimal values of those SDPs coincide. Furthermore,

The conditions in this theorem are sometimes called the **Slater conditions** for strong duality. There are other (somewhat more technical) conditions that guarantee that a primal/dual pair share their optimal value as well.

a) if the primal problem is strictly feasible then the optimal value is attained in the dual problem, and

b) if the dual problem is strictly feasible then the optimal value is attained in the primal problem.

Since the proof of this result relies on some facts about convex sets that we have not explicitly introduced in the main body of this text, we leave it to Appendix B.3. However, it is worth presenting some examples to illustrate what the various parts of the theorem mean and why they are important. To start, it is worthwhile to demonstrate what we mean when we say that strict feasibility implies that the optimal value is "attained" by the other problem in a primal/dual pair.

Example 3.C.6

An SDP That Does Not Attain Its Optimal Value

Show that no feasible point of the following SDP attains its optimal value:

$$\begin{aligned} \text{minimize:} \quad & x \\ \text{subject to:} \quad & \begin{bmatrix} x & 1 \\ 1 & y \end{bmatrix} \succeq O \\ & x, y \geq 0. \end{aligned}$$

Furthermore, explain why this phenomenon does not contradict strong duality (Theorem 3.C.2).

Be careful: even if the strong duality conditions of Theorem 3.C.2 hold (i.e., the primal/dual SDPs are strictly feasible) and the optimal value is attained, that does not mean that it is attained at a strictly feasible point.

Solution:

Recall from Theorem 2.2.7 that the matrix in this SDP is positive semidefinite if and only if $xy \geq 1$. In particular, this means that $(x,y) = (x, 1/x)$ is a feasible point of this SDP for all $x > 0$, so certainly the optimal value of this SDP cannot be bigger than 0. However, no feasible point has $x = 0$, since we would then have $xy = 0 \not\geq 1$. It follows that the optimal value of this SDP is 0, but no feasible point attains that value—they just get arbitrarily close to it.

The fact that the optimal value of this SDP is not attained does not contradict Theorem 3.C.2 since the dual of this SDP must not be strictly feasible. To verify this claim, we compute the dual SDP to have the following form (after simplifying somewhat):

$$\text{maximize:} \quad \text{Re}(z)$$
$$\text{subject to:} \quad O \preceq \begin{bmatrix} x & -z \\ -\bar{z} & y \end{bmatrix} \preceq \begin{bmatrix} 2 & 0 \\ 0 & 0 \end{bmatrix}$$

Recall from Theorem 2.2.4 that positive definite matrices have strictly positive diagonal entries.

The constraint in the above SDP forces $y = 0$, so there is no positive *definite* matrix that satisfies it (i.e., this dual SDP is not *strictly* feasible).

There are also conditions other than those of Theorem 3.C.2 that can be used to guarantee that the optimal value of an SDP is attained or its dual has the same optimal value. For example, the Extreme Value Theorem from real analysis says that every continuous function on a closed and bounded set necessarily attains it maximum and minimum values. Since the objective function of every SDP is linear (and thus continuous), and the feasible set of every SDP is closed, we obtain the following criterion:

> ! If the feasible region of an SDP is non-empty and bounded, its optimal value is attained.

On the other hand, notice that the feasible region of the SDP from Example 3.C.6 is unbounded, since we can make x and y as large as we like in it (see Figure 3.8).

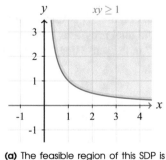

(a) The feasible region of this SDP is unbounded.

(b) The optimal value of this SDP is 0, but it is not attained by any point in the feasible region.

Figure 3.8: The feasible region of the semidefinite program from Example 3.C.6. The fact that the optimal value of this SDP is not attained corresponds to the fact that its feasible region gets arbitrarily close to the line $x = 0$ but does not contain any point on it.

In the previous example, even though the dual SDP was not strictly feasible, the primal SDP was, so the optimal values of both problems were still forced to equal each other by Theorem 3.C.2 (despite the optimal value not actually being attained in the primal problem). We now present an example that shows that it is also possible for *neither* SDP to be strictly feasible, and thus for their optimal values to differ.

Example 3.C.7

A Primal/Dual SDP Pair with Unequal Optimal Values

Show that the following primal/dual SDPs, which optimize over the variables $X \in \mathcal{M}_3^S$ in the primal and $y, z \in \mathbb{R}$ in the dual, have different optimal values:

Primal

maximize: $-x_{2,2}$

subject to: $2x_{1,3} + x_{2,2} = 1$

$x_{3,3} = 0$

$X \succeq O$

Dual

minimize: y

subject to: $\begin{bmatrix} 0 & 0 & y \\ 0 & y+1 & 0 \\ y & 0 & z \end{bmatrix} \succeq O$

Verify on your own that these problems are indeed duals of each other.

Furthermore, explain why this phenomenon does not contradict strong duality (Theorem 3.C.2).

Solution:

In the primal problem, the fact that $x_{3,3} = 0$ and $X \succeq O$ forces $x_{1,3} = 0$ as well. The first constraint then says that $x_{2,2} = 1$, so the optimal value of the primal problem is -1.

Recall from Exercise 2.2.11 that if a diagonal entry of a PSD matrix equals 0 then so does that entire row and column.

In the dual problem, the fact that the top-left entry of the 3×3 PSD matrix equals 0 forces $y = 0$, so the optimal value of this dual problem equals 0.

Since these two optimal values are not equal to each other, we know from Theorem 3.C.2 that neither problem is strictly feasible. Indeed, the primal problem is not strictly feasible since the constraint $x_{3,3} = 0$ ensures that the entries in the final row and column of X all equal 0 (so X cannot be invertible), and the dual problem is similarly not strictly feasible since the 3×3 PSD matrix must have every entry in its first row and column equal to 0.

In spite of semidefinite programs like those presented in the previous two examples, most real-world SDPs (i.e., ones that are not extremely "cooked up") are strictly feasible and we can thus make use of strong duality. That is, for most SDPs it is the case that Figure 3.7 is somewhat misleading—it is not just the case that the primal and dual programs bound each other, but rather they bound each other "tightly" in the sense that their optimal values coincide (see Figure 3.9).

Recall that strong duality holds (i.e., the optimal values are attained and equal in a primal/dual pair) for every linear program.

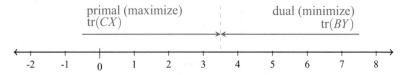

Figure 3.9: Strong duality says that, for many semidefinite programs, the objective function of the primal (maximization) problem can be increased to the exact same value that the objective function of the dual (minimization) problem can be decreased to.

Furthermore, showing that strict feasibility holds for an SDP is usually quite straightforward. For example, to see that the SDP from Example 3.C.5 is strictly feasible, we just need to observe that we can choose X and Y to each be suitably large multiples of the identity matrix.

Example 3.C.8

Using Strong Duality to Solve an SDP

Show that the following semidefinite program in the variables $Y \in \mathcal{M}_m^H$, $Z \in \mathcal{M}_n^H$, and $X \in \mathcal{M}_{m,n}(\mathbb{C})$ computes the operator norm of the matrix $A \in \mathcal{M}_{m,n}(\mathbb{C})$:

$$
\begin{aligned}
\text{maximize:} \quad & \mathrm{Re}\big(\mathrm{tr}(AX^*)\big) \\
\text{subject to:} \quad & \begin{bmatrix} Y & -X \\ -X^* & Z \end{bmatrix} \succeq O \\
& \mathrm{tr}(Y) + \mathrm{tr}(Z) \le 2
\end{aligned}
$$

Solution:

Recall from Example 3.C.3 that this is the dual of the SDP (3.C.3) that computes $\|A\|$. It follows that the optimal value of this SDP is certainly bounded above by $\|A\|$, so there are two ways we could proceed:

- we could find a feasible point of this SDP that attains the conjectured optimal value $\|A\|$, and then note that the true optimal value must be $\|A\|$ by weak duality, or

We find a feasible point attaining this optimal value in Exercise 3.C.8.

- we could show that the primal SDP (3.C.3) is strictly feasible, so this dual SDP must attain its optimal value, which is the same as the optimal value $\|A\|$ of the primal SDP by strong duality.

We opt for the latter method—we show that the primal SDP (3.C.3) is strictly feasible. To this end, we just note that if x is really, really large (in particular, larger than $\|A\|$) then $x > 0$ and

When proving strong duality, it is often useful to just let the variables be very big. Try not to get hung up on how big is big enough to make everything positive definite, as long as it is clear that there *is* a big enough choice that works.

$$
\begin{bmatrix} xI_m & A \\ A^* & xI_n \end{bmatrix} \succ O,
$$

so the primal SDP (3.C.3) is strictly feasible and strong duality holds.

We do not need to, but we can also show that the dual SDP that we started with is strictly feasible by choosing

$$
Y = I_m/m, \quad Z = I_n/n, \quad \text{and} \quad X = O.
$$

Remark 3.C.3

Unbounded and Feasibility SDPs

The corresponding table for linear programs is identical, except with a blank in the "infeasible/solvable" cells. There exists an infeasible SDP with solvable dual (see Exercise 3.C.4), but no such pair of LPs exists.

Just like linear programs, semidefinite programs can be **infeasible** (i.e., have an empty feasible region) or **unbounded** (i.e., have an objective function that can be made arbitrarily large). If a maximization problem is unbounded then we say that its optimal value is ∞, and if it is infeasible then we say that its optimal value is $-\infty$ (and of course these two quantities are swapped for minimization problems).

Weak duality immediately tells us that if an SDP is unbounded then its dual must be infeasible, which leads to the following possible infeasible/unbounded/solvable (i.e., finite optimal value) pairings that primal and dual problems can share:

		Primal problem		
		Infeasible	Solvable	Unbounded
Dual	Infeasible	✓	✓	✓
	Solvable	✓	✓	.
	Unbounded	✓	.	.

Allowing optimal values of $\pm\infty$ like this is particularly useful when considering **feasibility SDPs**—semidefinite programs in which we are only interested in whether or not a feasible point exists. For example, suppose we wanted to know whether or not there was a way to fill in the missing entries in the matrix

These entries can indeed be filled in to make this matrix PSD—see Exercise 3.C.7.

$$\begin{bmatrix} 3 & * & * & -2 & * \\ * & 3 & * & -2 & * \\ * & * & 3 & * & * \\ -2 & -2 & * & 3 & -2 \\ * & * & * & -2 & 3 \end{bmatrix}$$

so as to make it positive semidefinite. This isn't really an optimization problem per se, but we can nonetheless write it as the following SDP:

maximize: 0
subject to: $x_{j,j} = 3$ for all $1 \le j \le 5$
$x_{i,j} = -2$ for $\{i,j\} = \{1,4\}, \{2,4\}, \{4,5\}$
$X \succeq O$

If such a PSD matrix exists, this SDP has optimal value 0, otherwise it is infeasible and thus has optimal value $-\infty$.

Exercises

solutions to starred exercises on page 483

∗∗3.C.1 We now prove some of the basic properties of the Loewner partial order. Suppose $A, B, C \in \mathcal{M}_n^H$.

(a) Show that $A \succeq A$.
(b) Show that if $A \succeq B$ and $B \succeq A$ then $A = B$.
(c) Show that if $A \succeq B$ and $B \succeq C$ then $A \succeq C$.
(d) Show that if $A \succeq B$ then $A + C \succeq B + C$.
(e) Show that $A \succeq B$ if and only if $-A \preceq -B$.
(f) Show that $A \succeq B$ implies $PAP^* \succeq PBP^*$ for all $P \in \mathcal{M}_{m,n}(\mathbb{C})$.

3.C.2 Determine which of the following statements are true and which are false.

∗(a) Every linear program is a semidefinite program.
(b) Every semidefinite program is a linear program.
∗(c) If A and B are Hermitian matrices for which $A \succeq B$ then $\text{tr}(A) \ge \text{tr}(B)$.
(d) If an SDP is unbounded then its dual must be infeasible.
∗(e) If an SDP is infeasible then its dual must be unbounded.

∗∗3.C.3 Suppose $A \in \mathcal{M}_n^H$ and $c \in \mathbb{R}$. Let $\lambda_{max}(A)$ and $\lambda_{min}(A)$ denote the maximal and minimal eigenvalue, respectively, of A.

(a) Show that $A \preceq cI$ if and only if $\lambda_{max}(A) \le c$.
(b) Use part (a) to construct a semidefinite program that computes $\lambda_{max}(A)$.
(c) Construct a semidefinite program that computes $\lambda_{min}(A)$.

∗∗ 3.C.4 In this problem, we show that there are primal/dual pairs of semidefinite programs in which one problem is infeasible and the other is solvable. Consider the following SDP involving the variables $x, y \in \mathbb{R}$:

maximize: 0
subject to: $\begin{bmatrix} x+y & 0 \\ 0 & x \end{bmatrix} \preceq \begin{bmatrix} 0 & 1 \\ 1 & 0 \end{bmatrix}$
$x + y \le 0$
$x \ge 0$

(a) Show that this SDP is infeasible.
(b) Construct the dual of this semidefinite program and show that it is feasible and has an optimal value of 0.

∗∗3.C.5 Suppose $A, B \in \mathcal{M}_n^H$ are positive semidefinite and $A \succeq B$.

(a) Provide an example to show that it is *not* necessarily the case that $A^2 \succeq B^2$.

(b) Show that, in spite of part (a), it *is* the case that $\operatorname{tr}(A^2) \geq \operatorname{tr}(B^2)$.
[Hint: Factor a difference of squares.]

3.C.6 Suppose $A, B \in \mathcal{M}_n^H$ are positive semidefinite and $A \succeq B$.

(a) Show that if A is positive *definite* then $\sqrt{A} \succeq \sqrt{B}$.
[Hint: Use the fact that $\sqrt{B}\sqrt{A}^{-1}$ and $A^{-1/4}\sqrt{B}A^{-1/4}$ are similar to each other, where $A^{-1/4} = \sqrt{\sqrt{A}}^{-1}$.]

(b) Use the techniques from Section 2.D.3 to show that $\sqrt{A} \succeq \sqrt{B}$ even when A is just positive *semi*definite. You may use the fact that the principal square root function is continuous on the set of positive semidefinite matrices.

∗∗3.C.7 Use computer software to solve the SDP from Remark 3.C.3 and thus fill in the missing entries in the matrix

$$\begin{bmatrix} 3 & * & * & -2 & * \\ * & 3 & * & -2 & * \\ * & * & 3 & * & * \\ -2 & -2 & * & 3 & -2 \\ * & * & * & -2 & 3 \end{bmatrix}$$

so as to make it positive semidefinite.

∗∗3.C.8 Find a feasible point of the semidefinite program from Example 3.C.8 that attains its optimal value $\|A\|$.

∗∗3.C.9 Let $A \in \mathcal{M}_{m,n}(\mathbb{C})$. Show that the following semidefinite program in the variable $X \in \mathcal{M}_n^H$ computes $\|A\|_F^2$:

minimize: $\operatorname{tr}(X)$
subject to: $\begin{bmatrix} I_m & A \\ A^* & X \end{bmatrix} \succeq O$
$X \succeq O$

[Hint: Either use duality and mimic Example 3.C.5, or use the Schur complement from Section 2.B.1.]

3.C.10 Let $A \in \mathcal{M}_{m,n}(\mathbb{C})$. Show that the following semidefinite program in the variables $X, Y \in \mathcal{M}_n^H$ computes $\sigma_1^4 + \cdots + \sigma_r^4$, where $\sigma_1, \ldots, \sigma_r$ are the non-zero singular values of A:

minimize: $\operatorname{tr}(Y)$
subject to: $\begin{bmatrix} I_m & A \\ A^* & X \end{bmatrix}, \begin{bmatrix} I_n & X \\ X & Y \end{bmatrix} \succeq O$
$X, Y \succeq O$

[Hint: Solve Exercise 3.C.9 first, which computes $\sigma_1^2 + \cdots + \sigma_r^2$, and maybe make use of Exercise 3.C.5.]

[Side note: A similar method can be used to construct semidefinite programs that compute $\sigma_1^p + \cdots + \sigma_r^p$ whenever p is an integer power of 2.]

[Side note: The quantity $\left(\sigma_1^p + \cdots + \sigma_r^p\right)^{1/p}$ is sometimes called the **Schatten p-norm** of A.]

3.C.11 Let $A \in \mathcal{M}_n^H$ be positive semidefinite. Show that the following semidefinite program in the variable $X \in \mathcal{M}_n(\mathbb{C})$ computes $\operatorname{tr}(\sqrt{A})$ (i.e., the sum of the square roots of the eigenvalues of A):

maximize: $\operatorname{Re}(\operatorname{tr}(X))$
subject to: $\begin{bmatrix} I & X \\ X^* & A \end{bmatrix} \succeq O$

[Hint: Use duality or Exercises 2.2.18 and 3.C.6.]

∗∗3.C.12 Let $C \in \mathcal{M}_n^H$ and let k be a positive integer such that $1 \leq k \leq n$. Consider the following semidefinite program in the variable $X \in \mathcal{M}_n^H$:

maximize: $\operatorname{tr}(CX)$
subject to: $\operatorname{tr}(X) = k$
$X \preceq I$
$X \succeq O$

(a) Construct the dual of this semidefinite program.

(b) Show that the optimal value of this semidefinite program is the sum of the k largest eigenvalues of C.
[Hint: Try to "eyeball" optimal solutions for the primal and dual problems.]

3.C.13 Let $A \in \mathcal{M}_{m,n}(\mathbb{C})$. Construct a semidefinite program for computing the sum of the k largest singular values of A.

[Side note: This sum is called the **Ky Fan k-norm** of A.]

[Hint: The results of Exercises 2.3.14 and 3.C.12 may be helpful.]

∗∗3.C.14 Recall that a linear map $\Phi : \mathcal{M}_n \to \mathcal{M}_m$ is called **decomposable** if there exist completely positive linear maps $\Psi_1, \Psi_2 : \mathcal{M}_n \to \mathcal{M}_m$ such that $\Phi = \Psi_1 + T \circ \Psi_2$, where T is the transpose map.

Construct a semidefinite program that determines whether or not a given matrix-valued linear map Φ is decomposable.

∗∗3.C.15 Use computer software and the semidefinite program from Exercise 3.C.14 to show that the map $\Phi : \mathcal{M}_2 \to \mathcal{M}_4$ from Exercise 3.A.21 is decomposable.

3.C.16 Use computer software and the semidefinite program from Exercise 3.C.14 to show that the Choi map Φ from Theorem 3.A.7 is not decomposable.

[Side note: We already demonstrated this claim "directly" in the proof of that theorem.]

∗∗3.C.17   Use computer software and the semidefinite program from Exercise 3.C.14 to show that the map $\Phi : \mathcal{M}_2 \to \mathcal{M}_4$ from Exercise 3.A.22 is not decomposable.

[Side note: This map Φ is positive, so it serves as an example to show that Theorem 3.A.8 does not hold when $mn = 8$.]

∗∗3.C.18   Recall from Section 3.B.3 that a biquadratic form is a function $q : \mathbb{R}^m \times \mathbb{R}^n \to \mathbb{R}$ that can be written in the form $q(\mathbf{x}, \mathbf{y}) = \mathbf{x}^T \Phi(\mathbf{y}\mathbf{y}^T)\mathbf{x}$ for some linear map $\Phi : \mathcal{M}_n(\mathbb{R}) \to \mathcal{M}_m(\mathbb{R})$.

Construct a semidefinite program that determines whether or not q can be written as a sum of squares of bilinear forms.

[Hint: Instead of checking $q(\mathbf{x}, \mathbf{y}) = \mathbf{x}^T \Phi(\mathbf{y}\mathbf{y}^T)\mathbf{x}$ for all $\mathbf{x}$ and $\mathbf{y}$, it suffices to choose finitely many vectors $\{\mathbf{x}_i\}$ and $\{\mathbf{y}_j\}$ so that the sets $\{\mathbf{x}_i\mathbf{x}_i^T\}$ and $\{\mathbf{y}_j\mathbf{y}_j^T\}$ span the set of symmetric matrices.]

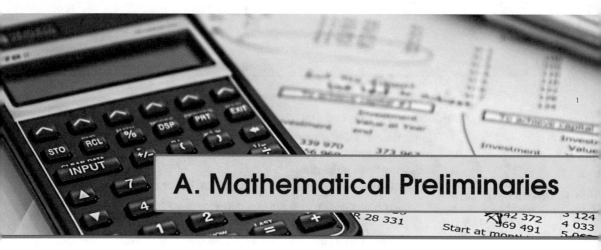

A. Mathematical Preliminaries

In this appendix, we present some of the miscellaneous bits of mathematical knowledge that are not topics of advanced linear algebra themselves, but are nevertheless useful and might be missing from the reader's toolbox. We also review some basic linear algebra from an introductory course that the reader may have forgotten.

A.1 Review of Introductory Linear Algebra

Here we review some of the basics of linear algebra that we expect the reader to be familiar with throughout the main text. We present some of the key results of introductory linear algebra, but we do not present any proofs or much context. For a more thorough presentation of these results and concepts, the reader is directed to an introductory linear algebra textbook like [Joh20].

A.1.1 Systems of Linear Equations

One of the first objects typically explored in introductory linear algebra is a system of linear equations (or a **linear system** for short), which is a collection of 1 or more linear equations (i.e., equations in which variables can only be added to each other and/or multiplied by scalars) in the same variables, like

$$\begin{aligned} y + 3z &= 3 \\ 2x + y - z &= 1 \\ x + y + z &= 2. \end{aligned} \qquad \text{(A.1.1)}$$

Linear systems can have zero, one, or infinitely many solutions, and one particularly useful method of finding these solutions (if they exist) starts by placing the coefficients of the linear system into a rectangular array called a **matrix**. For example, the matrix associated with the linear system (A.1.1) is

> The rows of a matrix represent the equations in the linear system, and its columns represent the variables as well as the coefficients on the right-hand side.

$$\left[\begin{array}{ccc|c} 0 & 1 & 3 & 3 \\ 2 & 1 & -1 & 1 \\ 1 & 1 & 1 & 2 \end{array} \right]. \qquad \text{(A.1.2)}$$

We then use a method called **Gaussian elimination** or **row reduction**, which works by using one of the three following **elementary row operations** to simplify this matrix as much as possible:

Multiplication. Multiplying row j by a non-zero scalar $c \in \mathbb{R}$, which we denote by cR_j.

© Springer Nature Switzerland AG 2021
N. Johnston, *Advanced Linear and Matrix Algebra*,
https://doi.org/10.1007/978-3-030-52815-7

Swap. Swapping rows i and j, which we denote by $R_i \leftrightarrow R_j$.

Addition. Replacing row i by $(\text{row i}) + c(\text{row j})$, which we denote by $R_i + cR_j$.

In particular, we can use these three elementary row operations to put any matrix into **reduced row echelon form** (**RREF**), which means that it has the following three properties:

*Any matrix that has the first two of these three properties is said to be in (not-necessarily-reduced) **row echelon form**.*

- all rows consisting entirely of zeros are below the non-zero rows,
- in each non-zero row, the first non-zero entry (called the **leading entry**) is to the left of any leading entries below it, and
- each leading entry equals 1 and is the only non-zero entry in its column.

For example, we can put the matrix (A.1.2) into reduced row echelon form via the following sequence of elementary row operations:

Every matrix can be converted into one, and only one, reduced row echelon form. However, there may be many different sequences of row operations that get there.

$$\begin{bmatrix} 0 & 1 & 3 & 3 \\ 2 & 1 & -1 & 1 \\ 1 & 1 & 1 & 2 \end{bmatrix} \xrightarrow{R_1 \leftrightarrow R_3} \begin{bmatrix} 1 & 1 & 1 & 2 \\ 2 & 1 & -1 & 1 \\ 0 & 1 & 3 & 3 \end{bmatrix} \xrightarrow{R_2 - 2R_1} \begin{bmatrix} 1 & 1 & 1 & 2 \\ 0 & -1 & -3 & -3 \\ 0 & 1 & 3 & 3 \end{bmatrix}$$

$$\xrightarrow{-R_2} \begin{bmatrix} 1 & 1 & 1 & 2 \\ 0 & 1 & 3 & 3 \\ 0 & 1 & 3 & 3 \end{bmatrix} \xrightarrow[R_3 - R_2]{R_1 - R_2} \begin{bmatrix} 1 & 0 & -2 & -1 \\ 0 & 1 & 3 & 3 \\ 0 & 0 & 0 & 0 \end{bmatrix}.$$

One of the useful features of reduced row echelon form is that the solutions of the corresponding linear system can be read off from it directly. For example, if we interpret the reduced row echelon form above as a linear system, the bottom row simply says $0x + 0y + 0z = 0$ (so we ignore it), the second row says that $y + 3z = 3$, and the top row says that $x - 2z = -1$. If we just move the "z" term in each of these equations over to the other side, we see that every solution of this linear system has $x = 2z - 1$ and $y = 3 - 3z$, where z is arbitrary (we thus call z a **free variable** and x and y **leading variables**).

A.1.2   Matrices as Linear Transformations

One of the central features of linear algebra is that there is a one-to-one correspondence between matrices and linear transformations. That is, every $m \times n$ matrix $A \in \mathcal{M}_{m,n}$ can be thought of as a function that sends $\mathbf{x} \in \mathbb{R}^n$ to the vector $A\mathbf{x} \in \mathbb{R}^m$. Conversely, every linear transformation $T : \mathbb{R}^n \to \mathbb{R}^m$ (i.e., function with the property that $T(\mathbf{x} + c\mathbf{y}) = T(\mathbf{x}) + cT(\mathbf{y})$ for all $\mathbf{x}, \mathbf{y} \in \mathbb{R}^n$ and $c \in \mathbb{R}$) can be represented by a matrix—there is a unique matrix $A \in \mathcal{M}_{m,n}$ with the property that $A\mathbf{x} = T(\mathbf{x})$ for all $\mathbf{x} \in \mathbb{R}^n$. We thus think of matrices and linear transformations as the "same thing".

In fact, vectors and matrices do not even need to have real entries. Their entries can come from any "field" (see the upcoming Appendix A.4).

Linear transformations are special for the fact that they are determined completely by how they act on the **standard basis vectors** $\mathbf{e}_1, \mathbf{e}_2, \ldots, \mathbf{e}_n$, which are the vectors with all entries equal to 0, except for a single entry equal to 1 in the location indicated by the subscript (e.g., in $\mathbb{R}^3$ there are three standard basis vectors: $\mathbf{e}_1 = (1,0,0)$, $\mathbf{e}_2 = (0,1,0)$, and $\mathbf{e}_3 = (0,0,1)$). In particular, $A\mathbf{e}_1, A\mathbf{e}_2, \ldots, A\mathbf{e}_n$ are exactly the n columns of A, and those n vectors form the sides of the parallelogram/ parallelepiped/hyperparallelepiped that the unit square/cube/ hypercube is mapped to by A (see Figure A.1).

A linear transformation is also completely determined by how it acts on any other basis of $\mathbb{R}^n$.

In particular, linear transformations act "uniformly" in the sense that they send a unit square/cube/hypercube grid to a parallelogram/parallelepiped/hyperparallelepiped grid without distorting any particular region of space more than other regions of space.

Squares are mapped to parallelograms in $\mathbb{R}^2$, cubes are mapped to parallelepipeds in $\mathbb{R}^3$, and so on.

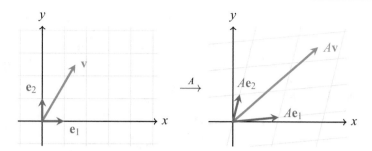

Figure A.1: A matrix $A \in \mathcal{M}_2$ acts as a linear transformation on $\mathbb{R}^2$ that transforms a square grid with sides $\mathbf{e}_1$ and $\mathbf{e}_2$ into a grid made up of parallelograms with sides $A\mathbf{e}_1$ and $A\mathbf{e}_2$ (i.e., the columns of A). Importantly, A preserves which cell of the grid each vector is in (in this case, $\mathbf{v}$ is in the 2nd square to the right, 3rd up, and $A\mathbf{v}$ is similarly in the 2nd parallelogram in the direction of $A\mathbf{e}_1$ and 3rd in the direction of $A\mathbf{e}_2$.

A.1.3 The Inverse of a Matrix

In the $n = 1$ case, the inverse of a 1×1 matrix (i.e., scalar) a is just $1/a$.

The **inverse** of a square matrix $A \in \mathcal{M}_n$ is a matrix $A^{-1} \in \mathcal{M}_n$ for which $AA^{-1} = A^{-1}A = I$ (the identity matrix). The inverse of a matrix is unique when it exists, but not all matrices have inverses (even in the $n = 1$ case, the scalar 0 does not have an inverse). The inverse of a matrix can be computed by using Gaussian elimination to row-reduce the block matrix $[\,A \mid I\,]$ into its reduced row echelon form $[\,I \mid A^{-1}\,]$. Furthermore, if the reduced row echelon form of $[\,A \mid I\,]$ has anything other than I in the left block, then A is not invertible.

We show in Remark 1.2.2 that one-sided inverses are not necessarily two-sided in the infinite-dimensional case.

A one-sided inverse of a matrix is automatically two-sided (that is, if either of the equations $AB = I$ or $BA = I$ holds then the other necessarily holds as well—we can deduce that $B = A^{-1}$ based on just one of the two defining equations). We can get some intuition for why this fact is true by thinking geometrically—if $AB = I$ then, as linear transformations, A simply undoes whatever B does to $\mathbb{R}^n$, and it perhaps seems believable that B similarly undoes whatever A does (see Figure A.2).

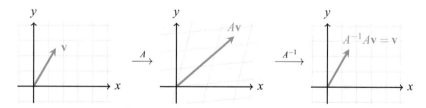

Figure A.2: As a linear transformation, A^{-1} undoes what A does to vectors in $\mathbb{R}^n$. That is, $A^{-1}A\mathbf{v} = \mathbf{v}$ for all $\mathbf{v} \in \mathbb{R}^n$ (and $AA^{-1}\mathbf{v} = \mathbf{v}$ too).

Row reducing a matrix is equivalent to multiplication on the left by an invertible matrix in the sense that $B \in \mathcal{M}_{m,n}$ can be obtained from $A \in \mathcal{M}_{m,n}$ via a sequence of elementary row operations if and only if there is an invertible matrix $P \in \mathcal{M}_m$ such that $B = PA$. The fact that every matrix can be row-reduced to a (unique) matrix in reduced row echelon form is thus equivalent to the following fact:

> (!) For every $A \in \mathcal{M}_{m,n}$, there exists an invertible $P \in \mathcal{M}_m$ such that $A = PR$, where $R \in \mathcal{M}_{m,n}$ is the RREF of A.

It is also useful to be familiar with the numerous different ways in which invertible matrices can be characterized:

Theorem A.1.1

**The Invertible
Matrix Theorem**

Suppose $A \in \mathcal{M}_n$. The following are equivalent:

a) A is invertible.
b) A^T is invertible.
c) The reduced row echelon form of A is I.
d) The linear system $A\mathbf{x} = \mathbf{b}$ has a solution for all $\mathbf{b} \in \mathbb{R}^n$.
e) The linear system $A\mathbf{x} = \mathbf{b}$ has a unique solution for all $\mathbf{b} \in \mathbb{R}^n$.
f) The linear system $A\mathbf{x} = \mathbf{0}$ has a unique solution ($\mathbf{x} = \mathbf{0}$).
g) The columns of A are linearly independent.
h) The columns of A span $\mathbb{R}^n$.
i) The columns of A form a basis of $\mathbb{R}^n$.

The final four characterizations here concern rank, determinants, and eigenvalues, which are the subjects of the next three subsections.

j) The rows of A are linearly independent.
k) The rows of A span $\mathbb{R}^n$.
l) The rows of A form a basis of $\mathbb{R}^n$.
m) $\text{rank}(A) = n$ (i.e., $\text{range}(A) = \mathbb{R}^n$).
n) $\text{nullity}(A) = 0$ (i.e., $\text{null}(A) = \{\mathbf{0}\}$).
o) $\det(A) \neq 0$.
p) All eigenvalues of A are non-zero.

A.1.4   Range, Rank, Null Space, and Nullity

The **range** of a matrix $A \in \mathcal{M}_{m,n}$ is the subspace of $\mathbb{R}^m$ consisting of all possible output vectors of A (when we think of it as a linear transformation), and its **null space** is the subspace of $\mathbb{R}^n$ consisting of all vectors that are sent to the zero vector:

$$\text{range}(A) \stackrel{\text{def}}{=} \{A\mathbf{v} : \mathbf{v} \in \mathbb{R}^n\} \quad \text{and} \quad \text{null}(A) \stackrel{\text{def}}{=} \{\mathbf{v} \in \mathbb{R}^n : A\mathbf{v} = \mathbf{0}\}.$$

The **rank** and **nullity** of A are the dimensions of its range and null space, respectively. The following theorem summarizes many of the important properties of the range, null space, rank, and nullity that are typically encountered in introductory linear algebra courses.

Theorem A.1.2

**Properties of Range,
Rank, Null Space
and Nullity**

Suppose $A \in \mathcal{M}_{m,n}$ has columns $\mathbf{v}_1, \mathbf{v}_2, \ldots, \mathbf{v}_n$. Then:

a) $\text{range}(A) = \text{span}(\mathbf{v}_1, \mathbf{v}_2, \ldots, \mathbf{v}_n)$.
b) $\text{range}(AA^*) = \text{range}(A)$.
c) $\text{null}(A^*A) = \text{null}(A)$.
d) $\text{rank}(AA^*) = \text{rank}(A^*A) = \text{rank}(A^*) = \text{rank}(A)$.

Property (e) is sometimes called the "rank-nullity theorem".

e) $\text{rank}(A) + \text{nullity}(A) = n$.
f) If $B \in \mathcal{M}_{n,p}$ then $\text{range}(AB) \subseteq \text{range}(A)$.
g) If $B \in \mathcal{M}_{m,n}$ then $\text{rank}(A+B) \leq \text{rank}(A) + \text{rank}(B)$.
h) If $B \in \mathcal{M}_{n,p}$ then $\text{rank}(AB) \leq \min\{\text{rank}(A), \text{rank}(B)\}$.
i) If $B \in \mathcal{M}_{n,p}$ has $\text{rank}(B) = n$ then $\text{rank}(AB) = \text{rank}(A)$.
j) $\text{rank}(A)$ equals that number of non-zero rows in any row echelon form of A.

We saw in Theorem A.1.1 that square matrices with maximal rank are exactly the ones that are invertible. Intuitively, we can think of the rank of a matrix as a rough measure of "how close to invertible" it is, so matrices with small rank are in some sense the "least invertible" matrices out there. Of particular interest are matrices with rank 1, which are exactly the ones that can be written in the form $\mathbf{v}\mathbf{w}^T$, where $\mathbf{v}$ and $\mathbf{w}$ are non-zero column vectors. More generally, a rank-r matrix can be written as a sum of r matrices of this form, but not fewer:

Theorem A.1.3	Suppose $A \in \mathcal{M}_{m,n}$. Then the smallest integer r for which there exist sets

Theorem A.1.3

Rank-One Sum Decomposition

This theorem works just fine if we replace $\mathbb{R}$ by an arbitrary field $\mathbb{F}$ (see Appendix A.4).

Suppose $A \in \mathcal{M}_{m,n}$. Then the smallest integer r for which there exist sets of vectors $\{\mathbf{v}_j\}_{j=1}^r \subset \mathbb{R}^m$ and $\{\mathbf{w}_j\}_{j=1}^r \subset \mathbb{R}^n$ with

$$A = \sum_{j=1}^r \mathbf{v}_j \mathbf{w}_j^T$$

is exactly $r = \text{rank}(A)$. Furthermore, the sets $\{\mathbf{v}_j\}_{j=1}^r$ and $\{\mathbf{w}_j\}_{j=1}^r$ can be chosen to be linearly independent.

A.1.5 Determinants and Permutations

The **determinant** of a matrix $A \in \mathcal{M}_n$, which we denote by $\det(A)$, is the area (or volume, or hypervolume, depending on the dimension n) of the output of the unit square/cube/hypercube after it is acted upon by A (see Figure A.3). In other words, it measures how much A expands space when acting as a linear transformation—it is the ratio

Actually, the determinant of a matrix can also be negative—this happens if A flips the orientation of space (i.e., reflects $\mathbb{R}^n$ through some $(n-1)$-dimensional hyperplane).

$$\frac{\text{volume of output region}}{\text{volume of input region}}.$$

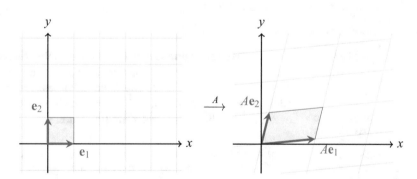

Figure A.3: A 2×2 matrix A stretches the unit square (with sides $\mathbf{e}_1$ and $\mathbf{e}_2$) into a parallelogram with sides $A\mathbf{e}_1$ and $A\mathbf{e}_2$ (the columns of A). The determinant of A is the area of this parallelogram.

This geometric interpretation of the determinant leads immediately to the fact that it is **multiplicative**—stretching space by the product of two matrices (i.e., the composition of two linear transformations) has the same effect as stretching by the first one and then stretching by the second one:

Similarly, $\det(I) = 1$ since the identity matrix does not stretch or squish space at all.

> **!** If $A, B \in \mathcal{M}_n$ then $\det(AB) = \det(A)\det(B)$.

There are several methods for actually computing the determinant of a matrix, but for our purposes the most useful method requires some brief background knowledge of permutations. A **permutation** is a function $\sigma : \{1, 2, \ldots, n\} \to \{1, 2, \ldots, n\}$ with the property that $\sigma(i) \neq \sigma(j)$ whenever $i \neq j$

(i.e., σ can be thought of as shuffling around the integers $1, 2, \ldots, n$). A **transposition** is a permutation with two distinct values i, j such that $\sigma(i) = j$ and $\sigma(j) = i$, and $\sigma(k) = k$ for all $k \neq i, j$ (i.e., σ swaps two integers and leaves the rest alone). We denote the set of all permutations acting on $\{1, 2, \ldots, n\}$ by S_n.

S_n is often called the **symmetric group**.

Every permutation can be written as a composition of transpositions. We define the **sign** of a permutation σ to be the quantity

$$\mathrm{sgn}(\sigma) \stackrel{\mathrm{def}}{=} (-1)^k, \text{ if } \sigma \text{ can be written as a composition of } k \text{ transpositions.}$$

Importantly, even though every permutation can be written as a composition of transpositions in numerous different ways, the number of transpositions used always has the same parity (i.e., it is either always even or always odd), so the sign of a permutation is well-defined.

With these details out of the way, we can now present a formula for computing the determinant. Note that we state this formula as a theorem, but many sources actually use it as the *definition* of the determinant:

Theorem A.1.4

Determinants via Permutations

Suppose $A \in \mathcal{M}_n$. Then $\det(A) = \displaystyle\sum_{\sigma \in S_n} \mathrm{sgn}(\sigma) a_{\sigma(1),1} a_{\sigma(2),2} \cdots a_{\sigma(n),n}$.

This theorem is typically not used for actual calculations in practice, as much faster methods of computing the determinant are known. However, this formula is useful for the fact that we can use it to quickly derive several useful properties of the determinant. For example, swapping two columns of a matrix has the effect of swapping the sign of each permutation in the sum in Theorem A.1.4, which leads to the following fact:

We can replace columns with rows in both of these facts, since $\det(A^T) = \det(A)$ for all $A \in \mathcal{M}_n$.

> (!) If $B \in \mathcal{M}_n$ is obtained from $A \in \mathcal{M}_n$ by swapping two of its columns then $\det(B) = -\det(A)$.

Similarly, the determinant is **multilinear** in the columns of the matrix it acts on: it acts linearly on each column individually, as long as all other columns are fixed. That is, much like we can "split up" linear transformations over vector addition and scalar multiplication, we can similarly split up the determinant over vector addition and scalar multiplication in a single column of a matrix:

These properties of the determinant can be derived from its geometric interpretation as well, but it's somewhat messy to do so.

> (!) For all matrices $A = [\, \mathbf{a}_1 \mid \cdots \mid \mathbf{a}_n \,] \in \mathcal{M}_n$, all $\mathbf{v}, \mathbf{w} \in \mathbb{R}^n$, and all scalars $c \in \mathbb{R}$, it is the case that
>
> $$\det([\, \mathbf{a}_1 \mid \cdots \mid \mathbf{v} + c\mathbf{w} \mid \cdots \mid \mathbf{a}_n \,])$$
> $$= \det([\, \mathbf{a}_1 \mid \cdots \mid \mathbf{v} \mid \cdots \mid \mathbf{a}_n \,]) + c \cdot \det([\, \mathbf{a}_1 \mid \cdots \mid \mathbf{w} \mid \cdots \mid \mathbf{a}_n \,]).$$

Finally, if $A \in \mathcal{M}_n$ is upper triangular then the only permutation σ for which $a_{\sigma(1),1} a_{\sigma(2),2} \cdots a_{\sigma(n),n} \neq 0$ is the identity permutation (i.e., the permutation for which $\sigma(j) = j$ for all $1 \leq j \leq n$), since every other permutation results in at least one factor in the product $a_{\sigma(1),1} a_{\sigma(2),2} \cdots a_{\sigma(n),n}$ coming from the strictly lower triangular portion of A. This observation gives us the following simple formula for the determinant of an upper triangular matrix:

This same result holds for lower triangular matrices as well.

> (!) If $A \in \mathcal{M}_n$ is upper triangular then $\det(A) = a_{1,1} \cdot a_{2,2} \cdots a_{n,n}$.

A similar argument shows the slightly more general fact that if A is *block upper triangular* with diagonal blocks $A_1, A_2, \ldots, A_k$ then $\det(A) = \displaystyle\prod_{j=1}^{k} \det(A_j)$.

A.1.6 Eigenvalues and Eigenvectors

If we allowed $\mathbf{v} = \mathbf{0}$ as an eigenvector then every scalar λ would be an eigenvalue corresponding to it.

If $A \in \mathcal{M}_n$, $v \neq \mathbf{0}$ is a vector, λ is a scalar, and $A\mathbf{v} = \lambda \mathbf{v}$, then we say that $\mathbf{v}$ is an **eigenvector** of A with corresponding **eigenvalue** λ. Geometrically, this means that A stretches $\mathbf{v}$ by a factor of λ, but does not rotate it at all (see Figure A.4). The set of all eigenvectors corresponding to a particular eigenvalue (together with the zero vector) is always a subspace of $\mathbb{R}^n$, which we call the **eigenspace** corresponding to that eigenvalue. The dimension of an eigenspace is called the **geometric multiplicity** of the corresponding eigenvalue.

This matrix *does* change the direction of any vector that is not on one of the two lines displayed here.

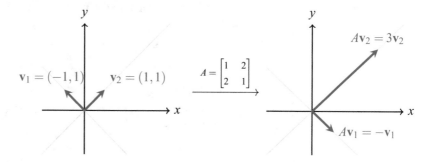

Figure A.4: Matrices do not change the line on which any of their eigenvectors lie, but rather just scale them by the corresponding eigenvalue. The matrix displayed here has eigenvectors $\mathbf{v}_1 = (-1, 1)$ and $\mathbf{v}_2 = (1, 1)$ with corresponding eigenvalues -1 and 3, respectively.

The standard method of computing a matrix's eigenvalues is to construct its **characteristic polynomial** $p_A(\lambda) = \det(A - \lambda I)$. This function p_A really is a polynomial in λ by virtue of the permutation formula for the determinant (i.e., Theorem A.1.4). Furthermore, the degree of p_A is n (the size of A), so it has at most n real roots and exactly n complex roots counting multiplicity (see the upcoming Theorem A.3.1). These roots are the eigenvalues of A, and the eigenvectors $\mathbf{v}$ that they correspond to can be found by solving the linear system $(A - \lambda I)\mathbf{v} = \mathbf{0}$ for each eigenvalue λ.

Example A.1.1

Computing the Eigenvalues and Eigenvectors of a Matrix

Compute all of the eigenvalues and corresponding eigenvectors of the matrix $A = \begin{bmatrix} 1 & 2 \\ 5 & 4 \end{bmatrix}$.

Solution:

To find the eigenvalues of A, we first compute the characteristic polynomial $p_A(\lambda) = \det(A - \lambda I)$:

For a 2×2 matrix, the determinant is simply

$$\det\left(\begin{bmatrix} a & b \\ c & d \end{bmatrix}\right) = ad - bc.$$

$$p_A(\lambda) = \det(A - \lambda I) = \det\left(\begin{bmatrix} 1-\lambda & 2 \\ 5 & 4-\lambda \end{bmatrix}\right)$$

$$= (1-\lambda)(4-\lambda) - 10 = \lambda^2 - 5\lambda - 6.$$

Setting this determinant equal to 0 then gives

$$\lambda^2 - 5\lambda - 6 = 0 \iff (\lambda + 1)(\lambda - 6) = 0$$

$$\iff \lambda = -1 \quad \text{or} \quad \lambda = 6,$$

so the eigenvalues of A are $\lambda = -1$ and $\lambda = 6$. To find the eigenvectors corresponding to these eigenvalues, we solve the linear systems $(A + I)\mathbf{v} = \mathbf{0}$ and $(A - 6I)\mathbf{v} = \mathbf{0}$, respectively:

$\lambda = -1$: In this case, we want to solve the linear system $(A - \lambda I)\mathbf{v} = (A + I)\mathbf{v} = \mathbf{0}$, which we can write explicitly as follows:

$$2v_1 + 2v_2 = 0$$
$$5v_1 + 5v_2 = 0.$$

To solve this linear system, we use Gaussian elimination as usual:

$$\left[\begin{array}{cc|c} 2 & 2 & 0 \\ 5 & 5 & 0 \end{array}\right] \xrightarrow{R_2 - \frac{5}{2}R_1} \left[\begin{array}{cc|c} 2 & 2 & 0 \\ 0 & 0 & 0 \end{array}\right].$$

It follows that v_2 is a free variable and v_1 is a leading variable with $v_1 = -v_2$. The eigenvectors corresponding to the eigenvalue $\lambda = -1$ are thus the non-zero vectors of the form $\mathbf{v} = (-v_2, v_2) = v_2(-1, 1)$.

$\lambda = 6$: Similarly, we now want to solve the linear system $(A - \lambda I)\mathbf{v} = (A - 6I)\mathbf{v} = \mathbf{0}$, which we can do as follows:

$$\left[\begin{array}{cc|c} -5 & 2 & 0 \\ 5 & -2 & 0 \end{array}\right] \xrightarrow{R_2 + R_1} \left[\begin{array}{cc|c} -5 & 2 & 0 \\ 0 & 0 & 0 \end{array}\right].$$

By multiplying $(2/5, 1)$ by 5, we could also say that the eigenvectors here are the multiples of $(2, 5)$, which is a slightly cleaner answer.

It follows that v_2 is a free variable and v_1 is a leading variable with $v_1 = 2v_2/5$. The eigenvectors corresponding to the eigenvalue $\lambda = 6$ are thus the non-zero vectors of the form $\mathbf{v} = (2v_2/5, v_2) = v_2(2/5, 1)$.

The multiplicity of an eigenvalue λ as a root of the characteristic polynomial is called its **algebraic multiplicity**, and the sum of all algebraic multiplicities of eigenvalues of an $n \times n$ matrix is no greater than n (and it exactly equals n if we consider complex eigenvalues). The following remarkable (and non-obvious) fact guarantees that the sum of geometric multiplicities is similarly no larger than n:

! The geometric multiplicity of an eigenvalue is never larger than its algebraic multiplicity.

A.1.7   Diagonalization

One of the primary reasons that eigenvalues and eigenvectors are of interest is that they let us **diagonalize** a matrix. That is, they give us a way of decomposing a matrix $A \in \mathcal{M}_n$ into the form $A = PDP^{-1}$, where P is invertible and D is diagonal. If the entries of P and D can be chosen to be real, we say that A is **diagonalizable over** $\mathbb{R}$. However, some real matrices A can only be diagonalized if we allow P and D to have complex entries (see the upcoming discussion of complex numbers in Appendix A.3). In that case, we say that A is **diagonalizable over** $\mathbb{C}$ (but not over $\mathbb{R}$).

Theorem A.1.5

Characterization of Diagonalizability

Suppose $A \in \mathcal{M}_n$. The following are equivalent:

a) A is diagonalizable over $\mathbb{R}$ (or $\mathbb{C}$).

b) There exists a basis of $\mathbb{R}^n$ (or $\mathbb{C}^n$) consisting of eigenvectors of A.

c) The sum of geometric multiplicities of the real (or complex) eigenvalues of A is n.

Furthermore, in any diagonalization $A = PDP^{-1}$, the eigenvalues of A are the diagonal entries of D and the corresponding eigenvectors are the columns of P in the same order.

To get a feeling for why diagonalizations are useful, notice that computing a large power of a matrix directly is quite cumbersome, as matrix multiplication itself is an onerous process, and repeating it numerous times only makes it worse. However, once we have diagonalized a matrix we can compute an arbitrary power of it via just two matrix multiplications, since

$$A^k = \underbrace{\left(PD\overbrace{P^{-1}}\right)\left(PD\overbrace{P^{-1}}\right)\left(PDP^{-1}\right)}_{k\,\text{times}} \cdots = PD^kP^{-1},$$

and D^k is trivial to compute (for diagonal matrices, matrix multiplication is the same as entrywise multiplication).

Example A.1.2

Diagonalizing a Matrix

Diagonalize the matrix $A = \begin{bmatrix} 1 & 2 \\ 5 & 4 \end{bmatrix}$ and then compute A^{314}.

Solution:

We showed in Example A.1.1 that this matrix has eigenvalues $\lambda_1 = -1$ and $\lambda_2 = 6$ corresponding to the eigenvectors $\mathbf{v}_1 = (-1, 1)$ and $\mathbf{v}_2 = (2, 5)$, respectively. Following the suggestion of Theorem A.1.5, we stick these eigenvalues along the diagonal of a diagonal matrix D, and the corresponding eigenvectors as columns into a matrix P in the same order:

We could have also chosen $\mathbf{v}_2 = (2/5, 1)$, but our choice here is prettier. Which multiple of each eigenvector we choose does not matter.

$$D = \begin{bmatrix} \lambda_1 & 0 \\ 0 & \lambda_2 \end{bmatrix} = \begin{bmatrix} -1 & 0 \\ 0 & 6 \end{bmatrix} \quad \text{and} \quad P = \begin{bmatrix} \mathbf{v}_1 \mid \mathbf{v}_2 \end{bmatrix} = \begin{bmatrix} -1 & 2 \\ 1 & 5 \end{bmatrix}.$$

It is straightforward to check that P is invertible, so Theorem A.1.5 tells us that A is diagonalized by this D and P.

To compute A^{314}, we first compute P^{-1} to be

The inverse of a 2×2 matrix is simply

$$\begin{bmatrix} a & b \\ c & d \end{bmatrix}^{-1} =$$

$$\frac{1}{\det(A)} \begin{bmatrix} d & -b \\ -c & a \end{bmatrix}.$$

$$P^{-1} = \frac{1}{7} \begin{bmatrix} -5 & 2 \\ 1 & 1 \end{bmatrix}.$$

We can then compute powers of A via powers of the diagonal matrix D in this diagonalization:

Since 314 is even,
$(-1)^{314} = 1$.

$$A^{314} = PD^{314}P^{-1} = \frac{1}{7}\begin{bmatrix} -1 & 2 \\ 1 & 5 \end{bmatrix}\begin{bmatrix} (-1)^{314} & 0 \\ 0 & 6^{314} \end{bmatrix}\begin{bmatrix} -5 & 2 \\ 1 & 1 \end{bmatrix}$$

$$= \frac{1}{7}\begin{bmatrix} 5+2\cdot6^{314} & -2+2\cdot6^{314} \\ -5+5\cdot6^{314} & 2+5\cdot6^{314} \end{bmatrix}.$$

We close this section with a reminder of a useful connection between diagonalizability of a matrix and the multiplicities of its eigenvalues:

(!) If a matrix is diagonalizable then, for each of its eigenvalues, the algebraic and geometric multiplicities coincide.

A.2   Polynomials and Beyond

A (single-variable) **polynomial** is a function $f : \mathbb{R} \to \mathbb{R}$ of the form

$$f(x) = a_p x^p + a_{p-1}x^{p-1} + \cdots + a_2 x^2 + a_1 x + a_0,$$

It's also OK to consider polynomials $f : \mathbb{C} \to \mathbb{C}$ with coefficients in $\mathbb{C}$, or even polynomials $f : \mathbb{F} \to \mathbb{F}$, where $\mathbb{F}$ is any field (see Appendix A.4).

where $a_0, a_1, a_2, \ldots, a_{p-1}, a_p \in \mathbb{R}$ are constants (called the **coefficients** of f). The highest power of x appearing in the polynomial is called its **degree** (so, for example, the polynomial f above has degree p).

More generally, a **multivariate polynomial** is a function $f : \mathbb{R}^n \to \mathbb{R}$ that can be written as a linear combination of products of powers of the n input variables. That is, there exist scalars $\{a_{j_1,\ldots,j_n}\}$, only finitely many of which are non-zero, such that

$$f(x_1,\ldots,x_n) = \sum_{j_1,\ldots,j_n} a_{j_1,\ldots,j_n} x_1^{j_1} \cdots x_n^{j_n} \quad \text{for all} \quad x_1,\ldots,x_n \in \mathbb{R}.$$

The degree of a multivariate polynomial is the largest sum $j_1 + \cdots + j_n$ of exponents in any of its terms. For example, the polynomials

$$f(x,y) = 3x^2y - 2y^2 + xy + x - 3 \quad \text{and} \quad g(x,y,z) = 4x^3yz^4 + 2xyz^2 - 4y$$

have degrees equal to $2+1 = 3$ and $3+1+4 = 8$, respectively.

A.2.1   Monomials, Binomials and Multinomials

Two of the simplest types of polynomials are **monomials**, which consist of just a single term (like $3x^2$ or $-4x^2y^3z$), and **binomials**, which consist of two terms added together (like $3x^2 + 4x$ or $2xy^2 - 7x^3y$). If we take a power of a binomial like $(x+y)^p$ then it is often useful to know what the coefficient of each term in the resulting polynomial is once everything is expanded out. For example, it is straightforward to check that

In other words, a binomial is a sum of two monomials.

$$(x+y)^2 = x^2 + 2xy + y^2 \quad \text{and}$$
$$(x+y)^3 = x^3 + 3x^2y + 3xy^2 + y^3.$$

More generally, we can notice that when we compute

$$(x+y)^p = \underbrace{(x+y)(x+y)\cdots(x+y)}_{p \text{ times}}$$

by repeatedly applying distributivity of multiplication over addition, we get 2^p terms in total—we can imagine constructing these terms by choosing either x or y in each copy of $(x+y)$, so we can make a choice between 2 objects p times.

However, when we expand $(x+y)^p$ in this way, many of the resulting terms are duplicates of each other due to commutativity of multiplication. For example, $x^2y = xyx = yx^2$. To count the number of times that each term is repeated, we notice that for each integer $0 \le k \le p$, the number of times that $x^{p-k}y^k$ is repeated equals the number of ways that we can choose y from k of the p factors of the form $(x+y)$ (and thus x from the other $p-k$ factors). This quantity is denoted by $\binom{p}{k}$ and is given by the formula

p! is called "p factorial".

$$\binom{p}{k} = \frac{p!}{k!\,(p-k)!}, \quad \text{where} \quad p! = p \cdot (p-1) \cdots 3 \cdot 2 \cdot 1.$$

This quantity $\binom{p}{k}$ is called a **binomial coefficient** (due to this connection with binomials) and is read as "p choose k". The following theorem states our observations about powers of binomials and binomial coefficients a bit more formally.

Theorem A.2.1

Binomial Theorem

Suppose $p \ge 0$ is an integer and $x, y \in \mathbb{R}$. Then

$$(x+y)^p = \sum_{k=0}^{p} \binom{p}{k} x^{p-k} y^k.$$

Example A.2.1

Using the Binomial Theorem

Expand the polynomial $(x+2y)^4$.

Solution:

We start by computing the binomial coefficients that we will need:

Binomial coefficients are symmetric: $\binom{p}{k} = \binom{p}{p-k}$.

$$\binom{4}{0} = 1, \quad \binom{4}{1} = 4, \quad \binom{4}{2} = 6, \quad \binom{4}{3} = 4, \quad \binom{4}{4} = 1.$$

Now that we have these quantities, we can just plug them into the binomial theorem (but we are careful to replace y in the theorem with $2y$):

As a strange edge case, we note that $0! = 1$.

$$\begin{aligned}
(x+2y)^4 &= \sum_{k=0}^{4} \binom{4}{k} x^{4-k}(2y)^k \\
&= x^4(2y)^0 + 4x^3(2y)^1 + 6x^2(2y)^2 + 4x^1(2y)^3 + x^0(2y)^4 \\
&= x^4 + 8x^3y + 24x^2y^2 + 32xy^3 + 16y^4.
\end{aligned}$$

More generally, we can use a similar technique to come up with an explicit formula for the coefficients of powers of *any* polynomials—not just binomials. In particular, when expanding out an expression like

$$(x_1 + x_2 + \cdots + x_n)^p = \underbrace{(x_1 + x_2 + \cdots + x_n)\cdots(x_1 + x_2 + \cdots + x_n)}_{p \text{ times}},$$

we find that the number of times that a particular term $x_1^{k_1} x_2^{k_2} \cdots x_n^{k_n}$ occurs equals the number of ways that we can choose x_1 from the p factors a total of k_1 times, x_2 from the p factors a total of k_2 times, and so on. This quantity is called a **multinomial coefficient**, and it is given by

Each of the p factors must have *some* x_j chosen from it, so $k_1 + k_2 + \cdots + k_n = p$.

$$\binom{p}{k_1, k_2, \ldots, k_n} = \frac{p!}{k_1! k_2! \cdots k_n!}.$$

These observations lead to the following generalization of the binomial theorem to multinomials (i.e., polynomials in general).

Theorem A.2.2

Multinomial Theorem

Suppose $p \geq 0$ and $n \geq 1$ are integers and $x_1, x_2, \ldots, x_n \in \mathbb{R}$. Then

$$(x_1 + x_2 + \cdots + x_n)^p = \sum_{k_1 + \cdots + k_n = p} \binom{p}{k_1, k_2, \ldots, k_n} x_1^{k_1} x_2^{k_2} \cdots x_n^{k_n}.$$

Example A.2.2

Using the Multinomial Theorem

Expand the polynomial $(x + 2y + 3z)^3$.

Solution:
We start by computing the multinomial coefficients that we will need:

$$\binom{3}{1,1,1} = 6, \quad \binom{3}{3,0,0} = \binom{3}{0,3,0} = \binom{3}{0,0,3} = 1, \quad \text{and}$$

$$\binom{3}{0,1,2} = \binom{3}{1,0,2} = \binom{3}{1,2,0}$$

$$= \binom{3}{0,2,1} = \binom{3}{2,0,1} = \binom{3}{2,1,0} = 3.$$

Expanding out $(x_1 + \cdots + x_n)^p$ results in $\binom{n+p-1}{p}$ terms in general (this fact can be proved using the method of Remark 3.1.2). In this example we have $n = p = 3$, so we get $\binom{5}{3} = 10$ terms.

Now that we have these quantities, we can just plug them into the multinomial theorem (but we replace y in the theorem with $2y$ and z with $3z$):

$$(x + 2y + 3z)^3 = \sum_{k+\ell+m=3} \binom{3}{k, \ell, m} x^k (2y)^\ell (3z)^m$$

$$= 6x^1 (2y)^1 (3z)^1$$

$$+ x^3 (2y)^0 (3z)^0 + x^0 (2y)^3 (3z)^0 + x^0 (2y)^0 (3z)^3$$

$$+ 3x^0 (2y)^1 (3z)^2 + 3x^1 (2y)^0 (3z)^2 + 3x^1 (2y)^2 (3z)^0$$

$$+ 3x^0 (2y)^2 (3z)^1 + 3x^2 (2y)^0 (3z)^1 + 3x^2 (2y)^1 (3z)^0$$

$$= 36xyz + x^3 + 8y^3 + 27z^3$$

$$+ 54yz^2 + 27xz^2 + 12xy^2 + 36y^2 z + 9x^2 z + 6x^2 y.$$

A.2.2   Taylor Polynomials and Taylor Series

Since we know so much about polynomials, it is often easier to approximate a function via a polynomial and then analyze that polynomial than it is to directly analyze the function that we are actually interested in. The degree-p polynomial $T_p : \mathbb{R} \to \mathbb{R}$ that best approximates a function $f : \mathbb{R} \to \mathbb{R}$ near some value $x = a$

is called its **degree-p Taylor polynomial**, and it has the form

*We sometimes say that these Taylor polynomials are **centered** at a.*

$$T_p(x) = \sum_{n=0}^{p} \frac{f^{(n)}(a)}{n!}(x-a)^n$$

$$= f(a) + f'(a)(x-a) + \frac{f''(a)}{2}(x-a)^2 + \cdots + \frac{f^{(p)}(a)}{p!}(x-a)^p,$$

Be somewhat careful with the $n=0$ term: $(x-a)^0 = 1$ for all x.

as long as f has p derivatives (i.e., $f^{(p)}$ exists).

For example, the lowest-degree polynomials that best approximate $f(x) = \sin(x)$ near $x = 0$ are

$$T_1(x) = T_2(x) = x,$$

$$T_3(x) = T_4(x) = x - \frac{x^3}{3!},$$

$$T_5(x) = T_6(x) = x - \frac{x^3}{3!} + \frac{x^5}{5!}, \quad \text{and}$$

$$T_7(x) = T_8(x) = x - \frac{x^3}{3!} + \frac{x^5}{5!} - \frac{x^7}{7!},$$

which are graphed in Figure A.5. In particular, notice that T_1 is simply the tangent line at the point $(0,0)$, and T_3, T_5, and T_7 provide better and better approximations of $f(x) = \sin(x)$.

In general, the graph of T_0 is a horizontal line going through $(a, f(a))$ and the graph of T_1 is the tangent line going through $(a, f(a))$.

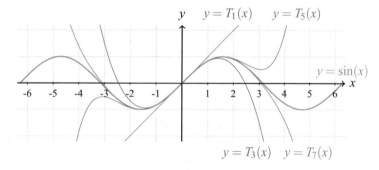

Figure A.5: The graphs of the first few Taylor polynomials of $f(x) = \sin(x)$. As the degree of the Taylor polynomial increases, the approximation gets better.

To get a rough feeling for why Taylor polynomials provide the best approximation of a function near a point, notice that the first p derivatives of $T_p(x)$ and of $f(x)$ agree with each other at $x = a$, so the local behavior of these functions is very similar. The following theorem pins this idea down more precisely and says that the difference between $f(x)$ and $T_p(x)$ behaves like $(x-a)^{p+1}$, which is smaller than any degree-p polynomial when x is sufficiently close to a:

Theorem A.2.3

Taylor's Theorem

Suppose $f : \mathbb{R} \to \mathbb{R}$ is a $(p+1)$-times differentiable function. For all x and a, there is a number c between x and a such that

$$f(x) - T_p(x) = \frac{f^{(p+1)}(c)}{(p+1)!}(x-a)^{p+1}.$$

A **Taylor series** is what we get if we consider the limit of the Taylor polynomials as p goes to infinity. The above theorem tells us that, as long as

$f^{(p+1)}(x)$ does not get too large near $x = a$, this limit converges to $f(x)$ near $x = a$, so we have

$$f(x) = \lim_{p \to \infty} T_p(x) = \sum_{n=0}^{\infty} \frac{f^{(n)}(a)}{n!}(x-a)^n$$

$$= f(a) + f'(a)(x-a) + \frac{f''(a)}{2}(x-a)^2 + \frac{f'''(a)}{3!}(x-a)^3 + \cdots$$

For example, e^x, $\cos(x)$, and $\sin(x)$ have the following Taylor series representations, which converge for all $x \in \mathbb{R}$:

$$e^x = \sum_{n=0}^{\infty} \frac{x^n}{n!} = 1 + x + \frac{x^2}{2} + \frac{x^3}{3!} + \frac{x^4}{4!} + \frac{x^5}{5!} + \cdots,$$

$$\cos(x) = \sum_{n=0}^{\infty} \frac{(-1)^n x^{2n}}{(2n)!} = 1 - \frac{x^2}{2} + \frac{x^4}{4!} - \frac{x^6}{6!} + \frac{x^8}{8!} - \cdots, \quad \text{and}$$

$$\sin(x) = \sum_{n=0}^{\infty} \frac{(-1)^n x^{2n+1}}{(2n+1)!} = x - \frac{x^3}{3!} + \frac{x^5}{5!} - \frac{x^7}{7!} + \frac{x^9}{9!} - \cdots$$

A.3 Complex Numbers

Many tasks in advanced linear algebra are based on the concept of eigenvalues and eigenvectors, and thus require us to be able to find roots of polynomials (after all, the eigenvalues of a matrix are exactly the roots of its characteristic polynomial). Because many polynomials do not have real roots, much of linear algebra works out much more cleanly if we instead work with "complex" numbers—a more general type of number with the property that every polynomial has complex roots (see the upcoming Theorem A.3.1).

To construct the complex numbers, we start by letting i be an object with the property that $i^2 = -1$. It is clear that i cannot be a real number, but we nonetheless think of it like a number anyway, as we will see that we can manipulate it much like we manipulate real numbers. We call any real scalar multiple of i like $2i$ or $-(7/3)i$ an **imaginary number**, and they obey the same laws of arithmetic that we might expect them to (e.g., $2i + 3i = 5i$ and $(3i)^2 = 3^2 i^2 = -9$).

> The term "imaginary" number is absolutely awful. These numbers are no more make-believe than real numbers are—they are both purely mathematical constructions and they are both useful.

We then let $\mathbb{C}$, the set of **complex numbers**, be the set

$$\mathbb{C} \stackrel{\text{def}}{=} \{a + bi : a, b \in \mathbb{R}\}$$

in which addition and multiplication work exactly as they do for real numbers, as long as we keep in mind that $i^2 = -1$. We call a the **real part** of $a + bi$, and it is sometimes convenient to denote it by $\text{Re}(a + bi) = a$. We similarly call b its **imaginary part** and denote it by $\text{Im}(a + bi) = b$.

Remark A.3.1

Yes, We Can Do That

It might seem extremely strange at first that we can just define a new number i and start doing arithmetic with it. However, this is perfectly fine, and we do this type of thing all the time—one of the beautiful things about mathematics is that we can define whatever we like. However, for that definition to actually be *useful*, it should mesh well with other definitions and objects that we use.

Complex numbers are useful because they let us do certain things that we cannot do in the real numbers (such as find roots of all polynomials), and furthermore they do not break any of the usual laws of arithmetic. That is, we still have all of the properties like $ab = ba$, $a+b = b+a$, and $a(b+c) = ab + ac$ for all $a, b, c \in \mathbb{C}$ that we would expect "numbers" to have. That is, $\mathbb{C}$ is a "field" (just like $\mathbb{R}$)—see Appendix A.4.

By way of contrast, suppose that we tried to do something similar to add a new number that lets us divide by zero. If we let ε be a number with the property that $\varepsilon \times 0 = 1$ (i.e., we are thinking of ε as $1/0$ much like we think of i as $\sqrt{-1}$), then we have

$$1 = \varepsilon \times 0 = \varepsilon \times (0+0) = (\varepsilon \times 0) + (\varepsilon \times 0) = 1+1 = 2.$$

We thus cannot work with such a number without breaking at least one of the usual laws of arithmetic.

As mentioned earlier, some polynomials like $p(x) = x^2 - 2x + 2$ do not have real roots. However, one of the most remarkable theorems concerning polynomials says that every polynomial, as long as it is not a constant function, has complex roots.

For example, $p(x) = x^2 - 2x + 2$ has roots $1 \pm i$.

Theorem A.3.1
Fundamental Theorem of Algebra

Every non-constant polynomial has at least one complex root.

Equivalently, the fundamental theorem of algebra tells us that every polynomial can be factored as a product of linear terms, as long as we allow the roots/factors to be complex numbers. That is, every degree-n polynomial p can be written in the form

$$p(x) = a(x - r_1)(x - r_2) \cdots (x - r_n),$$

where a is the coefficient of x^n in p and $r_1, r_2, \ldots, r_n$ are the (potentially complex, and not necessarily distinct) roots of p.

The proof of this theorem is outside the scope of this book, so we direct the interested reader to a book like [FR97] for various proofs and a discussion of the types of techniques needed to establish it.

A.3.1 Basic Arithmetic and Geometry

For the most part, arithmetic involving complex numbers works simply how we might expect it to. For example, to add two complex numbers together we just add up their real and imaginary parts: $(a+bi) + (c+di) = (a+c) + (b+d)i$, which is hopefully not surprising (it is completely analogous to how we can group and add real numbers and vectors). For example,

$$(3+7i) + (2-4i) = (3+2) + (7-4)i = 5+3i.$$

Similarly, to multiply two complex numbers together we just distribute parentheses like we do when we multiply real numbers together, and we make use of the fact that $i^2 = -1$:

Here we are making use of the fact that multiplication distributes over addition.

$$(a+bi)(c+di) = ac + bci + adi + bdi^2 = (ac - bd) + (ad + bc)i.$$

For example,

$$(3+7i)(4+2i) = (12-14)+(6+28)i = -2+34i.$$

Much like we think of $\mathbb{R}$ as a line, we can think of $\mathbb{C}$ as a plane, called the **complex plane**. The set of real numbers takes the place of the x-axis (which we call the real axis) and the set of imaginary numbers takes the place of the y-axis (which we called the imaginary axis), so that the number $a+bi$ has coordinates (a,b) on that plane, as in Figure A.6.

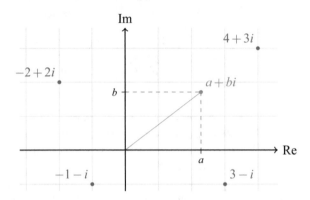

Figure A.6: The complex plane is a representation of the set $\mathbb{C}$ of complex numbers.

We can thus think of $\mathbb{C}$ much like we think of $\mathbb{R}^2$ (i.e., we think of the complex number $a+bi \in \mathbb{C}$ as the vector $(a,b) \in \mathbb{R}^2$), but with a multiplication operation that we do not have on $\mathbb{R}^2$. With this in mind, we define the **magnitude** of a complex number $a+bi$ to be the quantity

The magnitude of a real number is simply its absolute value: $|a| = \sqrt{a^2}$.

$$|a+bi| \overset{\text{def}}{=} \sqrt{a^2+b^2},$$

which is simply the length of the associated vector (a,b) (i.e., it is the distance between $a+bi$ and the origin in the complex plane).

A.3.2 The Complex Conjugate

One particularly important operation on complex numbers that does *not* have any natural analog on the set of real numbers is **complex conjugation**, which negates the imaginary part of a complex number and leaves its real part alone. We denote this operation by putting a horizontal bar over the complex number it is being applied to so that, for example,

$$\overline{3+4i} = 3-4i, \qquad \overline{5-2i} = 5+2i, \qquad \overline{3i} = -3i, \qquad \text{and} \qquad \overline{7} = 7.$$

Geometrically, complex conjugation corresponds to reflecting a number in the complex plane through the real axis, as in Figure A.7.

Notice that applying the complex conjugate twice simply undoes it: $\overline{\overline{a+bi}} = \overline{a-bi} = a+bi$. After all, reflecting a number or vector twice returns it to where it started.

Algebraically, complex conjugation is useful since many other common operations involving complex numbers can be expressed in terms of it. For example:

- The magnitude of a complex number $z = a+bi$ can be written in terms of the product of z with its complex conjugate: $|z|^2 = z\bar{z}$, since

$$z\bar{z} = (a+bi)\overline{(a+bi)} = (a+bi)(a-bi)$$
$$= a^2+b^2 = |a+bi|^2 = |z|^2.$$

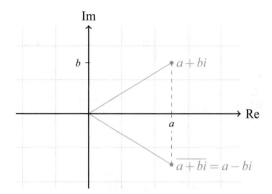

Figure A.7: Complex conjugation reflects a complex number through the real axis.

- The previous point tells us that we can multiply any complex number by another one (its complex conjugate) to get a real number. We can make use of this trick to come up with a method of dividing by complex numbers:

In the first step here, we just cleverly multiply by 1 so as to make the denominator real.

$$\frac{a+bi}{c+di} = \left(\frac{a+bi}{c+di}\right)\left(\frac{c-di}{c-di}\right)$$
$$= \frac{(ac+bd)+(bc-ad)i}{c^2+d^2} = \left(\frac{ac+bd}{c^2+d^2}\right)+\left(\frac{bc-ad}{c^2+d^2}\right)i.$$

- The real and imaginary parts of a complex number $z = a+bi$ can be computed via

$$\mathrm{Re}(z) = \frac{z+\bar{z}}{2} \quad \text{and} \quad \mathrm{Im}(z) = \frac{z-\bar{z}}{2i},$$

since

$$\frac{z+\bar{z}}{2} = \frac{(a+bi)+(a-bi)}{2} = \frac{2a}{2} = a = \mathrm{Re}(z) \quad \text{and}$$
$$\frac{z-\bar{z}}{2i} = \frac{(a+bi)-(a-bi)}{2i} = \frac{2bi}{2i} = b = \mathrm{Im}(z).$$

A.3.3 Euler's Formula and Polar Form

Since we can think of complex numbers as points in the complex plane, we can specify them via their length and direction rather than via their real and imaginary parts. In particular, we can write every complex number in the form $z = |z|u$, where $|z|$ is the magnitude of z and u is a number on the unit circle in the complex plane.

By recalling that every point on the unit circle in $\mathbb{R}^2$ has coordinates of the form $(\cos(\theta), \sin(\theta))$ for some $\theta \in [0, 2\pi)$, we see that every point on the unit circle in the complex plane can be written in the form $\cos(\theta) + i\sin(\theta)$, as illustrated in Figure A.8.

The notation $\theta \in [0, 2\pi)$ means that θ is between 0 (inclusive) and 2π (non-inclusive).

It follows that we can write every complex number in the form $z = |z|(\cos(\theta) + i\sin(\theta))$. However, we can simplify this expression somewhat

It is worth noting that $|\cos(\theta)+i\sin(\theta)|^2 = \cos^2(\theta)+\sin^2(\theta)=1$, so these numbers really are on the unit circle.

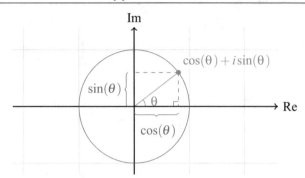

Figure A.8: Every number on the unit circle in the complex plane can be written in the form $\cos(\theta)+i\sin(\theta)$ for some $\theta \in [0,2\pi)$.

by using the following Taylor series that we saw in Appendix A.2.2:

$$e^x = 1+x+\frac{x^2}{2}+\frac{x^3}{3!}+\frac{x^4}{4!}+\frac{x^5}{5!}+\cdots,$$

$$\cos(x) = 1 \quad -\frac{x^2}{2!} \quad +\frac{x^4}{4!} \quad -\cdots, \quad \text{and}$$

$$\sin(x) = \quad x \quad -\frac{x^3}{3!} \quad +\frac{x^5}{5!} \quad \cdots.$$

Just like these Taylor series converge for all $x \in \mathbb{R}$, they also converge for all $x \in \mathbb{C}$.

In particular, if we plug $x = i\theta$ into the Taylor series for e^x, then we see that

$$e^{i\theta} = 1+i\theta-\frac{\theta^2}{2}-i\frac{\theta^3}{3!}+\frac{\theta^4}{4!}+i\frac{\theta^5}{5!}-\cdots$$
$$= \left(1-\frac{\theta^2}{2}+\frac{\theta^4}{4!}-\cdots\right)+i\left(\theta-\frac{\theta^3}{3!}+\frac{\theta^5}{5!}+\cdots\right),$$

which equals $\cos(\theta)+i\sin(\theta)$. We have thus proved the remarkable fact, called **Euler's formula**, that

In other words, $\mathrm{Re}(e^{i\theta}) = \cos(\theta)$ and $\mathrm{Im}(e^{i\theta}) = \sin(\theta)$.

$$e^{i\theta} = \cos(\theta)+i\sin(\theta) \quad \text{for all} \quad \theta \in [0,2\pi).$$

By making use of Euler's formula, we see that we can write every complex number $z \in \mathbb{C}$ in the form $z = re^{i\theta}$, where r is the magnitude of z (i.e., $r = |z|$) and θ is the angle that z makes with the positive real axis (see Figure A.9). This is called the **polar form** of z, and we can convert back and forth between the polar form $z = re^{i\theta}$ and its **Cartesian form** $z = a+bi$ via the formulas

In the formula for θ, $\mathrm{sign}(b) = \pm 1$, depending on whether b is positive or negative. If $b < 0$ then we get $-\pi < \theta < 0$, which we can put in the interval $[0,2\pi)$ by adding 2π to it.

$$a = r\cos(\theta) \qquad\qquad r = \sqrt{a^2+b^2}$$

$$b = r\sin(\theta) \qquad\qquad \theta = \mathrm{sign}(b)\arccos\left(\frac{a}{\sqrt{a^2+b^2}}\right).$$

There is no simple way to "directly" add two complex numbers that are in polar form, but multiplication is quite straightforward: $(r_1e^{i\theta_1})(r_2e^{i\theta_2}) = (r_1r_2)e^{i(\theta_1+\theta_2)}$. We can thus think of complex numbers as stretched rotations—multiplying by $re^{i\theta}$ stretches numbers in the complex plane by a factor of r and rotates them counter-clockwise by an angle of θ.

Because the polar form of complex numbers works so well with multiplication, we can use it to easily compute powers and roots. Indeed, repeatedly

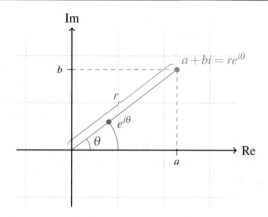

Figure A.9: Every complex number can be written in Cartesian form $a+bi$ and also in polar form $re^{i\theta}$.

You should try to convince yourself that raising any of these numbers to the n-th power results in $re^{i\theta}$. Use the fact that $e^{2\pi i} = e^{0i} = 1$.

multiplying a complex number in polar form by itself gives $(re^{i\theta})^n = r^n e^{in\theta}$ for all positive integers $n \geq 1$. We thus see that every non-zero complex number has at least n distinct n-th roots (and in fact, *exactly* n distinct n-th roots). In particular, the n roots of $z = re^{i\theta}$ are

$$r^{1/n}e^{i\theta/n}, \quad r^{1/n}e^{i(\theta+2\pi)/n}, \quad r^{1/n}e^{i(\theta+4\pi)/n}, \quad \ldots, \quad r^{1/n}e^{i(\theta+2(n-1)\pi)/n}.$$

Example A.3.1

Computing Complex Roots

Compute all 3 cube roots of the complex number $z = i$.

Solution:

We start by writing z in its polar form $z = e^{\pi i/2}$, from which we see that its 3 cube roots are

$$e^{\pi i/6}, \quad e^{5\pi i/6}, \quad \text{and} \quad e^{9\pi i/6}.$$

We can convert these three cube roots into Cartesian coordinates if we want to, though this is perhaps not necessary:

$$e^{\pi i/6} = (\sqrt{3}+i)/2, \quad e^{5\pi i/6} = (-\sqrt{3}+i)/2, \quad \text{and} \quad e^{9\pi i/6} = -i.$$

Geometrically, the cube root $e^{\pi i/6} = (\sqrt{3}+i)/2$ of i lies on the unit circle and has an angle one-third as large as that of i, and the other two cube roots are evenly spaced around the unit circle:

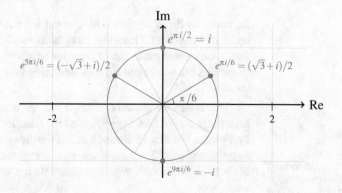

This definition of principal roots requires us to make sure that $\theta \in [0, 2\pi)$.

Among the n distinct n-th roots of a complex number $z = re^{i\theta}$, we call

$r^{1/n}e^{i\theta/n}$ its **principal n-th root**, which we denote by $z^{1/n}$. The principal root of a complex number is the one with the smallest angle so that, for example, if z is a positive real number (i.e., $\theta = 0$) then its principal roots are positive real numbers as well. Similarly, the principal square root of a complex number is the one in the upper half of the complex plane (for example, the principal square root of -1 is i, not $-i$), and we showed in Example A.3.1 that the principal cube root of $z = e^{\pi i/2} = i$ is $z^{1/3} = e^{\pi i/6} = (\sqrt{3}+i)/2$.

Remark A.3.2 **Complex Numbers as 2×2 Matrices**	Complex numbers can actually be represented by real matrices. If we define the 2×2 matrix $$J = \begin{bmatrix} 0 & -1 \\ 1 & 0 \end{bmatrix},$$ then it is straightforward to verify that $J^2 = -I$ (recall that we think of the identity matrix I as the "matrix version" of the number 1). It follows that addition and multiplication of the matrices of the form $$aI + bJ = \begin{bmatrix} a & -b \\ b & a \end{bmatrix}$$

work exactly the same as they do for complex numbers. We can thus think of the complex number $a + bi$ as "the same thing" as the 2×2 matrix $aI + bJ$.

In the language of Section 1.3.1, the set of 2×2 matrices of this special form is "isomorphic" to $\mathbb{C}$. However, they are not just isomorphic as vector spaces, but even as fields (i.e., the field operations of Definition A.4.1 are preserved).

This representation of complex numbers perhaps makes it clearer why they act like rotations when multiplying by them: we can rewrite $aI + bJ$ as

$$aI + bJ = \begin{bmatrix} a & -b \\ b & a \end{bmatrix} = \sqrt{a^2+b^2}\begin{bmatrix} \cos(\theta) & -\sin(\theta) \\ \sin(\theta) & \cos(\theta) \end{bmatrix},$$

where we recognize $\sqrt{a^2+b^2}$ as the magnitude of the complex number $a + bi$, and the matrix on the right as the one that rotates $\mathbb{R}^2$ counterclockwise by an angle of θ (here, θ is chosen so that $\cos(\theta) = a/\sqrt{a^2+b^2}$ and $\sin(\theta) = b/\sqrt{a^2+b^2}$).

A.4   Fields

Introductory linear algebra is usually carried out via vectors and matrices made up of real numbers. However, most of its results and methods carry over just fine if we instead make use of complex numbers, for example (in fact, much of linear algebra works out *better* when using complex numbers rather than real numbers).

This raises the question of what sets of "numbers" we can make use of in linear algebra. In fact, it even raises the question of what a "number" *is* in the first place. We saw in Remark A.3.2 that we can think of complex numbers as certain special 2×2 matrices, so it seems natural to wonder why we think of them as "numbers" even though we do not think of general matrices in the same way.

The upcoming definition generalizes $\mathbb{R}$ in much the same way that Definition 1.1.1 generalizes $\mathbb{R}^n$.

This appendix answers these questions. We say that a **field** is a set in which addition and multiplication behave in the same way that they do in the set of real or complex numbers, so it is reasonable to think of it as being made up of "numbers" or "scalars".

<table>
<tr><td>

Definition A.4.1

Field

</td><td>

Let $\mathbb{F}$ be a set with two operations called **addition** and **multiplication**. We write the addition of $a \in \mathbb{F}$ and $b \in \mathbb{F}$ as $a+b$, and the multiplication of a and b as ab.

If the following conditions hold for all $a, b, c \in \mathbb{F}$ then $\mathbb{F}$ is called a **field**:

</td></tr>
</table>

a) $a + b \in \mathbb{F}$ (closure under addition)
b) $a + b = b + a$ (commutativity of addition)
c) $(a+b) + c = a + (b+c)$ (associativity of addition)
d) There is a **zero element** $0 \in \mathbb{F}$ such that $0 + a = a$.
e) There is an **additive inverse** $-a \in \mathbb{F}$ such that $a + (-a) = 0$.

> **Notice that the first five properties concern addition, the next five properties concern multiplication, and the final property combines them.**

f) $ab \in \mathbb{F}$ (closure under multiplication)
g) $ab = ba$ (commutativity of multiplication)
h) $(ab)c = a(bc)$ (associativity of multiplication)
i) There is a **unit element** $1 \in \mathbb{F}$ such that $1a = a$.
j) If $a \neq 0$, there is a **multiplicative inverse** $1/a \in \mathbb{F}$ such that $a(1/a) = 1$.
k) $a(b+c) = (ab) + (bc)$ (distributivity)

The sets $\mathbb{R}$ and $\mathbb{C}$ of real and complex numbers are of course fields (we defined fields specifically so as to mimic $\mathbb{R}$ and $\mathbb{C}$, after all). Similarly, it is straightforward to show that the set $\mathbb{Q}$ of **rational numbers**—real numbers of the form p/q where p and $q \neq 0$ are integers—is a field. After all, if $a = p_1/q_1$ and $b = p_2/q_2$ are rational then so are $a + b = (p_1 q_2 + p_2 q_1)/(q_1 q_2)$ and $ab = (p_1 p_2)/(q_1 q_2)$, and all other field properties follow immediately from the corresponding facts about real numbers.

> **This is analogous to how we only need to prove closure to show that a subspace is a vector space (Theorem 1.1.2)—all other properties are inherited from the parent structure.**

Another less obvious example of a field is the set $\mathbb{Z}_2$ of numbers with arithmetic modulo 2. That is, $\mathbb{Z}_2$ is simply the set $\{0, 1\}$, but with the understanding that addition and multiplication of these numbers work as follows:

$$0 + 0 = 0 \qquad 0 + 1 = 1 \qquad 1 + 0 = 1 \qquad 1 + 1 = 0, \quad \text{and}$$
$$0 \times 0 = 0 \qquad 0 \times 1 = 0 \qquad 1 \times 0 = 0 \qquad 1 \times 1 = 1.$$

In other words, these operations work just as they normally do, with the exception that $1 + 1 = 0$ instead of $1 + 1 = 2$ (since there is no "2" in $\mathbb{Z}_2$). Phrased differently, we can think of this as a simplified form of arithmetic that only keeps track of whether or not a number is even or odd (with 0 for even and 1 for odd). All field properties are straightforward to show for $\mathbb{Z}_2$, as long as we understand that $-1 = 1$.

More generally, if p is a prime number then the set $\mathbb{Z}_p = \{0, 1, 2, \ldots, p-1\}$ with arithmetic modulo p is a field. For example, addition and multiplication in $\mathbb{Z}_5 = \{0, 1, 2, 3, 4\}$ work as follows:

+	0	1	2	3	4
0	0	1	2	3	4
1	1	2	3	4	0
2	2	3	4	0	1
3	3	4	0	1	2
4	4	0	1	2	3

×	0	1	2	3	4
0	0	0	0	0	0
1	0	1	2	3	4
2	0	2	4	1	3
3	0	3	1	4	2
4	0	4	3	2	1

Again, all of the field properties of $\mathbb{Z}_p$ are straightforward to show, with the exception of property (j)—the existence of multiplicative inverses. The standard way to prove this property is to use a theorem called **Bézout's identity**, which says that, for all $0 \leq a < p$, we can find integers x and y such that $ax + py = 1$. We can rearrange this equation to get $ax = 1 - py$, so that we can choose $1/a = x$. For example, in $\mathbb{Z}_5$ we have $1/1 = 1$, $1/2 = 3$, $1/3 = 2$, and $1/4 = 4$ (with $1/2 = 3$, for example, corresponding to the fact that the integer equation $2x + 5y = 1$ can be solved when $x = 3$).

Property (j) is why we need p to be prime. For example, $\mathbb{Z}_4$ is not a field since 2 does not have a multiplicative inverse.

A.5 Convexity

In linear algebra, the sets of vectors that arise most frequently are subspaces (which are closed under linear combinations) and the functions that arise most frequently are linear transformations (which preserve linear combinations). In this appendix, we briefly discuss the concept of **convexity**, which is a slight weakening of these linearity properties that can be thought of as requiring the set or graph of the function to "not bend inward" rather than "be flat".

A.5.1 Convex Sets

By a "real vector space" we mean a vector space over $\mathbb{R}$.

Roughly speaking, a convex subset S of a real vector space V is one for which every line segment between two points in S is also in S. The following definition pins down this idea algebraically:

Definition A.5.1

Convex Set

Suppose V is a real vector space. A subset $S \subseteq V$ is called **convex** if

$$(1-t)\mathbf{v} + t\mathbf{w} \in S \quad \text{whenever} \quad \mathbf{v}, \mathbf{w} \in S \quad \text{and} \quad 0 \leq t \leq 1.$$

If we required $c\mathbf{v} + d\mathbf{w} \in S$ for *all* $c, d \in \mathbb{R}$ then S would be a subspace. The restriction that the coefficients are non-negative and add up to 1 makes convex sets more general.

To convince ourselves that this definition captures the geometric idea involving line segments between two points, we observe that if $t = 0$ then $(1-t)\mathbf{v} + t\mathbf{w} = \mathbf{v}$, if $t = 1$ then $(1-t)\mathbf{v} + t\mathbf{w} = \mathbf{w}$, and as t increases from 0 to 1 the quantity $(1-t)\mathbf{v} + t\mathbf{w}$ travels along the line segment from $\mathbf{v}$ to $\mathbf{w}$ (see Figure A.10).

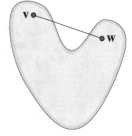

(a) The line segment between two vectors in a convex set is traced out by different values of t in Definition A.5.1.

(b) A non-convex set is one for which there is a line segment between two points in the set that leaves the set.

Figure A.10: A set is convex (a) if all line segments between two points in that set are contained within it, and it is non-convex (b) otherwise.

Another way of thinking about convex sets is as ones with no holes or sides that bend inward—their sides can be flat or bend outward only. Some standard examples of convex sets include

- Any interval in $\mathbb{R}$.
- Any subspace of a real vector space.
- Any quadrant in $\mathbb{R}^2$ or orthant in $\mathbb{R}^n$.
- The set of positive (semi)definite matrices in $\mathcal{M}_n^H$ (see Section 2.2).

Recall that even though the entries of Hermitian matrices can be complex, $\mathcal{M}_n^H$ is a real vector space (see Example 1.1.5).

The following theorem pins down a very intuitive, yet extremely powerful, idea—if two convex sets are disjoint (i.e., do not contain any vectors in common) then we must be able to fit a line/plane/hyperplane (i.e., a flat surface that cuts $\mathcal{V}$ into two halves) between them.

The way that we formalize this idea is based on linear forms (see Section 1.3), which are linear transformations from $\mathcal{V}$ to $\mathbb{R}$. We also make use of open sets, which are sets of vectors that do not include any of their boundary points (standard examples include any subinterval of $\mathbb{R}$ that does not include its endpoints, and the set of positive *definite* matrices in $\mathcal{M}_n^H$).

Theorem A.5.1

Separating Hyperplane Theorem

Suppose $\mathcal{V}$ is a finite-dimensional real vector space and $S, T \subseteq \mathcal{V}$ are disjoint convex subsets of $\mathcal{V}$, and T is open. Then there is a linear form $f : \mathcal{V} \to \mathbb{R}$ and a scalar $c \in \mathbb{R}$ such that

$$f(\mathbf{y}) > c \geq f(\mathbf{x}) \quad \text{for all} \quad \mathbf{x} \in S \quad \text{and} \quad \mathbf{y} \in T.$$

*If $c = 0$ then this separating hyperplane is a subspace of $\mathcal{V}$ (it is null(f)). Otherwise, it is a "shifted" subspace (which are typically called **affine spaces**).*

In particular, the line/plane/hyperplane that separates the convex sets S and T in this theorem is $\{\mathbf{v} \in \mathcal{V} : f(\mathbf{v}) = c\}$. This separating hyperplane may intersect the boundary of S and/or T, but because T is open we know that it does not intersect T itself, which guarantees the strict inequality $f(\mathbf{w}) > c$ (see Figure A.11).

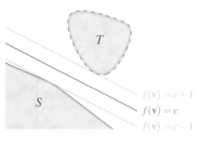
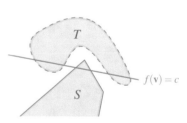

(a) Two convex sets and a separating hyperplane $f(\mathbf{v}) = c$. Because there is a gap between the sets, there are multiple separating hyperplanes.

(b) If one of the sets is non-convex, there may be no separating hyperplane between the sets, even if they are disjoint.

Figure A.11: The separating hyperplane theorem (Theorem A.5.1) tells us (a) that we can squeeze a hyperplane between any two disjoint convex sets. This is (b) not necessarily possible if one of the sets is non-convex.

While the separating hyperplane theorem itself perhaps looks a bit abstract, it can be made more concrete by choosing a particular vector space $\mathcal{V}$ and thinking about how we can represent linear forms acting on $\mathcal{V}$. For example, if $\mathcal{V} = \mathbb{R}^n$ then Theorem 1.3.3 tells us that for every linear form $f : \mathbb{R}^n \to \mathbb{R}$ there exists a vector $\mathbf{v} \in \mathbb{R}^n$ such that $f(\mathbf{x}) = \mathbf{v} \cdot \mathbf{x}$ for all $\mathbf{x} \in \mathbb{R}^n$. It follows that, in this setting, if S and T are convex and T is open then there exists a vector $\mathbf{v} \in \mathbb{R}^n$ and a scalar $c \in \mathbb{R}$ such that

In fact, $\mathbf{v}$ is exactly the unique (up to scaling) vector that is orthogonal to the separating hyperplane.

$$\mathbf{v} \cdot \mathbf{y} > c \geq \mathbf{v} \cdot \mathbf{x} \quad \text{for all} \quad \mathbf{x} \in S \quad \text{and} \quad \mathbf{y} \in T.$$

Similarly, we know from Exercise 1.3.7 that for every linear form $f : \mathcal{M}_n \to \mathbb{R}$, there exists a matrix $A \in \mathcal{M}_n$ such that $f(X) = \text{tr}(AX)$ for all $X \in \mathcal{M}_n$. The separating hyperplane theorem then tells us (under the usual hypotheses on S and T) that

$$\text{tr}(AY) > c \geq \text{tr}(AX) \quad \text{for all} \quad X \in S \quad \text{and} \quad Y \in T.$$

A.5.2   Convex Functions

Convex functions generalize linear forms in much the same way that convex sets generalize subspaces—they weaken the linearity (i.e., "flat") condition in favor of a convexity (i.e., "bend out") condition.

Definition A.5.2
Convex Function

Suppose $\mathcal{V}$ is a real vector space and $S \subseteq \mathcal{V}$ is a convex set. We say that a function $f : S \to \mathbb{R}$ is **convex** if

$$f\big((1-t)\mathbf{v}+t\mathbf{w}\big) \leq (1-t)f(\mathbf{v})+tf(\mathbf{w}) \quad \text{whenever} \quad \mathbf{v}, \mathbf{w} \in S$$
$$\text{and} \quad 0 \leq t \leq 1.$$

In the case when $\mathcal{V} = S = \mathbb{R}$, the above definition can be interpreted very naturally in terms of the function's graph: f is convex if and only if every line segment between two points on its graph lies above the graph itself. Equivalently, f is convex if and only if the region above its graph is a convex set (see Figure A.12).

The region above a function's graph is called its **epigraph**.

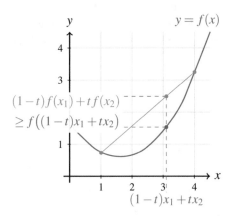
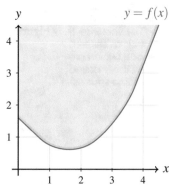

(a) Line segments between two points on the graph of a convex function always lie above the graph. Here, $x_1 = 1$, $x_2 = 4$, and $t = 0.7$.

(b) The region above the graph of a convex function is always a convex set.

Figure A.12: A function is convex if and only if (a) every line segment between two points on its graph lies above its graph, if and only if (b) the region above its graph is convex.

Some standard examples of convex functions include:

- The function $f(x) = e^x$ on $\mathbb{R}$.
- Given any real $p \geq 1$, the function $f(x) = x^p$ on $[0, \infty)$.
- Any linear form on a real vector space.
- Any norm (see Section 1.D) on a real vector space, since the triangle inequality and absolute homogeneity tell us that

$$\|(1-t)\mathbf{v}+t\mathbf{w}\| \leq \|(1-t)\mathbf{v}\| + \|t\mathbf{w}\| = (1-t)\|\mathbf{v}\|+t\|\mathbf{w}\|.$$

Some introductory calculus courses refer to convex and concave functions as **concave up** and **concave down**, respectively.

If f is a convex function then $-f$ is called a **concave** function. Geometrically, a concave function is one whose graph bends the opposite way of a convex function—a line segment between two points on its graph is always *below* the graph, and the region *below* its graph is a convex set (see Figure A.13). Examples of concave functions include

- Given any real $b > 1$, the base-b logarithm $f(x) = \log_b(x)$ on $(0, \infty)$.
- Given any real $0 < p \le 1$, the function $f(x) = x^p$ on $[0, \infty)$. For example, if $p = 1/2$ then we see that $f(x) = \sqrt{x}$ is concave.
- Any linear form on a real vector space. In fact, linear forms are exactly the functions that are both convex and concave.

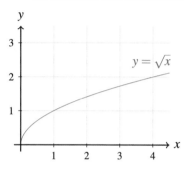

(a) The square root function is concave.

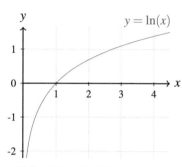

(b) The natural logarithm is concave.

Figure A.13: A function is concave if and only if every line segment between two points on its graph lies below its graph, if and only if the region below its graph is convex.

The following theorem provides what is probably the simplest method of checking whether or not a function of one real variable is convex, concave, or neither.

Theorem A.5.2

Convexity of Twice-Differentiable Functions

Suppose $f : (a, b) \to \mathbb{R}$ is twice differentiable and f'' is continuous. Then:

a) f is convex if and only if $f''(x) \ge 0$ for all $x \in (a, b)$, and

b) f is concave if and only if $f''(x) \le 0$ for all $x \in (a, b)$.

Similarly, a function is **strictly convex** if $f((1-t)\mathbf{v} + t\mathbf{w}) < (1-t)f(\mathbf{v}) + tf(\mathbf{w})$ whenever $\mathbf{v} \ne \mathbf{w}$ and $0 < t < 1$. Notice the strict inequalities.

For example, we can see that the function $f(x) = x^4$ is convex on $\mathbb{R}$ since $f''(x) = 12x^2 \ge 0$, whereas the natural logarithm $g(x) = \ln(x)$ is concave on $(0, \infty)$ since $g''(x) = -1/x^2 \le 0$. In fact, the second derivative of $g(x) = \ln(x)$ is *strictly* negative, which tells us that any line between two points on its graph is *strictly* below its graph—it only touches the graph at the endpoints of the line segment. We call such a function **strictly concave**.

One of the most useful applications of the fact that the logarithm is (strictly) concave is the following inequality that relates the arithmetic and geometric means of a set of n real numbers:

Theorem A.5.3

Arithmetic Mean–Geometric Mean (AM–GM) Inequality

If $x_1, x_2, \ldots, x_n \ge 0$ are real numbers then

$$\sqrt[n]{x_1 x_2 \cdots x_n} \le \frac{x_1 + x_2 + \cdots + x_n}{n}.$$

Furthermore, equality holds if and only if $x_1 = x_2 = \cdots = x_n$.

When $n = 2$, the AM–GM inequality just says that if $x, y \ge 0$ then

$\sqrt{xy} \leq (x+y)/2$, which can be proved "directly" by multiplying out the parenthesized term in the inequality $(\sqrt{x} - \sqrt{y})^2 \geq 0$. To see why it holds when $n \geq 3$, we use concavity of the natural logarithm:

This argument only works if each x_j is *strictly* positive, but that's okay because the AM–GM inequality is trivial if $x_j = 0$ for some $1 \leq j \leq n$.

$$\ln\left(\frac{x_1 + x_2 + \cdots + x_n}{n}\right) \geq \frac{1}{n}\sum_{j=1}^{n}\ln(x_j) = \sum_{j=1}^{n}\ln(x_j^{1/n}) = \ln\left(\sqrt[n]{x_1 x_2 \cdots x_n}\right).$$

Exponentiating both sides (i.e., raising e to the power of both sides of this inequality) gives exactly the AM–GM inequality. The equality condition of the AM–GM inequality follows from the fact that the logarithm is *strictly* convex.

B. Additional Proofs

In this appendix, we prove some of the technical results that we made use of throughout the main body of the textbook, but whose proofs are messy enough (or unenlightening enough, or simply not "linear algebra-y" enough) that they are hidden away here.

Equivalence of Norms

We mentioned near the start of Section 1.D that all norms on a finite-dimensional vector space are equivalent to each other. That is, the statement of Theorem 1.D.1 told us that, given any two norms $\|\cdot\|_a$ and $\|\cdot\|_b$ on $\mathcal{V}$, there exist real scalars $c, C > 0$ such that

$$c\|\mathbf{v}\|_a \leq \|\mathbf{v}\|_b \leq C\|\mathbf{v}\|_a \quad \text{for all} \quad \mathbf{v} \in \mathcal{V}.$$

We now prove this theorem.

Proof of Theorem 1.D.1. We first note that Exercise 1.D.28 tells us that it suffices to prove this theorem in $\mathbb{F}^n$ (where $\mathbb{F} = \mathbb{R}$ or $\mathbb{F} = \mathbb{C}$). Furthermore, transitivity of equivalence of norms (Exercise 1.D.27) tells us that it suffices just to show that every norm on $\mathbb{F}^n$ is equivalent to the usual vector length (2-norm) $\|\cdot\|$, so the remainder of the proof is devoted to showing why this is the case. In particular, we will show that any norm $\|\cdot\|_a$ is equivalent to the usual 2-norm $\|\cdot\|$.

The 2-norm is
$$\|\mathbf{v}\| = \sqrt{|v_1|^2 + \cdots + |v_n|^2}.$$

To see that there is a constant $C > 0$ such that $\|\mathbf{v}\|_a \leq C\|\mathbf{v}\|$ for all $\mathbf{v} \in \mathbb{F}^n$, we notice that

Recall that $\mathbf{e}_1, \ldots, \mathbf{e}_n$ denote the standard basis vectors in $\mathbb{F}^n$.

$$\begin{aligned}
\|\mathbf{v}\|_a &= \|v_1\mathbf{e}_1 + \cdots + v_n\mathbf{e}_n\|_a && \text{(\textbf{v} in standard basis)}\\
&\leq |v_1|\|\mathbf{e}_1\|_a + \cdots + |v_n|\|\mathbf{e}_n\|_a && \text{(triangle ineq. for } \|\cdot\|_a)\\
&\leq \sqrt{|v_1|^2 + \cdots + |v_n|^2}\sqrt{\|\mathbf{e}_1\|_a^2 + \cdots + \|\mathbf{e}_n\|_a^2} && \text{(Cauchy–Schwarz ineq.)}\\
&= \|\mathbf{v}\|\sqrt{\|\mathbf{e}_1\|_a^2 + \cdots + \|\mathbf{e}_n\|_a^2}. && \text{(formula for } \|\mathbf{v}\|)
\end{aligned}$$

We can thus choose

This might look like a weird choice for C, but all that matters is that it does not depend on the choice of **v**.

$$C = \sqrt{\|\mathbf{e}_1\|_a^2 + \cdots + \|\mathbf{e}_n\|_a^2}.$$

To see that there is similarly a constant $c > 0$ such that $c\|\mathbf{v}\| \leq \|\mathbf{v}\|_a$ for all $\mathbf{v} \in \mathcal{V}$, we need to put together three key observations. First, observe that it suffices to show that $c \leq \|\mathbf{v}\|_a$ whenever $\mathbf{v}$ is such that $\|\mathbf{v}\| = 1$, since otherwise we can just scale $\mathbf{v}$ so that this is the case and both sides of the desired inequality scale appropriately.

Second, we notice that the function $\|\cdot\|_a$ is continuous on $\mathbb{F}^n$, since the inequality $\|\mathbf{v}\|_a \leq C\|\mathbf{v}\|$ tells us that if a $\lim_{k \to \infty} \mathbf{v}_k = \mathbf{v}$ then

Some details about limits and continuity can be found in Section 2.D.2.

$$\lim_{k \to \infty} \|\mathbf{v}_k - \mathbf{v}\| = 0$$

$$\implies \lim_{k \to \infty} \|\mathbf{v}_k - \mathbf{v}\|_a = 0 \qquad \left(\lim_{k \to \infty} \|\mathbf{v}_k - \mathbf{v}\|_a \leq C \lim_{k \to \infty} \|\mathbf{v}_k - \mathbf{v}\| = 0\right)$$

$$\implies \lim_{k \to \infty} \|\mathbf{v}_k\|_a = \|\mathbf{v}\|_a. \qquad \text{(reverse triangle ineq.: Exercise 1.D.6)}$$

Third, we define

$$S^n = \left\{\mathbf{v} \in \mathbb{F}^n : \|\mathbf{v}\| = 1\right\} \subset \mathbb{F}^n,$$

which we can think of as the unit circle (or sphere, or hypersphere, depending on n), as illustrated in Figure B.14, and we notice that this set is closed and bounded.

*A closed and bounded subset of $\mathbb{F}^n$ is typically called **compact**.*

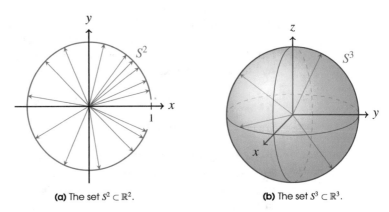

(a) The set $S^2 \subset \mathbb{R}^2$.   **(b)** The set $S^3 \subset \mathbb{R}^3$.

Figure B.14: The set $S^n \subset \mathbb{F}^n$ of unit vectors can be thought of as the unit circle, or sphere, or hypersphere, depending on the dimension.

Continuous functions also attain their maximum value on a closed and bounded set, but we only need the minimum value.

To put all of this together, we recall that the Extreme Value Theorem from analysis says that every continuous function on a closed and bounded subset of $\mathbb{F}^n$ attains its minimum value on that set (i.e., it does not just *approach* some minimum value). It follows that the norm $\|\cdot\|_a$ attains a minimum value (which we call c) on S^n, and since each vector in S^n is non-zero we know that $c > 0$. We thus conclude that $c \leq \|\mathbf{v}\|_a$ whenever $\mathbf{v} \in S^n$, which is exactly what we wanted to show. ∎

B.2   Details of the Jordan Decomposition

We presented the Jordan decomposition of a matrix in Section 2.4 and presented most of the results needed to demonstrate how to actually construct it, but we glossed over some of the details that guaranteed that our method of construction always works. We now fill in those details.

The first result that we present here demonstrates the claim that we made in the middle of the proof of Theorem 2.4.5.

Theorem B.2.1

Similarity of Block Triangular Matrices

If two matrices $A \in \mathcal{M}_m(\mathbb{C})$ and $C \in \mathcal{M}_n(\mathbb{C})$ do not have any eigenvalues in common then, for all $B \in \mathcal{M}_{m,n}(\mathbb{C})$, the following block matrices are similar:

$$\begin{bmatrix} A & B \\ O & C \end{bmatrix} \quad \text{and} \quad \begin{bmatrix} A & O \\ O & C \end{bmatrix}.$$

The m and n in the statement of this theorem are not the same as in Theorem 2.4.5. In the proof of that theorem, they were called d_1 and d_2, respectively (which are just any positive integers).

Proof. Notice that if $X \in \mathcal{M}_{m,n}(\mathbb{C})$ and

$$P = \begin{bmatrix} I & X \\ O & I \end{bmatrix} \quad \text{then} \quad P^{-1} = \begin{bmatrix} I & -X \\ O & I \end{bmatrix},$$

and

$$P^{-1}\begin{bmatrix} A & O \\ O & C \end{bmatrix}P = \begin{bmatrix} I & -X \\ O & I \end{bmatrix}\begin{bmatrix} A & O \\ O & C \end{bmatrix}\begin{bmatrix} I & X \\ O & I \end{bmatrix} = \begin{bmatrix} A & AX - XC \\ O & C \end{bmatrix}.$$

We provide another proof that such an X exists in Exercise 3.1.8.

It thus suffices to show that there exists a matrix $X \in \mathcal{M}_{m,n}(\mathbb{C})$ such that $AX - XC = B$. Since $B \in \mathcal{M}_{m,n}(\mathbb{C})$ is arbitrary, this is equivalent to showing that the linear transformation $T : \mathcal{M}_{m,n}(\mathbb{C}) \to \mathcal{M}_{m,n}(\mathbb{C})$ defined by $T(X) = AX - XC$ is invertible, which we do by showing that $T(X) = O$ implies $X = O$.

To this end, suppose that $T(X) = O$, so $AX = XC$. Simply using associativity of matrix multiplication then shows that this implies

$$A^2X = A(AX) = A(XC) = (AX)C = (XC)C = XC^2, \quad \text{so}$$
$$A^3X = A(A^2X) = A(XC^2) = (AX)C^2 = (XC)C^2 = XC^3,$$

and so on so that $A^kX = XC^k$ for all $k \geq 1$. Taking linear combinations of these equations then shows that $p(A)X = Xp(C)$ for any polynomial, so if we choose $p = p_A$ to be the characteristic polynomial of A then we see that $p_A(A)X = Xp_A(C)$. Since the Cayley–Hamilton theorem (Theorem 2.1.3) tells us that $p_A(A) = O$, it follows that $Xp_A(C) = O$. Our next goal is to show that $p_A(C)$ is invertible, which will complete the proof since we can then multiply the equation $Xp_A(C) = O$ on the right by $p_A(C)^{-1}$ to get $X = O$.

Well, if we denote the eigenvalues of A (listed according to algebraic multiplicity) by $\lambda_1, \lambda_2, \ldots, \lambda_m$, then

$$p_A(C) = (C - \lambda_1 I)(C - \lambda_2 I)\cdots(C - \lambda_m I).$$

Notice that, for each $1 \leq j \leq m$, the factor $C - \lambda_j I$ is invertible since λ_j is not an eigenvalue of C. It follows that their product $p_A(C)$ is invertible as well, as desired. ∎

The next result that we present fills in a gap in the proof of Theorem 2.4.5 (the Jordan decomposition) itself. In our method of constructing this decomposition, we needed a certain set that we built to be linearly independent. We now verify that it is.

Theorem B.2.2

Linear Independence of Jordan Chains

The set B_k from the proof of Theorem 2.4.1 is linearly independent.

Proof. Throughout this proof, we use the same notation as in the proof of Theorem 2.4.1, but before proving anything specific about B_k, we prove the following claim:

Claim: If S_1, S_2 are subspaces of $\mathbb{C}^n$ and $\mathbf{x} \in \mathbb{C}^n$ is a vector for which $\mathbf{x} \notin \text{span}(S_1 \cup S_2)$, then $\text{span}(\{\mathbf{x}\} \cup S_1) \cap \text{span}(S_2) = \text{span}(S_1) \cap \text{span}(S_2)$.

Note that $\text{span}(S_1 \cup S_2) = S_1 + S_2$ is the sum of S_1 and S_2, which was introduced in Exercise 1.1.19.

To see why this holds, notice that if $\{\mathbf{y}_1, \ldots, \mathbf{y}_s\}$ and $\{\mathbf{z}_1, \ldots, \mathbf{z}_t\}$ are bases of S_1 and S_2, respectively, then we can write every member of $\text{span}(\{\mathbf{x}\} \cup S_1) \cap \text{span}(S_2)$ in the form

$$b\mathbf{x} + \sum_{i=1}^{s} c_i \mathbf{y}_i = \sum_{i=1}^{t} d_i \mathbf{z}_i.$$

Moving the $c_i \mathbf{y}_i$ terms over to the right-hand side shows that $b\mathbf{x} \in \text{span}(S_1 \cup S_2)$. Since $\mathbf{x} \notin \text{span}(S_1 \cup S_2)$, this implies $b = 0$, so this member of $\text{span}(\{\mathbf{x}\} \cup S_1) \cap \text{span}(S_2)$ is in fact in $\text{span}(S_1) \cap \text{span}(S_2)$.

With that out of the way, we now inductively prove that B_k is linearly independent, assuming that B_{k+1} is linearly independent (with the base case of $B_\mu = \{\}$ trivially being linearly independent). Recall that C_k is constructed specifically to be linearly independent. For convenience, we start by giving names to the members of C_k:

$$B_{k+1} = \{\mathbf{v}_1, \mathbf{v}_2, \ldots, \mathbf{v}_p\} \quad \text{and} \quad C_k \backslash B_{k+1} = \{\mathbf{w}_1, \mathbf{w}_2, \ldots, \mathbf{w}_q\}.$$

Now suppose that some linear combination of the members of B_k equals zero:

$$\sum_{i=1}^{p} c_i \mathbf{v}_i + \sum_{i=1}^{q} \sum_{j=0}^{k-1} d_{i,j}(A - \lambda I)^j \mathbf{w}_i = \mathbf{0}. \tag{B.2.1}$$

Our goal is to show that this implies $c_i = d_{i,j} = 0$ for all i and j. To this end, notice that if we multiply the linear combination (B.2.1) on the left by $(A - \lambda I)^{k-1}$ then we get

$$
\begin{aligned}
\mathbf{0} &= (A - \lambda I)^{k-1}\left(\sum_{i=1}^{p} c_i \mathbf{v}_i\right) + \sum_{i=1}^{q} \sum_{j=0}^{k-1} d_{i,j}(A - \lambda I)^{j+k-1}\mathbf{w}_i \\
&= (A - \lambda I)^{k-1}\left(\sum_{i=1}^{p} c_i \mathbf{v}_i + \sum_{i=1}^{q} d_{i,0}\mathbf{w}_i\right),
\end{aligned} \tag{B.2.2}
$$

where the second equality comes from the fact that $(A - \lambda I)^{j+k-1}\mathbf{w}_i = \mathbf{0}$ whenever $j \geq 1$, since $C_k \backslash B_{k+1} \subseteq \text{null}\big((A - \lambda I)^k\big) \subseteq \text{null}\big((A - \lambda I)^{j+k-1}\big)$.

In other words, we have shown that $\sum_{i=1}^{p} c_i \mathbf{v}_i + \sum_{i=1}^{q} d_{i,0}\mathbf{w}_i \in \text{null}\big((A - \lambda I)^{k-1}\big)$, which we claim implies $d_{i,0} = 0$ for all $1 \leq i \leq q$. To see why this is the case, we repeatedly use the Claim from the start of this proof with $\mathbf{x}$ ranging over the members of $C_k \backslash B_{k+1}$ (i.e., the vectors that we added one at a time in Step 2 of the proof of Theorem 2.4.1), $S_1 = \text{span}(B_{k+1})$, and $S_2 = \text{null}\big((A - \lambda I)^{k-1}\big)$ to see that

$$
\begin{aligned}
\text{span}(C_k) \cap \text{null}\big((A - \lambda I)^{k-1}\big) &= \text{span}(B_{k+1}) \cap \text{null}\big((A - \lambda I)^{k-1}\big) \\
&\subseteq \text{span}(B_{k+1}).
\end{aligned} \tag{B.2.3}
$$

It follows that $\sum_{i=1}^{p} c_i \mathbf{v}_i + \sum_{i=1}^{q} d_{i,0}\mathbf{w}_i \in \text{span}(B_{k+1})$. Since C_k is linearly independent, this then tells us that $d_{i,0} = 0$ for all i.

We now repeat the above argument, but instead of multiplying the linear combination (B.2.1) on the left by $(A - \lambda I)^{k-1}$, we multiply it on the left by smaller powers of $A - \lambda I$. For example, if we multiply on the left by $(A - \lambda I)^{k-2}$ then we get

$$0 = (A - \lambda I)^{k-2} \left(\sum_{i=1}^{p} c_i \mathbf{v}_i + \sum_{i=1}^{q} d_{i,0} \mathbf{w}_i \right) + (A - \lambda I)^{k-1} \left(\sum_{i=1}^{q} d_{i,1} \mathbf{w}_i \right)$$

$$= (A - \lambda I)^{k-2} \left(\sum_{i=1}^{p} c_i \mathbf{v}_i \right) + (A - \lambda I)^{k-1} \left(\sum_{i=1}^{q} d_{i,1} \mathbf{w}_i \right),$$

where the first equality comes from the fact that $(A - \lambda I)^{j+k-2} \mathbf{w}_i = \mathbf{0}$ whenever $j \geq 2$, and the second equality following from the fact that we already showed that $d_{i,0} = 0$ for all $1 \leq i \leq q$.

In other words, we have shown that $\sum_{i=1}^{p} c_i \mathbf{v}_i + (A - \lambda I) \sum_{i=1}^{q} d_{i,1} \mathbf{w}_i \in \mathrm{null}\big((A - \lambda I)^{k-2}\big)$. If we notice that $\mathrm{null}\big((A - \lambda I)^{k-2}\big) \subseteq \mathrm{null}\big((A - \lambda I)^{k-1}\big)$ then it follows that $\sum_{i=1}^{p} c_i \mathbf{v}_i + \sum_{i=1}^{q} d_{i,1} \mathbf{w}_i \in \mathrm{null}\big((A - \lambda I)^{k-1}\big)$, so repeating the same argument from earlier shows that $d_{i,1} = 0$ for all $1 \leq i \leq q$.

Repeating this argument shows that $d_{i,j} = 0$ for all i and j, so the linear combination (B.2.1) simply says that $\sum_{i=1}^{p} c_i \mathbf{v}_i = 0$. Since B_{k+1} is linearly independent, this immediately implies $c_i = 0$ for all $1 \leq i \leq p$, which finally shows that B_k is linearly independent as well. ∎

The final proof of this subsection shows that the way we defined analytic functions of matrices (Definition 2.4.4) is actually well-defined. That is, we now show that this definition leads to the formula of Theorem 2.4.6 *regardless* of which value a we choose to center the Taylor series at, whereas we originally just proved that theorem in the case when $a = \lambda$. Note that in the statement of this theorem and its proof, we let $N_n \in M_k(\mathbb{C})$ denote the matrix with ones on its n-th superdiagonal and zeros elsewhere, just as in the proof of Theorem 2.4.6.

Theorem B.2.3	Suppose $J_k(\lambda) \in M_k(\mathbb{C})$ is a Jordan block and $f : \mathbb{C} \to \mathbb{C}$ is analytic on some open set D containing λ. Then for all $a \in D$ we have
Functions of Jordan Blocks	$$\sum_{n=0}^{\infty} \frac{f^{(n)}(a)}{n!} \big(J_k(\lambda) - aI \big)^n = \sum_{j=0}^{k-1} \frac{f^{(j)}(\lambda)}{j!} N_j.$$

The left-hand side of the concluding line of this theorem is just $f(J_k(\lambda))$, and the right-hand side is just the large matrix formula given by Theorem 2.4.6.

Proof. Since $J_k(\lambda) = \lambda I + N_1$, we can rewrite the sum on the left as

$$\sum_{n=0}^{\infty} \frac{f^{(n)}(a)}{n!} \big(\lambda I + N_1 - aI \big)^n = \sum_{n=0}^{\infty} \frac{f^{(n)}(a)}{n!} \big((\lambda - a)I + N_1 \big)^n.$$

Using the binomial theorem (Theorem A.2.1) then shows that this sum equals

$$\sum_{n=0}^{\infty} \frac{f^{(n)}(a)}{n!} \left(\sum_{j=0}^{k} \binom{n}{j} (\lambda - a)^{n-j} N_j \right).$$

Swapping the order of these sums and using the fact that $\binom{n}{j} = n!/((n-j)!j!)$ then puts it into the form

We can swap the order of summation here because the sum over n converges

$$\sum_{j=0}^{k} \frac{1}{j!} \left(\sum_{n=0}^{\infty} \frac{f^{(n)}(a)}{(n-j)!} (\lambda - a)^{n-j} \right) N_j. \tag{B.2.4}$$

Since f is analytic we know that $f(\lambda) = \sum_{n=0}^{\infty} \frac{f^{(n)}(a)}{n!}(\lambda - a)^n$. Furthermore, $f^{(j)}$ must also be analytic, and replacing f by $f^{(j)}$ in this Taylor series shows that $f^{(j)}(\lambda) = \sum_{n=0}^{\infty} \frac{f^{(n)}(a)}{(n-j)!}(\lambda - a)^{n-j}$. Substituting this expression into Equation (B.2.4) shows that

Keep in mind that
$N_k = O$, so we can
discard the last term
in this sum.

$$\sum_{j=0}^{k} \frac{1}{j!} \left(\sum_{n=0}^{\infty} \frac{f^{(n)}(a)}{(n-j)!}(\lambda - a)^{n-j} \right) N_j = \sum_{j=0}^{k} \frac{f^{(j)}(\lambda)}{j!} N_j,$$

which completes the proof.   ∎

B.3   Strong Duality for Semidefinite Programs

When discussing the duality properties of semidefinite programs in Section 3.C.3, one of the main results that we presented gave some conditions under which strong duality holds (this was Theorem 3.C.2). We now prove this theorem by making use of the properties of convex sets that were outlined in Appendix A.5.1.

Throughout this section, we use the same notation and terminology as in Section 3.C.3. In particular, $A \succeq B$ means that $A - B$ is positive semidefinite, $A \succ B$ means that $A - B$ is positive definite, and a primal/dual pair of semidefinite programs is a pair of optimization problems of the following form, where $B \in \mathcal{M}_m^H, C \in \mathcal{M}_n^H$, and $\Phi : \mathcal{M}_n^H \to \mathcal{M}_m^H$ is linear:

$X \in \mathcal{M}_n^H$ and $Y \in \mathcal{M}_m^H$
are matrix variables.

Primal	**Dual**
maximize: $\mathrm{tr}(CX)$	minimize: $\mathrm{tr}(BY)$
subject to: $\Phi(X) \preceq B$	subject to: $\Phi^*(Y) \succeq C$
$X \succeq O$	$Y \succeq O$

We start by proving a helper theorem that does most of the heavy lifting for strong duality. We note that we use the closely-related Exercises 2.2.20 and 2.2.21 in this proof, which tell us that $A \succeq O$ if and only if $\mathrm{tr}(AB) \geq 0$ whenever $B \succeq O$, as well as some other closely-related variants of this fact involving positive *definite* matrices.

Theorem B.3.1

Farkas Lemma

Suppose $B \in \mathcal{M}_m^H$ and $\Phi : \mathcal{M}_n^H \to \mathcal{M}_m^H$ is linear. The following are equivalent:

a) There does not exist $O \preceq X \in \mathcal{M}_n^H$ such that $\Phi(X) \prec B$.

b) There exists a non-zero $O \preceq Y \in \mathcal{M}_m^H$ such that $\Phi^*(Y) \succeq O$ and $\mathrm{tr}(BY) \leq 0$.

Proof. To see that (b) implies (a), suppose that Y is as described in (b). Then if a matrix X as described in (a) *did* exist, we would have

In particular, the first
and second
inequalities here use
Exercise 2.2.20.

$$0 < \mathrm{tr}\big((B - \Phi(X))Y\big) \qquad \text{(since } Y \neq O, Y \succeq O, B - \Phi(X) \succ O)$$
$$= \mathrm{tr}(BY) - \mathrm{tr}(X\Phi^*(Y)) \qquad \text{(linearity of trace, definition of adjoint)}$$
$$\leq 0, \qquad \text{(since } X \succeq O, \Phi^*(Y) \succeq O, \mathrm{tr}(BY) \leq 0)$$

which is impossible, so such an X does not exist.

For the opposite implication, we notice that if (a) holds then the set

$$S = \{B - \Phi(X) : X \in \mathcal{M}_n^H \text{ is positive semidefinite}\}$$

is disjoint from the set T of positive definite matrices in $\mathcal{M}_m^H$. Since T is open and both of these sets are convex, the separating hyperplane theorem (Theorem A.5.1) tells us that there is a linear form $f : \mathcal{M}_m^H \to \mathbb{R}$ and a scalar $c \in \mathbb{R}$ such that

$$f(Z) > c \geq f(B - \Phi(X))$$

for all $O \preceq X \in \mathcal{M}_n^H$ and all $O \prec Z \in \mathcal{M}_m^H$. We know from Exercise 1.3.8 that there exists a matrix $Y \in \mathcal{M}_m^H$ such that $f(A) = \operatorname{tr}(AY)$ for all $A \in \mathcal{M}_m^H$, so the above string of inequalities can be written in the form

$$\operatorname{tr}(ZY) > c \geq \operatorname{tr}\big((B - \Phi(X))Y\big) \qquad \text{(separating hyperplane)}$$
$$= \operatorname{tr}(BY) - \operatorname{tr}\big(X\Phi^*(Y)\big) \quad \text{(linearity of trace, definition of adjoint)}$$

for all $O \preceq X \in \mathcal{M}_n^H$ and all $O \prec Z \in \mathcal{M}_m^H$.

It follows that $Y \neq O$ (otherwise we would have $0 > c \geq 0$, which is impossible). We also know from Exercise 2.2.21(b) that the inequality $\operatorname{tr}(ZY) > c$ for all $Z \succ O$ implies $Y \succeq O$, and furthermore that we can choose $c = 0$. It follows that $0 = c \geq \operatorname{tr}(BY) - \operatorname{tr}\big(X\Phi^*(Y)\big)$, so $\operatorname{tr}\big(X\Phi^*(Y)\big) \geq \operatorname{tr}(BY)$ for all $X \succeq O$. Using Exercise 2.2.21(a) then shows that $\Phi^*(Y) \succeq O$ and that $\operatorname{tr}(BY) \leq 0$, which completes the proof. ∎

With the Farkas Lemma under our belt, we can now prove the main result of this section, which we originally stated as Theorem 3.C.2, but we re-state here for ease of reference.

Theorem B.3.2	Suppose that both problems in a primal/dual pair of SDPs are feasible, and at least one of them is strictly feasible. Then the optimal values of those SDPs coincide. Furthermore,
Strong Duality for Semidefinite Programs	a) if the primal problem is strictly feasible then the optimal value is attained in the dual problem, and
	b) if the dual problem is strictly feasible then the optimal value is attained in the primal problem.

Proof. We just prove part (a) of the theorem, since part (b) then follows from swapping the roles of the primal and dual problems.

Let α be the optimal value of the primal problem (which is not necessarily attained). If we define $\widetilde{\Phi} : \mathcal{M}_n^H \to \mathcal{M}_{m+1}^H$ and $B \in \mathcal{M}_{m+1}^H$ by

$$\widetilde{\Phi}(X) = \begin{bmatrix} -\operatorname{tr}(CX) & 0 \\ 0 & \Phi(X) \end{bmatrix} \quad \text{and} \quad \widetilde{B} = \begin{bmatrix} -\alpha & 0 \\ 0 & B \end{bmatrix},$$

then there does not exist $O \preceq X \in \mathcal{M}_n^H$ such that $\widetilde{\Phi}(X) \prec \widetilde{B}$ (since otherwise we would have $\Phi(X) \prec B$ and $\operatorname{tr}(CX) > \alpha$, which would mean that the objective function of the primal problem can be made larger than its optimal value α). Applying Theorem B.3.1 to $\widetilde{B}$ and $\widetilde{\Phi}$ then tells us that there exists a non-zero $O \preceq \widetilde{Y} \in \mathcal{M}_{m+1}^H$ such that $\widetilde{\Phi}^*(\widetilde{Y}) \succeq O$ and $\operatorname{tr}(\widetilde{B}\widetilde{Y}) \leq 0$. If we write

$$\widetilde{Y} = \begin{bmatrix} y & * \\ * & Y \end{bmatrix} \quad \text{then} \quad \widetilde{\Phi}^*(\widetilde{Y}) = \Phi^*(Y) - yC.$$

We use asterisks ($*$) to denote entries of $\widetilde{Y}$ that are irrelevant to us.

The inequality $\mathrm{tr}(\widetilde{B}\widetilde{Y}) \leq 0$ then tells us that $\mathrm{tr}(BY) \leq y\alpha$, and the constraint $\widetilde{\Phi}^*(\widetilde{Y}) \succeq O$ tells us that $\Phi^*(Y) \succeq yC$. If $y \neq 0$ then it follows that Y/y is a feasible point of the dual problem that produces a value in the objective function no larger than α (and thus necessarily equal to α, by weak duality), which is exactly what we wanted to find. All that remains is to show that $y \neq 0$, so that we know that Y/y exists.

Keep in mind that $\widetilde{Y} \succeq O$ implies $Y \succeq O$ and $y \geq 0$.

To pin down this final detail, we notice that if $y = 0$ then $\Phi^*(Y) = \widetilde{\Phi}^*(\widetilde{Y}) \succeq O$ and $\mathrm{tr}(BY) = \mathrm{tr}(\widetilde{B}\widetilde{Y}) \leq 0$. Applying Theorem B.3.1 (to B and Φ this time) shows that the primal problem is not strictly feasible, which is a contradiction that completes the proof. ∎

If $y = 0$, we know that $Y \neq O$ since $\widetilde{Y} \neq O$.

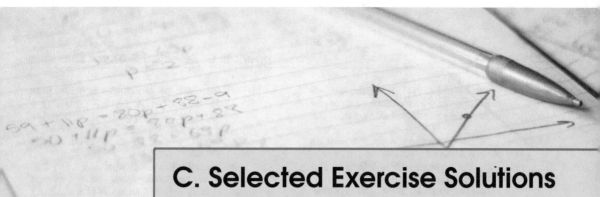

C. Selected Exercise Solutions

Section 1.1: Vector Spaces and Subspaces

1.1.1 (a) Not a subspace. For example, if $c = -1$ and $\mathbf{v} = (1,1)$ then $c\mathbf{v} = (-1,-1)$ is not in the set.

(c) Not a subspace. For example, I and $-I$ are both invertible, but $I + (-I) = O$, which is not invertible.

(e) Not a subspace. For example, $f(x) = x$ is in the set, but $2f$ is not since $2f(2) = 4 \neq 2$.

(g) Is a subspace. If f and g are even functions and $c \in \mathbb{R}$ then $f + g$ and cf are even functions too, since $(f+g)(x) = f(x) + g(x) = f(-x) + g(-x) = (f+g)(-x)$ and $(cf)(x) = cf(x) = cf(-x) = (cf)(-x)$.

(i) Is a subspace. If f and g are functions in this set, then $(f+g)''(x) - 2(f+g)(x) = (f''(x) - 2f(x)) + (g''(x) - 2g(x)) = 0 + 0 = 0$. Similarly, if $c \in \mathbb{R}$ then $(cf)''(x) - 2(cf)(x) = c(f''(x) - 2f(x)) = c0 = 0$.

(k) Is a subspace. If A and B both have trace 0 then $\text{tr}(cA) = c\text{tr}(A) = 0$ and $\text{tr}(A+B) = \text{tr}(A) + \text{tr}(B) = 0$ as well.

1.1.2 (a) Linearly independent, because every set containing exactly one non-zero vector is linearly independent (if there is only one vector $\mathbf{v}$ then the only way for $c_1\mathbf{v} = \mathbf{0}$ to hold is if $c_1 = 0$).

(c) Linearly dependent, since $\sin^2(x) + \cos^2(x) = 1$.

(e) Linearly dependent, since the angle-sum trigonometric identities tell us that

$$\sin(x+1) = \cos(1)\sin(x) + \sin(1)\cos(x).$$

We have thus written $\sin(x+1)$ as a linear combination of $\sin(x)$ and $\cos(x)$ (since $\cos(1)$ and $\sin(1)$ are just scalars).

(g) Linearly independent. To see this, consider the equation

$$c_1e^x + c_2xe^x + c_3x^2e^x = 0.$$

This equation has to be true no matter what value of x we plug in. Plugging in $x = 0$ tells us that $c_1 = 0$. Plugging in $x = 1$ (and using $c_1 = 0$) tells us that $c_2e + c_3e = 0$, so $c_2 + c_3 = 0$. Plugging in $x = 2$ (and using $c_1 = 0$) tells us that $2c_2e^2 + 4c_3e^2 = 0$, so $c_2 + 2c_3 = 0$. This is a system of two linear equations in two unknowns, and it is straightforward to check that its only solution is $c_2 = c_3 = 0$. Since we already showed that $c_1 = 0$ too, the functions are linearly independent.

1.1.3 (a) Not a basis, since it does not span all of $\mathcal{M}_2$. For example,

$$\begin{bmatrix} 1 & 0 \\ 0 & 0 \end{bmatrix} \notin \text{span}\left(\begin{bmatrix} 1 & 0 \\ 0 & 1 \end{bmatrix}, \begin{bmatrix} 1 & 1 \\ 0 & 1 \end{bmatrix}, \begin{bmatrix} 1 & 0 \\ 1 & 1 \end{bmatrix} \right),$$

since any linear combination of the three matrices on the right has both diagonal entries equal to each other.

(c) Is a basis. To see this, we note that

$$c_1 \begin{bmatrix} 1 & 0 \\ 0 & 1 \end{bmatrix} + c_2 \begin{bmatrix} 1 & 1 \\ 0 & 1 \end{bmatrix}$$
$$+ c_3 \begin{bmatrix} 1 & 0 \\ 1 & 1 \end{bmatrix} + c_4 \begin{bmatrix} 1 & 1 \\ 1 & 0 \end{bmatrix}$$
$$= \begin{bmatrix} c_1 + c_2 + c_3 + c_4 & c_2 + c_4 \\ c_3 + c_4 & c_1 + c_2 + c_3 \end{bmatrix}.$$

To make this equal to a given matrix

$$\begin{bmatrix} a & b \\ c & d \end{bmatrix},$$

we solve the linear system to get the unique solution $c_1 = 2a - b - c - d$, $c_2 = -a + b + d$, $c_3 = -a + c + d$, $c_4 = a - d$. This shows that this set of matrices spans all of $\mathcal{M}_2$. Furthermore, since this same linear system has a unique solution when $a = b = c = d = 0$, we see that this set is linearly independent, so it is a basis.

(e) Not a basis, since the set is linearly dependent. For example,

$$x^2 - x = (x^2 + 1) - (x + 1).$$

(g) Is a basis. We can use the same technique as in Example 1.1.15 to see that this set is linearly independent. Specifically, we want to know if there exist (not all zero) scalars $c_0, c_1, c_2, \ldots, c_n$ such that

$$c_0 + c_1(x-1) + c_2(x-1)^2 + \cdots + c_n(x-1)^n = 0.$$

Plugging in $x = 1$ shows that $c_0 = 0$, and then taking derivatives and plugging in $x = 1$ shows that $c_1 = c_2 = \cdots = c_n = 0$ as well, so the set is linearly independent. The fact that this set spans $\mathcal{P}$ follows either from the fact that we can build a Taylor series for polynomials centered at $x = 1$, or from the fact that in a linear combination of the form we can first choose c_n to give us whatever coefficient of x^n we want, then choose c_{n-1} to give us whatever coefficient of x^{n-1} we want, and so on.

1.1.4   **(b)** True. Since $\mathcal{V}$ must contain a zero vector $\mathbf{0}$, we can just define $\mathcal{W} = \{\mathbf{0}\}$.

     (d) False. For example, let $\mathcal{V} = \mathbb{R}^2$ and let $B = \{(1,0)\}, C = \{(2,0)\}$.

     (f) False. For example, if $\mathcal{V} = \mathcal{P}$ and $B = \{1, x, x^2, \ldots\}$ then $\mathrm{span}(B) = \mathcal{P}$, but there is no finite subset of B whose span is all of $\mathcal{P}$.

     (h) False. The set $\{\mathbf{0}\}$ is linearly dependent (but every other single-vector set is indeed linearly independent).

1.1.5 Property (i) of vector spaces tells us that $(-c)\mathbf{v} = (-1)(c\mathbf{v})$, and Theorem 1.1.1(b) tells us that $(-1)(c\mathbf{v}) = -(c\mathbf{v})$.

1.1.7 All 10 of the defining properties from Definition 1.1.1 are completely straightforward and follow immediately from the corresponding properties of $\mathbb{F}$. For example $\mathbf{v} + \mathbf{w} = \mathbf{w} + \mathbf{v}$ (property (b)) for all $\mathbf{v}, \mathbf{w} \in \mathbb{F}^{\mathbb{N}}$ because $v_j + w_j = w_j + v_j$ for all $j \in \mathbb{N}$.

1.1.8 If $A, B \in \mathcal{M}_n^S$ and $c \in \mathbb{F}$ then $(A + B)^T = A^T + B^T = A + B$, so $A + B \in \mathcal{M}_n^S$ and $(cA)^T = cA^T = cA$, so $cA \in \mathcal{M}_n^S$.

1.1.10   **(a)** $\mathcal{C}$ is a subspace of $\mathcal{F}$ because if f, g are continuous and $c \in \mathbb{R}$ then $f + g$ and cf are also continuous (both of these facts are typically proved in calculus courses).

     (b) $\mathcal{D}$ is a subspace of $\mathcal{F}$ because if f, g are differentiable and $c \in \mathbb{R}$ then $f + g$ and cf are also differentiable. In particular, $(f + g)' = f' + g'$ and $(cf)' = cf'$ (both of these facts are typically proved in calculus courses).

1.1.11 If S is closed under linear combinations then $c\mathbf{v} \in S$ and $\mathbf{v} + \mathbf{w} \in S$ whenever $c \in \mathbb{F}$ and $\mathbf{v}, \mathbf{w} \in S$ simply because $c\mathbf{v}$ and $\mathbf{v} + \mathbf{w}$ are both linear combinations of members of S, so S is a subspace of $\mathcal{V}$.

Conversely, if S is a subspace of $\mathcal{V}$ and $\mathbf{v}_1, \ldots, \mathbf{v}_k \in S$ then $c_j \mathbf{v}_j \in S$ for each $1 \leq j \leq k$. Then repeatedly using closure under vector addition gives $c_1 \mathbf{v}_1 + c_2 \mathbf{v}_2 \in S$, so $(c_1 \mathbf{v}_1 + c_2 \mathbf{v}_2) + c_3 \mathbf{v}_3 \in S$, and so on to $c_1 \mathbf{v}_1 + c_2 \mathbf{v}_2 + \cdots + c_k \mathbf{v}_k \in S$.

1.1.13 The fact that $\{E_{1,1}, E_{1,2}, \ldots, E_{m,n}\}$ spans $\mathcal{M}_{m,n}$ is clear: every matrix $A \in \mathcal{M}_{m,n}$ can be written in the form

$$A = \sum_{i=1}^{m} \sum_{j=1}^{n} a_{i,j} E_{i,j}.$$

Linear independence of $\{E_{1,1}, E_{1,2}, \ldots, E_{m,n}\}$ follows from the fact that if

$$\sum_{i=1}^{m} \sum_{j=1}^{n} c_{i,j} E_{i,j} = O$$

then

$$\begin{bmatrix} c_{1,1} & c_{1,2} & \cdots & c_{1,n} \\ c_{2,1} & c_{2,2} & \cdots & c_{2,n} \\ \vdots & \vdots & \ddots & \vdots \\ c_{m,1} & c_{m,2} & \cdots & c_{m,n} \end{bmatrix} = O,$$

so $c_{i,j} = 0$ for all i, j.

1.1.14   **(a)** We check the two closure properties from Theorem 1.1.2: if $f, g \in \mathcal{P}^O$ then $(f + g)(-x) = f(-x) + g(-x) = -f(x) - g(x) = -(f + g)(x)$, so $f + g \in \mathcal{P}^O$ too, and if $c \in \mathbb{R}$ then $(cf)(-x) = cf(-x) = -cf(x)$, so $cf \in \mathcal{P}^O$ too.

     (b) We first notice that $\{x, x^3, x^5, \ldots\} \subset \mathcal{P}^O$. This set is linearly independent since it is a subset of the linearly independent set $\{1, x, x^2, x^3, \ldots\}$ from Example 1.1.16. To see that it spans $\mathcal{P}^O$, we notice that if

$$f(x) = a_0 + a_1 x + a_2 x^2 + a_3 x^3 + \cdots \in \mathcal{P}^O$$

then $f(x) - f(-x) = 2f(x)$, so

$$2f(x) = f(x) - f(-x)$$
$$= (a_0 + a_1 x + a_2 x^2 + a_3 x^3 + \cdots)$$
$$- (a_0 - a_1 x + a_2 x^2 - a_3 x^3 + \cdots)$$
$$= 2(a_1 x + a_3 x^3 + a_5 x^5 + \cdots).$$

It follows that $f(x) = a_1 x + a_3 x^3 + a_5 x^5 + \cdots \in \mathrm{span}(x, x^3, x^5, \ldots)$, so $\{x, x^3, x^5, \ldots\}$ is indeed a basis of $\mathcal{P}^O$.

1.1.17   **(a)** We have to show that the following two properties hold for $S_1 \cap S_2$: (a) if $\mathbf{v}, \mathbf{w} \in S_1 \cap S_2$ then $\mathbf{v} + \mathbf{w} \in S_1 \cap S_2$ as well, and (b) if $\mathbf{v} \in S_1 \cap S_2$ and $c \in \mathbb{F}$ then $c\mathbf{v} \in S_1 \cap S_2$.

     For property (a), we note that $\mathbf{v} + \mathbf{w} \in S_1$ (since S_1 is a subspace) and $\mathbf{v} + \mathbf{w} \in S_2$ (since S_2 is a subspace). It follows that $\mathbf{v} + \mathbf{w} \in S_1 \cap S_2$ as well.

     Property (b) is similar: we note that $c\mathbf{v} \in S_1$ (since S_1 is a subspace) and $c\mathbf{v} \in S_2$ (since S_2 is a subspace). It follows that $c\mathbf{v} \in S_1 \cap S_2$ as well.

     (b) If $\mathcal{V} = \mathbb{R}^2$ and $S_1 = \{(x, 0) : x \in \mathbb{R}\}$, $S_2 = \{(0, y) : y \in \mathbb{R}\}$ then $S_1 \cup S_2$ is the set of points with at least one of their coordinates equal to 0. It is not a subspace since $(1, 0), (0, 1) \in S_1 \cup S_2$ but $(1, 0) + (0, 1) = (1, 1) \notin S_1 \cup S_2$.

1.1.19   **(a)** $S_1 + S_2$ consists of all vectors of the form $(x, 0, 0) + (0, y, 0) = (x, y, 0)$, which is the xy-plane.

     (b) We claim that $\mathcal{M}_2^S + \mathcal{M}_2^{sS}$ is all of $\mathcal{M}_2$. To verify this claim, we need to show that we can write every 2×2 matrix as a sum of a symmetric and a skew-symmetric matrix. This can be done as follows:

$$\begin{bmatrix} a & b \\ c & d \end{bmatrix} = \frac{1}{2} \begin{bmatrix} 2a & b+c \\ b+c & 2d \end{bmatrix} + \frac{1}{2} \begin{bmatrix} 0 & b-c \\ c-b & 0 \end{bmatrix}$$

[Note: We discuss a generalization of this fact in Example 1.B.3 and Remark 1.B.1.]

     (c) We check that the two properties of Theorem 1.1.2 are satisfied by $S_1 + S_2$. For property (a), notice that if $\mathbf{v}, \mathbf{w} \in S_1 + S_2$ then there exist $\mathbf{v}_1, \mathbf{w}_1 \in S_1$ and $\mathbf{v}_2, \mathbf{w}_2 \in S_2$ such that $\mathbf{v} = \mathbf{v}_1 + \mathbf{v}_2$ and $\mathbf{w} = \mathbf{w}_1 + \mathbf{w}_2$. Then

$$\mathbf{v} + \mathbf{w} = (\mathbf{v}_1 + \mathbf{v}_2) + (\mathbf{w}_1 + \mathbf{w}_2)$$
$$= (\mathbf{v}_1 + \mathbf{w}_1) + (\mathbf{v}_2 + \mathbf{w}_2) \in S_1 + S_2.$$

Similarly, if $c \in \mathbb{F}$ then

$$c\mathbf{v} = c(\mathbf{v}_1 + \mathbf{v}_2) = c\mathbf{v}_1 + c\mathbf{v}_2 \in S_1 + S_2,$$

since $c\mathbf{v}_1 \in S_1$ and $c\mathbf{v}_2 \in C_2$.

1.1.20 If every vector in $\mathcal{V}$ can be written as a linear combination of the members of B, then B spans $\mathcal{V}$ by definition, so we only need to show linear independence of B. To this end, suppose that $\mathbf{v} \in \mathcal{V}$ can be written as a linear combination of the members of B in exactly one way:

$$\mathbf{v} = c_1\mathbf{v}_1 + c_2\mathbf{v}_2 + \cdots + c_k\mathbf{v}_k$$

for some $\mathbf{v}_1, \mathbf{v}_2, \ldots, \mathbf{v}_k \in B$. If B were linearly dependent, then there must be a non-zero linear combination of the form

$$\mathbf{0} = d_1\mathbf{w}_1 + d_2\mathbf{w}_2 + \cdots + d_m\mathbf{w}_m$$

for some $\mathbf{w}_1, \mathbf{w}_2, \ldots, \mathbf{w}_m \in B$. By adding these two linear combinations, we see that

$$\mathbf{v} = c_1\mathbf{v}_1 + \cdots + c_k\mathbf{v}_k + d_1\mathbf{w}_1 + \cdots + d_m\mathbf{w}_m.$$

Since not all of the d_j's are zero, this is a different linear combination that gives $\mathbf{v}$, which contradicts uniqueness. We thus conclude that B must in fact be linearly independent.

1.1.21 (a) Since $f_1, f_2, \ldots, f_n$ are linearly dependent, it follows that there exist scalars $c_1, c_2, \ldots, c_n$ such that

$$c_1 f_1(x) + c_2 f_2(x) + \cdots + c_n f_n(x) = 0$$

for all $x \in \mathbb{R}$. Taking the derivative of this equation gives

$$c_1 f_1'(x) + c_2 f_2'(x) + \cdots + c_n f_n'(x) = 0,$$

taking the derivative again gives

$$c_1 f_1''(x) + c_2 f_2''(x) + \cdots + c_n f_n''(x) = 0,$$

and so on. We thus see that

$$c_1 \begin{bmatrix} f_1(x) \\ f_1'(x) \\ f_1''(x) \\ \vdots \\ f_1^{(n-1)}(x) \end{bmatrix} + \cdots + c_n \begin{bmatrix} f_n(x) \\ f_n'(x) \\ f_n''(x) \\ \vdots \\ f_n^{(n-1)}(x) \end{bmatrix} = \begin{bmatrix} 0 \\ 0 \\ 0 \\ \vdots \\ 0 \end{bmatrix}$$

for all $x \in \mathbb{R}$ as well. This is equivalent to saying that the columns of $W(x)$ are linearly dependent for all x, which is equivalent to $W(x)$ not being invertible for all x, which is equivalent to $\det(W(x)) = 0$ for all x.

(b) It is straightforward to compute $W(x)$:

$$W(x) = \begin{bmatrix} x & \ln(x) & \sin(x) \\ 1 & 1/x & \cos(x) \\ 0 & -1/x^2 & -\sin(x) \end{bmatrix}.$$

Then

$$\begin{aligned} \det(W(x)) &= (-\sin(x)) + 0 - \sin(x)/x^2 \\ &\quad + \cos(x)/x + \ln(x)\sin(x) - 0 \\ &= \sin(x)(\ln(x) - 1 - 1/x^2) + \cos(x)/x. \end{aligned}$$

To prove linear independence, we just need to find a particular value of x such that the above expression is non-zero. Almost any choice of x works: one possibility is $x = \pi$, which gives $\det(W(x)) = -1/\pi$.

(c) Notice that the k-th derivative of x^k is the scalar function $k!$, and every higher-order derivative of x^k is 0. It follows that $W(x)$ is an upper triangular matrix with $1, 1!, 2!, 3!, \ldots$ on its diagonal, so $\det(W(x)) = 1 \cdot 1! \cdot 2! \cdot 3! \cdots n! \neq 0$.

(d) We start by showing that $\det(W(x)) = 0$ for all x. First, we note that $f_2'(x) = 2|x|$ for all x. To see this, split it into three cases: If $x > 0$ then $f_2(x) = x^2$, which has derivative $2x = 2|x|$. If $x < 0$ then $f_2(x) = -x^2$, which has derivative $-2x = 2|x|$. If $x = 0$ then we have to use the limit definition of a derivative

$$\begin{aligned} f_2'(0) &= \lim_{h \to 0} \frac{f_2(h) - f_2(0)}{h} \\ &= \lim_{h \to 0} \frac{h|h| - 0}{h} = \lim_{h \to 0} |h| = 0, \end{aligned}$$

which also equals $2|x|$ (since $x = 0$). With that out of the way, we can compute

$$\begin{aligned} \det(W(x)) &= \det\left(\begin{bmatrix} x^2 & x|x| \\ 2x & 2|x| \end{bmatrix} \right) \\ &= 2x^2|x| - 2x^2|x| = 0 \end{aligned}$$

for all $x \in \mathbb{R}$, as desired.

Now we will show that f_1 and f_2 are linearly independent. To see this, suppose there exist constants c_1, c_2 such that

$$c_1 x^2 + c_2 x|x| = 0$$

for all x. Plugging in $x = 1$ gives the equation $c_1 + c_2 = 0$, while plugging in $x = -1$ gives the equation $c_1 - c_2 = 0$. It is straightforward to solve this system of two equations to get $c_1 = c_2 = 0$, which implies that x^2 and $x|x|$ are linearly independent.

1.1.22 The Wronskian of $\{e^{a_1 x}, e^{a_2 x}, \ldots, e^{a_n x}\}$ (see Exercise 1.1.21) is

$$\det\left(\begin{bmatrix} e^{a_1 x} & e^{a_2 x} & \cdots & e^{a_n x} \\ a_1 e^{a_1 x} & a_2 e^{a_2 x} & \cdots & a_n e^{a_n x} \\ a_1^2 e^{a_1 x} & a_2^2 e^{a_2 x} & \cdots & a_n^2 e^{a_n x} \\ \vdots & \vdots & \ddots & \vdots \\ a_1^{n-1} e^{a_1 x} & a_2^{n-1} e^{a_2 x} & \cdots & a_n^{n-1} e^{a_n x} \end{bmatrix} \right) =$$

$$e^{a_1 x} e^{a_2 x} \cdots e^{a_n x} \det\left(\begin{bmatrix} 1 & 1 & \cdots & 1 \\ a_1 & a_2 & \cdots & a_n \\ a_1^2 & a_2^2 & \cdots & a_n^2 \\ \vdots & \vdots & \ddots & \vdots \\ a_1^{n-1} & a_2^{n-1} & \cdots & a_n^{n-1} \end{bmatrix} \right),$$

with the equality coming from using multilinearity of the determinant to pull $e^{a_1 x}$ out of the first column of the matrix, $e^{a_2 x}$ out of the second column, and so on. Since the matrix on the right is the transpose of a Vandermonde matrix, we know that it is invertible and thus has non-zero determinant. Since $e^{a_1 x} e^{a_2 x} \cdots e^{a_n x} \neq 0$ as well, we conclude that the Wronskian is non-zero, so this set of functions is linearly independent.

Section 1.2: Coordinates and Linear Transformations

1.2.1 (a) $[\mathbf{v}]_B = (1,5)/2$　　　(c) $[\mathbf{v}]_B = (1,4,2,3)$

1.2.2 (b) One basis is $\{E_{i,i} : 1 \leq i \leq n\} \cup \{E_{i,j} + E_{j,i} : 1 \leq i < j \leq n\}$. This set contains $n + n(n-1)/2 = n(n+1)/2$ matrices, so that is the dimension of $\mathcal{V}$.

(d) One basis is $\{E_{i,j} - E_{j,i} : 1 \leq i < j \leq n\}$. This set contains $n(n-1)/2$ matrices, so that is the dimension of $\mathcal{V}$.

(f) One basis is $\{E_{j,j} : 1 \leq j \leq n\} \cup \{E_{j,k} + E_{k,j} : 1 \leq k < j \leq n\} \cup \{iE_{j,k} - iE_{k,j} : 1 \leq j < k \leq n\}$. This set contains $n + n(n-1)/2 + n(n-1)/2 = n^2$ matrices, so that is the dimension of $\mathcal{V}$.

(h) One basis is $\{(x-3), (x-3)^2, (x-3)^3\}$, so $\dim(\mathcal{V}) = 3$.

(j) To find a basis, we plug an arbitrary polynomial $f(x) = a_0 + a_1 x + a_2 x^2 + a_3 x^3$ into the equation that defines the vector space. If we do this, we get

$$(a_0 + a_1 x + a_2 x^2 + a_3 x^3)$$
$$- (a_1 x + 2a_2 x^2 + 3a_3 x^3) = 0,$$

so matching up powers of x shows that $a_0 = 0$, $-a_2 = 0$, and $-2a_3 = 0$. It follows that the only polynomials in $\mathcal{P}^3$ satisfying the indicated equation are those of the form $a_1 x$, so a basis is $\{x\}$ and $\mathcal{V}$ is 1-dimensional.

(l) $\mathcal{V}$ is 3-dimensional, since a basis is $\{e^x, e^{2x}, e^{3x}\}$ itself. It spans $\mathcal{V}$ by definition, so we just need to show that it is a linearly independent set. This can be done by plugging in 3 particular x values in the equation $c_1 e^x + c_2 e^{2x} + c_3 e^{3x}$ and then showing that the only solution is $c_1 = c_2 = c_3 = 0$, but we just note that it follows immediately from Exercise 1.1.22.

1.2.3 (b) Is a linear transformation, since $T(cf(x)) = cf(2x-1) = cT(f(x))$ and $T(f(x) + g(x)) = f(2x-1) + g(2x-1) = T(f(x)) + T(g(x))$.

(d) Is a linear transformation, since $R_B(A+C) = (A+C)B = AB + CB = R_B(A) + R_B(C)$ and $R_B(cA) = (cA)B = c(AB) = cR_B(A)$.

(f) Not a linear transformation. For example, $T(iA) = (iA)^* = -iA \neq iA = iT(A)$.

1.2.4 (a) True. In fact, this is part of the definition of a vector space being finite-dimensional.

(c) False. $\mathcal{P}^3$ is 4-dimensional (in general, $\mathcal{P}^p$ is $(p+1)$-dimensional).

(e) False. Its only basis is $\{\}$, which has 0 vectors, so $\dim(\mathcal{V}) = 0$.

(g) False. It is a basis of $\mathcal{P}$ (the vector space of polynomials).

(i) True. In fact, it is its own inverse since $(A^T)^T = A$ for all $A \in \mathcal{M}_{m,n}$.

1.2.5 (b) $P_{C \leftarrow B} = \dfrac{1}{9} \begin{bmatrix} -1 & 5 & -2 \\ 4 & -2 & -1 \\ 2 & -1 & 4 \end{bmatrix}$

(d) $P_{C \leftarrow B} = \begin{bmatrix} 1 & 1 & 1 & 1 \\ 0 & 1 & 0 & 1 \\ 0 & 0 & 1 & 1 \\ 1 & 1 & 1 & 0 \end{bmatrix}$

1.2.6 (b) Since

$$T(ax^2 + bx + c) = a(3x+1)^2 + b(3x+1) + c$$
$$= 9ax^2 + (6a+3b)x + (a+b+c),$$

we conclude that

$$[T]_{D \leftarrow B} = \begin{bmatrix} 9 & 0 & 0 \\ 6 & 3 & 0 \\ 1 & 1 & 1 \end{bmatrix}.$$

(d) We return to this problem much later, in Theorem 3.1.7. For now, we can solve it directly to get

$$[T]_{D \leftarrow B} = \begin{bmatrix} 0 & -2 & 2 & 0 \\ -3 & -3 & 0 & 2 \\ 3 & 0 & 3 & -2 \\ 0 & 3 & -3 & 0 \end{bmatrix}.$$

1.2.8 We start by showing that this set is linearly independent. To see why this is the case, consider the matrix equation

$$\sum_{i=1}^{n} \sum_{j=i}^{n} c_{i,j} S_{i,j} = O.$$

The top row of the matrix on the left equals $(\sum_{j=1}^{n} c_{1,j}, c_{1,2}, c_{1,3}, \ldots, c_{1,n})$, so we conclude that $c_{1,2} = c_{1,3} = \cdots = c_{1,n} = 0$ and thus $c_{1,1} = 0$ as well. A similar argument with the second row, third row, and so on then shows that $c_{i,j} = 0$ for all i and j, so B is linearly independent.

We could use a similar argument to show that B spans $\mathcal{M}_n^S$, but instead we just recall from Exercise 1.2.2 that $\dim(\mathcal{M}_n^S) = n(n+1)/2$, which is exactly the number of matrices in B. It follows from the upcoming Exercise 1.2.27(a) that B is a basis of $\mathcal{M}_n^S$.

1.2.10 (b) Let $B = \{e^x \sin(2x), e^x \cos(2x)\}$ be a basis of the vector space $\mathcal{V} = \text{span}(B)$ and let $D : \mathcal{V} \to \mathcal{V}$ be the derivative map. We start by computing $[D]_B$:

$$D(e^x \sin(2x)) = e^x \sin(2x) + 2e^x \cos(2x),$$
$$D(e^x \cos(2x)) = e^x \cos(2x) - 2e^x \sin(2x).$$

It follows that

$$[D]_B = \begin{bmatrix} 1 & -2 \\ 2 & 1 \end{bmatrix}.$$

The inverse of this matrix is

$$[D^{-1}]_B = \frac{1}{5} \begin{bmatrix} 1 & 2 \\ -2 & 1 \end{bmatrix}.$$

Since the coordinate vector of $e^x \sin(2x)$ is $(1,0)$, we then know that the coordinate vector of $\int e^x \sin(2x)\, dx$ is

$$\frac{1}{5} \begin{bmatrix} 1 & 2 \\ -2 & 1 \end{bmatrix} \begin{bmatrix} 1 \\ 0 \end{bmatrix} = \frac{1}{5} \begin{bmatrix} 1 \\ -2 \end{bmatrix}.$$

It follows that $\int e^x \sin(2x)\, dx = e^x \sin(2x)/5 - 2e^x \cos(2x)/5 + C$.

1.2.12 We prove this in the upcoming Definition 3.1.3, and the text immediately following it.

1.2.13 If $A^T = \lambda A$ then $a_{j,i} = \lambda a_{i,j}$ for all i,j, so $a_{i,j} = \lambda(\lambda a_{i,j}) = \lambda^2 a_{i,j}$ for all i,j. If $A \neq O$ then this implies $\lambda = \pm 1$. The fact that the eigenspaces are $\mathcal{M}_n^S$ and $\mathcal{M}_n^{sS}$ follows directly from the defining properties $A^T = A$ and $A^T = -A$ of these spaces.

1.2.15 (a) $\text{range}(T) = \mathcal{P}^2$, $\text{null}(T) = \{\mathbf{0}\}$.
(b) The only eigenvalue of T is $\lambda = 1$, and the corresponding eigenspace is $\mathcal{P}^0$ (the constant functions).
(c) One square root is the transformation $S: \mathcal{P}^2 \to \mathcal{P}^2$ given by $S(f(x)) = f(x+1/2)$.

1.2.18 $[D]_B$ can be (complex) diagonalized as $[D]_B = PDP^{-1}$ via

$$P = \begin{bmatrix} 1 & 1 \\ -i & i \end{bmatrix}, \quad D = \begin{bmatrix} i & 0 \\ 0 & -i \end{bmatrix}.$$

It follows that

$$[D]_B^{1/2} = PD^{1/2}P^{-1}$$
$$= \frac{1}{\sqrt{2}} P \begin{bmatrix} 1+i & 0 \\ 0 & 1-i \end{bmatrix} P^{-1}$$
$$= \frac{1}{\sqrt{2}} \begin{bmatrix} 1 & -1 \\ 1 & 1 \end{bmatrix},$$

which is the same square root we found in Example 1.2.19.

1.2.21 It then follows from Exercise 1.2.20(c) that if $\{\mathbf{v}_1, \mathbf{v}_2, \ldots, \mathbf{v}_n\}$ is a basis of $\mathcal{V}$ then $\{T(\mathbf{v}_1), T(\mathbf{v}_2), \ldots, T(\mathbf{v}_n)\}$ is a basis of $\mathcal{W}$, so $\dim(\mathcal{W}) = n = \dim(\mathcal{V})$.

1.2.22 (a) Write $B = \{\mathbf{v}_1, \mathbf{v}_2, \ldots, \mathbf{v}_n\}$ and give names to the entries of $[\mathbf{v}]_B$ and $[\mathbf{w}]_B$: $[\mathbf{v}]_B = (c_1, c_2, \ldots, c_n)$ and $[\mathbf{w}]_B = (d_1, d_2, \ldots, d_n)$, so that

$$\mathbf{v} = c_1\mathbf{v}_1 + c_2\mathbf{v}_2 + \cdots + c_n\mathbf{v}_n$$
$$\mathbf{w} = d_1\mathbf{v}_1 + d_2\mathbf{v}_2 + \cdots + d_n\mathbf{v}_n.$$

By adding these equations we see that

$$\mathbf{v} + \mathbf{w} = (c_1 + d_1)\mathbf{v}_1 + \cdots + (c_n + d_n)\mathbf{v}_n,$$

which means that $[\mathbf{v}+\mathbf{w}]_B = (c_1+d_1, c_2+d_2, \ldots, c_n+d_n)$, which is the same as $[\mathbf{v}]_B + [\mathbf{w}]_B$.
(b) Using the same notation as in part (a), we have

$$c\mathbf{v} = (cc_1)\mathbf{v}_1 + (cc_2)\mathbf{v}_2 + \cdots + (cc_n)\mathbf{v}_n,$$

which means that $[c\mathbf{v}]_B = (cc_1, cc_2, \ldots, cc_k)$, which is the same as $c[\mathbf{v}]_B$.
(c) It is clear that if $\mathbf{v} = \mathbf{w}$ then $[\mathbf{v}]_B = [\mathbf{w}]_B$. For the converse, we note that parts (a) and (b) tell us that if $[\mathbf{v}]_B = [\mathbf{w}]_B$ then $[\mathbf{v} - \mathbf{w}]_B = \mathbf{0}$, so $\mathbf{v} - \mathbf{w} = 0\mathbf{v}_1 + \cdots + 0\mathbf{v}_n = \mathbf{0}$, so $\mathbf{v} = \mathbf{w}$.

1.2.23 (a) By Exercise 1.2.22, we know that

$$c_1\mathbf{v}_1 + c_2\mathbf{v}_2 + \cdots + c_m\mathbf{v}_m = \mathbf{0}$$

if and only if

$$c_1[\mathbf{v}_1]_B + c_2[\mathbf{v}_2]_B + \cdots + c_m[\mathbf{v}_m]_B$$
$$= [c_1\mathbf{v}_1 + c_2\mathbf{v}_2 + \cdots + c_m\mathbf{v}_m]_B = [\mathbf{0}]_B = \mathbf{0}.$$

In particular, this means that $\{\mathbf{v}_1, \mathbf{v}_2, \ldots, \mathbf{v}_m\}$ is linearly independent if and only if $c_1 = c_2 = \cdots = c_m = 0$ is the unique solution to these equa-

(b) Suppose $\mathbf{v} \in \mathcal{V}$. Then Exercise 1.2.22 tells us that

$$c_1\mathbf{v}_1 + c_2\mathbf{v}_2 + \cdots + c_m\mathbf{v}_m = \mathbf{v}$$

if and only if

$$c_1[\mathbf{v}_1]_B + c_2[\mathbf{v}_2]_B + \cdots + c_m[\mathbf{v}_m]_B$$
$$= [c_1\mathbf{v}_1 + c_2\mathbf{v}_2 + \cdots + c_m\mathbf{v}_m]_B = [\mathbf{v}]_B.$$

In particular, we can find $c_1, c_2, \ldots, c_m$ to solve the first equation if and only if we can find $c_1, c_2, \ldots, c_m$ to solve the second equation, so $\mathbf{v} \in \text{span}(\mathbf{v}_1, \mathbf{v}_2, \ldots, \mathbf{v}_m)$ if and only if $[\mathbf{v}]_B \in \text{span}([\mathbf{v}_1]_B, [\mathbf{v}_2]_B, \ldots, [\mathbf{v}_m]_B)$ and thus $\text{span}(\mathbf{v}_1, \mathbf{v}_2, \ldots, \mathbf{v}_m) = \mathcal{V}$ if and only if $\text{span}([\mathbf{v}_1]_B, [\mathbf{v}_2]_B, \ldots, [\mathbf{v}_m]_B) = \mathbb{F}^n$.
(c) This follows immediately from combining parts (a) and (b).

1.2.24 (a) If $\mathbf{v}, \mathbf{w} \in \text{range}(T)$ and $c \in \mathbb{F}$ then there exist $\mathbf{x}, \mathbf{y} \in \mathcal{V}$ such that $T(\mathbf{x}) = \mathbf{v}$ and $T(\mathbf{y}) = \mathbf{w}$. Then $T(\mathbf{x} + c\mathbf{y}) = \mathbf{v} + c\mathbf{w}$, so $\mathbf{v} + c\mathbf{w} \in \text{range}(T)$ too, so $\text{range}(T)$ is a subspace of $\mathbf{w}$.
(b) If $\mathbf{v}, \mathbf{w} \in \text{null}(T)$ and $c \in \mathbb{F}$ then $T(\mathbf{v} + c\mathbf{w}) = T(\mathbf{v}) + cT(\mathbf{w}) = \mathbf{0} + c\mathbf{0} = \mathbf{0}$, so $\mathbf{v} + c\mathbf{w} \in \text{null}(T)$ too, so $\text{null}(T)$ is a subspace of $\mathcal{V}$.

1.2.25 (a) We just note that $\mathbf{w} \in \text{range}(T)$ if and only if there exists $\mathbf{v} \in \mathcal{V}$ such that $T(\mathbf{v}) = \mathbf{w}$, if and only if there exists $[\mathbf{v}]_B$ such that $[T]_{D \leftarrow B}[\mathbf{v}]_B = [\mathbf{w}]_D$, if and only if $[\mathbf{w}]_D \in \text{range}([T]_{D \leftarrow B})$.
(b) Similar to part (a), $\mathbf{v} \in \text{null}(T)$ if and only if $T(\mathbf{v}) = \mathbf{0}$, if and only if $[T]_{D \leftarrow B}[\mathbf{v}]_B = [\mathbf{0}]_B = \mathbf{0}$, if and only if $[\mathbf{v}]_B \in \text{null}([T]_{D \leftarrow B})$.
(c) Using methods like in part (a), we can show that if $\mathbf{w}_1, \ldots, \mathbf{w}_n$ is a basis of $\text{range}(T)$ then $[\mathbf{w}_1]_D, \ldots, [\mathbf{w}_n]_D$ is a basis of $\text{range}([T]_{D \leftarrow B})$, so $\text{rank}(T) = \dim(\text{range}(T)) = \dim(\text{range}([T]_{D \leftarrow B})) = \text{rank}([T]_{D \leftarrow B})$.
(d) Using methods like in part (b), we can show that if $\mathbf{v}_1, \ldots, \mathbf{v}_n$ is a basis of $\text{null}(T)$ then $[\mathbf{v}_1]_B, \ldots, [\mathbf{v}_n]_B$ is a basis of $\text{null}([T]_{D \leftarrow B})$, so $\text{nullity}(T) = \dim(\text{null}(T)) = \dim(\text{null}([T]_{D \leftarrow B})) = \text{nullity}([T]_{D \leftarrow B})$.

1.2.27 (a) The "only if" implication is trivial since a basis of $\mathcal{V}$ must, by definition, be linearly independent. Conversely, we note from Exercise 1.2.26(a) that, since B is linearly independent, we can add 0 or more vectors to B to create a basis of $\mathcal{V}$. However, B already has $\dim(\mathcal{V})$ vectors, so the only possibility is that B becomes a basis when we add 0 vectors to it—i.e., B itself is already a basis of $\mathcal{V}$.
(b) Again, the "only if" implication is trivial since a basis of $\mathcal{V}$ must span $\mathcal{V}$. Conversely, Exercise 1.2.26(b) tells us that we can remove 0 or more vectors from B to create a basis of $\mathcal{V}$. However, we know that all bases of $\mathcal{V}$ contain $\dim(\mathcal{V})$ vectors, and B already contains exactly this many vectors, so the only possibility is that B becomes a basis when we remove 0 vectors from it—i.e., B itself is already a basis of $\mathcal{V}$.

1.2.28 (a) All 10 vector space properties from Definition 1.1.1 are straightforward, so we do not show them all here. Property (a) just says that the sum of two linear transformations is again a linear transformation, for example.

    (b) $\dim(\mathcal{L}(V,W)) = mn$, which can be seen by noting that if $\{\mathbf{v}_1,\ldots,\mathbf{v}_n\}$ and $\{\mathbf{w}_1,\ldots,\mathbf{w}_m\}$ are bases of V and W, respectively, then the mn linear transformations defined by

$$T_{i,j}(\mathbf{v}_k) = \begin{cases} \mathbf{w}_i & \text{if } j=k, \\ \mathbf{0} & \text{otherwise} \end{cases}$$

form a basis of $\mathcal{L}(V,W)$. In fact, the standard matrices of these linear transformations make up the standard basis of $\mathcal{M}_{m,n}$: $[T_{i,j}]_{D\leftarrow B} = E_{i,j}$ for all i,j.

1.2.30 $[T]_B = I_n$ if and only if the j-th column of $[T]_B$ equals $\mathbf{e}_j$ for each j. If we write $B = \{\mathbf{v}_1,\mathbf{v}_2,\ldots,\mathbf{v}_n\}$ then we see that $[T]_B = I_n$ if and only if $[T(\mathbf{v}_j)]_B = \mathbf{e}_j$ for all j, which is equivalent to $T(\mathbf{v}_j) = \mathbf{v}_j$ for all j (i.e., $T = I_V$).

1.2.31 (a) If B is a basis of W then by Exercise 1.2.26 that we can extend B to a basis $C \supseteq B$ of V. However, since $\dim(V) = \dim(W)$ we know that B and C have the same number of vectors, so $B = C$, so $V = W$.

    (b) Let $V = c_{00}$ from Example 1.1.10 and let W be the subspace of V with the property that the first entry of every member of $\mathbf{w}$ equals 0. Then $\dim(V) = \dim(W) = \infty$ but $V \neq W$.

1.2.32 We check the two properties of Definition 1.2.4:

    (a) $R\big((v_1,v_2,\ldots) + (w_1,w_2,\ldots)\big) = R(v_1 + w_1, v_2 + w_2,\ldots) = (0, v_1 + w_1, v_2 + w_2,\ldots) = (0, v_1, v_2,\ldots) + (0, w_1, w_2,\ldots) = R(v_1, v_2,\ldots) + R(w_1, w_2,\ldots)$,

and (b) $R\big(c(v_1,v_2,\ldots)\big) = R(cv_1, cv_2,\ldots) = (0, cv_1, cv_2,\ldots) = c(0, v_1, v_2,\ldots) = cR(v_1, v_2,\ldots)$.

1.2.34 It is clear that $\dim(\mathcal{P}^2(\mathbb{Z}_2)) \leq 3$ since $\{1, x, x^2\}$ spans $\mathcal{P}^2(\mathbb{Z}_2)$ (just like it did in the real case). However, it is not linearly independent since the two polynomials $f(x) = x$ and $f(x) = x^2$ are the same on $\mathbb{Z}_2$ (i.e., they provide the same output for all inputs). The set $\{1, x\}$ is indeed a basis of $\mathcal{P}^2(\mathbb{Z}_2)$, so its dimension is 2.

1.2.35 Suppose $P \in \mathcal{M}_{m,n}$ is any matrix such that $P[\mathbf{v}]_B = [T(\mathbf{v})]_D$ for all $\mathbf{v} \in V$. For every $1 \leq j \leq n$, if $\mathbf{v} = \mathbf{v}_j$ then we see that $[\mathbf{v}]_B = \mathbf{e}_j$ (the j-th standard basis vector in $\mathbb{R}^n$), so $P[\mathbf{v}]_B = P\mathbf{e}_j$ is the j-th column of P. On the other hand, it is also the case that $P[\mathbf{v}]_B = [T(\mathbf{v})]_D = [T(\mathbf{v}_j)]_D$. It follows that the j-th column of P is $[T(\mathbf{v}_j)]_D$ for each $1 \leq j \leq n$, so $P = [T]_{D\leftarrow B}$, which shows uniqueness of $[T]_{D\leftarrow B}$.

1.2.36 Suppose that a set C has $m < n$ vectors, which we call $\mathbf{v}_1, \mathbf{v}_2,\ldots,\mathbf{v}_m$. To see that C does not span V, we want to show that there exists $\mathbf{x} \in V$ such that the equation $c_1\mathbf{v}_1 + \cdots + c_m\mathbf{v}_m = \mathbf{x}$ does not have a solution. This equation is equivalent to

$$c_1[\mathbf{v}_1]_B + \cdots + c_m[\mathbf{v}_m]_B = [\mathbf{x}]_B,$$

which is a system of n linear equations in m variables. Since $m < n$, this is a "tall and skinny" linear system, so applying Gaussian elimination to the augmented matrix form $[\,A \mid [\mathbf{x}]_B\,]$ of this linear system results in a row echelon form $[\,R \mid [\mathbf{y}]_B\,]$ where R has at least one zero row at the bottom. If $\mathbf{x}$ is chosen so that the bottom entry of $[\mathbf{y}]_B$ is non-zero then this linear system has no solution, as desired.

Section 1.3: Isomorphisms and Linear Forms

1.3.1 (a) Isomorphic, since they are finite-dimensional with the same dimension (6).

    (c) Isomorphic, since they both have dimension 9.

    (e) Not isomorphic, since V is finite-dimensional but W is infinite-dimensional.

1.3.2 (a) Not a linear form. For example, $f(\mathbf{e}_1) = f(-\mathbf{e}_1) = 1$, but $f(\mathbf{e}_1 - \mathbf{e}_1) = T(\mathbf{0}) = 0 \neq 2 = f(\mathbf{e}_1) + f(-\mathbf{e}_1)$.

    (c) Not a linear form, since it does not output scalars. It is a linear *transformation*, however.

    (e) Is a linear form, since $g(f_1 + f_2) = (f_1 + f_2)'(3) = f_1'(3) + f_2'(3) = g(f_1) + g(f_2)$ by linearity of the derivative, and similarly $g(cf) = cg(f)$.

1.3.3 (a) Not an inner product, since it is not conjugate symmetric. For example, $\langle \mathbf{e}_1, \mathbf{e}_2\rangle = 1$ but $\langle \mathbf{e}_2, \mathbf{e}_1\rangle = 0$.

    (d) Not an inner product, since it is not even linear in its second argument. For example, $\langle I, I\rangle = \text{tr}(I + I) = 2n$, but $\langle I, 2I\rangle = 3n \neq 2\langle I, I\rangle$.

    (f) Is an inner product. All three properties can be proved in a manner analogous to Example 1.3.17.

1.3.4 (a) True. This property of isomorphisms was stated explicitly in the text, and is proved in Exercise 1.3.6.

    (c) False. This statement is only true if V and W are finite-dimensional.

    (e) False. It is conjugate linear, not linear.

    (g) False. For example, if $f(x) = x^2$ then $E_1(f) = 1^2 = 1$ and $E_2(f) = 2^2 = 4$, but $E_3(f) = 3^2 = 9$, so $E_1(f) + E_2(f) \neq E_3(f)$.

1.3.6 (a) This follows immediately from things that we learned in Section 1.2.4. Linearity of T^{-1} is baked right into the definition of its existence, and it is invertible since T is its inverse.

(b) $S \circ T$ is linear (even when S and T are not necessarily invertible linear transformations) since $(S \circ T)(\mathbf{v}+c\mathbf{w}) = S(T(\mathbf{v}+c\mathbf{w})) = S(T(\mathbf{v})+cT(\mathbf{w})) = S(T(\mathbf{v}))+cS(T(\mathbf{w})) = (S \circ T)(\mathbf{v})+c(S \circ T)(\mathbf{w})$. Furthermore, $S \circ T$ is invertible since $T^{-1} \circ S^{-1}$ is its inverse.

1.3.7 If we use the standard basis of $\mathcal{M}_{m,n}(\mathbb{F})$, then we know from Theorem 1.3.3 that there are scalars $\{a_{i,j}\}$ such that $f(X) = \sum_{i,j} a_{i,j}x_{i,j}$. If we let A be the matrix with (i,j)-entry equal to $a_{i,j}$, we get $\text{tr}(AX) = \sum_{i,j} a_{i,j}x_{i,j}$ as well.

1.3.8 Just repeat the argument of Exercise 1.3.7 with the bases of $\mathcal{M}_n^S(\mathbb{F})$ and $\mathcal{M}_n^H$ that we presented in the solution to Exercise 1.2.2.

1.3.9 Just recall that inner products are linear in their second entry, so $\langle \mathbf{v}, \mathbf{0} \rangle = \langle \mathbf{v}, 0\mathbf{v} \rangle = 0\langle \mathbf{v}, \mathbf{v} \rangle = 0$.

1.3.10 Recall that

$$\|A\|_F = \sqrt{\sum_{i=1}^m \sum_{j=1}^n |a_{i,j}|^2}$$

which is clearly unchanged if we reorder the entries of A (i.e., $\|A\|_F = \|A^T\|_F$) and is also unchanged if we take the complex conjugate of some or all entries (so $\|A^*\|_F = \|A^T\|_F$).

1.3.12 The result follows from expanding out the norm in terms of the inner product: $\|\mathbf{v}+\mathbf{w}\|^2 = \langle \mathbf{v}+\mathbf{w}, \mathbf{v}+\mathbf{w} \rangle = \langle \mathbf{v}, \mathbf{v} \rangle + \langle \mathbf{v}, \mathbf{w} \rangle + \langle \mathbf{w}, \mathbf{v} \rangle + \langle \mathbf{w}, \mathbf{w} \rangle = \|\mathbf{v}\|^2 + 0 + 0 + \|\mathbf{w}\|^2$.

1.3.13 We just compute

$$\|\mathbf{v}+\mathbf{w}\|^2 + \|\mathbf{v}-\mathbf{w}\|^2$$
$$= \langle \mathbf{v}+\mathbf{w}, \mathbf{v}+\mathbf{w} \rangle + \langle \mathbf{v}-\mathbf{w}, \mathbf{v}-\mathbf{w} \rangle$$
$$= (\langle \mathbf{v}, \mathbf{v} \rangle + \langle \mathbf{v}, \mathbf{w} \rangle + \langle \mathbf{w}, \mathbf{v} \rangle + \langle \mathbf{w}, \mathbf{w} \rangle)$$
$$+ (\langle \mathbf{v}, \mathbf{v} \rangle - \langle \mathbf{v}, \mathbf{w} \rangle - \langle \mathbf{w}, \mathbf{v} \rangle + \langle \mathbf{w}, \mathbf{w} \rangle)$$
$$= 2\langle \mathbf{v}, \mathbf{v} \rangle + 2\langle \mathbf{w}, \mathbf{w} \rangle$$
$$= 2\|\mathbf{v}\|^2 + 2\|\mathbf{w}\|^2.$$

1.3.14 (a) Expanding the norm in terms of the inner product gives

$$\|\mathbf{v}+\mathbf{w}\|^2 - \|\mathbf{v}-\mathbf{w}\|^2$$
$$= \langle \mathbf{v}+\mathbf{w}, \mathbf{v}+\mathbf{w} \rangle - \langle \mathbf{v}-\mathbf{w}, \mathbf{v}-\mathbf{w} \rangle$$
$$= \langle \mathbf{v}, \mathbf{v} \rangle + 2\langle \mathbf{v}, \mathbf{w} \rangle + \langle \mathbf{w}, \mathbf{w} \rangle$$
$$- \langle \mathbf{v}, \mathbf{v} \rangle + 2\langle \mathbf{v}, \mathbf{w} \rangle - \langle \mathbf{w}, \mathbf{w} \rangle,$$
$$= 4\langle \mathbf{v}, \mathbf{w} \rangle,$$

from which the desired equality follows.

(b) This follows via the same method as in part (a). All that changes is that the algebra is uglier, and we have to be careful to not forget the complex conjugate in the property $\langle \mathbf{w}, \mathbf{v} \rangle = \overline{\langle \mathbf{v}, \mathbf{w} \rangle}$.

1.3.16 (a) If f is alternating then for all $\mathbf{v}, \mathbf{w} \in \mathcal{V}$ we have $0 = f(\mathbf{v}+\mathbf{w}, \mathbf{v}+\mathbf{w}) = f(\mathbf{v}, \mathbf{v}) + f(\mathbf{v}, \mathbf{w}) + f(\mathbf{w}, \mathbf{v}) + f(\mathbf{w}, \mathbf{w}) = f(\mathbf{v}, \mathbf{w}) + f(\mathbf{w}, \mathbf{v})$, so $f(\mathbf{v}, \mathbf{w}) = -f(\mathbf{w}, \mathbf{v})$, which means that f is skew-symmetric. Conversely, if f is skew-symmetric then choosing $\mathbf{v} = \mathbf{w}$ tells us that $f(\mathbf{v}, \mathbf{v}) = -f(\mathbf{v}, \mathbf{v})$, so $f(\mathbf{v}, \mathbf{v}) = 0$ for all $\mathbf{v} \in \mathcal{V}$, so f is alternating.

(b) It is still true that alternating implies skew-symmetric, but the converse fails because $f(\mathbf{v}, \mathbf{v}) = -f(\mathbf{v}, \mathbf{v})$ no longer implies $f(\mathbf{v}, \mathbf{v}) = 0$ (since $1+1 = 0$ implies $-1 = 1$ in this field).

1.3.18 (a) We could prove this directly by mimicking the proof of Theorem 1.3.5, replacing the transposes by conjugate transposes. Instead, we prove it by denoting the vectors in the basis B by $B = \{\mathbf{v}_1, \ldots, \mathbf{v}_m\}$ and defining a new *bilinear* function $g : \mathcal{V} \times \mathcal{W} \to \mathbb{C}$ by

$$g(\mathbf{v}_j, \mathbf{w}) = f(\mathbf{v}_j, \mathbf{w}) \text{ for all } \mathbf{v}_j \in B, \mathbf{w} \in \mathcal{W}$$

and extending to all $\mathbf{v} \in \mathcal{V}$ via linearity. That is, if $\mathbf{v} = c_1\mathbf{v}_1 + \cdots + c_m\mathbf{v}_m$ (i.e., $[\mathbf{v}]_B = (\mathbf{v}_1, \ldots, \mathbf{v}_m)$) then

$$g(\mathbf{v}, \mathbf{w}) = g(c_1\mathbf{v}_1 + \cdots + c_m\mathbf{v}_m, \mathbf{w})$$
$$= c_1 g(\mathbf{v}_1, \mathbf{w}) + \cdots + c_m g(\mathbf{v}_m, \mathbf{w}_m)$$
$$= c_1 f(\mathbf{v}_1, \mathbf{w}) + \cdots + c_m f(\mathbf{v}_m, \mathbf{w}_m)$$
$$= f(\overline{c_1}\mathbf{v}_1 + \cdots + \overline{c_m}\mathbf{v}_m, \mathbf{w}).$$

Then Theorem 1.3.5 tells us that we can write g in the form

$$g(\mathbf{v}, \mathbf{w}) = [\mathbf{v}]_B^T A [\mathbf{w}]_C.$$

When combined with the definition of g above, we see that

$$f(\mathbf{v}, \mathbf{w}) = g(\overline{c_1}\mathbf{v}_1 + \cdots + \overline{c_m}\mathbf{v}_m, \mathbf{w})$$
$$= (\overline{[\mathbf{v}]_B})^T A [\mathbf{w}]_C = [\mathbf{v}]_B^* A [\mathbf{w}]_C,$$

as desired.

(b) If $A = A^*$ (i.e., A is Hermitian) then

$$f(\mathbf{v}, \mathbf{w}) = [\mathbf{v}]_B^* A [\mathbf{w}]_B = [\mathbf{v}]_B^* A^* [\mathbf{w}]_B$$
$$= ([\mathbf{w}]_B^* A [\mathbf{v}]_B)^* = \overline{f(\mathbf{w}, \mathbf{v})},$$

with the final equality following from the fact that $[\mathbf{w}]_B^* A [\mathbf{v}]_B = f(\mathbf{v}, \mathbf{v})$ is a scalar and thus equal to its own transpose.

In the other direction, if $f(\mathbf{v}, \mathbf{w}) = \overline{f(\mathbf{w}, \mathbf{v})}$ for all $\mathbf{v}$ and $\mathbf{w}$ then in particular this holds if $[\mathbf{v}]_B = \mathbf{e}_i$ and $[\mathbf{w}]_B = \mathbf{e}_j$, so

$$a_{i,j} = \mathbf{e}_i^* A \mathbf{e}_j = [\mathbf{v}]_B^* A [\mathbf{w}]_B = f(\mathbf{v}, \mathbf{w})$$
$$= \overline{f(\mathbf{w}, \mathbf{v})} = ([\mathbf{w}]_B^* A [\mathbf{v}]_B)^* = (\mathbf{e}_j^* A \mathbf{e}_i)^* = \overline{a_{j,i}}.$$

Since this equality holds for all i and j, we conclude that A is Hermitian.

(c) Part (b) showed the equivalence of A being Hermitian and f being conjugate symmetric, so we just need to show that positive definiteness of f (i.e., $f(\mathbf{v}, \mathbf{v}) \geq 0$ with equality if and only if $f = \mathbf{0}$) is equivalent to $\mathbf{v}^* A \mathbf{v} \geq 0$ with equality if and only if $\mathbf{v} = \mathbf{0}$. This equivalence follows immediately from recalling that $f(\mathbf{v}, \mathbf{v}) = [\mathbf{v}]_B^* A [\mathbf{v}]_B$.

1.3.21 $g_x(y) = 1 + xy + x^2 y^2 + \cdots + x^p y^p$.

1.3.22 We note that $\dim((\mathcal{P}^p)^*) = \dim(\mathcal{P}^p) = p+1$, which is the size of the proposed basis, so we just need to show that it is linearly independent. To this end, we note that if

$$d_0 E_{c_0} + d_1 E_{c_1} + \cdots + d_p E_{c_p} = 0$$

then

$$d_0 f(c_0) + d_1 f(c_1) + \cdots + d_p f(c_p) = 0$$

for all $f \in \mathcal{P}^p$. However, if we choose f to be the non-zero polynomial with roots at each of $c_1, \ldots, c_p$ (but not c_0, since f has degree p and thus at most p roots) then this tells us that $d_0 f(c_0) = 0$, so $d_0 = 0$. A similar argument with polynomials having roots at all except for one of the c_j's shows that $d_0 = d_1 = \cdots = d_p = 0$, which gives linear independence.

1.3.24 Suppose that

$$c_1 T(\mathbf{v}_1) \cdots + c_n T(\mathbf{v}_n) = \mathbf{0}.$$

By linearity of T, this implies $T(c_1 \mathbf{v}_1 + \cdots + c_n \mathbf{v}_n) = \mathbf{0}$, so $\phi_{c_1 \mathbf{v}_1 + \cdots + c_n \mathbf{v}_n} = \mathbf{0}$, which implies

$$\phi_{c_1 \mathbf{v}_1 + \cdots + c_n \mathbf{v}_n}(f) = f(c_1 \mathbf{v}_1 + \cdots + c_n \mathbf{v}_n)$$
$$= c_1 f(\mathbf{v}_1) + \cdots + c_n f(\mathbf{v}_n) = 0$$

for all $f \in \mathcal{V}^*$. Since B is linearly independent, we can choose f so that $f(\mathbf{v}_1) = 1$ and $f(\mathbf{v}_2) = \cdots = f(\mathbf{v}_n) = 0$, which implies $c_1 = 0$. A similar argument (involving different choices of f) shows that $c_2 = \cdots = c_n = 0$ as well, so C is linearly independent.

1.3.25 We just check the three defining properties of inner products, each of which follows from the corresponding properties of the inner product on $\mathcal{W}$. (a) $\langle \mathbf{v}_1, \mathbf{v}_2 \rangle_{\mathcal{V}} = \langle T(\mathbf{v}_1), T(\mathbf{v}_2) \rangle_{\mathcal{W}} = \overline{\langle T(\mathbf{v}_2), T(\mathbf{v}_1) \rangle_{\mathcal{W}}} = \overline{\langle \mathbf{v}_2, \mathbf{v}_1 \rangle_{\mathcal{V}}}$. (b) $\langle \mathbf{v}_1, \mathbf{v}_2 + c\mathbf{v}_3 \rangle_{\mathcal{V}} = \langle T(\mathbf{v}_1), T(\mathbf{v}_2 + c\mathbf{v}_3) \rangle_{\mathcal{W}} = \langle T(\mathbf{v}_1), T(\mathbf{v}_2) + cT(\mathbf{v}_3) \rangle_{\mathcal{W}} = \langle T(\mathbf{v}_1), T(\mathbf{v}_2) \rangle_{\mathcal{W}} + c \langle T(\mathbf{v}_1), T(\mathbf{v}_3) \rangle_{\mathcal{W}} = \langle \mathbf{v}_1, \mathbf{v}_2 \rangle_{\mathcal{V}} + c \langle \mathbf{v}_1, \mathbf{v}_3 \rangle_{\mathcal{V}}$. (c) $\langle \mathbf{v}_1, \mathbf{v}_1 \rangle_{\mathcal{V}} = \langle T(\mathbf{v}_1), T(\mathbf{v}_1) \rangle_{\mathcal{W}} \geq 0$, with equality if and only if $T(\mathbf{v}_1) = \mathbf{0}$, which happens if and only if $\mathbf{v}_1 = \mathbf{0}$. As a side note, the equality condition of property (c) is the only place where we used the fact that T is an isomorphism (i.e., invertible).

1.3.27 If $j_k = j_\ell$ then the condition $a_{j_1, \ldots, j_k, \ldots, j_\ell, \ldots, j_p} = -a_{j_1, \ldots, j_\ell, \ldots, j_k, \ldots, j_p}$ tells us that $a_{j_1, \ldots, j_k, \ldots, j_\ell, \ldots, j_p} = 0$ whenever two or more of the subscripts are equal to each other. If all of the subscripts are distinct from each other (i.e., there is a permutation $\sigma : \{1, 2, \ldots, p\} \to \{1, 2, \ldots, p\}$ such that $\sigma(k) = j_k$ for all $1 \leq k \leq p$) then we see that $a_{\sigma(1), \ldots, \sigma(p)} = \text{sgn}(\sigma) a_{1,2,\ldots,p}$, where $\text{sgn}(\sigma)$ is the sign of σ (see Appendix A.1.5). It follows that A is completely determined by the value of $a_{1,2,\ldots,p}$, so it is unique up to scaling.

Section 1.4: Orthogonality and Adjoints

1.4.1 (a) Not an orthonormal basis, since $(0, 1, 1)$ does not have length 1.

(c) Not an orthonormal basis, since B contains 3 vectors living in a 2-dimensional subspace, and thus it is necessarily linearly dependent. Alternatively, we could check that the vectors are not mutually orthogonal.

(e) Is an orthonormal basis, since all three vectors have length 1 and they are mutually orthogonal:

$$(1,1,1,1) \cdot (0,1,-1,0) = 0$$
$$(1,1,1,1) \cdot (1,0,0,-1) = 0$$
$$(0,1,-1,0) \cdot (1,0,0,-1) = 0.$$

(We ignored some scalars since they do not affect whether or not the dot product equals 0.)

1.4.2 (b) The standard basis $\{(1,0),(0,1)\}$ works (the Gram–Schmidt process could also be used, but the resulting basis will be hideous).

(d) If we start with the basis $\{\sin(x), \cos(x)\}$ and perform the Gram–Schmidt process, we get the basis $\{\frac{1}{\sqrt{\pi}} \sin(x), \frac{1}{\sqrt{\pi}} \cos(x)\}$. (In other words, we can check that sin and cos are already orthogonal in this inner product, and we just need to normalize them to have length 1.)

(f) We refer to the three matrices as A_1, A_2, and A_3, respectively, and then we divide A_1 by its Frobenius norm to get B_1:

$$B_1 = \frac{1}{\sqrt{1^2 + 1^2 + 1^2 + 1^2}} A_1 = \frac{1}{2} \begin{bmatrix} 1 & 1 \\ 1 & 1 \end{bmatrix}.$$

Then

$$B_2 = A_2 - \langle B_1, A_2 \rangle B_1$$
$$= \begin{bmatrix} 1 & 0 \\ 0 & 1 \end{bmatrix} - \frac{1}{2} \begin{bmatrix} 1 & 1 \\ 1 & 1 \end{bmatrix} = \frac{1}{2} \begin{bmatrix} 1 & -1 \\ -1 & 1 \end{bmatrix}.$$

Since B_2 already has Frobenius norm 1, we do not need to normalize it. Similarly,

$$B_3 = A_3 - \langle B_1, A_3 \rangle B_1 - \langle B_2, A_3 \rangle B_2$$
$$= \begin{bmatrix} 0 & 2 \\ 1 & -1 \end{bmatrix} - \frac{1}{2} \begin{bmatrix} 1 & 1 \\ 1 & 1 \end{bmatrix} - (-2) \frac{1}{2} \begin{bmatrix} 1 & -1 \\ -1 & 1 \end{bmatrix}$$
$$= \frac{1}{2} \begin{bmatrix} 1 & 1 \\ -1 & -1 \end{bmatrix}.$$

Again, since B_3 already has Frobenius norm 1, we do not need to normalize it. We thus conclude that $\{B_1, B_2, B_3\}$ is an orthonormal basis of $\text{span}(\{A_1, A_2, A_3\})$.

1.4.3 (a) $T^*(\mathbf{w}) = (w_1 + w_2, w_2 + w_3)$.

(c) $T^*(\mathbf{w}) = (2w_1 - w_2, 2w_1 - w_2)$. Note that this can be found by constructing an orthonormal basis of $\mathbb{R}^2$ with respect to this weird inner product, but it is likely easier to just construct T^* directly via its definition.

1.4.4 (a) $\begin{bmatrix} 1 & 0 \\ 0 & 0 \end{bmatrix}$

(c) $\frac{1}{3} \begin{bmatrix} 2 & 1 & -1 \\ 1 & 2 & 1 \\ -1 & 1 & 2 \end{bmatrix}$

1.4.5 (a) True. This follows from Theorem 1.4.8.

(c) False. If $A = A^T$ and $B = B^T$ then $(AB)^T = B^T A^T = BA$, which in general does not equal AB. For an explicit counter-example, you can choose

$$A = \begin{bmatrix} 1 & 0 \\ 0 & -1 \end{bmatrix}, \quad B = \begin{bmatrix} 0 & 1 \\ 1 & 0 \end{bmatrix}.$$

(e) False. For example, $U = V = I$ are unitary, but $U + V = 2I$ is not.

(g) True. On any inner product space $\mathcal{V}$, $I_\mathcal{V}$ satisfies $I_\mathcal{V}^2 = I_\mathcal{V}$ and $I_\mathcal{V}^* = I_\mathcal{V}$.

1.4.8 If U and V are unitary then $(UV)^*(UV) = V^* U^* U V = V^* V = I$, so UV is also unitary.

1.4.10 If λ is an eigenvalue of U corresponding to an eigenvector $\mathbf{v}$ then $U\mathbf{v} = \lambda\mathbf{v}$. Unitary matrices preserve length, so $\|\mathbf{v}\| = \|U\mathbf{v}\| = \|\lambda\mathbf{v}\| = |\lambda|\|\mathbf{v}\|$. Dividing both sides of this equation by $\|\mathbf{v}\|$ (which is OK since eigenvectors are non-zero) gives $1 = |\lambda|$.

1.4.11 Since $U^* U = I$, we have $1 = \det(I) = \det(U^* U) = \det(U^*)\det(U) = \overline{\det(U)}\det(U) = |\det(U)|^2$. Taking the square root of both sides of this equation gives $|\det(U)| = 1$.

1.4.12 We know from Exercise 1.4.10 that the eigenvalues (diagonal entries) of U must all have magnitude 1. Since the columns of U each have norm 1, it follows that the off-diagonal entries of U must each be 0.

1.4.15 Direct computation shows that the (i,j)-entry of $F^* F$ equals $\frac{1}{n}\sum_{k=0}^{n-1}\omega^{(j-i)k}$. If $i = j$ (i.e., $j - i = 0$) then $\omega^{(j-i)k} = 1$, so the (i,i)-entry of $F^* F$ equals 1. If $i \neq j$ then $\omega^{(j-i)k} \neq 1$ is an n-th root of unity, so we claim that this sum equals 0 (and thus $F^* F = I$, so F is unitary).

To see why $\sum_{k=0}^{n-1}\omega^{(j-i)k} = 0$ when $i \neq j$, we use a standard formula for summing a geometric series:

$$\sum_{k=0}^{n-1}\omega^{(j-i)k} = \frac{1-\omega^{(j-i)n}}{1-\omega^{j-i}} = 0,$$

with the final equality following from the fact that $\omega^{(j-i)n} = 1$.

1.4.17 (a) For the "if" direction we note that $(A\mathbf{x}) \cdot \mathbf{y} = (A\mathbf{x})^T\mathbf{y} = \mathbf{x}^T A^T \mathbf{y}$ and $\mathbf{x} \cdot (A^T\mathbf{y}) = \mathbf{x}^T A^T\mathbf{y}$, which are the same. For the "only if" direction, we can either let $\mathbf{x}$ and $\mathbf{y}$ range over all standard basis vectors to see that the (i,j)-entry of A equals the (j,i)-entry of B, or use the "if" direction together with uniqueness of the adjoint (Theorem 1.4.8).

(b) Almost identical to part (a), but just recall that the complex dot product has a complex conjugate in it, so $(A\mathbf{x}) \cdot \mathbf{y} = (A\mathbf{x})^*\mathbf{y} = \mathbf{x}^* A^*\mathbf{y}$ and $\mathbf{x} \cdot (A^*\mathbf{y}) = \mathbf{x}^* A^*\mathbf{y}$, which are equal to each other.

1.4.18 This follows directly from the fact that if $B \{\mathbf{v}_1, \dots, \mathbf{v}_n\}$ then $[\mathbf{v}_j]_E = \mathbf{v}_j$ for all $1 \leq j \leq n$, so $P_{E \leftarrow B} = [\mathbf{v}_1 \mid \mathbf{v}_2 \mid \cdots \mid \mathbf{v}_n]$, which is unitary if and only if its columns (i.e., the members of B) form an orthonormal basis of $\mathbb{F}^n$.

1.4.19 We mimic the proof of Theorem 1.4.9. To see that (a) implies (c), suppose $B^* B = C^* C$. Then for all $\mathbf{v} \in \mathbb{F}^n$ we have

$$\|B\mathbf{v}\| = \sqrt{(B\mathbf{v}) \cdot (B\mathbf{v})} = \sqrt{\mathbf{v}^* B^* B\mathbf{v}}$$
$$= \sqrt{\mathbf{v}^* C^* C\mathbf{v}} = \sqrt{(C\mathbf{v}) \cdot (C\mathbf{v})} = \|C\mathbf{v}\|.$$

For the implication (c) $\implies$ (b), note that if $\|B\mathbf{v}\|^2 = \|C\mathbf{v}\|^2$ for all $\mathbf{v} \in \mathbb{F}^n$ then $(B\mathbf{v}) \cdot (B\mathbf{v}) = (C\mathbf{v}) \cdot (C\mathbf{v})$. If $\mathbf{x}, \mathbf{y} \in \mathbb{F}^n$ then this tells us (by choosing $\mathbf{v} = \mathbf{x} + \mathbf{y}$) that $(B(\mathbf{x}+\mathbf{y})) \cdot (B(\mathbf{x}+\mathbf{y})) = (C(\mathbf{x}+\mathbf{y})) \cdot (C(\mathbf{x}+\mathbf{y}))$. Expanding this dot product on both the left and right then gives

$$(B\mathbf{x}) \cdot (B\mathbf{x}) + 2\mathrm{Re}((B\mathbf{x}) \cdot (B\mathbf{y})) + (B\mathbf{y}) \cdot (B\mathbf{y})$$
$$= (C\mathbf{x}) \cdot (C\mathbf{x}) + 2\mathrm{Re}((C\mathbf{x}) \cdot (C\mathbf{y})) + (C\mathbf{y}) \cdot (C\mathbf{y}).$$

By then using the facts that $(B\mathbf{x}) \cdot (B\mathbf{x}) = (C\mathbf{x}) \cdot (C\mathbf{x})$ and $(B\mathbf{y}) \cdot (B\mathbf{y}) = (C\mathbf{y}) \cdot (C\mathbf{y})$, we can simplify the above equation to the form

$$\mathrm{Re}((B\mathbf{x}) \cdot (B\mathbf{y})) = \mathrm{Re}((C\mathbf{x}) \cdot (C\mathbf{y})).$$

If $\mathbb{F} = \mathbb{R}$ then this implies $(B\mathbf{x}) \cdot (B\mathbf{y}) = (C\mathbf{x}) \cdot (C\mathbf{y})$ for all $\mathbf{x}, \mathbf{y} \in \mathbb{F}^n$, as desired. If instead $\mathbb{F} = \mathbb{C}$ then we can repeat the above argument with $\mathbf{v} = \mathbf{x} + i\mathbf{y}$ to see that

$$\mathrm{Im}((B\mathbf{x}) \cdot (B\mathbf{y})) = \mathrm{Im}((C\mathbf{x}) \cdot (C\mathbf{y})),$$

so in this case we have $(B\mathbf{x}) \cdot (B\mathbf{y}) = (C\mathbf{x}) \cdot (C\mathbf{y})$ for all $\mathbf{x}, \mathbf{y} \in \mathbb{F}^n$ too, establishing (b).

Finally, to see that (b) $\implies$ (a), note that if we rearrange $(B\mathbf{v}) \cdot (B\mathbf{w}) = (C\mathbf{v}) \cdot (C\mathbf{w})$ slightly, we get

$$((B^* B - C^* C)\mathbf{v}) \cdot \mathbf{w} = 0 \quad \text{for all} \quad \mathbf{v}, \mathbf{w} \in \mathbb{F}^n.$$

If we choose $\mathbf{w} = (B^* B - C^* C)\mathbf{v}$ then this implies $\|(B^* B - C^* C)\mathbf{v}\|^2 = 0$ for all $\mathbf{v} \in \mathbb{F}^n$, so $(B^* B - C^* C)\mathbf{v} = 0$ for all $\mathbf{v} \in \mathbb{F}^n$. This in turn implies $B^* B - C^* C = O$, so $B^* B = C^* C$, which completes the proof.

1.4.20 Since B is mutually orthogonal and thus linearly independent, Exercise 1.2.26(a) tells us that there is a (not necessarily orthonormal) basis D of $\mathcal{V}$ such that $B \subseteq D$. Applying the Gram–Schmidt process to this bases D results in an orthonormal basis C of $\mathcal{V}$ that also contains B (since the vectors from B are already mutually orthogonal and normalized, the Gram–Schmidt process does not affect them).

1.4.22 Pick orthonormal bases B and C of $\mathcal{V}$ and $\mathcal{W}$, respectively, so that $\mathrm{rank}(T) = \mathrm{rank}([T]_{C \leftarrow B}) = \mathrm{rank}([T]_{C \leftarrow B}^*) = \mathrm{rank}([T^*]_{B \leftarrow C}) = \mathrm{rank}(T^*)$. In the second equality, we used the fact that $\mathrm{rank}(A^*) = \mathrm{rank}(A)$ for all *matrices* A, which is typically proved in introduced linear algebra courses.

1.4.23 If R and S are two adjoints of T, then

$$\langle \mathbf{v}, R(\mathbf{w}) \rangle = \langle T(\mathbf{v}), \mathbf{w} \rangle = \langle \mathbf{v}, S(\mathbf{w}) \rangle \quad \text{for all} \quad \mathbf{v}, \mathbf{w}.$$

Rearranging slightly gives

$$\langle \mathbf{v}, (R - S)(\mathbf{w}) \rangle \quad \text{for all} \quad \mathbf{v} \in \mathcal{V}, \mathbf{w} \in \mathcal{W}.$$

Exercise 1.4.27 then shows that $R - S = O$, so $R = S$, as desired.

1.4.24 (a) If $\mathbf{v}, \mathbf{w} \in \mathbb{F}^n$, E is the standard basis of $\mathbb{F}^n$, and B is any basis of $\mathbb{F}^n$, then

$$P_{B \leftarrow E}\mathbf{v} = [\mathbf{v}]_B \quad \text{and} \quad P_{B \leftarrow E}\mathbf{w} = [\mathbf{w}]_B$$

By plugging this fact into Theorem 1.4.3, we see that $\langle \cdot, \cdot \rangle$ is an inner product if and only if it has the form

$$\langle \mathbf{v}, \mathbf{w} \rangle = [\mathbf{v}]_B \cdot [\mathbf{w}]_B$$
$$= (P_{B \leftarrow E} \mathbf{v}) \cdot (P_{B \leftarrow E} \mathbf{w})$$
$$= \mathbf{v}^* (P_{B \leftarrow E}^* P_{B \leftarrow E}) \mathbf{w}.$$

Recalling that change-of-basis matrices are invertible and every invertible matrix is a change-of-basis matrix completes the proof.

(b) The matrix

$$P = \begin{bmatrix} 1 & 2 \\ 0 & 1 \end{bmatrix}$$

works.

(c) If P is not invertible then it may be the case that $\langle \mathbf{v}, \mathbf{v} \rangle = 0$ even if $\mathbf{v} \neq \mathbf{0}$, which violates the third defining property of inner products. In particular, it will be the case that $\langle \mathbf{v}, \mathbf{v} \rangle = 0$ whenever $\mathbf{v} \in \text{null}(P)$.

1.4.25 If $B = \{\mathbf{v}_1, \ldots, \mathbf{v}_n\}$ then set $Q = \begin{bmatrix} \mathbf{v}_1 \mid \cdots \mid \mathbf{v}_n \end{bmatrix} \in \mathcal{M}_n$, which is invertible since B is linearly independent. Then set $P = Q^{-1}$, so that the function

$$\langle \mathbf{v}, \mathbf{w} \rangle = \mathbf{v}^* (P^* P) \mathbf{w}$$

(which is an inner product by Exercise 1.4.24) satisfies

$$\langle \mathbf{v}_i, \mathbf{v}_j \rangle = \mathbf{v}_i^* (P^* P) \mathbf{v}_j = \mathbf{e}_i^* \mathbf{e}_j = \begin{cases} 1 & \text{if } i = j \\ 0 & \text{if } i \neq j, \end{cases}$$

with the second-to-last equality following from the fact that $Q \mathbf{e}_j = \mathbf{v}_j$, so $P \mathbf{v}_j = Q^{-1} \mathbf{v}_j = \mathbf{e}_j$ for all j.

1.4.27 The "if" direction is trivial. For the "only if" direction, choose $\mathbf{w} = T(\mathbf{v})$ to see that $\|T(\mathbf{v})\|^2 = \langle T(\mathbf{v}), T(\mathbf{v}) \rangle = 0$ for all $\mathbf{v} \in \mathcal{V}$. It follows that $T(\mathbf{v}) = \mathbf{0}$ for all $\mathbf{v} \in \mathcal{V}$, so $T = O$.

1.4.28 (a) The "if" direction is trivial. For the "only if" direction, choose $\mathbf{v} = \mathbf{x} + \mathbf{y}$ to see that $\langle T(\mathbf{x}) + T(\mathbf{y}), \mathbf{x} + \mathbf{y} \rangle = 0$ for all $\mathbf{x}, \mathbf{y} \in \mathcal{V}$, which can be expanded and simplified to $\langle T(\mathbf{x}), \mathbf{y} \rangle + \langle T(\mathbf{y}), \mathbf{x} \rangle = 0$, which can in turn be rearranged to $\langle T(\mathbf{x}), \mathbf{y} \rangle + \overline{\langle T^*(\mathbf{x}), \mathbf{y} \rangle} = 0$ for all $\mathbf{x}, \mathbf{y} \in \mathcal{V}$. If we perform the same calculation with $\mathbf{v} = \mathbf{x} + i\mathbf{y}$ instead, then we similarly learn that $i\langle T(\mathbf{x}), \mathbf{y} \rangle - i\overline{\langle T^*(\mathbf{x}), \mathbf{y} \rangle} = 0$, which can be multiplied by $-i$ to get $\langle T(\mathbf{x}), \mathbf{y} \rangle - \overline{\langle T^*(\mathbf{x}), \mathbf{y} \rangle} = 0$ for all $\mathbf{x}, \mathbf{y} \in \mathcal{V}$.
Adding the equations $\langle T(\mathbf{x}), \mathbf{y} \rangle + \overline{\langle T^*(\mathbf{x}), \mathbf{y} \rangle} = 0$ and $\langle T(\mathbf{x}), \mathbf{y} \rangle - \overline{\langle T^*(\mathbf{x}), \mathbf{y} \rangle} = 0$ together reveals that $\langle T(\mathbf{x}), \mathbf{y} \rangle = 0$ for all $\mathbf{x}, \mathbf{y} \in \mathcal{V}$, so Exercise 1.4.27 tells us that $T = O$.

(b) For the "if" direction, notice that if $T^* = -T$ then $\langle T(\mathbf{v}), \mathbf{v} \rangle = \langle \mathbf{v}, T^*(\mathbf{v}) \rangle = \langle \mathbf{v}, -T(\mathbf{v}) \rangle = -\langle \mathbf{v}, T(\mathbf{v}) \rangle = -\langle T(\mathbf{v}), \mathbf{v} \rangle$, so $\langle T(\mathbf{v}), \mathbf{v} \rangle = 0$ for all $\mathbf{v}$.
For the "only if" direction, choose $\mathbf{x} = \mathbf{v} + \mathbf{y}$ as in part (a) to see that $\langle (T + T^*)(\mathbf{x}), \mathbf{y} \rangle = \langle T(\mathbf{x}), \mathbf{y} \rangle + \langle T^*(\mathbf{x}), \mathbf{y} \rangle = 0$ for all $\mathbf{x}, \mathbf{y} \in \mathcal{V}$. Exercise 1.4.27 then tells us that $T + T^* = O$, so $T^* = -T$.

1.4.29 (a) If $P = P^*$ then

$$\langle P(\mathbf{v}), \mathbf{v} - P(\mathbf{v}) \rangle = \langle P^2(\mathbf{v}), \mathbf{v} - P(\mathbf{v}) \rangle$$
$$= \langle P(\mathbf{v}), P^*(\mathbf{v} - P(\mathbf{v})) \rangle$$
$$= \langle P(\mathbf{v}), P(\mathbf{v} - P(\mathbf{v})) \rangle$$
$$= \langle P(\mathbf{v}), P(\mathbf{v}) - P(\mathbf{v}) \rangle = 0,$$

so P is orthogonal.

(b) If P is orthogonal (i.e., $\langle P(\mathbf{v}), \mathbf{v} - P(\mathbf{v}) \rangle = 0$ for all $\mathbf{v} \in \mathcal{V}$) then $\langle (P - P^* \circ P)(\mathbf{v}), \mathbf{v} \rangle = 0$ for all $\mathbf{v} \in \mathcal{V}$, so Exercise 1.4.28 tells us that $P - P^* \circ P = O$, so $P = P^* \circ P$. Since $P^* \circ P$ is self-adjoint, it follows that P is as well.

1.4.30 (a) If the columns of A are linearly independent then $\text{rank}(A) = n$, and we know in general that $\text{rank}(A^* A) = \text{rank}(A)$, so $\text{rank}(A^* A) = n$ as well. Since $A^* A$ is an $n \times n$ matrix, this tells us that it is invertible.

(b) We first show that P is an orthogonal projection:

$$P^2 = A(A^* A)^{-1} (A^* A)(A^* A)^{-1} A^*$$
$$= A(A^* A)^{-1} A^* = P,$$

and

$$P^* = (A(A^* A)^{-1} A^*)^* = A(A^* A)^{-1} A^* = P.$$

By uniqueness of orthogonal projections, we thus just need to show that $\text{range}(P) = \text{range}(A)$. The key fact that lets us prove this is the general fact that $\text{range}(AB) \subseteq \text{range}(A)$ for all matrices A and B.
The fact that $\text{range}(A) \subseteq \text{range}(P)$ follows from noting that $PA = A(A^* A)^{-1} (A^* A) = A$, so $\text{range}(A) = \text{range}(PA) \subseteq \text{range}(P)$. Conversely, $P = A((A^* A)^{-1} A^*)$ immediately implies $\text{range}(P) \subseteq \text{range}(A)$ (by choosing $B = (A^* A)^{-1} A^*$ in the fact we mentioned earlier), so we are done.

1.4.31 If $\mathbf{v} \in \text{range}(P)$ then $P\mathbf{v} = \mathbf{v}$ (so $\mathbf{v}$ is an eigenvector with eigenvalue 1), and if $\mathbf{v} \in \text{null}(P)$ then $P\mathbf{v} = \mathbf{0}$ (so $\mathbf{v}$ is an eigenvector with eigenvalue 0). It follows that the geometric multiplicity of the eigenvalue 1 is at least $\text{rank}(P) = \dim(\text{range}(P))$, and the multiplicity of the eigenvalue 0 is at least $\text{nullity}(P) = \dim(\text{null}(P))$. Since $\text{rank}(P) + \text{nullity}(P) = n$ and the sum of multiplicities of eigenvalues of P cannot exceed n, we conclude that actually the multiplicities of the eigenvalues *equal* $\text{rank}(P)$ and $\text{nullity}(P)$, respectively. It then follows immediately from Theorem A.1.5 that P is diagonalizable and has the claimed form.

1.4.32 (a) Applying the Gram–Schmidt process to the standard basis $\{1, x, x^2\}$ of $\mathcal{P}^2[-c, c]$ produces the orthonormal basis

$$C = \left\{ \frac{1}{\sqrt{2c}}, \frac{\sqrt{3}}{\sqrt{2c^3}} x, \frac{\sqrt{5}}{\sqrt{8c^5}} (3x^2 - c^2) \right\}.$$

Constructing $P_c(e^x)$ via this basis (as we did in Example 1.4.18) gives the following polynomial (we use $\sinh(c) = (e^c - e^{-c})/2$ and $\cosh(c) = (e^c + e^{-c})/2$ to simplify things a bit):

$$P_c(e^x) = \frac{15}{2c^5}\left((c^2+3)\sinh(c) - 3c\cosh(c)\right)x^2$$
$$- \frac{3}{c^3}\left(\sinh(c) - c\cosh(c)\right)x$$
$$- \frac{3}{2c^3}\left((c^2+5)\sinh(c) - 5c\cosh(c)\right).$$

(b) Standard techniques from calculus like L'Hôpital's rule show that as $c \to 0$, the coefficients of x^2, x, and 1 in $P_c(e^x)$ above go to $1/2$, 1, and 1, respectively, so

$$\lim_{c \to 0^+} P_c(e_x) = \tfrac{1}{2}x^2 + x + 1,$$

which we recognize as the degree-2 Taylor polynomial of e^x at $x = 0$. This makes intuitive sense since $P_c(e^x)$ is the best approximation of e^x on the interval $[-c, c]$, whereas the Taylor polynomial is its best approximation at $x = 0$.

Section 1.5: Summary and Review

1.5.1 (a) False. This is not even true if B and C are bases of V (vector spaces have many bases).
(c) True. This follows from Theorem 1.2.9.
(e) False. This is true if V is finite-dimensional, but we showed in Exercise 1.2.33 that the "right shift" transformation on $\mathbb{C}^{\mathbb{N}}$ has no eigenvalues or eigenvectors.

1.5.2 (a) If $\{v_1, \ldots, v_k\}$ is a basis of $S_1 \cap S_2$ then we can extend it to bases $\{v_1, \ldots, v_k, w_1, \ldots, w_m\}$ and $\{v_1, \ldots, v_k, x_1, \ldots, x_n\}$ of S_1 and S_2, respectively, via Exercise 1.2.26. It is then straightforward to check that $\{v_1, \ldots, v_k, w_1, \ldots, w_m, x_1, \ldots, x_n\}$ is a basis of $S_1 + S_2$, so $\dim(S_1 + S_2) = k + m + n$, and $\dim(S_1) + \dim(S_2) - \dim(S_1 \cap S_2) = (k+m) + (k+n) - k = k + m + n$ too.

(b) Since $\dim(S_1 + S_2) \leq \dim(V)$, rearranging part (a) shows that $\dim(S_1 \cap S_2) = \dim(S_1) + \dim(S_2) - \dim(S_1 + S_2) \geq \dim(S_1) + \dim(S_2) - \dim(V)$.

1.5.4 One basis is $\{x^\ell y^k : \ell, k \geq 0, \ell + k \leq p\}$. To count the members of this basis, we note that for each choice of ℓ there are $p + 1 - \ell$ possible choices of k, so there are $(p+1) + p + (p-1) + \cdots + 2 + 1 = (p+1)(p+2)/2$ vectors in this basis, so $\dim(\mathcal{P}_2^p) = (p+1)(p+2)/2$.

1.5.6 We can solve this directly from the definition of T^*: $\langle T(X), Y \rangle = \langle AXB^*, Y \rangle = \mathrm{tr}\left((AXB^*)^*Y\right) = \mathrm{tr}(BX^*A^*Y) = \mathrm{tr}(X^*A^*YB) = \langle X, A^*YB \rangle = \langle X, T^*(Y) \rangle$ for all $X \in \mathcal{M}_n$ and $Y \in \mathcal{M}_m$. It follows that $T^*(Y) = A^*YB$.

Section 1.A: Extra Topic: More About the Trace

1.A.1 (b) False. For example, if $A = B = I \in \mathcal{M}_2$ then $\mathrm{tr}(AB) = 2$ but $\mathrm{tr}(A)\mathrm{tr}(B) = 4$.
(d) True. A and A^T have the same diagonal entries.

1.A.2 We just repeatedly use properties of the trace: $\mathrm{tr}(AB) = -\mathrm{tr}(A^T B^T) = -\mathrm{tr}((BA)^T) = -\mathrm{tr}(BA) = -\mathrm{tr}(AB)$, so $\mathrm{tr}(AB) = 0$. Note that this argument only works if $\mathbb{F}$ is a field in which $1 + 1 \neq 0$.

1.A.4 Recall from Exercise 1.4.31 that we can write

$$P = Q\begin{bmatrix} I_r & O \\ O & O \end{bmatrix}Q^{-1},$$

where $r = \mathrm{rank}(P)$. Since the trace is similarity-invariant, it follows that $\mathrm{tr}(P) = \mathrm{tr}(I_r) = r = \mathrm{rank}(P)$.

1.A.5 (a) If A is invertible then choosing $P = A^{-1}$ shows that

$$P(AB)P^{-1} = A^{-1}ABA = BA,$$

so AB and BA are similar. A similar argument works if B is invertible.

(b) If

$$A = \begin{bmatrix} 1 & 0 \\ 0 & 0 \end{bmatrix} \quad \text{and} \quad B = \begin{bmatrix} 0 & 1 \\ 0 & 0 \end{bmatrix}$$

then

$$AB = \begin{bmatrix} 0 & 1 \\ 0 & 0 \end{bmatrix} \quad \text{and} \quad BA = \begin{bmatrix} 0 & 0 \\ 0 & 0 \end{bmatrix},$$

which are not similar since they do not have the same rank ($\mathrm{rank}(A) = 1$ and $\mathrm{rank}(B) = 0$).

1.A.6 (a) If $W \in \mathcal{W}$ then $W = \sum_i c_i(A_iB_i - B_iA_i)$, which has $\mathrm{tr}(W) = \sum_i c_i(\mathrm{tr}(A_iB_i) - \mathrm{tr}(B_iA_i)) = 0$ by commutativity of the trace, so $W \in \mathcal{Z}$, which implies $\mathcal{W} \subseteq \mathcal{Z}$. The fact that $\mathcal{W}$ is a subspace follows from the fact that every span is a subspace, and the fact that $\mathcal{Z}$ is a subspace follows from the fact that it is the null space of a linear transformation (the trace).

(b) We claim that $\dim(\mathcal{Z}) = n^2 - 1$. One way to see this is to notice that $\mathrm{rank}(\mathrm{tr}) = 1$ (since its output is just 1-dimensional), so the rank-nullity theorem tells us that $\mathrm{nullity}(\mathrm{tr}) = \dim(\mathrm{null}(\mathrm{tr})) = \dim(\mathcal{Z}) = n^2 - 1$.

(c) If we let $A = E_{j,j+1}$ and $B = E_{j+1,j}$ (with $1 \leq j < n$), then we get $AB - BA = E_{j,j+1}E_{j+1,j} - E_{j+1,j}E_{j,j+1} = E_{j,j} - E_{j+1,j+1}$. Similarly, if $A = E_{i,k}$ and $B = E_{k,j}$ (with $i \neq j$) then we get $AB - BA = E_{i,k}E_{k,j} - E_{k,j}E_{i,k} = E_{i,j}$. It follows that $\mathcal{W}$ contains each of the $n^2 - 1$ matrices in the following set:

$$\{E_{j,j} - E_{j+1,j+1}\} \cup \{E_{i,j} : i \neq j\}.$$

Furthermore, it is straightforward to show that this set is linearly independent, so $\dim(\mathcal{W}) \geq n^2 - 1$. Since $\mathcal{W} \subseteq \mathcal{Z}$ and $\dim(\mathcal{Z}) = n^2 - 1$, it follows that $\dim(\mathcal{W}) = n^2 - 1$ as well, so $\mathcal{W} = \mathcal{Z}$.

1.A.7   (a) We note that $f(E_{j,j}) = 1$ for all $1 \leq j \leq n$ since $E_{j,j}$ is a rank-1 orthogonal projection. To see that $f(E_{j,k}) = 0$ whenever $1 \leq j \neq k \leq n$ (and thus show that f is the trace), we notice that $E_{j,j} + E_{j,k} + E_{k,j} + E_{k,k}$ is a rank-2 orthogonal projection, so applying f to it gives a value of 2. Linearity of f then says that $f(E_{j,k}) + f(E_{k,j}) = 0$ whenever $j \neq k$. A similar argument shows that $E_{j,j} + iE_{j,k} - iE_{k,j} + E_{k,k}$ is a rank-2 orthogonal projection, so linearity of f says that $f(E_{j,k}) - f(E_{k,j}) = 0$ whenever $j \neq k$. Putting these two facts together shows that $f(E_{j,k}) = f(E_{k,j}) = 0$, which completes the proof.

(b) Consider the linear form $f : \mathcal{M}_2(\mathbb{R}) \to \mathbb{R}$ defined by

$$f\left(\begin{bmatrix} a & b \\ c & d \end{bmatrix}\right) = a + b - c + d.$$

This linear form f coincides with the trace on all symmetric matrices (and thus all orthogonal projections $P \in \mathcal{M}_2(\mathbb{R})$), but not on all $A \in \mathcal{M}_2(\mathbb{R})$.

1.A.8   (a) First use the fact that $I = AA^{-1}$ and multiplicativity of the determinant to write

$$f_{A,B}(x) = \det(A + xB) = \det(A)\det\left(I + x(A^{-1}B)\right).$$

It then follows from Theorem 1.A.4 that $f'_{A,B}(0) = \det(A)\mathrm{tr}(A^{-1}B)$.

(b) The definition of the derivative says that

$$\frac{d}{dt}\det\left(A(t)\right) = \lim_{h \to 0} \frac{\det\left(A(t+h)\right) - \det\left(A(t)\right)}{h}.$$

Using Taylor's theorem (Theorem A.2.3) tells us that

$$A(t+h) = A(t) + h\frac{dA}{dt} + hP(h),$$

where $P(h)$ is a matrix whose entries depend on h in a way so that $\lim_{h \to 0} P(h) = O$. Substituting this expression for $A(t+h)$ into the earlier expression for the derivative of $\det\left(A(t)\right)$ shows that

$$\frac{d}{dt}\det\left(A(t)\right)$$
$$= \lim_{h \to 0} \frac{\det\left(A(t) + h\frac{dA}{dt} + hP(h)\right) - \det\left(A(t)\right)}{h}$$
$$= \lim_{h \to 0} \frac{\det\left(A(t) + h\frac{dA}{dt}\right) - \det\left(A(t)\right)}{h}$$
$$= f'_{A(t),dA/dt}(0),$$

where the second-to-last equality follows from the fact that adding $hP(h)$ inside the determinant, which is a polynomial in the entries of its argument, just adds lots of terms that have $h[P(h)]_{i,j}$ as a factor for various values of i and j. Since $\lim_{h \to 0} h[P(h)]_{i,j}/h = 0$ for all i and j, these terms make no contribution to the value of the limit.

(c) This follows immediately from choosing $B = dA/dt$ in part (a) and then using the result that we proved in part (b).

Section 1.B: Extra Topic: Direct Sum, Orthogonal Complement

1.B.1   (a) The orthogonal complement is the line going through the origin that is perpendicular to $(3,2)$. It thus has $\{(-2,3)\}$ as a basis.

(c) Everything is orthogonal to the zero vector, so the orthogonal complement is all of $\mathbb{R}^3$. One possible basis is just the standard basis $\{\mathbf{e}_1, \mathbf{e}_2, \mathbf{e}_3\}$.

(e) Be careful: all three of the vectors in this set lie on a common plane, so their orthogonal complement is 1-dimensional. The orthogonal complement can be found by solving the linear system $\mathbf{v} \cdot (1,2,3) = 0, \mathbf{v} \cdot (1,1,1) = 0, \mathbf{v} \cdot (3,2,1) = 0$, which has infinitely many solutions of the form $\mathbf{v} = v_3(1,-2,1)$, so $\{(1,-2,1)\}$ is a basis.

(g) We showed in Example 1.B.7 that $(\mathcal{M}_n^S)^\perp = \mathcal{M}_n^{sS}$. When $n = 2$, there is (up to scaling) only one skew-symmetric matrix, so one basis of this space is

$$\left\{\begin{bmatrix} 0 & 1 \\ -1 & 0 \end{bmatrix}\right\}.$$

1.B.2 In all parts of this solution, we refer to the given matrix as A, and the sets that we list are bases of the indicated subspaces.

(a) range(A): $\{(1,3)\}$
null(A^*): $\{(3,-1)\}$
range(A^*): $\{(1,2)\}$
null(A): $\{(-2,1)\}$
The fact that range$(A)^\perp = $ null(A^*) and null$(A)^\perp = $ range(A^*) can be verified by noting that the dimensions of each of the subspace pairs add up to the correct dimension (2 here) and that all vectors in one basis are orthogonal to all vectors in the other basis (e.g., $(1,3) \cdot (3,-1) = 0$).

(c) range(A): $\{(1,4,7), (2,5,8)\}$
null(A^*): $\{(1,-2,1)\}$
range(A^*): $\{(1,0,-1), (0,1,2)\}$
null(A): $\{(1,-2,1)\}$

1.B.3   (b) True. This just says that the only vector in $\mathcal{V}$ that is orthogonal to everything in $\mathcal{V}$ is $\mathbf{0}$.

(d) True. We know from Theorem 1.B.7 that the range of A and the null space of A^* are orthogonal complements of each other.

(f) False. $\mathbb{R}^2 \oplus \mathbb{R}^3$ has dimension 5, so it cannot have a basis consisting of 6 vectors. Instead, the standard basis of $\mathbb{R}^2 \oplus \mathbb{R}^3$ is $\{(\mathbf{e}_1, \mathbf{0}), (\mathbf{e}_2, \mathbf{0}), (\mathbf{0}, \mathbf{e}_1), (\mathbf{0}, \mathbf{e}_2), (\mathbf{0}, \mathbf{e}_3)\}$.

1.B.4 We need to show that $\operatorname{span}(\operatorname{span}(\mathbf{v}_1), \ldots, \operatorname{span}(\mathbf{v}_k)) = \mathcal{V}$ and

$$\operatorname{span}(\mathbf{v}_i) \cap \operatorname{span}\left(\bigcup_{j \neq i} \mathbf{v}_j\right) = \{\mathbf{0}\}$$

for all $1 \leq i \leq k$. The first property follows immediately from the fact that $\{\mathbf{v}_1, \ldots, \mathbf{v}_k\}$ is a basis of $\mathcal{V}$ so $\operatorname{span}(\mathbf{v}_1, \ldots, \mathbf{v}_k) = \mathcal{V}$, so $\operatorname{span}(\operatorname{span}(\mathbf{v}_1), \ldots, \operatorname{span}(\mathbf{v}_k)) = \mathcal{V}$ too. To see the other property, suppose $\mathbf{w} \in \operatorname{span}(\mathbf{v}_i) \cap \operatorname{span}(\bigcup_{j \neq i} \mathbf{v}_j)$ for some $1 \leq i \leq k$. Then there exist scalars $c_1, c_2, \ldots, c_k$ such that $\mathbf{w} = c_i \mathbf{v}_i = \sum_{j \neq i} c_j \mathbf{v}_j$, which (since $\{\mathbf{v}_1, \mathbf{v}_2, \ldots, \mathbf{v}_k\}$ is a basis and thus linearly independent) implies $c_1 = c_2 = \cdots = c_k = 0$, so $\mathbf{w} = \mathbf{0}$.

1.B.5 If $\mathbf{v} \in \mathcal{S}_1 \cap \mathcal{S}_2$ when we can write $\mathbf{v} = c_1 \mathbf{v}_1 + \cdots + c_k \mathbf{v}_k$ and $\mathbf{v} = d_1 \mathbf{w}_1 + \cdots + d_m \mathbf{w}_m$ for some $\mathbf{v}_1, \ldots, \mathbf{v}_k \in B$, $\mathbf{w}_1, \ldots, \mathbf{w}_m \in C$, and scalars $c_1, \ldots, c_k, d_1, \ldots, d_m$. By subtracting these two equations for $\mathbf{v}$ from each other, we see that

$$c_1 \mathbf{v}_1 + \cdots + c_k \mathbf{v}_k - d_1 \mathbf{w}_1 - \cdots - d_m \mathbf{w}_m = \mathbf{0},$$

which (since $B \cup C$ is linearly independent) implies $c_1 = \cdots = c_k = 0$ and $d_1 = \cdots = d_m = 0$, so $\mathbf{v} = \mathbf{0}$.

1.B.6 It suffices to show that $\operatorname{span}(\mathcal{S}_1 \cup \mathcal{S}_2) = \mathcal{V}$. To this end, recall that Theorem 1.B.2 says that if B and C are bases of $\mathcal{S}_1$ and $\mathcal{S}_2$, respectively, then (since (ii) holds) $B \cup C$ is linearly independent. Since $|B \cup C| = \dim(\mathcal{V})$, Exercise 1.2.27 implies that $B \cup C$ is a basis of $\mathcal{V}$, so using Theorem 1.B.2 again tells us that $\mathcal{V} = \mathcal{S}_1 \oplus \mathcal{S}_2$.

1.B.7 Notice that if P is any projection with $\operatorname{range}(P) = \mathcal{S}_1$ and $\operatorname{null}(P) = \mathcal{S}_2$, then we know from Example 1.B.8 that $\mathcal{V} = \mathcal{S}_1 \oplus \mathcal{S}_2$. If B and C are bases of $\mathcal{S}_1$ and $\mathcal{S}_2$, respectively, then Theorem 1.B.2 tells us that $B \cup C$ is a basis of $\mathcal{V}$ and thus P is completely determined by how it acts on the members of B and C. However, P being a projection tells us exactly how it behaves on those two sets: $P\mathbf{v} = \mathbf{v}$ for all $\mathbf{v} \in B$ and $P\mathbf{w} = \mathbf{0}$ for all $\mathbf{w} \in C$, so P is completely determined.

1.B.8 (a) Since $\operatorname{range}(A) \oplus \operatorname{null}(B^*) = \mathbb{F}^m$, we know that if $B^* A \mathbf{v} = \mathbf{0}$ then $A \mathbf{v} = \mathbf{0}$ (since $A\mathbf{v} \in \operatorname{range}(A)$ so the only way $A\mathbf{v}$ can be in $\operatorname{null}(B^*)$ too is if $A\mathbf{v} = \mathbf{0}$). Since A has linearly independent columns, this further implies $\mathbf{v} = \mathbf{0}$, so the linear system $B^* A \mathbf{v} = \mathbf{0}$ in fact has a unique solution. Since $B^* A$ is square, it must be invertible.

(b) We first show that P is a projection:

$$P^2 = A(B^* A)^{-1}(B^* A)(B^* A)^{-1} B^*$$
$$= A(B^* A)^{-1} B^* = P.$$

By uniqueness of projections (Exercise 1.B.7), we now just need to show that $\operatorname{range}(P) = \operatorname{range}(A)$ and $\operatorname{null}(P) = \operatorname{null}(B^*)$. These facts both follow quickly from the facts that $\operatorname{range}(AC) \subseteq \operatorname{range}(A)$ and $\operatorname{null}(A) \subseteq \operatorname{null}(CA)$ for all matrices A and C.

More specifically, these facts immediately tell us that $\operatorname{range}(P) \subseteq \operatorname{range}(A)$ and $\operatorname{null}(B^*) \subseteq \operatorname{null}(P)$. To see the opposite inclusions, notice that $PA = A$ so $\operatorname{range}(A) \subseteq \operatorname{range}(P)$, and $B^* P = B^*$ so $\operatorname{null}(P) \subseteq \operatorname{null}(B^*)$.

1.B.10 To see that the orthogonal complement is indeed a subspace, we need to check that it is non-empty (which it is, since it contains $\mathbf{0}$) and verify that the two properties of Theorem 1.1.2 hold. For property (a), suppose $\mathbf{v}_1, \mathbf{v}_2 \in B^\perp$. Then for all $\mathbf{w} \in B$ we have $\langle \mathbf{v}_1 + \mathbf{v}_2, \mathbf{w} \rangle = \langle \mathbf{v}_1, \mathbf{w} \rangle + \langle \mathbf{v}_2, \mathbf{w} \rangle = 0 + 0 = 0$, so $\mathbf{v}_1 + \mathbf{v}_2 \in B^\perp$. For property (b), suppose $\mathbf{v} \in B^\perp$ and $c \in \mathbb{F}$. Then for all $\mathbf{w} \in B$ we have $\langle c\mathbf{v}, \mathbf{w} \rangle = \overline{c}\langle \mathbf{v}, \mathbf{w} \rangle = \overline{c}0 = 0$, so $c\mathbf{v} \in B^\perp$.

1.B.14 Since $B^\perp = (\operatorname{span}(B))^\perp$ (by Exercise 1.B.13), it suffices to prove that $(\mathcal{S}^\perp)^\perp = \mathcal{S}$ when $\mathcal{S}$ is a subspace of $\mathcal{V}$. To see this, just recall from Theorem 1.B.6 that $\mathcal{S} \oplus \mathcal{S}^\perp = \mathcal{V}$. However, it is also the case that $(\mathcal{S}^\perp)^\perp \oplus \mathcal{S}^\perp = \mathcal{V}$, so $\mathcal{S} = (\mathcal{S}^\perp)^\perp$ by Exercise 1.B.12.

1.B.15 We already know from part (a) that $\operatorname{range}(T^*)^\perp = \operatorname{null}(T^*)$ for all T. If we replace T by T^* throughout that equation, we see that $\operatorname{range}(T^*)^\perp = \operatorname{null}(T)$. Taking the orthogonal complement of both sides (and using the fact that $(\mathcal{S}^\perp)^\perp = \mathcal{S}$ for all subspaces $\mathcal{S}$, thanks to Exercise 1.B.14) we then see $\operatorname{range}(T^*) = \operatorname{null}(T)^\perp$, as desired.

1.B.16 Since $\operatorname{rank}(T) = \operatorname{rank}(T^*)$ (i.e., $\dim(\operatorname{range}(T)) = \dim(\operatorname{range}(T^*))$, we just need to show that $\mathbf{v} = \mathbf{0}$ is the only solution to $S(\mathbf{v}) = \mathbf{0}$. To this end, notice that if $\mathbf{v} \in \operatorname{range}(T^*)$ is such that $S(\mathbf{v}) = \mathbf{0}$ then $\mathbf{v} \in \operatorname{null}(T)$ too. However, since $\operatorname{range}(T^*) = \operatorname{null}(T)^\perp$, we know that the only such $\mathbf{v}$ is $\mathbf{v} = \mathbf{0}$, which is exactly what we wanted to show.

1.B.17 Unfortunately, we cannot quite use Theorem 1.B.6, since that result only applies to finite-dimensional inner product spaces, but we can use the same idea. First notice that $\langle f, g \rangle = \int_{-1}^{1} f(x)g(x)\, dx = 0$ for all $f \in \mathcal{P}^{\mathrm{E}}[-1, 1]$ and $g \in \mathcal{P}^{\mathrm{O}}[-1, 1]$. This follows immediately from the fact that if f is even and g is odd then fg is odd and thus has integral (across any interval that is symmetric through the origin) equal to 0. It follows that $(\mathcal{P}^{\mathrm{E}}[-1, 1])^\perp \supseteq \mathcal{P}^{\mathrm{O}}[-1, 1]$. To see that equality holds, suppose $g \in (\mathcal{P}^{\mathrm{E}}[-1, 1])^\perp$ and write $g = g^{\mathrm{E}} + g^{\mathrm{O}}$, where $g^{\mathrm{E}} \in \mathcal{P}^{\mathrm{E}}[-1, 1]$ and $g^{\mathrm{O}} \in \mathcal{P}^{\mathrm{O}}[-1, 1]$ (which we know we can do thanks to Example 1.B.2). If $\langle f, g \rangle = 0$ for all $f \in \mathcal{P}^{\mathrm{E}}[-1, 1]$ then $\langle f, g^{\mathrm{E}} \rangle = \langle f, g^{\mathrm{O}} \rangle = 0$, so $\langle f, g^{\mathrm{E}} \rangle = 0$ (since $\langle f, g^{\mathrm{O}} \rangle = 0$ when f is even and g^{O} is odd), which implies (by choosing $f = g^{\mathrm{E}}$) that $g^{\mathrm{E}} = 0$, so $g \in \mathcal{P}^{\mathrm{O}}[-1, 1]$.

1.B.18 (a) Suppose $f \in S^{\perp}$ so that $\langle f,g \rangle = 0$ for all $g \in S$. Notice that the function $g(x) = xf(x)$ satisfies $g(0) = 0$, so $g \in S$, so $0 = \langle f,g \rangle = \int_0^1 x(f(x))^2 \, dx$. Since $x(f(x))^2$ is continuous and non-negative on the interval $[0,1]$, the only way this integral can equal 0 is if $x(f(x))^2 = 0$ for all x, so (by continuity of f) $f(x) = 0$ for all $0 \le x \le 1$ (i.e., $f = 0$).

(b) It follows from part (a) that $(S^{\perp})^{\perp} = \{\mathbf{0}\}^{\perp} = C[0,1] \ne S$.

1.B.19 Most of the 10 properties from Definition 1.1.1 follow straightforwardly from the corresponding properties of V and W, so we just note that the zero vector in $V \oplus W$ is $(\mathbf{0},\mathbf{0})$, and $-(\mathbf{v},\mathbf{w}) = (-\mathbf{v},\mathbf{w})$ for all $\mathbf{v} \in V$ and $\mathbf{w} \in W$.

1.B.20 Suppose

$$\sum_{i=1}^{k} c_i(\mathbf{v}_i,\mathbf{0}) + \sum_{j=1}^{m} d_j(\mathbf{0},\mathbf{w}_j) = (\mathbf{0},\mathbf{0})$$

for some $\{(\mathbf{v}_i,\mathbf{0})\} \subseteq B'$ (i.e., $\{\mathbf{v}_i\} \subseteq B$), $\{(\mathbf{0},\mathbf{w}_j)\} \subseteq C'$ (i.e., $\{\mathbf{w}_j\} \subseteq C$), and scalars $c_1,\ldots,c_k, d_1,\ldots,d_m$. This implies $c_1\mathbf{v}_1 + \cdots + c_k\mathbf{v}_k = \mathbf{0}$, which (since B is linearly independent) implies $c_1 = \cdots = c_k = 0$. A similar argument via linear independence of C shows that $d_1 = \cdots = d_m = 0$, which implies $B' \cup C'$ is linearly independent as well.

1.B.21 (a) If $(\mathbf{v}_1,\mathbf{0}),(\mathbf{v}_2,\mathbf{0}) \in V'$ (i.e., $\mathbf{v}_1,\mathbf{v}_2 \in V$) and c is a scalar then $(\mathbf{v}_1,\mathbf{0}) + c(\mathbf{v}_2,\mathbf{0}) = (\mathbf{v}_1 + c\mathbf{v}_2,\mathbf{0}) \in V'$ as well, since $\mathbf{v}_1 + c\mathbf{v}_2 \in V$. It follows that V' is a subspace of $V \oplus W$, and a similar argument works for W'.

(b) The function $T: V \to V'$ defined by $T(\mathbf{v}) = (\mathbf{v},\mathbf{0})$ is clearly an isomorphism.

(c) We need to show that $V' \cap W' = \{(\mathbf{0},\mathbf{0})\}$ and $\operatorname{span}(V',W') = V \oplus W$. For the first property, suppose $(\mathbf{v},\mathbf{w}) \in V' \cap W'$. Then $\mathbf{w} = \mathbf{0}$ (since $(\mathbf{v},\mathbf{w}) \in V'$) and $\mathbf{v} = \mathbf{0}$ (since $(\mathbf{v},\mathbf{w}) \in W'$), so $(\mathbf{v},\mathbf{w}) = (\mathbf{0},\mathbf{0})$, as desired. For the second property, just notice that every $(\mathbf{v},\mathbf{w}) \in V \oplus W$ can be written in the form $(\mathbf{v},\mathbf{w}) = (\mathbf{v},\mathbf{0}) + (\mathbf{0},\mathbf{w})$, where $(\mathbf{v},\mathbf{0}) \in V'$ and $(\mathbf{0},\mathbf{w}) \in W'$.

Section 1.C: Extra Topic: The QR Decomposition

1.C.1 (a) This matrix has QR decomposition UT, where

$$U = \frac{1}{5}\begin{bmatrix} 3 & 4 \\ 4 & -3 \end{bmatrix} \quad \text{and} \quad T = \begin{bmatrix} 5 & 4 \\ 0 & 2 \end{bmatrix}.$$

(c) This QR decomposition is not unique, so yours may differ:

$$U = \frac{1}{3}\begin{bmatrix} 2 & 1 & -2 \\ 1 & 2 & 2 \\ -2 & 2 & -1 \end{bmatrix} \quad \text{and} \quad T = \begin{bmatrix} 9 & 4 \\ 0 & 1 \\ 0 & 0 \end{bmatrix}.$$

(e) $U = \frac{1}{5}\begin{bmatrix} 2 & -1 & 2 & 4 \\ 1 & 2 & -4 & 2 \\ 4 & -2 & -1 & -2 \\ 2 & 4 & 2 & -1 \end{bmatrix}$ and

$$T = \begin{bmatrix} 5 & 3 & 2 & 1 \\ 0 & 1 & 1 & 2 \\ 0 & 0 & 1 & 2 \\ 0 & 0 & 0 & 4 \end{bmatrix}.$$

1.C.2 (a) $\mathbf{x} = (-1,1)/\sqrt{2}$.   (c) $\mathbf{x} = (-2,-1,2)$.

1.C.3 (a) True. This follows immediately from Theorem 1.C.1.

(c) False. Almost any matrix's QR decomposition provides a counter-example. See the matrix from Exercise 1.C.6, for example.

1.C.5 Suppose $A = U_1 T_1 = U_2 T_2$ are two QR decompositions of A. We then define $X = U_1^* U_2 = T_1 T_2^{-1}$ and note that since T_1 and T_2^{-1} are upper triangular with positive real diagonal entries, so is X (see Exercise 1.C.4). On the other hand, X is both unitary and upper triangular, so Exercise 1.4.12 tells us that it is diagonal and its diagonal entries have magnitude equal to 1. The only positive real number with magnitude 1 is 1 itself, so $X = I$, which implies $U_1 = U_2$ and $T_1 = T_2$.

1.C.7 (a) If we write $A = [B \mid C]$ has QR decomposition $A = U[T \mid D]$ where B and T are $m \times m$, B is invertible, and T is upper triangular, then $B = UT$. Since B is invertible, we know from Exercise 1.C.5 that this is its *unique* QR decomposition, so the U and T in the QR decomposition of A are unique. To see that D is also unique, we just observe that $D = U^*C$.

(b) We gave one QR decomposition for a particular 3×2 matrix in the solution to Exercise 1.C.1(c), and another one can be obtained simply by negating the final column of U:

$$U = \frac{1}{3}\begin{bmatrix} 2 & 1 & 2 \\ 1 & 2 & -2 \\ -2 & 2 & 1 \end{bmatrix} \quad \text{and} \quad T = \begin{bmatrix} 9 & 4 \\ 0 & 1 \\ 0 & 0 \end{bmatrix}.$$

Section 1.D: Extra Topic: Norms and Isometries

1.D.1 (a) Is a norm. This is the 1-norm of $P\mathbf{v}$, where P is the diagonal matrix with 1 and 3 on its diagonal. In general, $\|P\mathbf{v}\|_p$ is a norm whenever P is invertible (and it is in this case).

(c) Not a norm. For example, if $\mathbf{v} = (1,-1)$ then

$$\|\mathbf{v}\| = \left|v_1^3 + v_1^2 v_2 + v_1 v_2^2 + v_2^3\right|^{1/3}$$
$$= |1 - 1 + 1 - 1|^{1/3} = 0.$$

Since $\mathbf{v} \neq \mathbf{0}$, it follows that $\|\cdot\|$ is not a norm.

(e) Not a norm. For example, if $p(x) = x$ then $\|p\| = 1$ and $\|2p\| = 1 \neq 2$.

1.D.2 (b) No, not induced by an inner product. For example,

$$2\|\mathbf{e}_1\|^2 + 2\|\mathbf{e}_2\|^2 = 1 + 3^{2/3}, \quad \text{but}$$
$$\|\mathbf{e}_1 + \mathbf{e}_2\|^2 + \|\mathbf{e}_1 - \mathbf{e}_2\|^2 = 4^{2/3} + 4^{2/3}.$$

Since these numbers are not equal, the parallelogram law does not hold, so the norm is not induced by an inner product.

1.D.3 (a) True. All three of the defining properties of norms are straightforward to prove. For example, the triangle inequality for $c\|\cdot\|$ follows from the triangle inequality for $\|\cdot\|$:

$$c\|\mathbf{v} + \mathbf{w}\| \leq c(\|\mathbf{v}\| + \|\mathbf{w}\|) = c\|\mathbf{v}\| + c\|\mathbf{w}\|.$$

(c) False. This is not even true for unitary matrices (e.g., $I + I$ is not a unitary matrix/isometry).

(e) False. For example, if $\mathcal{V}$ and $\mathcal{W}$ have different dimensions then T cannot be invertible. However, we show in Exercise 1.D.21 that this statement is true if $\mathcal{V} = \mathcal{W}$ is finite-dimensional.

1.D.6 The triangle inequality tells us that $\|\mathbf{x} + \mathbf{y}\| \leq \|\mathbf{x}\| + \|\mathbf{y}\|$ for all $\mathbf{x}, \mathbf{y} \in \mathcal{V}$. If we choose $\mathbf{x} = \mathbf{v} - \mathbf{w}$ and $\mathbf{y} = \mathbf{w}$, then this says that

$$\|(\mathbf{v} - \mathbf{w}) + \mathbf{w}\| \leq \|\mathbf{v} - \mathbf{w}\| + \|\mathbf{w}\|.$$

After simplifying and rearranging, this becomes $\|\mathbf{v}\| - \|\mathbf{w}\| \leq \|\mathbf{v} - \mathbf{w}\|$, as desired.

1.D.7 First, choose an index k such that $\|\mathbf{v}\|_\infty = |v_k|$. Then

$$\|\mathbf{v}\|_p = \sqrt[p]{|v_1|^p + \cdots + |v_n|^p}$$
$$\geq \sqrt[p]{|v_k|^p} = |v_k| = \|\mathbf{v}\|_\infty.$$

1.D.8 (a) If $|v| < 1$ then multiplying $|v|$ by itself decreases it. By multiplying it by itself more and more, we can make it decrease to as close to 0 as we like.

[Note: To make this more rigorous, you can note that for every sufficiently small $\varepsilon > 0$, you can choose $p \geq \log(\varepsilon)/\log(|v|)$, which is ≥ 1 when $0 < \varepsilon \leq |v|$, so that $|v|^p \leq \varepsilon$.]

(b) If $\|\mathbf{v}\|_\infty = 1$ then we know from Exercise 1.D.7 that $\|\mathbf{v}\|_p \geq 1$. Furthermore, $|v_j| \leq 1$ for each $1 \leq j \leq n$, so $|v_j|^p \leq 1$ for all $p \geq 1$ as well. Then

$$\|\mathbf{v}\|_p = \sqrt[p]{|v_1|^p + \cdots + |v_n|^p}$$
$$\leq \sqrt[p]{1 + \cdots + 1} \leq \sqrt[p]{n}.$$

Since $\lim_{p \to \infty} \sqrt[p]{n} = 1$, the squeeze theorem for limits tells us that $\lim_{p \to \infty} \|\mathbf{v}\|_p = 1$ too.

(c) We already know that the desired claim holds whenever $\|\mathbf{v}\|_\infty = 1$, so for other vectors we just note that

$$\lim_{p \to \infty} \|c\mathbf{v}\|_p = |c| \lim_{p \to \infty} \|\mathbf{v}\|_p = |c|\|\mathbf{v}\|_\infty = \|c\mathbf{v}\|_\infty.$$

1.D.9 This inequality can be proved "directly":

$$|\mathbf{v} \cdot \mathbf{w}| = \left|\sum_{j=1}^n \overline{v_j} w_j\right|$$
$$\leq \sum_{j=1}^n |v_j||w_j| \leq \sum_{j=1}^n |v_j|\|\mathbf{w}\|_\infty = \|\mathbf{v}\|_1\|\mathbf{w}\|_\infty,$$

where the final inequality follows from the fact that $|w_j| \leq \|\mathbf{w}\|_\infty$ for all j (straight from the definition of that norm).

1.D.10 (a) For the "if" direction, just notice that $\|c\mathbf{w} + \mathbf{w}\|_p = \|(c+1)\mathbf{w}\|_p = (c+1)\|\mathbf{w}\|_p = \|c\mathbf{w}\|_p + \|\mathbf{w}\|_p$. For the "only if" direction, we refer back to the proof of Minkowski's inequality (Theorem 1.D.2). There were only two places in that proof where an inequality was introduced: once when using the triangle inequality on $\mathbb{C}$ and once when using convexity of the function $f(x) = x^p$.

From equality in the triangle inequality we see that, for each $1 \leq j \leq n$, we have (using the notation established in that proof) $y_j = 0$ or $x_j = c_j y_j$ for some $0 < c_j \in \mathbb{R}$. Furthermore, since $f(x) = x^p$ is *strictly* convex whenever $p > 1$, we conclude that the only way the corresponding inequality can be equality is if $|x_j| = |y_j|$ for all $1 \leq j \leq n$. Since $\|\mathbf{x}\|_p = \|\mathbf{y}\|_p = 1$, this implies $\mathbf{x} = \mathbf{y}$, so $\mathbf{v}$ and $\mathbf{w}$ are non-negative multiples of $\mathbf{x} = \mathbf{y}$ and thus of each other.

(b) The "if" direction is straightforward, so we focus on the "only if" direction. If $\|\mathbf{v} + \mathbf{w}\|_1 = \|\mathbf{v}\|_1 + \|\mathbf{w}\|_1$ then

$$\sum_j |v_j + w_j| = \sum_j (|v_j| + |w_j|).$$

Since $|v_j + w_j| \leq |v_j| + |w_j|$ for all j, the above equality implies $|v_j + w_j| = |v_j| + |w_j|$ for all j. This equation holds if and only if v_j and w_j lie on the same closed ray starting at the origin in the complex plane (i.e., either $w_j = 0$ or $v_j = c_j w_j$ for some $0 < c_j \in \mathbb{R}$).

1.D.12 (a) If $\mathbf{v} = \mathbf{e}_1$ is the first standard basis vector then $\|\mathbf{v}\|_p = \|\mathbf{v}\|_q = 1$.

(b) If $\mathbf{v} = (1,1,\ldots,1)$ then $\|\mathbf{v}\|_p = n^{1/p}$ and $\|\mathbf{v}\|_q = n^{1/q}$, so

$$\|\mathbf{v}\|_p = \left(n^{\frac{1}{p} - \frac{1}{q}}\right)\|\mathbf{v}\|_q,$$

as desired.

1.D.13 We already proved the triangle inequality for this norm as Theorem 1.D.7, so we just need to show the remaining two defining properties of norms. First,

$$\|cf\|_p = \left(\int_a^b |cf(x)|^p \, dx\right)^{1/p}$$
$$= \left(|c|^p \int_a^b |f(x)|^p \, dx\right)^{1/p}$$
$$= |c| \left(\int_a^b |f(x)|^p \, dx\right)^{1/p} = |c|\|f\|_p.$$

Second, $\|f\|_p \geq 0$ for all $f \in C[a,b]$ because integrating a non-negative function gives a non-negative answer. Furthermore, $\|f\|_p = 0$ implies f is the zero function since otherwise $f(x) > 0$ for some $x \in [a,b]$ and thus $f(x) > 0$ on some subinterval of $[a,b]$ by continuity of f, so the integral and thus $\|f\|_p$ would both have to be strictly positive as well.

1.D.14 We just mimic the proof of Hölder's inequality for vectors in $\mathbb{C}^n$. Without loss of generality, we just need to prove the theorem in the case when $\|f\|_p = \|g\|_q = 1$. By Young's inequality, we know that

$$|f(x)g(x)| \leq \frac{|f(x)|^p}{p} + \frac{|g(x)|^q}{q}$$

for all $x \in [a,b]$. Integrating then gives

$$\int_a^b |f(x)g(x)|\, dx \leq \int_a^b \frac{|f(x)|^p}{p}\, dx + \int_a^b \frac{|g(x)|^q}{q}\, dx$$
$$= \frac{\|f\|_p^p}{p} + \frac{\|g\|_q^q}{q} = \frac{1}{p} + \frac{1}{q} = 1,$$

as desired.

1.D.16 First, we compute

$$\langle \mathbf{v}, \mathbf{v} \rangle = \frac{1}{4} \sum_{k=0}^3 \frac{1}{i^k} \|\mathbf{v} + i^k \mathbf{v}\|_\mathcal{V}^2 = \|\mathbf{v}\|_\mathcal{V}^2,$$

which is clearly non-negative with equality if and only if $\mathbf{v} = \mathbf{0}$. Similarly,

$$\langle \mathbf{v}, \mathbf{w} \rangle = \frac{1}{4} \sum_{k=0}^3 \frac{1}{i^k} \|\mathbf{v} + i^k \mathbf{w}\|_\mathcal{V}^2$$
$$= \frac{1}{4} \sum_{k=0}^3 \frac{1}{i^k} \|\overline{i^k}\mathbf{v} + \mathbf{w}\|_\mathcal{V}^2$$
$$= \frac{1}{4} \sum_{k=0}^3 \overline{i^k} \|\mathbf{w} + \overline{i^k}\mathbf{v}\|_\mathcal{V}^2 = \overline{\langle \mathbf{w}, \mathbf{v} \rangle}.$$

All that remains is to show that $\langle \mathbf{v}, \mathbf{w} + c\mathbf{x} \rangle = \langle \mathbf{v}, \mathbf{w} \rangle + c\langle \mathbf{v}, \mathbf{x} \rangle$ for all $\mathbf{v}, \mathbf{w}, \mathbf{x} \in \mathcal{V}$ and all $c \in \mathbb{C}$. The fact that $\langle \mathbf{v}, \mathbf{w} + \mathbf{x} \rangle = \langle \mathbf{v}, \mathbf{w} \rangle + \langle \mathbf{v}, \mathbf{x} \rangle$ for all $\mathbf{v}, \mathbf{w}, \mathbf{x} \in \mathcal{V}$ can be proved in a manner identical to that given in the proof of Theorem 1.D.8, so we just need to show that $\langle \mathbf{v}, c\mathbf{w} \rangle = c\langle \mathbf{v}, \mathbf{w} \rangle$ for all $\mathbf{v}, \mathbf{w} \in \mathcal{V}$ and $c \in \mathbb{C}$. As suggested by the hint, we first notice that

$$\langle \mathbf{v}, i\mathbf{w} \rangle = \frac{1}{4} \sum_{k=0}^3 \frac{1}{i^k} \|\mathbf{v} + i^{k+1}\mathbf{w}\|_\mathcal{V}^2$$
$$= \frac{1}{4} \sum_{k=0}^3 \frac{1}{i^{k-1}} \|\mathbf{v} + i\mathbf{w}\|_\mathcal{V}^2$$
$$= i\frac{1}{4} \sum_{k=0}^3 \frac{1}{i^k} \|\mathbf{v} + i\mathbf{w}\|_\mathcal{V}^2 = i\langle \mathbf{v}, \mathbf{w} \rangle.$$

When we combine this observation with the fact that this inner product reduces to exactly the one from the proof of Theorem 1.D.8 when $\mathbf{v}$ and $\mathbf{w}$ are real, we see that $\langle \mathbf{v}, c\mathbf{w} \rangle = c\langle \mathbf{v}, \mathbf{w} \rangle$ simply by splitting everything into their real and imaginary parts.

1.D.20 If $\dim(\mathcal{V}) > \dim(\mathcal{W})$ and $\{\mathbf{v}_1, \ldots, \mathbf{v}_n\}$ is a basis of $\mathcal{V}$, then $\{T(\mathbf{v}_1), \ldots, T(\mathbf{v}_n)\}$ is necessarily linearly dependent (since it contains more than $\dim(\mathcal{W})$ vectors). There thus exist scalars $c_1, \ldots, c_n$, not all equal to 0, such that

$$T(c_1\mathbf{v}_1 + \cdots + c_n\mathbf{v}_n) = c_1 T(\mathbf{v}_1) + \cdots + c_n T(\mathbf{v}_n)$$
$$= \mathbf{0}.$$

In particular, this means that there is a non-zero vector (which has non-zero norm) $c_1\mathbf{v}_1 + \cdots + c_n\mathbf{v}_n \in \mathcal{V}$ that gets sent to the zero vector (which has norm 0), so T is not an isometry.

1.D.22 We prove this theorem by showing the chain of implications (b) $\implies$ (a) $\implies$ (c) $\implies$ (b), and mimic the proof of Theorem 1.4.9.
To see that (b) implies (a), suppose $T^* \circ T = I_\mathcal{V}$. Then for all $\mathbf{v} \in \mathcal{V}$ we have

$$\|T(\mathbf{v})\|_\mathcal{W}^2 = \langle T(\mathbf{v}), T(\mathbf{v}) \rangle = \langle \mathbf{v}, (T^* \circ T)(\mathbf{v}) \rangle$$
$$= \langle \mathbf{v}, \mathbf{v} \rangle = \|\mathbf{v}\|_\mathcal{V}^2,$$

so T is an isometry.
For (a) $\implies$ (c), note that if T is an isometry then $\|T(\mathbf{v})\|_\mathcal{W}^2 = \|\mathbf{v}\|_\mathcal{V}^2$ for all $\mathbf{v} \in \mathcal{V}$, so expanding these quantities in terms of the inner product (like we did above) shows that

$$\langle T(\mathbf{v}), T(\mathbf{v}) \rangle = \langle \mathbf{v}, \mathbf{v} \rangle \quad \text{for all} \quad \mathbf{v} \in \mathcal{V}.$$

Well, if $\mathbf{x}, \mathbf{y} \in \mathcal{V}$ then this tells us (by choosing $\mathbf{v} = \mathbf{x} + \mathbf{y}$) that

$$\langle T(\mathbf{x}+\mathbf{y}), T(\mathbf{x}+\mathbf{y}) \rangle = \langle \mathbf{x}+\mathbf{y}, \mathbf{x}+\mathbf{y} \rangle.$$

Expanding the inner product on both sides of the above equation then gives

$$\langle T(\mathbf{x}), T(\mathbf{x}) \rangle + 2\text{Re}\big(\langle T(\mathbf{x}), T(\mathbf{y}) \rangle\big) + \langle T(\mathbf{y}), T(\mathbf{y}) \rangle$$
$$= \langle \mathbf{x}, \mathbf{x} \rangle + 2\text{Re}\big(\langle \mathbf{x}, \mathbf{y} \rangle\big) + \langle \mathbf{y}, \mathbf{y} \rangle.$$

By then using the fact that $\langle T(\mathbf{x}), T(\mathbf{x}) \rangle = \langle \mathbf{x}, \mathbf{x} \rangle$ and $\langle T(\mathbf{y}), T(\mathbf{y}) \rangle = \langle \mathbf{y}, \mathbf{y} \rangle$, we can simplify the above equation to the form

$$\text{Re}\big(\langle T(\mathbf{x}), T(\mathbf{y}) \rangle\big) = \text{Re}\big(\langle \mathbf{x}, \mathbf{y} \rangle\big).$$

If $\mathcal{V}$ is a vector space over $\mathbb{R}$, then this implies $\langle T(\mathbf{x}), T(\mathbf{y}) \rangle = \langle \mathbf{x}, \mathbf{y} \rangle$ for all $\mathbf{x}, \mathbf{y} \in \mathcal{V}$. If instead $\mathcal{V}$ is a vector space over $\mathbb{C}$ then we can repeat the above argument with $\mathbf{v} = \mathbf{x} + i\mathbf{y}$ to see that

$$\text{Im}\big(\langle T(\mathbf{x}), T(\mathbf{y}) \rangle\big) = \text{Im}\big(\langle \mathbf{x}, \mathbf{y} \rangle\big),$$

so in this case we have $\langle T(\mathbf{x}), T(\mathbf{y}) \rangle = \langle \mathbf{x}, \mathbf{y} \rangle$ for all $\mathbf{x}, \mathbf{y} \in \mathcal{V}$ too, establishing (c).
Finally, to see that (c) $\implies$ (b), note that if we rearrange the equation $\langle T(\mathbf{x}), T(\mathbf{y}) \rangle = \langle \mathbf{x}, \mathbf{y} \rangle$ slightly, we get

$$\langle \mathbf{x}, (T^* \circ T - I_\mathcal{V})(\mathbf{y}) \rangle = 0 \quad \text{for all} \quad \mathbf{x}, \mathbf{y} \in \mathcal{V}.$$

Well, if we choose $\mathbf{x} = (T^* \circ T - I_\mathcal{V})(\mathbf{y})$ then this implies

$$\|(T^* \circ T - I_\mathcal{V})(\mathbf{y})\|_\mathcal{V}^2 = 0 \quad \text{for all} \quad \mathbf{y} \in \mathcal{V}.$$

This implies $(T^* \circ T - I_\mathcal{V})(\mathbf{y}) = 0$, so $(T^* \circ T)(\mathbf{y}) = \mathbf{y}$ for all $\mathbf{y} \in \mathcal{V}$, which means exactly that $T^* \circ T = I_\mathcal{V}$, thus completing the proof.

1.D.24 For the "if" direction, we note (as was noted in the proof of Theorem 1.D.10) that any P with the specified form just permutes the entries of $\mathbf{v}$ and possibly multiplies them by a number with absolute value 1, and such an operation leaves $\|\mathbf{v}\|_1 = \sum_j |v_j|$ unchanged.

For the "only if" direction, suppose $P = \begin{bmatrix} \mathbf{p}_1 & \mathbf{p}_2 & \cdots & \mathbf{p}_n \end{bmatrix}$ is an isometry of the 1-norm. Then $P\mathbf{e}_j = \mathbf{p}_j$ for all j, so

$$\|\mathbf{p}_j\|_1 = \|P\mathbf{e}_j\|_1 = \|\mathbf{e}_j\|_1 = 1 \quad \text{for all} \quad 1 \le j \le n.$$

Similarly, $P(\mathbf{e}_j + \mathbf{e}_k) = \mathbf{p}_j + \mathbf{p}_k$ for all j, k, so

$$\|\mathbf{p}_j + \mathbf{p}_k\|_1 = \|P(\mathbf{e}_j + \mathbf{e}_k)\|_1 = \|\mathbf{e}_j + \mathbf{e}_k\|_1 = 2$$

for all j and k. We know from the triangle inequality (or equivalently from Exercise 1.D.10(b)) that the above equality can only hold if there exist non-negative real constants $c_{i,j,k} \ge 0$ such that, for each i, j, k, it is the case that either $p_{i,j} = c_{i,j,k} p_{i,k}$ or $p_{i,k} = 0$.
However, we can repeat this argument with the fact that $P(\mathbf{e}_j - \mathbf{e}_k) = \mathbf{p}_j - \mathbf{p}_k$ for all j, k to see that

$$\|\mathbf{p}_j - \mathbf{p}_k\|_1 = \|P(\mathbf{e}_j - \mathbf{e}_k)\|_1 = \|\mathbf{e}_j - \mathbf{e}_k\|_1 = 2$$

for all j and k as well. Then by using Exercise 1.D.10(b) again, we see that there exist non-negative real constants $d_{i,j,k} \ge 0$ such that, for each i, j, k, it is the case that either $p_{i,j} = -d_{i,j,k} p_{i,k}$ or $p_{i,k} = 0$.
Since each $c_{i,j,k}$ and $d_{i,j,k}$ is non-negative, it follows that if $p_{i,k} \ne 0$ then $p_{i,j} = 0$ for all $j \ne k$. In other words, each row of P contains at most one non-zero entry (and each row must indeed contain at least one non-zero entry since P is invertible by Exercise 1.D.21).
Every row thus has exactly one non-zero entry. By using (again) the fact that isometries must be invertible, it follows that each of the non-zero entries must occur in a distinct column (otherwise there would be a zero column). The fact that each non-zero entry has absolute value 1 follows from simply noting that P must preserve the 1-norm of each standard basis vector $\mathbf{e}_j$.

1.D.25 Instead of just noting that $\max\limits_{1 \le i \le n} \{|p_{i,j} \pm p_{i,k}|\} = 1$ for all $1 \le j, k \le n$, we need to observe that $\max_{1 \le i \le n} \{|p_{i,j} + z p_{i,k}|\} = 1$ whenever $z \in \mathbb{C}$ is such that $|z| = 1$. The rest of the proof follows with no extra changes needed.

1.D.26 Notice that if $p_n(x) = x^n$ then

$$\|p_n\|_1 = \int_0^1 x^n \, dx = \frac{1}{n+1} \quad \text{and}$$

$$\|p_n\|_\infty = \max_{0 \le x \le 1} \{x^n\} = 1.$$

In particular, this means that there does not exist a constant $C > 0$ such that $1 = \|p_n\|_\infty \le C\|p_n\|_1 = C/(n+1)$ for all $n \ge 1$, since we would need $C \ge n+1$ for all $n \ge 1$.

1.D.27 Since $\|\cdot\|_a$ and $\|\cdot\|_b$ are equivalent, there exist scalars $c, C > 0$ such that

$$c\|\mathbf{v}\|_a \le \|\mathbf{v}\|_b \le C\|\mathbf{v}\|_a \quad \text{for all} \quad \mathbf{v} \in \mathcal{V}.$$

Similarly, equivalence of $\|\cdot\|_b$ and $\|\cdot\|_c$ tells us that there exist scalars $d, D > 0$ such that

$$d\|\mathbf{v}\|_b \le \|\mathbf{v}\|_c \le D\|\mathbf{v}\|_b \quad \text{for all} \quad \mathbf{v} \in \mathcal{V}.$$

Basic algebraic manipulations of these inequalities show that

$$cd\|\mathbf{v}\|_a \le \|\mathbf{v}\|_c \le CD\|\mathbf{v}\|_a \quad \text{for all} \quad \mathbf{v} \in \mathcal{V},$$

so $\|\cdot\|_a$ and $\|\cdot\|_c$ are equivalent too.

1.D.28 Both directions of this claim follow just from noticing that all three defining properties of $\|\cdot\|_{\mathcal{V}}$ follow immediately from the three corresponding properties of $\|\cdot\|_{\mathbb{F}^n}$ (e.g., if we know that $\|\cdot\|_{\mathbb{F}^n}$ is a norm then we can argue that $\|\mathbf{v}\| = 0$ implies $\|[\mathbf{v}]_B\|_{\mathbb{F}^n} = 0$, so $[\mathbf{v}]_B = \mathbf{0}$, so $\mathbf{v} = \mathbf{0}$).
More generally, we recall that the function that sends a vector $\mathbf{v} \in \mathcal{V}$ to its coordinate vector $[\mathbf{v}]_B \in \mathbb{F}^n$ is an isomorphism. It is straightforward to check that if $T : \mathcal{V} \to \mathcal{W}$ is an isomorphism then the function defined by $\|\mathbf{v}\| = \|T(\mathbf{v})\|_{\mathcal{W}}$ is a norm on $\mathcal{V}$ (compare with the analogous statement for inner products given in Exercise 1.3.25).

Section 2.1: The Schur and Spectral Decompositions

2.1.1 (a) Normal.
  (c) Unitary and normal.
  (e) Hermitian, skew-Hermitian, and normal.
  (g) Unitary, Hermitian, and normal.

2.1.2 (a) Not normal, since

$$A^*A = \begin{bmatrix} 5 & 1 \\ 1 & 10 \end{bmatrix} \ne \begin{bmatrix} 5 & -1 \\ -1 & 10 \end{bmatrix} = AA^*.$$

  (c) Is normal, since

$$A^*A = \begin{bmatrix} 2 & 0 \\ 0 & 2 \end{bmatrix} = AA^*.$$

  (e) Is normal (all Hermitian matrices are).
  (g) Not normal.

2.1.3 (a) The eigenvalues of this matrix are 3 and 4, with corresponding eigenvectors $(1,1)/\sqrt{2}$ and $(3,2)/\sqrt{13}$, respectively (only one eigenvalue/eigenvector pair is needed). If we choose to Schur triangularize via $\lambda = 3$ then we get

$$U = \frac{1}{\sqrt{2}} \begin{bmatrix} 1 & 1 \\ 1 & -1 \end{bmatrix} \quad \text{and} \quad T = \begin{bmatrix} 3 & 5 \\ 0 & 4 \end{bmatrix},$$

whereas if we Schur triangularize via $\lambda = 4$ then we get

$$U = \frac{1}{\sqrt{13}} \begin{bmatrix} 3 & 2 \\ 2 & -3 \end{bmatrix} \quad \text{and} \quad T = \begin{bmatrix} 4 & 5 \\ 0 & 3 \end{bmatrix}.$$

These are both valid Schur triangularizations of A (and there are others too).

2.1.4 In all parts of this question, we call the given matrix A.

(a) $A = UDU^*$, where

$$D = \begin{bmatrix} 5 & 0 \\ 0 & 1 \end{bmatrix} \quad \text{and} \quad U = \frac{1}{\sqrt{2}} \begin{bmatrix} 1 & 1 \\ 1 & -1 \end{bmatrix}.$$

(c) $A = UDU^*$, where

$$D = \begin{bmatrix} 1 & 0 \\ 0 & -1 \end{bmatrix} \quad \text{and} \quad U = \frac{1}{\sqrt{2}} \begin{bmatrix} 1 & -1 \\ i & i \end{bmatrix}.$$

(e) $A = UDU^*$, where

$$D = \frac{1}{2} \begin{bmatrix} 1+\sqrt{3}i & 0 & 0 \\ 0 & 1-\sqrt{3}i & 0 \\ 0 & 0 & 4 \end{bmatrix} \quad \text{and}$$

$$U = \frac{1}{2\sqrt{3}} \begin{bmatrix} 2 & 2 & -2 \\ -1+\sqrt{3}i & -1-\sqrt{3}i & -2 \\ -1-\sqrt{3}i & -1+\sqrt{3}i & -2 \end{bmatrix}.$$

2.1.5 (b) False. For example, the matrices

$$A = \begin{bmatrix} 1 & 0 \\ 0 & 0 \end{bmatrix} \quad \text{and } B = \begin{bmatrix} 0 & 1 \\ -1 & 0 \end{bmatrix}$$

are normal, but $A + B$ is not.

(d) False. This was shown in part (b) above.

(f) False. By the real spectral decomposition, we know that such a decomposition of A is possible if and only if A is *symmetric*. There are (real) normal matrices that are not symmetric (and they require a complex D and/or U).

(h) False. For a counter-example, just pick any (non-diagonal) upper triangular matrix with real diagonal entries. However, this becomes true if you add in the restriction that A is *normal* and has real eigenvalues.

2.1.7 Recall that $p_A(\lambda) = \lambda^2 - \text{tr}(A)\lambda + \det(A)$. By the Cayley–Hamilton theorem, it follows that

$$p_A(A) = A^2 - \text{tr}(A)A + \det(A)I = O.$$

Multiplying this equation through by A^{-1} shows that $\det(A)A^{-1} = \text{tr}(A)I - A$, which we can rearrange as

$$A^{-1} = \frac{1}{\det(A)} \begin{bmatrix} d & -b \\ -c & a \end{bmatrix}.$$

The reader may have seen this formula when learning introductory linear algebra.

2.1.9 (a) The characteristic polynomial of A is $p_A(\lambda) = \lambda^3 - 3\lambda^2 + 4$, so the Cayley–Hamilton theorem tells us that $A^3 - 3A^2 + 4I = O$. Moving I to one side and then factoring A out of the other side then gives $I = A(\frac{3}{4}A - \frac{1}{4}A^2)$. It follows that $A^{-1} = \frac{3}{4}A - \frac{1}{4}A^2$.

(b) From part (a) we know that $A^3 - 3A^2 + 4I = O$. Multiplying through by A gives $A^4 - 3A^3 + 4A = O$, which can be solved for A to get $A = \frac{1}{4}(3A^3 - A^4)$. Plugging this into the formula $A^{-1} = \frac{3}{4}A - \frac{1}{4}A^2$ gives us what we want: $A^{-1} = \frac{9}{16}A^3 - \frac{3}{4}A^4 - \frac{1}{4}A^2$.

2.1.12 If A has Schur triangularization $A = UTU^*$, then cyclic commutativity of the trace shows that $\|A\|_F = \|T\|_F$. Since the diagonal entries of T are the eigenvalues of A, we have

$$\|T\|_F = \sqrt{\sum_{j=1}^{n} |\lambda_j|^2 + \sum_{i<j} |t_{i,j}|^2}.$$

It follows that

$$\|A\|_F = \sqrt{\sum_{j=1}^{n} |\lambda_j|^2}$$

if and only if T can actually be chosen to be diagonal, which we know from the spectral decomposition happens if and only if A is normal.

2.1.13 Just apply Exercise 1.4.19 with $B = A$ and $C = A^*$. In particular, the equivalence of conditions (a) and (c) gives us what we want.

2.1.14 If $A^* \in \text{span}\{I, A, A^2, A^3, \ldots\}$ then A^* commutes with A (and thus A is normal) since A commutes with each of its powers.

On the other hand, if A is normal then it has a spectral decomposition $A = UDU^*$, and $A^* = U\overline{D}U^*$. Then let p be the interpolating polynomial with the property that $p(\lambda_j) = \overline{\lambda}_j$ for all $1 \leq j \leq n$ (some of the eigenvalues of A might be repeated, but that is OK because the eigenvalues of A^* are then repeated as well, so we do not run into a problem with trying to set $p(\lambda_j)$ to equal two different values). Then $p(A) = A^*$, so $A^* \in \text{span}\{I, A, A^2, A^3, \ldots\}$. In particular, this tells us that if A has k distinct eigenvalues then $A^* \in \text{span}\{I, A, A^2, A^3, \ldots, A^{k-1}\}$.

2.1.17 In all cases, we write A in a spectral decomposition $A = UDU^*$.

(a) A is Hermitian if and only if $A^* = (UDU^*)^* = UD^*U^*$ equals $A = UDU^*$. Multiplying on the left by U^* and the right by U shows that this happens if and only if $D^* = D$, if and only if the entries of D (i.e., the eigenvalues of A) are all real.

(b) The same as part (a), but noting that $D^* = -D$ if and only if its entries (i.e., the eigenvalues of A) are all imaginary.

(c) A is unitary if and only if $I = A^*A = (UDU^*)^*(UDU^*) = UD^*DU^*$. Multiplying on the left by U^* and the right by U shows that A is unitary if and only if $D^*D = I$, which (since D is diagonal) is equivalent to $|d_{j,j}|^2 = 1$ for all $1 \leq j \leq n$.

(d) If A is not normal then we can let it be triangular (but not diagonal) with whatever eigenvalues (i.e., eigenvalues) we like. Such a matrix is not normal (see Exercise 2.1.16) and thus is not Hermitian, skew-Hermitian, or unitary.

2.1.19 (a) Use the spectral decomposition to write $A = UDU^*$, where D is diagonal with the eigenvalues of A along its diagonal. If we recall that rank is similarity-invariant then we see that $\text{rank}(A) = \text{rank}(D)$, and the rank of a diagonal matrix equals the number of non-zero entries that it has (i.e., the $\text{rank}(D)$ equals the number of non-zero eigenvalues of A).

(b) Any non-zero upper triangular matrix with all diagonal entries (i.e., eigenvalues) equal to 0 works. They have non-zero rank, but no non-zero eigenvalues.

2.1.20 **(a)** We already proved this in the proof of Theorem 2.1.6. Since A is real and symmetric, we have

$$\lambda \|\mathbf{v}\|^2 = \lambda \mathbf{v}^* \mathbf{v} = \mathbf{v}^* A \mathbf{v} = \mathbf{v}^* A^* \mathbf{v}$$
$$= (\mathbf{v}^* A \mathbf{v})^* = (\lambda \mathbf{v}^* \mathbf{v})^* = \overline{\lambda} \|\mathbf{v}\|^2,$$

which implies $\lambda = \overline{\lambda}$ (since every eigenvector $\mathbf{v}$ is, by definition, non-zero), so λ is real.

(b) Let λ be a (necessarily real) eigenvalue of A with corresponding eigenvector $\mathbf{v} \in \mathbb{C}^n$. Then

$$A\overline{\mathbf{v}} = \overline{A\mathbf{v}} = \overline{\lambda \mathbf{v}} = \lambda \overline{\mathbf{v}},$$

so $\overline{\mathbf{v}}$ is also an eigenvector corresponding to λ. Since linear combinations of eigenvectors (corresponding to the same eigenvalue) are still eigenvectors, we conclude that $\mathrm{Re}(\mathbf{v}) = (\mathbf{v} + \overline{\mathbf{v}})/2$ is a real eigenvector of A corresponding to the eigenvalue λ.

(c) We proceed by induction on n (the size of A) and note that the $n = 1$ base case is trivial since every 1×1 real symmetric matrix is real diagonal. For the inductive step, let λ be a real eigenvalue of A with corresponding real eigenvector $\mathbf{v} \in \mathbb{R}^n$. By using the Gram–Schmidt process we can find a unitary matrix $V \in \mathcal{M}_n(\mathbb{R})$ with $\mathbf{v}$ as its first column:

$$V = \big[\, \mathbf{v} \mid V_2 \, \big],$$

where $V_2 \in \mathcal{M}_{n,n-1}(\mathbb{R})$ satisfies $V_2^T \mathbf{v} = \mathbf{0}$ (since V is unitary, $\mathbf{v}$ is orthogonal to every column of V_2). Then direct computation shows that

$$V^T A V = \big[\, \mathbf{v} \mid V_2 \, \big]^T A \big[\, \mathbf{v} \mid V_2 \, \big]$$
$$= \begin{bmatrix} \mathbf{v}^T \\ V_2^T \end{bmatrix} \big[A\mathbf{v} \mid AV_2 \big] = \begin{bmatrix} \mathbf{v}^T A\mathbf{v} & \mathbf{v}^T AV_2 \\ \lambda V_2^T \mathbf{v} & V_2^T AV_2 \end{bmatrix}$$
$$= \begin{bmatrix} \lambda & \mathbf{0}^T \\ \mathbf{0} & V_2^T AV_2 \end{bmatrix}.$$

We now apply the inductive hypothesis—since $V_2^T AV_2$ is an $(n-1) \times (n-1)$ symmetric matrix, there exists a unitary matrix $U_2 \in \mathcal{M}_{n-1}(\mathbb{R})$ and a diagonal $D_2 \in \mathcal{M}_{n-1}(\mathbb{R})$ such that $V_2^T AV_2 = U_2 D_2 U_2^T$. It follows that

$$V^T A V = \begin{bmatrix} \lambda & \mathbf{0}^T \\ \mathbf{0} & U_2 D_2 U_2^T \end{bmatrix}$$
$$= \begin{bmatrix} 1 & \mathbf{0}^T \\ \mathbf{0} & U_2 \end{bmatrix} \begin{bmatrix} \lambda & \mathbf{0}^T \\ \mathbf{0} & D_2 \end{bmatrix} \begin{bmatrix} 1 & \mathbf{0}^T \\ \mathbf{0} & U_2^T \end{bmatrix}.$$

By multiplying on the left by V and on the right by V^T, we see that $A = UDU^T$, where

$$U = V \begin{bmatrix} 1 & \mathbf{0}^T \\ \mathbf{0} & U_2 \end{bmatrix} \quad \text{and} \quad D = \begin{bmatrix} \lambda & \mathbf{0}^T \\ \mathbf{0} & D_2 \end{bmatrix}.$$

Since U is unitary and D is diagonal, this completes the inductive step and the proof.

2.1.23 **(a)** See solution to Exercise 1.4.29(a).

(b) Orthogonality of P tells us that $\langle P(\mathbf{v}), \mathbf{v} - P(\mathbf{v}) \rangle = 0$, so $\langle T(\mathbf{v}), \mathbf{v} \rangle = 0$ for all $\mathbf{v} \in \mathcal{V}$. Exercise 1.4.28 then tells us that $T^* = -T$.

(c) If B is an orthonormal basis of $\mathcal{V}$ then $[T]_B^T = -[T]_B$, so $\mathrm{tr}([T]_B) = \mathrm{tr}([T]_B^T) = -\mathrm{tr}([T]_B)$. It follows that $\mathrm{tr}([P]_B - [P]_B^T[P]_B) = \mathrm{tr}([T]_B) = 0$, so $\mathrm{tr}([P]_B) = \mathrm{tr}([P]_B^T[P]_B) = \|[P]_B^T[P]_B\|_F^2$. Since $\mathrm{tr}([P]_B)$ equals the sum of the eigenvalues of $[P]_B$, all of which are 0 or 1, it follows from Exercise 2.1.12 that $[P]_B$ is normal and thus has a spectral decomposition $[P]_B = UDU^*$. The fact that $[P]_B^T = [P]_B^* = [P]_B$ and thus $P^* = P$ follows from again recalling that the diagonal entries of D are all 0 or 1 (and thus real).

2.1.26 If A has distinct eigenvalues, we can just notice that if $\mathbf{v}$ is an eigenvector of A (i.e., $A\mathbf{v} = \lambda\mathbf{v}$ for some λ) then $AB\mathbf{v} = BA\mathbf{v} = \lambda B\mathbf{v}$, so $B\mathbf{v}$ is also an eigenvector of A corresponding to the same eigenvalue. However, the eigenvalues of A being distinct means that its eigenspaces are 1-dimensional, so $B\mathbf{v}$ must in fact be a multiple of $\mathbf{v}$: $B\mathbf{v} = \mu\mathbf{v}$ for some scalar μ, which means exactly that $\mathbf{v}$ is an eigenvector of B as well. If the eigenvalues of A are *not* distinct, we instead proceed by induction on n (the size of the matrices). The base case $n = 1$ is trivial, and for the inductive step we suppose that the result holds for matrices of size $(n-1) \times (n-1)$ and smaller. Let $\{\mathbf{v}_1, \ldots, \mathbf{v}_k\}$ be an orthonormal basis of any one of the eigenspaces $\mathcal{S}$ of A. If we let $V_1 = \big[\, \mathbf{v}_1 \mid \cdots \mid \mathbf{v}_k \, \big]$ then we can extend V_1 to a unitary matrix $V = \big[\, V_1 \mid V_2 \, \big]$. Furthermore, $B\mathbf{v}_j \in \mathcal{S}$ for all $1 \le j \le k$, by the same argument used in the previous paragraph, and straightforward calculation shows that

$$V^* A V = \begin{bmatrix} V_1^* AV_1 & V_1^* AV_2 \\ O & V_2^* AV_2 \end{bmatrix} \quad \text{and}$$

$$V^* B V = \begin{bmatrix} V_1^* BV_1 & V_1^* BV_2 \\ O & V_2^* BV_2 \end{bmatrix}.$$

Since the columns of V_1 form an orthonormal basis of an eigenspace of A, we have $V_1^* AV_1 = \lambda I_k$, where λ is the corresponding eigenvalue, so $V_1^* AV_1$ and $V_1^* BV_1$ commute. By the inductive hypothesis, $V_1^* AV_1$ and $V_1^* BV_1$ share a common eigenvector $\mathbf{x} \in \mathbb{C}^k$, and it follows that $V_1\mathbf{x}$ is a common eigenvector of each of A and B.

Section 2.2: Positive Semidefiniteness

2.2.1 **(a)** Positive definite, since its eigenvalues equal 1.

(c) Not positive semidefinite, since it is not even square.

(e) Not positive semidefinite, since it is not even Hermitian.

(g) Positive definite, since it is strictly diagonally dominant.

2.2.2 **(a)** $D(1,2)$ and $D(-1,3)$.

(c) $D(1,0)$, $D(2,0)$, and $D(3,0)$. Note that these discs are really just points located at 1, 2, and 3 in the complex plane, so the eigenvalues of this matrix must be 1, 2, and 3 (which we could see directly since it is diagonal).

(e) $D(1,3)$, $D(3,3)$, and another (redundant) copy of $D(1,3)$.

2.2.3 (a) $\dfrac{1}{\sqrt{2}} \begin{bmatrix} 1 & -1 \\ -1 & 1 \end{bmatrix}$

(c) $\begin{bmatrix} 1 & 0 & 0 \\ 0 & \sqrt{2} & 0 \\ 0 & 0 & \sqrt{3} \end{bmatrix}$

(e) $\dfrac{1}{2\sqrt{3}} \begin{bmatrix} 1+\sqrt{3} & 2 & 1-\sqrt{3} \\ 2 & 4 & 2 \\ 1-\sqrt{3} & 2 & 1+\sqrt{3} \end{bmatrix}$

2.2.4 (a) This matrix is positive semidefinite so we can just set $U = I$ and choose P to be itself.

(d) Since this matrix is invertible, its polar decomposition is unique:
$$U = \frac{1}{3} \begin{bmatrix} 1 & -2 & 2 \\ -2 & 1 & 2 \\ 2 & 2 & 1 \end{bmatrix}, \quad P = \begin{bmatrix} 2 & 1 & 0 \\ 1 & 3 & 1 \\ 0 & 1 & 1 \end{bmatrix}.$$

2.2.5 (a) False. The matrix B from Equation (2.2.1) is a counter-example.

(c) True. The identity matrix is Hermitian and has all eigenvalues equal to 1.

(e) False. For example, let
$$A = \begin{bmatrix} 2 & 1 \\ 1 & 1/2 \end{bmatrix} \quad \text{and} \quad B = \begin{bmatrix} 1/2 & 1 \\ 1 & 2 \end{bmatrix},$$
which are both positive semidefinite. Then
$$AB = \begin{bmatrix} 2 & 4 \\ 1 & 2 \end{bmatrix},$$
which is not even Hermitian, so it cannot be positive semidefinite.

(g) False. A counter-example to this claim was provided in Remark 2.2.2.

(i) False. I has many non-PSD square roots (e.g., any diagonal matrix with ± 1) entries on its diagonal.

2.2.6 (a) $x \geq 0$, since a diagonal matrix is PSD if and only if its diagonal entries are non-negative.

(c) $x = 0$. The matrix (which we will call A) is clearly PSD if $x = 0$, and if $x \neq 0$ then we note from Exercise 2.2.11 that if A is PSD and has a diagonal entry equal to 0 then every entry in that row and column must equal 0 too.
More explicitly, we can see that A is not PSD by letting $\mathbf{v} = (0, -1, v_3)$ and computing $\mathbf{v}^*A\mathbf{v} = 1 - 2v_3 x$, which is negative as long as we choose v_3 large enough.

2.2.7 If A is PSD then we can write $A = UDU^*$, where U is unitary and D is diagonal with non-negative real diagonal entries. However, since A is also unitary, its eigenvalues (i.e., the diagonal entries of D) must lie on the unit circle in the complex plane. The only non-negative real number on the unit circle is 1, so $D = I$, so $A = UDU^* = UU^* = I$.

2.2.11 First note that $a_{j,i} = \overline{a_{i,j}}$ since PSD matrices are Hermitian, so it suffices to show that $a_{i,j} = 0$ for all i. To this end, let $\mathbf{v} \in \mathbb{F}^n$ be a vector with $v_j = -1$ and all entries except for v_i and v_j equal to 0. Then $\mathbf{v}^*A\mathbf{v} = a_{j,j} - 2\mathrm{Re}(v_i a_{i,j})$. If it were the case that $a_{i,j} \neq 0$, then we could choose v_i to be a sufficiently large multiple of $\overline{a_{i,j}}$ so that $\mathbf{v}^*A\mathbf{v} < 0$. Since this contradicts positive semidefiniteness of A, we conclude that $a_{i,j} = 0$.

2.2.13 The "only if" direction follows immediately from Theorem 2.2.4. For the "if" direction, note that for any vector $\mathbf{v}$ (of suitable size that we partition in the same way as A) we have
$$\mathbf{v}^*A\mathbf{v} = \begin{bmatrix} \mathbf{v}_1^* & \cdots & \mathbf{v}_n^* \end{bmatrix} \begin{bmatrix} A_1 & \cdots & O \\ \vdots & \ddots & \vdots \\ O & \cdots & A_n \end{bmatrix} \begin{bmatrix} \mathbf{v}_1 \\ \vdots \\ \mathbf{v}_n \end{bmatrix}$$
$$= \mathbf{v}_1^*A_1\mathbf{v}_1 + \cdots + \mathbf{v}_n^*A_n\mathbf{v}_n.$$
Since $A_1, A_2, \ldots, A_n$ are positive (semi)definite it follows that each term in this sum is non-negative (or strictly positive, as appropriate), so the sum is as well, so A is positive (semi)definite.

2.2.14 This follows immediately from Theorem 2.2.10, which says that $A^*A = B^*B$ if and only if there exists a unitary matrix U such that $B = UA$. If we choose $B = A^*$ then we see that $A^*A = AA^*$ (i.e., A is normal) if and only if $A^* = UA$.

2.2.16 (a) For the "if" direction just note that any real linear combination of PSD (self-adjoint) matrices is self-adjoint. For the "only" direction, note that if $A = UDU^*$ is a spectral decomposition of A then we can define $P = UD_+U^*$ and $N = -UD_-U^*$, where D_+ and D_- are the diagonal matrices containing the strictly positive and negative entries of D, respectively. Then P and N are each positive semidefinite and $P - N = UD_+U^* + UD_-U^* = U(D_+ + D_-)U^* = UDU^* = A$.
[Side note: The P and N constructed in this way are called the **positive semidefinite part** and **negative semidefinite part** of A, respectively.]

(b) We know from Remark 1.B.1 that we can write A as a linear combination of the two Hermitian matrices $A + A^*$ and $iA - iA^*$:
$$A = \frac{1}{2}(A + A^*) + \frac{1}{2i}(iA - iA^*).$$
Applying the result of part (a) to these 2 Hermitian matrices writes A as a linear combination of 4 positive semidefinite matrices.

(c) If $\mathbb{F} = \mathbb{R}$ then every PSD matrix is symmetric, and the set of symmetric matrices is a vector space. It follows that there is no way to write a non-symmetric matrix as a (real) linear combination of (real) PSD matrices.

2.2.19 (a) Since A is PSD, we know that $a_{j,j} = \mathbf{e}_j^*A\mathbf{e}_j \geq 0$ for all $1 \leq j \leq n$. Adding up these non-negative diagonal entries shows that $\mathrm{tr}(A) \geq 0$.

(b) Write $A = D^*D$ and $B = E^*E$ (which we can do since A and B are PSD). Then cyclic commutativity of the trace shows that $\operatorname{tr}(AB) = \operatorname{tr}(D^*DE^*E) = \operatorname{tr}(ED^*DE^*) = \operatorname{tr}((DE^*)^*(DE^*))$. Since $(DE^*)^*(DE^*)$ is PSD, it follows from part (a) that this quantity is non-negative.

(c) One example that works is

$$A = \begin{bmatrix} 1 & 1 \\ 1 & 1 \end{bmatrix}, B = \begin{bmatrix} 1 & -2 \\ -2 & 4 \end{bmatrix},$$

$$C = \begin{bmatrix} 4 & -2 \\ -2 & 1 \end{bmatrix}.$$

It is straightforward to verify that each of A, B, and C are positive semidefinite, but $\operatorname{tr}(ABC) = -4$.

2.2.20 (a) The "only if" direction is exactly Exercise 2.2.19(b). For the "if" direction, note that if A is not positive semidefinite then it has a strictly negative eigenvalue $\lambda < 0$ with a corresponding eigenvector $\mathbf{v}$. If we let $B = \mathbf{v}\mathbf{v}^*$ then $\operatorname{tr}(AB) = \operatorname{tr}(A\mathbf{v}\mathbf{v}^*) = \mathbf{v}^*A\mathbf{v} = \mathbf{v}^*(\lambda\mathbf{v}) = \lambda\|\mathbf{v}\|^2 < 0$.

(b) In light of part (a), we just need to show that if A is positive *semi*definite but not positive definite, then there is a PSD B such that $\operatorname{tr}(AB) = 0$. To this end, we just let $\mathbf{v}$ be an eigenvector of A corresponding to the eigenvalue 0 of A and set $B = \mathbf{v}\mathbf{v}^*$. Then $\operatorname{tr}(AB) = \operatorname{tr}(A\mathbf{v}\mathbf{v}^*) = \mathbf{v}^*A\mathbf{v} = \mathbf{v}^*(0\mathbf{v}) = 0$.

2.2.21 (a) The fact that $c \le 0$ follows simply from choosing $B = O$. If A were not positive semidefinite then it has a strictly negative eigenvalue $\lambda < 0$ with a corresponding eigenvector $\mathbf{v}$. If we let $x \ge 0$ be a real number and $B = I + x\mathbf{v}\mathbf{v}^*$ then

$$\operatorname{tr}(AB) = \operatorname{tr}(A(I + x\mathbf{v}\mathbf{v}^*)) = \operatorname{tr}(A) + x\lambda\|\mathbf{v}\|^2,$$

which can be made arbitrarily large and negative (in particular, more negative than c) by choosing x sufficiently large.

(b) To see that $c \le 0$, choose $B = \varepsilon I$ so that $\operatorname{tr}(AB) = \varepsilon \operatorname{tr}(A)$, which tends to 0^+ as $\varepsilon \to 0^+$. Positive semidefiniteness of A then follows via the same argument as in the proof of part (a) (notice that the matrix $B = I + x\mathbf{v}\mathbf{v}^*$ from that proof is positive definite).

2.2.22 (a) Recall from Theorem A.1.2 that $\operatorname{rank}(A^*A) = \operatorname{rank}(A)$. Furthermore, if A^*A has spectral decomposition $A^*A = UDU^*$ then $\operatorname{rank}(A^*A)$ equals the number of non-zero diagonal entries of D, which equals the number of non-zero diagonal entries of $\sqrt{D}$, which equals $\operatorname{rank}(|A|) = \operatorname{rank}(\sqrt{A^*A})$.

(b) Just recall that $\|A\|_F = \sqrt{\operatorname{tr}(A^*A)}$, so $\||A|\|_F = \sqrt{\operatorname{tr}(|A|^2)} = \sqrt{\operatorname{tr}(A^*A)}$ too.

(c) We compute $\||A|\mathbf{v}\|^2 = (|A|\mathbf{v}) \cdot (|A|\mathbf{v}) = \mathbf{v}^*|A|^2\mathbf{v} = \mathbf{v}^*A^*A\mathbf{v} = (A\mathbf{v}) \cdot (A\mathbf{v}) = \|A\mathbf{v}\|^2$ for all $\mathbf{v} \in \mathbb{F}^n$.

2.2.23 To see that (a) $\implies$ (b), let $\mathbf{v}$ be an eigenvector of A with corresponding eigenvalue λ. Then $A\mathbf{v} = \lambda\mathbf{v}$, and multiplying this equation on the left by $\mathbf{v}^*$ shows that $\mathbf{v}^*A\mathbf{v} = \lambda\mathbf{v}^*\mathbf{v} = \lambda\|\mathbf{v}\|^2$. Since A is positive definite, we know that $\mathbf{v}^*A\mathbf{v} > 0$, so it follows that $\lambda > 0$ too.

To see that (b) $\implies$ (d), we just apply the spectral decomposition theorem (either the complex Theorem 2.1.4 or the real Theorem 2.1.6, as appropriate) to A.

To see why (d) $\implies$ (c), let $\sqrt{D}$ be the diagonal matrix that is obtained by taking the (strictly positive) square root of the diagonal entries of D, and define $B = \sqrt{D}U^*$. Then B is invertible since it is the product of two invertible matrices, and $B^*B = (\sqrt{D}U^*)^*(\sqrt{D}U^*) = U\sqrt{D}^*\sqrt{D}U^* = UDU^* = A$. Finally, to see that (c) $\implies$ (a), we let $\mathbf{v} \in \mathbb{F}^n$ be any non-zero and we note that

$$\mathbf{v}^*A\mathbf{v} = \mathbf{v}^*B^*B\mathbf{v} = (B\mathbf{v})^*(B\mathbf{v}) = \|B\mathbf{v}\|^2 > 0,$$

with the final inequality being strict because B is invertible so $B\mathbf{v} \ne \mathbf{0}$ whenever $\mathbf{v} \ne \mathbf{0}$.

2.2.24 In all parts of this question, we prove the statement for positive definiteness. For semidefiniteness, just make the inequalities not strict.

(a) If A and B are self-adjoint then so is $A + B$, and if $\mathbf{v}^*A\mathbf{v} > 0$ and $\mathbf{v}^*B\mathbf{v} > 0$ for all $\mathbf{v} \in \mathbb{F}^n$ then $\mathbf{v}^*(A+B)\mathbf{v} = \mathbf{v}^*A\mathbf{v} + \mathbf{v}^*B\mathbf{v} > 0$ for all $\mathbf{v} \in \mathbb{F}^n$ too.

(b) If A is self-adjoint then so is cA (recall c is real), and $\mathbf{v}^*(cA)\mathbf{v} = c(\mathbf{v}^*A\mathbf{v}) > 0$ for all $\mathbf{v} \in \mathbb{F}^n$ whenever $c > 0$ and A is positive definite.

(c) $(A^T)^* = (A^*)^T$, so A^T is self-adjoint whenever A is, and A^T always has the same eigenvalues as A, so if A is positive definite (i.e., has positive eigenvalues) then so is A^T.

2.2.25 (a) We use characterization (c) of positive semidefiniteness from Theorem 2.2.1. If we let $\{\mathbf{v}_1, \mathbf{v}_2, \ldots, \mathbf{v}_n\}$ be the columns of B (i.e., $B = [\,\mathbf{v}_1 \mid \mathbf{v}_2 \mid \cdots \mid \mathbf{v}_n\,]$) then we have

$$B^*B = \begin{bmatrix} \mathbf{v}_1^* \\ \mathbf{v}_2^* \\ \vdots \\ \mathbf{v}_n^* \end{bmatrix} [\,\mathbf{v}_1 \mid \mathbf{v}_2 \mid \cdots \mid \mathbf{v}_n\,]$$

$$= \begin{bmatrix} \mathbf{v}_1^*\mathbf{v}_1 & \mathbf{v}_1^*\mathbf{v}_2 & \cdots & \mathbf{v}_1^*\mathbf{v}_n \\ \mathbf{v}_2^*\mathbf{v}_1 & \mathbf{v}_2^*\mathbf{v}_2 & \cdots & \mathbf{v}_2^*\mathbf{v}_n \\ \vdots & \vdots & \ddots & \vdots \\ \mathbf{v}_n^*\mathbf{v}_1 & \mathbf{v}_n^*\mathbf{v}_2 & \cdots & \mathbf{v}_n^*\mathbf{v}_n \end{bmatrix}.$$

In particular, it follows that $A = B^*B$ (i.e., A is positive semidefinite) if and only if $a_{i,j} = \mathbf{v}_i^*\mathbf{v}_j = \mathbf{v}_i \cdot \mathbf{v}_j$ for all $1 \le i, j \le n$.

(b) The set $\{\mathbf{v}_1, \mathbf{v}_2, \ldots, \mathbf{v}_n\}$ is linearly independent if and only if the matrix B from our proof of part (a) is invertible, if and only if $\operatorname{rank}(B^*B) = \operatorname{rank}(B) = n$, if and only if B^*B is invertible, if and only if B^*B is positive definite.

2.2.28 (a) This follows from multilinearity of the determinant: multiplying one of the columns of a matrix multiplies its determinant by the same amount.

(b) Since A^*A is positive semidefinite, its eigenvalues $\lambda_1, \lambda_2, \ldots, \lambda_n$ are non-negative, so we can apply the AM–GM inequality to them to get

$$\det(A^*A) = \lambda_1\lambda_2\cdots\lambda_n \le \left(\frac{1}{n}\sum_{j=1}^{n}\lambda_j\right)^n$$

$$= \left(\frac{1}{n}\mathrm{tr}(A^*A)\right)^n = \left(\frac{1}{n}\|A\|_F^2\right)^n$$

$$= \left(\frac{1}{n}\sum_{j=1}^{n}\|\mathbf{a}_j\|^2\right)^n = 1^n = 1.$$

(c) We just recall that the determinant is multiplicative, so $\det(A^*A) = \det(A^*)\det(A) = |\det(A)|^2$, so $|\det(A)|^2 \le 1$, so $|\det(A)| \le 1$.

(d) Equality is attained if and only if the columns of A form a mutually orthogonal set. The reason for this is that equality is attained in the AM–GM inequality if and only if $\lambda_1 = \lambda_2 = \ldots = \lambda_n$, which happens if and only if A^*A is the identity matrix, so A must be unitary in part (b) above. However, the argument in part (b) relied on having already scaled A so that its columns have length 1— after "un-scaling" the columns, we see that any matrix with orthogonal columns also attains equality.

Section 2.3: The Singular Value Decomposition

2.3.1 In all parts of this solution, we refer to the given matrix as A and its SVD as $A = U\Sigma V^*$.

(a) $U = \dfrac{1}{\sqrt{2}}\begin{bmatrix} 1 & 1 \\ -1 & 1 \end{bmatrix}$, $\Sigma = \begin{bmatrix} 4 & 0 \\ 0 & 2 \end{bmatrix}$, $V = \dfrac{1}{\sqrt{2}}\begin{bmatrix} -1 & 1 \\ 1 & 1 \end{bmatrix}$

(c) $U = \dfrac{1}{\sqrt{6}}\begin{bmatrix} \sqrt{2} & \sqrt{3} & -1 \\ \sqrt{2} & 0 & 2 \\ -\sqrt{2} & \sqrt{3} & 1 \end{bmatrix}$, $\Sigma = \begin{bmatrix} \sqrt{6} & 0 \\ 0 & 2 \\ 0 & 0 \end{bmatrix}$,

$V = \dfrac{1}{\sqrt{2}}\begin{bmatrix} 1 & -1 \\ 1 & 1 \end{bmatrix}$

(e) $U = \dfrac{1}{3}\begin{bmatrix} 1 & 2 & 2 \\ -2 & -1 & 2 \\ -2 & 2 & -1 \end{bmatrix}$, $\Sigma = \begin{bmatrix} 6\sqrt{2} & 0 & 0 \\ 0 & 3 & 0 \\ 0 & 0 & 0 \end{bmatrix}$,

$V = \dfrac{1}{\sqrt{2}}\begin{bmatrix} 1 & 0 & 1 \\ 0 & \sqrt{2} & 0 \\ -1 & 0 & 1 \end{bmatrix}$

2.3.2 (a) 4   (c) $\sqrt{6}$ (e) $6\sqrt{2}$

2.3.3 (a) range(A): $\{(1,0),(0,1)\}$,
null(A^*): $\{\}$,
range(A^*): $\{(1,0),(0,1)\}$, and
null(A): $\{\}$.
(c) range(A): $\{(1,1,-1)/\sqrt{3},(1,0,1)/\sqrt{2}\}$,
null(A^*): $\{(-1,2,1)/\sqrt{6}\}$,
range(A^*): $\{(1,0),(0,1)\}$, and
null(A): $\{\}$.
(e) range(A): $\{(1,-2,-2)/3,(2,-1,2)/3\}$,
null(A^*): $\{(2,2,-1)/3\}$,
range(A^*): $\{(1,0,-1)/\sqrt{2},(0,1,0)\}$, and
null(A): $\{(1,0,1)/\sqrt{2}\}$.

2.3.4 (a) False. This statement is true if A is normal by Theorem 2.3.4, but is false in general.
(c) True. We can write $A = U\Sigma V^*$, so $A^*A = V\Sigma^*\Sigma V^*$, whose singular values are the diagonal entries of $\Sigma^*\Sigma$, which are the squares of the singular values of A.
(e) False. For example, if $A = I$ then $\|A^*A\|_F = \sqrt{n}$, but $\|A\|_F^2 = n$.

(g) False. All we can say in general is that $A^2 = UDV^*UDV^*$. This does not simply any further, since we cannot cancel out the V^*U in the middle.
(i) True. If A has SVD $A = U\Sigma V^*$ then $A^T = \overline{V}\Sigma^T U^T$ is also an SVD, and Σ and Σ^T have the same diagonal entries.

2.3.5 Use the singular value decomposition to write $A = U\Sigma V^*$. If each singular value is 1 then $\Sigma = I$, so $A = UV^*$ is unitary. Conversely, there were a singular value σ unequal to 1 then there would be a corresponding normalized left- and right-singular vectors $\mathbf{u}$ and $\mathbf{v}$, respectively, for which $A\mathbf{v} = \sigma\mathbf{u}$, so $\|A\mathbf{v}\| = \|\sigma\mathbf{u}\| = |\sigma|\|\mathbf{u}\| = \sigma \ne 1 = \|\mathbf{v}\|$, so A is not unitary by Theorem 1.4.9.

2.3.6 For the "only if" direction, notice that A has singular value decomposition

$$A = U\begin{bmatrix} D & O \\ O & O \end{bmatrix}V^*,$$

where D is an $r \times r$ diagonal matrix with non-zero diagonal entries (so in particular, D is invertible). Then

$$A = U\begin{bmatrix} I_r & O \\ O & O \end{bmatrix}Q, \quad \text{where} \quad Q = \begin{bmatrix} D & O \\ O & I \end{bmatrix}V^*$$

is invertible. For the "if" direction, just use the fact that rank$(I_r) = r$ and multiplying one the left or right by an invertible matrix does not change rank.

2.3.7 If $A = U\Sigma V^*$ is a singular value decomposition then $|\det(A)| = |\det(U\Sigma V^*)| = |\det(U)\det(\Sigma)\det(V^*)| = |\det(\Sigma)| = \sigma_1\sigma_2\cdots\sigma_n$, where the second-to-last equality follows from the fact that if U is unitary then $|\det(U)| = 1$ (see Exercise 1.4.11).

2.3.10 Submultiplicativity of the operator norm tells us that $\|I\| = \|AA^{-1}\| \le \|A\|\|A^{-1}\|$, so dividing through by $\|A\|$ gives $\|A^{-1}\| \ge 1/\|A\|$. To see that equality does not always hold, consider

$$A = \begin{bmatrix} 1 & 0 \\ 0 & 2 \end{bmatrix},$$

which has $\|A\| = 2$ and $\|A^{-1}\| = 1$.

2.3.12 (a) If $A = U\Sigma V^*$ is an SVD (with the diagonal entries of Σ being $\sigma_1 \geq \sigma_2 \geq \cdots$) then $\|AB\|_F^2 = \operatorname{tr}((U\Sigma V^*B)^*(U\Sigma V^*B)) = \operatorname{tr}(B^*V\Sigma^*\Sigma V^*B) = \operatorname{tr}(\Sigma^*\Sigma V^*BB^*V) \leq \sigma_1^2 \operatorname{tr}(V^*BB^*V) = \sigma_1^2 \operatorname{tr}(BB^*) = \sigma_1^2\|B\|_F^2 = \|A\|^2\|B\|_F^2$, where the inequality in the middle comes from the fact that multiplying V^*BB^*V on the left by $\Sigma^*\Sigma$ multiplies its j-th diagonal entry by σ_j^2, which is no larger than σ_1^2.

(b) This follows from part (a) and the fact that
$$\|A\| = \sigma_1 \leq \sqrt{\sigma_1^2 + \cdots + \sigma_{\min\{m,n\}}^2} = \|A\|_F.$$

2.3.13 The Cauchy–Schwarz inequality tells us that if $\|\mathbf{v}\| = 1$ then $|\mathbf{v}^*A\mathbf{w}| \leq \|\mathbf{v}\|\|A\mathbf{w}\| = \|A\mathbf{w}\|$. Furthermore, equality is attained when $\mathbf{v}$ is parallel to $A\mathbf{w}$. It follows that the given maximization over $\mathbf{v}$ and $\mathbf{w}$ is equal to
$$\max_{\mathbf{w}\in\mathbb{F}^n}\{\|A\mathbf{w}\| : \|\mathbf{w}\| = 1\}$$
which (by definition) equals $\|A\|$.

2.3.15 If $\mathbf{v} \in \mathbb{C}^m$ and $\mathbf{w} \in \mathbb{C}^n$ are unit vectors then
$$\begin{bmatrix} \mathbf{v}^* & \mathbf{w}^* \end{bmatrix}\begin{bmatrix} cI_m & A \\ A^* & cI_n \end{bmatrix}\begin{bmatrix} \mathbf{v} \\ \mathbf{w} \end{bmatrix}$$
$$= c(\|\mathbf{v}\|^2 + \|\mathbf{w}\|^2) + 2\operatorname{Re}(\mathbf{v}^*A\mathbf{w}).$$

This quantity is always non-negative (i.e., the block matrix is positive semidefinite) if and only if $\operatorname{Re}(\mathbf{v}^*A\mathbf{w}) \leq 1$, which is equivalent to $|\mathbf{v}^*A\mathbf{w}| \leq 1$ (since we can multiply $\mathbf{w}$ by some $e^{i\theta}$ so that $\operatorname{Re}(\mathbf{v}^*A\mathbf{w}) = |\mathbf{v}^*A\mathbf{w}|$), which is equivalent to $\|A\| \leq 1$ by Exercise 2.3.13.

2.3.16 All three properties follow quickly from the definition of $\|A\|$ and the corresponding properties of the norm on $\mathbb{F}^n$. Property (a):
$$\|cA\| = \max_{\mathbf{v}\in\mathbb{F}^n}\{\|cA\mathbf{v}\| : \|\mathbf{v}\| \leq 1\}$$
$$= \max_{\mathbf{v}\in\mathbb{F}^n}\{|c|\|A\mathbf{v}\| : \|\mathbf{v}\| \leq 1\}$$
$$= |c|\max_{\mathbf{v}\in\mathbb{F}^n}\{\|A\mathbf{v}\| : \|\mathbf{v}\| \leq 1\} = |c|\|A\|.$$

Property (b):
$$\|A+B\| = \max_{\mathbf{v}\in\mathbb{F}^n}\{\|(A+B)\mathbf{v}\| : \|\mathbf{v}\| \leq 1\}$$
$$= \max_{\mathbf{v}\in\mathbb{F}^n}\{\|A\mathbf{v} + B\mathbf{v}\| : \|\mathbf{v}\| \leq 1\}$$
$$\leq \max_{\mathbf{v}\in\mathbb{F}^n}\{\|A\mathbf{v}\| + \|B\mathbf{v}\| : \|\mathbf{v}\| \leq 1\}$$
$$\leq \max_{\mathbf{v}\in\mathbb{F}^n}\{\|A\mathbf{v}\| : \|\mathbf{v}\| \leq 1\}$$
$$\qquad + \max_{\mathbf{v}\in\mathbb{F}^n}\{\|A\mathbf{v}\| : \|\mathbf{v}\| \leq 1\}$$
$$= \|A\| + \|B\|,$$

where the final inequality comes from the fact that there is more freedom in two separate maximizations than there is in a single maximization. For property (c), the fact that $\|A\| \geq 0$ follows simply from the fact that it involves maximizing a bunch of non-negative quantities, and $\|A\| = 0$ if and only if $A = O$ since $\|A\| = 0$ implies $\|A\mathbf{v}\| = 0$ for all $\mathbf{v}$, which implies $A = O$.

2.3.17 (a) Suppose A has SVD $A = U\Sigma V^*$. If we let $B = UV^*$ then B is unitary and thus has $\|UV^*\| = 1$ (by Exercise 2.3.5, for example), so the given maximization is at least as large as $\langle A, B\rangle = \operatorname{tr}(V\Sigma U^*UV^*) = \operatorname{tr}(\Sigma) = \|A\|_{\operatorname{tr}}$. To show the opposite inequality, note that if B is *any* matrix for which $\|B\| \leq 1$ then $|\langle A, B\rangle| = |\langle U\Sigma V^*, B\rangle| = |\langle \Sigma, U^*BV\rangle| = |\sum_j \sigma_j[U^*BV]_{j,j}| \leq \sum_j \sigma_j|\mathbf{u}_j^*B\mathbf{v}_j| \leq \sum_j \sigma_j\|B\| \leq \sum_j \sigma_j = \|A\|_{\operatorname{tr}}$, where we referred to the j-th columns of U and V as $\mathbf{u}_j$ and $\mathbf{v}_j$, respectively, and we used Exercise 2.3.13 to see that $|\mathbf{u}_j^*B\mathbf{v}_j| \leq \|B\|$.

(b) Property (a) follows quickly from the fact that if $A = U\Sigma V^*$ is a singular value decomposition then so is $cA = U(c\Sigma)V^*$. Property (c) follows from the fact that singular values are non-negative, and if all singular values equal 0 then the matrix they come from is $UOV^* = O$. Finally, for property (b) we use part (a) above:
$$\|A+B\| = \max_{C\in\mathcal{M}_{m,n}}\{|\langle A+B, C\rangle| : \|C\| \leq 1\}$$
$$\leq \max_{C\in\mathcal{M}_{m,n}}\{|\langle A, C\rangle| : \|C\| \leq 1\}$$
$$+ \max_{C\in\mathcal{M}_{m,n}}\{|\langle B, C\rangle| : \|C\| \leq 1\}$$
$$= \|A\| + \|B\|.$$

2.3.19 Just notice that A and UAV have the same singular values since if $A = U_2\Sigma V_2^*$ is a singular value decomposition then so is $UAV = UU_2\Sigma V_2^*V = (UU_2)\Sigma(V^*V_2)^*$.

2.3.20 Recall from Exercise 2.1.12 that A is normal if and only if
$$\|A\|_F = \sqrt{\sum_{j=1}^n |\lambda_j|^2}.$$
Since
$$\|A\|_F = \sqrt{\sum_{j=1}^n \sigma_j^2}$$
by definition, we conclude that if A is not normal then these two sums do not coincide, so at least one of the terms in the sum must not coincide: $\sigma_j \neq |\lambda_j|$ for some $1 \leq j \leq n$.

2.3.25 (a) Since $P \neq O$, there exists some non-zero vector $\mathbf{v}$ in its range. Then $P\mathbf{v} = \mathbf{v}$, so $\|P\mathbf{v}\|/\|\mathbf{v}\| = 1$, so $\|P\| \geq 1$.

(b) We know from Theorem 1.4.13 that $\|P\mathbf{v}\| \leq \|\mathbf{v}\|$ for all $\mathbf{v}$, so $\|P\| \leq 1$. When combined with part (a), this means that $\|P\| = 1$.

(c) We know from Exercise 1.4.31 that every eigenvalue of P equals 0 or 1, so it has a Schur triangularization $P = UTU^*$, where
$$T = \begin{bmatrix} A & B \\ O & I+C \end{bmatrix},$$
where A and C are *strictly* upper triangular (i.e., we are just saying that the diagonal entries of T are all 0 in the top-left block and all 1 in the bottom-right block). If $\mathbf{u}_j$ is the j-th column of U then $\|\mathbf{u}_j\| = 1$ and $\|P\mathbf{u}_j\| = \|UTU^*\mathbf{u}_j\| = \|UT\mathbf{e}_j\|$, which is the norm of the j-th column of T. Since $\|P\| = 1$, it follows that the j-th column of T cannot have norm bigger than 1, which implies $B = O$ and $C = O$.

To see that $A = O$ (and thus complete the proof), note that $P^2 = P$ implies $T^2 = T$, which in turn implies $A^2 = A$, so $A^k = A$ for all $k \geq 1$. However, the diagonal of A consists entirely of zeros, so the first diagonal above the main diagonal in A^2 consists of zeros, the diagonal above that one in A^3 consists of zeros, and so on. Since these powers of A all equal A itself, we conclude that $A = O$.

2.3.26 (a) One simple example is

$$A = \begin{bmatrix} 0 & 1+i \\ 1+i & 1 \end{bmatrix},$$

which has

$$A^*A - AA^* = \begin{bmatrix} 0 & -2i \\ 2i & 0 \end{bmatrix},$$

so A is complex symmetric but not normal.

(b) Since A^*A is positive semidefinite, its spectral decomposition has the form $A^*A = VDV^*$ for some real diagonal D and unitary V. Then

$$B^*B = (V^TAV)^*(V^TAV) = V^*A^*AV = D,$$

so B is real (and diagonal and entrywise non-negative, but we do not need those properties). Furthermore, B is complex symmetric (for *all* unitary matrices V) since

$$B^T = (V^TAV)^T = V^TA^TV = V^TAV = B.$$

(c) To see that B_R is real we note that symmetry of B ensures that the (i, j)-entry of B_R is $(b_{i,j} + \overline{b_{i,j}})/2 = \text{Re}(b_{i,j})$ (a similar calculation shows that the (i, j)-entry of B_I is $\text{Im}(b_{i,j})$). Since B_R and B_I are clearly Hermitian, they must be symmetric. To see that they commute, we compute

$$B^*B = (B_R + iB_I)^*(B_R + iB_I)$$
$$= B_R^2 + B_I^2 + i(B_RB_I - B_IB_R).$$

Since B^*B, B_R, and B_I are all real, this implies $B_RB_I - B_IB_R = O$, so B_R and B_I commute.

(d) Since B_R and B_I are real symmetric and commute, we know by (the "Side note" underneath of) Exercise 2.1.28 that there exists a unitary matrix $W \in M_n(\mathbb{R})$ such that each of W^TB_RW and W^TB_IW are diagonal. Since $B = B_R + iB_I$, we conclude that

$$W^TBW = W^T(B_R + iB_I)W$$
$$= (W^TB_RW) + (W^TB_IW)$$

is diagonal too.

(e) By part (d), we know that if $U = VW$ then

$$U^TAU = W^T(V^TAV)W = W^TBW$$

is diagonal. It does not necessarily have non-negative (or even real) entries on its diagonal, but this can be fixed by multiplying U on the right by a suitable diagonal unitary matrix, which can be used to adjust the complex phases of the diagonal matrix as we like.

Section 2.4: The Jordan Decomposition

2.4.1 (a) $\begin{bmatrix} 1 & 1 \\ 0 & 1 \end{bmatrix}$

(c) $\begin{bmatrix} 2 & 1 \\ 0 & 2 \end{bmatrix}$

(e) $\begin{bmatrix} 1 & 0 & 0 \\ 0 & 2 & 1 \\ 0 & 0 & 2 \end{bmatrix}$

(g) $\begin{bmatrix} 3 & 0 & 0 \\ 0 & 3 & 1 \\ 0 & 0 & 3 \end{bmatrix}$

(e) Similar, since they both have

$$\begin{bmatrix} -1 & 0 & 0 \\ 0 & 3 & 1 \\ 0 & 0 & 3 \end{bmatrix}$$

as their Jordan canonical form.

2.4.2 We already computed the matrix J in the Jordan decomposition $A = PJP^{-1}$ in Exercise 2.4.1, so here we just present a matrix P that completes the decomposition. Note that this matrix is not unique, so your answer may differ.

(a) $\begin{bmatrix} 2 & 2 \\ 2 & 1 \end{bmatrix}$

(c) $\begin{bmatrix} 1 & 0 \\ -2 & -1 \end{bmatrix}$

(e) $\begin{bmatrix} 0 & 1 & 1 \\ 2 & 1 & 2 \\ 1 & 1 & 2 \end{bmatrix}$

(g) $\begin{bmatrix} 1 & -2 & -1 \\ 1 & 1 & 0 \\ 2 & -1 & 0 \end{bmatrix}$

2.4.3 (a) Not similar, since their traces are not the same (8 and 10).

(c) Not similar, since their determinants are not the same (0 and -2).

2.4.4 (a) $\begin{bmatrix} \sqrt{2} & 0 \\ \sqrt{2}/4 & \sqrt{2} \end{bmatrix}$

(e) $\begin{bmatrix} e & -2e & -e \\ 0 & 3e & e \\ 0 & -4e & -e \end{bmatrix}$

2.4.5 (a) False. Any matrix with a Jordan block of size 2×2 or larger (i.e., almost any matrix from this section) serves as a counter-example.

(c) False. The Jordan canonical forms J_1 and J_2 are only unique up to re-ordering of their Jordan blocks (so we can shift the diagonal blocks of J_1 around to get J_2).

(e) True. If A and B are diagonalizable then their Jordan canonical forms are diagonal and so Theorem 2.4.3 tells us they are similar if those 1×1 Jordan blocks (i.e., their eigenvalues) coincide.

(g) True. Since the function $f(x) = e^x$ is analytic on all of $\mathbb{C}$, Theorems 2.4.6 and 2.4.7 tell us that this sum converges for all A (and furthermore we can compute it via the Jordan decomposition).

2.4.8 We can choose C to be the standard basis of $\mathbb{R}^2$ and $T(\mathbf{v}) = A\mathbf{v}$. Then it is straightforward to check that $[T]_C = A$. All that is left to do is find a basis D such that $[T]_D = B$. To do so, we find a Jordan decomposition of A and B, which are actually diagonalizations in this case. In particular, if we define

$$P_1 = \begin{bmatrix} 1 & 2 \\ -1 & 5 \end{bmatrix} \text{ and } P_2 = \begin{bmatrix} 1 & 1 \\ -1 & -2 \end{bmatrix} \text{ then}$$

$$P_1^{-1}AP_1 = \begin{bmatrix} -1 & 0 \\ 0 & 6 \end{bmatrix} = P_2^{-1}BP_2.$$

Rearranging gives $P_2P_1^{-1}AP_1P_2^{-1} = B$. In other words, $P_1P_2^{-1}$ is the change-of-basis matrix from D to C. Since C is the standard basis, the columns of $P_1P_2^{-1}$ are the basis vectors of D. Now we can just compute:

$$P_1P_2^{-1} = \begin{bmatrix} 3 & 1 \\ 11 & 6 \end{bmatrix}, \quad \text{so} \quad D = \left\{ \begin{bmatrix} 3 \\ 11 \end{bmatrix}, \begin{bmatrix} 1 \\ 6 \end{bmatrix} \right\}.$$

2.4.11 Since $e^{\operatorname{tr}(A)} \neq 0$, this follows immediately from Exercise 2.4.10.

2.4.14 Since $\sin^2(x) + \cos^2(x) = 1$ for all $x \in \mathbb{C}$, the function $f(x) = \sin^2(x) + \cos^2(x)$ is certainly analytic (it is constant). Furthermore, $f'(x) = 0$ for all $x \in \mathbb{C}$ and more generally $f^{(k)}(x) = 0$ for all $x \in \mathbb{C}$ and integers $k \geq 1$. It follows from Theorem 2.4.6 that $f(J_k(\lambda)) = I$ for all Jordan blocks $J_k(\lambda)$, and Theorem 2.4.7 then tells us that $f(A) = I$ for all $A \in M_n(\mathbb{C})$.

2.4.16 (a) This follows directly from the definition of matrix multiplication:

$$[A^2]_{i,j} = \sum_{\ell=1}^{k} a_{i,\ell}a_{\ell,j},$$

which equals 0 unless $\ell \geq i+1$ and $j \geq \ell+1 \geq i+2$. A similar argument via induction shows that $[A^k]_{i,j} = 0$ unless $j \geq i+k$.

(b) This is just the $k = n$ case of part (a)—A^n only has n superdiagonals, and since they are all 0 we know that $A^n = O$.

2.4.17 (a) This follows directly from the definition of matrix multiplication:

$$[N_1^2]_{i,j} = \sum_{\ell=1}^{k} [N_1]_{i,\ell}[N_1]_{\ell,j},$$

which equals 1 if $\ell = i+1$ and $j = \ell+1 = i+2$, and equals 0 otherwise. In other words, $N_1^2 = N_2$, and a similar argument via induction shows that $N_1^n = N_n$ whenever $1 \leq n < k$. The fact that $N_1^k = O$ follows from Exercise 2.4.16.

(b) Simply notice that N_n has $k - n$ of the standard basis vectors as its columns so its rank is $\max\{k - n, 0\}$, so its nullity is $k - \max\{k-n,0\} = \min\{k,n\}$.

Section 2.5: Summary and Review

2.5.1 (a) All five decompositions apply to this matrix.

(c) Schur triangularization, singular value decomposition, and Jordan decomposition (it does not have a linearly independent set of eigenvectors, so we cannot diagonalize or apply the spectral decomposition).

(e) Diagonalization, Schur triangularization, singular value decomposition, and Jordan decomposition (it is not normal, so we cannot apply the spectral decomposition).

(g) All five decompositions apply to this matrix.

2.5.2 (a) True. This is the statement of Corollary 2.1.5.

(c) True. This fact was stated as part of Theorem 2.2.12.

2.5.4 For the "only if" direction, use the spectral decomposition to write $A = UDU^*$ so $A^* = UD^*U^* = U\overline{D}U^*$. It follows that A and A^* have the same eigenspaces and the corresponding eigenvalues are just complex conjugates of each other, as claimed.

For the "if" direction, note that we can use Schur triangularization to write $A = UTU^*$, where the leftmost column of U is $\mathbf{v}$ and we denote the other columns of U by $\mathbf{u}_2, \dots, \mathbf{u}_n$. Then $A\mathbf{v} = t_{1,1}\mathbf{v}$ (i.e., the eigenvalue of A corresponding to $\mathbf{v}$ is $t_{1,1}$), but $A^*\mathbf{v} = (UT^*U)\mathbf{v} = UT^*\mathbf{e}_1 = \overline{t_{1,1}}\mathbf{v} + \overline{t_{1,2}}\mathbf{u}_2 + \cdots + \overline{t_{1,n}}\mathbf{u}_n$. The only way that this equals $\overline{t_{1,1}}\mathbf{v}$ is if $t_{1,2} = \cdots = t_{1,n} = 0$, so the only non-zero entry in the first row of T is its $(1,1)$-entry.

It follows that the second column of U must also be an eigenvector of A, and then we can repeat the same argument as in the previous paragraph to show that the only non-zero entry in the second row of T is its $(2,2)$-entry. Repeating in this way shows that $k = n$ and T is diagonal, so $A = UTU^*$ is normal.

2.5.7 (a) $\operatorname{tr}(A)$ is the sum of the diagonal entries of A, each of which is 0 or 1, so $\operatorname{tr}(A) \leq 1 + 1 + \cdots + 1 = n$.

(b) The fact that $\det(A)$ is an integer follows from formulas like Theorem A.1.4 and the fact that each entry of A is an integer. Since $\det(A) > 0$ (by positive definiteness of A), it follows that $\det(A) \geq 1$.

(c) The AM–GM inequality tells us that $\sqrt[n]{\det(A)} = \sqrt[n]{\lambda_1 \cdots \lambda_n} \le (\lambda_1 + \cdots + \lambda_n)/n = \text{tr}(A)/n$. Since $\text{tr}(A) \le n$ by part (a), we conclude that $\det(A) \le 1$.

(d) Parts (b) and (c) tell us that $\det(A) = 1$, so equality holds in the AM–GM inequality in part (c). By the equality condition of the AM–GM inequality, we conclude that $\lambda_1 = \ldots = \lambda_n$, so they all equal 1 and thus A has spectral decomposition $A = UDU^*$ with $D = I$, so $A = UIU^* = UU^* = I$.

Section 2.A: Extra Topic: Quadratic Forms and Conic Sections

2.A.1   (a) Positive definite.
        (c) Indefinite.
        (e) Positive definite.
        (g) Positive semidefinite.
        (i) Positive semidefinite.

2.A.2   (a) Ellipse.
        (c) Ellipse.
        (e) Hyperbola.
        (g) Hyperbola.

2.A.3   (a) Ellipsoid.
        (c) Hyperboloid of one sheet.
        (e) Elliptical cylinder.

(g) Ellipsoid.

2.A.4   (a) False. Quadratic forms arise from bilinear forms in the sense that if $f(\mathbf{v}, \mathbf{w})$ is a bilinear form then $q(\mathbf{v}) = f(\mathbf{v}, \mathbf{v})$ is a quadratic form, but they are not the same thing. For example, bilinear forms act on two vectors, whereas quadratic forms act on just one.
        (c) True. We stated this fact near the start of Section 2.A.1.
        (e) True. This is the quadratic form associated with the 1×1 matrix $A = [1]$.

Section 2.B: Extra Topic: Schur Complements and Cholesky

2.B.2   (a) The Schur complement is

$$S = \frac{1}{5} \begin{bmatrix} 7 & -1 \\ -1 & 3 \end{bmatrix},$$

which has $\det(S) = 4/5$ and is positive definite. It follows that the original matrix has determinant $\det(A)\det(S) = 5(4/5) = 4$ and is also positive definite (since its top-left block is).

2.B.3   (a) True. More generally, the Schur complement of the top-left $n \times n$ block of an $(m+n) \times (m+n)$ matrix is an $m \times m$ matrix.
        (c) False. Only positive semidefinite matrices have a Cholesky decomposition.

2.B.4 There are many possibilities—we can choose the $(1,2)$- and $(1,3)$-entries of the upper triangular matrix to be anything sufficiently small that we want, and then adjust the $(2,2)$- and $(2,3)$-entries accordingly (as suggested by Remark 2.B.1). For example, if $0 \le a \le 1$ is a real number and we define

$$T = \begin{bmatrix} 0 & \sqrt{a} & \sqrt{a} \\ 0 & \sqrt{1-a} & \sqrt{1-a} \\ 0 & 0 & 1 \end{bmatrix}$$

then

$$T^*T = \begin{bmatrix} 0 & 0 & 0 \\ 0 & 1 & 1 \\ 0 & 1 & 2 \end{bmatrix}.$$

2.B.7 This follows immediately from Theorem 2.B.1 and the following facts: the upper and lower triangular matrices in that theorem are invertible due to having all diagonal entries equal to 1, $\text{rank}(XY) = \text{rank}(X)$ whenever Y is invertible, and the rank of a block diagonal matrix is the sum of the ranks of its diagonal blocks.

2.B.8   (a) The following matrices work:

$$U = \begin{bmatrix} I & BD^{-1} \\ O & I \end{bmatrix}, \quad L = \begin{bmatrix} I & O \\ D^{-1}C & I \end{bmatrix}.$$

        (b) The argument is the exact same as in Theorem 2.B.2: $\det(U) = \det(L) = 1$, so $\det(Q)$ equals the determinant of the middle block diagonal matrix, which equals $\det(D)\det(S)$.
        (c) Q is invertible if and only if $\det(Q) \ne 0$, if and only if $\det(S) \ne 0$, if and only if S is invertible. The inverse of Q can be computed using the same method of Theorem 2.B.3 to be

$$\begin{bmatrix} I & O \\ -D^{-1}C & I \end{bmatrix} \begin{bmatrix} S^{-1} & O \\ O & D^{-1} \end{bmatrix} \begin{bmatrix} I & -BD^{-1} \\ O & I \end{bmatrix}.$$

        (d) The proof is almost identical to the one provided for Theorem 2.B.4.

2.B.11 We need to transform $\det(AB - \lambda I_m)$ into a form that lets us use Sylvester's determinant identity (Exercise 2.B.9). Well,

$$\det(AB - \lambda I_m) = (-1)^m \det(\lambda I_m - AB)$$
$$= (-\lambda)^m \det(I_m + (-A/\lambda)B)$$
$$= (-\lambda)^m \det(I_n + B(-A/\lambda))$$
$$= ((-\lambda)^m/\lambda^n) \det(\lambda I_n - BA)$$
$$= (-\lambda)^{m-n} \det(BA - \lambda I_n),$$

which is what we wanted to show.

2.B.12 In the 1×1 case, it is clear that if $A = [a]$ then the Cholesky decomposition $A = [\sqrt{a}]^*[\sqrt{a}]$ is unique. For the inductive step, we assume that the Cholesky decompositions of $(n-1) \times (n-1)$ matrices are unique. If $a_{1,1} = 0$ (i.e., Case 1 of the proof of Theorem 2.B.5) then solving

$$A = \begin{bmatrix} 0 & \mathbf{0}^T \\ \mathbf{0} & A_{2,2} \end{bmatrix} = [\mathbf{x} \mid B]^*[\mathbf{x} \mid B]$$

for $\mathbf{x}$ and B reveals that $\|\mathbf{x}\|^2 = \mathbf{x}^*\mathbf{x} = 0$, so $\mathbf{x} = \mathbf{0}$, so the only way for this to be a Cholesky decomposition of A is if $B = T$, where $A_{2,2} = T^*T$ is a Cholesky decomposition of $A_{2,2}$ (which is unique by the inductive hypothesis).

On the other hand, if $a_{1,1} \neq 0$ (i.e., Case 1 of the proof of Theorem 2.B.5) then the Schur complement S has unique Cholesky decomposition $S = T^*T$, so we know that the Cholesky decomposition of A must be of the form

$$A = \begin{bmatrix} \sqrt{a_{1,1}} & \mathbf{x}^* \\ \mathbf{0} & T \end{bmatrix}^* \begin{bmatrix} \sqrt{a_{1,1}} & \mathbf{x}^* \\ \mathbf{0} & T \end{bmatrix} = \begin{bmatrix} a_{1,1} & \mathbf{a}_{2,1}^* \\ \mathbf{a}_{2,1} & A_{2,2} \end{bmatrix},$$

where $\mathbf{x} \in \mathbb{F}^{n-1}$ is some unknown column vector. Just performing the matrix multiplication reveals that it must be the case that $\mathbf{x} = \mathbf{a}_{2,1}/\sqrt{a_{1,1}}$, so the Cholesky decomposition of A is unique in this case well.

Section 2.C: Extra Topic: Applications of the SVD

2.C.1 In all cases, we simply compute $\mathbf{x} = A^\dagger\mathbf{b}$.
   (a) No solution, closest is $\mathbf{x} = (1,2)/10$.
   (c) Infinitely many solutions, with the smallest being $\mathbf{x} = (-5,-1,3)$.

2.C.2 (a) $y = 3x - 2$ (c) $y = (5/2)x - 4$

2.C.3 (a) $\begin{bmatrix} 2 & 2 \\ 2 & 2 \end{bmatrix}$ (c) $\dfrac{1}{5}\begin{bmatrix} 4 & 2 & 0 \\ 2 & 1 & 0 \\ -10 & -5 & 0 \end{bmatrix}$

2.C.4 (a) True. More generally, the pseudoinverse of a diagonal matrix is obtained by taking the reciprocals of its non-zero diagonal entries.
   (c) False. Theorem 2.C.1 shows that $\text{range}(A^\dagger) = \text{range}(A^*)$, which typically does not equal $\text{range}(A)$.
   (e) False. The pseudoinverse finds one such vector, but there may be many of them. For example, if the linear system $A\mathbf{x} = \mathbf{b}$ has infinitely many solutions then there are infinitely many vectors that minimize $\|A\mathbf{x} - \mathbf{b}\|$ (i.e., make it equal to 0).

2.C.5 We plug the 4 given data points into the equation $z = ax + by + c$ to get 4 linear equations in the 3 variables a, b, and c. This linear system has no solution, but we can compute the least squares solution $\mathbf{x} = A^\dagger\mathbf{b} = (1,2,1)$, so the plane of best fit is $z = x + 2y + 1$.

2.C.6 We plug the 3 given data points into the equation $y = c_1 \sin(x) + c_2 \cos(x)$ to get 3 linear equations in the 2 variables c_1 and c_2. This linear system has no solution, but we can compute the least squares solution $\mathbf{x} = A^\dagger\mathbf{b} = (1,-1/2)$, so the curve of best fit is $y = \sin(x) - \cos(x)/2$.

2.C.7 These parts can both be proved directly in a manner analogous to the proofs of parts (a) and (c) given in the text. However, a quicker way is to notice that part (d) follows from part (c), since Exercise 1.B.9 tells us that $I - A^\dagger A$ is the orthogonal projection onto $\text{range}(A^*)^\perp$, which equals $\text{null}(A)$ by Theorem 1.B.7.
   Part (b) follows from part (a) via a similar argument: $I - AA^\dagger$ is the orthogonal projection onto $\text{range}(A)^\perp$, which equals $\text{null}(A^*)$ by Theorem 1.B.7. The fact that it is also the orthogonal projection onto $\text{null}(A^\dagger)$ follows from swapping the roles of A and $A^\dagger$ (and using the fact that $(A^\dagger)^\dagger = A$) in part (d).

2.C.9 (a) If A has linearly independent columns then $\text{rank}(A) = n$, so A has n non-zero singular values. It follows that if $A = U\Sigma V^*$ is a singular value decomposition then $(A^*A)^{-1}A^* = (V\Sigma^*\Sigma V^*)^{-1}V\Sigma^*U^* = V(\Sigma^*\Sigma)^{-1}V^*V\Sigma^*U^* = V(\Sigma^*\Sigma)^{-1}\Sigma^*U^*$. We now note that $\Sigma^*\Sigma$ is indeed invertible, since it is an $n \times n$ diagonal matrix with n non-zero diagonal entries. Furthermore, $(\Sigma^*\Sigma)^{-1}\Sigma^* = \Sigma^\dagger$ since the inverse of a diagonal matrix is the diagonal matrix with the reciprocal diagonal entries. It follows that $(A^*A)^{-1}A^* = V\Sigma^\dagger U^* = A^\dagger$, as claimed.
   (b) Almost identical to part (a), but noting instead that $\text{rank}(A) = m$ so $\Sigma\Sigma^*$ is an $m \times m$ invertible diagonal matrix.

Section 2.D: Extra Topic: Continuity and Matrix Analysis

2.D.2   (a) True. $\|A\|_F = \sqrt{\operatorname{tr}(A^*A)}$ is continuous since it can be written as a composition of three continuous functions (the square root, the trace, and multiplication by A^*).

2.D.3 This follows from continuity of singular values. If $f(A) = \sigma_{r+1}$ is the $(r+1)$-th largest singular value of A then f is continuous and $f(A_k) = 0$ for all k. It follows that $f\left(\lim_{k\to\infty} A_k\right) = \lim_{k\to\infty} f(A_k) = 0$ too, so $\operatorname{rank}\left(\lim_{k\to\infty} A_k\right) \leq r$.

2.D.5 Just notice that if A is positive semidefinite then $A_k = A + \frac{1}{k}I$ is positive definite for all integers $k \geq 1$, and $\lim_{k\to\infty} A_k = A$.

2.D.7 For any $A \in \mathcal{M}_{m,n}$, A^*A is positive semidefinite with $m \geq \operatorname{rank}(A) = \operatorname{rank}(A^*A)$, so the Cholesky decomposition (Theorem 2.B.5) tells us that there exists an upper triangular matrix $T \in \mathcal{M}_{m,n}$ with non-negative real diagonal entries such that $A^*A = T^*T$. Applying Theorem 2.2.10 then tells us that there exists a unitary matrix $U \in \mathcal{M}_m$ such that $A = UT$, as desired.

Section 3.1: The Kronecker Product

3.1.1   (a) $\left[\begin{array}{cc|cc} -1 & 2 & -2 & 4 \\ 0 & -3 & 0 & -6 \\ \hline -3 & 6 & 0 & 0 \\ 0 & -9 & 0 & 0 \end{array}\right]$

     (c) $\left[\begin{array}{ccc} 2 & -3 & 1 \\ \hline 4 & -6 & 2 \\ \hline 6 & -9 & 3 \end{array}\right]$

3.1.2   (a) $\begin{bmatrix} a & 0 \\ 0 & b \end{bmatrix}$ (c) $\begin{bmatrix} a & b \\ c & d \end{bmatrix}$

3.1.3   (a) True. In general, if A is $m \times n$ and B is $p \times q$ then $A \otimes B$ is $mp \times nq$.

     (c) False. If $A^T = -A$ and $B^T = -B$ then $(A \otimes B)^T = A^T \otimes B^T = (-A) \otimes (-B) = A \otimes B$. That is, $A \otimes B$ is *symmetric*, not skew-symmetric.

3.1.4 Most randomly-chosen matrices serve as counterexamples here. For example, if

$$A = \begin{bmatrix} 1 & 0 & 0 \\ 0 & 0 & 0 \end{bmatrix} \quad \text{and} \quad B = \begin{bmatrix} 0 & 0 \\ 0 & 1 \\ 0 & 0 \end{bmatrix}$$

then $\operatorname{tr}(A \otimes B) = 1$ and $\operatorname{tr}(B \otimes A) = 0$. This does not contradict Theorem 3.1.3 or Corollary 3.1.9 since those results require A and B to be square.

3.1.6 If $A = U_1\Sigma_1V_1^*$ and $B = U_2\Sigma_2V_2^*$ are singular value decompositions then $A^\dagger = V_1\Sigma_1^\dagger U_1^*$ and $B^\dagger = V_2\Sigma_2^\dagger U_2^*$. Then

$$A^\dagger \otimes B^\dagger = (V_1\Sigma_1^\dagger U_1^*) \otimes (V_2\Sigma_2^\dagger U_2^*)$$
$$= (V_1 \otimes V_2)(\Sigma_1^\dagger \otimes \Sigma_2^\dagger)(U_1 \otimes U_2)^*$$
$$= (V_1 \otimes V_2)(\Sigma_1 \otimes \Sigma_2)^\dagger (U_1 \otimes U_2)^* = (A \otimes B)^\dagger,$$

where the second-to-last equality comes from recalling that the pseudoinverse of a diagonal matrix is just obtained by taking the reciprocal of its non-zero entries (and leaving its zero entries alone).

3.1.8   (a) This follows immediately from taking the vectorization of both sides of the equation $AX + XB = C$ and using Theorem 3.1.7.

     (b) We know from Exercise 3.1.7 that the eigenvalues of $A \otimes I + I \otimes B^T$ are the sums of the eigenvalues of A and B^T (which has the same eigenvalues as B). It follows that the equation $AX + XB = C$ has a unique solution if and only if $A \otimes I + I \otimes B^T$ is invertible (i.e., has no 0 eigenvalues), if and only if A and $-B$ do not have any eigenvalues in common.

3.1.9   (a) If $A = U_1T_1U_1^*$ and $B = U_2T_2U_2^*$ are Schur triangularizations then so is

$$A \otimes B = (U_1 \otimes U_2)(T_1 \otimes T_2)(U_1 \otimes U_2)^*.$$

It is straightforward to check that the diagonal entries of $T_1 \otimes T_2$ are exactly the products of the diagonal entries of T_1 and T_2, so the eigenvalues of $A \otimes B$ are exactly the products of the eigenvalues of A and B.

     (b) We know from part (a) that if A has eigenvalues $\lambda_1, \ldots, \lambda_m$ and B has eigenvalues $\mu_1, \ldots, \mu_n$ then $A \otimes B$ has eigenvalues $\lambda_1\mu_1, \ldots, \lambda_m\mu_n$. Since $\det(A) = \lambda_1 \cdots \lambda_m$ and $\det(B) = \mu_1 \cdots \mu_n$, we then have $\det(A \otimes B) = (\lambda_1\mu_1) \cdots (\lambda_m\mu_n) = \lambda_1^n \cdots \lambda_m^n \mu_1^m \cdots \mu_n^m = \det(A)^n \det(B)^m$.

3.1.11   (a) Just apply Theorem 3.1.6 to the rank-one sum decomposition of $\operatorname{mat}(\mathbf{x})$ (i.e., Theorem A.1.3).

     (b) Instead apply Theorem 3.1.6 to the *orthogonal* rank-one sum decomposition (i.e., Theorem 2.3.3).

3.1.13 We just notice that $\mathbb{Z}_2^2$ contains only 3 non-zero vectors: $(1,0)$, $(0,1)$, and $(1,1)$, so tensor powers of these vectors cannot possibly span a space that is larger than 3-dimensional, but $\mathcal{S}_2^3$ is 4-dimensional. Explicitly, the vector $(0,0,0,1,0,1,1,0) \in \mathcal{S}_2^3$ cannot be written in the form $c_1(1,0)^{\otimes 3} + c_2(0,1)^{\otimes 3} + c_3(1,1)^{\otimes 3}$.

3.1.14 If $\sigma \in S_p$ is any permutation with $\operatorname{sgn}(\sigma) = -1$ then $W_{\sigma^{-1}}\mathbf{v} = \mathbf{v}$ and $W_\sigma\mathbf{w} = -\mathbf{w}$. It follows that $\mathbf{v} \cdot \mathbf{w} = \mathbf{v} \cdot (-W_\sigma\mathbf{w}) = -(W_\sigma^*\mathbf{v}) \cdot \mathbf{w} = -(W_{\sigma^{-1}}\mathbf{v}) \cdot \mathbf{w} = -(\mathbf{v} \cdot \mathbf{w})$, where the second-to-last equality uses the fact that W_σ is unitary so $W_\sigma^* = W_\sigma^{-1} = W_{\sigma^{-1}}$. It follows that $\mathbf{v} \cdot \mathbf{w} = 0$.

3.1.15 (a) Just notice that $P^T = \sum_{\sigma \in S_p} W_\sigma^T/p! = \sum_{\sigma \in S_p} W_{\sigma^{-1}}/p! = \sum_{\sigma \in S_p} W_\sigma/p! = P$, with the second equality following from the fact that W_σ is unitary so $W_\sigma^T = W_\sigma^{-1} = W_{\sigma^{-1}}$, and the third equality coming from the fact that changing the order in which we sum over S_p does not change the sum itself.

(b) We compute

$$P^2 = \Big(\sum_{\sigma \in S_p} W_\sigma/p!\Big)\Big(\sum_{\tau \in S_p} W_\tau/p!\Big)$$
$$= \sum_{\sigma,\tau \in S_p} W_{\sigma \circ \tau}/(p!)^2$$
$$= \sum_{\sigma \in S_p} p! W_\sigma/(p!)^2 = P,$$

with the third equality following from the fact that summing $W_{\sigma \circ \tau}$ over all σ and all τ in S_p just sums each W_σ a total of $|S_p| = p!$ times (once for each value of $\tau \in S_p$).

3.1.16 By looking at the blocks in the equation $\sum_{j=1}^k \mathbf{v}_j \otimes \mathbf{w}_j = \mathbf{0}$, we see that $\sum_{j=1}^k [\mathbf{v}_j]_i \mathbf{w}_j = \mathbf{0}$ for all $1 \le i \le m$. By linear independence, this implies $[\mathbf{v}_j]_i = 0$ for each i and j, so $\mathbf{v}_j = \mathbf{0}$ for each j.

3.1.17 Since $B \otimes C$ consists of mn vectors, it suffices via Exercise 1.2.27(a) to show that it is linearly independent. To this end, write $B = \{\mathbf{v}_1,\ldots,\mathbf{v}_m\}$ and $C = \{\mathbf{w}_1,\ldots,\mathbf{w}_n\}$ and suppose

$$\sum_{i=1}^m \sum_{j=1}^n c_{i,j} \mathbf{v}_i \otimes \mathbf{w}_j = \sum_{j=1}^n \Big(\sum_{i=1}^m c_{i,j} \mathbf{v}_i\Big) \otimes \mathbf{w}_j = \mathbf{0}.$$

We then know from Exercise 3.1.16 that (since C is linearly independent) $\sum_{i=1}^m c_{i,j} \mathbf{v}_i = \mathbf{0}$ for each $1 \le j \le n$. By linear independence of B, this implies $c_{i,j} = 0$ for each i and j, so $B \otimes C$ is linearly independent too.

3.1.18 (a) This follows almost immediately from Definition 3.1.3(b), which tells us that the columns of $W_{m,n}$ are (in some order) $\{\mathbf{e}_j \otimes \mathbf{e}_i\}_{i,j}$, which are the standard basis vectors in $\mathbb{F}^{mn}$. In other words, $W_{m,n}$ is obtained from the identity matrix by permuting its columns in some way.

(b) Since the columns of $W_{m,n}$ are standard basis vectors, the dot products of its columns with each other equal 1 (the dot product of a column with itself) or 0 (the dot product of a column with another column). This means exactly that $W_{m,n}^* W_{m,n} = I$, so $W_{m,n}$ is unitary.

(c) This follows immediately from Definition 3.1.3(c) and the fact that $E_{i,j}^* = E_{j,i}$.

3.1.20 (a) The (i,j)-block of $A \otimes B$ is $a_{i,j}B$, so the (k,ℓ)-block within the (i,j)-block of $(A \otimes B) \otimes C$ is $a_{i,j}b_{k,\ell}C$. On the other hand, the (k,ℓ)-block of $B \otimes C$ is $b_{k,\ell}C$, so the (k,ℓ)-block within the (i,j)-block of $A \otimes (B \otimes C)$ is $a_{i,j}b_{k,\ell}C$.

(b) The (i,j)-block of $(A+B) \otimes C$ is $(a_{i,j}+b_{i,j})C$, which equals the (i,j)-block of $A \otimes C + B \otimes C$, which is $a_{i,j}C + b_{i,j}C$.

(c) The (i,j)-block of these matrices equal $(ca_{i,j})B$, $a_{i,j}(cB)$, $c(a_{i,j}B)$, respectively, which are all equal.

3.1.21 (a) If A^{-1} and B^{-1} exist then we just use Theorem 3.1.2(a) to see that $(A \otimes B)(A^{-1} \otimes B^{-1}) = (AA^{-1}) \otimes (BB^{-1}) = I \otimes I = I$, so $(A \otimes B)^{-1} = A^{-1} \otimes B^{-1}$. If $(A \otimes B)^{-1}$ exists then we can use Theorem 3.1.3(a) to see that A and B have no zero eigenvalues (otherwise $A \otimes B$ would too), so A^{-1} and B^{-1} exist.

(b) The (j,i)-block of $(A \otimes B)^T$ is the (i,j)-block of $A \otimes B^T$, which equals $a_{i,j}B^T$, which is also the (j,i)-block of $A^T \otimes B^T$.

(c) Just apply part (b) above (part (c) of the theorem) to the complex conjugated matrices $\overline{A}$ and $\overline{B}$.

3.1.22 (a) If A and B are upper triangular then the (i,j)-block of $A \otimes B$ is $a_{i,j}B$. If $i > j$ then this entire block equals O since A is upper triangular. If $i = j$ then this block is upper triangular since B is. If A and B are lower triangular then just apply this result to A^T and B^T.

(b) Use part (a) and the fact that diagonal matrices are both upper *lower* triangular.

(c) If A and B are normal then $(A \otimes B)^*(A \otimes B) = (A^*A) \otimes (B^*B) = (AA^*) \otimes (BB^*) = (A \otimes B)(A \otimes B)^*$.

(d) If $A^*A = B^*B = I$ then $(A \otimes B)^*(A \otimes B) = (A^*A) \otimes (B^*B) = I \otimes I = I$.

(e) If $A^T = A$ and $B^T = B$ then $(A \otimes B)^T = A^T \otimes B^T = A \otimes B$. Similarly, if $A^* = A$ and $B^* = B$ then $(A \otimes B)^* = A^* \otimes B^* = A \otimes B$.

(f) If the eigenvalues of A and B are non-negative then so are the eigenvalues of $A \otimes B$, by Theorem 3.1.3.

3.1.23 In all parts of this exercise, we use the fact that if $A = U_1 \Sigma_1 V_1^*$ and $B = U_2 \Sigma_2 V_2^*$ are singular value decompositions then so is $A \otimes B = (U_1 \otimes U_2)(\Sigma_1 \otimes \Sigma_2)(V_1 \otimes V_2)^*$.

(a) If σ is a diagonal entry of Σ_1 and τ is a diagonal entry of Σ_2 then $\sigma\tau$ is a diagonal entry of $\Sigma_1 \otimes \Sigma_2$.

(b) rank$(A \otimes B)$ equals the number of non-zero entries of $\Sigma_1 \otimes \Sigma_2$, which equals the product of the number of non-zero entries of Σ_1 and Σ_2 (i.e., rank(A)rank(B)).

(c) If $A\mathbf{v} \in \text{range}(A)$ and $B\mathbf{w} \in \text{range}(B)$ then $(A\mathbf{v}) \otimes (B\mathbf{w}) = (A \otimes B)(\mathbf{v} \otimes \mathbf{w}) \in \text{range}(A \otimes B)$. Since range$(A \otimes B)$ is a subspace, this shows that "$\supseteq$" inclusion. For the opposite inclusion, recall from Theorem 2.3.2 that range$(A \otimes B)$ is spanned by the rank$(A \otimes B)$ columns of $U_1 \otimes U_2$ that correspond to non-zero diagonal entries of $\Sigma_1 \otimes \Sigma_2$, which are all of the form $\mathbf{u}_1 \otimes \mathbf{u}_2$ for some $\mathbf{u}_1 \text{range}(A)$ and $\mathbf{u}_2 \in \text{range}(B)$.

(d) Similar to part (c) (i.e., part (d) of the Theorem), but using columns of $V_1 \otimes V_2$ instead of $U_1 \otimes U_2$.

(e) $\|A \otimes B\| = \|A\|\|B\|$ since we know from part (a) of this theorem that the largest singular value of $A \otimes B$ is the product of the largest singular values of A and B. To see that $\|A \otimes B\|_F = \|A\|_F \|B\|_F$, just note that $\|A \otimes B\|_F = \sqrt{\text{tr}((A \otimes B)^*(A \otimes B))} = \sqrt{\text{tr}((A^*A) \otimes (B^*B))} = \sqrt{\text{tr}(A^*A)}\sqrt{\text{tr}(B^*B)} = \|A\|_F \|B\|_F$.

3.1.24 For property (a), let $P = \sum_{\sigma \in S_p} \text{sgn}(\sigma) W_\sigma / p!$ and note that $P^2 = P = P^T$ via an argument almost identical to that of Exercise 3.1.15. It thus suffices to show that $\text{range}(P) = \mathcal{A}_n^p$. To this end, notice that for all $\tau \in S_p$ we have

$$\text{sgn}(\tau) W_\tau P = \frac{1}{p!} \sum_{\sigma \in S_p} \text{sgn}(\tau) \text{sgn}(\sigma) W_\tau W_\sigma$$

$$= \frac{1}{p!} \sum_{\sigma \in S_p} \text{sgn}(\tau \circ \sigma) W_{\tau \circ \sigma} = P.$$

It follows that everything in $\text{range}(P)$ is unchanged by $\text{sgn}(\tau) W_\tau$ (for all $\tau \in S_p$), so $\text{range}(P) \subseteq \mathcal{A}_n^p$. To prove the opposite inclusion, we just notice that if $\mathbf{v} \in \mathcal{A}_n^p$ then

$$P\mathbf{v} = \frac{1}{p!} \sum_{\sigma \in S_p} \text{sgn}(\sigma) W_\sigma \mathbf{v} = \frac{1}{p!} \sum_{\sigma \in S_p} \mathbf{v} = \mathbf{v},$$

so $\mathbf{v} \in \text{range}(P)$ and thus $\mathcal{A}_n^p \subseteq \text{range}(P)$, so P is a projection onto $\mathcal{A}_n^p$ as claimed.

To prove property (c), we first notice that the columns of the projection P from part (a) have the form

$$P(\mathbf{e}_{j_1} \otimes \mathbf{e}_{j_2} \otimes \cdots \otimes \mathbf{e}_{j_p}) =$$

$$\frac{1}{p!} \sum_{\sigma \in S_p} \text{sgn}(\sigma) W_\sigma(\mathbf{e}_{j_1} \otimes \mathbf{e}_{j_2} \otimes \cdots \otimes \mathbf{e}_{j_p}),$$

where $1 \le j_1, j_2, \ldots, j_p \le n$. To turn this set of vectors into a basis of $\text{range}(P) = \mathcal{A}_n^p$, we omit the columns that are equal to each other or equal to $\mathbf{0}$ by only considering the columns for which $1 \le j_1 < j_2 < \cdots < j_p \le n$. If $\mathbb{F} = \mathbb{R}$ or $\mathbb{F} = \mathbb{C}$ (so we have an inner product to work with) then these remaining vectors are mutually orthogonal and thus form an orthogonal basis of $\text{range}(P)$, and otherwise they are linearly independent (and thus form a basis of $\text{range}(P)$) since the coordinates of their non-zero entries in the standard basis form disjoint subsets of $\{1, 2, \ldots, n^p\}$. If we multiply these vectors each by $p!$ then they form the basis described in the statement of the theorem.

To demonstrate property (b), we simply notice that the basis from part (c) of the theorem contains as many vectors as there are sets of p distinct numbers $\{j_1, j_2, \ldots, j_p\} \subseteq \{1, 2, \ldots, n\}$, which equals $\binom{n}{p}$.

3.1.26 (a) We can scale $\mathbf{v}$ and $\mathbf{w}$ freely, since if we replace $\mathbf{v}$ and $\mathbf{w}$ by $c\mathbf{v}$ and $d\mathbf{w}$ (where $c, d > 0$) then the inequality becomes

$$cd|\mathbf{v} \cdot \mathbf{w}| = |(c\mathbf{v}) \cdot (d\mathbf{w})|$$

$$\le \|c\mathbf{v}\|_p \|d\mathbf{w}\|_q = cd \|\mathbf{v}\|_p \|\mathbf{w}\|_q,$$

which is equivalent to the original inequality that we want to prove. We thus choose $c = 1/\|\mathbf{v}\|_p$ and $d = 1/\|\mathbf{w}\|_q$.

(b) We simply notice that we can rearrange the condition $1/p + 1/q = 1$ to the form $1/(q - 1) = p - 1$. Then dividing the left inequality above by $|v_j|$ shows that it is equivalent to $|w_j| \le |v_j|^{p-1}$, whereas dividing the right inequality by $|w_j|$ and then raising it to the power $1/(q - 1) = p - 1$ shows that it is equivalent to $|v_j|^{p-1} \le |w_j|$. It is thus clear that at least one of these two inequalities must hold (they simply point in opposite directions of each other), as claimed.

As a minor technicality, we should note that the above argument only works if $p, q > 1$ and each $v_j, w_j \ne 0$. If $p = 1$ then $|v_j w_j| \le |v_j|^p$ (since $\|\mathbf{w}\|_q = 1$ so $|w_j| \le 1$). If $q = 1$ then $|v_j w_j| \le |w_j|^q$. If $v_j = 0$ or $w_j = 0$ then both inequalities hold trivially.

(c) Part (b) implies that $|v_j w_j| \le |v_j|^p + |w_j|^q$ for each $1 \le j \le n$, so

$$|\mathbf{v} \cdot \mathbf{w}| = \left| \sum_{j=1}^n \overline{v_j} w_j \right| \le \sum_{j=1}^n |v_j w_j|$$

$$\le \sum_{j=1}^n |v_j|^p + \sum_{j=1}^n |w_j|^q$$

$$= \|\mathbf{v}\|_p^p + \|\mathbf{w}\|_q^q = 2.$$

(d) Notice that the inequality above does not depend on the dimension n at all. It follows that if we pick a positive integer k, replacing $\mathbf{v}$ and $\mathbf{w}$ by $\mathbf{v}^{\otimes k}$ and $\mathbf{w}^{\otimes k}$ respectively shows that

$$2 \ge |\mathbf{v}^{\otimes k} \cdot \mathbf{w}^{\otimes k}| = |\mathbf{v} \cdot \mathbf{w}|^k,$$

with the final equality coming from Exercise 3.1.25. Taking the k-th root of this inequality shows us that $|\mathbf{v} \cdot \mathbf{w}| \le \sqrt[k]{2}$ for all integers $k \ge 1$. Since $\lim_{k \to \infty} \sqrt[k]{2} = 1$, it follows that $|\mathbf{v} \cdot \mathbf{w}| \le 1$, which completes the proof.

Section 3.2: Multilinear Transformations

3.2.1 (a) This is *not* a multilinear transformation since, for example, if $\mathbf{w} = (1,0)$ then the function $S(\mathbf{v}) = T(\mathbf{v}, \mathbf{w}) = (v_1 + v_2, 1)$ is not a linear transformation (e.g., it does not satisfy $S(\mathbf{0}) = \mathbf{0}$).

(c) Yes, this is a multilinear transformation (we mentioned that the Kronecker product is bilinear in the main text).

(e) Yes, this is a multilinear transformation.

3.2.2 (b) $\begin{bmatrix} 0 & 2 & 1 & 0 & 3 & 0 \\ 1 & 0 & 0 & 1 & 1 & 0 \\ -1 & 0 & 0 & 0 & 0 & 1 \end{bmatrix}$

(c) $\begin{bmatrix} 1 & 0 & 0 & 0 \\ 0 & 1 & 0 & 0 \\ 0 & 0 & 1 & 0 \\ 0 & 0 & 0 & 1 \end{bmatrix}$

3.2.3 (a) False. The type of a multilinear transformation does not say the dimensions of the vector spaces, but rather how many of them there are. D has type $(2, 0)$.

(c) False. In the standard block matrix, the rows index the output space and the columns index the different input spaces, so it has size $mp \times mn^2p$.

3.2.8 range$(C) = \mathbb{R}^3$. To see why this is the case, recall that $C(\mathbf{v}, \mathbf{w})$ is always orthogonal to each of $\mathbf{v}$ and $\mathbf{w}$. It follows that if we are given $\mathbf{x} \in \mathbb{R}^3$ and we want to find $\mathbf{v}$ and $\mathbf{w}$ so that $C(\mathbf{v}, \mathbf{w}) = \mathbf{x}$, we can just choose $\mathbf{v}$ and $\mathbf{w}$ to span the plane orthogonal to $\mathbf{x}$ and then rescale $\mathbf{v}$ and $\mathbf{w}$ appropriately.

3.2.9 If the standard array has k non-zero entries then so does the standard block matrix A. We can thus write A as a sum of k terms of the form $\mathbf{e}_i \mathbf{e}_j^T$, where $\mathbf{e}_i \in \mathbb{F}^{d_{\mathcal{W}}}$ and $\mathbf{e}_j \in \mathbb{F}^{d_1} \otimes \cdots \otimes \mathbb{F}^{d_p}$ are standard basis vectors. Since each standard basis vector in $\mathbb{F}^{d_1} \otimes \cdots \otimes \mathbb{F}^{d_p}$ is an elementary tensor, it follows from Theorem 3.2.4 that rank$(T) \leq k$.

3.2.10 The bound rank$(T_\times) \leq mnp$ comes from the standard formula for matrix multiplication in the exact same way as in Example 3.2.15 (equivalently, the standard array of $T_\times$ has exactly mnp non-zero entries, so Exercise 3.2.9 gives this bound).
The bound rank$(T_\times) \geq mp$ comes from noting that the standard block matrix A of $T_\times$ (with respect to the standard basis) is an $mp \times mn^2p$ matrix with full rank mp (it is straightforward to check that there are standard block matrices that multiply to $E_{i,j}$ for each i, j). It follows that rank$(T_\times) \geq$ rank$(A) = mp$.

3.2.11 (a) By definition, rank(f) is the least r such that we can write $f(\mathbf{v}, \mathbf{w}) = \sum_{j=1}^r f_j(\mathbf{v})g_j(\mathbf{w})$ for some linear forms $\{f_j\}$ and $\{g_j\}$. If we represent each of these linear forms as row vectors via $f_j(\mathbf{v}) = \mathbf{x}_j^T[\mathbf{v}]_B$ and $g_j(\mathbf{w}) = \mathbf{y}_j^T[\mathbf{w}]_C$ then we can write this sum in the form $f(\mathbf{v}, \mathbf{w}) = [\mathbf{v}]_B^T\left(\sum_{j=1}^r \mathbf{x}_j\mathbf{y}_j^T\right)[\mathbf{w}]_C$. Since $f(\mathbf{v}, \mathbf{w}) = [\mathbf{v}]_B^TA[\mathbf{w}]_C$, Theorem A.1.3 tells us that $r = $ rank(A) as well.

(b) By definition, $\|f\| = \max\{|f(\mathbf{v}, \mathbf{w})| : \|\mathbf{v}\| = \|\mathbf{w}\| = 1\}$. If we represent f via Theorem 1.3.5 and use the fact that $\|\mathbf{v}\| = \|[\mathbf{v}]_B\|$ whenever B is an orthonormal basis, we see that $\|f\| = \max\{[\mathbf{v}]_B^TA[\mathbf{w}]_C : \|[\mathbf{v}]_B\| = \|[\mathbf{w}]_C\| = 1\}$. It follows from Exercise 2.3.13 that this quantity equals $\|A\|$.

3.2.13 (a) One such decomposition is $C(\mathbf{v}, \mathbf{w}) = \sum_{j=1}^6 f_{1,j}(\mathbf{v})f_{2,j}(\mathbf{w})\mathbf{x}_j$, where

$$f_{1,1}(\mathbf{v}) = v_2 \qquad f_{1,2}(\mathbf{v}) = v_3$$
$$f_{1,3}(\mathbf{v}) = v_3 \qquad f_{1,4}(\mathbf{v}) = v_1$$
$$f_{1,5}(\mathbf{v}) = v_1 \qquad f_{1,6}(\mathbf{v}) = v_2$$
$$f_{2,1}(\mathbf{w}) = w_3 \qquad f_{2,2}(\mathbf{w}) = w_2$$
$$f_{2,3}(\mathbf{w}) = w_1 \qquad f_{2,4}(\mathbf{w}) = w_3$$
$$f_{2,5}(\mathbf{w}) = w_2 \qquad f_{2,6}(\mathbf{w}) = w_1$$
$$\mathbf{x}_1 = \mathbf{e}_1 \qquad \mathbf{x}_2 = -\mathbf{e}_1$$
$$\mathbf{x}_3 = \mathbf{e}_2 \qquad \mathbf{x}_4 = -\mathbf{e}_2$$
$$\mathbf{x}_5 = \mathbf{e}_3 \qquad \mathbf{x}_6 = -\mathbf{e}_3.$$

(b) We just perform the computation directly:

$$\sum_{j=1}^5 f_{1,j}(\mathbf{v})f_{2,j}(\mathbf{w})\mathbf{x}_j$$
$$= v_1(w_2 + w_3)(\mathbf{e}_1 + \mathbf{e}_3) - (v_1 + v_3)w_2\mathbf{e}_1$$
$$\quad - v_2(w_1 + w_3)(\mathbf{e}_2 + \mathbf{e}_3)$$
$$\quad + (v_2 + v_3)w_1\mathbf{e}_2 + (v_2 - v_1)w_3(\mathbf{e}_1 + \mathbf{e}_2 + \mathbf{e}_3)$$
$$= \big(v_1(w_2 + w_3) - (v_1 + v_3)w_2 + (v_2 - v_1)w_3\big)\mathbf{e}_1$$
$$\quad + \big(-v_2(w_1 + w_3) + (v_2 + v_3)w_1 + (v_2 - v_1)w_3\big)\mathbf{e}_2$$
$$\quad + \big(v_1(w_2 + w_3) - v_2(w_1 + w_3) + (v_2 - v_1)w_3\big)\mathbf{e}_3$$
$$= (v_2w_3 - v_3w_2)\mathbf{e}_1 + (v_3w_1 - v_1w_3)\mathbf{e}_2$$
$$\quad + (v_1w_2 - v_2w_1)\mathbf{e}_3 = C(\mathbf{v}, \mathbf{w}).$$

3.2.16 Recall from Exercise 3.2.13 that if $C : \mathbb{R}^3 \times \mathbb{R}^3 \to \mathbb{R}^3$ is the cross product then rank$(C) = 5$. An optimal rank sum decomposition of the standard block matrix of C thus makes use of sets consisting of 5 vectors in $\mathbb{R}^3$, which cannot possibly be linearly independent. A similar argument works with the matrix multiplication transformation $T_\times : \mathcal{M}_2 \times \mathcal{M}_2 \to \mathcal{M}_2$, which has rank 7 and vectors living in a 4-dimensional space.

Section 3.3: The Tensor Product

3.3.1 (a) Just recall that when $p = 2$ we have rank$(\mathbf{v}) = $ rank$(\text{mat}(\mathbf{v}))$ for all $\mathbf{v}$. If $\mathbf{v} = \mathbf{e}_1 \otimes \mathbf{e}_1 + \mathbf{e}_2 \otimes \mathbf{e}_2$ then mat$(\mathbf{v}) = I$, which has rank 2, so rank$(\mathbf{v}) = 2$.

(c) This vector clearly has rank at most 3, and it has rank at *least* 3 (and thus exactly 3) since it has $\mathbf{e}_1 \otimes \mathbf{e}_1 + \mathbf{e}_2 \otimes \mathbf{e}_5 + \mathbf{e}_3 \otimes \mathbf{e}_9$ as a flattening, which has rank-3 matricization.

3.3.2 Equation (3.3.3) itself shows that rank$(\mathbf{v}) \leq 3$, so we just need to prove the other inequality. If we mimic Example 3.3.4, we find that the set $S_\mathbf{v}$ from Theorem 3.3.6 has the form

$$S_\mathbf{v} = \{(a, b, b, -a) \mid a, b \in \mathbb{R}\}.$$

It is clear that dim$(S_\mathbf{v}) = 2$, and we can see that the only elementary tensor in $S_\mathbf{v}$ is $(0, 0, 0, 0)$, since the matricization of $(a, b, b, -a)$ is

$$\begin{bmatrix} a & b \\ b & -a \end{bmatrix},$$

which has determinant $-a^2 - b^2$ and thus rank 2 whenever $(a,b) \neq (0,0)$. It follows that $\mathcal{S}_\mathbf{v}$ does not have a basis consisting of elementary tensors, so Theorem 3.3.6 tells us that $\mathrm{rank}(\mathbf{v}) > \dim(\mathcal{S}_\mathbf{v}) = 2$, so $\mathrm{rank}(\mathbf{v}) = 3$.

This argument breaks down if we work over the complex numbers since the above matrix can be made rank 1 by choosing $a = 1$, $b = i$, for example, and $\mathcal{S}_\mathbf{v}$ is spanned by the elementary tensors $(1,i,i,-1)$ and $(1,-i,-i,-1)$.

3.3.3   (a) True. This is simply a consequence of $\mathbb{C}$ being 1-dimensional, so $\dim(\mathbb{C} \otimes \mathbb{C}) = \dim(\mathbb{C})\dim(\mathbb{C}) = 1$. Since all vector spaces of the same (finite) dimension over the same field are isomorphic, we conclude that $\mathbb{C} \otimes \mathbb{C} \cong \mathbb{C}$.

  (c) True. This follows from the fact that $\mathrm{rank}(\mathbf{v}) = \mathrm{rank}(\mathrm{mat}(\mathbf{v}))$, where $\mathrm{mat}(\mathbf{v}) \in \mathcal{M}_{m,n}(\mathbb{F})$.

  (e) True. This follows from every real tensor rank decomposition being a complex one, since $\mathbb{R} \subseteq \mathbb{C}$.

3.3.4 If there were two such linear transformations $S_1, S_2 : V \otimes W \to X$ then we would have $(S_1 - S_2)(\mathbf{v} \otimes \mathbf{w}) = \mathbf{0}$ for all $\mathbf{v} \in V$ and $\mathbf{w} \in W$. Since $V \otimes W$ is spanned by elementary tensors, it follows that $(S_1 - S_2)(\mathbf{x}) = \mathbf{0}$ for all $\mathbf{x} \in V \otimes W$, so $S_1 - S_2$ is the zero linear transformation and $S_1 = S_2$.

3.3.5 For property (a), we just note that it is straightforward to see that every function of the form $f(x) = Ax^2 + Bx + C$ can be written as a linear combination of elementary tensors—Ax^2, Bx, and C are themselves elementary tensors. For (b) and (c), we compute

$$\big((A+cB) \otimes f\big)(x) = (A+cB)f(x)$$
$$= Af(x) + cBf(x)$$
$$= (A \otimes f)(x) + c(B \otimes f)(x)$$

for all x, so $(A + cB) \otimes f = (A \otimes f) + c(B \otimes f)$ for all $A, B \in \mathcal{M}_2$, $f \in \mathcal{P}^2$, and $c \in \mathbb{F}$. The fact that $A \otimes (f + cg) = (A \otimes f) + c(A \otimes g)$ is similar. For (d), we just define $S(A) = T(A,1)$, $S(Ax) = T(A,x)$, and $S(Ax^2) = T(A,x^2)$ for all $A \in \mathcal{M}_2$ and extend via linearity.

3.3.6 Suppose that we could write $f(x,y) = e^{xy}$ as a sum of elementary tensors: $e^{xy} = f_1(x)g_1(y) + \cdots + f_k(x)g_k(y)$ for some $f_1, g_1, \ldots, f_k, g_k \in \mathcal{C}$. Choosing $y = 1, 2, \ldots, k, k+1$ then gives us the system of equations

$$e^x = f_1(x)g_1(1) + \cdots + f_k(x)g_k(1)$$
$$e^{2x} = f_1(x)g_1(2) + \cdots + f_k(x)g_k(2)$$
$$\vdots$$
$$e^{kx} = f_1(x)g_1(k) + \cdots + f_k(x)g_k(k)$$
$$e^{(k+1)x} = f_1(x)g_1(k+1) + \cdots + f_k(x)g_k(k+1).$$

We know from Exercise 1.1.22 that the set of functions $\{e^x, e^{2x}, \ldots, e^{kx}, e^{(k+1)x}\}$ is linearly independent. However, this contradicts the above linear system, which says that $\{e^x, e^{2x}, \ldots, e^{kx}, e^{(k+1)x}\}$ is contained within the (k-dimensional) span of the functions $f_1, \ldots, f_k$.

3.3.10 Let $T : V \times W \to W \otimes V$ be the bilinear transformation defined by $T(\mathbf{v}, \mathbf{w}) = \mathbf{w} \otimes \mathbf{v}$. By the universal property, there exists a linear transformation $S : V \otimes W \to W \otimes V$ with $S(\mathbf{v} \otimes \mathbf{w}) = \mathbf{w} \otimes \mathbf{v}$. A similar argument shows that there exists a linear transformation that sends $\mathbf{w} \otimes \mathbf{v}$ to $\mathbf{v} \otimes \mathbf{w}$. Since elementary tensors span the entire tensor product space, it follows that S is invertible and thus an isomorphism.

3.3.12 Suppose

$$\mathbf{v} = \lim_{k \to \infty} \left(\mathbf{v}_k^{(1)} \otimes \mathbf{v}_k^{(2)} \otimes \cdots \otimes \mathbf{v}_k^{(p)} \right),$$

where we absorb scalars among these Kronecker factors so that $\|\mathbf{v}_k^{(j)}\| = 1$ for all k and all $j \geq 2$. The length of a vector (i.e., its norm induced by the dot product) is continuous in its entries, so

$$\|\mathbf{v}\| = \lim_{k \to \infty} \left\| \mathbf{v}_k^{(1)} \otimes \mathbf{v}_k^{(2)} \otimes \cdots \otimes \mathbf{v}_k^{(p)} \right\| = \lim_{k \to \infty} \left\| \mathbf{v}_k^{(1)} \right\|.$$

In particular, this implies that the vectors $\{\mathbf{v}_k^{(j)}\}$ have bounded norm, so the Bolzano–Weierstrass theorem tells us that there is a sequence $k_1, k_2, k_3, \ldots$ with the property that

$$\lim_{\ell \to \infty} \mathbf{v}_{k_\ell}^{(1)}$$

exists. We can then find a subsequence $k_{\ell_1}, k_{\ell_2}, k_{\ell_3}, \ldots$ of *that* with the property that

$$\lim_{m \to \infty} \mathbf{v}_{k_{\ell_m}}^{(2)}$$

exist too. Continuing in this way gives us a subsequence with the property that

$$\lim_{k \to \infty} \mathbf{v}_k^{(j)}$$

exists for each $1 \leq j \leq p$ (note that we have relabeled this subsequence just in terms of a single subscript k for simplicity). It follows that

$$\mathbf{v} = \lim_{k \to \infty} \left(\mathbf{v}_k^{(1)} \otimes \mathbf{v}_k^{(2)} \otimes \cdots \otimes \mathbf{v}_k^{(p)} \right) =$$
$$\left(\lim_{k \to \infty} \mathbf{v}_k^{(1)} \right) \otimes \left(\lim_{k \to \infty} \mathbf{v}_k^{(2)} \right) \otimes \cdots \otimes \left(\lim_{k \to \infty} \mathbf{v}_k^{(p)} \right),$$

since all of these limits exist (after all, the Kronecker product is made up of entrywise products of vector entries and is thus continuous in those entries). The decomposition on the right shows that $\mathbf{v}$ has tensor rank 1.

Section 3.4: Summary and Review

3.4.1 (a) True. This was part of Theorem 3.1.4.

(c) True. This follows from Theorem 3.1.3, for example.

(e) False. The Kronecker (and tensor) product is not commutative.

(g) False. We gave an example at the end of Section 3.3.3 with real tensor rank 3 but complex tensor rank 2.

3.4.2 We need to perform $n(n-1)/2$ column swaps to obtain $W_{n,n}$ from I (one for each column $e_i \otimes e_j$ with $i < j$), so $\det(W_{n,n}) = (-1)^{n(n-1)/2}$. Equivalently, $\det(W_{n,n}) = -1$ if $n \equiv 2$ or 3 (mod 4), and $\det(W_{n,n}) = 1$ otherwise.

Section 3.A: Extra Topic: Matrix-Valued Linear Maps

3.A.1 (a) True. In fact, we know from Theorem 3.A.6 that it is completely positive.

(c) True. If X is positive semidefinite then so are $\Phi(X)$ and $\Psi(X)$, so $(\Phi + \Psi)(X) = \Phi(X) + \Psi(X)$ is as well.

3.A.2 We mimic the proof of Theorem 3.A.6: we construct C_Ψ and then let the A_i's be the matricizations of its scaled eigenvectors. We recall from Example 3.A.8 that $C_\Psi = I - W_{3,3}$, and applying the spectral decomposition to this matrix shows that $C_\Psi = \sum_{i=1}^{3} v_i v_i^T$, where

$$v_1 = (0,1,0,-1,0,0,0,0,0),$$
$$v_2 = (0,0,1,0,0,0,-1,0,0), \quad \text{and}$$
$$v_3 = (0,0,0,0,0,1,0,-1,0).$$

It follows that $\Psi(X) = \sum_{i=1}^{3} A_i X A_i^T$, where $A_i = \text{mat}(v_i)$ for all i, so

$$A_1 = \begin{bmatrix} 0 & 1 & 0 \\ -1 & 0 & 0 \\ 0 & 0 & 0 \end{bmatrix}, \quad A_2 = \begin{bmatrix} 0 & 0 & 1 \\ 0 & 0 & 0 \\ -1 & 0 & 0 \end{bmatrix}, \quad \text{and}$$

$$A_3 = \begin{bmatrix} 0 & 0 & 0 \\ 0 & 0 & 1 \\ 0 & -1 & 0 \end{bmatrix}.$$

3.A.4 Direct computation shows that $(\Phi_C \otimes I_2)(xx^*)$ equals

$$\begin{bmatrix} 2 & 0 & -1 & 0 & 0 & -1 \\ 0 & 0 & 0 & 0 & 0 & 0 \\ -1 & 0 & 1 & 0 & 0 & -1 \\ 0 & 0 & 0 & 1 & 0 & 0 \\ 0 & 0 & 0 & 0 & 1 & 0 \\ -1 & 0 & -1 & 0 & 0 & 1 \end{bmatrix},$$

which is not positive semidefinite (it has $1 - \sqrt{3} \approx -0.7321$ as an eigenvalue), so Φ_C is not 2-positive.

3.A.5 Recall that $\|X\|_F = \|\text{vec}(X)\|$ for all matrices X, so $\|\Phi(X)\|_F = \|X\|_F$ for all X if and only if $\|[\Phi]\text{vec}(X)\| = \|\text{vec}(X)\|$ for all $\text{vec}(X)$. It follows from Theorem 1.4.9 that this is equivalent to $[\Phi]$ being unitary.

3.A.7 By Exercise 2.2.20, Φ is positive if and only if $\text{tr}(\Phi(X)Y) \geq 0$ for all positive semidefinite X and Y. Using the definition of the adjoint Φ^* shows that this is equivalent to $\text{tr}(X\Phi^*(Y)) \geq 0$ for all PSD X and Y, which (also via Exercise 2.2.20) is equivalent to Φ^* being positive.

3.A.8 If $C_\Phi = \sum_{i,j} \Phi(E_{i,j}) \otimes E_{i,j}$ then $\text{tr}_2(C_\Phi) = \sum_{i,j} \Phi(E_{i,j}) \otimes tr(E_{i,j}) = \sum_i \Phi(E_{i,i}) = \Phi(I)$, which demonstrates the equivalence of (i) and (ii). The equivalence of (i) and (iii) just follows from choosing $X = I$ in (iii).

3.A.12 (a) If Φ is transpose-preserving and X is symmetric then $\Phi(X)^T = \Phi(X^T) = \Phi(X)$.

(b) For example, consider the linear map $\Phi : \mathcal{M}_2 \to \mathcal{M}_2$ defined by

$$\Phi\left(\begin{bmatrix} a & b \\ c & d \end{bmatrix}\right) = \begin{bmatrix} a & b \\ (b+c)/2 & d \end{bmatrix}.$$

If the input is symmetric (i.e., $b = c$) then so is the output, but $\Phi(E_{1,2}) \neq \Phi(E_{2,1})^T$, so Φ is not transpose-preserving.

3.A.13 (a) Recall from Exercise 2.2.16 that every self-adjoint $A \in \mathcal{M}_n$ can be written as a linear combination of two PSD matrices: $A = P - N$. Then $\Phi(A) = \Phi(P) - \Phi(N)$, and since P and N are PSD, so are $\Phi(P)$ and $\Phi(N)$, so $\Phi(A)$ is also self-adjoint (this argument works even if $\mathbb{F} = \mathbb{R}$). If $\mathbb{F} = \mathbb{C}$ then this means that Φ is Hermiticity-preserving, so it follows from Theorem 3.A.2 that Φ is also adjoint-preserving.

(b) The same map from the solution to Exercise 3.A.12(b) works here. It sends PSD matrices to PSD matrices (in fact, it acts as the identity map on all symmetric matrices), but is not transpose-preserving.

3.A.14 We know from Theorem 3.A.1 that $T \circ \Phi = \Phi \circ T$ (i.e., Φ is transpose-preserving) if and only if $C_\Phi = C_\Phi^T$. The additional requirement that $\Phi = \Phi \circ T$ tells us (in light of Exercise 3.A.10) that $C_\Phi = (I \otimes T)(C_\Phi) = \Gamma(C_\Phi)$.

As a side note, observe this condition is also equivalent to $C_\Phi = C_\Phi^T = \mathsf{T}(C_\Phi)$, since $\Gamma(C_\Phi) = \Gamma(C_\Phi^T) = \mathsf{T}(C_\Phi)$.

3.A.15 If Φ is bisymmetric and decomposable then $\Phi = T \circ \Phi = \Phi \circ T = \Psi_1 + T \circ \Psi_2$ for some completely positive Ψ_1, Ψ_2. Then

$$2\Phi = \Phi + (T \circ \Phi)$$
$$= (\Psi_1 + \Psi_2) + T \circ (\Psi_1 + \Psi_2),$$

so we can choose $\Psi = (\Psi_1 + \Psi_2)/2$. Conversely, if $\Phi = \Psi + T \circ \Psi$ for some completely positive Ψ then Φ is clearly decomposable, and it is bisymmetric because

$$T \circ \Phi = T \circ (\Psi + T \circ \Psi) = T \circ \Psi + \Psi = \Phi,$$

and

$$\Phi \circ T = (\Psi + T \circ \Psi) \circ T = \Psi \circ T + T \circ \Psi \circ T$$
$$= T \circ (T \circ \Psi \circ T) + (T \circ \Psi \circ T) = T \circ \Psi + \Psi$$
$$= \Phi,$$

where we used the fact that $T \circ \Psi \circ T = \Psi$ thanks to Ψ being real and CP (and thus transpose-preserving).

3.A.16 Recall that $\{E_{i,j} \otimes E_{k,\ell}\}$ is a basis of $\mathcal{M}_{mp,nq}$, and $\Phi \otimes \Psi$ must satisfy

$$(\Phi \otimes \Psi)(E_{i,j} \otimes E_{k,\ell}) = \Phi(E_{i,j}) \otimes \Psi(E_{k,\ell})$$

for all i, j, k, and ℓ. Since linear maps are determined by how they act on a basis, this shows that $\Phi \otimes \Psi$, if it exists, is unique. Well-definedness (i.e., existence of $\Phi \otimes \Psi$) comes from reversing this argument— if we define $\Phi \otimes \Psi$ by $(\Phi \otimes \Psi)(E_{i,j} \otimes E_{k,\ell}) = \Phi(E_{i,j}) \otimes \Psi(E_{k,\ell})$ then linearity shows that $\Phi \otimes \Psi$ also satisfies $(\Phi \otimes \Psi)(A \otimes B) = \Phi(A) \otimes \Psi(B)$ for all A and B.

3.A.18 If $X \in \mathcal{M}_n \otimes \mathcal{M}_k$ is positive semidefinite then we can construct a PSD matrix $\tilde{X} \in \mathcal{M}_n \otimes \mathcal{M}_{k+1}$ simply by padding it with n rows and columns of zeros without affecting positive semidefiniteness. Then $(\Phi \otimes I_k)(X) = (\Phi \otimes I_{k+1})(\tilde{X})$ is positive semidefinite by $(k+1)$-positivity of Φ, and since X was arbitrary this shows that Φ is k-positive.

3.A.19 We need to show that $\Phi(X)$ is positive *semi*definite whenever X is positive semidefinite. To this end, notice that every PSD matrix X can be written as a limit of positive definite matrices $X_1, X_2, X_3, \ldots$: $X = \lim_{k \to \infty} X_k$, where $X_k = X + I/k$. Linearity (and thus continuity) of Φ then tells us that

$$\Phi(X) = \Phi\left(\lim_{k \to \infty} X_k\right) = \lim_{k \to \infty} \Phi(X_k),$$

which is positive semidefinite since each $\Phi(X_k)$ is positive definite (recall that the coefficients of characteristic polynomials are continuous, so the roots of a characteristic polynomial cannot jump from positive to negative in a limit).

3.A.22 (a) If X is Hermitian (so $c = \bar{b}$ and a and d are real) then the determinants of the top-left 1×1, 2×2, and 3×3 blocks of $\Phi(X)$ are

$$4a - 4\text{Re}(b) + 3d, \quad 4a^2 - 4|b|^2 + 6ad, \quad \text{and}$$
$$8a^2d + 12ad^2 - 4a|b|^2 + 4\text{Re}(b)|b|^2 - 11d|b|^2,$$

respectively. If X is positive definite then $|\text{Re}(b)| \leq |b| < \sqrt{ad} \leq (a+d)/2$, which easily shows that the 1×1 and 2×2 determinants are positive. For the 3×3 determinant, after we use $|b|^2 < ad$ three times and $\text{Re}(b) > -\sqrt{ad}$ once, we see that

$$8a^2d + 12ad^2 - 4a|b|^2 + 4\text{Re}(b)|b|^2 - 11d|b|^2$$
$$> 4a^2d + ad^2 - 4(ad)^{3/2}.$$

This quantity is non-negative since the inequality $4a^2d + ad^2 \geq 4(ad)^{3/2}$ is equivalent to $(4a^2d + ad^2)^2 \geq 16a^3d^3$, which is equivalent to $a^2d^2(d - 4a)^2 \geq 0$.

(b) After simplifying and factoring as suggested by this hint, we find

$$\det(\Phi(X)) = 2\det(X)p(X), \quad \text{where}$$
$$p(X) = 16a^2 + 30ad + 9d^2$$
$$- 8a\text{Re}(b) - 12\text{Re}(b)d - 8|b|^2.$$

If we use the facts that $|b|^2 < ad$ and $\text{Re}(b) < (a+d)/2$ then we see that

$$p(X) > 12a^2 + 12ad + 3d^2 > 0,$$

so $\det(\Phi(X)) > 0$ as well.

(c) Sylvester's Criterion tells us that $\Phi(X)$ is positive definite whenever X is positive definite. Exercise 3.A.19 then says that Φ is positive.

Section 3.B: Extra Topic: Homogeneous Polynomials

3.B.1 (a) Since this is a quadratic form, we can represent it via a symmetric matrix

$$3x^2 + 3y^2 - 2xy = \begin{bmatrix} x & y \end{bmatrix} \begin{bmatrix} 3 & -1 \\ -1 & 3 \end{bmatrix} \begin{bmatrix} x \\ y \end{bmatrix}$$

and then apply the spectral decomposition to this matrix to get the sum-of-squares form $(x+y)^2 + 2(x-y)^2$.

(c) We mimic Example 3.B.1: we try to write $2x^3 - 9x^2y + 3xy^2 - y^3 = c_1x^3 + c_2y^3 + c_3(x+y)^3 + c_4(x-y)^3$ and solve for c_1, c_2, c_3, c_4 to get the decomposition $x^3 + 2y^3 - (x+y)^3 + 2(x-y)^3$.

(e) We again mimic Example 3.B.1, but this time we are working in a 10-dimensional space so we use the basis $\{x^3, y^3, z^3, (x+y)^3, (x-y)^3, (y+z)^3, (y-z)^3, (z+x)^3, (z-x)^3, (x+y+z)^3\}$. Solving the resulting linear system gives the decomposition $6(x^2y + y^2z + z^2x) = (x+y)^3 - (x-y)^3 + (y+z)^3 - (y-z)^3 + (z+x)^3 - (z-x)^3 - 2x^3 - 2y^3 - 2z^3$.

(g) This time we use the basis $\{x^4, y^4, (x+y)^4, (x-y)^4, (x+2y)^4\}$. Solving the resulting linear system gives the decomposition $x^4 + (x+y)^4 + (x-y)^4 - (x+2y)^4 + 4y^4$.

3.B.2 (a) One possibility is $(2,3,3,5) = (1,1)^{\otimes 2} + (1,2)^{\otimes 2}$.

(c) $(1,0)^{\otimes 3} + (0,1)^{\otimes 3} - (1,1)^{\otimes 3} + (1,-1)^{\otimes 3}$

3.B.3 (a) False. The degree of g might be strictly less than that of f. For example, if $f(x_1,x_2) = x_1x_2 + x_2^2$ then dividing through by x_2^2 gives the dehomogenization $g(x) = x + 1$.

(c) True. This follows from Theorem 3.B.2.

(e) True. This follows from the real spectral decomposition (see Corollary 2.A.2).

3.B.5 We just observe that in any 4th power of a linear form

$$(ax + by)^4,$$

if the coefficient of the x^4 and y^4 terms equal 0 then $a = b = 0$ as well, so no quartic form without x^4 or y^4 terms can be written as a sum of 4th powers of linear forms.

3.B.6 For both parts of this question, we just need to show that $\operatorname{rank}^S(\mathbf{w}) \leq \operatorname{rank}(\mathbf{w})$, since the other inequality is trivial.

(a) $\mathbf{w} \in \mathcal{S}_n^2$ if and only if $\operatorname{mat}(\mathbf{w})$ is a symmetric matrix. Applying the real spectral decomposition shows that

$$\operatorname{mat}(\mathbf{w}) = \sum_{i=1}^{r} \lambda_i \mathbf{v}_i \mathbf{v}_i^T,$$

where $r = \operatorname{rank}(\operatorname{mat}(\mathbf{w})) = \operatorname{rank}(\mathbf{w})$. It follows that $\mathbf{w} = \sum_{i=1}^{r} \lambda_i \mathbf{v}_i \otimes \mathbf{v}_i$, so $\operatorname{rank}^S(\mathbf{w}) \leq r = \operatorname{rank}(\mathbf{w})$.

(b) The Takagi factorization (Exercise 2.3.26) says that we can write $\operatorname{mat}(\mathbf{w}) = UDU^T$, where U is unitary and D is diagonal (with the singular values of $\operatorname{mat}(\mathbf{w})$ on its diagonal). If $r = \operatorname{rank}(\operatorname{mat}(\mathbf{w})) = \operatorname{rank}(\mathbf{w})$ and the first r columns of U are $\mathbf{v}_1, \ldots, \mathbf{v}_r$, then

$$\operatorname{mat}(\mathbf{w}) = \sum_{i=1}^{r} d_{i,i} \mathbf{v}_i \mathbf{v}_i^T,$$

so $\mathbf{w} = \sum_{i=1}^{r} d_{i,i} \mathbf{v}_i \otimes \mathbf{v}_i$, which tells us that $\operatorname{rank}^S(\mathbf{w}) \leq r = \operatorname{rank}(\mathbf{w})$, as desired.

3.B.7 (a) Apply the AM–GM inequality to the 4 quantities w^4, x^2y^2, y^2z^2, and z^2x^2.

(b) If we could write f as a sum of squares of quadratic forms, those quadratic forms must contain no x^2, y^2, or z^2 terms (or else f would have an x^4, y^4, or z^4 term, respectively). It then follows that they also have no wx, wy, or wz terms (or else f would have an $a w^2x^2$, w^2y^2, or w^2z^2 term, respectively) and thus no way to create the cross term $-4wxyz$ in f, which is a contradiction.

3.B.8 Recall from Remark 3.B.2 that

$$f(x,y,z) = \frac{g(x,y,z)^2 + g(y,z,x)^2 + g(z,x,y)^2}{x^2 + y^2 + z^2}$$
$$+ \frac{h(x,y,z)^2 + h(y,z,x)^2 + h(z,x,y)^2}{2(x^2 + y^2 + z^2)}.$$

If we multiply and divide the right-hand side by $x^2 + y^2 + z^2$ and then multiply out the numerators, we get $f(x,y,z)$ as a sum of 18 squares of rational functions. For example, the first three terms in this sum of squares are of the form

$$\frac{(x^2 + y^2 + z^2)g(x,y,z)^2}{(x^2 + y^2 + z^2)^2} = \left(\frac{xg(x,y,z)}{x^2 + y^2 + z^2} \right)^2$$
$$+ \left(\frac{yg(x,y,z)}{x^2 + y^2 + z^2} \right)^2 + \left(\frac{zg(x,y,z)}{x^2 + y^2 + z^2} \right)^2.$$

3.B.9 Recall from Theorem 1.3.6 that we can write every quadrilinear form f in the form

$$f(\mathbf{x},\mathbf{y},\mathbf{z},\mathbf{w}) = \sum_{i,j,k,\ell} b_{i,j,k,\ell} x_i z_j y_k w_\ell.$$

If we plug $\mathbf{z} = \mathbf{x}$ and $\mathbf{w} = \mathbf{y}$ into this equation then we see that

$$f(\mathbf{x},\mathbf{y},\mathbf{x},\mathbf{y}) = \sum_{i,j,k,\ell} b_{i,j,k,\ell} x_i x_j y_k y_\ell.$$

If we define $a_{i,j;k,\ell} = (b_{i,j,k,\ell} + b_{j,i,k,\ell} + b_{i,j,\ell,k} + b_{j,i,\ell,k})/4$ then we get

$$f(\mathbf{x},\mathbf{y},\mathbf{x},\mathbf{y}) = \sum_{i \leq j, k \leq \ell} a_{i,j;k,\ell} x_i x_j y_k y_\ell = q(\mathbf{x},\mathbf{y}),$$

and the converse is similar.

3.B.10 We think of $q(\mathbf{x},\mathbf{y})$ as a (quadratic) polynomial in y_3, keeping x_1, x_2, x_3, y_1, and y_2 fixed. That is, we write $q(\mathbf{x},\mathbf{y}) = ay_3^2 + by_3 + c$, where a, b, and c are complicated expressions that depend on x_1, x_2, x_3, y_1, and y_2:

$$a = x_2^2 + x_3^2,$$
$$b = -2x_1x_3y_1 - 2x_2x_3y_2, \quad \text{and}$$
$$c = x_1^2y_1^2 + x_1^2y_2^2 + x_2^2y_2^2 + x_3^2y_1^2 - 2x_1x_2y_1y_2.$$

If $a = 0$ then $x_2 = x_3 = 0$, so $q(\mathbf{x},\mathbf{y}) = x_1^2(y_1^2 + y_2^2) \geq 0$, as desired, so we assume from now on that $a > 0$. Our goal is to show that q has at most one real root, since that implies it is never negative. To this end, we note that after performing a laborious calculation to expand out the discriminant $b^2 - 4ac$ of q, we get the following sum of squares decomposition:

$$b^2 - 4ac = -4(abx - b^2y)^2$$
$$- 4(aby - c^2x)^2 - 4(acy - bcx)^2.$$

It follows that $b^2 - 4ac \leq 0$, so q has at most one real root and is thus positive semidefinite.

3.B.11 We check the 3 defining properties of Definition 1.3.6. Properties (a) and (b) are both immediate, and property (c) comes from computing

$$\langle f, f \rangle = \sum_{k_1 + \cdots + k_n = p} \frac{|a_{k_1,k_2,\ldots,k_n}|^2}{\binom{p}{k_1,k_2,\ldots,k_n}},$$

which is clearly non-negative and equals zero if and only if $a_{k_1,k_2,\ldots,k_n} = 0$ for all $k_1, k_2, \ldots, k_n$ (i.e., if and only if $f = 0$).

Section 3.C: Extra Topic: Semidefinite Programming

3.C.1   (a) This is equivalent to $A - A = O$ being positive semidefinite, which is clear.

     (b) If $A - B$ and $B - A = -(A - B)$ are both positive semidefinite then all eigenvalues of $A - B$ must be ≥ 0 and ≤ 0, so they must all equal 0. Since $A - B$ is Hermitian, this implies that it equals O (by the spectral decomposition, for example), so $A = B$.

     (c) If $A - B$ and $B - C$ are both PSD then so is $(A - B) + (B - C) = A - C$.

     (d) If $A - B$ is PSD then so is $(A + C) - (B + C) = A - B$.

     (e) This is equivalent to the (obviously true) statement that $A - B$ is PSD if and only if $-B + A$ is PSD.

     (f) If $A - B$ is PSD then so is $P(A - B)P^* = PAP^* - PBP^*$, by Theorem 2.2.3(d).

3.C.2   (a) True. We showed how to represent a linear program as a semidefinite program at the start of this section.

     (c) True. If $A - B$ is PSD then $\text{tr}(A - B) \geq 0$, so $\text{tr}(A) \geq \text{tr}(B)$.

     (e) False. This is not even true for linear programs.

3.C.3   (a) Suppose A has spectral decomposition $A = UDU^*$. Then $A \preceq cI$ if and only if $cI - A \succeq O$, if and only if $U(cI - D)U^* \succeq O$, if and only if $cI - D \succeq O$, if and only if $c \geq \lambda_{\max}(A)$.

     (b) There are many possibilities, including the following (where $c \in \mathbb{R}$ is the variable):

$$\begin{aligned} \text{minimize:} &\quad c \\ \text{subject to:} &\quad cI \succeq A \end{aligned}$$

     (c) By an argument like that from part (a), $A \succeq cI$ if and only if $\lambda_{\min}(A) \geq c$. It follows that the following SDP has optimal value $\lambda_{\min}(A)$:

$$\begin{aligned} \text{maximize:} &\quad c \\ \text{subject to:} &\quad cI \preceq A \end{aligned}$$

3.C.4   (a) The constraints of this SDP force

$$\begin{bmatrix} -x - y & 1 \\ 1 & -x \end{bmatrix} \succeq O,$$

which is not possible since $-x \leq 0$ and PSD matrices cannot have negative diagonal entries (furthermore, we showed in Exercise 2.2.11 that if a diagonal entry of a PSD matrix equals 0 then so does every entry in that row and column, which rules out the $x = 0$ case).

     (b) The dual SDP has the following form (where $X \in \mathcal{M}_3^H$ is a matrix variable):

$$\begin{aligned} \text{minimize:} &\quad 2\text{Re}(x_{1,2}) \\ \text{subject to:} &\quad x_{1,1} + x_{2,2} + x_{3,3} \geq 0 \\ &\quad x_{1,1} + x_{3,3} = 0 \\ &\quad X \succeq O \end{aligned}$$

This SDP is feasible with optimal value 0 since the feasible points are exactly the matrices X with $x_{2,2} \geq 0$ and all other entries equal to 0.

3.C.5   (a) One possibility is

$$A = \begin{bmatrix} 2 & 0 \\ 0 & 3 \end{bmatrix} \quad \text{and} \quad B = \begin{bmatrix} 1 & -1 \\ -1 & 2 \end{bmatrix}.$$

Then A and B are both positive semidefinite, and so is

$$A - B = \begin{bmatrix} 1 & 1 \\ 1 & 1 \end{bmatrix}.$$

However,

$$A^2 - B^2 = \begin{bmatrix} 2 & 3 \\ 3 & 4 \end{bmatrix},$$

which is not positive semidefinite (it has determinant -1).

     (c) Notice that both $A - B$ and $A + B$ are positive semidefinite (the former by assumption, the latter since A and B are positive semidefinite). By Exercise 2.2.19, it follows that

$$\text{tr}\big((A - B)(A + B)\big) \geq 0.$$

Expanding out this trace then shows that

$$0 \leq \text{tr}(A^2) - \text{tr}(BA) + \text{tr}(AB) - \text{tr}(B^2).$$

Since $\text{tr}(AB) = \text{tr}(BA)$, it follows that $\text{tr}(A^2) - \text{tr}(B^2) \geq 0$, so $\text{tr}(A^2) \geq \text{tr}(B^2)$, as desired.

3.C.7 One matrix that works is

$$\begin{bmatrix} 3 & 2 & 1 & -2 & 1 \\ 2 & 3 & -1 & -2 & 1 \\ 1 & -1 & 3 & 1 & -1 \\ -2 & -2 & 1 & 3 & -2 \\ 1 & 1 & -1 & -2 & 3 \end{bmatrix}.$$

3.C.8 If A has orthogonal rank-one sum decomposition $A = \sum_{j=1}^r \sigma_j \mathbf{u}_j \mathbf{v}_j^*$ then we choose $X = \mathbf{u}_1 \mathbf{v}_1^*$, $Y = \mathbf{u}_1 \mathbf{u}_1^*$, and $Z = \mathbf{v}_1 \mathbf{v}_1^*$. It is straightforward to then check that $\text{tr}(Y) = \text{tr}(Z) = 1$, $\text{Re}(\text{tr}(AX^*)) = \text{Re}(\text{tr}(\sigma_1(\mathbf{v}_1^* \mathbf{v}_1)(\mathbf{u}_1^* \mathbf{u}_1))) = \sigma_1 = \|A\|$, and

$$\begin{bmatrix} Y & -X \\ -X^* & Z \end{bmatrix} = \begin{bmatrix} \mathbf{u}_1 \mathbf{u}_1^* & -\mathbf{u}_1 \mathbf{v}_1^* \\ -\mathbf{v}_1 \mathbf{u}_1^* & \mathbf{v}_1 \mathbf{v}_1^* \end{bmatrix}$$
$$= \begin{bmatrix} \mathbf{u}_1 \\ -\mathbf{v}_1 \end{bmatrix} \begin{bmatrix} \mathbf{u}_1 \\ -\mathbf{v}_1 \end{bmatrix}^* \succeq O.$$

3.C.9 We could use duality and mimic Example 3.C.5, but instead we recall from Theorem 2.B.4 that

$$\begin{bmatrix} I & A \\ A^* & X \end{bmatrix} \succeq O$$

if and only if $X - A^*A \succeq O$. It follows that, for any feasible X, we have $\text{tr}(X) \geq \text{tr}(A^*A) = \|A\|_F^2$, which shows that the optimal value of the semidefinite program is at least as large as $\|A\|_F^2$.

To show that there exists a feasible X that attains the bound of $\|A\|_F^2$, let $A = U\Sigma V^*$ be a singular value decomposition of A. Then we define $X = V\Sigma^*\Sigma V^*$, which has $\text{tr}(X) = \text{tr}(V\Sigma^*\Sigma V^*) = \text{tr}(\Sigma^*\Sigma) = \|A\|_F^2$. Furthermore, X is feasible since it is positive semidefinite and so is

$$\begin{bmatrix} I & A \\ A^* & X \end{bmatrix} = \begin{bmatrix} I & U\Sigma V^* \\ V\Sigma^*U^* & V\Sigma^*\Sigma V^* \end{bmatrix}$$
$$= \begin{bmatrix} U \\ V\Sigma^* \end{bmatrix} \begin{bmatrix} U \\ V\Sigma^* \end{bmatrix}^*.$$

3.C.12 (a) The dual of this SDP can be written in the following form, where $Y \in \mathcal{M}_n^H$ and $z \in \mathbb{R}$ are variables:

$$\begin{aligned}
\text{minimize:} \quad & \text{tr}(Y) + kz \\
\text{subject to:} \quad & Y + zI \succeq C \\
& Y \succeq O
\end{aligned}$$

(b) If X is the orthogonal projection onto the span of the eigenspaces corresponding to the k largest eigenvalues of C then X is a feasible point of the primal problem of this SDP and $\text{tr}(CX)$ is the sum of those k largest eigenvalues, as claimed. On the other hand, if C has spectral decomposition $C = \sum_{j=1}^n \lambda_j \mathbf{v}_j \mathbf{v}_j^*$ then we can choose $z = \lambda_{k+1}$ (where $\lambda_1 \geq \cdots \geq \lambda_n$ are the eigenvalues of C) and $Y = \sum_{j=1}^k (\lambda_j - z) \mathbf{v}_j \mathbf{v}_j^*$. It is straightforward to check that $Y \succeq O$ and $Y + zI \succeq C$, so Y is feasible. Furthermore, $\text{tr}(Y) + kz = \sum_{j=1}^k \lambda_j - kz + kz = \sum_{j=1}^k \lambda_j$, as desired.

3.C.14 If $\Phi = \Psi_1 + T \circ \Psi_2$ then $C_\Phi = C_{\Psi_1} + T(C_{\Psi_2})$, so Φ is decomposable if and only if there exist PSD matrices X and Y such that $C_\Phi = X + T(Y)$. It follows that the optimal value of the following SDP is 0 if Φ is decomposable, and it is $-\infty$ otherwise (this is a feasibility SDP—see Remark 3.C.3):

$$\begin{aligned}
\text{maximize:} \quad & 0 \\
\text{subject to:} \quad & X + T(Y) = C_\Phi \\
& X, Y \succeq O
\end{aligned}$$

3.C.15 We can write $\Phi = \Psi_1 + T \circ \Psi_2$, where $2C_{\Psi_1}$ and $2C_{\Psi_2}$ are the (PSD) matrices

$$\begin{bmatrix}
3 & 0 & 0 & 0 & 0 & -2 & -1 & 0 \\
0 & 1 & 0 & 0 & 0 & 0 & 0 & -1 \\
0 & 0 & 2 & 0 & 0 & 0 & 0 & -2 \\
0 & 0 & 0 & 2 & -2 & 0 & 0 & 0 \\
0 & 0 & 0 & -2 & 2 & 0 & 0 & 0 \\
-2 & 0 & 0 & 0 & 0 & 2 & 0 & 0 \\
-1 & 0 & 0 & 0 & 0 & 0 & 1 & 0 \\
0 & -1 & -2 & 0 & 0 & 0 & 0 & 3
\end{bmatrix}$$

and

$$\begin{bmatrix}
1 & 0 & 0 & 0 & 0 & 0 & -1 & 0 \\
0 & 3 & 0 & 0 & -2 & 0 & 0 & -1 \\
0 & 0 & 2 & 0 & 0 & -2 & 0 & 0 \\
0 & 0 & 0 & 2 & 0 & 0 & -2 & 0 \\
0 & -2 & 0 & 0 & 2 & 0 & 0 & 0 \\
0 & 0 & -2 & 0 & 0 & 2 & 0 & 0 \\
-1 & 0 & 0 & -2 & 0 & 0 & 3 & 0 \\
0 & -1 & 0 & 0 & 0 & 0 & 0 & 1
\end{bmatrix}$$

respectively.

3.C.17 (a) The Choi matrix of Φ is

$$\begin{bmatrix}
4 & -2 & -2 & 2 & 0 & 0 & 0 & 0 \\
-2 & 3 & 0 & 0 & 0 & 0 & 0 & 0 \\
-2 & 0 & 2 & 0 & 0 & 1 & 0 & 0 \\
2 & 0 & 0 & 0 & 0 & 0 & 0 & -2 \\
0 & 0 & 0 & 0 & 0 & 0 & 0 & -2 \\
0 & 0 & 1 & 0 & 0 & 2 & 0 & -1 \\
0 & 0 & 0 & 0 & 0 & 0 & 4 & 0 \\
0 & 0 & 0 & 0 & -2 & -1 & 0 & 2
\end{bmatrix}$$

Using the same dual SDP from the solution to Exercise 3.C.16, we find that the matrix Z given by

$$\begin{bmatrix}
2 & 0 & 0 & -28 & 0 & 0 & 0 & 0 \\
0 & 16 & 0 & 0 & 0 & 0 & 0 & 0 \\
0 & 0 & 49 & 0 & 0 & -49 & 0 & 0 \\
-28 & 0 & 0 & 392 & 0 & 0 & 0 & 0 \\
0 & 0 & 0 & 0 & 392 & 0 & 0 & 28 \\
0 & 0 & -49 & 0 & 0 & 49 & 0 & 0 \\
0 & 0 & 0 & 0 & 0 & 0 & 16 & 0 \\
0 & 0 & 0 & 0 & 28 & 0 & 0 & 2
\end{bmatrix}$$

is feasible and has $\text{tr}(C_\Phi Z) = -2$. Scaling Z up by an arbitrary positive scalar shows that the optimal value of this SDP is $-\infty$ and thus the primal problem is infeasible and so Φ is not decomposable.

3.C.18 Recall from Theorem 3.B.3 that q is a sum of squares of bilinear forms if and only if we can choose Φ to be decomposable and bisymmetric. Let $\{\mathbf{x}_i \mathbf{x}_i^T\}$ and $\{\mathbf{y}_j \mathbf{y}_j^T\}$ be the rank-1 bases of $\mathcal{M}_m^S$ and $\mathcal{M}_n^S$, respectively, described by Exercise 1.2.8. It follows that the following SDP determines whether or not q is a sum of squares, since it determines whether or not there is a decomposable bisymmetric map Φ that represents q:

$$\begin{aligned}
\text{maximize:} \quad & 0 \\
\text{subject to:} \quad & \mathbf{x}_i^T \Phi(\mathbf{y}_j \mathbf{y}_j^T) \mathbf{x}_i = q(\mathbf{x}_i, \mathbf{y}_j) \text{ for all } i, j \\
& \Phi = T \circ \Phi \\
& \Phi = \Phi \circ T \\
& \Phi = \Psi_1 + T \circ \Psi_2 \\
& C_{\Psi_1}, C_{\Psi_2} \succeq O
\end{aligned}$$

[Side note: We can actually choose $\Psi_1 = \Psi_2$ to make this SDP slightly simpler, thanks to Exercise 3.A.15.]

Bibliography

[Aga85] S.S. Agaian, *Hadamard Matrices and Their Applications* (Springer, 1985)

[AM57] A.A. Albert, B. Muckenhoupt, On matrices of trace zeros. Mich. Math. J. **4**(1), 1–3 (1957)

[Art27] E. Artin, Über die zerlegung definiter funktionen in quadrate. Abhandlungen aus dem Mathematischen Seminar der Universität Hamburg **5**(1), 100–115 (1927)

[BBP95] P. Borwein, J. Borwein, S. Plouffe, The inverse symbolic calculator (1995), http://wayback.cecm.sfu.ca/projects/ISC/

[BV09] S.P. Boyd, L. Vandenberghe, *Convex Optimization* (Cambridge University Press, 2009)

[Cal73] A.P. Calderón, A note on biquadratic forms. Linear Algebra Appl. **7**(2), 175–177 (1973)

[Cho75] M.-D. Choi, Positive semidefinite biquadratic forms. Linear Algebra Appl. **12**(2), 95–100 (1975)

[Chv83] V. Chvatal, *Linear Programming* (Freeman, W. H, 1983)

[CL92] S. Chang, C.-K. Li, Certain isometries on $\mathbb{R}^n$. Linear Algebra Appl. **165**, 251–265 (1992)

[CVX12] CVX Research, Inc. CVX: Matlab software for disciplined convex programming, version 2.0 (August 2012), http://cvxr.com/cvx

[DCAB18] S. Diamond, E. Chu, A. Agrawal, S. Boyd, CVXPY: A Python-embedded modeling language for convex optimization (2018), http://www.cvxpy.org

[FR97] B. Fine, G. Rosenberger, *The Fundamental Theorem of Algebra* (Undergraduate Texts in Mathematics (Springer, New York, 1997)

[Hil88] D. Hilbert, Ueber die darstellung definiter formen als summe von formenquadraten. Mathematische Annalen **32**(3), 342–350 (1888)

[HJ12] R.A. Horn, C.R. Johnson, *Matrix Analysis* (Cambridge University Press, 2012)

[Hor06] K.J. Horadam, *Hadamard Matrices and Their Applications* (Princeton University Press, 2006)

[Joh20] N. Johnston, *Introduction to Linear and Matrix Algebra* (Undergraduate Texts in Mathematics (Springer, New York, 2020)

[KB09] T.G. Kolda, B.W. Bader, Tensor decompositions and applications. SIAM Rev. **51**(3), 455–500 (September 2009)

[Lan12] J.M. Landsberg, *Tensors: Geometry and Applications* (American Mathematical Society, 2012)

[LP01] C.-K. Li, S. Pierce, Linear preserver problems. Am. Math. Mon. **108**, 591–605 (2001)

[LS94] C.-K. Li, W. So, Isometries of ℓ_p norm. Am. Math. Mon. **101**, 452–453 (1994)

[LT92] C.-K. Li, N.-K. Tsing, Linear preserver problems: a brief introduction and some special techniques. Linear Algebra Appl. **162**, 217–235 (1992)

[Mir60] L. Mirsky, Symmetric gauge functions and unitarily invariant norms. Q. J. Math. **11**, 50–59 (1960)

[MO15] M. Miller, R. Olkiewicz, Topology of the cone of positive maps on qubit systems. J. Phys. A Math. Theor. **48**, (2015)

[Shi18] Y. Shitov, A counterexample to Comon's conjecture. SIAM J. Appl. Algebra Geom. **2**(3), 428–443 (2018)

[Stø63] E. Størmer, Positive linear maps of operator algebras. Acta Mathematica **110**(1), 233–278 (1963)

[Sto10] A. Stothers, *On the Complexity of Matrix Multiplication*. Ph.D. thesis, University of Edinburgh (2010)

[TB97] L.N. Trefethen, D. Bau, *Numerical Linear Algebra* (SIAM, 1997)

[Wor76] S.L. Woronowicz, Positive maps of low dimensional matrix algebras. Rep. Math. Phys. **10**, 165–183 (1976)

[YYT16] Y. Yang, D.H. Leung, W.-S. Tang, All 2-positive linear maps from $\mathcal{M}_3(\mathbb{C})$ to $\mathcal{M}_3(\mathbb{C})$ are decomposable. Linear Algebra Appl. **503**, 233–247 (2016)

Index

© Springer Nature Switzerland AG 2021
N. Johnston, *Advanced Linear and Matrix Algebra*,
https://doi.org/10.1007/978-3-030-52815-7

Symbol Index

n the United States
& Taylor Publisher Services